ELECTRICAL CIRCUIT ANALYSIS

K. MAHADEVAN
Professor
Electrical and Electronics Engineering
PSNA College of Engineering and Technology
Dindigul, Tamil Nadu

C. CHITRA
Professor
Electronics and Communication Engineering
PSNA College of Engineering and Technology
Dindigul, Tamil Nadu

PHI Learning Private Limited
Delhi-110092
2015

₹ 550.00

ELECTRICAL CIRCUIT ANALYSIS
K. Mahadevan and C. Chitra

ISBN-978-81-203-5047-2

Published by Asoke K. Ghosh, PHI Learning Private Limited, Rimjhim House, 111, Patparganj Industrial Estate, Delhi-110092 and Printed by Mohan Makhijani at Rekha Printers Private Limited, New Delhi-110020.

Dedicated to

OUR PARENTS

Contents

Preface

Electrical circuit analysis is the foundation of the technologies of electrical, electronics, communication and computer engineering. Hence it becomes an essential course for the specialized branches of engineering viz. electrical and electronics engineering, electronics and communication engineering, instrumentation and control engineering, computer science and engineering, biomedical engineering etc. which deals electron and its related applications.

The book is primarily intended for the beginners who are pursuing undergraduate study in the above branches in various universities. As 'Electrical Circuit Analysis' or 'Circuit theory' being the very first course in their graduation study, more emphasize need to be given on explaining the basic concepts. Hence, the book aims at presenting the concepts of electrical circuits in lucid language with easy-to-understand approach and is done based on the classroom experience of the authors.

The presentation is very simple to facilitate self study to the students. A better way to understand the various aspects of electrical circuits is to solve many problems. Keeping this point in view a large number of solved and unsolved problems have been included which are selected from previously asked end semester examination question papers of various universities across the country. Detailed explanation of each step makes it easy for the students to understand complicated problems. Most of the mathematical steps have been explained in the book leaving only a few intermediate simple steps to the students to work on. The chapters are arranged logically in a proper sequence so that successive topics build upon earlier topics, which is important in learning subjects like this one. Each chapter is supported with necessary illustrations. Contents of the book are spread over 14 chapters and the organization is as follows.

The book begins with explaining the fundamentals of DC circuits, its associated symbols and units in Chapter 1. One of the very fundamental laws in electrical engineering, Ohm's law, is introduced in this chapter. Network reduction techniques, conversion of sources are also given in this chapter. Chapter 2 is devoted for studying Kirchhoff's laws. Mesh analysis and nodal analysis techniques derived from Kirchhoff's laws are presented here with sufficient examples. Introduction to AC fundamentals are given in Chapter 3. Discussion about *j* operator and its significance is also found in this chapter. The steady state analysis of single phase circuits such as *RL*, *RC* and *RLC* are done in Chapter 4. Mesh analysis and nodal analysis techniques as regard to AC circuits is covered in Chapter 5 with examples. In Chapter 6, detailed analysis of various network theorems applicable to both DC and AC circuits are presented. A comprehensive description of three phase systems of both balanced and unbalanced types are presented in Chapter 7. Measurement of real power, reactive power and power factor are studied as well in Chapter 7. Chapter 8 deals the topic on *magnetic circuits* and the applications of Faraday's laws of electromagnetic induction. A thorough study has been carried out on the topic *coupled circuits* in Chapter 9. Tuned circuits have also been discussed in this chapter. The phenomenon 'resonance' and its behavior with respect to series and parallel circuits are presented in Chapter 10. Chapters 11 and 12 respectively explain the transient behavior of DC circuits and AC circuits with worked problems. Both classical and laplace transformation methods are employed for solving all varieties of problems discussed in these two chapters. The two-port networks along with various parameters and their inter-relationships are discussed in Chapter 13. In Chapter 14, the topic *graph theory* and its application to simple DC circuits are given. This chapter includes the concepts of duality with examples as well.

Last section of each chapter is included with practice problems along with answers. Short questions with answers are given at the end of each chapter to help the students to understand the basic laws, theorems and some key terms involved in learning the subject. To ensure that the students get the gist of the subject matter clearly, annotations are given in boxes and also in equations at appropriate places.

K. Mahadevan
C. Chitra

Acknowledgements

Foremost we would like to express our sincere gratitude to Dr. C. Thangaraj, former Vice-Chancellor, Kalasalingam University, an inspirational leader, who initiated us to bring out a book of this kind.

Our unbounded thanks are due to the Directors of PSNA College of Engineering and Technology, Rtn. MPHF. R.S.K. Raguraam, Dr. R.S.K. Laksshmana Prabu and Mr. R.S.K. Sukumaran for their encouragement and support in preparing this book.

We are thankful to our friends Dr. S. Kannan and Dr. N. Karpagam for their help and support during the preparation of the text.

We are also very thankful to our colleague Dr. V. Rajasekaran who has helped us in proof reading the manuscript.

We are grateful to the editorial and production staff of PHI Learning Private Limited for having shown interest in publishing this edition.

K. Mahadevan
C. Chitra

CHAPTER 1

Introduction to DC Circuits

1.1 INTERNATIONAL SYSTEM OF UNITS

The International System of Units (*Le Système international d'unités*), usually abbreviated as *SI units*, is the most widely used system of measurement. Seven basic units in the SI system are listed with their symbols in Table 1.1. Other units, *derived SI units*, use combinations of basic units and there are many of them. For example, velocity is derived as metres per second (m/s) and force is as kilogram metre (kg m).

Table 1.1 Basic SI Units

Quantity	*Unit*	*Symbol*
length	metre	m
mass	kilogram	kg
time	second	s
electric current	ampere	A
thermodynamic resistance	kelvin	K
luminous intensity	candela	cd
amount of substance	mole	mol

The *metre* is the length equal to 16,50,763.73 wavelengths of the orange line in the spectrum of an internationally specified krypton discharge lamp.

The *kilogram* is the mass of a platinum-iridium cylinder preserved at the International Bureau of Weights and Measures at Sevres, near Paris.

The *second* is the interval of time taken by 9,19,26,31,770 cycles of radiation corresponding to the transition of the Caesium-133 atom.

The *ampere* is that current which, if maintained in two straight infinitely long parallel conductors of negligible cross-section and placed 1 metre apart in vacuum, would produce a force of 2×10^{-7} newtons per metre length between these conductors.

The *Kelvin* is 1/273.16 of the thermodynamic temperature of the triple point of water.

The *candela* is the luminous intensity in the perpendicular direction of a surface of 1/6,00,000 square metre of a black body at the temperature of freezing under standard atmospheric surface.

The *mole* is the amount of substance of a system which contains as many elementary entities as there are atoms in 0.012 kg of the carbon isotype. When the mole is used, it is essential to specify the elementary entities under consideration. These may be atoms, molecules, ions, electrons or other particles or groups of particles.

SI units may be made larger or smaller by using prefixes which denote multiplication or division by a particular amount. The most common multiples are listed in Table 1.2.

Table 1.2 Multiple and Sub-multiple Units

Prefix	*Name*	*Meaning*	
M	mega	multiply by 1 000 000	(i.e. $\times 10^6$)
k	kilo	multiply by 1 000	(i.e. $\times 10^3$)
m	milli	divide by 1 000	(i.e. $\times 10^{-3}$)
μ	micro	divide by 1 000 000	(i.e. $\times 10^{-6}$)
n	nano	divide by 1 000 000 000	(i.e. $\times 10^{-9}$)
p	pico	divide by 1 000 000 000 000	(i.e. $\times 10^{-12}$)

The above multiple and sub-multiple units are applicable to all units such as ohm, volt, ampere, henry, farad, and so on.

1.2 NETWORK TERMINOLOGY

Network

A network or circuit is a closed path consists of source of power (battery, generator, etc.), the conductors to carry the current and the loads (lamp, motor, fan, etc.).

Parameters

The various elements of an electric circuit are called its parameters (resistance, inductance, capacitance etc.) and these may be lumped or distributed.

Network element

Any individual element which can be connected to some other element is called network element. Network elements are classified into four groups, viz.

1. Active or passive
2. Bilateral or unilateral

3. Linear or nonlinear
4. Lumped or distributed

Active element Elements which supply power or energy to the circuit are known as active elements. Typical examples are voltage source and current source.

Passive element Elements which store or dissipate energy are called passive elements. Resistor, inductor and capacitor are the three basic passive elements. Resistor dissipates energy, whereas inductor and capacitor can store energy.

Bilateral element In bilateral elements, the voltage–current relationship does not change with the change in current direction. Resistor is an example of bilateral element.

Unilateral element In unilateral elements the voltage–current relationship changes with the change in current direction. Examples of unilateral elements are junction diodes and transistors.

Linear element Elements which obey the principle of superposition are called linear elements (e.g. resistor).

Nonlinear element Elements which do not obey the principle of superposition are called nonlinear elements (e.g. junction diode, transistor).

Lumped element Elements which are separated are known as lumped elements (e.g. resistor, inductor and capacitor).

Distributed element Elements which are not separated are known as distributed elements. For example, in transmission lines, resistors, inductors and capacitors are distributed throughout the length.

Node

Node refers to a point where two or more elements meet. In a circuit if more than two elements meet at a node, then the node is called major or principal node. In Fig. 1.1, B and E are called major nodes.

Loop

Any closed path of a circuit is called loop. Thus, in Fig. 1.1, ABEFA, BCDEB and ABCDEFA are the loops.

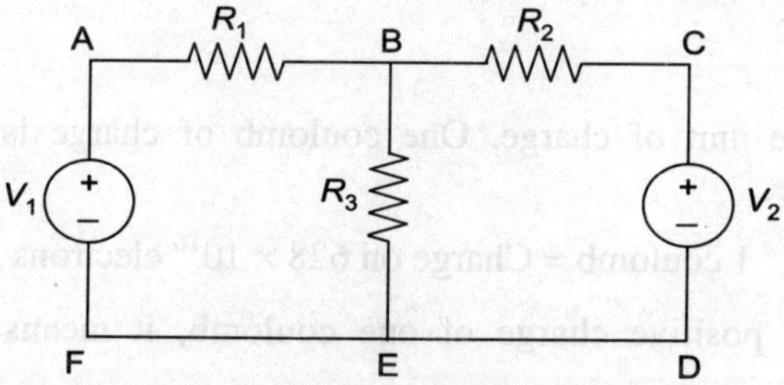

Fig. 1.1 Illustration of mesh and loop.

Mesh

Mesh is a loop that contains no loops inside it. For instance, in Fig. 1.1, ABCDEFA is a loop, whereas ABEFA and BCDEB are meshes.

1.3 LUMPED CIRCUIT VERSUS DISTRIBUTED CIRCUIT

The circuits which are made up of lumped elements are called lumped circuits. Resistor, inductor, capacitor, voltage source and current source are examples of lumped elements. In lumped circuits, the differential equations obtained for the circuit using KCL and/or KVL (will be introduced in Chapter 2) involve only one variable, that is, time. The entire circuit is assumed at a single point in space, or equivalently, that the travel time of the electric signal is negligible.

A circuit is said to be a distributed one, if it has elements which are not separated. Transmission lines, coaxial cables, hollow waveguides, handset antennas for wireless communication are some of the examples for distributed circuits. The voltages and currents in such a circuit are functions of position as well as time. A lumped circuit is easy to understand and analyse compared to a distributed circuit. A distributed circuit may be approximated by considering it being made up of a number of lumped circuits.

1.4 QUANTITY OF ELECTRICITY

All matters (solid, liquid, gas) consist of atoms and all atoms consist of *protons*, *neutrons* and *electrons*. The electron has negative charge, proton has positive charge and neutron has no charge. When the number of protons and the number of electrons in an atom are equal, then the resultant charge is zero. As a result, the atom becomes electrically neutral. If an atom loses electrons, it is positively charged. Similarly, if an atom is supplied with electrons, it is negatively charged. Electrons have very small mass. So, they are more mobile than protons. Whereas protons are powerfully held in the nucleus and cannot be removed or detached. If an electric pressure (voltage) is applied across the body, there is a tendency for electrons to move in a particular direction. This movement of free electrons constitutes an electric current flow.

1.5 CHARGE

Coulomb is used as the unit of charge. One coulomb of charge is equal to the charge on 628×10^{16} electrons, i.e.

$$1 \text{ coulomb} = \text{Charge on } 628 \times 10^{16} \text{ electrons}$$

Thus, if a body has a positive charge of one coulomb, it means that it has a deficit of 628×10^{16} electrons.

1.6 CURRENT

Current is defined as the rate of flow of charge. If a charge q passes through a conductor in time t (the rate of flow of charge is constant), then current $I = q/t$. The unit of current is coulomb per second or ampere. If the rate of flow of charge is not constant, the instantaneous value of current is given by

$$i = \frac{dq}{dt}$$

1.7 ELECTRIC POTENTIAL

If one joule work is done to move a charge of one coulomb between two points, then the potential of the first point is said to be 1 V with respect to the second point.

$$v = \frac{dW}{dq}$$

It is generally measured between two points and its unit is volt.

1.8 CIRCUIT ELEMENTS

1.8.1 Resistance

The property of any material which opposes the flow of electric current through it is called resistance. Its symbol is shown in Fig. 1.2, and is denoted by letter 'R'. Its unit is ohms (Ω).

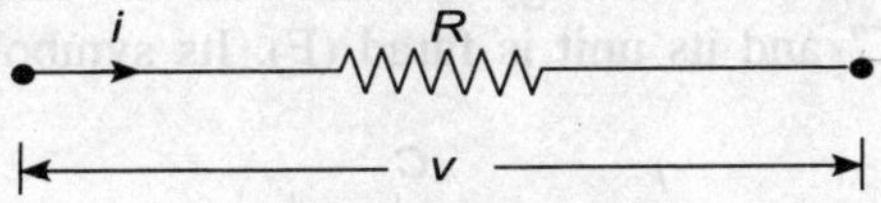

Fig. 1.2 Resistance.

By Ohm's law, the voltage across the two terminals of the conductor is written as $V = IR$.

If a current of I amperes flows through the resistor, it dissipates energy in the form of heat. The power absorbed by the resistor is given by

$$P = VI = I^2 R = \frac{V^2}{R} \text{ watts}$$

The power absorbed by the resistor is converted into heat (energy) during a period of time t, and is given by

$$W = Pdt = \int_0^t I^2 R dt = I^2 Rt$$

$$= VI \times t \text{ joules}$$

1.8.2 Inductance

Inductance is the element in which energy is stored in the form of electromagnetic field. The inductance is denoted by '*L*' and its unit is henry (H). The symbol of inductance is shown in Fig. 1.3.

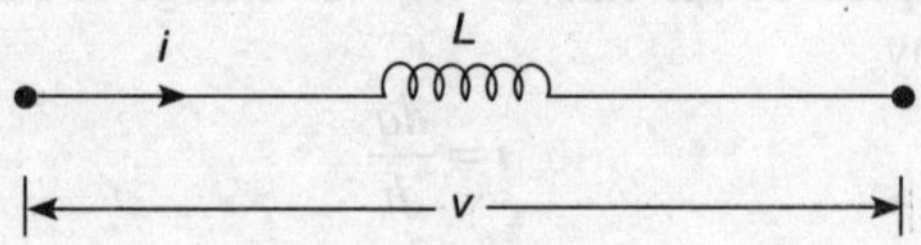

Fig. 1.3 Inductance.

In inductance, the voltage across it is proportional to the rate of change of current through it.

i.e. $$v \propto \frac{di}{dt}$$

or $$v = L\frac{di}{dt}$$

Assuming that previously no current was flowing through the inductance and if a current of i amperes flows through it, the energy stored during a time interval of t is obtained as:

$$W = \int_0^t v\,i\,dt = \int_0^t \left(L\frac{di}{dt}\right) i \cdot dt = L\int_0^i i \cdot di = \frac{1}{2}Li^2 \text{ joules}$$

1.8.3 Capacitance

Capacitance is the element in which energy is stored in the form of electrostatic field. The capacitance is denoted by '*C*' and its unit is farad (F). Its symbol is shown in Fig. 1.4.

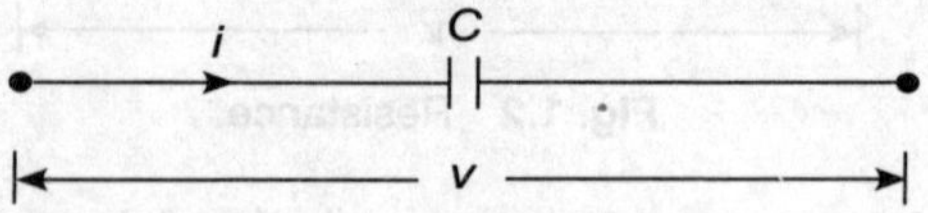

Fig. 1.4 Capacitance.

In capacitance, charge q, is directly proportional to voltage v across it.

i.e. $$q \propto v$$

or $$q = Cv$$

Differentiating,

$$\frac{dq}{dt} = C\frac{dv}{dt}$$

or $$idt = Cdv \qquad \left(\because i = \frac{dq}{dt}\right) \qquad \text{[from Section 1.6]}$$

Integrating both sides, we get

$$\int i\,dt = C\int dv$$

or

$$v = \frac{1}{C}\int i\,dt$$

Assuming no initial charge on the capacitor, if current *i* flows for a period of *t* seconds, the energy stored in the capacitor will be

$$W = \int_0^t vi\,dt = \int_0^v vC\,dv = C\int_0^v v\,dv = \frac{1}{2}Cv^2 \text{ joules}$$

Table 1.3 shows the voltage–current relationships of three basic elements.

Table 1.3 Voltage–Current Relationships of Basic Elements

Element	*Voltage*	*Current*
R	$V = IR$	$I = \dfrac{V}{R}$
L	$v = L\dfrac{di}{dt}$	$i = \dfrac{1}{L}\int v\,dt$
C	$v = \dfrac{1}{C}\int i\,dt$	$i = C\dfrac{dv}{dt}$

1.9 OHM'S LAW

The relationship between voltage (V), current (I) and resistance (R) in a DC circuit was first discovered by scientist George Simon Ohm. This relationship is called Ohm's law and may be stated as under.

The ratio of potential difference (V) between the ends of a conductor to the current (I) flowing between them is constant, provided the physical conditions (e.g. temperature) do not change.

i.e.

$$\frac{V}{I} = \text{constant} = R$$

where R is the resistance of the conductor between the two ends considered.

For example, if the voltage between two points of a resistor is V volts and the current flowing is I amperes, then V/I will be constant and is equal to R. If the voltage is increased/decreased, the current will also be increased/decreased accordingly; thus, the V/I ratio is constant. The unit of voltage is in volts, current in amperes and resistance in ohms.

Limitations of Ohm's law

1. Ohm's law is applicable for conductors at constant temperature.
2. It does not apply to non-metallic conductors (e.g. graphite).
3. Ohm's law does not apply to nonlinear devices (e.g. zener diode).

1.10 SPECIFIC RESISTANCE (RESISTIVITY)

Consider a conductor of length l m and area of cross-section a m^2 (Fig. 1.5). If R be its resistance, then

$$R \propto \frac{l}{a}$$

or

$$R = \rho \frac{l}{a} \tag{1.1}$$

ρ is the constant of proportionality and is known as the specific resistivity of the material.

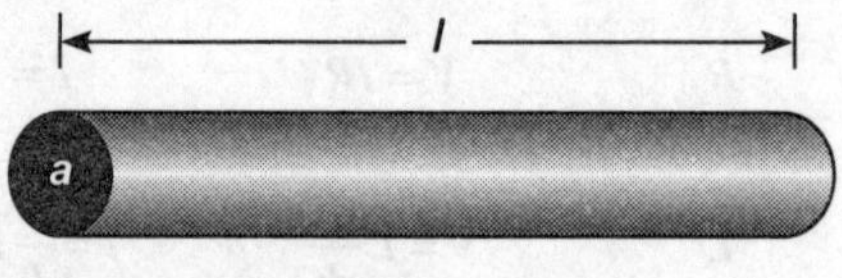

Fig. 1.5 Geometrical parameters of a conductor.

Substituting l = 1 m and a = 1 m^2 into Eq. (1.1), we get

$$R = \rho$$

From Eq. (1.1), we get

$$\rho = \frac{Ra}{l}$$

Its unit is $\dfrac{\Omega \text{m}^2}{\text{m}}$, i.e. Ω m.

EXAMPLE 1.1 Calculate the resistance of 500 m length of aluminium wire given that the diameter of the wire is 6 mm and the resistivity of aluminium is 0.028 μ Ω m.

Solution

Area of cross-section of wire, $a = \frac{\pi}{4} d^2 = \frac{\pi}{4} 6^2 \text{ mm}^2$

$= 9\pi \times 10^{-6} \text{ m}^2$

Length of wire, $= 500$ m

$\rho = 0.028\ \mu\ \Omega\text{m} = 0.028 \times 10^{-6}\ \Omega\text{m}$

Resistance, $R = \rho \frac{l}{a} = \frac{0.028 \times 10^{-6} \times 500}{9\pi \times 10^{-6}}$

$\mathbf{R = 0.4951\ \Omega}$

EXAMPLE 1.2 A conductor has a resistance of 5 Ω. What is the resistance of another conductor of the same material, which has half of the diameter and six times length of the first conductor?

Solution Let R_1, l_1 and a_1 are the resistance, length and cross-sectional area of the first conductor respectively. Similarly, R_2, l_2 and a_2 represent resistance, length and cross-sectional area of the second conductor respectively.

$$R_1 = \rho \frac{l_1}{a_1} \quad \text{and} \quad R_2 = \rho \frac{l_2}{a_2}$$

$$\frac{R_2}{R_1} = \frac{l_2/a_2}{l_1/a_1} = \left(\frac{l_2}{l_1}\right) \times \left(\frac{a_1}{a_2}\right)$$

$$a_1 = \pi \frac{d_1^2}{4},\; a_2 = \pi \frac{d_2^2}{4} \quad \text{and} \quad d_2 = \frac{1}{2} d_1$$

$$\frac{a_1}{a_2} = \frac{d_1^2}{d_2^2} = \left(\frac{d_1}{d_2}\right)^2 = \left(\frac{d_1}{d_1/2}\right)^2 = 4$$

Length of the second conductor, l_2, is given as six times that of l_1.

$$\therefore \qquad \frac{l_2}{l_1} = 6$$

$$\frac{R_2}{R_1} = \frac{l_2}{l_1} \times \frac{a_1}{a_2} = 6 \times 4 = 24$$

Therefore, $\qquad R_2 = 24R_1 = 24 \times 5 = \mathbf{120\ \Omega}$

1.11 CONDUCTANCE AND CONDUCTIVITY

Conductance is defined as *the reciprocal of resistance*. The resistance of a conductor refers to the opposition it offers to the flow of current through it. The conductance G refers to the inducement it offers to the flow of current.

$$G = \frac{1}{R} = \frac{1}{\rho}\left(\frac{a}{l}\right) \tag{1.2}$$

$$\sigma = \frac{1}{\rho}$$

where σ is called *conductivity* or *specific conductance* of a conductor. Its unit is Siemens (S) or mho (℧). From Eq. (1.2), it can be written as:

$$\sigma = \frac{Gl}{a}$$

1.12 OPEN-CIRCUIT AND SHORT-CIRCUIT

Figure 1.6 is an illustration of open-circuit. There is no load connected across terminals A and B. In this case, $R = \infty$. Then, current I through the circuit is given by Ohm's law as:

$$I = \frac{V}{R} = \frac{V}{\infty} = 0$$

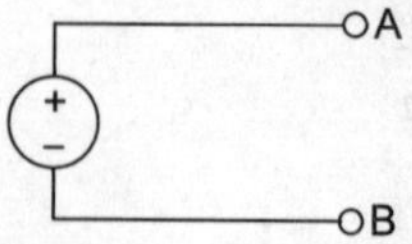

Fig. 1.6 An open-circuit.

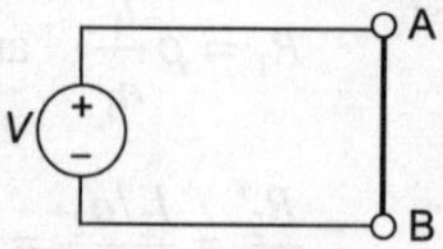

Fig. 1.7 A short-circuit.

Figure 1.7 is an illustration of short-circuit made across terminals A and B. Ideally, short-circuit measures $R = 0\ \Omega$, then the voltage drop across AB is given by

$$V_{AB} = I \times R = I \times 0 = 0 \text{ volts}$$

Hence, it is to be noted that current through open-circuit is zero amperes and voltage across short-circuit is zero volts.

1.13 VARIATION OF RESISTANCE WITH TEMPERATURE

The resistance of a material does not remain constant when temperature changes. It has been found that the resistance in some cases increases with the rise in temperature and decreases in some other cases as well. In the former case, the temperature coefficient of resistance is said to be positive, whereas in the latter case, it is negative. If R_0 be the resistance at 0°C and R_t be the resistance at t°C, we can write:

$$R_t = R_0(1 + \alpha t) \tag{1.3}$$

where α is called the *temperature coefficient of resistance.*

From Eq. (1.3), we get

$$\alpha = \frac{R_t - R_0}{tR_0} \tag{1.4}$$

The temperature coefficient of resistance is defined as *the increase in resistance per degree rise in temperature.*

In pure metals, resistance increases with the rise in temperature. Alloys like brass and manganin have negligible temperature coefficient and so their resistance value does not increase with the rise in temperature. When carbon, insulators and electrolytes are heated, there is a decrease in temperature. Thermistor is one which has high negative temperature coefficient of resistance.

The resistivity of a metallic conductor too increases with an increase in temperature and decreases when the temperature decreases. It has been found that the change in resistivity is proportional to the change in temperature over the normal operating range of temperature. Mathematically,

$$\rho_t = \rho_0(1 + \alpha_0 t)$$

where α_0 is the temperature coefficient of resistance at 0°C, ρ_0 is resistivity at 0°C, ρ_t is resistivity at t°C, and t is the change in temperature.

EXAMPLE 1.3 Find the resistance of aluminium wire at 40°C when it has a resistance of 100 Ω at 25°C. Take the temperature coefficient of aluminium at 0°C as 0.045 per °C.

Solution Let

$$t_1 = 25°\text{C} \quad \text{and} \quad R_{t1} = 100\ \Omega$$

$$t_2 = 40°\text{C} \quad \text{and} \quad R_{t2} = ?$$

$$R_t = R_0(1 + \alpha t) \qquad \text{[from Eq. 1.3]}$$

Hence,

$$R_{t1} = R_0(1 + \alpha t_1) \qquad \text{(i)}$$

and

$$R_{t2} = R_0(1 + \alpha t_2) \qquad \text{(ii)}$$

(i) ÷ (ii) gives

$$\frac{R_{t2}}{R_{t1}} = \frac{(1+\alpha t_2)}{(1+\alpha t_1)}$$

Therefore,

$$R_{t2} = R_{t1} \times \frac{(1+\alpha t_2)}{(1+\alpha t_1)}$$

$$R_{t2} = 100 \times \frac{(1+0.045\times 40)}{(1+0.045\times 25)} = \mathbf{131.76\ \Omega}$$

EXAMPLE 1.4 A coil of copper wire in a machine has a resistance of 150 Ω at 20°C. When the coil is connected to a 240 V supply, current taken is 1.25 A after four hours of run. Calculate the final temperature assuming the temperature coefficient of copper at 20°C to be 0.0039.

Solution Let

$$R_1 = 150\ \Omega \text{ and } t_1 = 20°\text{C}$$

$$R_2 = 240/1.25 = 192\ \Omega \text{ and } t_2 = ?$$

$$\alpha_1 = 0.0039$$

$$t_1 = \text{initial temperature} = 20°\text{C}$$

$$R_2 = R_1\,[1 + \alpha(t_2 - t_1)]$$

$$192 = 150\,[1 + 0.0039(t_2 - 20)]$$

Hence, the final temperature of copper wire after four hours of run, t_2 = **91.79°C**

EXAMPLE 1.5 A coil has a resistance of 18 Ω when its mean temperature is 20°C and of 20 Ω when its mean temperature is 50°C. Find its mean temperature rise when its resistance is 21 Ω and the surrounding temperature is 15°C.

Solution

$$\frac{R_{t2}}{R_{t1}} = \frac{(1+\alpha t_2)}{(1+\alpha t_1)}$$

$$\frac{20}{18} = \frac{(1+50\alpha)}{(1+20\alpha)}$$

$$1.11\ (1 + 20\alpha) = (1 + 50\alpha)$$

$$22.2\alpha - 50\alpha = 1 - 1.11$$

$$\alpha = 0.004$$

$$\frac{21}{20} = \frac{(1+0.004t)}{(1+0.004\times 50)}$$

Solving, $t = 65°C$

Therefore, the rise in temperature $= 65 - 15 = \mathbf{50°C}$

1.14 ELECTRICAL POWER

The rate at which work is done is called power and its unit is joule per second or watt. When one joule work is done in moving a charge of one coulomb in 1 s, the rate of work is one watt (or) joule/s. Hence, power in electric circuits is obtained as the product of voltage (V) and current (I).

$$P = VI \text{ watts}$$

Using Ohm's law, it could be written as:

$$P = I^2R \text{ or } \frac{V^2}{R} \text{ watts}$$

1.15 ELECTRICAL ENERGY

Electrical energy is the total work done in an electric circuit.

$$\text{Electrical energy} = \text{electrical power} \times \text{time}$$

$$= VIt = I^2Rt = \frac{V^2}{R}t$$

In practice, electrical energy is measured in kilowatt hour (kWh). One kWh is also called 1 unit.

$$\text{Energy in kWh} = \text{power in kW} \times \text{time in hours}$$

1.16 ENERGY SOURCES

Electrical energy supplied from any form of source is decided by the two quantities, namely voltage and current. Generally, the energy sources are operated by keeping one of them constant and allowing the other quantity to vary. If the voltage is maintained constant and the current is allowed to vary, the source is called *voltage source*. On the other hand, if the current is maintained constant and the voltage is allowed to vary, it is called as *current source*.

1.16.1 Independent sources

The energy source which maintains constant voltage across its terminals independent of the output current is called ideal voltage source. Figure 1.8(a) and (b) shows the representation of ideal voltage source and its voltage–current relationship, respectively. Similarly, the ideal current source delivers constant current irrespective of the voltage across its terminals. Figure 1.9(a) and (b) shows the representation of ideal current source and its voltage–current relationship, respectively.

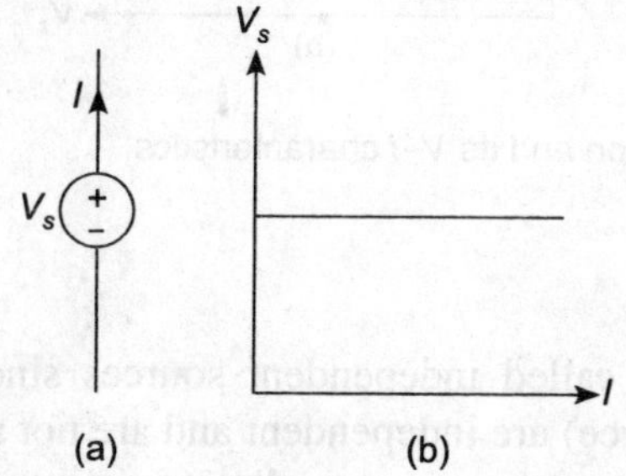

Fig. 1.8 Ideal voltage source representation and its *V–I* characteristics.

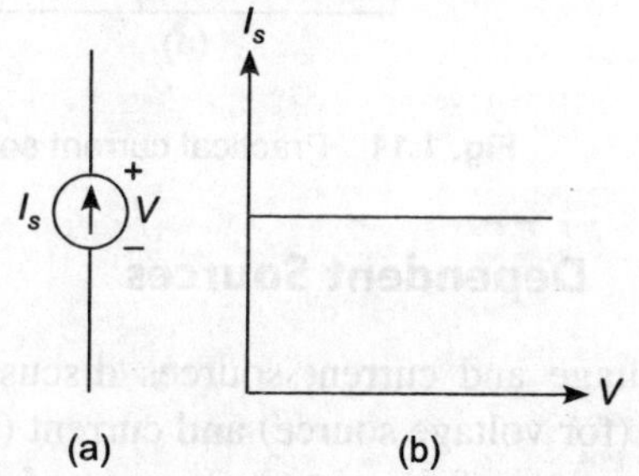

Fig. 1.9 Ideal current source representation and its *V–I* characteristics.

However, ideal voltage sources are never realisable in practice. They have their own internal resistance. This is represented by connecting it in series with the voltage source as shown in Fig. 1.10. The voltage across the source decreases with increasing load current and is due to the voltage drop in internal resistance.

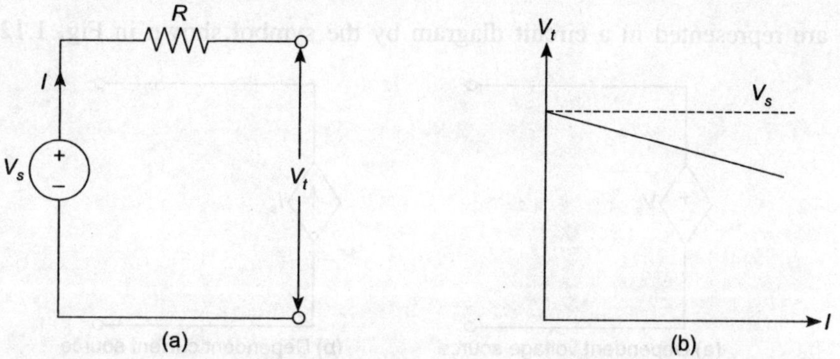

Fig. 1.10 Practical voltage source representation and its *V–I* characteristics.

Similarly, practical current source has its own internal resistance. This is represented by connecting it in parallel with the current source as shown in Fig. 1.11. In practical current source, the current delivered to the external circuit decreases with increasing load voltage and the reduction in current is due to its internal characteristics. The characteristics of practical voltage and current sources are shown in Fig. 1.10(b) and 1.11(b), respectively.

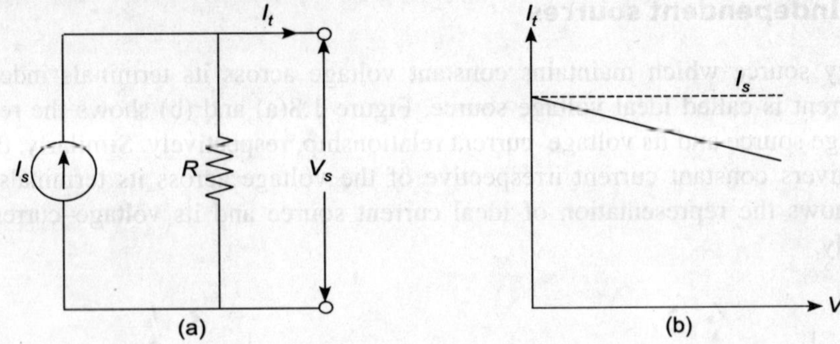

Fig. 1.11 Practical current source representation and its *V–I* characteristics.

1.16.2 Dependent Sources

The voltage and current sources discussed above are called independent sources since their voltage (for voltage source) and current (for current source) are independent and are not affected by other parts of the circuit. In case of dependent sources, the source voltage or current is *not fixed*, but is dependent on the voltage or current in some part of the same circuit. Such sources mainly occur in modelling of electronic components like transistors, op-amp and so on.

Dependent or controlled sources are of the following types:

1. Voltage-controlled voltage source (VCVS)
2. Current-controlled voltage source (CCVS)
3. Voltage-controlled current source (VCCS)
4. Current-controlled current source (CCCS)

These are represented in a circuit diagram by the symbol shown in Fig. 1.12(a) and (b).

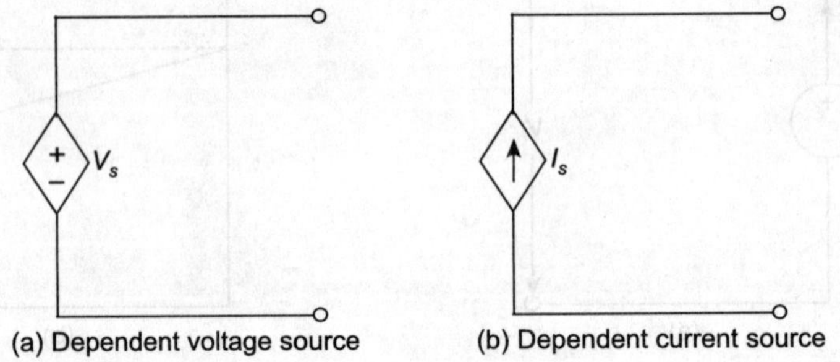

Fig. 1.12 Dependent sources.

1.17 RESISTANCES IN SERIES

In series circuit, the resistances are connected end to end and there is only one path for the current to flow as shown in Fig. 1.13. Therefore, all the resistances carry same current but the voltage drops across them are different depending on the ohmic value of resistances.

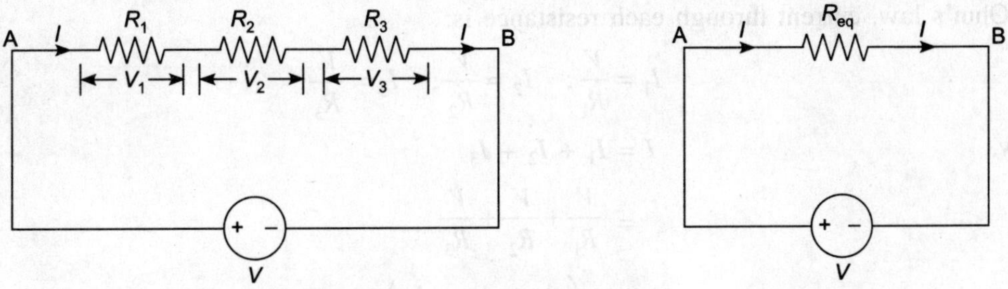

(a) Series circuit (b) Equivalent circuit

Fig. 1.13 Resistances in series.

Let R_1, R_2 and R_3 are the values of individual resistances and V_1, V_2 and V_3 are the respective voltage drops across them. By ohm's law, voltage across various resistances are:

$$V_1 = IR_1;\ V_2 = IR_2;\ V_3 = IR_3$$

Now,
$$V = V_1 + V_2 + V_3 = IR_1 + IR_2 + IR_3$$
$$= I(R_1 + R_2 + R_3)$$

or
$$\frac{V}{I} = R_1 + R_2 + R_3$$

Here, V/I is the equivalent resistance R_{eq} between points A and B.

Therefore,
$$R_{eq} = R_1 + R_2 + R_3$$

Hence, when a number of resistances are connected in series, the equivalent resistance is equal to the sum of individual resistances.

1.18 RESISTANCES IN PARALLEL

When several resistances are shunted across each other so that there are as many paths as the number of resistances, as shown in Fig. 1.14, the circuit is known as a parallel circuit. In parallel circuits, all resistances take up the same potential difference, but draw different currents depending on the ohmic value of resistances.

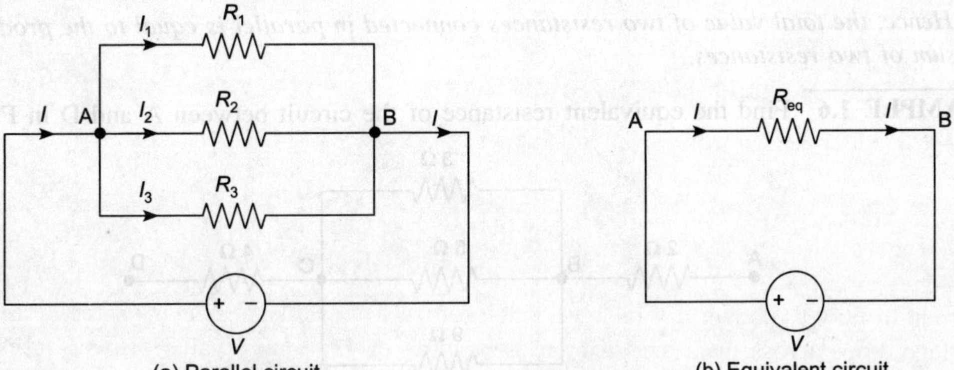

(a) Parallel circuit (b) Equivalent circuit

Fig. 1.14 Resistances in Parallel.

By Ohm's law, current through each resistance is:

$$I_1 = \frac{V}{R_1}; \quad I_2 = \frac{V}{R_2}; \quad I_3 = \frac{V}{R_3}$$

Now,

$$I = I_1 + I_2 + I_3$$

$$= \frac{V}{R_1} + \frac{V}{R_2} + \frac{V}{R_3}$$

$$I = V\left(\frac{1}{R_1} + \frac{1}{R_2} + \frac{1}{R_3}\right)$$

From Fig. 1.14(b), $I = \frac{V}{R_{eq}}$

$$\frac{V}{R_{eq}} = V\left(\frac{1}{R_1} + \frac{1}{R_2} + \frac{1}{R_3}\right)$$

Therefore,

$$\frac{1}{R_{eq}} = \frac{1}{R_1} + \frac{1}{R_2} + \frac{1}{R_3}$$

Hence, when a number of resistances are connected in parallel, the reciprocal of equivalent resistance is equal to the sum of reciprocals of individual resistances.

Special case of parallel circuit

In parallel circuits,

$$\frac{1}{R_{eq}} = \frac{1}{R_1} + \frac{1}{R_2} + \frac{1}{R_3} + \cdots + \frac{1}{R_n}$$

If there are only two resistances connected in parallel, their equivalent resistance will be

$$\frac{1}{R_{eq}} = \frac{1}{R_1} + \frac{1}{R_2} = \frac{R_2 + R_1}{R_1 R_2}$$

Therefore,

$$R_{eq} = \frac{R_1 R_2}{R_1 + R_2}$$

Hence, the total value of two resistances connected in parallel is equal to the product over the sum of two resistances.

EXAMPLE 1.6 Find the equivalent resistance of the circuit between A and D in Fig. 1.15.

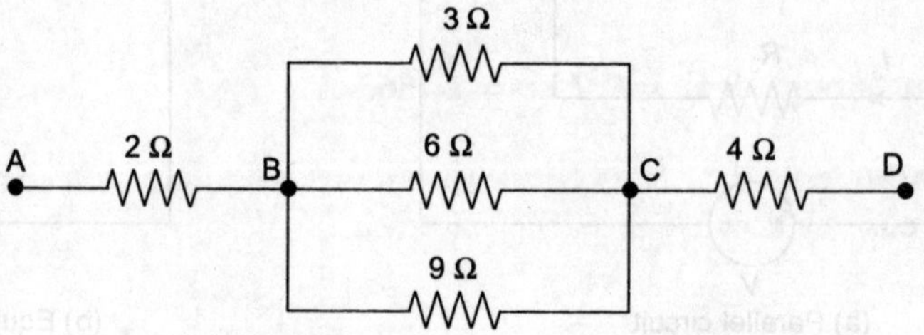

Fig. 1.15

Solution The parallel combination of 3 Ω, 6 Ω and 9 Ω gives,

$$\frac{1}{R}=\frac{1}{3}+\frac{1}{6}+\frac{1}{9}=\frac{11}{18}$$

Hence, $$R=\frac{18}{11}=1.64\ \Omega$$

Now, the circuit is equivalent to three resistances in series as shown in Fig. 1.16.

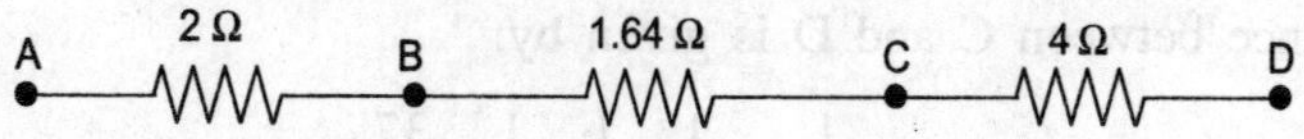

Fig. 1.16

Therefore, the resistance between terminals A and D, R_{AD} = 2 + 1.64 + 4 = **7.64 Ω.**

EXAMPLE 1.7 Three resistors 4 Ω, 6 Ω and 8 Ω are connected in parallel. If the total current taken is 5 A, find the current through each resistor.

Solution

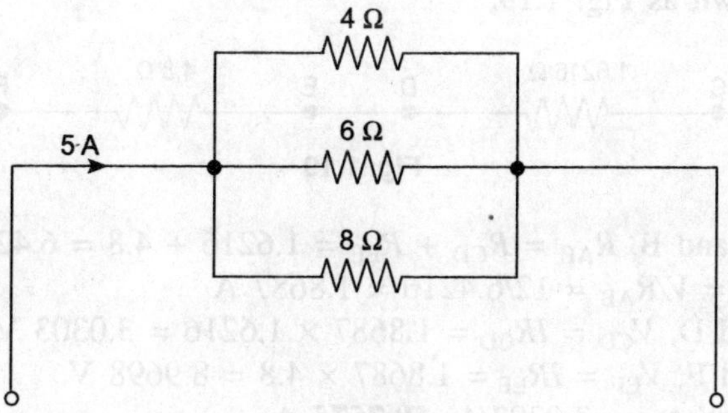

Fig. 1.17

$$\frac{1}{R_{eq}}=\frac{1}{4}+\frac{1}{6}+\frac{1}{8}=\frac{13}{24}$$

Hence, $$R_{eq}=\frac{24}{13}=1.85\ \Omega$$

Voltage across the parallel circuit = IR_{eq} = 5 × 1.85 = 9.25 V

So, Current through 4 Ω = 9.25/4 = **2.3125 A**

Current through 6 Ω = 9.25/6 = **1.5417 A**

Current through 8 Ω = 9.25/8 = **1.1563 A**

EXAMPLE 1.8 A battery of 12 V is connected across terminals AB of the circuit shown in Fig. 1.18. Find the current flowing in each resistance.

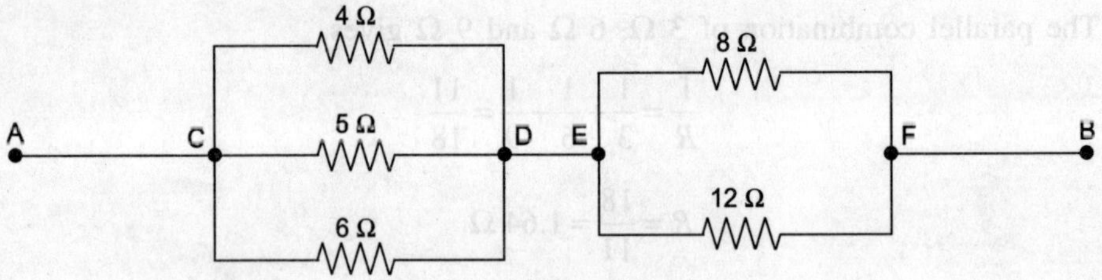

Fig. 1.18

Solution Resistance between C and D is given by:

$$\frac{1}{R_{CD}} = \frac{1}{4} + \frac{1}{5} + \frac{1}{6} = \frac{37}{60}$$

$$R_{CD} = 1.6216\ \Omega$$

Resistance between E and F, $R_{EF} = \frac{8 \times 12}{8 + 12} = 4.8\ \Omega$

The circuit is now drawn as Fig. 1.19,

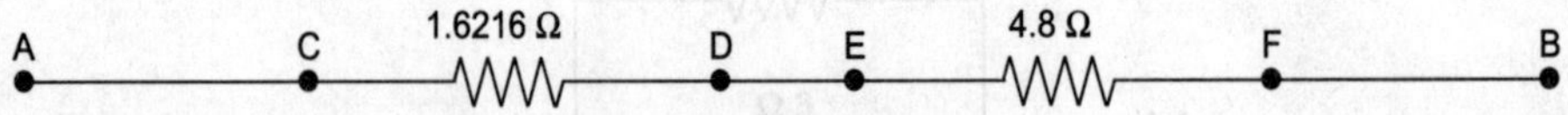

Fig. 1.19

Resistance between A and B, $R_{AB} = R_{CD} + R_{EF} = 1.6216 + 4.8 = 6.4216\ \Omega$
Total circuit current, $I = V/R_{AB} = 12/6.4216 = 1.8687$ A
Voltage between C and D, $V_{CD} = IR_{CD} = 1.8687 \times 1.6216 = 3.0303$ V
Voltage between E and F, $V_{EF} = IR_{EF} = 1.8687 \times 4.8 = 8.9698$ V
Current through 4 Ω resistor = 3.0303/4 = **0.7575 A**
Current through 5 Ω resistor = 3.0303/5 = **0.6061 A**
Current through 6 Ω resistor = 3.0303/6 = **0.5051 A**
Current through 8 Ω resistor = 8.9698/8 = **1.1212 A**
Current through 8 Ω resistor = 8.9698/12 = **0.7475 A**

EXAMPLE 1.9 Find the equivalent resistance between A and B in the circuit shown in Fig. 1.20.

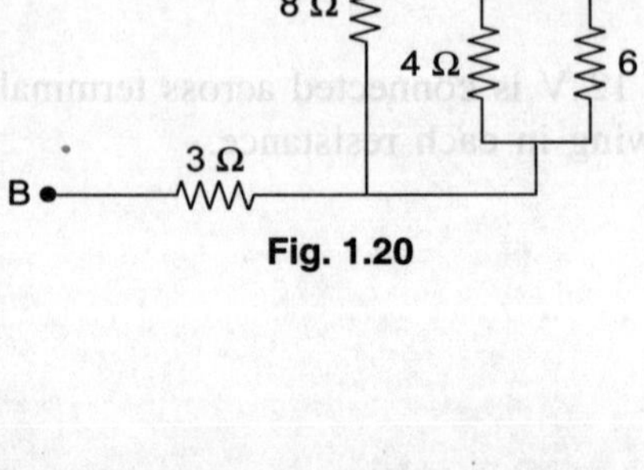

Fig. 1.20

Solution Since 6 Ω and 4 Ω are in parallel, the equivalent resistance is obtained as:

$$(6\|4) = \frac{6 \times 4}{6+4} = 2.4\ \Omega$$

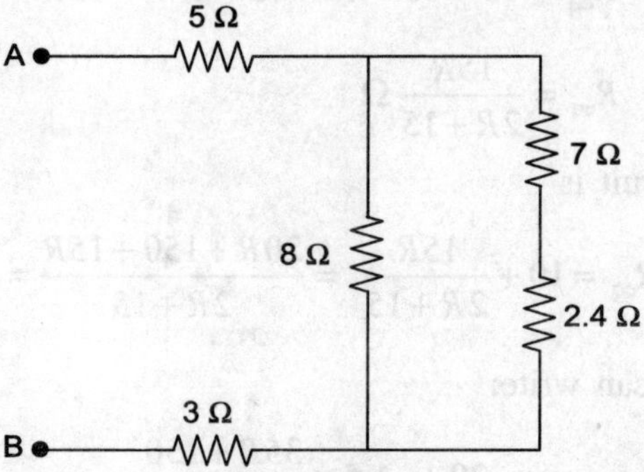

Fig. 1.21

Now, 7 Ω and 2.4 Ω are in series and this series resistance (7 + 2.4 = 9.4) is in parallel with 8 Ω. This gives

$$(9.4\|8) = \frac{9.4 \times 8}{9.4 \times 8} = 4.32\ \Omega$$

Now, the circuit is equivalent to three resistances in series as shown in Fig. 1.22.

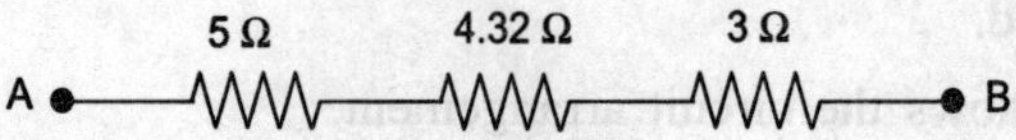

Fig. 1.22

Therefore, the resistance between terminals A and B, R_{AB} = 5 + 4.32 + 3 = **12.32 Ω.**

EXAMPLE 1.10 A resistance of 10 Ω is connected in series with two resistances each of 15 Ω, arranged in parallel. What resistance must be added across this parallel combination so that the total current taken will be 1.5 A with 20 V applied?

Solution

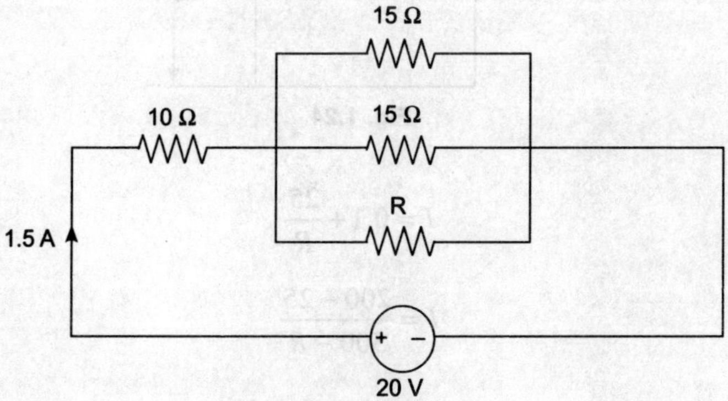

Fig. 1.23

In Fig. 1.23, let R Ω be connected with two 15 Ω resistances. The equivalent resistances of these three parallel resistances are

$$\frac{1}{R_{eq}} = \frac{1}{15} + \frac{1}{15} + \frac{1}{R} = \frac{2}{15} + \frac{1}{R} = \frac{2R+15}{15R}$$

$$R_{eq} = \frac{15R}{2R+15}\ \Omega$$

Total resistance of the circuit is

$$R_{eq} = 10 + R_{eq} = 10 + \frac{15R}{2R+15} = \frac{20R+150+15R}{2R+15} = \frac{35R+150}{2R+15}\ \Omega$$

Applying Ohm's law, we can write:

$$20 = 1.5 \times \frac{35R+150}{2R+15}$$

or
$$\frac{35R+150}{2R+15} = \frac{20}{1.5}$$

or
$$R = \mathbf{6\ \Omega}$$

EXAMPLE 1.11 A voltage of 200 V is applied to a tapped resistor of 500 Ω. Find the resistance between two tapping points connected to a circuit needing 0.1 A at 25 V. Calculate the total power consumed.

Solution Figure 1.24 shows the circuit arrangement.

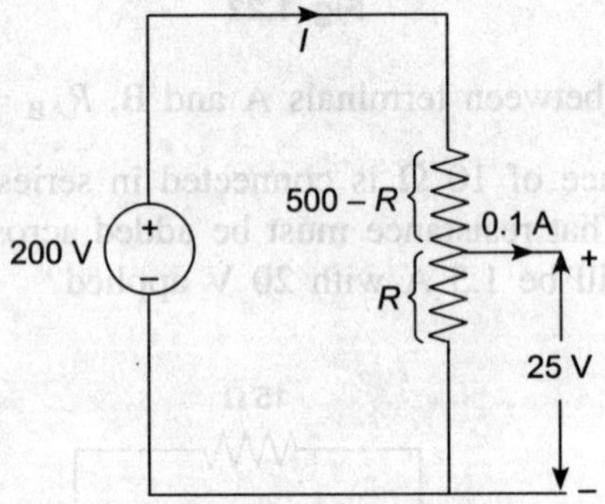

Fig. 1.24

Total current,
$$I = 0.1 + \frac{25}{R} \qquad \text{(i)}$$

Also,
$$I = \frac{200-25}{500-R} \qquad \text{(ii)}$$

Equating Eqs. (i) and (ii), we get

$$0.1 + \frac{25}{R} = \frac{200 - 25}{500 - R}$$

$$\frac{25 + 0.1R}{R} = \frac{175}{500 - R}$$

$$175R = (25 + 0.1R)\,(500 - R)$$

Simplifying, we get

$$0.1R^2 + 150R - 12500 = 0$$

Solving the quadratic equation, we get

$$R = 79.15\ \Omega \text{ or } -1579.15\ \Omega$$

Taking only the positive value, $R = 79.15\ \Omega$

Therefore, from Eq. (i), the current $I = 0.1 + \dfrac{25}{R}$

$$= 0.1 + \frac{25}{79.15} = 0.4159\text{ A.}$$

The total power consumed = 200 × 0.4159 = **83.18 W**

EXAMPLE 1.12 Find the equivalent resistance across A and B in the circuit shown in Fig. 1.25.

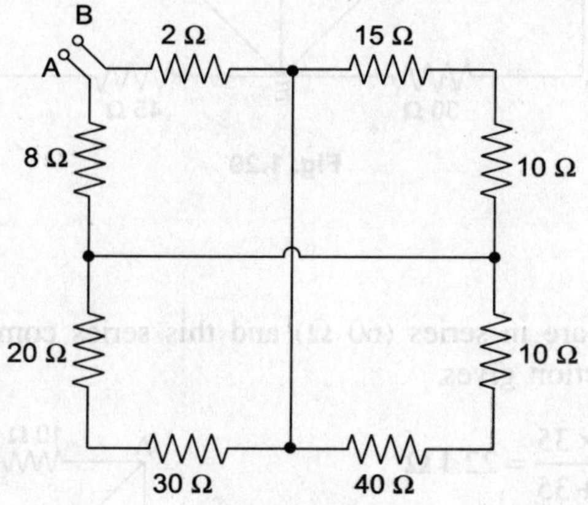

Fig. 1.25

Solution The two short-circuits in Fig. 1.25 are labelled C and D as shown in Fig. 1.26. The circuit is redrawn as shown in Fig. 1.27. Further simplification gives the circuit as shown in Fig. 1.28.

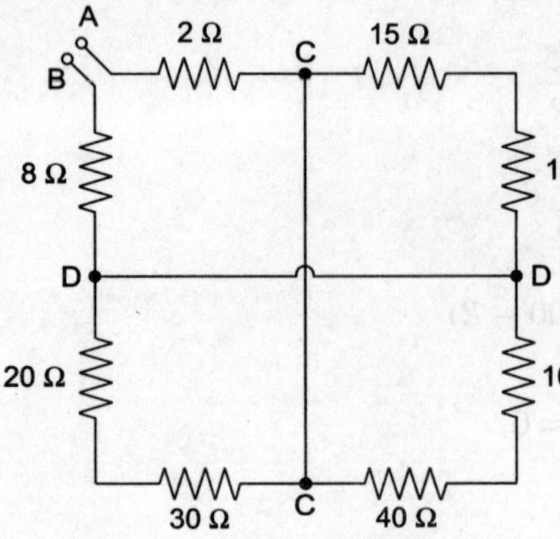

Fig. 1.26

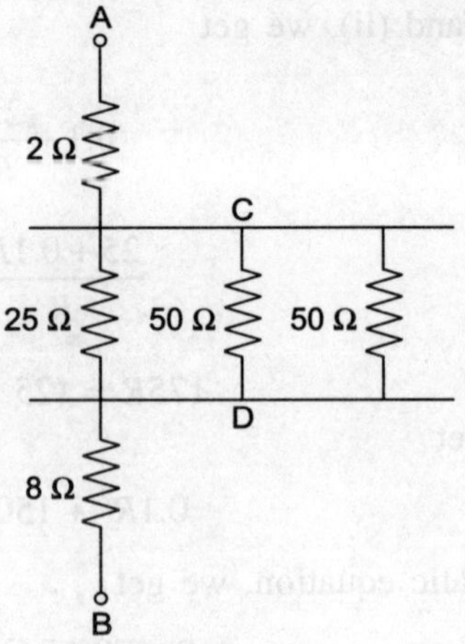

Fig. 1.27

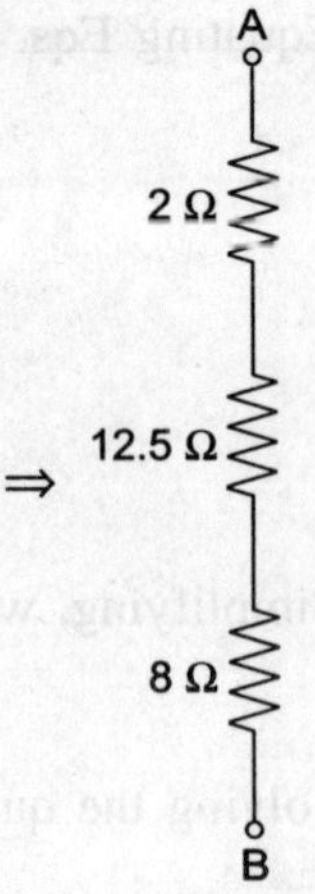

Fig. 1.28

From Fig. 1.28, resistance across A and B, $R_{AB} = 2 + 12.5 + 8 = \mathbf{22.5\ \Omega}$

EXAMPLE 1.13 Simplify the network given in Fig. 1.29 and find (i) R_{AD} (ii) R_{BE} and (iii) R_{CF}.

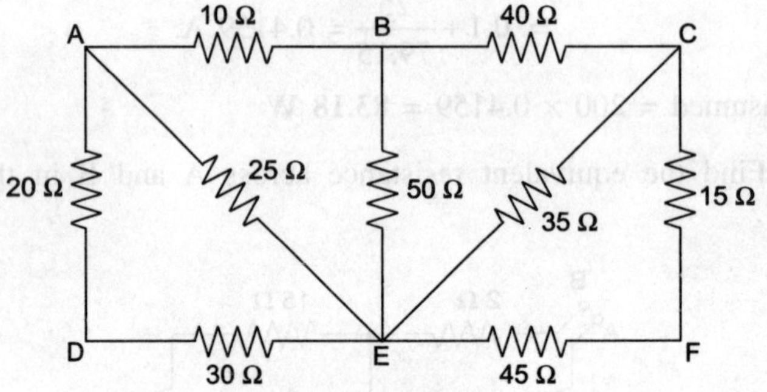

Fig. 1.29

Solution

(i) **To find R_{AD}**

15 Ω and 45 Ω are in series (60 Ω) and this series combination is in parallel with 35 Ω. This reduction gives

$$(60\|35) = \frac{60 \times 35}{60 + 35} = 22.1\ \Omega$$

as illustrated in Fig. 1.30.

22.1 Ω and 40 Ω are in series (62.1 Ω) and this series combination is in parallel with 50 Ω.

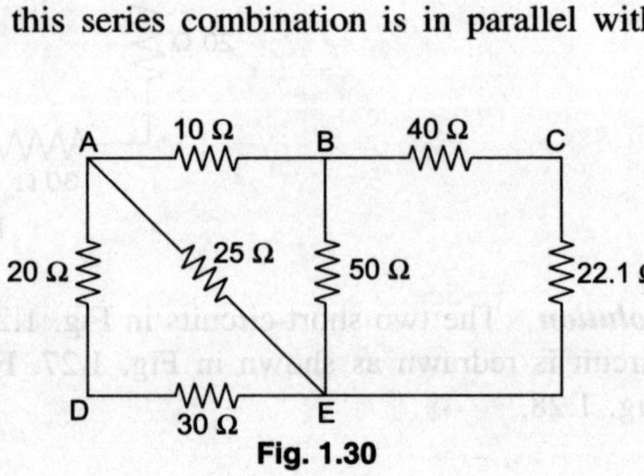

Fig. 1.30

This reduction gives

$$(62.1\|50) = \frac{62.1 \times 50}{62.1 + 50} = 27.7\ \Omega$$

as illustrated in Fig. 1.31.

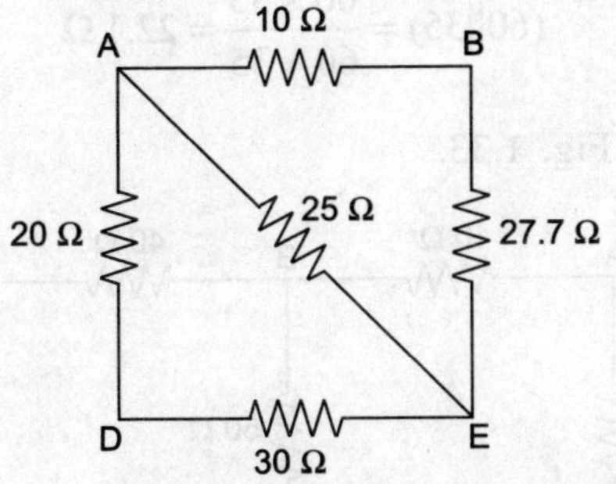

Fig. 1.31

27.7 Ω and 10 Ω are in series (37.7 Ω) and this series combination is in parallel with 25 Ω.

This reduction yields

$$(37.7\|25) = \frac{37.7 \times 25}{37.7 + 25} = 15.03\ \Omega$$

as illustrated in Fig. 1.32.

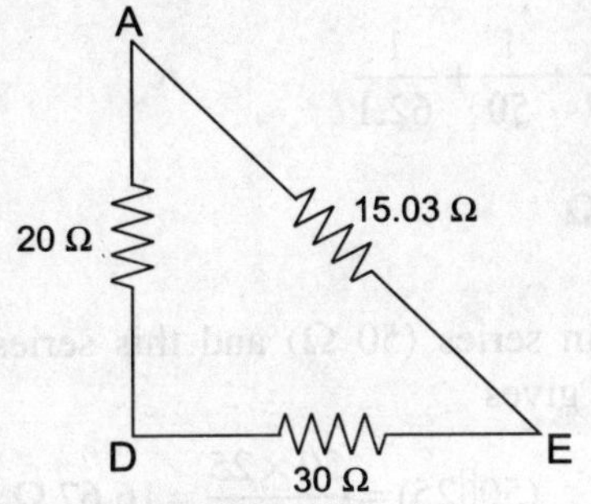

Fig. 1.32

Now 15.03 Ω and 30 Ω are in series (45.03 Ω) and this series combination is in parallel with 20 Ω.

It is finally reduced to $(45.03\|20) = \dfrac{45.03 \times 20}{45.03 + 20} = 13.85\ \Omega$

Hence, $R_{AD} = \mathbf{13.85\ \Omega}$

(ii) **To find R_{BE}**

20 Ω and 30 Ω are in series (50 Ω) and this series combination is in parallel with 25 Ω. This reduction gives

$$(50||25) = \frac{50 \times 25}{50 + 25} = 16.67\ \Omega$$

Similarly, 15 Ω and 45 Ω are in series (60 Ω) and this series combination is in parallel with 35 Ω. This simplification gives

$$(60||35) = \frac{60 \times 35}{60 + 35} = 22.1\ \Omega$$

This is illustrated in Fig. 1.33.

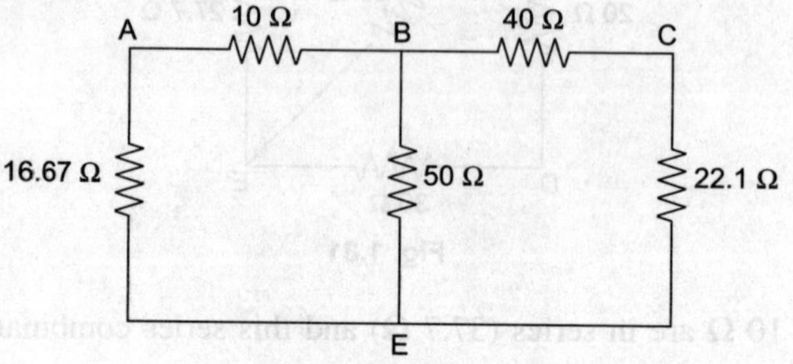

Fig. 1.33

10 Ω and 16.67 Ω are in series (26.67 Ω) and 40 Ω and 22.1 Ω are in series (62.1 Ω). The circuit is reduced to as shown in Fig. 1.34.

With reference to Fig. 1.34,

$$\frac{1}{R_{BE}} = \frac{1}{26.67} + \frac{1}{50} + \frac{1}{62.1}$$

Hence, R_{BE} = **13.58 Ω**

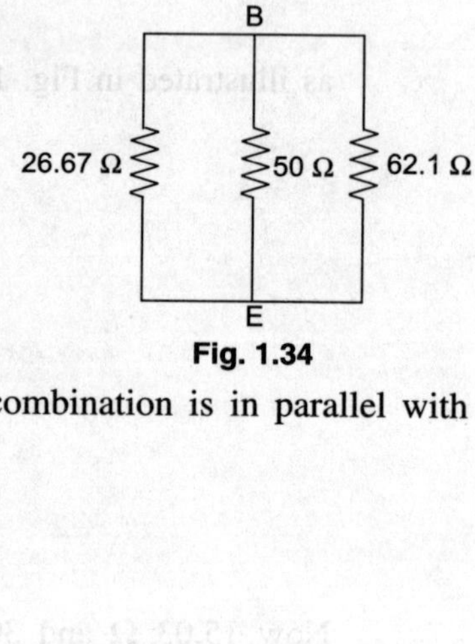

Fig. 1.34

(iii) **To find R_{CF}**

20 Ω and 30 Ω are in series (50 Ω) and this series combination is in parallel with 25 Ω. This reduction gives

$$(50||25) = \frac{50 \times 25}{50 + 25} = 16.67\ \Omega$$

This is illustrated in Fig. 1.35.

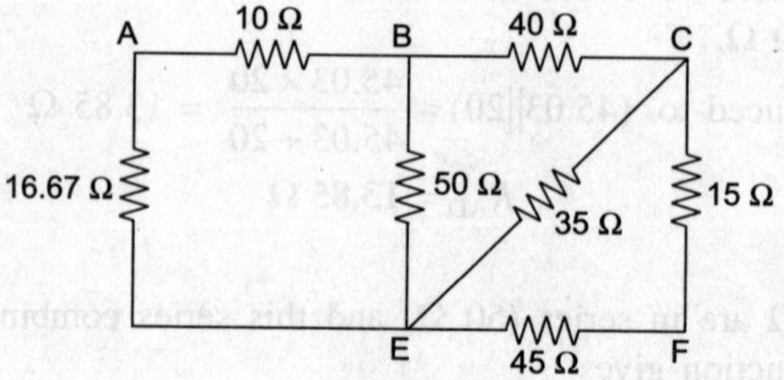

Fig. 1.35

16.67 Ω and 10 Ω are in series (26.67 Ω) and this series combination is in parallel with 50 Ω. This reduction gives

$$(26.67 \| 50) = \frac{26.67 \times 50}{26.67 + 50} = 17.39\ \Omega$$

This is illustrated in Fig. 1.36.

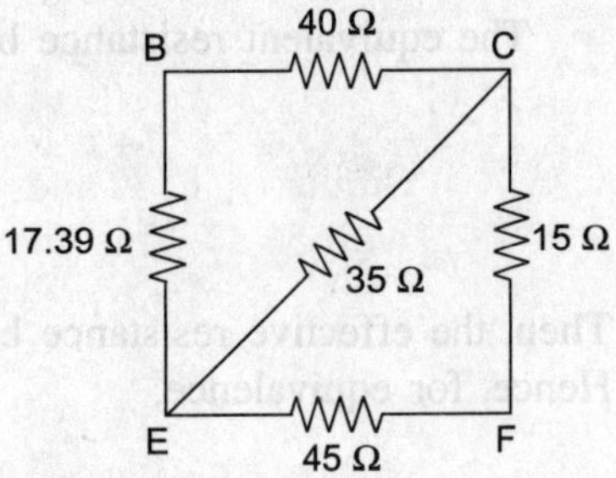

Fig. 1.36

17.39 Ω and 40 Ω are in series (57.39 Ω) and this series combination is in parallel with 35 Ω. This reduction gives

$$(57.39 \| 35) = \frac{57.39 \times 35}{57.39 + 35} = 21.74\ \Omega$$

as illustrated in Fig. 1.37.

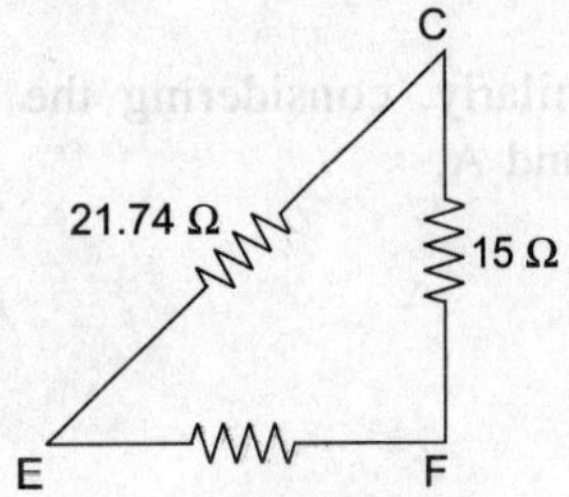

Fig. 1.37

21.74 Ω and 45 Ω are in series as, 21.74 + 45 = 66.74 Ω

$$(66.74 \| 15) = \frac{66.74 \times 15}{66.74 + 15} = 12.25\ \Omega$$

Therefore, R_{CF} = **12.25 Ω**

1.19 STAR–DELTA TRANSFORMATION

In some networks, resistances are interconnected to form complex networks and they are neither in series nor in parallel to solve easily. Series and parallel circuits are no longer useful under such conditions. Generally, two types of interconnections create such difficulties. They are *delta* (Δ connection) connection and *star* (Y connection) connection as shown in Figs. 1.38 and 1.39, respectively. The transformation of a *delta* into an equivalent *star* or vice-versa makes the circuit easily solved by applying further series–parallel circuit techniques.

1.19.1 Delta–Star Transformation

Consider three resistors R_{AB}, R_{BC} and R_{CA} connected in Δ to three terminals A, B and C as shown in Fig. 1.38. Let R_A, R_B and R_C be their Y equivalent original resistances (Fig. 1.39).

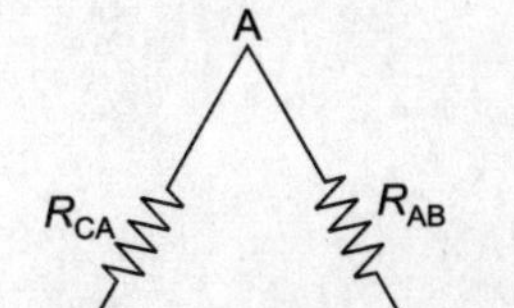

Fig. 1.38 Delta connection.

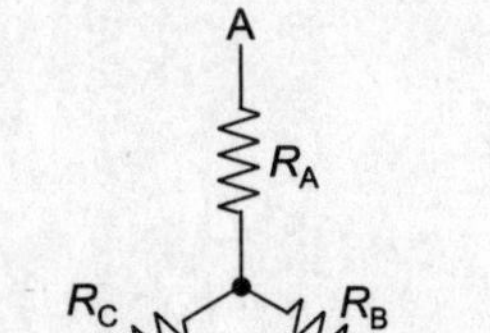

Fig. 1.39 Star connection.

The equivalent resistance between A and B in the Δ connection

$$= \frac{R_{AB}(R_{BC} + R_{CA})}{R_{AB} + (R_{BC} + R_{CA})}$$

Then, the effective resistance between A and B in the Y connection = $R_A + R_B$

Hence, for equivalence,

$$R_A + R_B = \frac{R_{AB}(R_{BC} + R_{CA})}{R_{AB} + R_{BC} + R_{CA}} \tag{1.5}$$

Similarly, considering the effective resistance between terminals B and C and terminals C and A,

$$R_B + R_C = \frac{R_{BC}(R_{CA} + R_{AB})}{R_{AB} + R_{BC} + R_{CA}} \tag{1.6}$$

and

$$R_C + R_A = \frac{R_{CA}(R_{AB} + R_{BC})}{R_{AB} + R_{BC} + R_{CA}} \tag{1.7}$$

Subtracting Eq. (1.6) from Eq. (1.7), we get

$$(R_C + R_A) - (R_B + R_C) = \frac{R_{CA}(R_{AB} + R_{BC}) - R_{BC}(R_{CA} + R_{AB})}{R_{AB} + R_{BC} + R_{CA}}$$

$$R_A - R_B = \frac{R_{AB}(R_{CA} - R_{BC})}{R_{AB} + R_{BC} + R_{CA}} \tag{1.8}$$

Adding Eqs. (1.5) and (1.8), we get

$$2R_A = \frac{2R_{AB} \cdot R_{CA}}{R_{AB} + R_{BC} + R_{CA}}$$

$$R_A = \frac{R_{AB} \cdot R_{CA}}{R_{AB} + R_{BC} + R_{CA}} \tag{1.9}$$

Similarly, R_B and R_C are obtained as:

$$R_B = \frac{R_{BC} \cdot R_{AB}}{R_{AB} + R_{BC} + R_{CA}} \tag{1.10}$$

and

$$R_C = \frac{R_{CA} \cdot R_{BC}}{R_{AB} + R_{BC} + R_{CA}} \tag{1.11}$$

This is required Δ–Y transformation.

1.19.2 Star–Delta Transformation

To convert a star into its equivalent delta, Eqs. (1.8, 1.9 and 1.10) developed in Δ to Y are required.

Taking products of Eqs. (1.9) and (1.10), we get

$$R_A R_B = \frac{R_{AB}^2 \cdot R_{CA} \cdot R_{BC}}{(R_{AB} + R_{BC} + R_{CA})^2} \tag{1.12}$$

Taking products of Eqs. (1.10) and (1.11), we get

$$R_B R_C = \frac{R_{BC}^2 \cdot R_{CA} \cdot R_{AB}}{(R_{AB} + R_{BC} + R_{CA})^2} \tag{1.13}$$

Taking products of Eqs. (1.11) and (1.9), we get

$$R_C R_A = \frac{R_{CA}^2 \cdot R_{AB} \cdot R_{BC}}{(R_{AB} + R_{BC} + R_{CA})^2} \tag{1.14}$$

Adding Eqs. (1.12), (1.13) and (1.14), we get

$$R_A R_B + R_B R_C + R_C R_A = \frac{R_{AB}^2 \cdot R_{CA} \cdot R_{BC} + R_{BC}^2 \cdot R_{CA} \cdot R_{AB} + R_{CA}^2 \cdot R_{AB} \cdot R_{BC}}{(R_{AB} + R_{BC} + R_{CA})^2}$$

$$= \frac{R_{AB} \cdot R_{CA} \cdot R_{BC}(R_{AB} + R_{BC} + R_{CA})}{(R_{AB} + R_{BC} + R_{CA})^2}$$

$$= \frac{R_{AB} \cdot R_{CA} \cdot R_{BC}}{R_{AB} + R_{BC} + R_{CA}} \tag{1.15}$$

Substituting $R_A = \dfrac{R_{AB} \cdot R_{CA}}{R_{AB} + R_{BC} + R_{CA}}$ from Eq. (1.9) into Eq. (1.15), we get

$$R_A R_B + R_B R_C + R_C R_A = R_A \cdot R_{BC}$$

Hence,

$$R_{BC} = \frac{R_A R_B + R_B R_C + R_C R_A}{R_A} \tag{1.16}$$

Similarly,

$$R_{CA} = \frac{R_A R_B + R_B R_C + R_C R_A}{R_B} \tag{1.17}$$

and

$$R_{AB} = \frac{R_A R_B + R_B R_C + R_C R_A}{R_C} \tag{1.18}$$

This is required Y–Δ transformation.

EXAMPLE 1.14 Find R_{AB} in Fig. 1.40 using star–delta transformation.

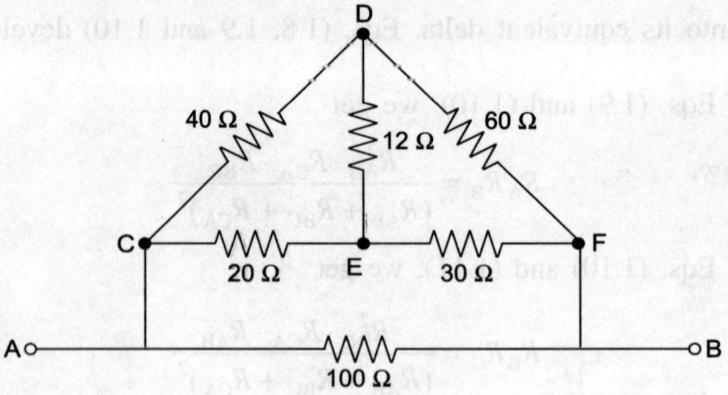

Figure 1.40

Solution The loop CDE forms one delta (Fig. 1.41) and its equivalent star connection (Fig. 1.42) is obtained as:

$$R_{CN} = \frac{40 \times 20}{40 + 12 + 20} = 11.11\ \Omega$$

$$R_{DN} = \frac{40 \times 12}{40 + 12 + 20} = 6.66\ \Omega$$

$$R_{EN} = \frac{12 \times 20}{40 + 12 + 20} = 3.33\ \Omega$$

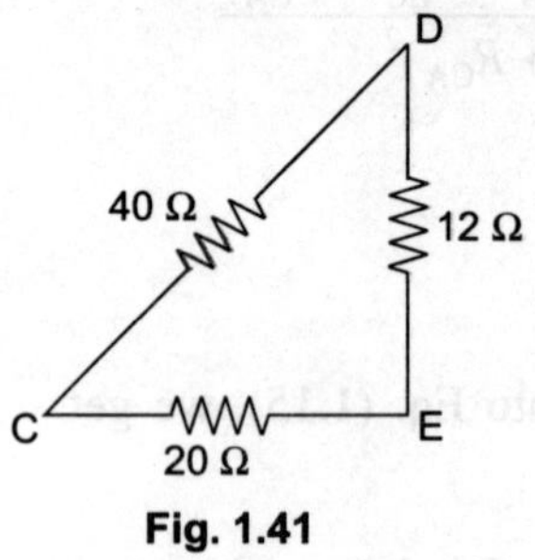

Fig. 1.41

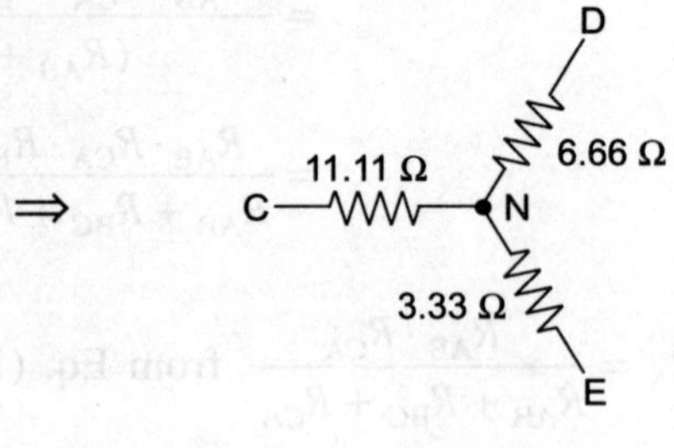

Fig. 1.42

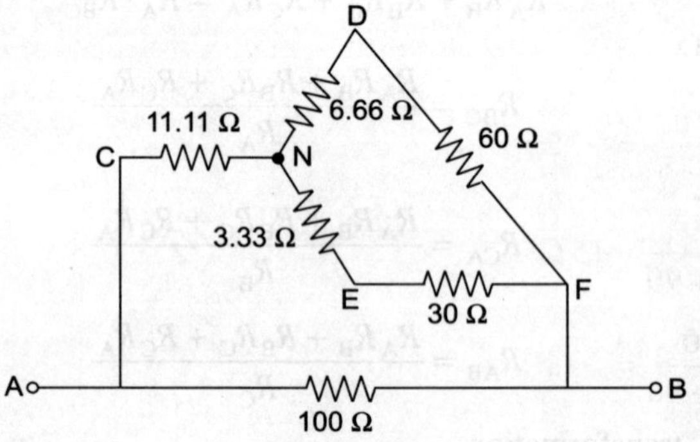

Fig. 1.43

Replacing the CDE delta by its equivalent star, the circuit is redrawn as in Fig. 1.43. Resistances 6.66 Ω and 60 Ω are in series as well as 3.33 Ω and 30 Ω are in series. These two series combinations form one parallel connection. Simplifying,

$$R_{NF} = \frac{66.66 \times 33.33}{66.66 + 33.33} = 22.22\ \Omega$$

The circuit is redrawn in Fig. 1.44.

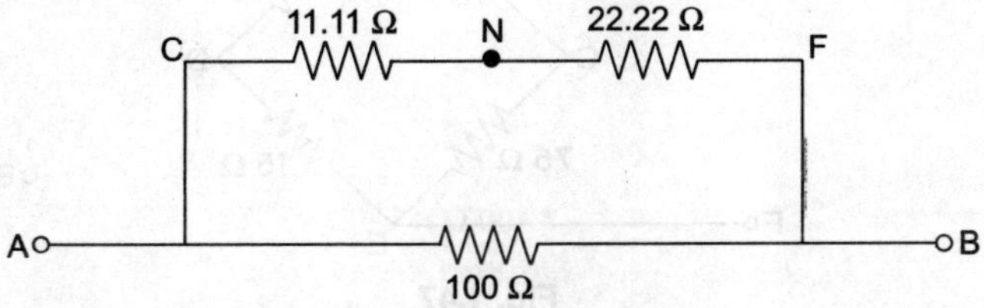

Fig. 1.44

Now, 11.11 Ω and 22.22 Ω are in series. The equivalent value is 33.33 Ω, and is in parallel with 100 Ω.

$$\therefore \qquad R_{AB} = \frac{33.33 \times 100}{33.33 + 100} = \mathbf{25\ \Omega}$$

EXAMPLE 1.15 Find R_{AF} in Fig. 1.45, using star–delta transformation technique.

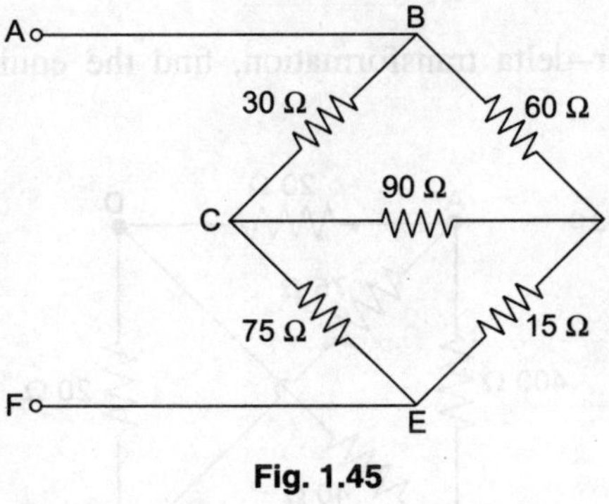

Fig. 1.45

Solution Delta BCD is converted into star (Fig. 1.46) and its values are obtained as:

$$R_{BN} = \frac{30 \times 60}{30 + 60 + 90} = 10\ \Omega$$

$$R_{CN} = \frac{30 \times 90}{30 + 60 + 90} = 15\ \Omega$$

$$R_{DN} = \frac{60 \times 90}{30 + 60 + 90} = 30\ \Omega$$

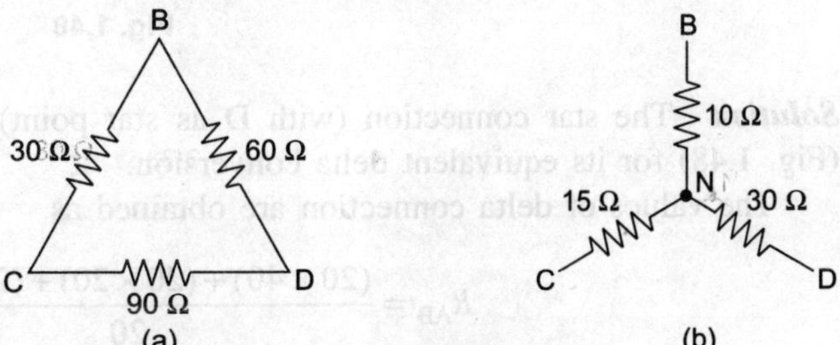

Fig. 1.46

Replacing the BCD delta by its equivalent star, the circuit is redrawn as shown in Fig. 1.47.

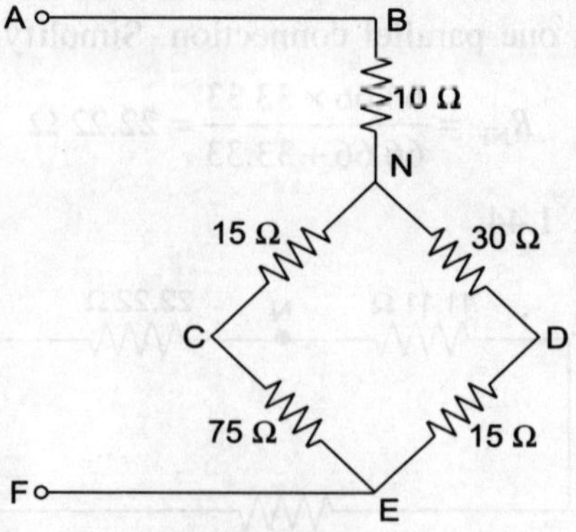

Fig. 1.47

15 Ω (branch CN) and 75 Ω are in series as well as 30 Ω and 15 Ω (branch DE) are in series. These two series combinations form one parallel connection. Simplifying, we get

$$R_{NE} = \frac{90 \times 45}{90 + 45} = 30\ \Omega$$

Therefore, the resistance between R and F is,

$$R_{AF} = 30 + 10 = \mathbf{40\ \Omega}$$

EXAMPLE 1.16 Using star–delta transformation, find the equivalent resistance between J and K.

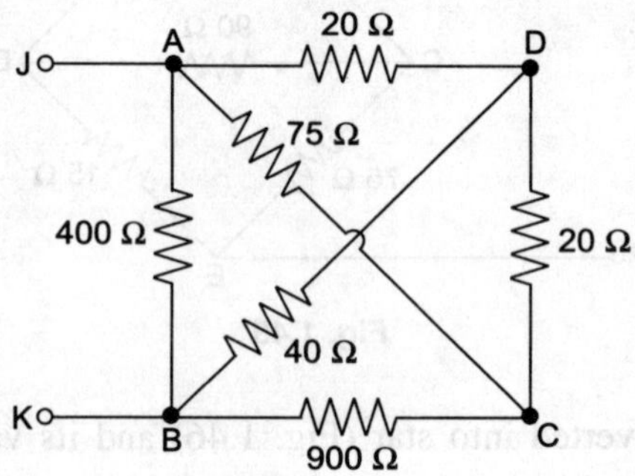

Fig. 1.48

Solution The star connection (with D as star point) is removed from the original network (Fig. 1.48) for its equivalent delta conversion.

The values of delta connection are obtained as

$$R_{AB} = \frac{(20 \times 40) + (20 \times 20) + (20 \times 40)}{20} = 100\ \Omega$$

$$R_{BC} = \frac{(20\times 40)+(20\times 20)+(20\times 40)}{20} = 100\ \Omega$$

$$R_{AC} = \frac{(20\times 40)+(20\times 20)+(20\times 40)}{40} = 50\ \Omega$$

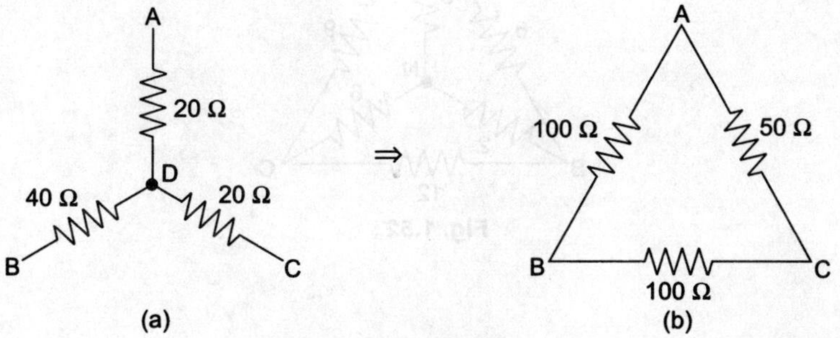

Fig. 1.49

The circuit in Fig. 1.48 is redrawn by fitting the equivalent delta into the original network.

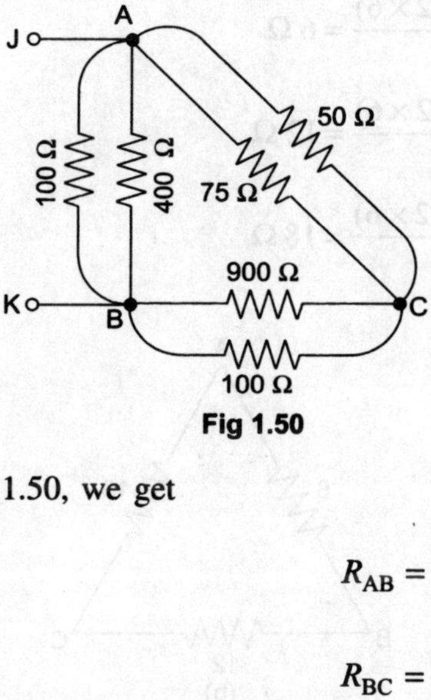

Fig 1.50

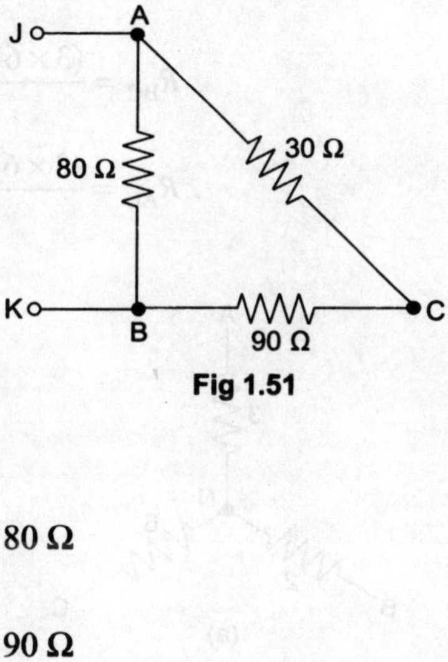

Fig 1.51

From Fig. 1.50, we get

$$R_{AB} = \frac{100\times 400}{100+400} = 80\ \Omega$$

$$R_{BC} = \frac{100\times 900}{100+900} = 90\ \Omega$$

$$R_{AC} = \frac{50\times 75}{50+75} = 30\ \Omega$$

90 Ω and 30 Ω are in series and this series combination is in parallel with 80 Ω (Fig. 1.51).

Therefore, $$R_{JK} = \frac{80\times(90+30)}{80+90+30} = \mathbf{48\ \Omega}$$

EXAMPLE 1.17 In the circuit of Fig. 1.52, find (i) R_{AB}, R_{BC} and R_{AC} and (ii) R_{AN}. All the values are in ohms.

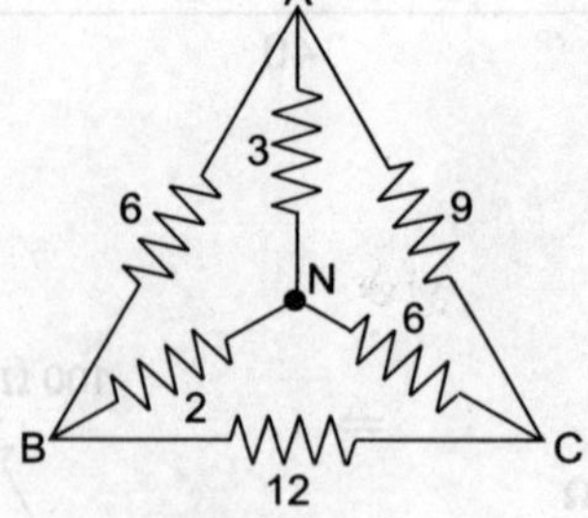

Fig. 1.52

Solution

(i) **To find R_{AB}, R_{BC} and R_{AC}**

The star connection (ABC with N as star point as shown in Fig. 1.53(a)) is removed from the original network and its equivalent delta values are obtained as:

$$R_{AB} = \frac{(3\times 6)+(3\times 2)+(2\times 6)}{6} = 6\ \Omega$$

$$R_{BC} = \frac{(3\times 6)+(3\times 2)+(2\times 6)}{3} = 12\ \Omega$$

$$R_{AC} = \frac{(3\times 6)+(3\times 2)+(2\times 6)}{2} = 18\ \Omega$$

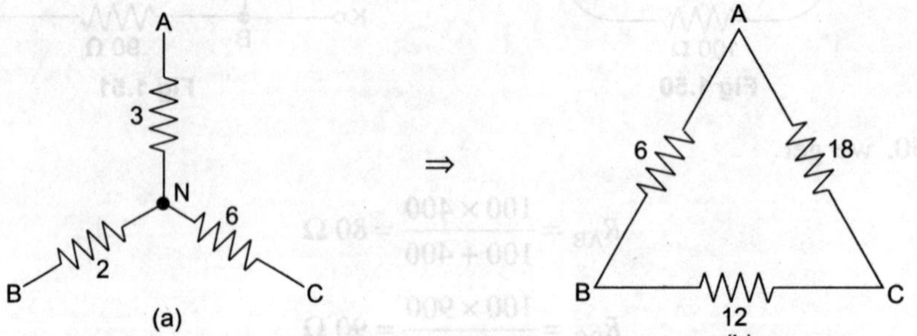

Fig. 1.53

The circuit in Fig. 1.52 is redrawn by fitting the delta equivalent obtained as shown in Fig. 1.54.

6 Ω and 6 Ω are in parallel in section AB. Similarly 12 Ω and 12 Ω are in parallel in section BC and 9 Ω and 18 Ω are in parallel in section AC.

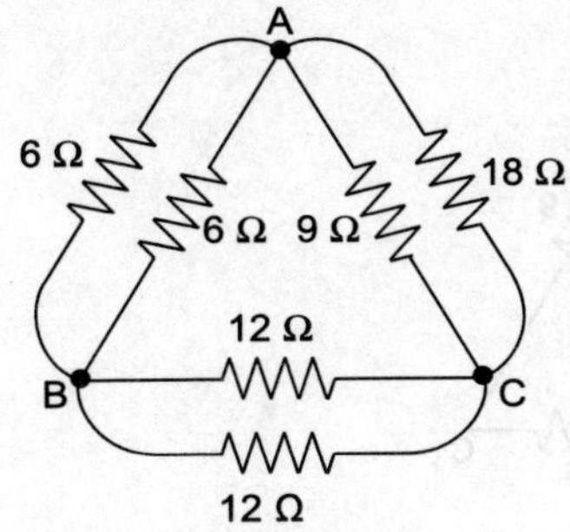

Fig. 1.54

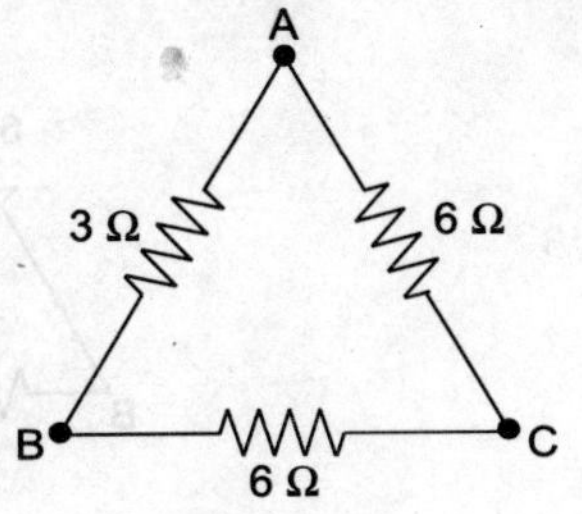

Fig. 1.55

$$R_{AB} = (3||12) = \frac{3 \times 12}{3+12} = \mathbf{2.4\ \Omega}$$

$$R_{BC} = (6||9) = \frac{6 \times 9}{6+9} = \mathbf{3.6\ \Omega}$$

$$R_{AC} = (6||9) = \frac{6 \times 9}{6+9} = \mathbf{3.6\ \Omega}$$

(ii) **To find R_{AN}**

To find the resistance R_{AN}, delta BNC comprising 2 Ω, 6 Ω and 12 Ω is converted into star as follows:

$$R_{NS} = \frac{2 \times 6}{2+6+12} = 0.6\ \Omega$$

$$R_{BS} = \frac{2 \times 12}{2+6+12} = 1.2\ \Omega$$

$$R_{CS} = \frac{6 \times 12}{2+6+12} = 3.6\ \Omega$$

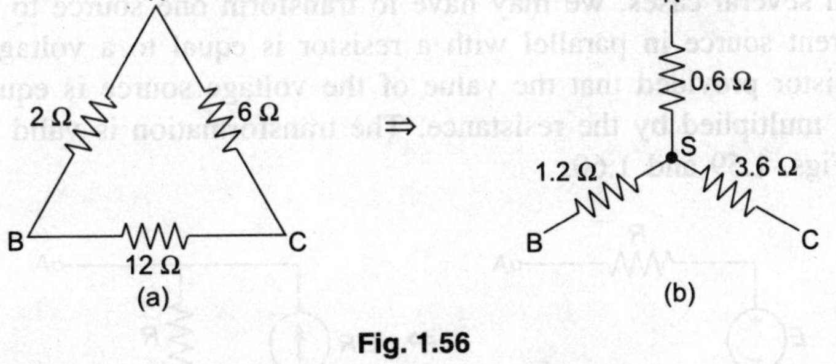

Fig. 1.56

The circuit in Fig. 1.52 is redrawn by fitting the equivalent star for the removed delta. This is illustrated in Fig. 1.57.

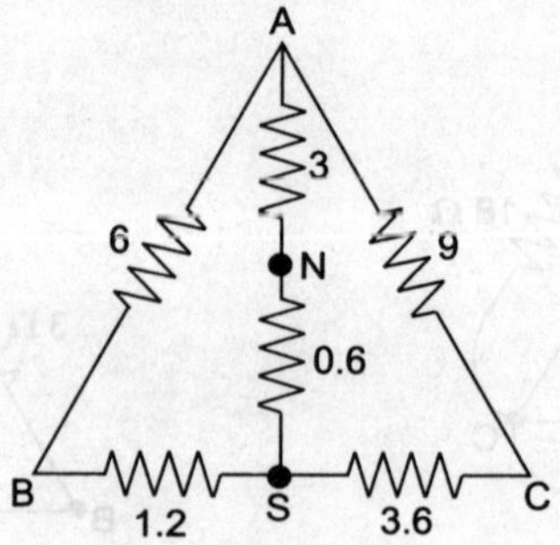

Fig. 1.57

The step-by-step reduction of network in Fig. 1.57 is shown in Fig. 1.58.

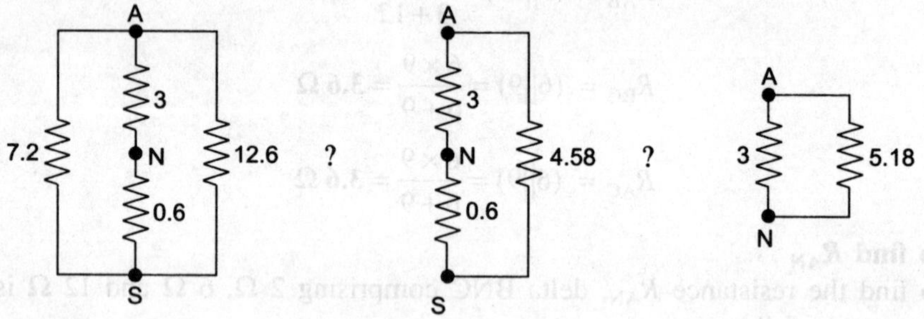

Fig. 1.58

From Fig. 1.57(c), we get

$$R_{AN} = (3\|5.18) = \frac{3 \times 5.18}{3 + 5.18} = \mathbf{1.9\ \Omega}$$

1.20 CONVERSION OF SOURCES

It can be seen that in some networks, it is easier if the circuit has only voltage sources or current sources. In several cases, we may have to transform one source to another. It can be proved that a current source in parallel with a resistor is equal to a voltage source in series with the same resistor provided that the value of the voltage source is equal to the value of the current source multiplied by the resistance. The transformation is valid in either direction and is shown in Figs. 1.59 and 1.60.

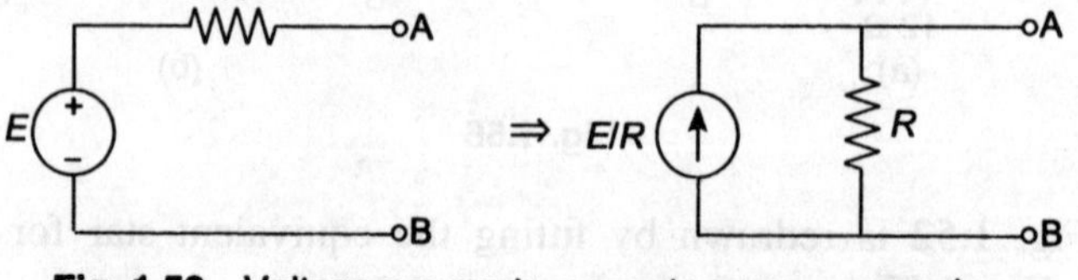

Fig. 1.59 Voltage source-to-current source conversion.

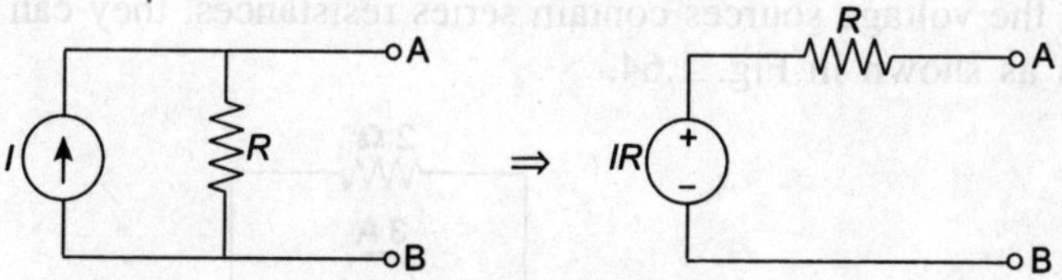

Fig. 1.60 Current source-to-voltage source conversion.

EXAMPLE 1.18 Reduce the network shown in Fig. 1.61 into a single source.

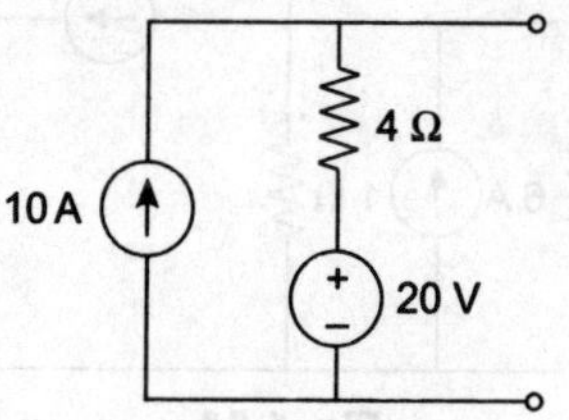

Fig. 1.61

Solution The reduction procedure is shown in Figs. 1.62(a) and (b). The 20 V voltage source is first converted into a current source. Then, the two current sources are combined to get a single current source.

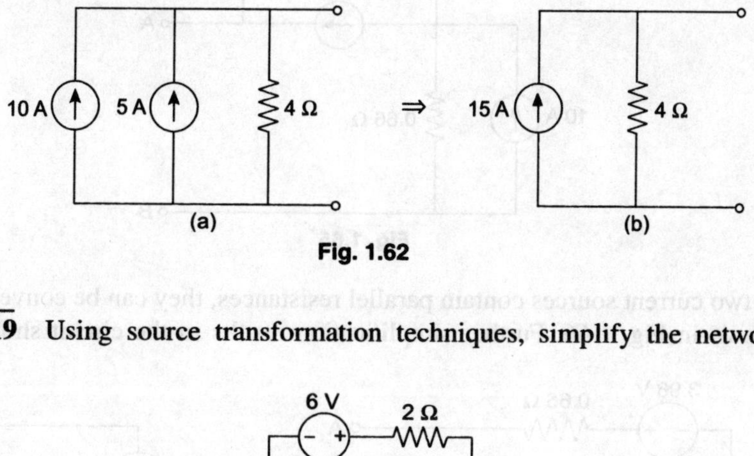

Fig. 1.62

EXAMPLE 1.19 Using source transformation techniques, simplify the network shown in Fig. 1.63.

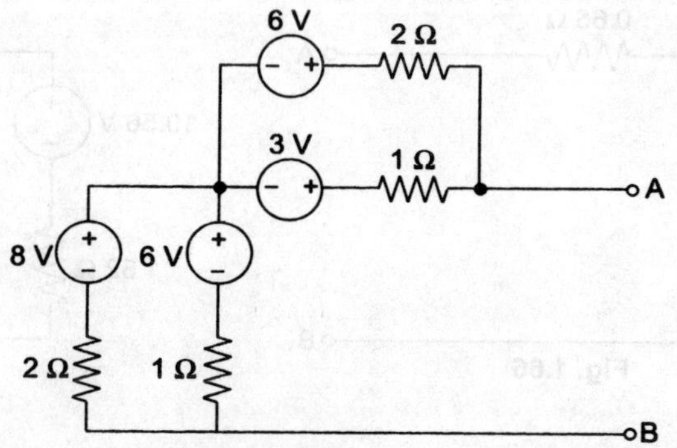

Fig. 1.63

Solution Since all the voltage sources contain series resistances, they can be easily converted into current sources as shown in Fig. 1.64.

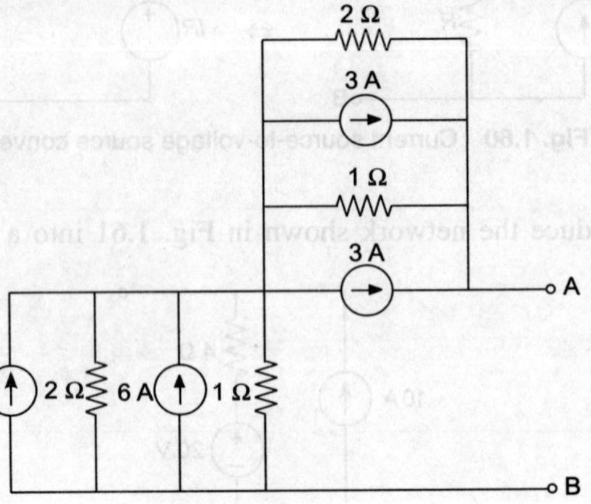

Fig. 1.64

Further simplification results in the circuit shown in Fig. 1.65.

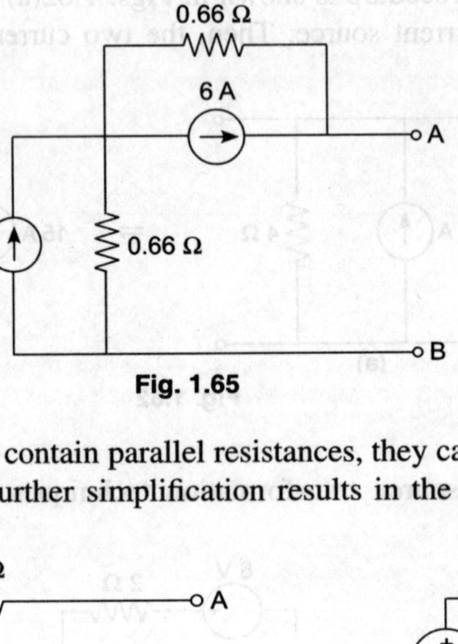

Fig. 1.65

Since the two current sources contain parallel resistances, they can be converted into voltage sources as shown in Fig. 1.66. Further simplification results in the circuit shown in Fig. 1.67.

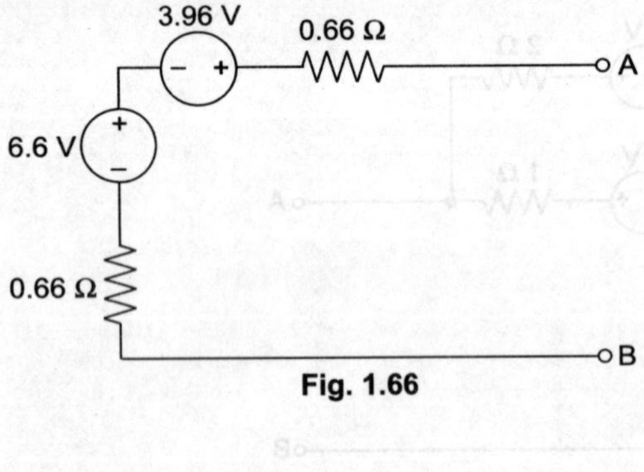

Fig. 1.66

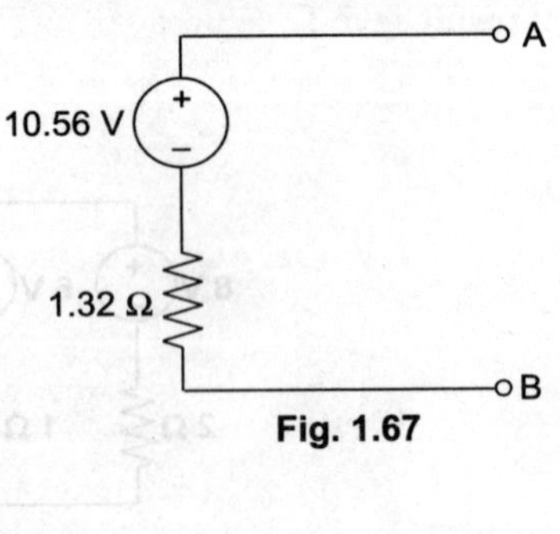

Fig. 1.67

EXAMPLE 1.20 Use source transformation to find I_0 in the circuit shown in Fig. 1.68.

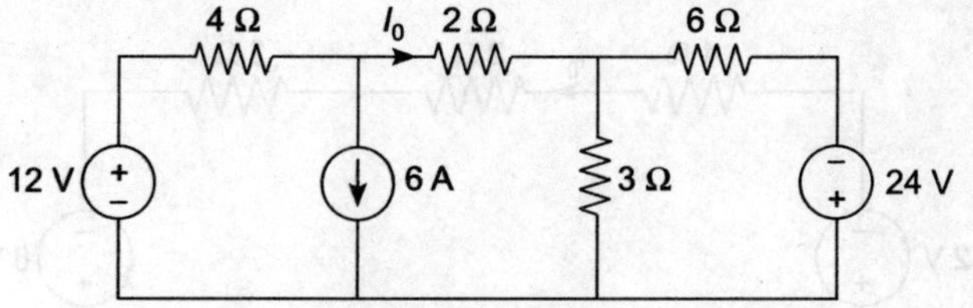

Fig. 1.68

Solution

Step 1 The two voltage sources are converted into current sources as shown in Fig. 1.69.

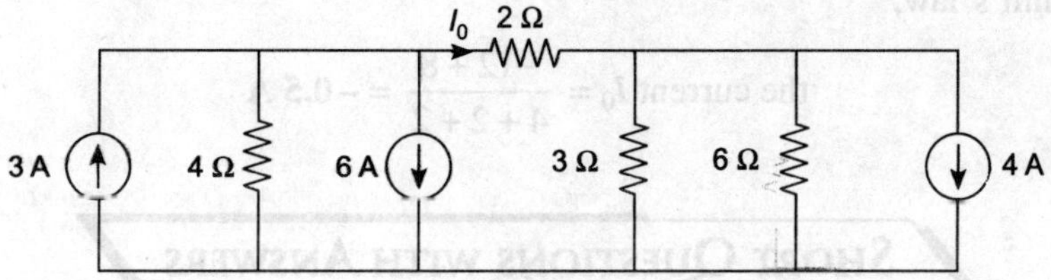

Fig. 1.69

Step 2 Current sources 3 A and 6 A are combined and made as a single current source of 3 A (Fig. 1.70).

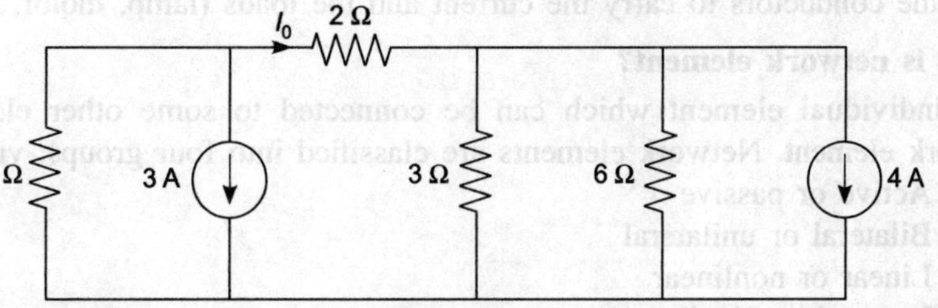

Fig. 1.70

Step 3 3 Ω and 6 Ω resistances are found in parallel; hence, they are simplified and the result is 2 Ω as shown in Fig. 1.71.

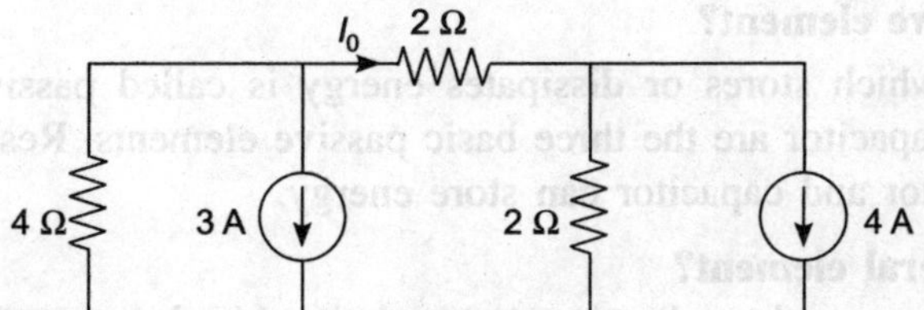

Fig. 1.71

Step 4 Both 3 A and 4 A current sources are converted into voltage sources as shown in Fig. 1.72.

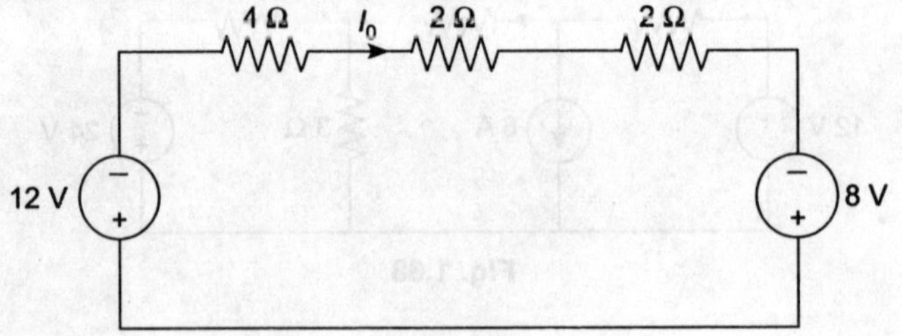

Fig. 1.72

Step 5 By Ohm's law,

$$\text{the current } I_0 = \frac{-12+8}{4+2+2} = \mathbf{-0.5\ A}$$

SHORT QUESTIONS WITH ANSWERS

1. **Define network.**

   A network or circuit is a closed path consists of source of power (battery, generator, etc.), the conductors to carry the current and the loads (lamp, motor, fan, etc.)

2. **What is network element?**

   Any individual element which can be connected to some other element is called network element. Network elements are classified into four groups, viz.
   (i) Active or passive
   (ii) Bilateral or unilateral
   (iii) Linear or nonlinear
   (iv) Lumped or distributed

3. **What is an active element?**

   Active element is the element which supplies power or energy to the network. Examples of active elements are voltage source and current source.

4. **What is passive element?**

   The element which stores or dissipates energy is called passive elements. Resistor, inductor and capacitor are the three basic passive elements. Resistor dissipates energy whereas inductor and capacitor can store energy.

5. **What is bilateral element?**

   In bilateral elements the voltage–current relationship does not change with the change in current direction. Resistor is an example of bilateral element.

6. What is a unilateral element?

In unilateral elements the voltage–current relationship changes with the change in current direction. Examples of unilateral elements are junction diodes, transistors and others.

7. Define linear element.

Elements which obey the principle of superposition are called linear elements (e.g. resistor).

8. Define nonlinear element.

Elements which do not obey the principle of superposition are called nonlinear elements (e.g. junction diode, transistor).

9. What is lumped element?

Elements which can be separated are known as lumped elements (e.g. resistor, inductor and capacitor).

10. What is distributed element?

Elements which cannot be separated are known as distributed elements. For example, in transmission lines, resistors, inductors and capacitors are distributed throughout the length.

11. Define node.

In a circuit the meeting point of two or more elements is called a node. If more than two elements meet at a node, then the particular node is called major or principal node.

12. Define loop.

Any closed path of a circuit is called loop.

13. Define mesh.

A mesh is the most elementary form of a loop and it cannot be further divided into loops.

14. What is lumped circuit?

The circuits which are made up of lumped elements are called lumped circuits. Resistor, inductor, capacitor, voltage source and current source are examples of lumped elements.

15. What is distributed circuit?

A circuit is said to be a distributed one, if it has elements which cannot be separated. Transmission lines, coaxial cables, hollow waveguides, handset antennas for wireless communication are some of the examples of distributed circuits.

16. Define charge.

The charge of an electron is so small that it is not convenient to select it as the unit of charge. In practice, *coulomb* is used as the unit of charge. One coulomb of charge is equal to the charge on 628×10^{16} electrons i.e.,

1 coulomb = charge on 628×10^{16} electrons.

Thus, when we say that a body has a positive charge of one coulomb, it means that it has a deficit of 628×10^{16} electrons.

17. Define current.

Current is defined as the rate of flow of charge. If charge q passes through a conductor in time t (the rate of flow of charge is constant), then current $I = q/t$. The unit of current is coulomb per second or ampere. If the rate of flow of charge is not constant, the instantaneous value of current is given by

$$i = \frac{dq}{dt}$$

18. Define electric potential.

If one Joule work is done to move a charge of one coulomb between two points, then the potential of first point is said as one volt with respect to the second point.

$$v = \frac{dW}{dq}$$

It is generally measured between two points and its unit is volt.

19. Define resistance.

The property of any material which opposes the flow of electric current through it is called resistance. Its unit is ohms (Ω).

20. Define inductance.

Inductance is the element in which energy is stored in the form of electromagnetic field. Its unit is henry (H).

21. Define capacitance.

Capacitance is the element in which energy is stored in the form of electrostatic field. The unit of capacitance is farad (F).

22. State Ohm's law.

The ratio of potential difference (V) between the ends of a conductor to the current (I) flowing between them is constant, provided the physical conditions (e.g. temperature) do not change, i.e.

$$\frac{V}{I} = \text{constant} = R$$

where R is the resistance of the conductor between the two ends considered.

23. What are the limitations of Ohm's law?

Ohm's law is true for metal conductors at *constant temperature*. If the temperature changes, the law is not applicable. It does not apply to all non-metallic conductors such as *silicon carbide*, nor to nonlinear devices such as *zener diode*.

24. Define specific resistivity.

The specific resistivity of a material is defined as the resistance (R) offered to current flow by a conductor of unit length (l) having unit area (a) of cross-section. The unit of resistivity is ohm-metre (Ω m).

Therefore, $$\rho = \frac{Ra}{l}$$

25. Define conductance and conductivity.

Conductance is defined as the reciprocal of resistance. The resistance of a conductor measures the opposition which it offers during the flow of current through it. Conductance G measures the inducement it offers to the flow of current.

$$G = \frac{1}{R} = \frac{1}{\rho}\left(\frac{a}{l}\right)$$

$$\sigma = \frac{1}{\rho}$$

where σ is called *conductivity* or *specific conductance* of a conductor. Its unit is Siemens (S), whereas its earlier unit was mho. As above, it can be written as:

$$\sigma = \frac{Gl}{a}$$

26. Define temperature coefficient of resistance.

The temperature coefficient of resistance is defined as the increase in resistance per degree rise in temperature.

27. Define electrical power.

The rate at which work is done is called power and its unit is joule per second or watt. When one joule work is done in moving a charge of one coulomb in one second, the rate of work is one watt (or) joule/second. Hence, power in electric circuits is obtained as the product of voltage (V) and current (I).

$$P = VI \text{ watts}$$

Using Ohm's law, it can be written as:

$$P = I^2R \quad \text{or} \quad \frac{V^2}{R} \text{ watts}$$

28. Define electric energy.

Electrical energy is the total work done in an electric circuit.

$$\text{Electrical energy} = \text{electrical power} \times \text{time}$$

$$= VIt = I^2Rt = \frac{V^2}{R}t$$

In practice, electrical energy is measured in kilowatt hour (kWh).

$$\text{Energy in kWh} = \text{power in kW} \times \text{time in hours}$$

One kilowatt hour (kWh) of electrical energy is expanded in a circuit if 1 kW (1000 watts) of power is supplied for one hour. One kWh is also called 1 unit.

29. Define voltage source.

In a source, if the voltage is maintained constant and the current is allowed to vary, the source is called voltage source.

30. Define current source.

If the current of a source is maintained constant and the voltage is allowed to vary, it is called as current source.

31. Define ideal voltage source.

Voltage source having zero source impedance is called ideal voltage source. Such voltage source gives a constant voltage V irrespective of current drawn from it.

32. Define ideal current source.

Current source having infinite source impedance is called ideal current source. Such current source gives a constant current I irrespective of voltage across it.

33. What is dependent source?

In case of dependent sources, the source voltage or current is *not fixed*, but is dependent on the voltage or current in some part of the same circuit. Such sources mainly occur in modelling of electronic components like transistors, op-amp and so on.

34. What are the types of dependent sources?

Dependent or controlled sources are of the following types:

Voltage-controlled voltage source (VCVS)
Current-controlled voltage source (CCVS)
Voltage-controlled current source (VCCS)
Current-controlled current source (CCCS)

35. Write the formula to find the equivalent resistance in a network when a number of resistances are connected in parallel.

If there are n number of resistances connected in parallel, then

$$\frac{1}{R_{eq}} = \frac{1}{R_1} + \frac{1}{R_2} + \frac{1}{R_3} + \cdots + \frac{1}{R_n}$$

where R_{eq} is the equivalent value of resistance.

36. Write the formula to find the equivalent resistance in a network when a number of resistances are connected in series.

If there are n number of resistances connected in series, then

$$R_{eq} = R_1 + R_2 + R_3 + \cdots + R_n$$

where R_{eq} is the equivalent value of resistance.

37. Give the equation to convert delta connected resistive network into a star one.

For the star connected elements,

$$R_A = \frac{R_{AB} \cdot R_{CA}}{R_{AB} + R_{BC} + R_{CA}}, \quad R_B = \frac{R_{AB} \cdot R_{BC}}{R_{AB} + R_{BC} + R_{CA}} \quad \text{and} \quad R_C = \frac{R_{BC} \cdot R_{CA}}{R_{AB} + R_{BC} + R_{CA}}$$

38. Give the equation to convert star connected resistive network into a delta one.

For the delta connected elements,

$$R_{AB} = \frac{R_A R_B + R_B R_C + R_C R_A}{R_C}$$

$$R_{BC} = \frac{R_A R_B + R_B R_C + R_C R_A}{R_A}$$

and

$$R_{CA} = \frac{R_A R_B + R_B R_C + R_C R_A}{R_B}$$

39. What is meant by source transformation?

Source transformation means converting a voltage source into current source or vice versa.

40. Explain how a voltage source in series with a source resistance can be converted into a current source?

Let a voltage source of E volts is having series resistance of R Ω. Then, the equivalent current source is made as E/R amperes with the resistance R in parallel with the current source as shown in Fig. S1.1.

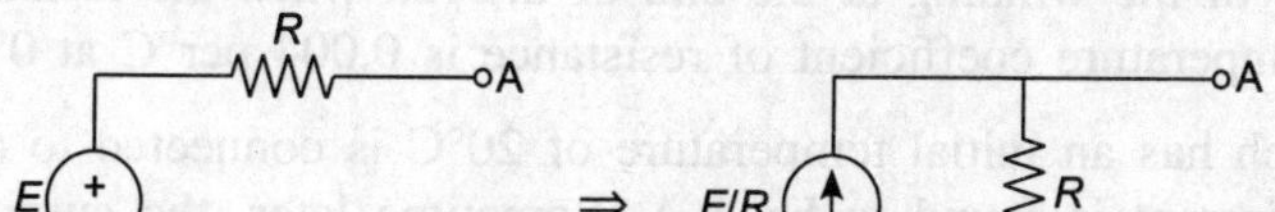

Fig. S1.1 Voltage source-to-current source conversion.

41. Explain how a current source in parallel with a source resistance can be converted into a voltage source?

Suppose a current source of I A is having parallel resistance of R Ω. Then, the equivalent current source is made as IR volts with the resistance R in series with the voltage source as shown in Fig. S1.2.

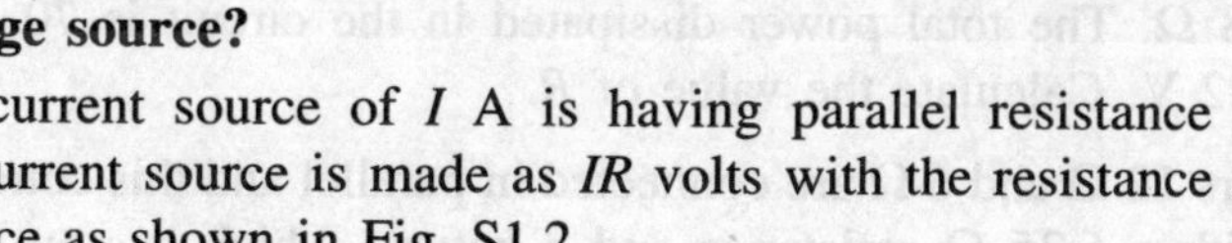

Fig. S1.2 Current source-to-voltage source conversion.

UNSOLVED PROBLEMS

1. A potential difference of 120 V exists between the ends of a wire of 200 cm. Its diameter is 0.15 mm. The current flowing is 2.4 A. Calculate the resistivity. [0.442×10^{-6} Ω m]

2. A bulb rated 110 V, 60 W is connected with another bulb rated 110 V, 100 W across 220 V mains. Calculate the resistance which should be shunted with the first bulb so that the bulbs may take their rated power. [302.5 Ω]

3. A 20 Ω resistor is connected in series with an unknown resistor to a 200 V supply mains. If the current drawn is 4 A, find the value of the unknown resistor and power in each resistor. [30 Ω, 320 W, 480 W]

4. A wire of length 1 m has a resistance of 2 Ω. What is the resistance of second wire, whose specific resistance is double that of the first, if the length of wire is 3 m and the diameter is double that of first? [3 Ω]

5. An aluminium wire 10 m long and 2 mm in diameter is connected in parallel to a copper wire 6 m long. A total current of 2 A is passed through the combination and it is found that current through the aluminium wire is 1.25 A. Calculate the diameter of copper wire. Specific resistance of copper is 1.6×10^{-6} Ω cm and that of aluminium is 2.6×10^{-6} Ω cm. [0.94 mm]

6. The field coil of a shunt motor has a resistance of 45 Ω at 20°C. Find the average temperature of the winding at the end of the run when the resistance is increased to 48.5 Ω. Temperature coefficient of resistance is 0.004 per°C at 0°C. [41°C]

7. A coil which has an initial temperature of 20°C is connected to a 180 V supply and the initial current is found to be 4 A. Sometime later, the current is found to have decreased to 3.4 A, the supply voltage being unchanged at 180 V. Determine the temperature rise of the coil. Assume the temperature coefficient of copper to be 0.0043 per°C at 0°C. [44.56°C]

8. Resistance *R* is connected in series with a parallel circuit comprising two resistors 12 Ω and 8 Ω. The total power dissipated in the circuit is 70 W when the applied voltage is 22 V. Calculate the value of *R*. [2.12 Ω]

9. Two resistors 12 Ω and 6 Ω are connected in parallel and this combination is connected in series with a 6.25 Ω resistance and a battery which has an internal resistance of 0.25 Ω. Determine the emf of the battery if the potential difference across 6 Ω resistance is 6 V. [15.75 volts]

10. Find the voltage across terminals A and B in the circuit shown in Fig. U1.1.

[18 V]

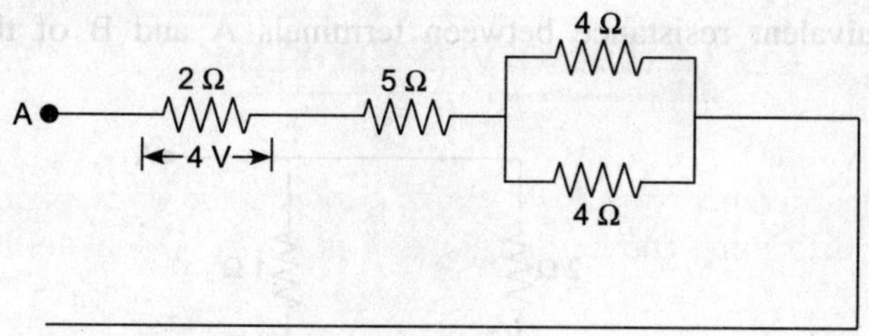

Fig. U1.1

11. Determine the total resistance of the circuit as seen by the voltage source shown in Fig. U1.2. [1.5 kΩ]

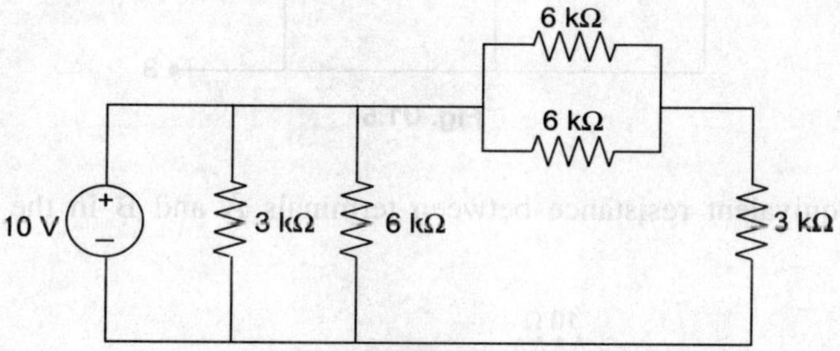

Fig. U1.2

12. Find the effective resistance across terminals AB of the circuit shown in Fig. U1.3. [3 Ω]

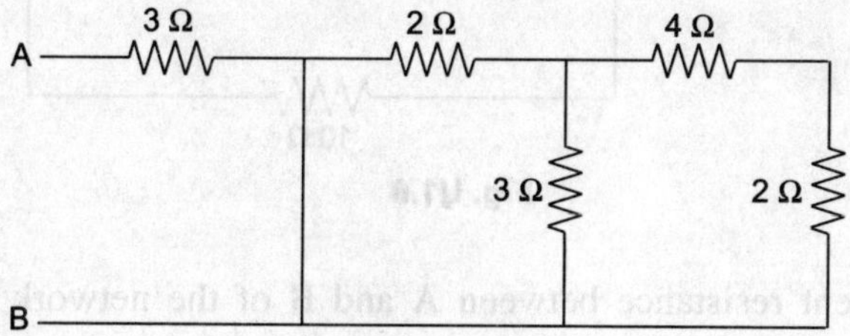

Fig. U1.3

13. Find the effective resistance across terminals AB of the circuit shown in Fig. U1.4. [7.63 Ω]

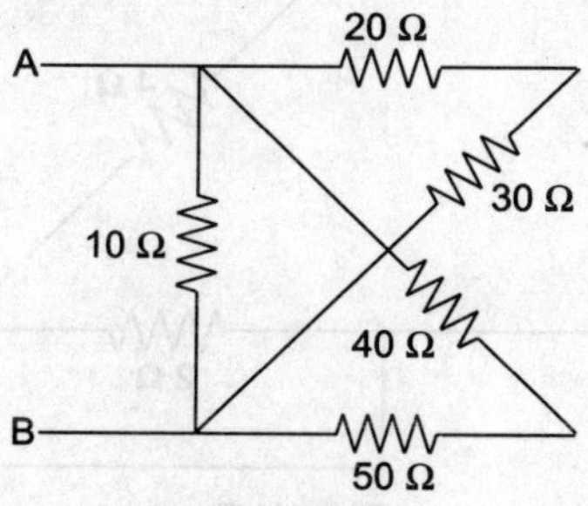

Fig. U1.4

14. Find the equivalent resistance between terminals A and B of the circuit shown in Fig. U1.5. [1.5 Ω]

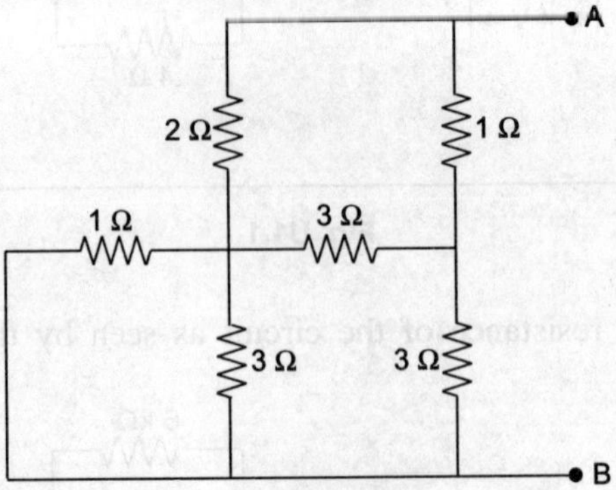

Fig. U1.5

15. Find the equivalent resistance between terminals A and B in the network shown in Fig. U1.6. [10 Ω]

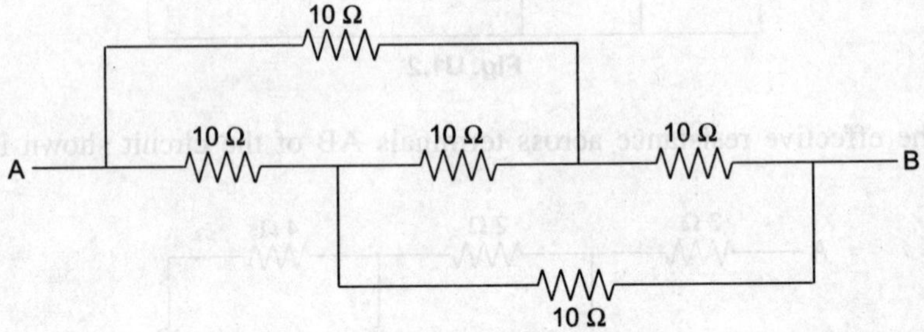

Fig. U1.6

16. Find the equivalent resistance between A and B of the network shown in Fig. U1.7. [4 Ω]

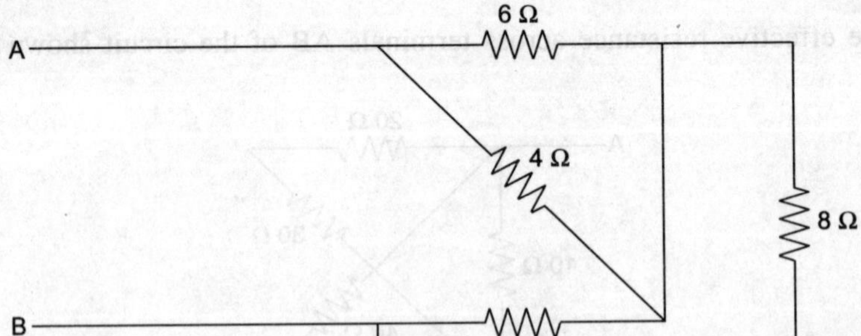

Fig. U1.7

17. Calculate the current supplied by the 30 V source in the network shown in Fig. U1.8. [28.3 A]

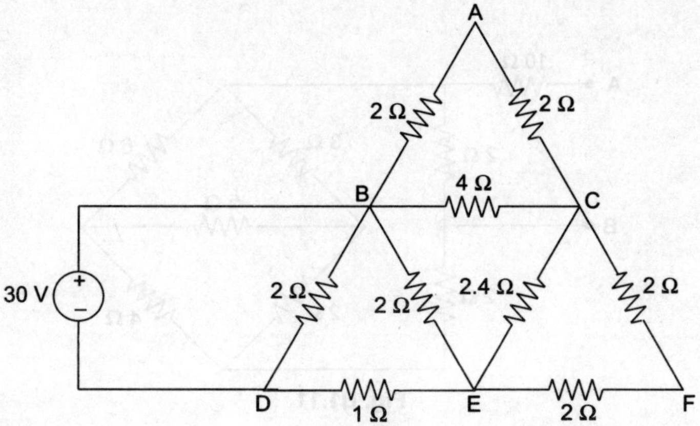

Fig. U1.8

18. Twelve resistances each of 1 Ω are connected to form twelve edges of a cube as in Fig. U1.9. Find the effective resistance between A and B. [5/6 Ω]

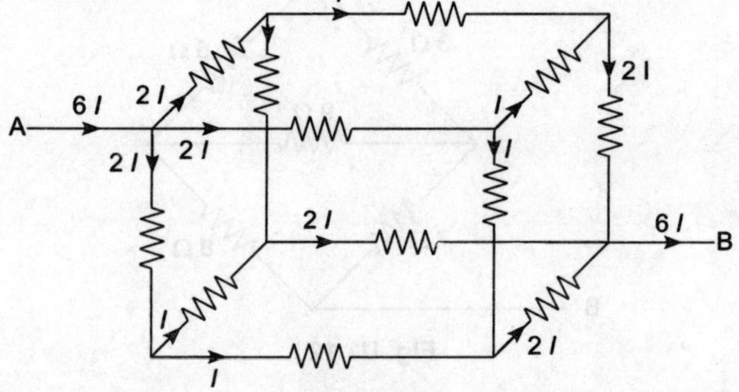

Fig. U1.9

19. Using star–delta conversion, determine the equivalent resistance between A and B in the network shown in Fig. U1.10. [11.54 Ω]

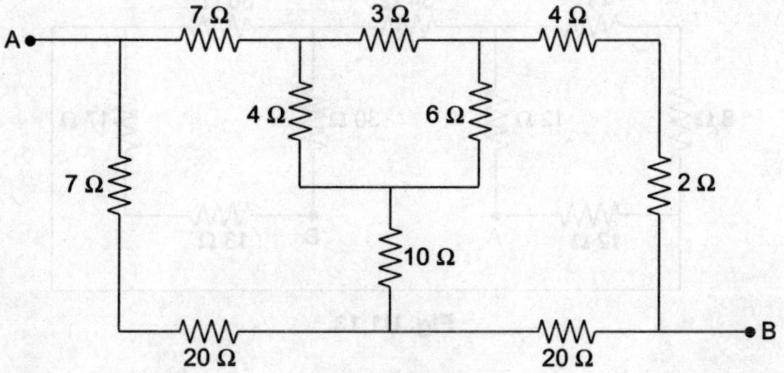

Fig. U1.10

20. Find the equivalent resistance between terminals A and B of the circuit shown in Fig. U1.11. [11.4545 Ω]

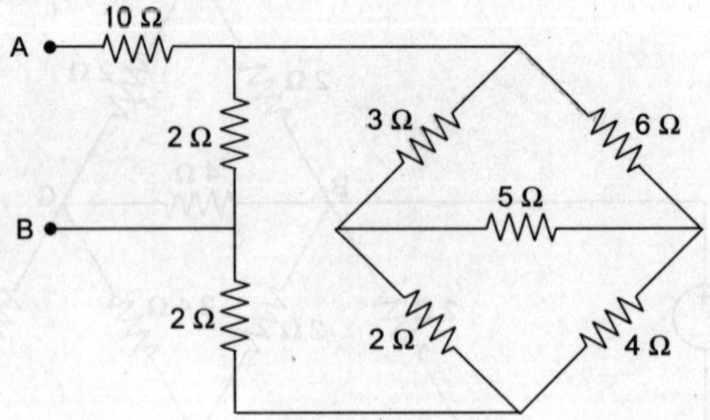

Fig. U1.11

21. Using star–delta conversion, determine the equivalent resistance between A and B in the network shown in Fig. U1.12. [5.79 Ω]

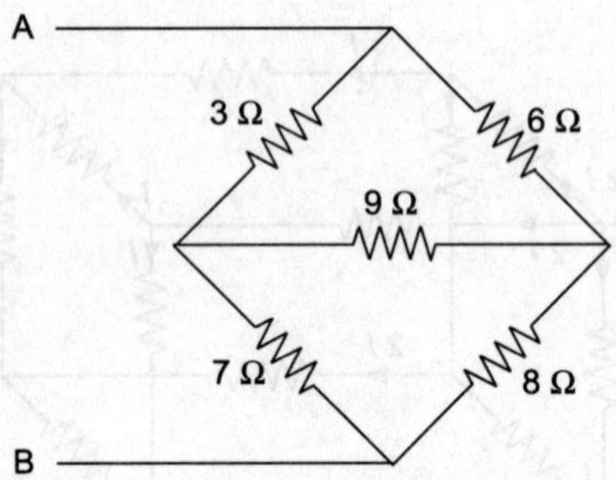

Fig. U1.12

22. Using star–delta conversion, determine the equivalent resistance between A and B in the network shown in Fig. U1.13. [24.84 Ω]

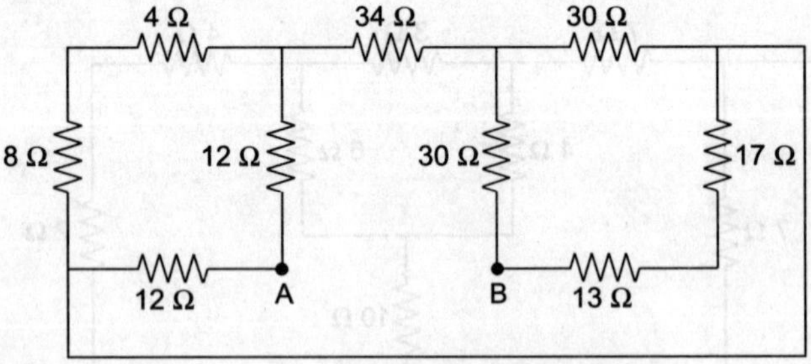

Fig. U1.13

23. In the circuit shown in Fig. U1.14, find the value of resistance *R*. [9.72 Ω]

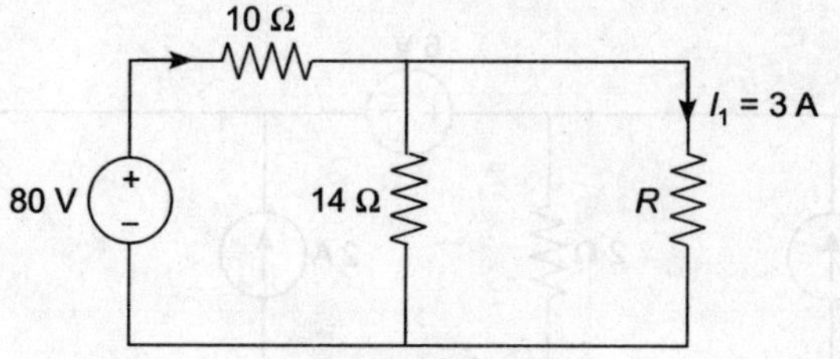

Fig. U1.14

24. Find the value of *R* in the circuit shown in Fig. U1.15 when the current is zero in the branch CD. [40 Ω]

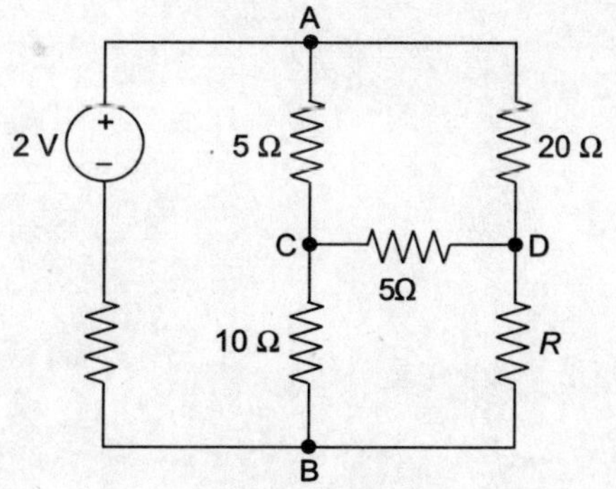

Fig. U1.15

25. Using source transformation, find the power delivered by the 50 V source in the given network as shown in Fig. U1.16. [274.2 W]

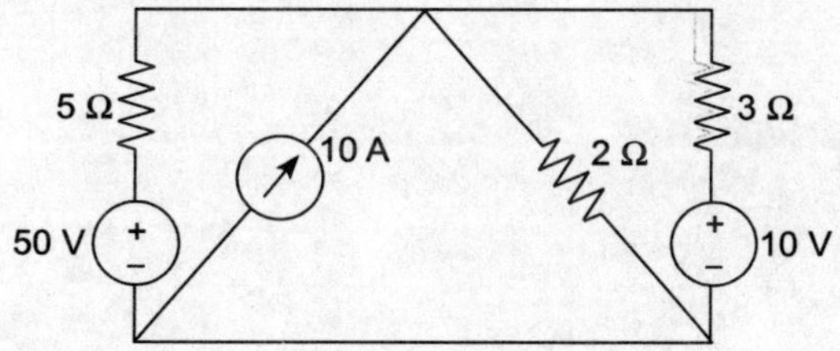

Fig. U1.16

26. Using source transformation, find the current flowing through the 4 Ω resistor in the given network as shown in Fig. U1.17. [1.33 A]

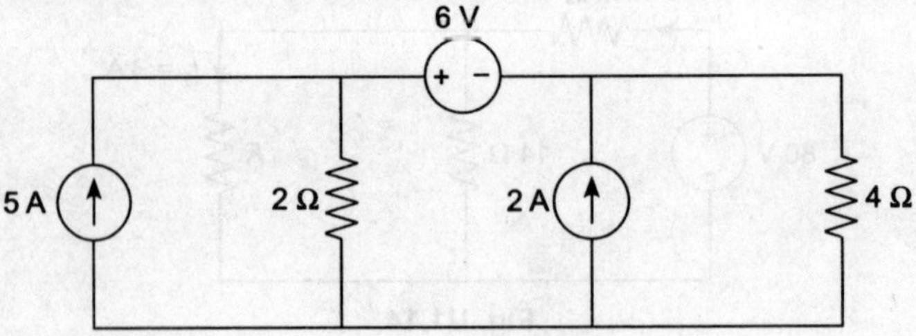

Fig. U1.17

CHAPTER 2

Kirchhoff's Laws

2.1 INTRODUCTION

Some networks are difficult to solve by applying Ohm's law. In such cases, Kirchhoff's laws are very useful to obtain the solution easily. The following sections discuss the two laws formulated by Gustav Kirchhoff in 1845.

2.2 KIRCHHOFF'S CURRENT LAW (KCL)

Kirchhoff's current law is stated as:

The algebraic sum of currents meeting at a point or junction is zero.

In Fig. 2.1, it has been shown that currents I_2, I_3 and I_6 are entering at point A, while currents I_1, I_4 and I_5 are leaving from point A. Applying KCL at point A, we can write:

$$I_1 + I_4 + I_5 + (-I_2) + (-I_3) + (-I_6) = 0$$

or

$$I_1 + I_4 + I_5 = I_2 + I_3 + I_6$$

So another form of KCL is that *the sum of currents entering at any point is equal to that leaving from that point.*

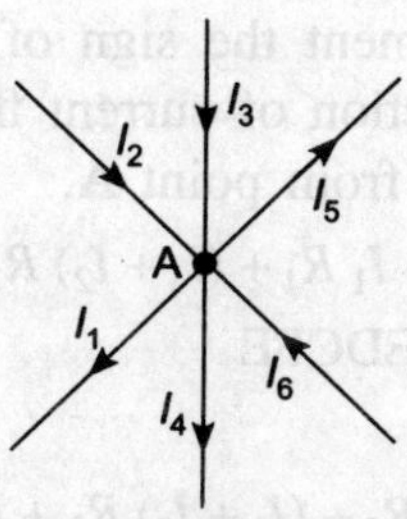

Fig. 2.1 Illustration of KCL.

2.3 KIRCHHOFF'S VOLTAGE LAW (KVL)

Kirchhoff's voltage law is stated as:

The algebraic sum of all the voltages taken around in a closed circuit is zero.

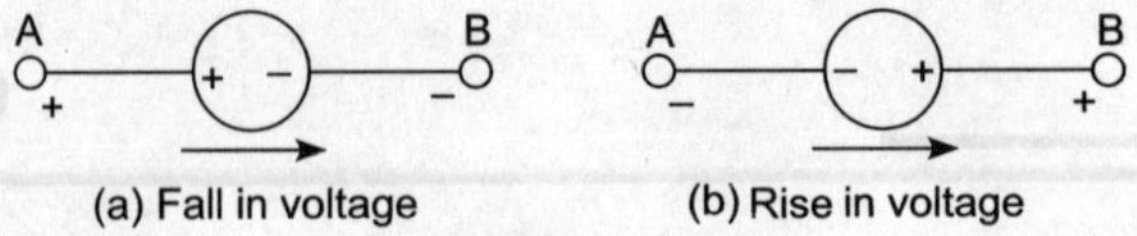

Fig. 2.2 Sign convention.

From Fig. 2.2, it is clear if the circuit is traced from higher voltage terminal to lower voltage terminal, there is a fall in voltage and vice versa.

Kirchhoff's voltage law is better understood by the following example. Consider the circuit shown in Fig. 2.3(a). The currents supplied by sources V_1 and V_2 are respectively I_1 and I_2 assumed in the directions shown. Directions of current flow in other branches are also marked accordingly. The direction of current flow is chosen arbitrarily, even if we assume wrong direction it will be reflected in the result by a negative sign.

Fig. 2.3(b) shows the voltage drops developed across each element as the result of current flowing through it. The current entering terminal of the passive element (resistor, inductor, capacitor) is marked with '+' sign and leaving terminal is marked with '–' sign. Conversely, current is assumed to be leaving from positive terminal and entering through negative terminal in active element (voltage source, current source).

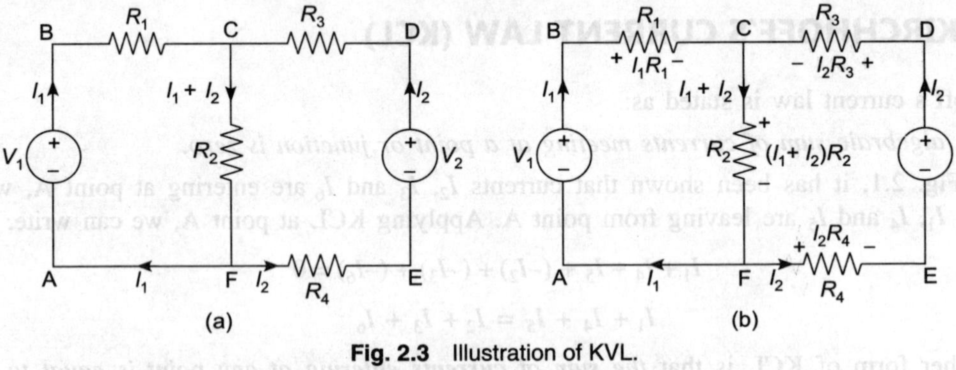

Fig. 2.3 Illustration of KVL.

Kirchhoff's Voltage law around loop ABCFA

To apply KVL, across each element the sign of voltage drop which is encountered first is taken irrespective of the direction of current flow.

Referring to Fig. 2.3(b), starting from point A,

$$-V_1 + I_1 R_1 + (I_1 + I_2) R_2 = 0$$

Kirchhoff's Voltage law around loop EDCFE

Starting from point E,

$$-V_2 + I_2 R_3 + (I_1 + I_2) R_2 + I_2 R_4 = 0$$

Kirchhoff's Voltage law around loop ABCDEFA

Starting from point A,

$$-V_1 + I_1 R_1 - I_2 R_3 + V_2 - I_2 R_4 = 0$$

If V_1, V_2, R_1, R_2, R_3 and R_4 are known, then two currents I_1 and I_2 can be obtained by solving any two of these three linear equations.

EXAMPLE 2.1 Calculate the current in 2 Ω resistor in Fig. 2.4.

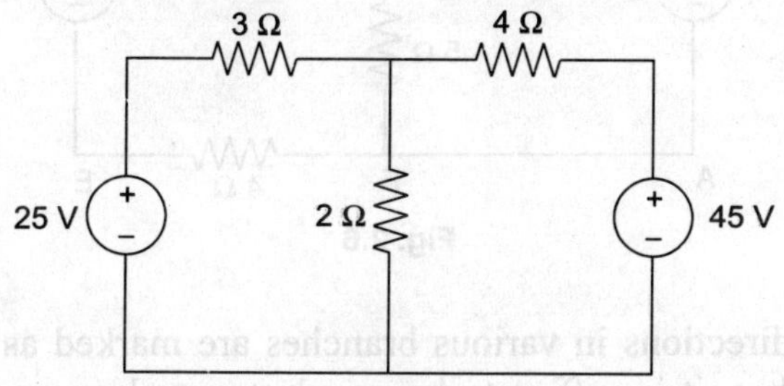

Fig. 2.4

Solution The current directions in various branches are marked as shown in Fig. 2.5. The voltage drops across each element are also shown in Fig. 2.5.

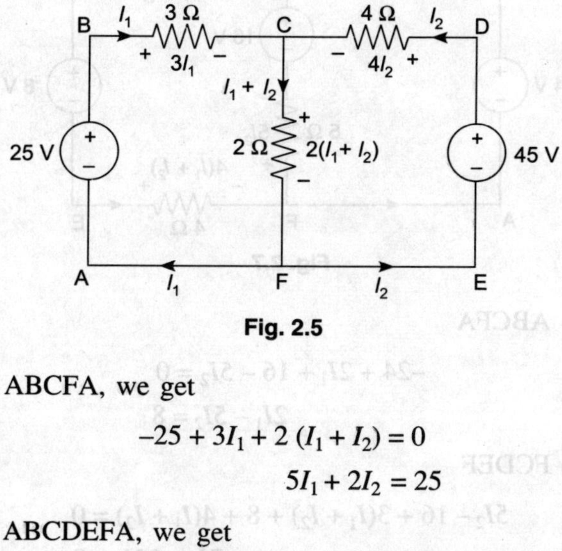

Fig. 2.5

Writing KVL for loop ABCFA, we get

$$-25 + 3I_1 + 2\,(I_1 + I_2) = 0$$

or
$$5I_1 + 2I_2 = 25 \quad \text{(i)}$$

Writing KVL for loop ABCDEFA, we get

$$-25 + 3I_1 - 4I_2 + 45 = 0$$

or
$$3I_1 - 4I_2 = -20 \quad \text{(ii)}$$

Solving Eqs. (i) and (ii), we get

$$I_1 = 2.31 \text{ A and } I_2 = 6.73 \text{ A}$$

Therefore, the current in 2 Ω resistor $= I_1 + I_2 =$ **9.04 A.**

EXAMPLE 2.2 For the circuit shown below (Fig. 2.6), find the current flowing in all branches.

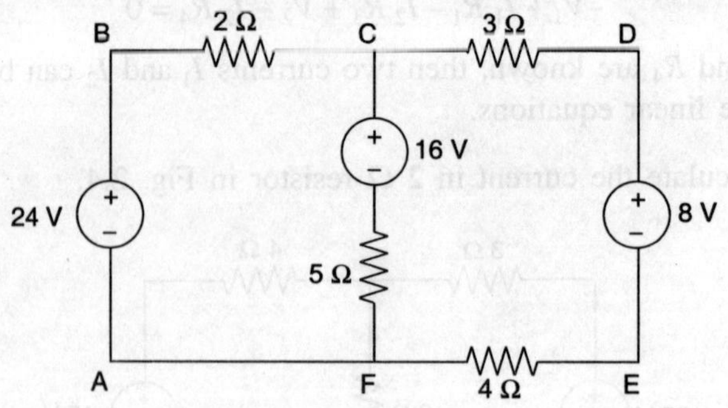

Fig. 2.6

Solution The current directions in various branches are marked as shown in Fig. 2.7. As there are only two meshes, it is suffice to have only two unknown currents I_1 and I_2, even though there are three voltage sources.

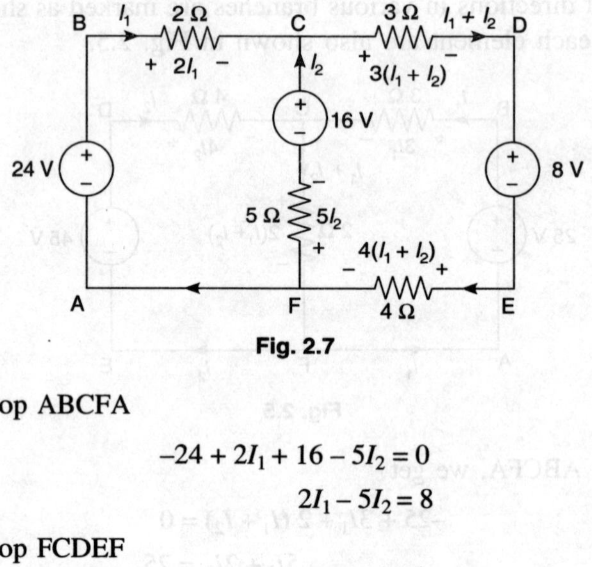

Fig. 2.7

Applying KVL for loop ABCFA

$$-24 + 2I_1 + 16 - 5I_2 = 0$$

or
$$2I_1 - 5I_2 = 8 \quad \text{(i)}$$

Applying KVL for loop FCDEF

$$5I_2 - 16 + 3(I_1 + I_2) + 8 + 4(I_1 + I_2) = 0$$

or
$$7I_1 + 12I_2 = 8 \quad \text{(ii)}$$

Solving Eqs. (i) and (ii), we get

$$I_1 = 2.31 \text{ A}$$

$$I_2 = -0.68 \text{ A}$$

and current in branch CDEF = $I_1 + I_2$ = **1.63 A**

EXAMPLE 2.3 Determine currents I_1, I_2 and I_3 in the network shown in Fig. 2.8.

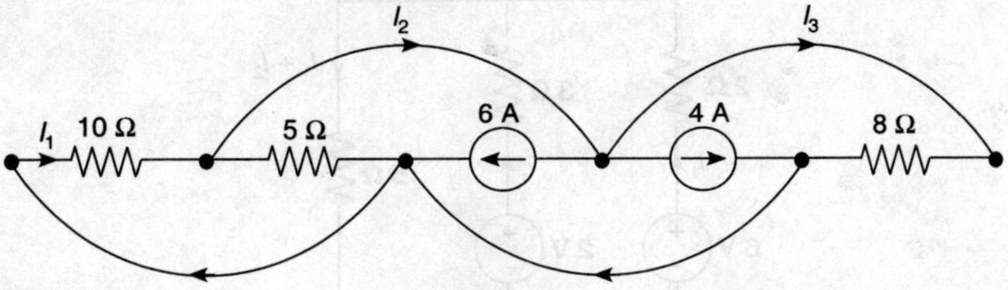

Fig. 2.8

Solution The current directions in various branches are marked by using KCL as shown in Fig. 2.9.

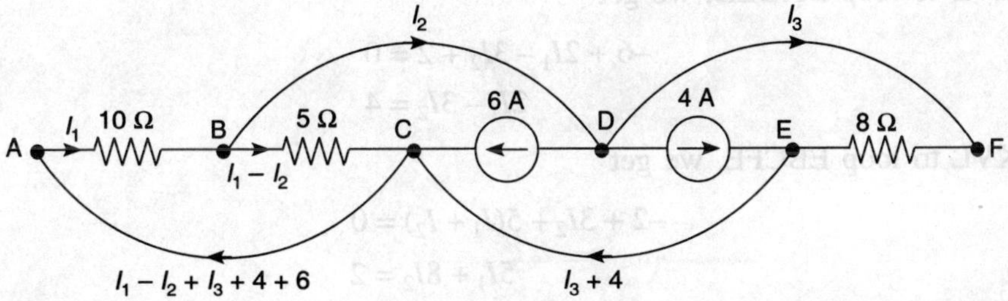

Fig. 2.9

Referring to Fig. 2.9,

$$I_1 = I_1 - I_2 + I_3 + 4 + 6$$

or
$$I_2 - I_3 = 10 \quad \text{(i)}$$

Applying KVL around loop ABDFECA, we get

$$10I_1 + 8I_3 = 0 \quad \text{(ii)}$$

Applying KVL around loop ABCA, we get

$$10I_1 + 5(I_1 - I_2) = 0 \quad \text{(iii)}$$

Solving Eqs. (i), (ii) and (iii), we get

$$\mathbf{I_1 = 2.3529\ A}$$

$$\mathbf{I_2 = 7.0588\ A}$$

$$\mathbf{I_3 = -2.9412\ A}$$

EXAMPLE 2.4 Two batteries of e.m.f.s of 6 V and 2 V having internal resistances of 2 Ω and 3 Ω, respectively are connected in parallel across a 5 Ω resistor. Calculate the current through each battery and the voltage across 5 Ω resistor.

Solution

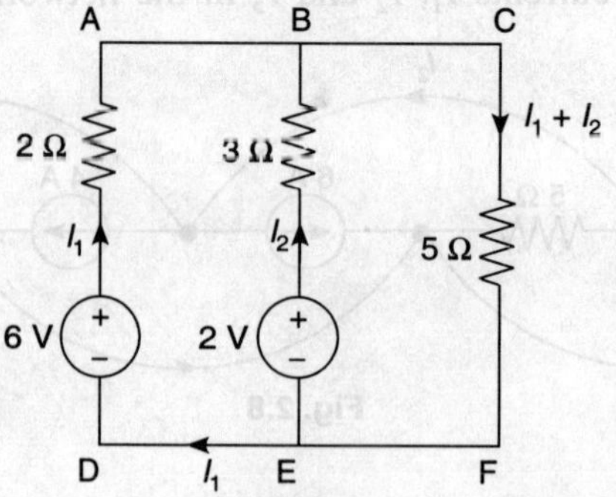

Fig. 2.10

Applying KVL to loop DABED, we get

$$-6 + 2I_1 - 3I_2 + 2 = 0$$

or $$2I_1 - 3I_2 = 4 \quad \text{(i)}$$

Applying KVL to loop EBCFE, we get

$$-2 + 3I_2 + 5(I_1 + I_2) = 0$$

or $$5I_1 + 8I_2 = 2 \quad \text{(ii)}$$

Solving Eqs. (i) and (ii), we get

$$I_1 = \mathbf{1.2258\ A}$$

$$I_2 = \mathbf{-0.5161\ A}$$

Voltage across 5 Ω resistor $= (I_1 + I_2) \times 5$

$= (1.2258 - 0.5161) \times 5 = \mathbf{3.5485\ V.}$

EXAMPLE 2.5 Find the current through 4 Ω resistance in the network shown in Fig. 2.11.

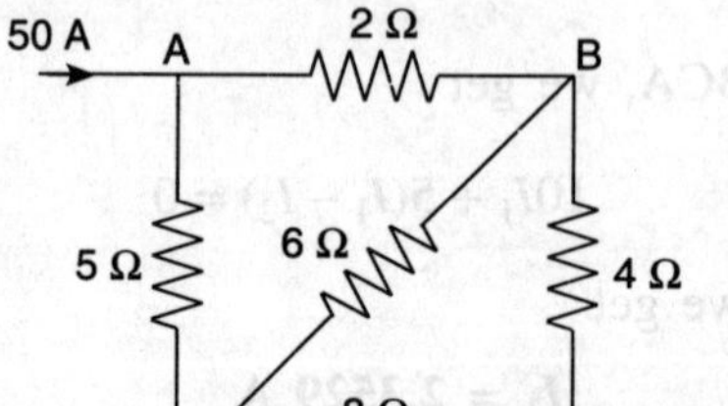

Fig. 2.11

Solution In Fig. 2.12, using KCL the currents flowing through various resistances are labelled on the network arbitrarily first assuming current through 5 Ω resistance as I_1 amperes.

Applying KVL around loop ABCA, we get

$$2(50 - I_1) - 6I_2 - 5I_1 = 0 \quad \text{(i)}$$

Applying KVL around loop CBDC, we get

$$6I_2 + 4(50 - I_1 + I_2) - 8(I_1 - I_2) = 0 \quad \text{(ii)}$$

Solving Eqs. (i) and (ii), we get

$$I_1 = 15.1515 \text{ A}$$

$$I_2 = -1.0101 \text{ A}$$

Therefore, the current in 4 Ω resistor = $50 - I_1 + I_2$ = **33.8384 A.**

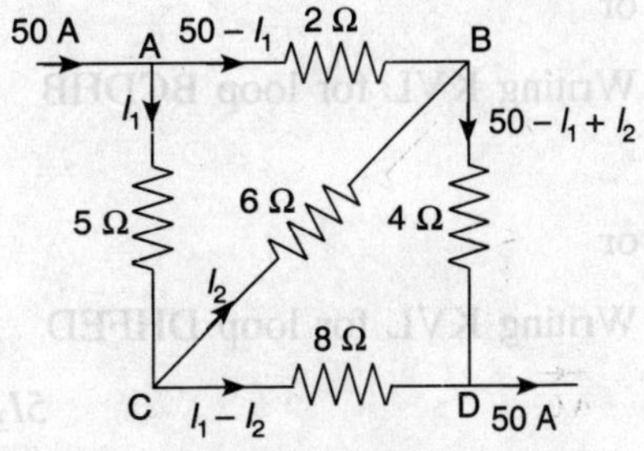

Fig. 2.12

EXAMPLE 2.6 Calculate the current in 5 Ω resistor shown in Fig. 2.13.

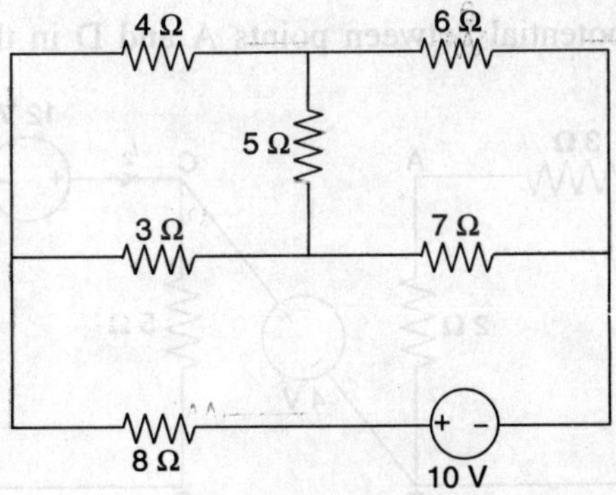

Fig. 2.13

Solution Mark the direction of currents in the branches as shown in Fig. 2.14.

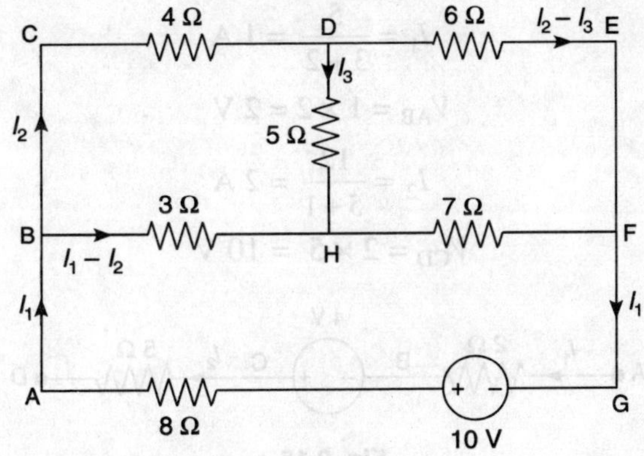

Fig. 2.14

Writing KVL for loop ABHFGA,

$$3(I_1 - I_2) + 7(I_1 - I_2 + I_3) - 10 + 8I_1 = 0$$

or $$18I_1 - 10I_2 + 7I_3 = 10 \quad \text{(i)}$$

Writing KVL for loop BCDHB

$$4I_2 + 5I_3 - 3(I_1 - I_2) = 0$$

or $$-3I_1 + 7I_2 + 5I_3 = 0 \quad \text{(ii)}$$

Writing KVL for loop DHFED

$$5I_3 - 6(I_2 - I_3) + 7(I_1 - I_2 + I_3) = 0$$

or $$7I_1 - 13I_2 + 18I_3 = 0 \quad \text{(iii)}$$

Solving Eqs. (i), (ii) and (iii), we get

$$I_3 = \mathbf{-0.0403\ A}$$

EXAMPLE 2.7 Find the potential between points A and D in the network shown below.

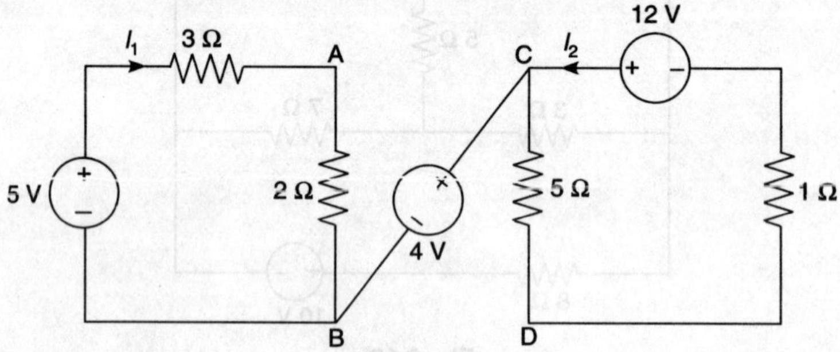

Fig. 2.15

Solution From Fig. 2.15,

$$I_1 = \frac{5}{3+2} = 1\text{ A}$$

Hence, $$V_{AB} = 1 \times 2 = 2\text{ V}$$

Similarly, $$I_2 = \frac{12}{5+1} = 2\text{ A}$$

Hence, $$V_{CD} = 2 \times 5 = 10\text{ V}$$

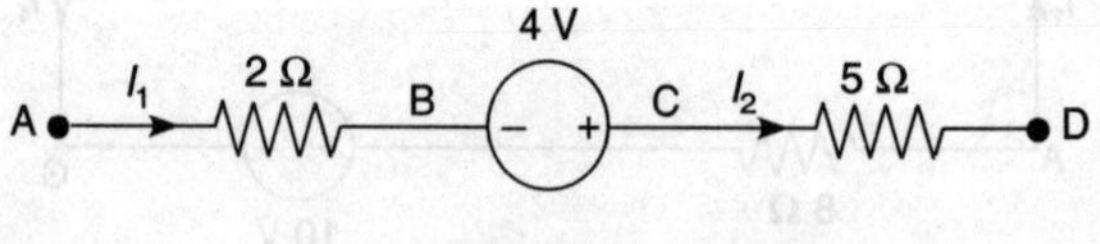

Fig. 2.16

From Fig. 2.16,

$$V_{AD} = V_{AB} - V_{BC} + V_{CD}$$
$$= +2 - 4 + 10 = \mathbf{8\ V}$$

2.4 CURRENT DIVISION IN PARALLEL RESISTORS

When two resistances are connected in parallel as shown in Fig. 2.17, the equivalent resistance is given by

$$R_T = \frac{R_1 R_2}{R_1 + R_2}$$

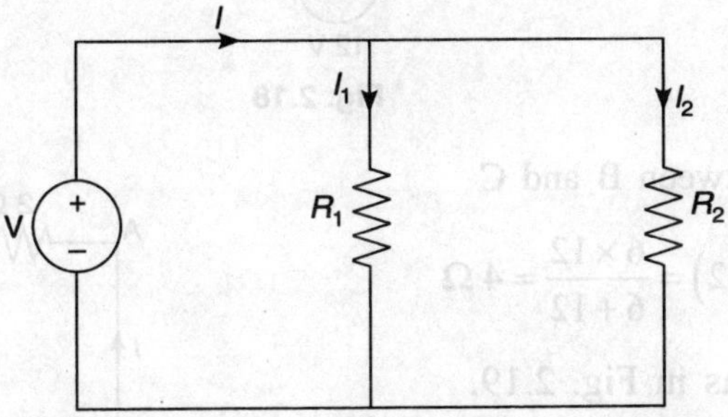

Fig. 2.17 Current division between two resistors.

The total current flows in the circuit is I amperes. Now, current through R_1 is given by

$$I = \frac{V}{R_1} = \frac{IR_T}{R_1}$$

$$= \frac{I}{R_1}\left(\frac{R_1 R_2}{R_1 + R_2}\right)$$

$$= I\left(\frac{R_2}{R_1 + R_2}\right)$$

Similarly, current through R_2 is given by

$$I_2 = \frac{V}{R_2} = \frac{IR_T}{R_2}$$

$$= \frac{I}{R_2}\left(\frac{R_1 R_2}{R_1 + R_2}\right) = I\left(\frac{R_1}{R_1 + R_2}\right)$$

Therefore, the current through a resistor (parallel combination of two resistors)

$$= \text{Total current} \times \left(\frac{\text{Resistance of opposite resistor}}{\text{Sum of resistances of two resistors}}\right)$$

EXAMPLE 2.8 Determine I, I_1 and I_2 in the network shown in Fig. 2.18.

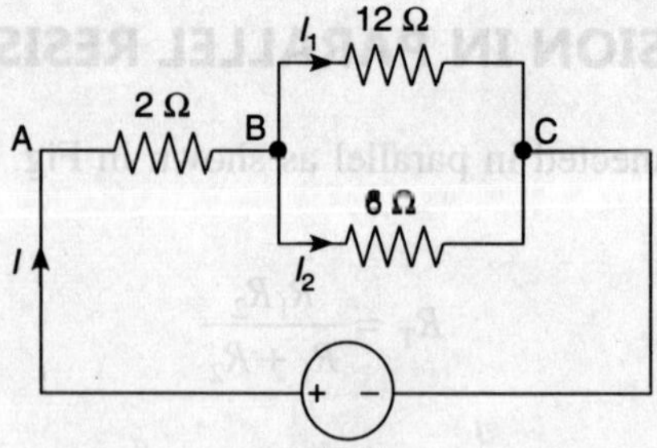

Fig. 2.18

Solution Resistance between B and C

$$R_{BC} = (6\|12) = \frac{6\times 12}{6+12} = 4\ \Omega$$

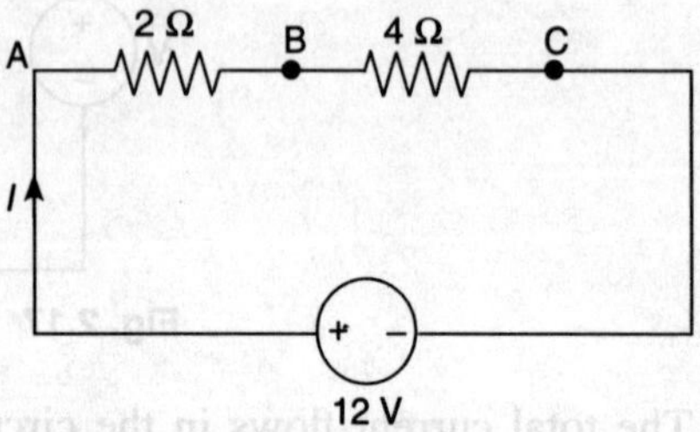

Fig. 2.19

The circuit is redrawn as in Fig. 2.19.

Now, the current supplied by the battery, $I = \dfrac{12}{2+4} = \mathbf{2\ A}$

Hence, the current through 6 Ω resistance is

$$I_1 = I\times\frac{12}{12+6} = 2\times\frac{12}{12+6} = \mathbf{\frac{4}{3}\ A}$$

Similarly, the current through 12 Ω resistance is

$$I_2 = I\times\frac{6}{12+6} = 2\times\frac{6}{12+6} = \mathbf{\frac{2}{3}\ A}$$

EXAMPLE 2.9 Calculate V_{ab} and I of the network shown in Fig. 2.20.

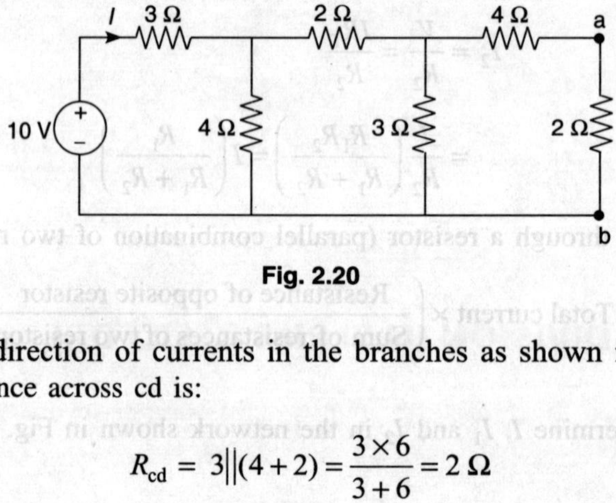

Fig. 2.20

Solution Mark the direction of currents in the branches as shown in Fig. 2.21(a).

Equivalent resistance across cd is:

$$R_{cd} = 3\|(4+2) = \frac{3\times 6}{3+6} = 2\ \Omega$$

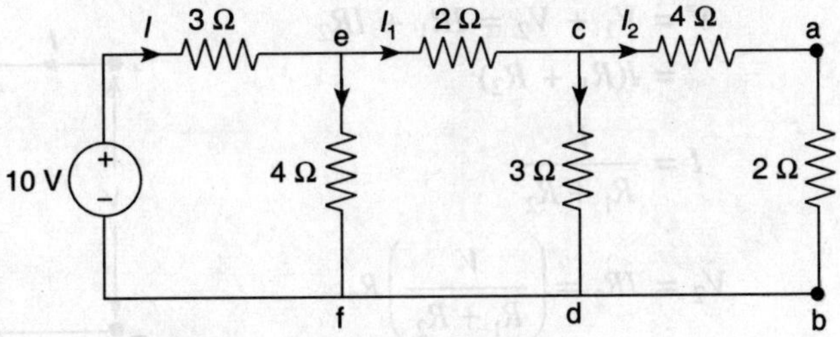

Fig. 2.21(a)

The reduced circuit is shown in Fig. 2.21(b)

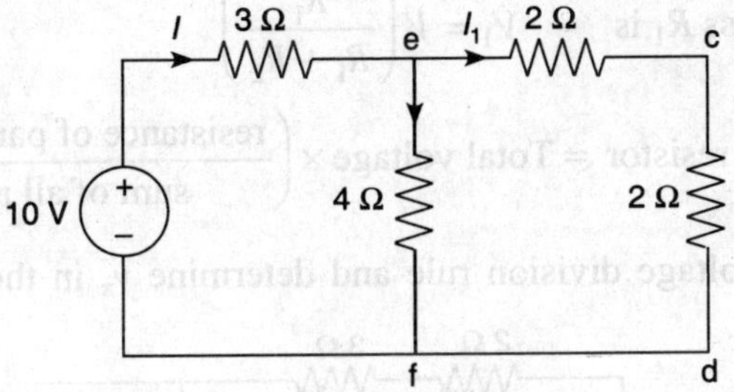

Fig. 2.21(b)

Equivalent resistance across ef is: $R_{ef} = 4\|(2+2) = \dfrac{4 \times 4}{4+4} = 2\ \Omega$

The reduced circuit is shown in Fig. 2.21(c)

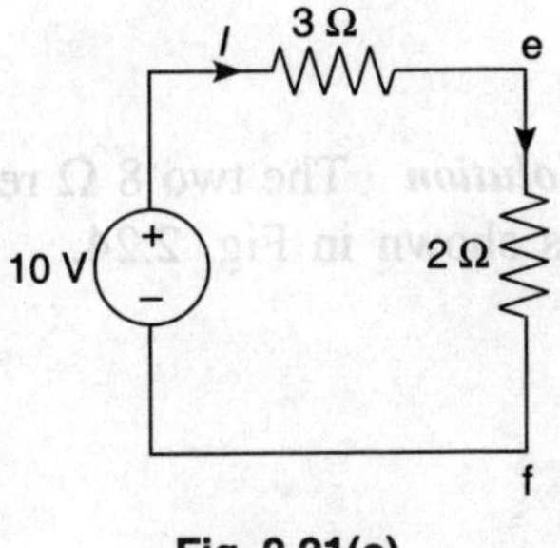

Fig. 2.21(c)

Therefore, $I = \dfrac{10}{3+2} = 2\text{ A}$

From Fig. 2.21(b), $I_1 = I \times \dfrac{4}{4+2+2} = 2 \times \dfrac{4}{8} = 1\text{ A}$

From Fig. 2.21(a), $I_2 = I_1 \times \dfrac{3}{3+4+2} = 1 \times \dfrac{3}{9} = \dfrac{1}{3}\text{ A}$

$$V_{ab} = 2 \times \frac{1}{3} = \frac{2}{3}\text{ V}$$

$$\boldsymbol{V_{ab} = \frac{2}{3}\text{ V}; \quad I = 2\text{ A}.}$$

2.5 VOLTAGE DIVISION IN SERIES RESISTORS

Voltage division is used to express the voltage across one of several series resistors in terms of the voltage across the combination. In Fig. 2.22, the voltage across R_2 is found using KVL and Ohm's law:

$$V = V_1 + V_2 = IR_1 + IR_2$$
$$= I(R_1 + R_2)$$

So, $$I = \frac{V}{R_1 + R_2}$$

Thus, $$V_2 = IR_2 = \left(\frac{V}{R_1 + R_2}\right)R_2$$

or $$V_2 = V\left(\frac{R_2}{R_1 + R_2}\right)$$

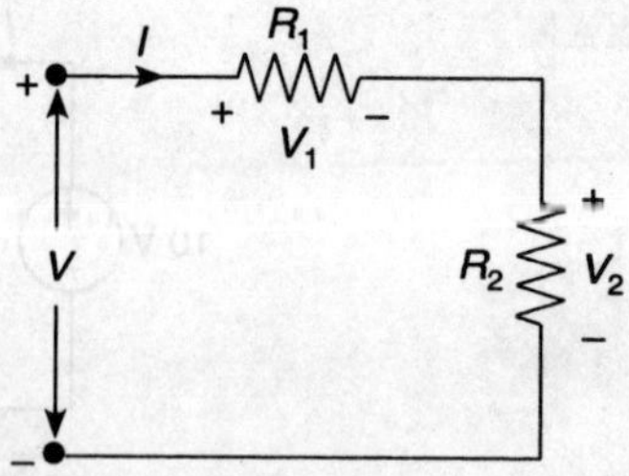

Fig. 2.22 Voltage division between two resistors.

Similarly, the voltage across R_1 is $V_1 = V\left(\frac{R_1}{R_1 + R_2}\right)$

∴ Voltage across a resistor = Total voltage × $\left(\frac{\text{resistance of particular resistor}}{\text{sum of all resistances}}\right)$

EXAMPLE 2.10 Use voltage division rule and determine v_x in the circuit of Fig. 2.23.

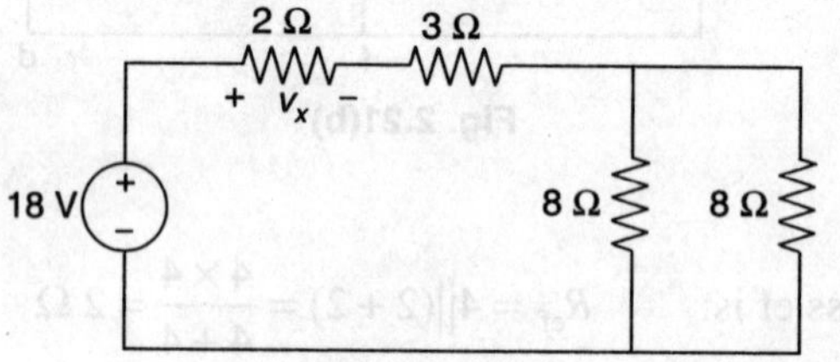

Fig. 2.23

Solution The two 8 Ω resistors are combined to have all the resistors in series connection as shown in Fig. 2.24.

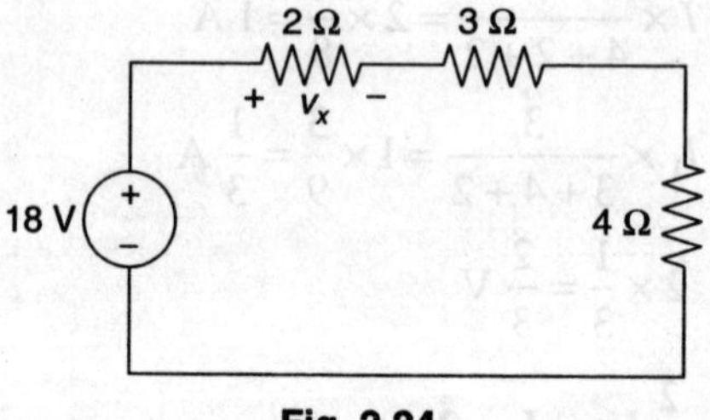

Fig. 2.24

$$v_x = 18 \times \frac{2}{2+3+4} = 4\text{ V}$$

EXAMPLE 2.11 Find the value of R, if the current in branch AD is zero.

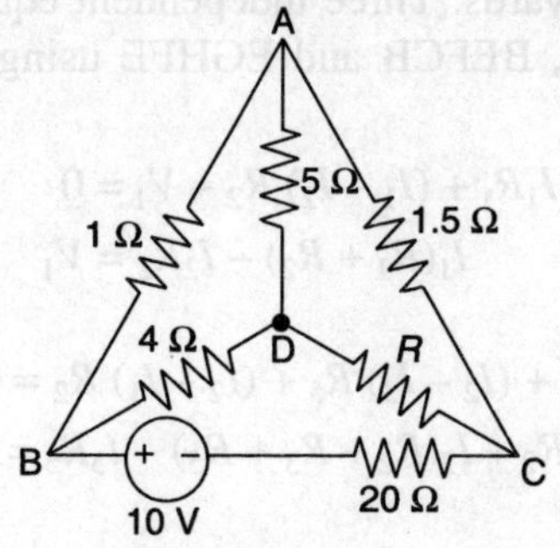

Fig. 2.25

Solution The circuit shown in Fig. 2.25 is redrawn as in Fig. 2.26.

Current through branch AD is zero means the voltages of V_{DC} and V_{AC} are equal.

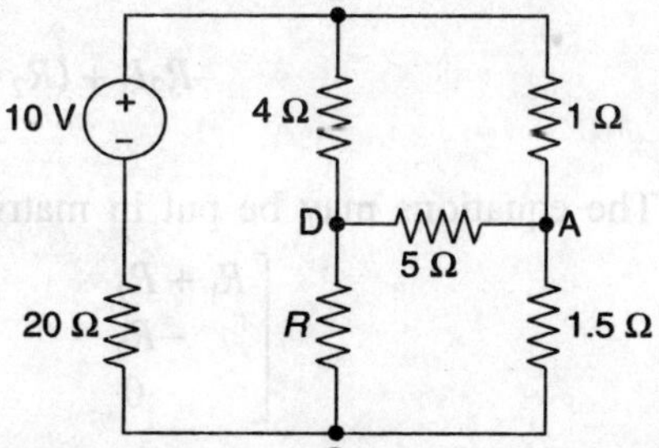

Fig. 2.26

i.e. $$V_{DC} = V_{AC}$$

or $$V_{BC} \times \frac{R}{4+R} = V_{BC} \times \frac{1.5}{1+1.5}$$

$$2.5R = 6 + 1.5R$$

Therefore, $$R = 6\ \Omega$$

2.6 MESH ANALYSIS

This method was first suggested by Maxwell. This method simplifies the solution of several networks. So far we have solved examples using Kirchhoff's basic laws, assuming branch currents to start with. In this method, we consider *loop* or *mesh* current instead of *branch* current. For each loop, an equation is written based on Kirchhoff's voltage law and there are as many equations as the number of loops. For example, let us consider a circuit with 5 resistors and 2 voltage sources as shown in Fig. 2.27. The three loop currents are marked as I_1, I_2 and I_3 and these are the currents flowing through resistors R_1, R_3 and R_4, respectively. The current directions are chosen arbitrarily (either clockwise or anticlockwise). The current flows through

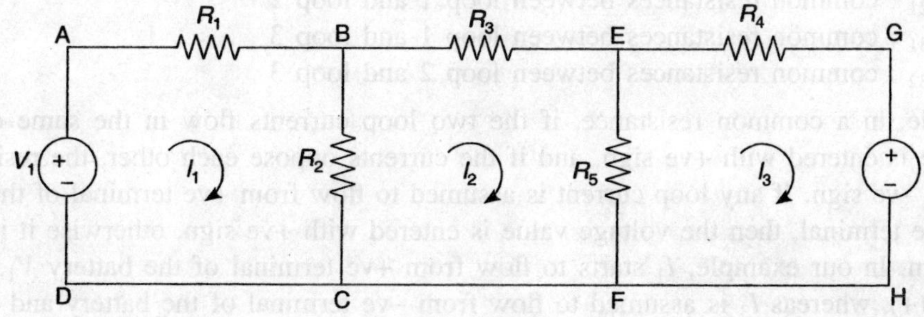

Fig. 2.27

R_2 is $(I_1 - I_2)$ downwards or $(I_2 - I_1)$ upwards. Similarly, the current flows through R_5 is $(I_2 - I_3)$ downwards or $(I_3 - I_2)$ upwards. Three independent equations can be formed in respect of the three closed loops ABCDA, BEFCB and EGHFE using Kirchhoff's voltage law.

Loop ABCDA

$$I_1R_1 + (I_1 - I_2)\,R_2 - V_1 = 0$$

or
$$I_1(R_1 + R_2) - I_2R_2 = V_1 \tag{2.1}$$

Loop BEFCB

$$I_2R_3 + (I_2 - I_3)\,R_5 + (I_2 - I_1)\,R_2 = 0$$

or
$$-I_1\,R_2 + I_2(R_2 + R_3 + R_5) - I_3R_5 = 0 \tag{2.2}$$

Loop EGHFE

$$I_3R_4 + V_2 + (I_3 - I_2)\,R_5 = 0$$

or
$$-I_2R_5 + I_3\,(R_4 + R_5) = -V_2 \tag{2.3}$$

Rearranging the equations,

$$(R_1 + R_2)\,I_1 - R_2I_2 = V_1$$
$$-R_2I_1 + (R_2 + R_3 + R_5)\,I_2 - R_5I_3 = 0$$
$$-R_5I_2 + (R_4 + R_5)\,I_3 = -V_2$$

The equations may be put in matrix form as:

$$\begin{bmatrix} R_1 + R_2 & -R_2 & 0 \\ -R_2 & R_2 + R_3 + R_5 & -R_5 \\ 0 & -R_5 & R_4 + R_5 \end{bmatrix} \begin{bmatrix} I_1 \\ I_2 \\ I_3 \end{bmatrix} = \begin{bmatrix} V_1 \\ 0 \\ -V_2 \end{bmatrix} \tag{2.4}$$

The above matrix equations can be solved for I_1, I_2 and I_3, when all the resistances and voltage source values are known. The above matrix is written in the form $[R]\,[I] = [V]$, where $[R]$ is the resistance matrix, $[I]$ is the current matrix and $[V]$ is the voltage matrix. If we observe the resistance matrix in Eq. (2.4), it is in the form:

$$\begin{bmatrix} R_{11} & R_{12} & R_{13} \\ R_{21} & R_{22} & R_{23} \\ R_{31} & R_{32} & R_{33} \end{bmatrix}$$

R_{11} : self resistances of loop 1
R_{22} : self resistances of loop 2
R_{33} : self resistances of loop 3
$R_{12} = R_{21}$: common resistances between loop 1 and loop 2
$R_{13} = R_{31}$: common resistances between loop 1 and loop 3
$R_{23} = R_{32}$: common resistances between loop 2 and loop 3

As a rule, in a common resistance, if the two loop currents flow in the same direction, its resistance is entered with +ve sign, and if the currents oppose each other, the resistance is entered with –ve sign. If any loop current is assumed to flow from +ve terminal of the voltage source to –ve terminal, then the voltage value is entered with +ve sign, otherwise it is entered with –ve sign. In our example, I_1 starts to flow from +ve terminal of the battery V_1, so V_1 is obtained as (+), whereas I_3 is assumed to flow from –ve terminal of the battery and hence V_2 is written as (–). Based on the above expression, we can write $[R]$ matrix merely by inspection without actually deriving the basic equations.

EXAMPLE 2.12 In the circuit shown in Fig. 2.28, find the current through 15 Ω resistor.

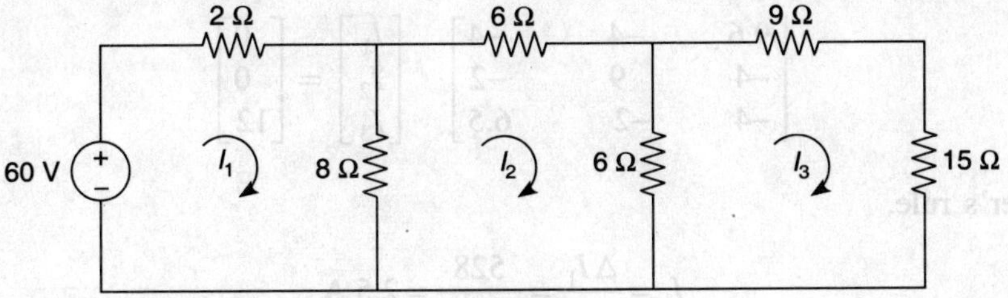

Fig. 2.28

Solution The matrix has to be written in the form:

$$\begin{bmatrix} R_{11} & R_{12} & R_{13} \\ R_{21} & R_{22} & R_{23} \\ R_{31} & R_{32} & R_{33} \end{bmatrix} \begin{bmatrix} I_1 \\ I_2 \\ I_3 \end{bmatrix} = \begin{bmatrix} V_1 \\ V_2 \\ V_3 \end{bmatrix}$$

$R_{11} = 2 + 8 = 10$	$R_{12} = R_{21} = -8$	('–' sign is because the directions of currents I_1 and I_2 through 8 Ω are in opposite)
$R_{22} = 8 + 6 + 6 = 20$	$R_{13} = R_{31} = 0$	(no resistance is available which is common for both 1st and 3rd loop)
$R_{33} = 6 + 9 + 15 = 30$	$R_{23} = R_{32} = -6$	('–' sign is because the directions of currents I_2 and I_3 through 6 Ω are in opposite)

So, we can write:

$$\begin{bmatrix} 10 & -8 & 0 \\ -8 & 20 & -6 \\ 0 & -6 & 30 \end{bmatrix} \begin{bmatrix} I_1 \\ I_2 \\ I_3 \end{bmatrix} = \begin{bmatrix} 60 \\ 0 \\ 0 \end{bmatrix}$$

The current through 15 Ω resistor, I_3 can be determined by applying Cramer's rule as follows:

$$\Delta = \begin{vmatrix} 10 & -8 & 0 \\ -8 & 20 & -6 \\ 0 & -6 & 30 \end{vmatrix} = 3720$$

$$\Delta I_3 = \begin{vmatrix} 10 & -8 & 60 \\ -8 & 20 & 0 \\ 0 & -6 & 0 \end{vmatrix} = 2880$$

Therefore,
$$I_3 = \frac{\Delta I_3}{\Delta} = \frac{2880}{3720} = \mathbf{0.77\ A}$$

EXAMPLE 2.13 Find currents I_1, I_2 and I_3 in the network shown in Fig. 2.29 by using mesh analysis.

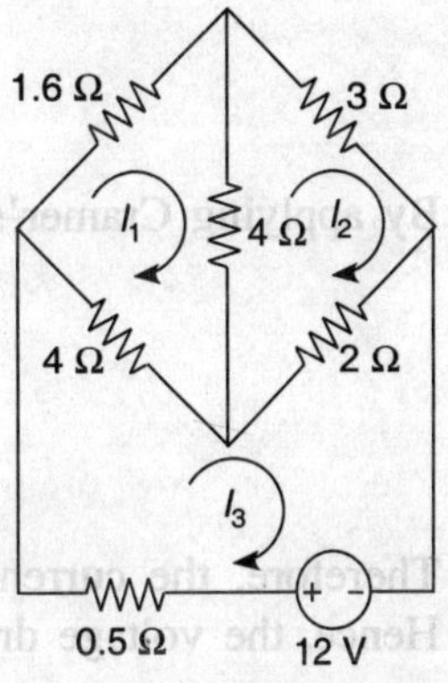

Fig. 2.29

Solution By inspection,

$$\begin{bmatrix} 9.6 & -4 & -4 \\ -4 & 9 & -2 \\ -4 & -2 & 6.5 \end{bmatrix} \begin{bmatrix} I_1 \\ I_2 \\ I_3 \end{bmatrix} = \begin{bmatrix} 0 \\ 0 \\ 12 \end{bmatrix}$$

Using Cramer's rule,

$$I_1 = \frac{\Delta I_1}{\Delta} = \frac{528}{211.2} = 2.5 \text{ A}$$

$$I_2 = \frac{\Delta I_2}{\Delta} = \frac{422.4}{211.2} = 2\text{A}$$

$$I_3 = \frac{\Delta I_3}{\Delta} = \frac{844.8}{211.2} = 4 \text{ A}$$

EXAMPLE 2.14 Find the voltage drop across 1.1 Ω resistor in the circuit of Fig. 2.30.

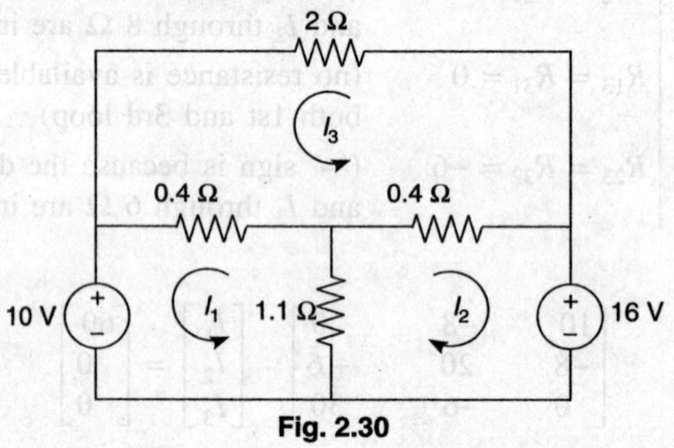

Fig. 2.30

Solution To find the voltage drop across 1.1 Ω resistor, we need to calculate the current flowing through it.

By inspection, we can write

$$\begin{bmatrix} 1.5 & 1.1 & -0.4 \\ 1.1 & 1.5 & 0.4 \\ -0.4 & 0.4 & 2.8 \end{bmatrix} \begin{bmatrix} I_1 \\ I_2 \\ I_3 \end{bmatrix} = \begin{bmatrix} -10 \\ -16 \\ 0 \end{bmatrix}$$

By applying Cramer's rule,

$$I_1 = \frac{\Delta I_1}{\Delta} = \frac{11.44}{2.08} = 5.5 \text{ A}$$

$$I_2 = \frac{\Delta I_2}{\Delta} = \frac{-32.24}{2.08} = -15.5 \text{ A}$$

Therefore, the current through 1.1 Ω resistor is 10 A in the direction from top to bottom. Hence, the voltage drop across it is

$$10 \times 1.1 = \mathbf{11\ V}$$

EXAMPLE 2.15 Use mesh analysis method and find the three mesh currents in the circuit shown in Fig. 2.31.

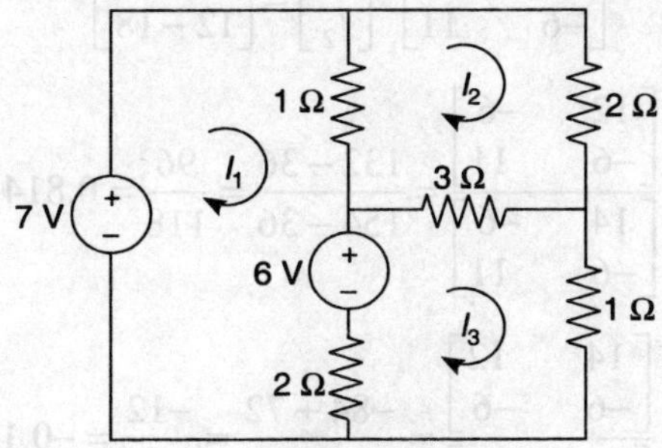

Fig. 2.31

Solution By inspection, we can write:

$$\begin{bmatrix} 3 & -1 & -2 \\ -1 & 6 & -3 \\ -2 & -3 & 6 \end{bmatrix} \begin{bmatrix} I_1 \\ I_2 \\ I_3 \end{bmatrix} = \begin{bmatrix} 7-6 \\ 0 \\ 6 \end{bmatrix}$$

By applying Cramer's rule,

$$I_1 = \frac{\Delta I_1}{\Delta} = \frac{117}{39} = \mathbf{3\ A}$$

$$I_2 = \frac{\Delta I_2}{\Delta} = \frac{78}{39} = \mathbf{2\ A}$$

$$I_3 = \frac{\Delta I_3}{\Delta} = \frac{117}{39} = \mathbf{3\ A}$$

EXAMPLE 2.16 In the circuit shown in Fig. 2.32,

(i) Use mesh method to find out the power delivered to 6 Ω resistor.

(ii) To what voltage should the 18 V battery be changed so that no power is delivered to the 6 Ω resistor?

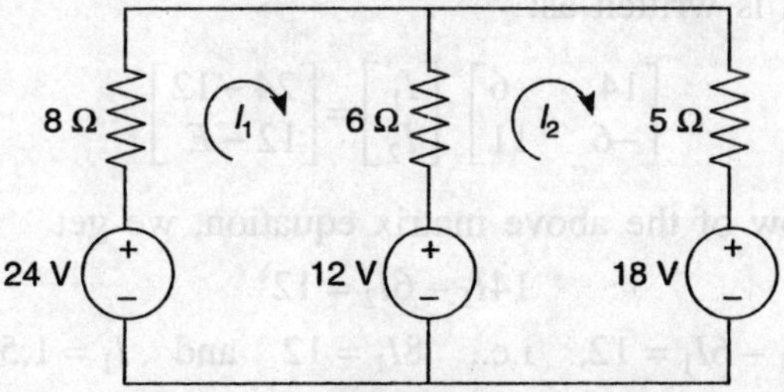

Fig. 2.32

Solution

(i) To find the value of power delivered to 6 Ω resistor

By inspection we can write

$$\begin{bmatrix} 14 & -6 \\ -6 & 11 \end{bmatrix} \begin{bmatrix} I_1 \\ I_2 \end{bmatrix} = \begin{bmatrix} 24-12 \\ 12-18 \end{bmatrix}$$

$$I_1 = \frac{\begin{bmatrix} 12 & -6 \\ -6 & 11 \end{bmatrix}}{\begin{bmatrix} 14 & -6 \\ -6 & 11 \end{bmatrix}} = \frac{132-36}{154-36} = \frac{96}{118} = 0.814 \text{ A}$$

$$I_2 = \frac{\begin{bmatrix} 14 & 12 \\ -6 & -6 \end{bmatrix}}{\begin{bmatrix} 14 & -6 \\ -6 & 11 \end{bmatrix}} = \frac{-84+72}{154-36} = \frac{-12}{118} = -0.1 \text{ A}$$

Hence, the current through 6 Ω resistor is 0.914 A.
Power delivered to 6 Ω resistor is $0.914^2 \times 6$ = **5.012 W**.

(ii) To find the replacement battery value for 18 V battery so that no power is delivered to 6 Ω resistor

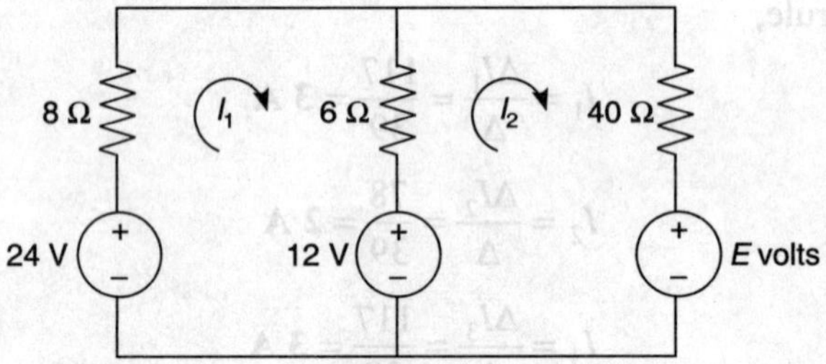

Fig. 2.33

Firstly, the 18 V battery is replaced by an unknown battery of E volts. Then, power delivered is zero means no current flows through the resistor or the two currents of equal value are flowing in the opposite directions. Therefore, in this case currents I_1 and I_2 which are flowing through 6 Ω are equal in magnitude. The matrix equation for this condition is written as:

$$\begin{bmatrix} 14 & -6 \\ -6 & 11 \end{bmatrix} \begin{bmatrix} I_1 \\ I_2 \end{bmatrix} = \begin{bmatrix} 24-12 \\ 12-E \end{bmatrix}$$

Using the first row of the above matrix equation, we get

$$14I_1 - 6I_2 = 12$$

$$14I_1 - 6I_1 = 12, \quad \text{i.e.,} \quad 8I_1 = 12 \quad \text{and} \quad I_1 = 1.5 \text{ A}$$

Substituting I_1 = 1.5 into the second row of matrix equation, we get

$$5I_1 = 12 - E \quad \text{and} \quad E = 4.5$$

Therefore, the value of 18 V has to be changed to **4.5 V** to make the power delivered to 6 Ω resistor be zero.

EXAMPLE 2.17 Find the current through 4 Ω resistance by mesh analysis method in the network shown in Fig. 2.34.

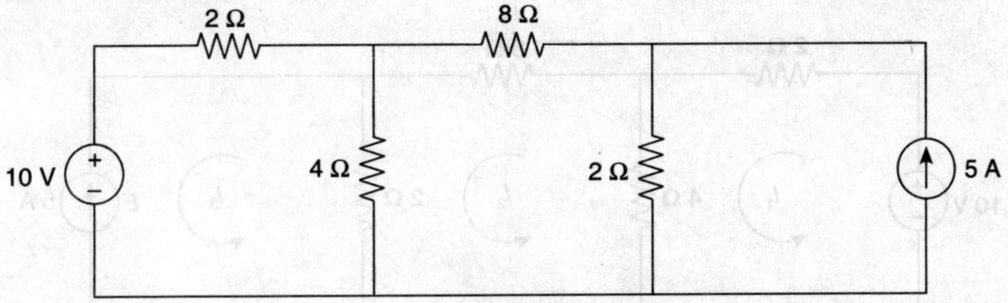

Fig. 2.34

Solution

Method I To start-with, the 5 A current source with its parallel resistance of 2 Ω shall be converted into an equivalent voltage source of 5 × 2 = 10 V and series resistance of 2 Ω. The redrawn circuit is shown in Fig. 2.35.

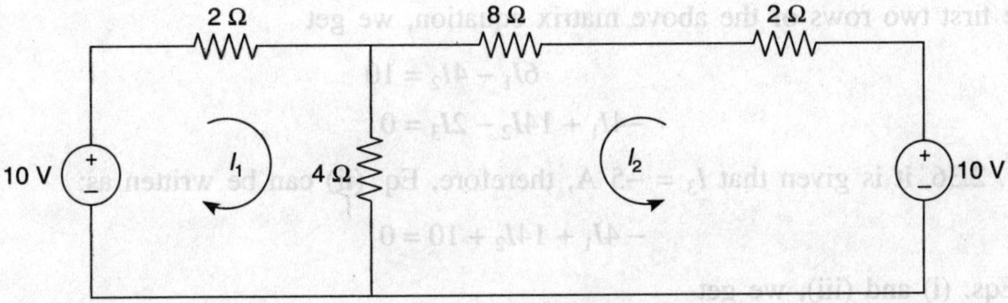

Fig. 2.35

Now the matrix equation is written as:

$$\begin{bmatrix} 6 & 4 \\ 4 & 14 \end{bmatrix} \begin{bmatrix} I_1 \\ I_2 \end{bmatrix} = \begin{bmatrix} 10 \\ 10 \end{bmatrix}$$

$$\Delta = \begin{vmatrix} 6 & 4 \\ 4 & 14 \end{vmatrix} = 84 - 16 = 68$$

$$\Delta I_1 = \begin{vmatrix} 10 & 4 \\ 10 & 14 \end{vmatrix} = 140 - 40 = 100$$

$$\Delta I_2 = \begin{vmatrix} 6 & 10 \\ 4 & 10 \end{vmatrix} = 60 - 40 = 20$$

$$I_1 = \frac{100}{68} = 1.47 \text{ A}; \quad I_2 = \frac{20}{68} = 0.29 \text{ A}$$

Hence, the current through 4 Ω resistor = $I_1 + I_2$ = 1.47 + 0.29 = **1.76 A.**

Method II This method does not use the source transformation technique. It directly applies the Maxwell's mesh method by assuming voltage E across the current source. And the matrix equation is written as usual as follows:

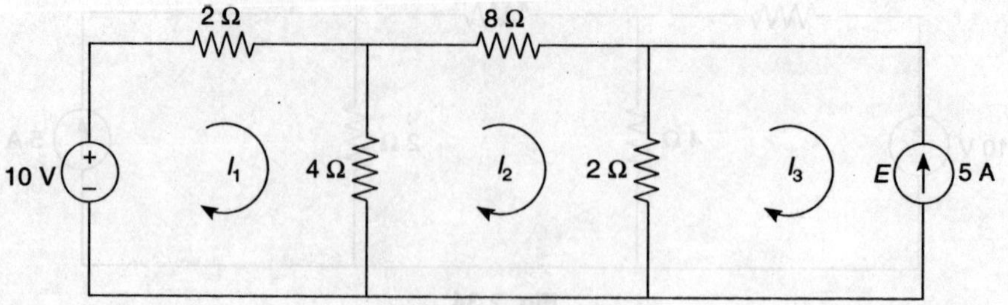

Fig. 2.36

$$\begin{bmatrix} 6 & -4 & 0 \\ -4 & 14 & -2 \\ 0 & -2 & 2 \end{bmatrix} \begin{bmatrix} I_1 \\ I_2 \\ I_3 \end{bmatrix} = \begin{bmatrix} 10 \\ 0 \\ -E \end{bmatrix}$$

Using the first two rows of the above matrix equation, we get

$$6I_1 - 4I_2 = 10 \quad \text{(i)}$$

$$-4I_1 + 14I_2 - 2I_3 = 0 \quad \text{(ii)}$$

From Fig. 2.36, it is given that $I_3 = -5$ A, therefore, Eq. (ii) can be written as:

$$-4I_1 + 14I_2 + 10 = 0 \quad \text{(iii)}$$

Solving Eqs. (i) and (iii), we get

$$I_1 = 1.47 \text{ A}$$

$$I_2 = -0.29 \text{ A}$$

Therefore, the current through 4 Ω resistor = $I_1 - I_2 = 1.47 - (-0.29) =$ **1.76 A.**

EXAMPLE 2.18 Find the current through 2 Ω resistance by mesh analysis method in the network shown in Fig. 2.37.

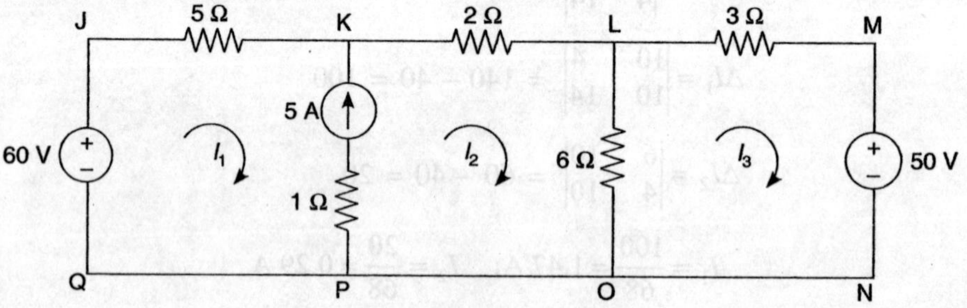

Fig. 2.37

Solution

Method I A fictitious voltage of E is assumed across the current source as shown in Fig. 2.38.

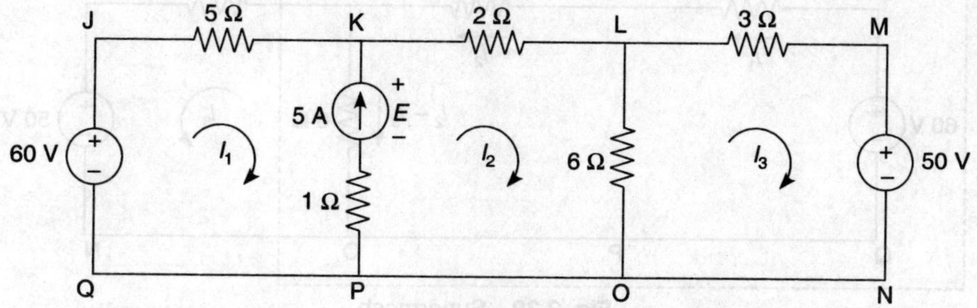

Fig. 2.38

$$\begin{bmatrix} 6 & -1 & 0 \\ -1 & 9 & -6 \\ 0 & -6 & 9 \end{bmatrix} \begin{bmatrix} I_1 \\ I_2 \\ I_3 \end{bmatrix} = \begin{bmatrix} 60 - E \\ E \\ -50 \end{bmatrix}$$

Matrix equation is expanded as:

$$6I_1 - I_2 = 60 - E \quad \text{(i)}$$

$$-I_1 + 9I_2 - 6I_3 = E \quad \text{(ii)}$$

$$-6I_2 + 9I_3 = -50 \quad \text{(iii)}$$

In branch KP, the current flow is 5 A, i.e.,

$$I_2 - I_1 = 5 \quad \text{(iv)}$$

Substituting $I_1 = I_2 - 5$ into Eq. (i), we get

$$5I_2 = 90 - E \quad \text{(v)}$$

Again, substituting $I_1 = I_2 - 5$ into Eq. (ii), we get

$$8I_2 - 6I_3 = E - 5 \quad \text{(vi)}$$

Equation (vi) × 3	⇒	$24I_2 - 18I_3$	=	$3E - 15$	
Equation (iii) × 2	⇒	$-12I_2 + 18I_3$	=	-100	
		$12I_2$	=	$3E - 115$	(vii)
Equation (vii) × 1	⇒	$12I_2$	=	$3E - 115$	
Equation (v) × 3	⇒	$15I_2$	=	$-3E + 270$	
		$27I_2$	=	155	(viii)

Solving, $\quad$ $I_2 = \mathbf{5.74\ A}$

Method II Since the current source is common for mesh 1 and mesh 2, this problem can be solved by *supermesh analysis*. We need three equations involving the three unknown mesh currents, I_1, I_2 and I_3. The first equation is written as:

$$I_2 - I_1 = 5 \quad \text{(i)}$$

The second equation is written by removing the current source and forming the supermesh as shown in Fig. 2.39.

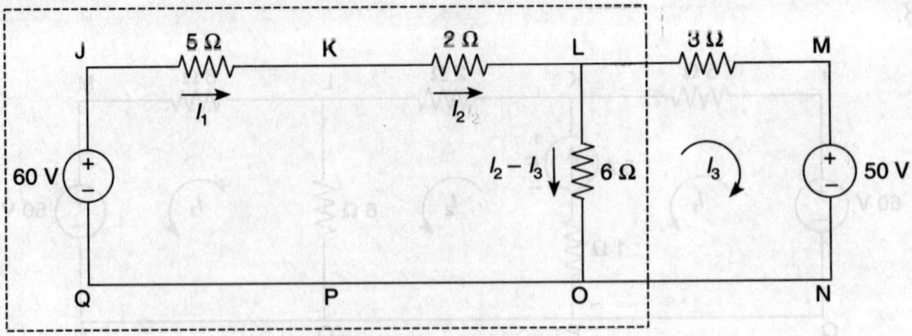

Fig. 2.39 Supermesh.

KVL is written for the supermesh as:

$$5I_1 + 2I_2 + 6(I_2 - I_3) - 60 = 0$$

$$5I_1 + 8I_2 - 6I_3 = 60 \tag{ii}$$

The third equation is written considering the mesh which does not involve the current source as shown dotted in Fig. 2.40.

$$3I_3 + 50 + 6(I_3 - I_2) = 0$$

$$9I_3 - 6I_2 = -50 \tag{iii}$$

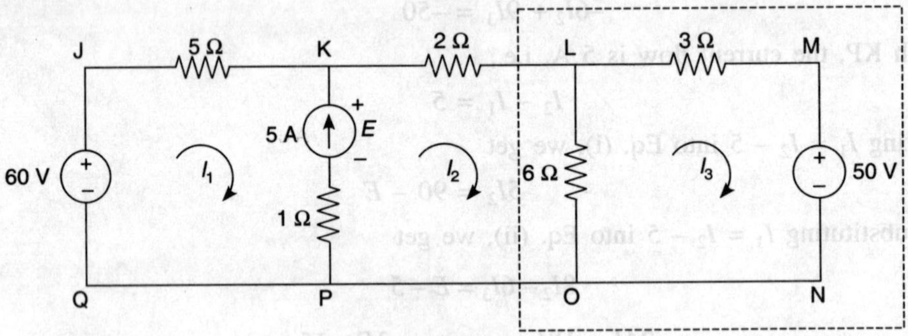

Fig. 2.40

Substituting I_1 value from Eq. (i) into Eq. (ii), we get

$$13I_2 - 6I_3 = 85 \tag{iv}$$

Equation (iv) × 3 ⇒ $39I_2 - 18I_3 = 255$

Equation (iii) × 2 ⇒ $-12I_2 + 18I_3 = -100$

$$27I_2 = 155 \tag{v}$$

$$I_2 = \frac{155}{27} = 5.74 \text{ A}$$

Hence, the current flowing through 2 Ω resistor is **5.74 A.**

EXAMPLE 2.19 Determine the loop currents for the circuit shown in Fig. 2.41 by using mesh analysis.

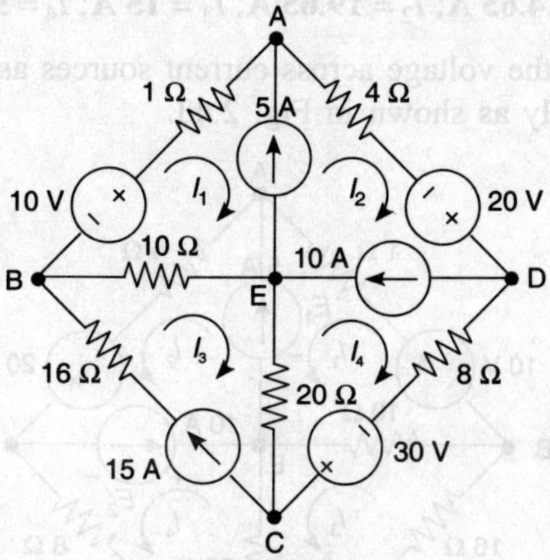

Fig. 2.41

Solution

Method I

Current in branch AE, $I_2 - I_1 = 5$ (i)

Current in branch BC, $I_3 = 15$ (ii)

Current in branch DE, $I_2 - I_4 = 10$ (iii)

Note: Supermesh cannot be formed using mesh 3 since the 15 A current source lies on the perimeter.

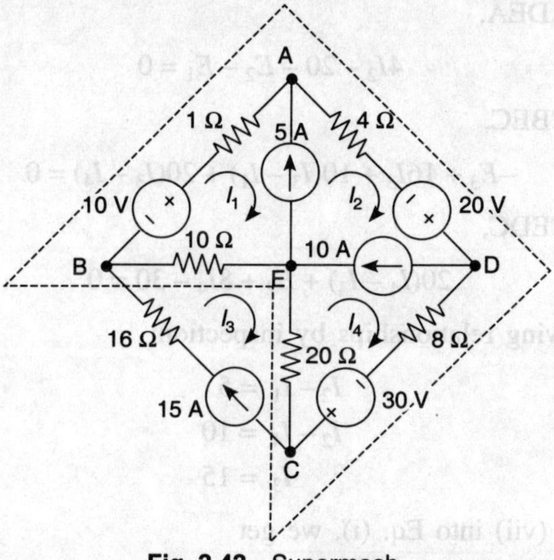

Fig. 2.42 Supermesh.

The supermesh is formed as shown in Fig. 2.42 and the equation is written as:

$$-10 + I_1 + 4I_2 - 20 + 8I_4 - 30 + 20(I_4 - I_3) + 10(I_1 - I_3) = 0$$

or

$$11I_1 + 4I_2 - 30I_3 + 28I_4 = 60 \quad \text{(iv)}$$

Solving these four equations we get,

$$I_1 = 14.65 \text{ A}; I_2 = 19.65 \text{ A}; I_3 = 15 \text{ A}; I_4 = 9.65 \text{ A}$$

Method II Let us assume the voltage across current sources as E_1, E_2 and E_3 for 5 A, 10 A and 15 A sources respectively as shown in Fig. 2.43.

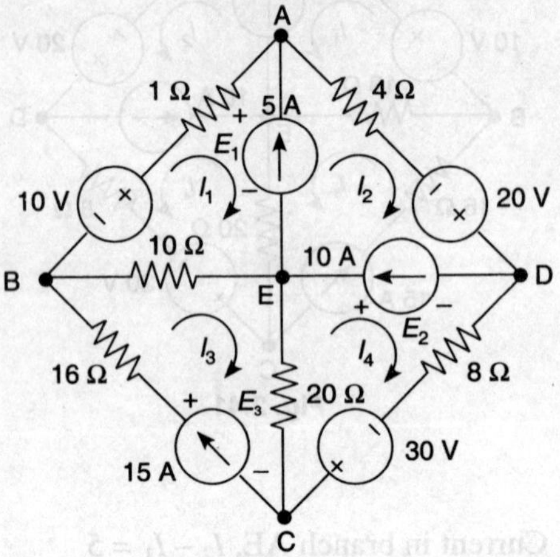

Fig. 2.43

Writing KVL for loop BAEB,

$$-10 + I_1 + E_1 + 10(I_1 - I_3) = 0 \quad \text{(i)}$$

Writing KVL for loop ADEA,

$$4I_2 - 20 - E_2 - E_1 = 0 \quad \text{(ii)}$$

Writing KVL for loop CBEC,

$$-E_3 + 16I_3 + 10(I_3 - I_1) + 20(I_3 - I_4) = 0 \quad \text{(iii)}$$

Writing KVL for loop CEDC,

$$20(I_4 - I_3) + E_2 + 8I_4 - 30 = 0 \quad \text{(iv)}$$

Also we write the following relationships by inspection:

$$I_2 - I_1 = 5 \quad \text{(v)}$$

$$I_2 - I_4 = 10 \quad \text{(vi)}$$

$$I_3 = 15 \quad \text{(vii)}$$

Substituting I_3 from Eq. (vii) into Eq. (i), we get

$$11I_1 + E_1 = 160 \quad \text{(viii)}$$

Substituting the value of I_2 from Eq. (v) into Eq. (ii), we get

$$4I_1 - E_1 - E_2 = 0 \quad \text{(ix)}$$

Substituting Eqs. (v), (vi) and (vii) into Eq. (iv), we get

$$28I_1 + E_2 = 470 \tag{x}$$

Solving Eqs. (viii), (ix) and (x), we get

$$\mathbf{I_1 = 14.65\ A;\ I_2 = 19.65\ A;\ I_3 = 15\ A;\ I_4 = 9.65\ A}$$

EXAMPLE 2.20 Using mesh analysis method, find the three mesh currents in the circuit shown in Fig. 2.44.

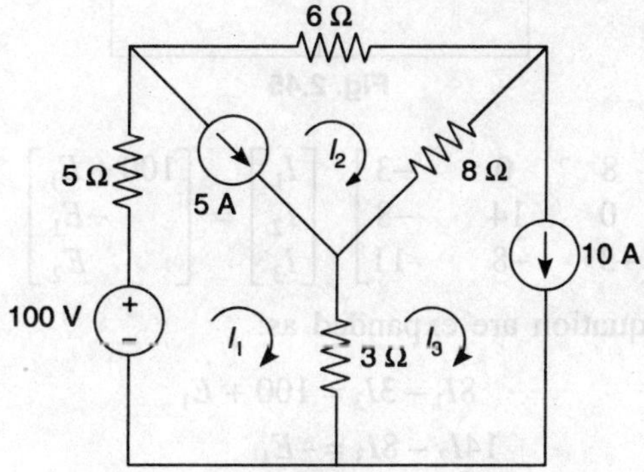

Fig. 2.44

Solution

Method I

$$I_3 = 10\ A \tag{i}$$

$$I_1 - I_2 = 5\ A \tag{ii}$$

Supermesh is formed using mesh 1 and mesh 2 and the equation is written as:

$$-100 + 5I_1 + 6I_2 + 8(I_2 - I_3) + 3(I_1 - I_3) = 0$$

or

$$8I_1 + 14I_2 - 11I_3 = 100 \tag{iii}$$

Substituting I_3 = 10 A into Eq. (iii), we get

$$8I_1 + 14I_2 = 210 \tag{iv}$$

Solving Eqs. (ii) and (iv), we get

$$I_1 = 12.72\ A$$

$$I_2 = 7.73\ A$$

and

$$I_3 = 10\ A$$

Note: Supermesh cannot be formed using mesh 3 since the 10 A current source lies on its perimeter.

Method II Let us assume the voltage across current sources as E_1 and E_2 for 5 A source and 10 A source respectively as shown in Fig. 2.45.

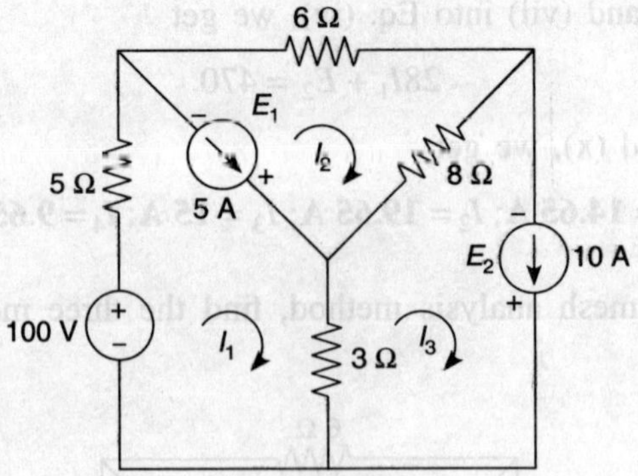

Fig. 2.45

$$\begin{bmatrix} 8 & 0 & -3 \\ 0 & 14 & -8 \\ -3 & -8 & 11 \end{bmatrix} \begin{bmatrix} I_1 \\ I_2 \\ I_3 \end{bmatrix} = \begin{bmatrix} 100 + E_1 \\ -E_1 \\ E_2 \end{bmatrix}$$

First two rows of matrix equation are expanded as:

$$8I_1 - 3I_3 = 100 + E_1 \quad \text{(i)}$$

$$14I_2 - 8I_3 = -E_1 \quad \text{(ii)}$$

Substituting I_3 = 10 into Eqs. (i) and (ii), we get

$$8I_1 = 130 + E_1 \quad \text{(iii)}$$

$$14I_2 = 80 - E_1 \quad \text{(iv)}$$

Substituting $I_1 - I_2 = 5$ into Eq. (iii) and solving with Eq. (iv), we get

$$I_1 = \mathbf{12.72\ A}$$

$$I_2 = \mathbf{7.73\ A}$$

and $$I_3 = \mathbf{10\ A}$$

2.7 NODAL ANALYSIS

Consider the circuit shown in Fig. 2.46. The branch currents in the circuit can be found by Kirchhoff's laws or Maxwell's mesh current method. There is another method, called *nodal analysis*, for determining the branch currents in the circuit. In this method, one of the nodes (a node is a point in a network where two or more circuit elements meet) is taken as the *reference node*. The potentials of all the points in the circuit are measured w.r.t. this reference node.

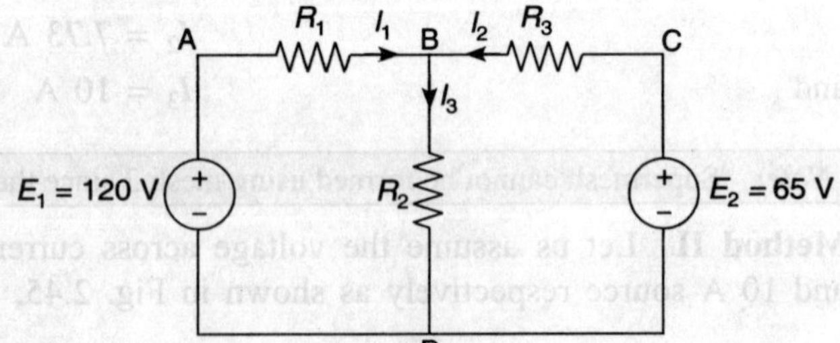

Fig. 2.46

In Fig. 2.46, A, B, C and D are four nodes and node D has been taken as the reference node. A glance at the circuit shows that voltages at nodes A and C w.r.t. the reference node are known. These are E_1 = 120 V and E_2 = 65 V. The only potential of node B w.r.t. D (call it V_B) is unknown. If this potential V_B can be found, each branch current can be determined because the voltage across each resistor will then be known. The voltage V_B can be found by applying Kirchhoff's current law at point B.

$$I_1 + I_2 = I_3 \tag{2.5}$$

In mesh ABDA, the voltage drop across R_1 is $E_1 - V_B$

$$\therefore \qquad I_1 = \frac{E_1 - V_B}{R_1}$$

In mesh CBDC, the voltage drop across R_3 is $E_2 - V_B$

$$\therefore \qquad I_2 = \frac{E_2 - V_B}{R_3}$$

Also, Current, $I_3 = \dfrac{V_B}{R_2}$

Putting the values of I_1, I_2 and I_3 into Eq. (2.5), we get

$$\frac{E_1 - V_B}{R_1} + \frac{E_2 - V_B}{R_3} = \frac{V_B}{R_2}$$

V_B is calculated from the above equation from which the branch currents are determined.

EXAMPLE 2.21 Find the currents in the various branches of the circuit shown in Fig. 2.47.

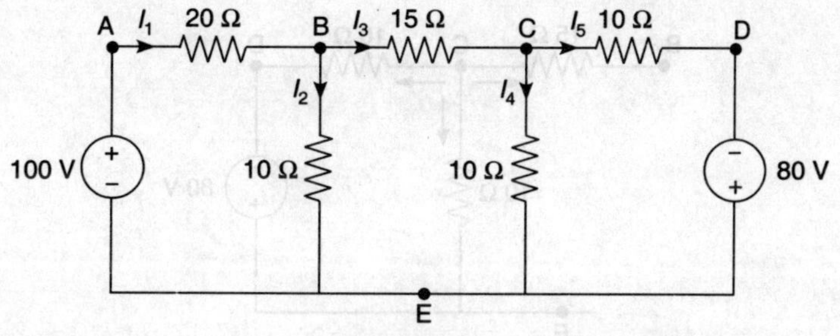

Fig. 2.47

Solution Node equations are written using Kirchhoff's current law. The node equation for a node is formed by equating the sum of currents leaving that node to the sum of currents entering that node. While writing the node equation for a node, it is assumed that all the currents are leaving that node irrespective of what is happening by these currents in other nodes. If current source exists in a particular node, the direction of its current is taken as such, i.e. +ve, if the current leaves that node and –ve, if the current enters.

Referring to Fig. 2.48, the node equation for node B is formed as:

$$\frac{V_B - V_A}{20} + \frac{V_B - V_E}{10} + \frac{V_B - V_C}{15} = 0 \quad \text{(i)}$$

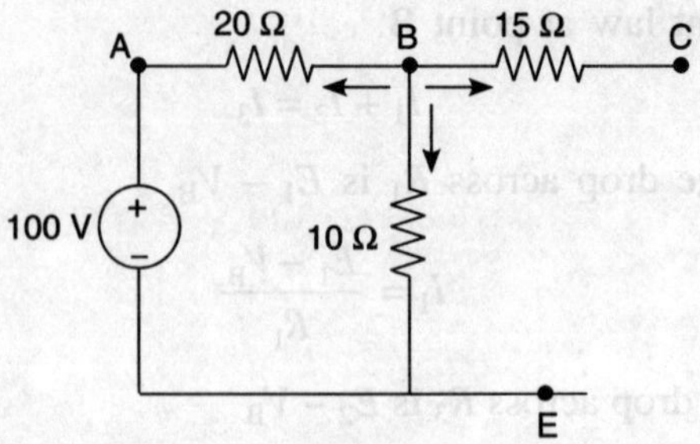

Fig. 2.48

Since E is taken as reference node, its voltage V_E is zero. The voltage at point A, w.r.t the reference node, V_A is 100 V. Substituting the values of V_A and V_E into Eq. (i), we get

$$\frac{V_B - 100}{20} + \frac{V_B}{10} + \frac{V_B - V_C}{15} = 0$$

or

$$13V_B - 4V_C = 300 \quad \text{(ii)}$$

Referring to Fig. 2.49, the node equation for node C is formed as:

$$\frac{V_C - V_B}{15} + \frac{V_C - V_E}{10} + \frac{V_C + 80}{10} = 0 \quad \text{(iii)}$$

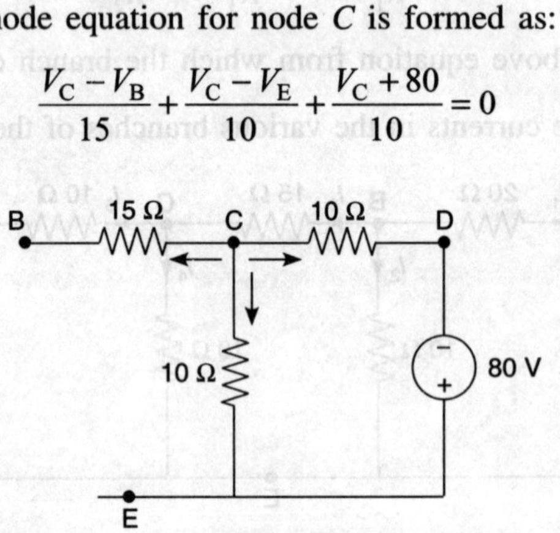

Fig. 2.49

Since $V_E = 0$, Eq. (iii) becomes

$$V_B - 4V_C = 120 \quad \text{(iv)}$$

Solving Eqs. (ii) and (iv), we get

$$V_B = 15 \text{ V}$$

and

$$V_C = -26.25 \text{ V}$$

From which,

$$\text{Current, } I_1 = \frac{100 - V_B}{20} = \frac{100-15}{20} = \mathbf{4.25\ A}$$

$$\text{Current, } I_2 = \frac{V_B}{10} = \frac{15}{10} = \mathbf{1.5\ A}$$

$$\text{Current, } I_3 = \frac{V_B - V_C}{15} = \frac{15-(-26.25)}{15} = \mathbf{2.75\ A}$$

$$\text{Current, } I_4 = \frac{V_C}{10} = \frac{-26.25}{10} = \mathbf{-2.625\ A}$$

$$\text{Current, } I_5 = \frac{V_C + 80}{10} = \frac{-26.25+80}{10} = \mathbf{5.375\ A}$$

EXAMPLE 2.22 For the circuit shown in Fig. 2.50, find (i) node voltages V_A and V_B (ii) currents in various branches.

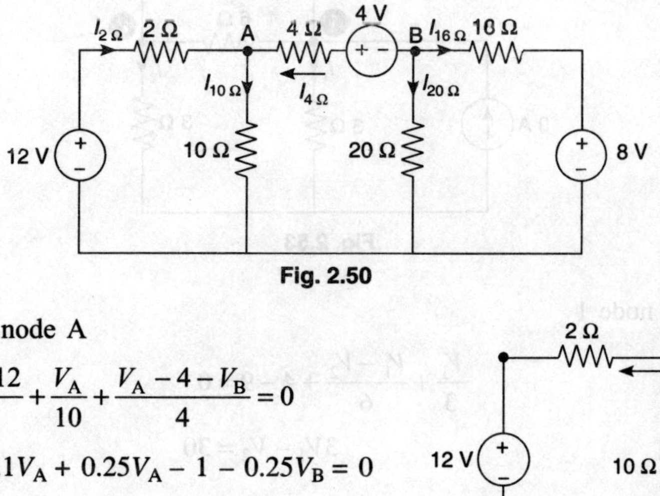

Fig. 2.50

Solution KCL at node A

$$\frac{V_A - 12}{2} + \frac{V_A}{10} + \frac{V_A - 4 - V_B}{4} = 0$$

or $\quad 0.5V_A - 6 + 0.1V_A + 0.25V_A - 1 - 0.25V_B = 0$

or $\quad 0.85V_A - 0.25V_B = 7 \qquad$ (i)

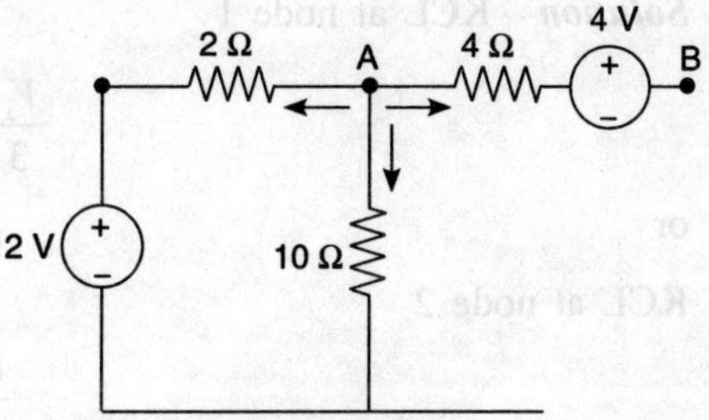

Fig. 2.51

KCL at node B

$$\frac{V_B + 4 - V_A}{4} + \frac{V_B}{20} + \frac{V_B - 8}{16} = 0$$

$$-0.25V_A + 0.3625V_B = -\ 0.5 \qquad \text{(ii)}$$

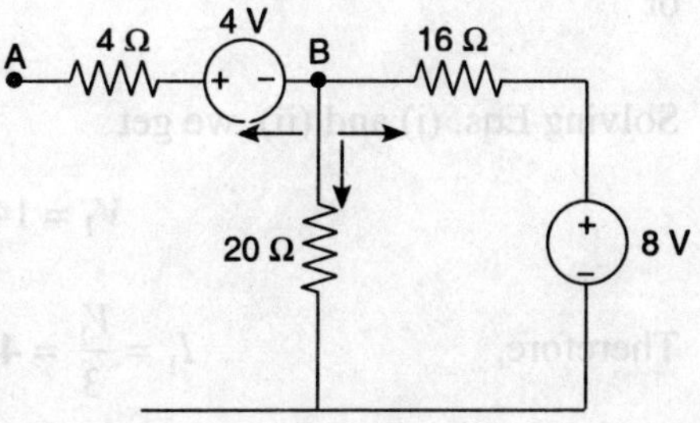

Fig. 2.52

Solving Eqs. (i) and (ii), we get

$$V_A = \mathbf{9.82\ V} \quad \text{and} \quad V_B = \mathbf{5.4\ V}$$

Therefore, current, $I_{2\,\Omega} = \dfrac{12 - V_A}{2} = \dfrac{12 - 9.82}{2} = \mathbf{1.09\ A}$

$$\text{Current, } I_{10\,\Omega} = \frac{V_A}{10} = \frac{9.82}{10} = \mathbf{0.982\ A}$$

$$\text{Current, } I_{4\,\Omega} = \frac{V_D + 4 - V_A}{4} = \frac{5.4 + 4 - 9.82}{4} = \mathbf{-0.105\ A}$$

$$\text{Current, } I_{20\,\Omega} = \frac{V_B}{20} = \frac{5.4}{20} = \mathbf{0.27\ A}$$

$$\text{Current, } I_{16\,\Omega} = \frac{8 - V_B}{16} = \frac{8 - 5.4}{16} = \mathbf{0.162\ A}$$

EXAMPLE 2.23 Find I_1 and I_2 in Fig. 2.53 with the help of nodal analysis.

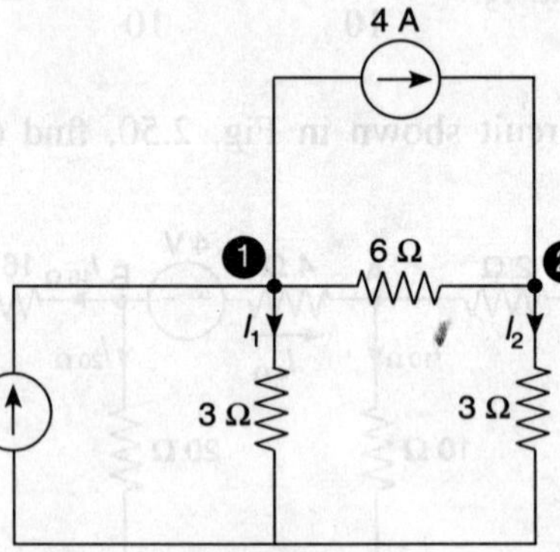

Fig. 2.53

Solution KCL at node 1

$$\frac{V_1}{3} + \frac{V_1 - V_2}{6} + 4 - 9 = 0$$

or $$3V_1 - V_2 = 30 \qquad \text{(i)}$$

KCL at node 2

$$\frac{V_2}{3} + \frac{V_2 - V_1}{6} - 4 = 0$$

or $$-V_1 + 3V_2 = 24 \qquad \text{(ii)}$$

Solving Eqs. (i) and (ii), we get

$$V_1 = 14.25\ \text{V} \quad \text{and} \quad V_2 = 12.75\ \text{V}$$

Therefore, $$I_1 = \frac{V_1}{3} = \mathbf{4.75\ A} \quad \text{and} \quad I_2 = \frac{V_2}{3} = \mathbf{4.25\ A}$$

EXAMPLE 2.24 Find I in the network shown in Fig. 2.54 by applying nodal analysis.

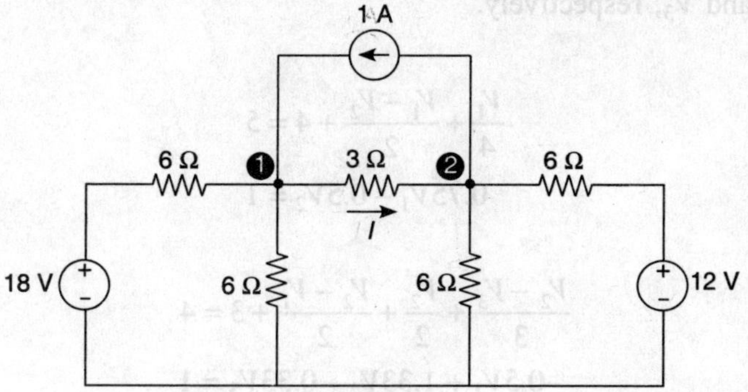

Fig. 2.54

Solution Let the voltages at nodes 1 and 2 are V_1 and V_2.
KCL at node 1

$$\frac{V_1-18}{6}+\frac{V_1}{6}+\frac{V_1-V_2}{3}-1=0$$

or
$$2V_1-V_2=12 \qquad \text{(i)}$$

KCL at node 2

$$\frac{V_2-12}{6}+\frac{V_2}{6}+\frac{V_2-V_1}{3}+1=0$$

or
$$4V_2-2V_1=6 \qquad \text{(ii)}$$

Solving Eqs. (i) and (ii), we get

$$V_1=9\text{ V} \quad \text{and} \quad V_2=6\text{ V}$$

Therefore,
$$I=\frac{V_1-V_2}{3}=\frac{9-6}{3}=\mathbf{1\ A}$$

EXAMPLE 2.25 Find the current through 5 Ω resistor using nodal analysis in the circuit shown in Fig. 2.55.

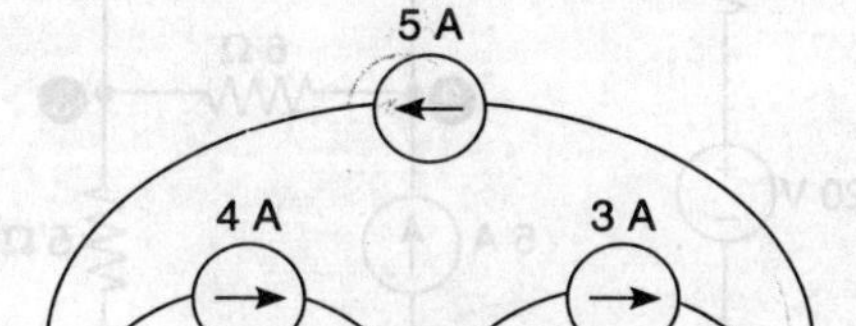

Fig. 2.55

Solution Taking node D as reference node, the unknown voltages at nodes A, B and C are assumed as V_1, V_2 and V_3, respectively.

KCL at node A

$$\frac{V_1}{4} + \frac{V_1 - V_2}{2} + 4 = 5$$

or

$$0.75V_1 - 0.5V_2 = 1 \qquad \text{(i)}$$

KCL at node B

$$\frac{V_2 - V_3}{3} + \frac{V_2}{2} + \frac{V_2 - V_1}{2} + 3 = 4$$

or

$$-0.5V_1 + 1.33V_2 - 0.33V_3 = 1 \qquad \text{(ii)}$$

KCL at node C

$$\frac{V_3}{5} + \frac{V_3 - V_2}{3} + 5 = 3$$

or

$$-0.33V_2 + 0.53V_3 = -2 \qquad \text{(iii)}$$

Solving Eqs. (i), (ii) and (iii) gives

$$V_1 = 1.6884 \text{ V}, V_2 = 0.5326 \text{ V}, V_3 = -3.442 \text{ V}$$

Therefore, the current through 5 Ω resistor $= \frac{V_3}{5} = \frac{-3.442}{5} = \mathbf{-0.6884\ A}$

EXAMPLE 2.26 Use nodal analysis to find the power dissipated in 6 Ω resistor in the circuit shown in Fig. 2.56.

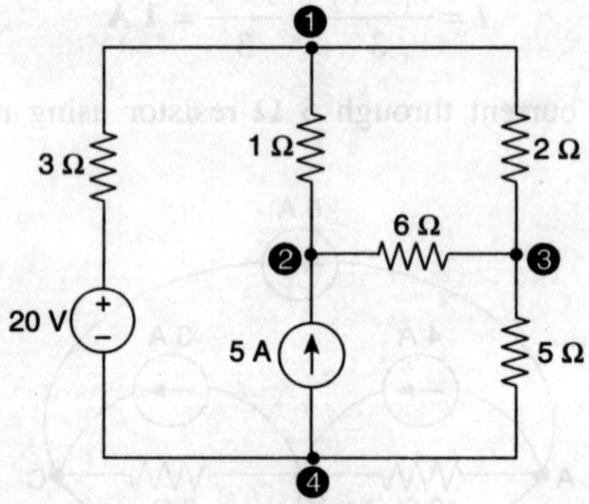

Fig. 2.56

Solution Taking node 4 as reference node, the unknown voltages at nodes 1, 2 and 3 are assumed as V_1, V_2 and V_3, respectively.

KCL at node 1

$$\frac{V_1 - 20}{3} + \frac{V_1 - V_2}{1} + \frac{V_1 - V_3}{2} = 0$$

or $$1.83V_1 - V_2 - 0.5V_3 = 6.67 \quad \text{(i)}$$

KCL at node 2

$$\frac{V_2 - V_1}{1} + \frac{V_2 - V_3}{6} = 5$$

or $$-V_1 - 1.167V_2 - 0.167V_3 = 5 \quad \text{(ii)}$$

KCL at node 3

$$\frac{V_3 - V_1}{2} + \frac{V_3 - V_2}{6} + \frac{V_3}{5} = 0$$

or $$-0.5V_1 - 0.167V_2 + 0.867V_3 = 0 \quad \text{(iii)}$$

Solving Eqs. (i), (ii) and (iii), we get

$$V_1 = 0.83\text{ V},\ V_2 = -4.92\text{ V},\ V_3 = -0.47\text{ V}$$

Hence, the current through 6 Ω resistor is $\frac{V_2 - V_3}{6} = \frac{-4.92 + 0.47}{6} = -0.74$ A.

The power dissipated in 6 Ω resistor = $(0.74)^2 \times 6$ = **3.29 W**.

EXAMPLE 2.27 Use nodal analysis and find the current through 5 Ω resistor in the circuit shown in Fig. 2.57.

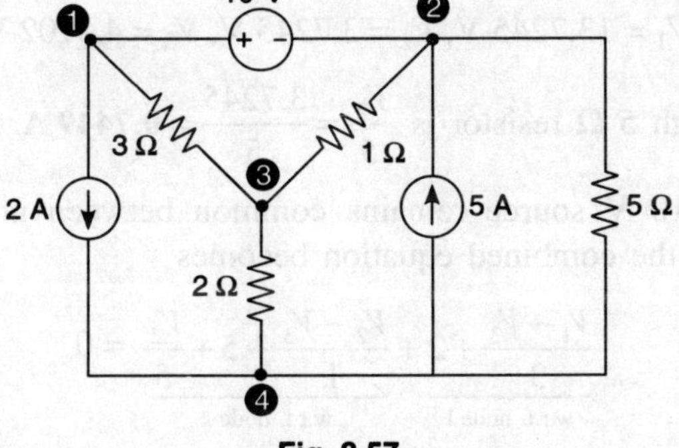

Fig. 2.57

Solution Node 4 is taken as reference node and the voltages at the other nodes are assumed as V_1, V_2 and V_3 at nodes 1, 2 and 3, respectively.

Method I Since it has a voltage source without a resistor in series with it, a current of I_x amperes is assumed to flow through it as shown in Fig. 2.58.

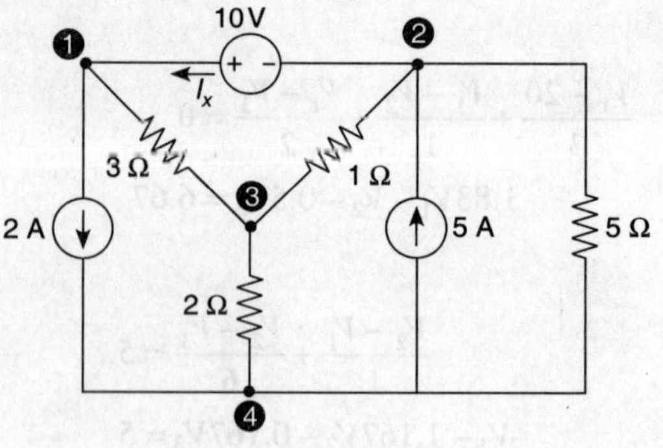

Fig. 2.58

KCL at the three nodes is formed as follows:

$$\frac{V_1 - V_3}{3} + 2 - I_x = 0$$

or $$0.33V_1 - 0.33V_3 = I_x - 2 \quad \text{(i)}$$

$$\frac{V_2 - V_3}{1} + \frac{V_2}{5} + I_x - 5 = 0$$

or $$1.2V_2 - V_3 = -I_x + 5 \quad \text{(ii)}$$

$$\frac{V_3}{2} + \frac{V_3 - V_1}{3} + \frac{V_3 - V_2}{1} = 0$$

or $$-0.33V_1 - V_2 + 1.83V_3 = 0 \quad \text{(iii)}$$

Substituting, $V_1 - V_2 = 10$ and solving the above, we get

$$V_1 = 13.7245\text{ V}, V_2 = 3.7245\text{ V}, V_3 = 4.5102\text{ V}$$

Hence, the current through 5 Ω resistor is $\frac{V_2}{5} = \frac{3.7245}{5} = \mathbf{0.7449\ A}$.

Method II Since the 10 V source remains common between nodes 1 and 2, *supernode* technique is applied and the combined equation becomes

$$\underbrace{\frac{V_1 - V_3}{3} + 2}_{\text{w.r.t. node 1}} + \underbrace{\frac{V_2 - V_3}{1} - 5 + \frac{V_2}{5}}_{\text{w.r.t. node 2}} = 0 \quad \text{(i)}$$

KCL at node 3 is written as:

$$\frac{V_3}{2} + \frac{V_3 - V_1}{3} + \frac{V_3 - V_2}{1} = 0 \quad \text{(ii)}$$

Also, $$V_1 - V_2 = 10 \quad \text{(iii)}$$

Solving Eqs. (i), (ii) and (iii), we get

$$V_1 = 13.7245 \text{ V}, V_2 = 3.7245 \text{ V}, V_3 = 4.5102 \text{ V}$$

Hence, the current through 5 Ω resistor is $\frac{V_2}{5} = \frac{3.7245}{5}$ = **0.7449 A**.

EXAMPLE 2.28 In the circuit shown in Fig. 2.59, find the voltage across 25 Ω resistor and the power supplied by 5 A source, by node analysis.

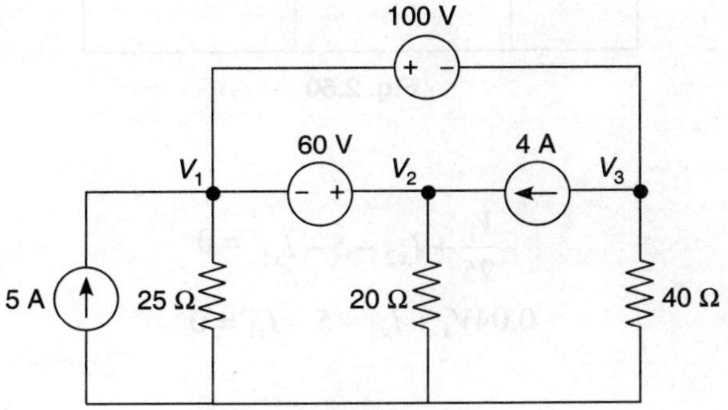

Fig. 2.59

Solution

Method I From Fig. 2.59,

$$V_1 - V_3 = 100 \quad \text{(i)}$$

and

$$V_2 - V_1 = 60 \quad \text{(ii)}$$

Since 100 V source is connected between node 1 and node 3 and 60 V source is connected between node 1 and node 2, the supernode equation is formed as:

$$\underbrace{-5 + \frac{V_1}{25}}_{\text{w.r.t. node 1}} + \underbrace{\frac{V_2}{20} - 4}_{\text{w.r.t. node 2}} + \underbrace{\frac{V_3}{40} + 4}_{\text{w.r.t. node 3}} = 0 \quad \text{(iii)}$$

Substituting Eqs. (i) and (ii) into Eq. (iii), we get

$$-5 + \frac{V_1}{25} + \frac{60 + V_1}{20} - 4 + \frac{V_1 - 100}{40} + 4 = 0$$

Solving,

$$V_1 = 39.13 \text{ V}$$

Hence, the power supplied by the 5 A source $= V_1 \times 5$

$$= 39.13 \times 5 = \mathbf{195.65\ W}$$

Method II In this method, two currents I_{x1} and I_{x2} are assumed to be flowing through the 100 V source and 60 V source, respectively as shown in Fig. 2.60.

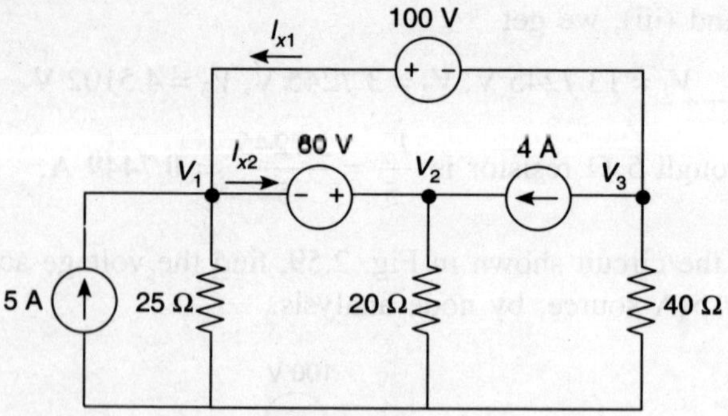

Fig. 2.60

KCL at node 1 is

$$\frac{V_1}{25} + I_{x2} - 5 - I_{x1} = 0$$

or $$0.04V_1 + I_{x2} - 5 - I_{x1} = 0 \quad \text{(i)}$$

KCL at node 2 is

$$\frac{V_2}{20} - I_{x2} - 4 = 0$$

or $$0.05V_2 - I_{x2} - 4 = 0 \quad \text{(ii)}$$

KCL at node 3 is

$$\frac{V_3}{40} + I_{x1} + 4 = 0$$

or $$0.025V_3 + I_{x1} + 4 = 0 \quad \text{(iii)}$$

Also, $$V_1 - V_3 = 100 \quad \text{(iv)}$$

and $$V_2 - V_1 = 60 \quad \text{(v)}$$

Substituting Eqs. (iv) and (v) into Eqs. (i), (ii) and (iii) and solving, we get

$$V_1 = 39.13 \text{ V}$$

Therefore, the power supplied by the 5 A source $= V_1 \times 5$

$= 39.13 \times 5$

$=$ **195.65 W.**

SHORT QUESTIONS WITH ANSWERS

1. State and explain Kirchhoff's laws.

Kirchhoff's current law is stated as:

The algebraic sum of currents meeting at a point or junction is zero.

In Fig. S2.1, it has been shown that currents I_2, I_3 and I_6 are entering at point A, while currents I_1, I_4 and I_5 are leaving from point A. Applying KCL at point A, we can write:

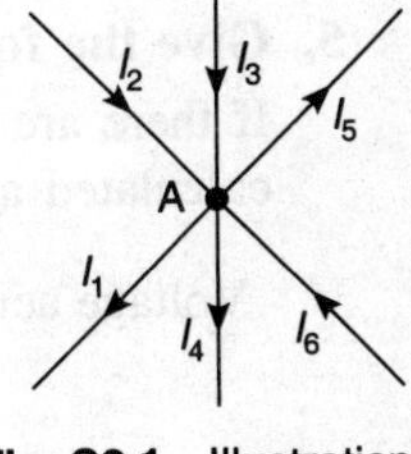

Fig. S2.1 Illustration of KCL.

$$I_1 + I_4 + I_5 + (-I_2) + (-I_3) + (-I_6) = 0$$

or
$$I_1 + I_4 + I_5 = I_2 + I_3 + I_6$$

So another form of KCL is that *the sum of currents entering at any point is equal to that leaving from that point.*

Kirchhoff's Voltage Law is stated as:

The algebraic sum of all the voltages taken around in a closed circuit or loop or mesh is zero.

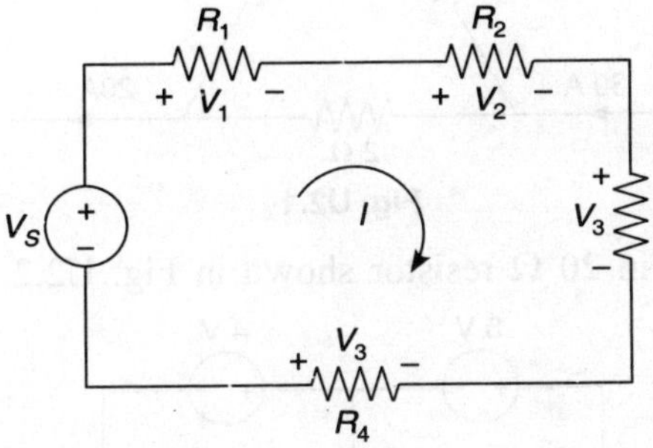

Fig. S2.2

Let the voltages across resistors R_1, R_2, R_3 and R_4 be V_1, V_2, V_3 and V_4, (Fig. S2.2), respectively Then, according to Kirchhoff's voltage law,

$$-V_S + V_1 + V_2 + V_3 + V_4 = 0$$

or
$$V_S = V_1 + V_2 + V_3 + V_4$$

2. When 'supermesh' analysis is used?

In mesh analysis method, if there exists a current source on the common branch of the two adjacent meshes, supermesh analysis is used.

3. When 'supernode' analysis is used?

In node analysis method, if there exists a voltage source between two nodes, supernode analysis is used.

4. Give the formula to find the current dividing into two parallel connected resistors.

Current through a resistor (parallel combination of TWO resistors)

$$= \text{Total current} \times \left(\frac{\text{Resistance of opposite resistor}}{\text{Sum of resistances of two resistors}} \right)$$

5. Give the formula to find the voltage dividing among series connected resistors.

If there are two resistors connected in series, then the voltage divided among them is calculated as follows:

$$\text{Voltage across a resistor} = \text{Total voltage} \times \left(\frac{\text{Resistance of particular resistor}}{\text{Sum of all resistances}} \right)$$

UNSOLVED PROBLEMS

1. Find the current flows through 1 Ω resistance in the circuit shown in Fig. U2.1.

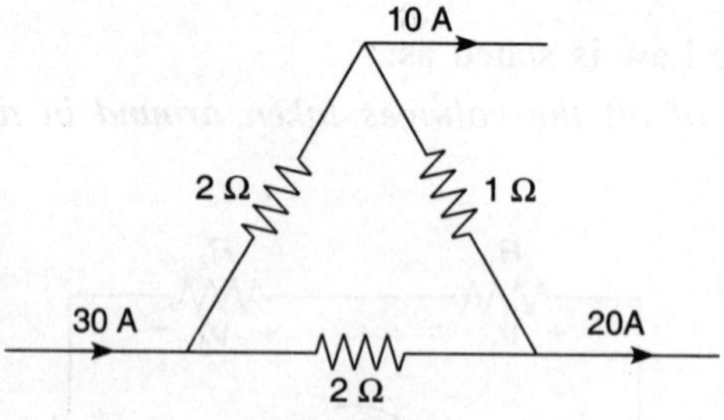

Fig. U2.1

[**Ans.** 4 A]

2. Calculate the current in 20 Ω resistor shown in Fig. U2.2.

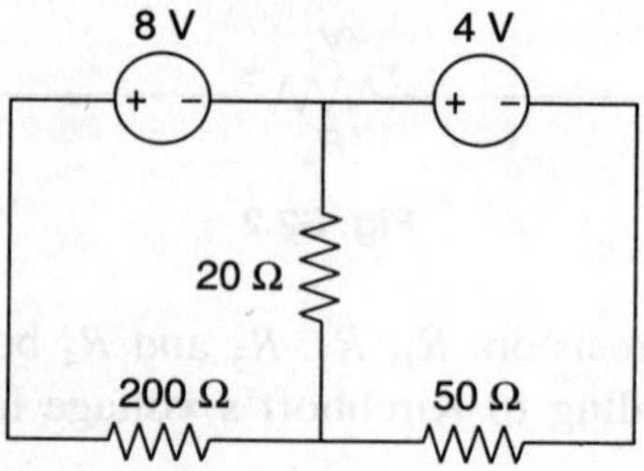

Fig. U2.2

[**Ans.** 26.67 mA]

3. A Wheatstone bridge ABCD has the following details: AB = 1000 Ω; BC = 100 Ω; CD = 450 Ω; DA = 5000 Ω. A galvanometer of resistance 500 Ω is connected between B and D. A 4.5 V battery of negligible resistance is connected between A and C with A positive. Find the magnitude and direction of current through the galvanometer.

[**Ans.** 37.391 μA, direction of flow is B to D]

4. Find V_{AD} in the network shown in Fig. U2.3.

[**Ans.** –15.2 V]

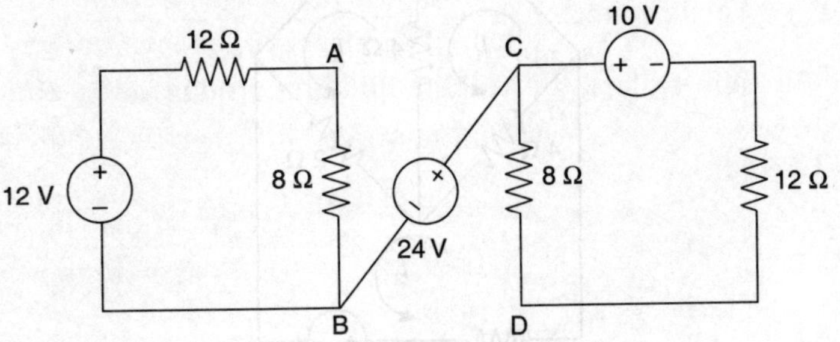

Fig. U2.3

5. Determine voltage V_{AB} in the circuit shown in Fig. U2.4.

[**Ans.** 3.37 V]

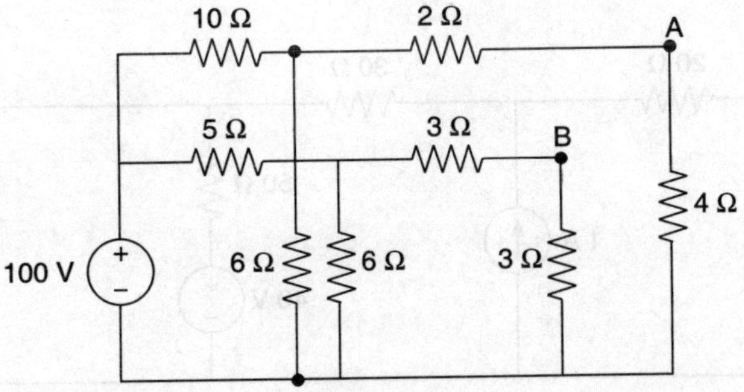

Fig. U2.4

6. Using mesh analysis method, find the current flowing through 2 Ω resistor in the circuit shown in Fig. U2.5.

[**Ans.** 1.7372 A]

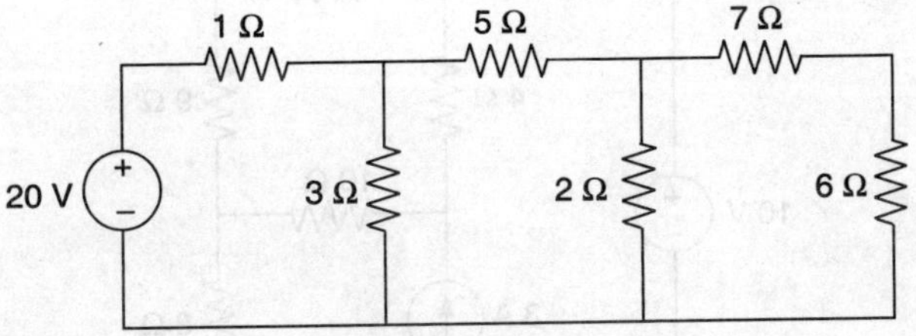

Fig. U2.5

7. Find currents I_1, I_2 and I_3 in the network shown in Fig. U2.6 by using mesh analysis.

[**Ans.** I_1 = 1.6815 A; I_2 = 1.2993 A; I_3 = 2.4841 A]

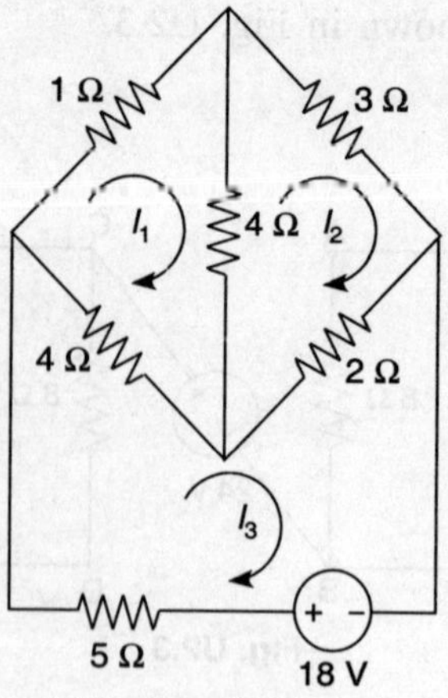

Fig. U2.6

8. By mesh analysis, find the current through 100 Ω resistor in the circuit shown in Fig. U2.7. **[Ans.** 0.48 A]

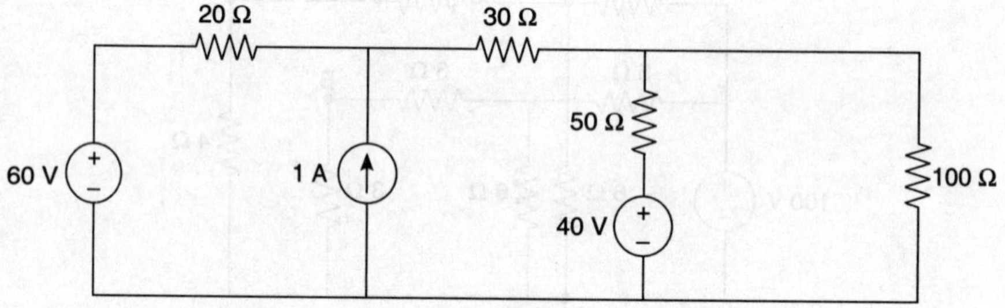

Fig. U2.7

9. By mesh analysis, find current *I* in the circuit shown in Fig. U2.8. **[Ans.** –1.9333 A]

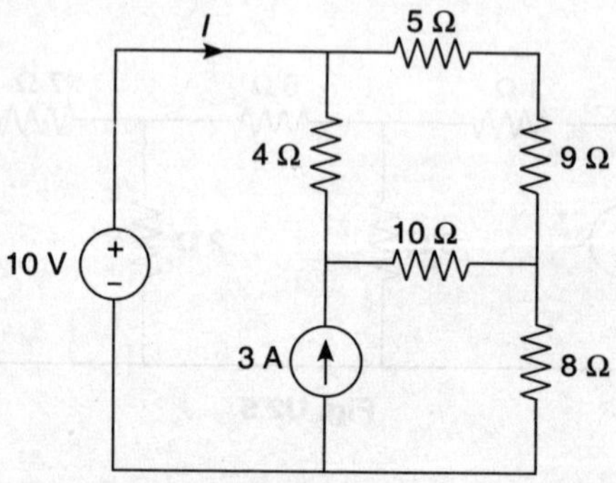

Fig. U2.8

10. Determine the mesh currents I_1 and I_2 for the given circuit shown in Fig. U2.9.
[**Ans.** I_1 = 3.7561 A, I_2 = 7.5366 A]

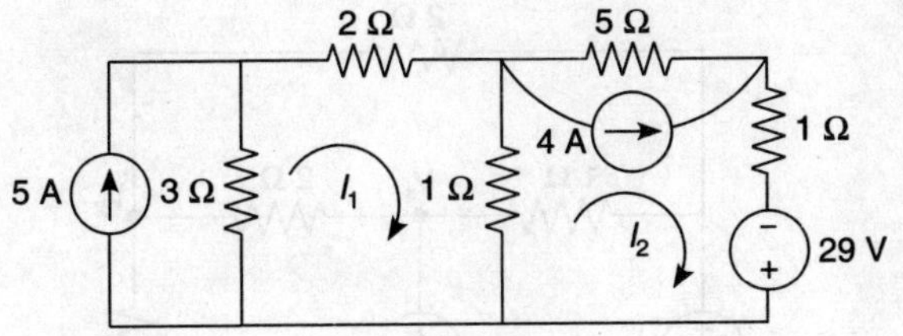

Fig. U2.9

11. By nodal analysis method, find voltages V_1, V_2 and V_3 in the network shown in Fig. U2.10.
[**Ans.** V_1 = 17.099 V; V_2 = 6.73 V; V_3 = 50.99 V]

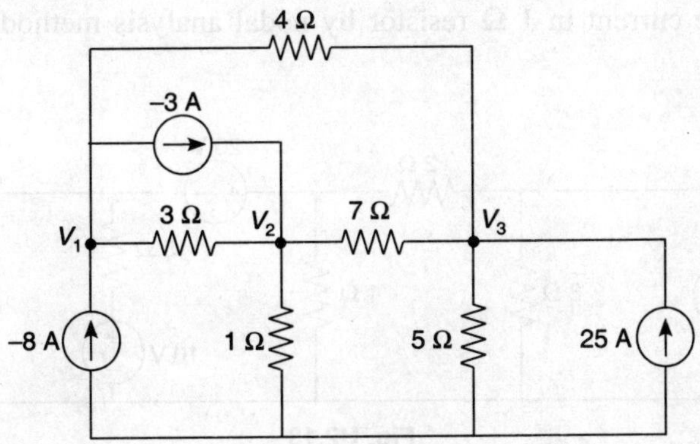

Fig. U2.10

12. Find the current flowing through resistance R_L using nodal analysis in the circuit shown in Fig. U2.11.
[**Ans.** 0 A]

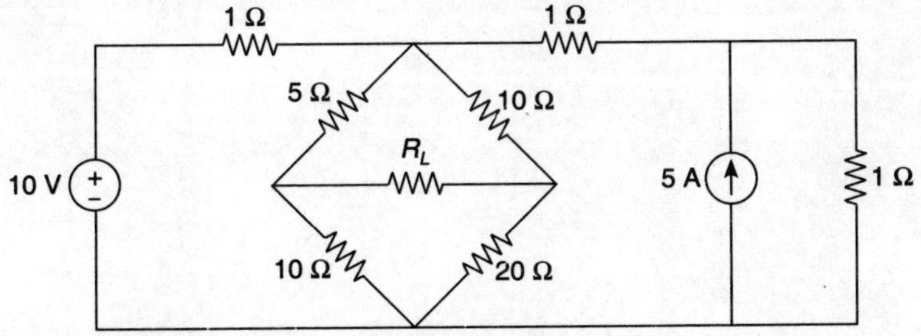

Fig. U2.11

13. By nodal analysis method, find V_1 and V_2 in the circuit shown in Fig. U2.12.

[**Ans.** $V_1 = 8.9231$ V; $V_2 = 6.7692$ V]

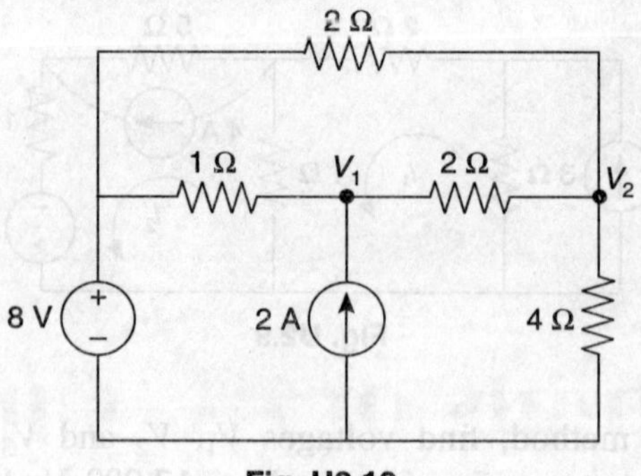

Fig. U2.12

14. Determine the current in 1 Ω resistor by nodal analysis method for the circuit shown in Fig. U2.13.

[**Ans.** 11.58 A]

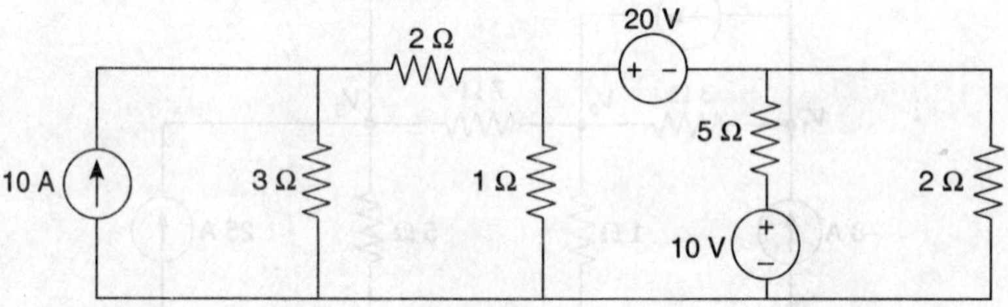

Fig. U2.13

CHAPTER 3

Fundamentals of AC Circuits

3.1 INTRODUCTION

The circuits we considered in the first two chapters are DC circuits in which DC currents and voltages are involved. This chapter discusses yet another form of electricity, *alternating quantity*. Alternating current/voltage is defined as the current/voltage that varies with respect to time, assuming positive and negative values. The wave shape of AC current or voltage values when plotted against time may be of rectangular, square, triangular or sinusoidal. However, we confine our discussion only to sinusoidal waveform because sine waveforms are easy to generate by generators at power stations and transmit by transmission lines to industry and for domestic use. Since AC quantity is time varying in nature, at any time it can be represented in three ways: (1) by its average value, (2) by its effective value and (3) by its peak value.

3.2 GENERATION OF ALTERNATING E.M.F.

According to Faraday's law of electromagnetic induction, whenever a conductor cuts magnetic field, e.m.f. is induced in the conductor. The magnitude of e.m.f. induced depends on the rate of cutting of magnetic flux. This voltage (e.m.f.) is generated either by rotating a coil inside a magnetic field, or by rotating the magnetic field while keeping the coil stationary. The voltage thus generated will be of sinusoidal form.

Let us consider a conductor placed in a uniform magnetic field as shown in Fig. 3.1. Fig. 3.2 shows the amplitude of e.m.f. induced with respect to time at each position of the conductor. When the conductor is rotated with an angular velocity ω rad/s, the cutting of flux by the conductor is not uniform. It depends on the position of the conductor.

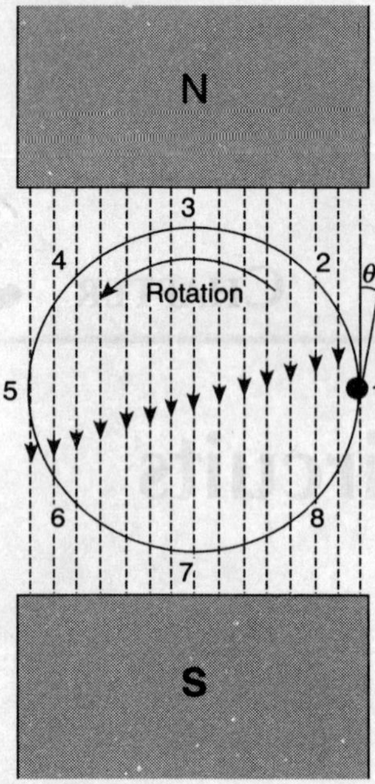

Fig. 3.1 Simple AC generator.

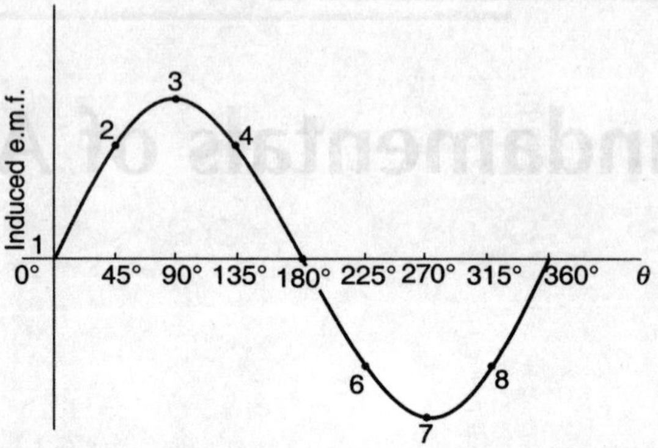

Fig. 3.2 Induced e.m.f. at each position of conductor.

When the conductor is at position 1, the rate of change of flux is zero because the conductor is seen moving parallel to the lines of flux. Hence, no e.m.f. is generated in the coil. When it is rotated by 90º in anticlockwise direction (position 3), the flux linkage is maximum, therefore, maximum possible e.m.f. is generated in the coil. Hence, the magnitude of e.m.f. generated depends on the following:

B = flux density of the magnetic field
l = active length of the conductor
v = linear velocity of coil
θ = angle between plane of the coil and axis of rotation

By Faraday's law of electromagnetic induction,

$$v = Blv \sin \theta \tag{3.1}$$

When θ = 90º, induced e.m.f. is of maximum value (V_m).

i.e. $$V_m = Blv$$

or $$v = V_m \sin \theta$$

The different positions of the conductor and the respective values of the induced e.m.f. are tabulated in Table 3.1.

Table 3.1 Induced e.m.f. against position of conductor

ωt	0	$\pi/4$	$\pi/2$	$3\pi/4$	π	$5\pi/4$	$3\pi/2$	$7\pi/4$	2π
θ	0°	45°	90°	135°	180°	225°	270°	315°	360°
v	0	$\frac{V_m}{\sqrt{2}}$	V_m	$\frac{V_m}{\sqrt{2}}$	0	$-\frac{V_m}{\sqrt{2}}$	$-V_m$	$-\frac{V_m}{\sqrt{2}}$	0

The graph between θ and v is shown in Fig. 3.3.

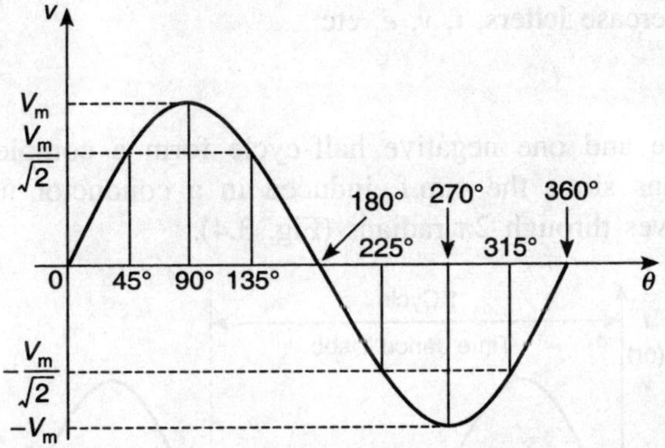

Fig. 3.3 Generation of voltage with respect to angle of rotation.

In one revolution of coil, the angle covered is 2π radian (360°) and the voltage wave completes one cycle. The time taken to complete one cycle is the time period T of the alternating voltage. Therefore,

$$\text{Angular velocity, } \omega = \frac{\text{angle covered}}{\text{time taken}} = \frac{2\pi}{T}$$

or

$$\omega = 2\pi f \qquad \left(\because f = \frac{1}{T}\right)$$

Hence, a sinusoidal alternating quantity can be represented in any of the forms given in Table 3.2.

Table 3.2 Representation of Alternating Quantity

Voltage, v	*Current, i*
$V_m \sin \theta$	$I_m \sin \theta$
$V_m \sin \omega t$	$I_m \sin \omega t$
$V_m \sin 2\pi ft$	$I_m \sin 2\pi ft$

3.3 TERMINOLOGY

Amplitude

The maximum magnitude of the quantity (voltage or current) is known as amplitude. The figure of an AC sine wave is obtained as a graphical representation of the magnitude plotted as a function of angular position (or equivalent time).

Instantaneous values

Instantaneous values are the values of the alternating quantities at any instant of time. They are represented by lowercase letters, *i*, *v*, *e*, etc.

Cycle

One positive half-cycle and one negative half-cycle form a complete cycle. One cycle is equivalent to 2π radians since the e.m.f. induced in a conductor undergoes one complete alternation when it moves through 2π radians (Fig. 3.4).

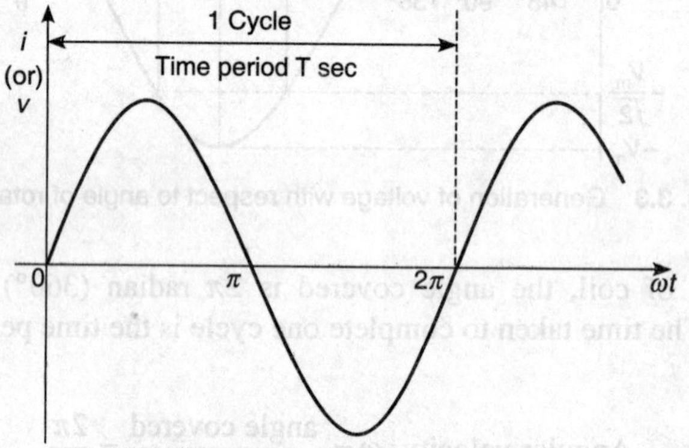

Fig. 3.4 Sine waveform

Frequency

The number of cycles completed by the AC quantity in one second is called frequency. It is expressed in hertz (Hz) or cycles per second.

As time period T is the time for one cycle, i.e. seconds/cycle and frequency is cycles/second, we can say that the frequency is reciprocal of the time period.

$$f = \frac{1}{T} \text{ Hz}$$

Angular frequency

The angle equivalent of one full-cycle is 360° or 2π radians. When 'f' cycles are covered in one second, the angle covered per second is $2\pi f$ radians. Then, angular frequency is $2\pi f$ rad/s. It is expressed in ω.

EXAMPLE 3.1 An alternating current is given by $i = 100 \sin 314t$. Find (i) the maximum value, (ii) the frequency, (iii) the time period, and (iv) the instantaneous value when time is 6 ms.

Solution Comparing the given equation of alternating current with the standard form $i = I_m \sin \omega t$, we get

(i) Maximum value, $I_m = \mathbf{100\ A}$

(ii) $\omega = 314$, therefore, $f = \dfrac{\omega}{2\pi} = \dfrac{314}{2\pi} = \mathbf{50\ Hz}$

(iii) time period, $T = \dfrac{1}{50} = \mathbf{0.02\ s}$

(iv) given $i = 100 \sin 314t$

Substituting $t = 6\text{ ms} = 6 \times 10^{-3}$ s, we get

$$i = 100 \sin (314 \times 6 \times 10^{-3})$$

$$= 100 \sin\left(314 \times 6 \times 10^{-3} \times \frac{180^\circ}{\pi}\right) = \mathbf{95.13\ A}$$

3.4 AVERAGE AND ROOT MEAN SQUARE VALUES

As the alternating quantity (current or voltage) varies instantaneously, it is always essential to represent their values along with the time at which it occurs. Alternately, there are other two forms of representation of AC quantities, viz. (i) average value and (ii) root mean square value.

3.4.1 Average Value

The mean value of all the instantaneous values of an alternating quantity over one cycle is called its average value:

i.e.
$$\text{Average value} = \frac{\text{area under the curve}}{\text{base}}$$

For symmetrical waves (a wave which has equal amplitudes of positive and negative half-cycles), the average value is zero. Hence, the average value is computed for one half-cycle only, i.e.

$$\text{Average value of a symmetrical wave} = \frac{\text{area of one half-cycle}}{\text{base length of one half-cycle}}$$

On the other hand, in unsymmetrical waves, the average value can be taken over one full-cycle:

$$\text{Average value of an unsymmetrical wave} = \frac{\text{area of one half-cycle}}{\text{base length of one full-cycle}}$$

3.4.2 Average Value of Sinusoidal Current

The instantaneous value of an alternating current is given by

$$i = I_m \sin \omega t$$

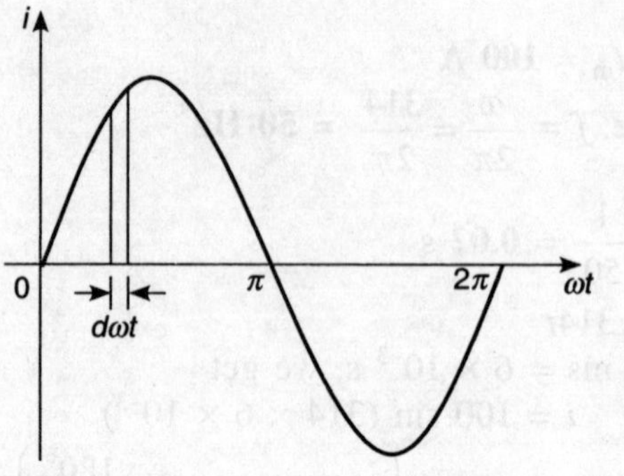

Fig. 3.5 Average value of an alternating current.

Let us consider an elementary strip of width $d(\omega t)$ in the first half-cycle of the wave as shown in Fig. 3.5.

$$\text{Area of the strip} = i d(\omega t)$$

$$\text{Area of half-cycle} = \int_0^{\pi} i d(\omega t) = \int_0^{\pi} I_m \sin \omega t \, d(\omega t) = I_m [-\cos \omega t]_0^{\pi}$$

$$= 2I_m$$

$$\therefore \quad \text{Average value, } I_{av} = \frac{\text{area of half-cycle}}{\text{base length of half-cycle}} = \frac{2I_m}{\pi}$$

3.4.3 Root Mean Square (RMS) Value or Effective Value

The RMS or effective value of an alternating current is that value of an alternating current that produces the same heating effect as that of a given DC current when passed through the same resistance for the same duration. Mathematically, this DC equivalent of AC current is given by,

$$I_{rms} = \sqrt{\frac{1}{T}\int_0^T i^2 dt}$$

= square root of the mean of the squares of the current

Similarly,

$$V_{rms} = \sqrt{\frac{1}{T}\int_0^T v^2 dt}$$

= square root of the mean of the squares of the voltage

3.4.4 RMS Value of Sinusoidal Current

The instantaneous value of an alternating current is given by

$$i = I_m \sin \omega t$$

Consider an elementary strip of width $d(\omega t)$ in the first half-cycle of the squared current wave in Fig 3.6. Let i^2 be the mid-ordinate of the strip.

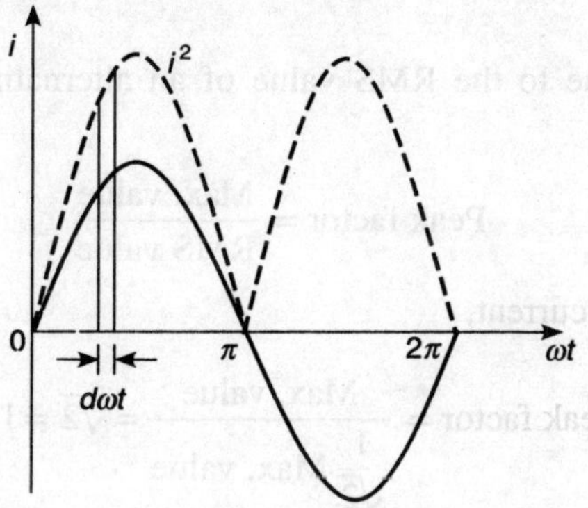

Fig. 3.6 RMS value of alternating current.

Area of the strip = $i^2 d(\omega t)$

Area of half-cycle of the squared wave $= \int_0^\pi i^2 d(\omega t) = \int_0^\pi I_m^2 \sin^2 \omega t \, d(\omega t)$

$$= I_m^2 \int_0^\pi \left(\frac{1 - \cos 2\omega t}{2} \right) d(\omega t)$$

$$= \frac{I_m^2}{2} \left[\omega t - \frac{\sin 2\omega t}{2} \right]_0^\pi = \frac{I_m^2}{2} \pi$$

Therefore, RMS value, $I_{rms} = \sqrt{\frac{\text{area of half-cycle squared wave}}{\text{base length of half-cycle}}} = \sqrt{\frac{\pi I_m^2/2}{\pi}} = \frac{I_m}{\sqrt{2}}$

3.5 FORM FACTOR AND PEAK FACTOR

There exists a relation among the peak value, average value and RMS value of an alternating quantity. The relationship is expressed by two factors, namely form factor and peak factor.

3.5.1 Form Factor

The ratio of RMS value to the average value of an alternating quantity is known as form factor,

i.e. $$\text{Form factor} = \frac{\text{RMS value}}{\text{average value}}$$

For a sinusoidal voltage or current,

$$\text{Form factor} = \frac{\frac{1}{\sqrt{2}}\text{Max. value}}{\frac{2}{\pi}\text{Max. value}} = \frac{\pi}{2\sqrt{2}} = 1.11$$

3.5.2 Peak Factor

The ratio of maximum value to the RMS value of an alternating quantity is known as peak factor, i.e.

$$\text{Peak factor} = \frac{\text{Max. value}}{\text{RMS value}}$$

For a sinusoidal voltage or current,

$$\text{Peak factor} = \frac{\text{Max. value}}{\frac{1}{\sqrt{2}}\text{Max. value}} = \sqrt{2} = 1.414$$

EXAMPLE 3.2 Find the average value of the half-rectified sine waveform shown in Fig. 3.7.

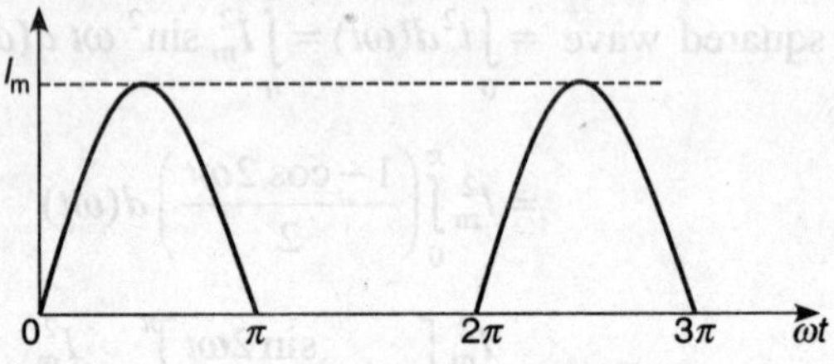

Fig. 3.7

Solution

$$\text{Average value} = \frac{\text{area under the curve}}{\text{base}}$$

Here, the area under the curve $= \int_0^{\pi} i\, d\omega t + \int_{\pi}^{2\pi} 0\, d\omega t$

$$= \int_0^{\pi} I_m \sin \omega t\, d\omega t = I_m[-\cos \omega t]_0^{\pi} = 2I_m$$

Hence, average value $= \dfrac{2I_m}{2\pi} = \dfrac{I_m}{\pi}$

EXAMPLE 3.3 Find the average value of the triangular wave shown in Fig. 3.8.

Solution

$$\text{Area under one complete cycle} = \frac{1}{2} \times 30 \times T.$$

$$\text{Average value} = \frac{1}{2} \times \frac{30 \times T}{T}$$

$$= \mathbf{15\ V}$$

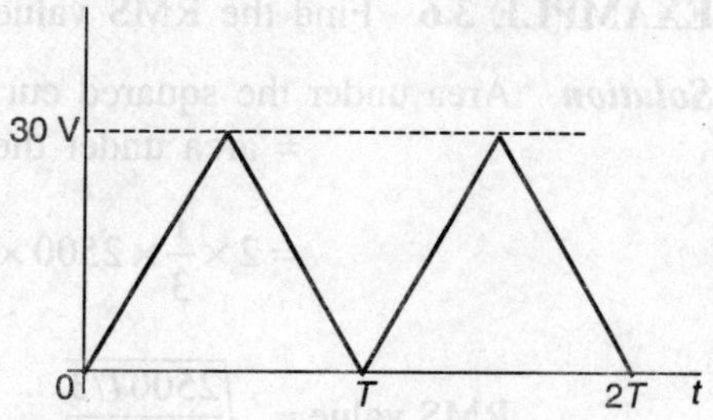

Fig. 3.8

EXAMPLE 3.4 Find the average value of the wave shown in Fig. 3.9.

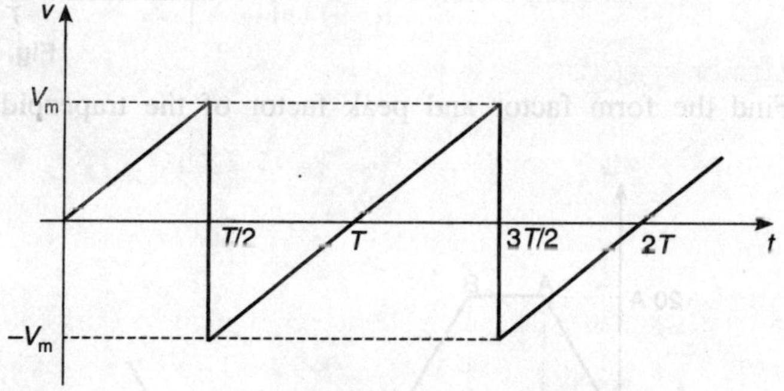

Fig. 3.9

Solution

$$\text{Area under half-cycle} = \frac{1}{2} \times V_m \times \frac{T}{2}.$$

$$\text{Average value} = \frac{1}{2} \times \frac{V_m \times T/2}{T/2} = \frac{V_m}{2}\ \mathbf{V}$$

EXAMPLE 3.5 Find the RMS value of the saw tooth wave shown in Fig. 3.10.

Solution Area under the squared curve = area under the parabola

$$= \frac{1}{3} \times I^2 \times T$$

$$\text{RMS value} = \sqrt{\frac{1}{3}\frac{I^2 T}{T}}$$

$$= \frac{I}{\sqrt{3}}\ \mathbf{A}$$

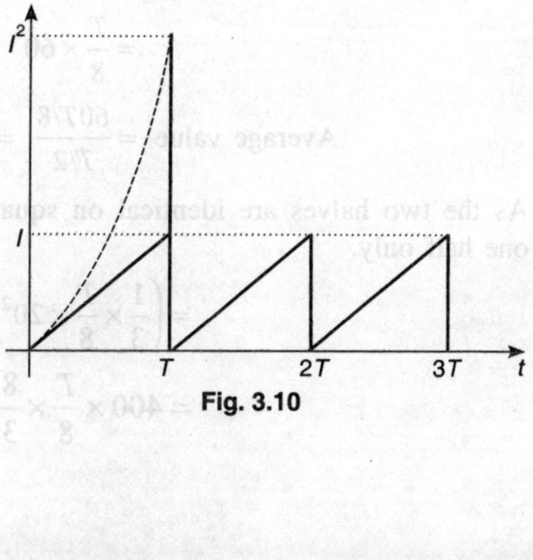

Fig. 3.10

EXAMPLE 3.6 Find the RMS value of the triangular wave shown in Fig. 3.11.

Solution Area under the squared curve

$$= \text{area under the parabola}$$

$$= 2 \times \frac{1}{3} \times 2500 \times \frac{T}{2}$$

$$\text{RMS value} = \sqrt{\frac{2500T/3}{T}}$$

$$= \frac{\mathbf{50}}{\sqrt{\mathbf{3}}}\ \mathbf{V}$$

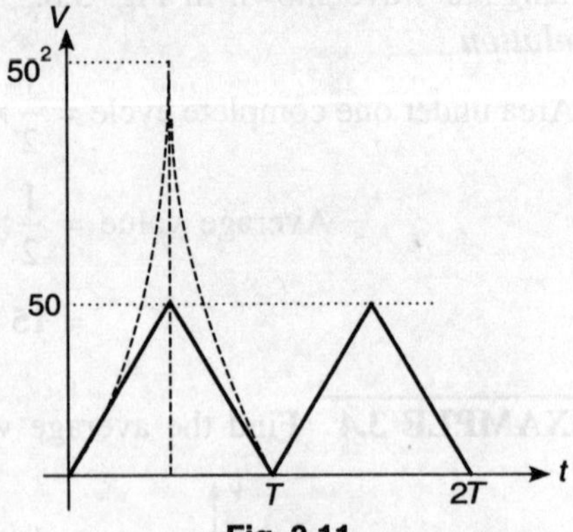

Fig. 3.11

EXAMPLE 3.7 Find the form factor and peak factor of the trapezoidal wave shown in Fig. 3.12.

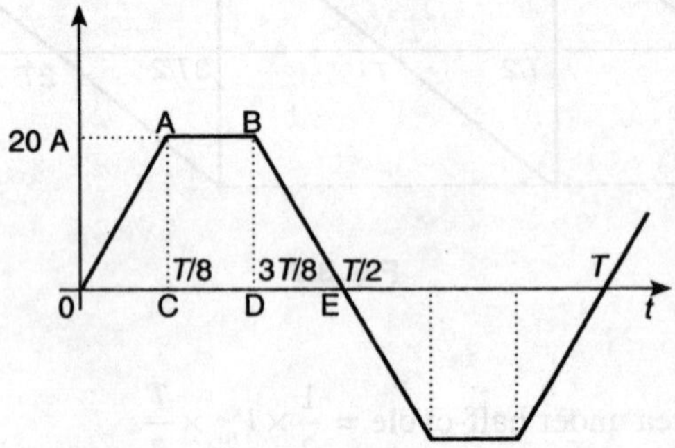

Fig. 3.12

Solution As the wave is symmetrical, the average value is calculated over one half-cycle.

Area under one half-cycle = Area of ΔOAC + Area of ΔBDE + Area of rectangle ABDC

$$= \left(\frac{1}{2} \times \frac{T}{8} \times 20\right) + \left(\frac{1}{2} \times \frac{T}{8} \times 20\right) + \left(\frac{T}{4} \times 20\right)$$

$$= \frac{T}{8} \times 60$$

$$\text{Average value } = \frac{60T/8}{T/2} = 15\ \text{A}$$

As the two halves are identical on squaring as well, the RMS value may be calculated for one half only.

$$= \left(\frac{1}{3} \times \frac{T}{8} \times 20^2\right) + \left(\frac{1}{3} \times \frac{T}{8} \times 20^2\right) + \left(\frac{T}{4} \times 20^2\right)$$

$$= 400 \times \frac{T}{8} \times \frac{8}{3}$$

$$\text{RMS value} = \sqrt{\frac{400 \times T/3}{T/2}} = \sqrt{\frac{800}{3}} = 16.33 \text{ A}$$

$$\text{Form factor} = \frac{\text{RMS value}}{\text{Average value}} = \frac{16.33}{15} = \mathbf{1.089}$$

$$\text{Peak factor} = \frac{\text{Max. value}}{\text{RMS value}} = \frac{20}{16.33} = \mathbf{1.225}$$

EXAMPLE 3.8 Find the RMS value of the waveform shown in Fig. 3.13.

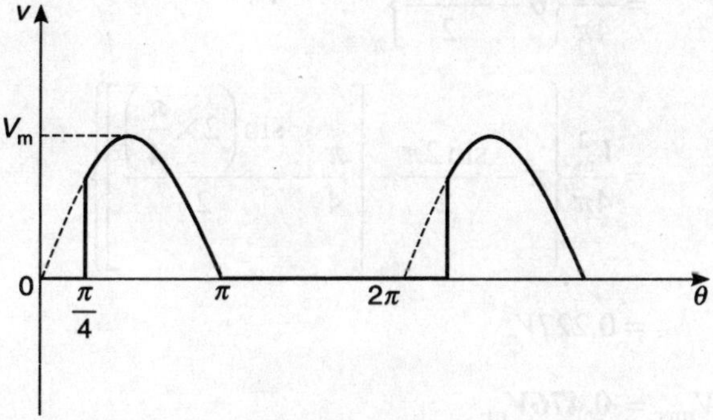

Fig. 3.13

Solution As the wave is unsymmetrical, the RMS value is calculated over one full-cycle.

The equations of the voltage waveform are given as (Fig. 3.14):

$$v = 0, \qquad 0 < \theta < \frac{\pi}{4}$$

$$v = V_m \sin\theta, \qquad \frac{\pi}{4} < \theta < \pi$$

$$v = 0, \qquad \pi < \theta < 2\pi$$

$$V_{rms} = \sqrt{\frac{1}{2\pi}\int_0^{2\pi} v^2 d\theta}$$

or

$$V_{rms}^2 = \frac{1}{2\pi}\int_0^{2\pi} v^2 d\theta$$

$$= \frac{1}{2\pi}\left\{\int_0^{\pi/4} v^2 d\theta + \int_{\pi/4}^{\pi} v^2 d\theta + \int_{\pi}^{2\pi} v^2 d\theta\right\}$$

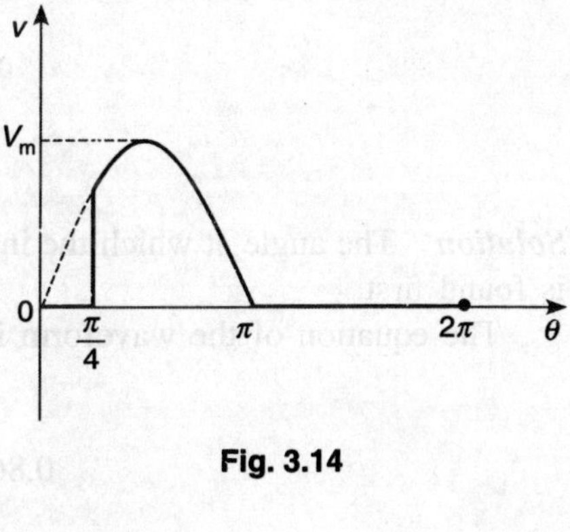

Fig. 3.14

$$=\frac{1}{2\pi}\left\{0+\int_{\pi/4}^{\pi} v^2\,d\theta+0\right\}$$

$$=\frac{1}{2\pi}\int_{\pi/4}^{\pi}(V_m\sin\theta)^2\,d\theta=\frac{1}{2\pi}\int_{\pi/4}^{\pi}(V_m^2\sin^2\theta)\,d\theta$$

$$=\frac{V_m^2}{2\pi}\int_{\pi/4}^{\pi}\left(\frac{1-\cos 2\theta}{2}\right)d\theta$$

$$=\frac{V_m^2}{4\pi}\left\{\theta-\frac{\sin 2\theta}{2}\right\}_{\pi/4}^{\pi}$$

$$=\frac{V_m^2}{4\pi}\left\{\pi-\frac{\sin 2\pi}{2}-\left[\frac{\pi}{4}-\frac{\sin\left(2\times\frac{\pi}{4}\right)}{2}\right]\right\}$$

$$=0.227V_m^2$$

Hence, $V_{rms} = \mathbf{0.476V_m}$.

EXAMPLE 3.9 Find the RMS value of the waveform shown in Fig. 3.15.

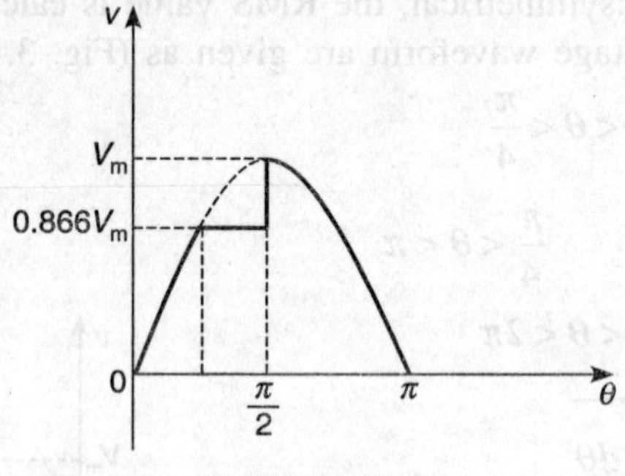

Fig. 3.15

Solution The angle at which the instantaneous value of the voltage becomes equal to $0.866V_m$ is found first.

The equation of the waveform is given by (Fig. 3.16),

$$v = V_m \sin\theta$$

$$0.866V_m = V_m \sin\theta$$

or $$\theta = 60^\circ\ (\pi/3\text{ rad})$$

The equations of the voltage waveform are given as:

$$v = V_m \sin\theta, \quad 0 < \theta < \pi/3$$
$$v = 0.866V_m, \quad \pi/3 < \theta < \pi/2$$
$$v = V_m \sin\theta, \quad \pi/2 < \theta < \pi$$

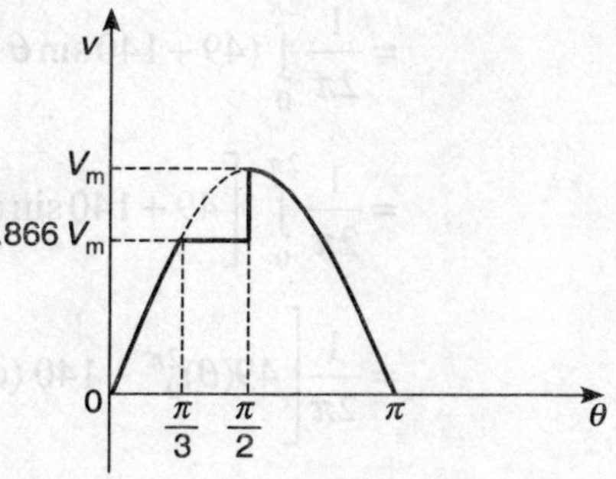

Fig. 3.16

So, $$V_{rms} = \sqrt{\frac{1}{\pi}\int_0^{\pi} v^2 d\theta}$$

or $$V_{rms}^2 = \frac{1}{\pi}\int_0^{\pi} v^2 d\theta$$

$$= \frac{1}{\pi}\left\{\int_0^{\pi/3} v^2 d\theta + \int_{\pi/3}^{\pi/2} v^2 d\theta + \int_{\pi/2}^{\pi} v^2 d\theta\right\}$$

$$= \frac{1}{\pi}\left\{\int_0^{\pi/3} (V_m \sin\theta)^2 d\theta + \int_{\pi/3}^{\pi/2} (0.866V_m)^2 d\theta + \int_{\pi/2}^{\pi} (V_m \sin\theta)^2 d\theta\right\}$$

$$= \frac{1}{\pi}\left\{\int_0^{\pi/3} V_m^2 \sin^2\theta\, d\theta + \int_{\pi/3}^{\pi/2} (0.866V_m)^2 d\theta + \int_{\pi/2}^{\pi} V_m^2 \sin^2\theta\, d\theta\right\}$$

$$= \frac{V_m^2}{\pi}\left\{\int_0^{\pi/3} \sin^2\theta\, d\theta + \int_{\pi/3}^{\pi/2} (0.866)^2 d\theta + \int_{\pi/2}^{\pi} \sin^2\theta\, d\theta\right\}$$

$$= \frac{V_m^2}{\pi}\left\{\int_0^{\pi/3} \left(\frac{1-\cos 2\theta}{2}\right) d\theta + \int_{\pi/3}^{\pi/2} (0.866)^2 d\theta + \int_{\pi/2}^{\pi} \left(\frac{1-\cos 2\theta}{2}\right) d\theta\right\}$$

$$= \frac{V_m^2}{\pi}\left\{\frac{1}{2}\left[\theta - \frac{\sin 2\theta}{2}\right]_0^{\pi/3} + (0.866)^2[\theta]_{\pi/3}^{\pi/2} + \frac{1}{2}\left[\theta - \frac{\sin 2\theta}{2}\right]_{\pi/2}^{\pi}\right\}$$

$$= 0.4726V_m^2$$

Hence, $$V_{rms} = \mathbf{0.687V_m}$$

EXAMPLE 3.10 Find the effective value of the resultant current that carries simultaneously a direct current of 7 A and a sinusoidal AC of peak value 10 A.

Solution The equation of the resultant current will be $i = (7 + 10 \sin\theta)$

So, $$I_{rms} = \sqrt{\frac{1}{2\pi}\int_0^{2\pi} i^2 d\theta}$$

or $$I_{rms}^2 = \frac{1}{2\pi}\int_0^{2\pi} i^2 d\theta$$

$$= \frac{1}{2\pi}\int_0^{2\pi} (7 + 10\sin\theta)^2 d\theta$$

$$= \frac{1}{2\pi}\int_0^{2\pi}(49 + 140\sin\theta + 100\sin^2\theta)d\theta$$

$$= \frac{1}{2\pi}\int_0^{2\pi}\left[49 + 140\sin\theta + 100\left(\frac{1-\cos 2\theta}{2}\right)\right]d\theta$$

$$= \frac{1}{2\pi}\left[49(\theta)_0^{2\pi} - 140\,(\cos\theta)_0^{2\pi} + \frac{100}{2}\left(\theta - \frac{\sin 2\theta}{2}\right)_0^{2\pi}\right]$$

$$= \frac{1}{2\pi}\left[49(2\pi - 0) - 140(\cos 2\pi - \cos 0) + 50\left(2\pi - 0 - \frac{\sin 4\pi}{2} + \sin 0\right)\right]$$

$$= \frac{1}{2\pi}[49(2\pi) - 0 + 50(2\pi)]$$

$$= \frac{2\pi}{2\pi}(99)$$

$$I_{rms}^2 = 99$$

Hence, I_{rms} = **9.95 A**

EXAMPLE 3.11 The equation of an alternating current is $i = 70.71 \sin 314.16t$ A. Determine the (i) maximum value, (ii) frequency, (iii) RMS value and (iv) average value.

Solution Comparing the given current equation with the standard form of $i = I_m \sin \omega t$, we obtain

(i) Maximum value, I_m = **70.71 A**

(ii) Angular frequency, ω = 314.16 rad/s

so, frequency, $f = \frac{\omega}{2\pi} = \frac{314.16}{2\pi}$ = **50 Hz**

(iii) RMS value, $I_{rms} = \frac{I_m}{\sqrt{2}} = \frac{70.71}{\sqrt{2}}$ = **50 A**

(iv) Average value, $I_{av} = \frac{2I_m}{\pi} = \frac{2 \times 70.71}{\pi}$ = **45.02 A**

3.6 IN-PHASE, OUT-OF-PHASE AND PHASE DIFFERENCE

When two alternating quantities of same frequency reach their zero values, maximum values and minimum values at the same instant, they are said to be *in-phase*. On the other hand, if they reach their zero values, maxima and minima at different instants, they are said to be *out-of-phase*.

For example, in Fig. 3.17, two current waveforms i_1 and i_2 of the same frequency are in phase to each other. In other words, the phase difference between the two is zero (ϕ = 0). Let, $i_1 = I_{m1} \sin \omega t$ and $i_2 = I_{m2} \sin \omega t$. They may be added to give a resultant waveform (Fig. 3.17) of the same frequency as follows:

$$i = (I_{m1} + I_{m2}) \sin \omega t$$

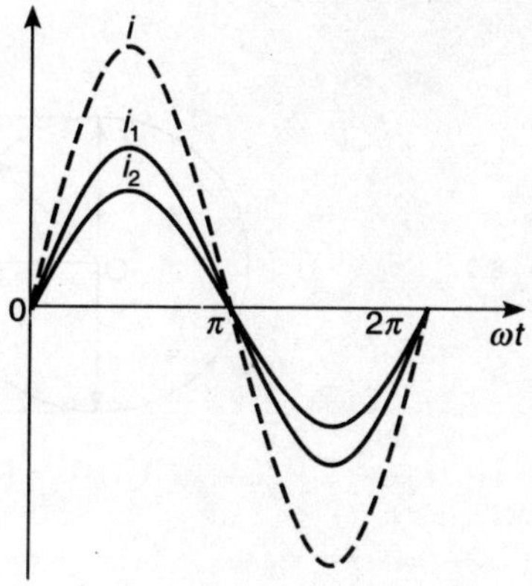

Fig. 3.17 Concept of in-phase quantities.

The current waveforms i_1 and i_2 shown in Fig. 3.18 are out-of-phase. They are said to have a phase difference of $\pi/3$ radians or 60°. Here, i_2 is said to be lagging behind i_1 by 60° as it reaches its maximum value 60° after i_1. Similarly, waveforms v_1 and v_2 shown in Fig. 3.19 are said to be out-of-phase. However, v_2 leads v_1 by 60° as it reaches its maximum value 60° before v_1.

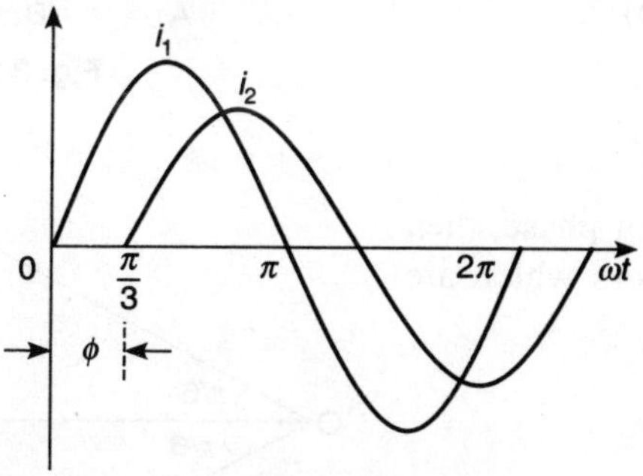

Fig. 3.18 Concept of out-of-phase (lagging).

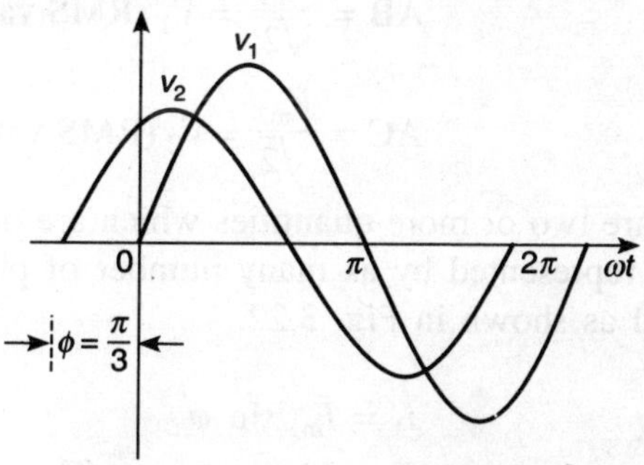

Fig. 3.19 Concept of out-of-phase (leading).

3.7 PHASOR REPRESENTATION

In analysing sinusoidal quantities, it is very easy to handle the equations or waveforms using phasor representation.

Consider a rotating vector (phasor) OP in Fig. 3.20. It is assumed that the vector is rotating in anticlockwise direction at an angular velocity of ω rad/s. At some point of time t, it reaches position OQ after traversing through an angle of ϕ rad. The vertical component QR is now found as OQ sin ϕ. The projection of the position of this phasor forms a sine wave as also shown in Fig. 3.20. This means that the vertical component varies sinusoidally. If OP is made equal to V_m, then the vertical component varies as $v = V_m \sin \phi$ (or) $V_m \sin \omega t$, where v represents instantaneous value. It is conventional to use the RMS value instead of the maximum value. Then, the vertical component QR will have to be scaled properly.

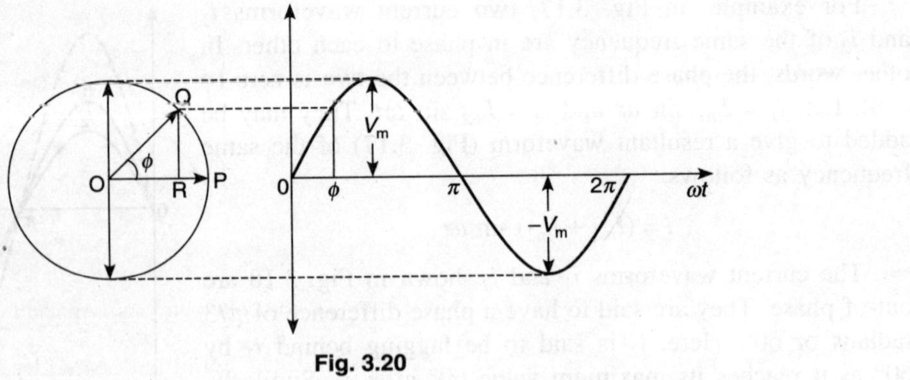

Fig. 3.20

Let there are two voltages v_1 and v_2 given by $v_1 = V_{m1} \sin \phi$ and $v_2 = V_{m2} \sin \phi$. If both are in phase with each other, they may be represented by two phasors as shown in Fig. 3.21.

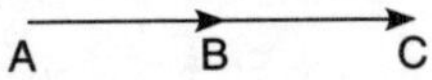

Fig. 3.21

$$AB = \frac{V_{m1}}{\sqrt{2}} = V_1 \text{ (RMS value)}$$

and

$$AC = \frac{V_{m2}}{\sqrt{2}} = V_2 \text{ (RMS value)}$$

If there are two or more quantities which are out of phase, then they are represented by as many number of phasors which are displaced as shown in Fig. 3.22.

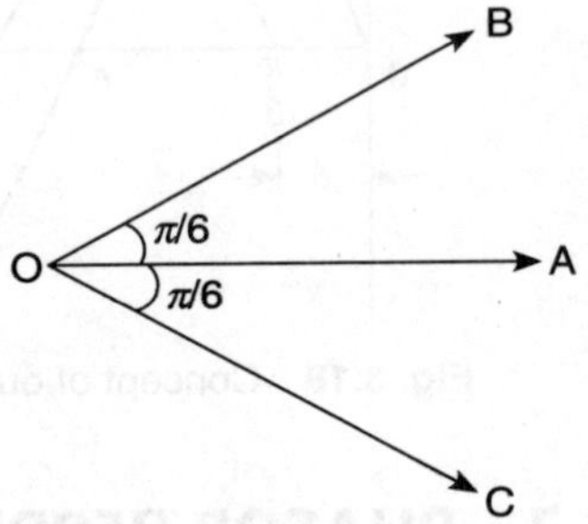

Fig. 3.22

Let

$$i_1 = I_{m1} \sin \phi$$

$$i_2 = I_{m2} \sin (\phi + \pi/6)$$

and

$$i_3 = I_{m3} \sin (\phi - \pi/6)$$

$$OA = \frac{I_{m1}}{\sqrt{2}} = I_1 \text{ ; } OB = \frac{I_{m2}}{\sqrt{2}} = I_2 \text{ ; } OC = \frac{I_{m3}}{\sqrt{2}} = I_3$$

3.8 PHASOR ALGEBRA AND OPERATOR *j*

Time-varying quantities can be represented on a conventional *X–Y* axis. Sinusoidal quantities such as current, voltage and impedance are all vectors having both magnitude and direction. Mathematical representation of any phasor is known as *phasor algebra*. They may be represented mathematically by two ways, polar form and rectangular form.

Consider point A shown in Fig. 3.23(a). It measures 3 units along *X* axis and 4 units along *Y* axis. The location of point A could be written as A (3, 4). But the vector position of 4 units is not well defined. So this information is written more precisely using operator *j* in complex form, i.e., $3 + j4$. Letter *j* is placed before the vertical axis component to mean it is turned by 90° (in anticlockwise direction) from the reference axis. Operator *j* is same as operator *i* in complex form. Generally, alternating quantities written using rectangular form are represented as $x + jy$, where *x* is real part and *y* is imaginary part.

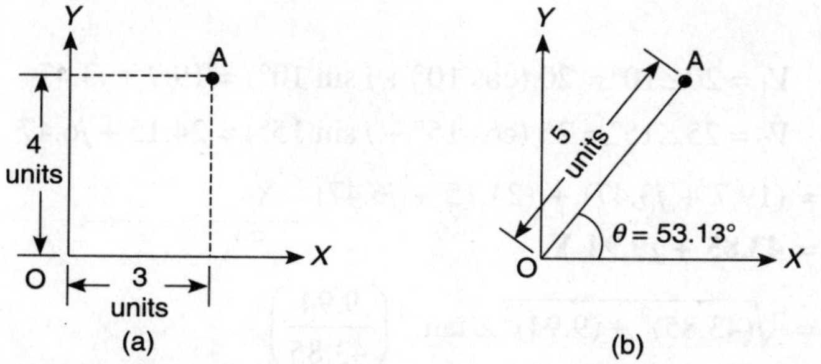

Fig. 3.23 Illustration of rectangular and polar form representation.

With reference to Fig. 3.23(b), the same point A could be located by knowing distance OA and angle θ measured from +ve X axis. Here, OA = $\sqrt{3^2+4^2}=5$ units, and the angle, θ = $\tan^{-1}(4/3)$ = 53.13°. Thus, in polar form, point A is written as $5\angle 53.13°$. In general, alternating quantities written using polar form is represented as $M\angle\theta$, where M is the magnitude of the quantity (RMS value) and θ is the phase angle (i.e. angle measured from +ve X axis). Angle θ is taken as positive if measured anticlockwise direction while negative if measured clockwise direction with respect to positive X axis.

In polar form, j represents $1\angle 90°$. For example, if current $I\angle 0°$ is shifted through 90°, it becomes $I\angle 90°$ and is written as jI. When jI vector is once more rotated through 90°, it becomes

$$\begin{aligned} j(jI) &= j^2 I \\ &= I\angle 180° \\ &= -1 \end{aligned}$$

So, j^2 is equal to –1, and $j = \sqrt{-1}$ becomes imaginary term.

Similarly, $\quad j^3 I = j(j^2 I) = I\angle 270° \quad$ and $\quad j^4 I = I\angle 360° = I.$

All the rules of algebraic manipulations used for complex numbers are applicable to phasors too.

Let two phasors be

$$A_1 = a + jb$$

and

$$A_2 = c + jd$$

then,

$$A_1 + A_2 = (a+jb) + (c+jd) = (a+c) + j(b+d)$$

$$A_1 - A_2 = (a+jb) - (c+jd) = (a-c) + j(b-d)$$

$$\begin{aligned} A_1 \times A_2 &= (a+jb)\times(c+jd) = ac + jad + jbc + j^2 bd \\ &= (ac - bd) + j(ad + bc) \end{aligned}$$

$$\frac{A_1}{A_2} = \frac{a+jb}{c+jd} = \frac{(a+jb)}{(c+jd)}\times\frac{(c-jd)}{(c-jd)} = \frac{ac+bd}{c^2+d^2} + j\frac{bc-ad}{c^2+d^2}$$

For addition and subtraction of phasor quantities, rectangular form is used; and for multiplication and division, polar form is used.

EXAMPLE 3.12 Two voltages are $V_1 = 20\angle 10°$ and $V_2 = 25\angle 15°$. Find (i) $V_1 + V_2$ and (ii) $V_1 - V_2$. Express the answers both in polar and rectangular forms.

Solution

$$V_1 = 20\angle 10^\circ = 20\,(\cos 10^\circ + j\sin 10^\circ) = 19.7 + j3.47$$

$$V_2 = 25\angle 15^\circ = 25\,(\cos 15^\circ + j\sin 15^\circ) = 24.15 + j6.47$$

(i) $V_1 + V_2 = (19.7 + j3.47) + (24.15 + j6.47)$

$$= \mathbf{43.85 + j9.94\ V}$$

$$= \sqrt{(43.85)^2 + (9.94)^2}\,\angle\tan^{-1}\left(\frac{9.94}{43.85}\right)$$

$$= \mathbf{44.96\angle 12.77^\circ\ V}$$

(ii) $V_1 - V_2 = (19.7 + j3.47) - (24.15 + j6.47)$

$$= \mathbf{-4.45 - j3\ V}$$

$$= \sqrt{(4.45)^2 + (3)^2}\,\angle\tan^{-1}\left(\frac{-3}{-4.45}\right)$$

$$= \mathbf{5.37\angle -146^\circ\ V}$$

EXAMPLE 3.13 Two phasors are $A = 5 + j4$ and $B = 3 + j2$. Find (i) AB (ii) A/B.

Solution

(i) $AB = (5 + j4)\,(3 + j2) = 15 + j10 + j12 + j^2 8$

$$= 15 + j22 - 8$$

$$= \mathbf{7 + j22}$$

$$= \sqrt{(7)^2 + (22)^2}\,\angle\tan^{-1}\left(\frac{22}{7}\right)$$

$$= \mathbf{23.09\angle 72.35^\circ}$$

(ii) $\dfrac{A}{B} = \dfrac{5 + j4}{3 + j2}$

$$= \frac{(5 + j4)}{(3 + j2)} \times \frac{(3 - j2)}{(3 - j2)}$$

$$= \frac{(15 - j10 + j12 - j^2 8)}{(3^2 - j^2 4)} = \frac{(15 + j2 + 8)}{(9 + 4)}$$

$$= \frac{23 + j2}{13} = \mathbf{1.77 + j0.154}$$

$$= \mathbf{1.77\angle 4.97^\circ\ V}$$

EXAMPLE 3.14 Two phasors are $A = 5\angle 20^\circ$ V and $B = 6\angle 30^\circ$ V. Find (i) AB and (ii) A/B.

Solution

(i) $AB = 5\angle 20^\circ \times 6\angle 30^\circ$

$$= (5 \times 6)\angle(20^\circ + 30^\circ)$$

$$= \mathbf{30\angle 50^\circ}$$

(ii) $$\frac{A}{B} = \frac{5\angle 20^\circ}{6\angle 30^\circ}$$

$$= \frac{5}{6}\angle(20^\circ - 30^\circ)$$

$$= \mathbf{0.8333\angle{-10^\circ}}$$

EXAMPLE 3.15 Find the resultant of the following instantaneous voltages.

$$v_1 = 20 \sin \omega t$$

$$v_2 = 10 \sin\left(\omega t + \frac{\pi}{4}\right)$$

$$v_3 = 30 \sin\left(\omega t - \frac{\pi}{6}\right)$$

Solution The given instantaneous voltages in standard sinusoidal forms are converted into RMS values in polar forms as below:

$$V_1 = \frac{20}{\sqrt{2}}\angle 0^\circ = 14.14\angle 0^\circ$$

$$V_2 = \frac{10}{\sqrt{2}}\angle 45^\circ = 7.07\angle 45^\circ$$

$$V_3 = \frac{30}{\sqrt{2}}\angle -30^\circ = 21.21\angle -30^\circ$$

Now the voltages in polar forms are converted into rectangular forms as shown below:

$$V_1 = 14.14\angle 0^\circ = 14.14 + j0$$

$$V_2 = 7.07\angle 45^\circ = 5 + j5$$

$$V_3 = 21.21\angle -30^\circ = 18.37 - j10.605$$

Then, the resultant voltage, $V = V_1 + V_2 + V_3$

$$= (14.14 + j0) + (5 + j5) + (18.37 - j10.605)$$

or $$V = 37.51 - j5.605$$

Converting back to polar form,

$$V = 37.93\angle -8.5^\circ$$

Converting into standard sinusoidal form,

$$\mathbf{v = 53.64 \sin(\omega t - 8.5^\circ)\ V}$$

Short Questions with Answers

1. Define amplitude of alternating quantity.

The maximum magnitude of the quantity (voltage or current) is known as amplitude.

2. Define instantaneous value.

Instantaneous values are the values of the alternating quantities at any instant of time. They are represented by lowercase letters, i, v, e, etc.

3. Define cycle.

One positive half-cycle and one negative half-cycle form a complete cycle. One cycle is equivalent to 2π radians.

4. Define frequency.

The number of cycles completed by the AC quantity in one second is called frequency. It is expressed in hertz (Hz) or cycles per second.

As time period T is the time for one cycle, i.e. seconds/cycle and frequency is cycles/second, we can say that the frequency is reciprocal of the time period.

$$f = \frac{1}{T}\ \text{Hz}$$

5. Define angular frequency.

The angle equivalent of one full-cycle is 360° or 2π rad. When 'f' cycles are covered in one second, the angle covered per second is $2\pi f$ rad. Then, angular frequency is $2\pi f$ rad/s. It is expressed in ω.

6. What are the different forms of representing AC voltage mathematically?

$$V_m \sin\theta \quad \text{(or)} \quad V_m \sin\omega t \quad \text{(or)} \quad V_m \sin 2\pi f t$$

7. What are the different forms of representing AC current mathematically?

$$I_m \sin\theta \quad \text{(or)} \quad I_m \sin\omega t \quad \text{(or)} \quad I_m \sin 2\pi f t$$

8. Define average value.

The arithmetical average of all the values of an alternating quantity over one cycle is called its average value, i.e.

$$\text{Average value} = \frac{\text{Area under the curve}}{\text{Base}}$$

(i) In case of symmetrical waves (a wave which has positive half-cycle exactly equal to negative half-cycle), the average value over one cycle is zero. However, the average value of positive or negative half is not zero. Hence, in case of symmetrical waves, average value means the average value of half-cycle.

$$\text{Average value of a symmetrical wave} = \frac{\text{Area of one half-cycle}}{\text{Base length of one half-cycle}}$$

(ii) In case of unsymmetrical waves, the average value is taken over the full-cycle.

$$\text{Average value of an unsymmetrical wave} = \frac{\text{Area over one half-cycle}}{\text{Base length of one full-cycle}}$$

9. What is the average value of a sinusoidal quantity?

$$\text{Average value of current, } I_{av} = \frac{2I_m}{\pi}$$

$$\text{Average value of voltage, } V_{av} = \frac{2V_m}{\pi}$$

10. Define RMS value.

The RMS (or) effective value of an alternating current is that steady current (DC) which when flowing through a given resistance for a given time produces the same amount of heat as produced by the alternating current when flowing through the same resistance for the same time.

11. What is the RMS value of a sinusoidal quantity?

$$\text{RMS value of current, } I_{rms} = \frac{I_m}{\sqrt{2}}$$

$$\text{RMS value of voltage, } V_{rms} = \frac{V_m}{\sqrt{2}}$$

12. Define form factor.

The ratio of RMS value to the average value of an alternating quantity is known as form factor, i.e.

$$\text{Form factor} = \frac{\text{RMS value}}{\text{Average value}}$$

13. What is the form factor for a sinusoidal voltage or current?

$$\text{Form factor} = \frac{\text{RMS value}}{\text{Average value}}$$

Hence,
$$\text{Form factor} = \frac{\frac{1}{\sqrt{2}}\text{Max. value}}{\frac{2}{\pi}\text{Max. value}} = \frac{\pi}{2\sqrt{2}} = 1.11$$

14. Define peak factor.

The ratio of maximum value to the RMS value of an alternating quantity is known as peak factor, i.e.

$$\text{Peak factor} = \frac{\text{Max. value}}{\text{RMS value}}$$

15. What is the peak factor for a sinusoidal voltage or current?

$$\text{Peak factor} = \frac{\text{Max. value}}{\text{RMS value}}$$

Hence,
$$\text{Peak factor} = \frac{\text{Max. value}}{\frac{1}{\sqrt{2}}\text{Max. value}} = \sqrt{2} = 1.414$$

16. What is a phasor?

Rotating vector is called phasor.

UNSOLVED PROBLEMS

1. An alternating quantity is given by i = 141.4 sin 314t. Find (a) the maximum value, (b) the frequency, (c) the time period and (d) the instantaneous value when time is 3 ms. **[Ans.** (a) 141.4 A; (b) 50 Hz; (c) 0.02 s; (d) 114.36 A]
2. An alternating current of frequency 60 Hz has a maximum value of 12 A. (a) Write down the equation of the instantaneous value, (b) find the value of current after 1/360 s and (c) find the time taken to reach 9.6 A for the first time.
[Ans. (a) 12 sin 377t; (b) 10.39 A; (c) 2.46 ms]
3. An alternating voltage is given by v = 10 sin 942t. Determine the time taken from t = 0 for the voltage to reach a value of 6 V for first and second time.
[Ans. 6.66 ms; 7.34 ms]
4. A 50 Hz sinusoidal current wave has a peak amplitude of 10 A. Find the rate of change of current in amperes per second at time (a) t = 0.0025 s, (b) t = 0.005 s and (c) t = 0.01 s after i = 0 and is increasing.
[Ans. (a) 2220 A/s; (b) 0 A/s; (c) –3140 A/s]
5. A wire carries simultaneously a sinusoidal current of amplitude 10 A superimposed on a DC current of 10 A. Find the RMS value of the resultant current.
[Ans. 12.25 A]
6. Determine the instantaneous values of waves A and B shown in Fig. U3.1 at 90° point.
[Ans. 10 V; 5.657 V]

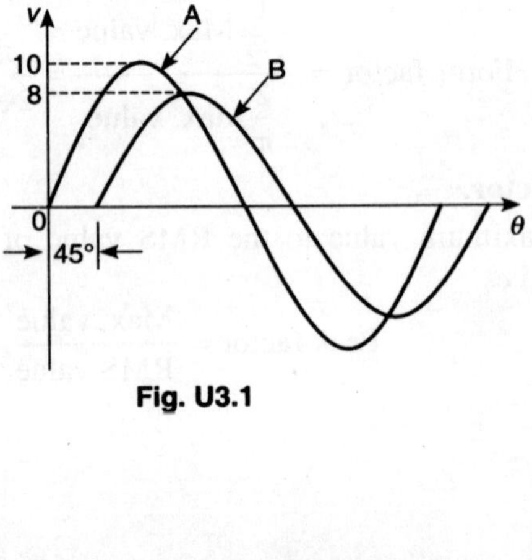

Fig. U3.1

7. Find the RMS and average value of the voltage waveform shown in Fig. U3.2.

[**Ans.** 25 V; 28.9 V]

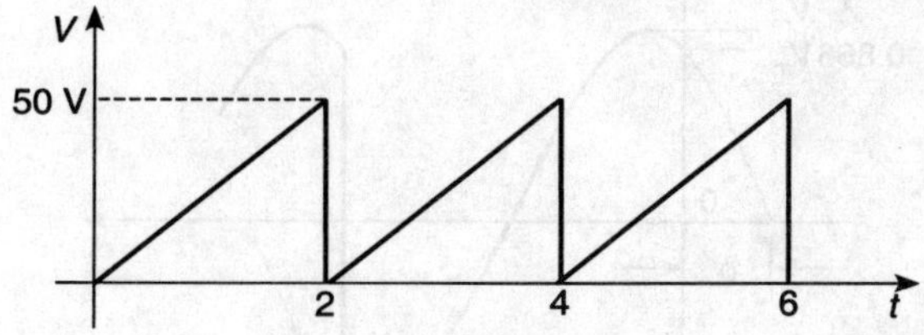

Fig. U3.2

8. Find the form factor and peak factor of the alternating quantity shown in Fig. U3.3.

[**Ans.** 1.088; 1.225]

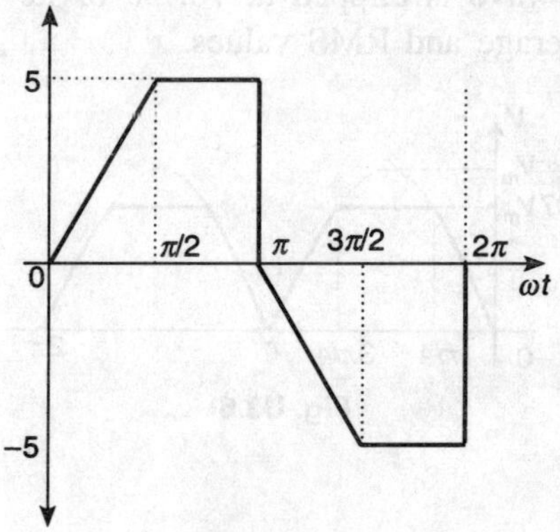

Fig. U3.3

9. Calculate the RMS value, the form and peak factors for a periodic voltage having the following values at equal intervals: 0, 5, 10, 20, 40, 60, 40, 20, 10, 5, 0, –5, –10, –20, etc. The values change from one to another suddenly. [**Ans.** 28 V; 1.33; 2.14]

10. Find the RMS and average value of the voltage waveform shown in Fig. U3.4.

[**Ans.** 10 V; 12.9 V]

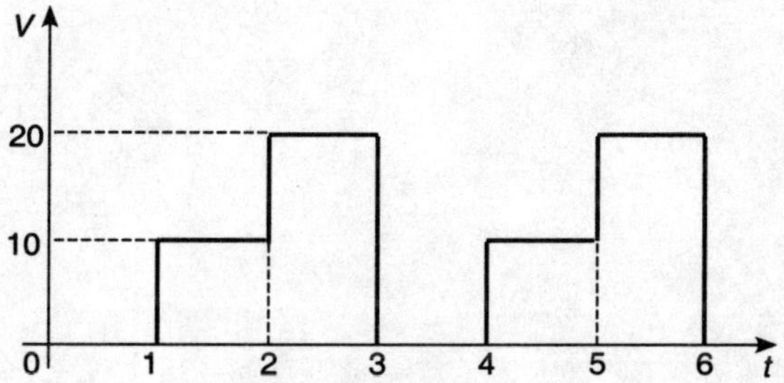

Fig. U3.4

11. Find the RMS value of the waveform shown in Fig. U3.5. [**Ans.** $0.776V_m$]

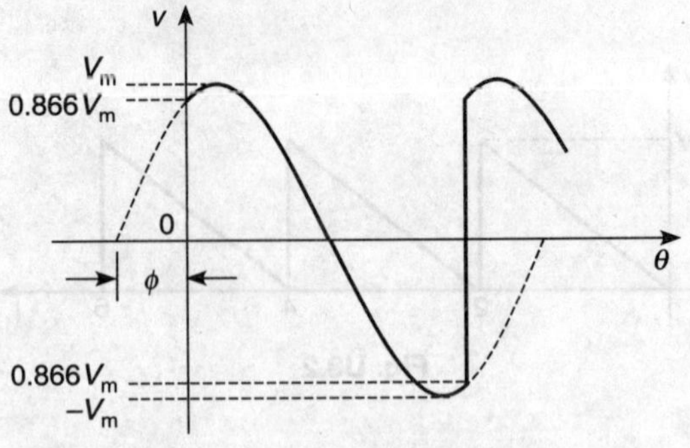

Fig. U3.5

12. A full wave rectified wave is clipped at 70.7% of its maximum value as shown in Fig. U3.6. Find its average and RMS values. [**Ans.** $0.54V_m$; $0.584V_m$]

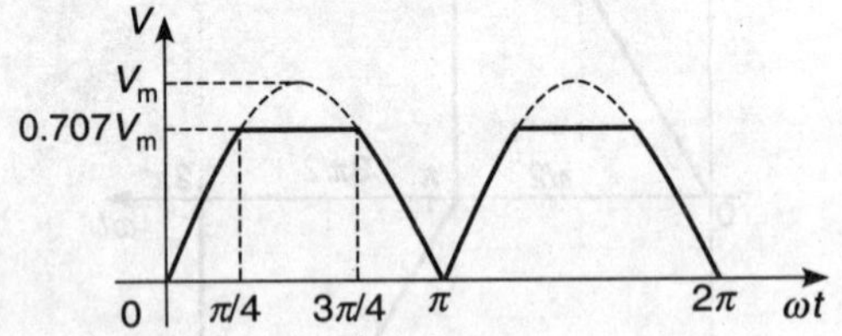

Fig. U3.6

CHAPTER 4

Single-Phase AC Circuits

4.1 ANALYSIS OF AC CIRCUITS

When a sinusoidal alternating voltage is applied to a circuit, the resulting alternating current is also sinusoidal and has the same frequency as that of the applied voltage. However, there is generally a phase difference between the applied voltage and the resulting current if the circuit contains inductances and/or capacitances. This chapter discusses the response of electric circuits to alternating current by passing alternating current through the basic circuit elements, namely resistor (R), inductor (L) and capacitor (C).

4.1.1 AC Circuit Containing Resistance Only

Consider a circuit containing a pure resistance of $R\,\Omega$ connected across an AC voltage of v as shown in Fig. 4.1.

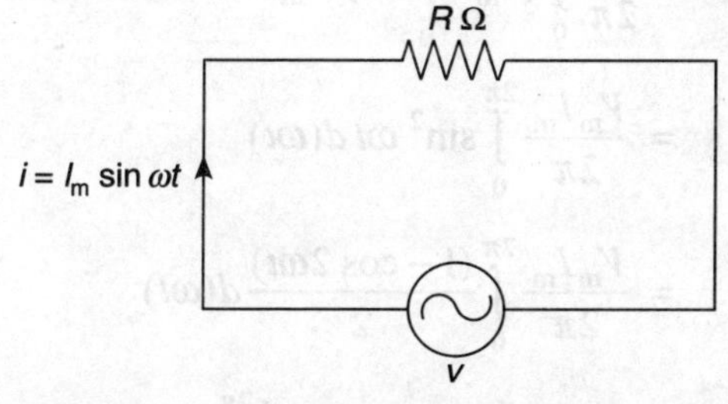

Fig. 4.1 AC circuit with pure resistance.

Let the alternating voltage be given by the equation:

$$v = V_{\text{m}} \sin \omega t \tag{4.1}$$

As a result of this voltage, an alternating current i will flow in the circuit. Then, the applied voltage,

$$v = Ri$$

or

$$V_m \sin \omega t = Ri$$

$$\therefore \quad i = \frac{V_m}{R} \sin \omega t \tag{4.2}$$

The value of i will be maximum (i.e. I_m) when $\sin \omega t$ is unity. Therefore, Eq. (4.2) becomes,

$$i = I_m \sin \omega t \tag{4.3}$$

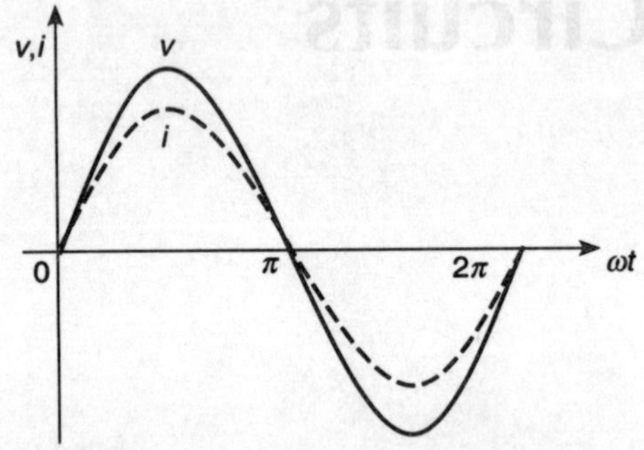

Fig. 4.2 Voltage and current waveforms for a pure resistive circuit.

Fig. 4.3 Voltage and current phasors in a pure resistive circuit.

If we put $t = 0$ in Eqs. (4.1) and (4.3), both voltage and current become zero. Now they are said to be in-phase. The waveform and phasor diagram are shown in Figs. 4.2 and 4.3, respectively. While drawing the phasor diagrams, RMS values are used. The ratio of voltage to current is opposition offered by the circuit element in AC circuit and in this case, $V_m/I_m = R$.

Power in a pure resistive circuit

In a pure resistive circuit, if $v = V_m \sin \omega t$, and $i = I_m \sin \omega t$, then the average power,

$$P = \frac{1}{2\pi} \int_0^{2\pi} (V_m \sin \omega t)(I_m \sin \omega t) d(\omega t)$$

$$= \frac{V_m I_m}{2\pi} \int_0^{2\pi} \sin^2 \omega t \, d(\omega t)$$

$$= \frac{V_m I_m}{2\pi} \int_0^{2\pi} \frac{(1 - \cos 2\omega t)}{2} d(\omega t)$$

$$= \frac{V_m I_m}{4\pi} \left(\omega t - \frac{\sin 2\omega t}{2} \right)_0^{2\pi}$$

$$= \frac{V_m I_m}{4\pi}\left(2\pi - 0 - \frac{\sin 4\pi}{2} + \frac{\sin 0}{2}\right)$$

$$= \frac{V_m I_m}{4\pi} 2\pi = \frac{V_m I_m}{2}$$

$$= \frac{V_m}{\sqrt{2}} \frac{I_m}{\sqrt{2}} = VI$$

where

$\frac{V_m}{\sqrt{2}} = V$ is the RMS value of source voltage.

$\frac{I_m}{\sqrt{2}} = I$ is the RMS value of current.

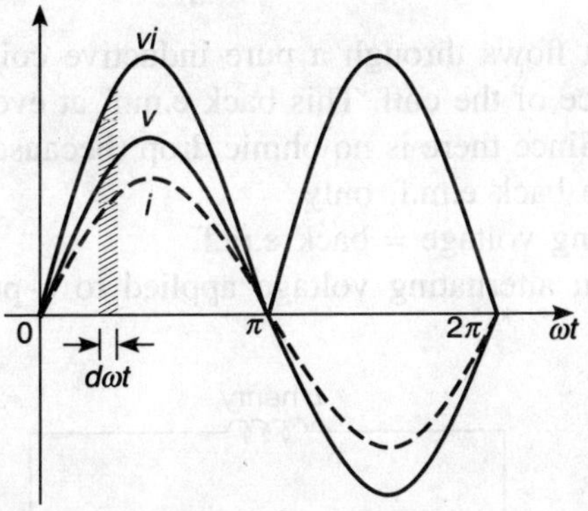

Fig. 4.4 Power curve for a pure resistive circuit.

Thus, like in DC circuits, in AC also, the power in a pure resistance is simply the product of RMS value of voltage and RMS value of current. Fig. 4.4 shows the power curve for a pure resistive circuit. It is clear that power is always positive, that is, the voltage source constantly supplies power to the circuit which is consumed by the circuit.

4.1.2 AC Circuit Containing Inductance Only

Property of Inductance

We know that whenever current passes through a conductor, a magnetic field is set up around the conductor. If the current changes with time, the magnetic flux produced by it also changes. This change in flux linkage causes an e.m.f. to be induced in the circuit. The induced e.m.f. is equal to the rate of change of flux linkages.

i.e. $$e = -\frac{d}{dt}\text{(flux changes)}$$

Flux multiplied by number of turns gives flux linkages, so,

$$e = -\frac{d}{dt}(N\psi) \tag{4.4}$$

This is called self-induced e.m.f. or back e.m.f. The inductance (L) is defined as the flux linkage produced per ampere of the current flowing through it. Its unit is henry (H).

i.e. $$\text{Inductance, } L = \frac{\text{Flux linkage}}{\text{Current}} = \frac{N\psi}{i}$$

Substituting for $N\psi = Li$ into Eq. (4.4), we get

$$e = -\frac{d}{dt}(Li)$$

To keep the current flowing in the circuit, the applied voltage should be equal and opposite to the induced e.m.f.

i.e. $$V = -e = L\frac{di}{dt}$$

When an alternating current flows through a pure inductive coil, a back e.m.f. (= $L\ di/dt$) is induced due to the inductance of the coil. This back e.m.f. at every instant opposes the change in current through the coil. Since there is no ohmic drop (because of no resistance), the applied voltage has to overcome the back e.m.f. only.

Hence, applied alternating voltage = back e.m.f.

Now, let us consider an alternating voltage applied to a pure inductance of L henry as shown in Fig. 4.5.

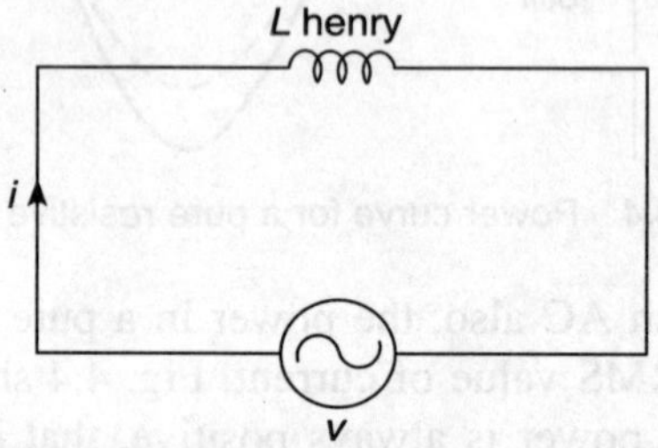

Fig. 4.5 AC circuit with pure inductance.

Let the equation of the applied voltage be:

$$v = V_m \sin \omega t \tag{4.5}$$

$$V_m \sin \omega t = L\frac{di}{dt} \quad \text{(as above)}$$

or $$di = \frac{V_m}{L} \sin \omega t\, dt$$

Integrating both sides, we get,

$$i = \frac{V_m}{L} \int \sin \omega t\, dt$$

$$= \frac{V_m}{\omega L}(-\cos \omega t)$$

or
$$i = \frac{V_m}{\omega L}\sin\left(\omega t - \frac{\pi}{2}\right) \tag{4.6}$$

The value of i will be maximum (i.e. I_m) when $\sin(\omega t - \pi/2)$ is unity. So,

$$I_m = V_m/\omega L$$

Therefore, Eq. (4.6) becomes

$$i = I_m \sin\left(\omega t - \frac{\pi}{2}\right) \quad \text{where } I_m = V_m/\omega L \tag{4.7}$$

It is clear from Eqs. (4.5) and (4.7) that current lags behind the voltage by $\pi/2$ radians or 90°. Hence, in a pure inductance, current lags behind the voltage by 90°. This fact is also shown in the waveform in Fig. 4.6 and phasor diagram in Fig. 4.7.

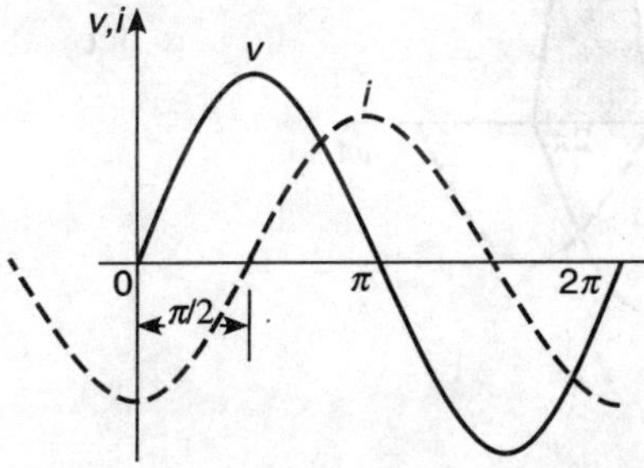

Fig. 4.6 Voltage and current waveforms for a pure inductive circuit.

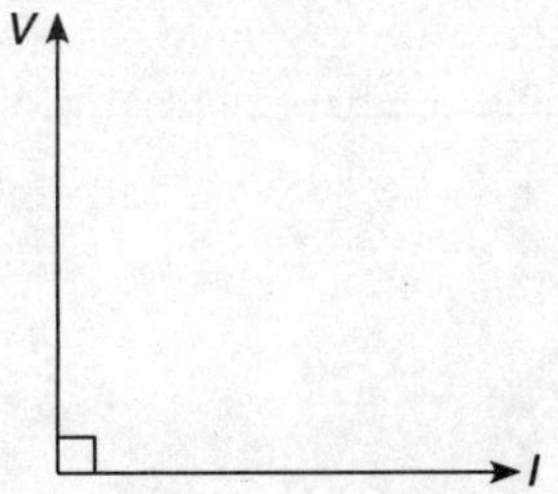

Fig. 4.7 Voltage and current phasors in a pure inductive circuit.

The ratio of voltage to current is opposition offered by the circuit element and in this case, it is $V_m/I_m = \omega L$. The quantity ωL is called *inductive reactance* X_L of the coil and is measured in Ω. Note that $X_L(= \omega L = 2\pi f L)$ will be in Ω if L is in henry and f in Hz.

Power in a pure inductive circuit

In a pure inductive circuit, $v = V_m \sin \omega t$ and $i = I_m \sin(\omega t - \pi/2)$. Then, the power consumed

$$P = \frac{1}{2\pi}\int_0^{2\pi} V_m \sin \omega t \, I_m \sin(\omega t - \pi/2) d(\omega t)$$

$$= \frac{V_m I_m}{2\pi}\int_0^{2\pi} \sin \omega t(-\cos \omega t) d(\omega t)$$

$$= \frac{V_m I_m}{2\pi}\int_0^{2\pi}\left(-\frac{\sin 2\omega t}{2}\right) d(\omega t)$$

$$= \frac{V_m I_m}{4\pi}\left(\frac{\cos 2\omega t}{2}\right)_0^{2\pi}$$

$$= \frac{V_m I_m}{8\pi}(\cos 4\pi - \cos 0)$$

$$= \frac{V_m I_m}{8\pi}(1 - 1)$$

$$= 0$$

Hence, the power absorbed in pure inductance is zero. Figure 4.8 shows the power curve for a pure inductive circuit. An examination of power curve over one full cycle shows that positive power equals the negative power. Hence, the resultant power over one full cycle is zero. In other words, *pure inductance consumes no power.*

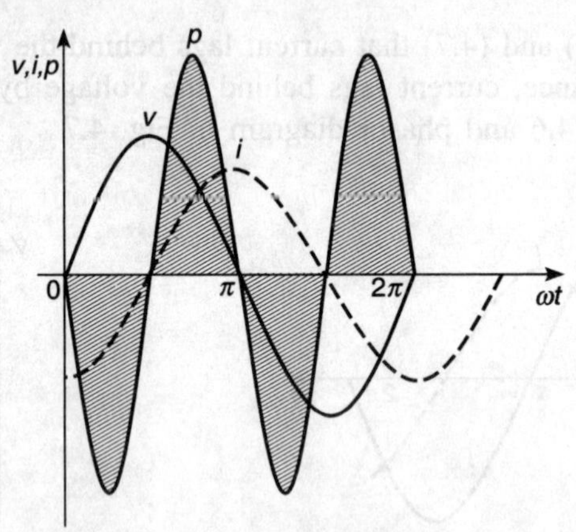

Fig. 4.8 Power curve for a pure inductive circuit.

4.1.3 AC Circuit Containing Capacitance Only

Property of Capacitance

A capacitor is formed by two metallic conductors or plates separated by an insulating medium and the property which permits it to accumulate charge is called capacitance. The potential difference v between the terminals of a capacitor is proportional to the charge q on it.

i.e. $$v \propto q$$

or $$v = \frac{q}{C}$$

Constant C is called the capacitance. q is in coulombs, v is in volts and C in coulombs/volt or farad. As

$$i = \frac{dq}{dt} = C\frac{dv}{dt}$$

then $$v = \frac{1}{C}\int i\,dt$$

Let us consider an AC voltage of $v = V_m \sin \omega t$ is applied to a capacitor of C farad as shown in Fig. 4.9.

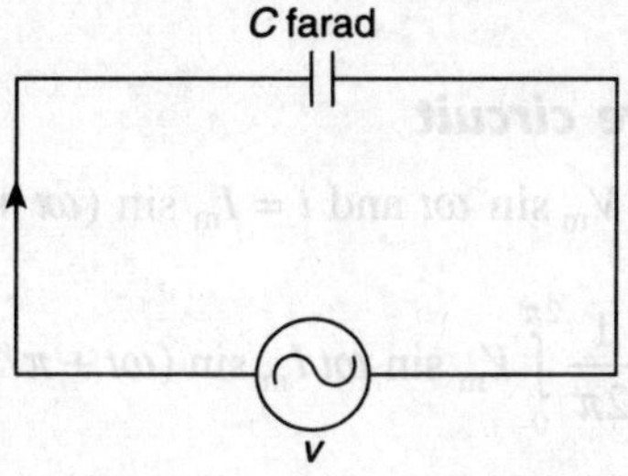

Fig. 4.9 AC circuit with pure capacitance.

As a result of this alternating voltage, alternating current will flow through the circuit. Let at any instant, i is the current and q is the charge on the plates.

Charge on capacitor,

$$q = Cv = CV_m \sin \omega t$$

Hence, circuit current, $i = \frac{d}{dt}(q) = \frac{d}{dt}(CV_m \sin \omega t) = \omega CV_m \cos \omega t$

or $$i = \omega CV_m \sin(\omega t + \pi/2) \qquad (4.8)$$

The value of i will be maximum (i.e. I_m) when $\sin(\omega t + \pi/2)$ is unity.

So, $$I_m = \omega CV_m$$

Substituting the value of $\omega CV_m = I_m$ into Eq. (4.8), we get

$$i = I_m \sin(\omega t + \pi/2) \qquad (4.9)$$

It is clear from Eqs. (4.9) that current leads the voltage by $\pi/2$ radians or 90°. Hence, in a pure capacitance, current leads the voltage by 90°. This fact is also shown in the waveform in Fig. 4.10 and phasor diagram in Fig. 4.11.

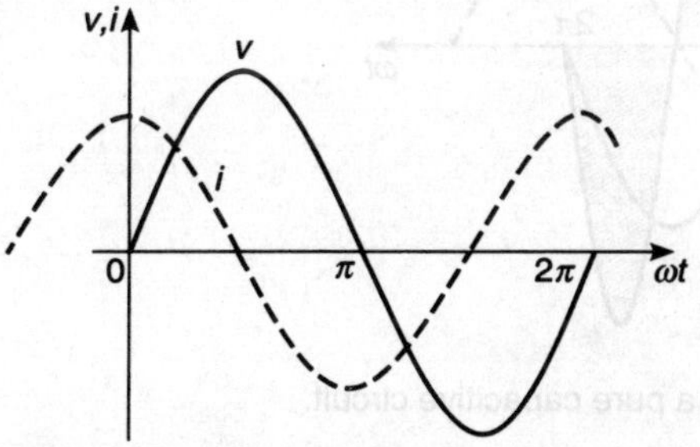

Fig. 4.10 Voltage and current waveforms for a pure capacitive circuit.

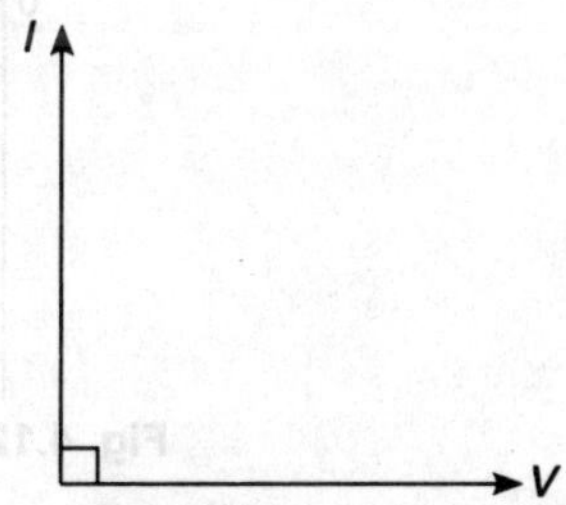

Fig. 4.11 Voltage and current phasors in a pure capacitive circuit.

The ratio of voltage to current is the opposition offered by the circuit element to the flow of current and in this case it is $V_m/I_m = 1/\omega C$. The quantity $1/\omega C$ is called *capacitive reactance* X_C of the capacitor and is measured in Ω. Note that X_c ($= 1/\omega C = 1/2\pi f C$) will be in Ω if C is in farad and f in Hz.

Power in a pure capacitive circuit

In a pure capacitive circuit, $v = V_m \sin \omega t$ and $i = I_m \sin (\omega t + \pi/2)$. Then, the power consumed,

$$P = \frac{1}{2\pi} \int_0^{2\pi} V_m \sin \omega t \, I_m \sin (\omega t + \pi/2) \, d(\omega t)$$

$$= \frac{V_m I_m}{2\pi} \int_0^{2\pi} \sin \omega t \cos \omega t \, d(\omega t)$$

$$= \frac{V_m I_m}{2\pi} \int_0^{2\pi} \frac{\sin 2\omega t}{2} d(\omega t)$$

$$= \frac{V_m I_m}{4\pi} \left(-\frac{\cos 2\omega t}{2} \right)_0^{2\pi}$$

$$= \frac{V_m I_m}{8\pi} (-\cos 4\pi + \cos 0)$$

$$= \frac{V_m I_m}{8\pi} (-1 + 1) = 0$$

Hence, the power absorbed in pure capacitance is zero.

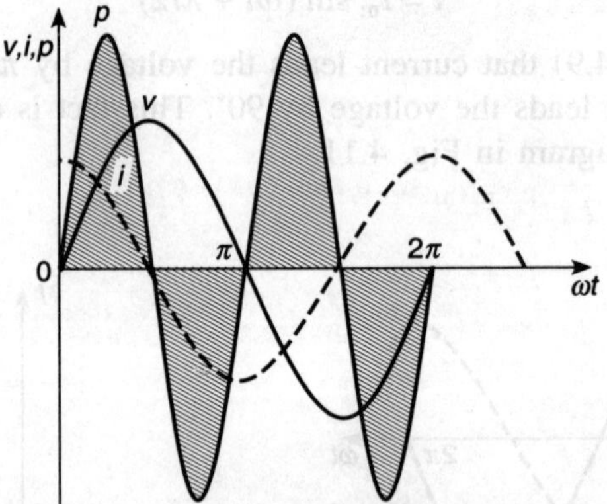

Fig. 4.12 Power curve for a pure capacitive circuit.

Fig. 4.12 shows the power curve for a pure capacitive circuit. The power curve is similar to that for a pure inductor because now current leads the voltage by 90°. It is clear that positive power is equal to the negative power over one full cycle. Hence, *net power absorbed by a capacitor is zero.*

4.1.4 *RL* Series Circuit

Consider the *RL* circuit in Fig. 4.13. Let the applied voltage be *V* volts and the current flowing is *I* amperes. V_R and V_L are the voltage drops across resistance *R* and inductance *L*, respectively. Fig. 4.14 shows the phasor diagram.

The voltage across resistor *R*, $V_R = IR$

The voltage across inductor *L*, $V_L = IX_L$

Therefore,

$$V = \sqrt{V_R^2 + V_L^2} = \sqrt{(IR)^2 + (IX_L)^2}$$
$$= I\sqrt{R^2 + X_L^2} = IZ$$

where $Z = \sqrt{R^2 + X_L^2}$ is called the impedance of the *RL* circuit. Using $Z = \sqrt{R^2 + X_L^2}$, the impedance triangle can be drawn and is shown in Fig. 4.15. From Fig. 4.15, ϕ is the angle by which the total current *I* is out of phase from the voltage *V*.

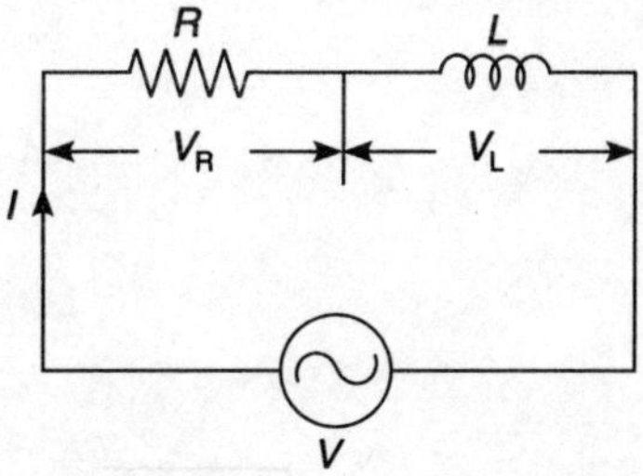

Fig. 4.13 *RL* series circuit.

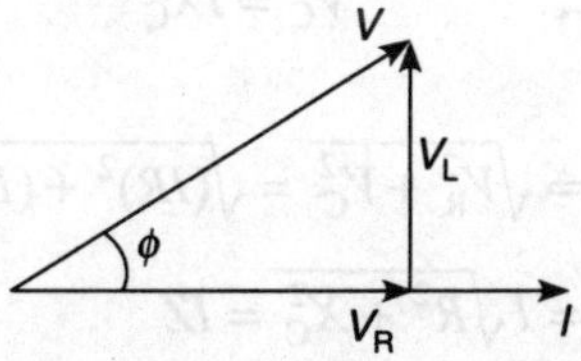

Fig. 4.14 Phasor diagram.

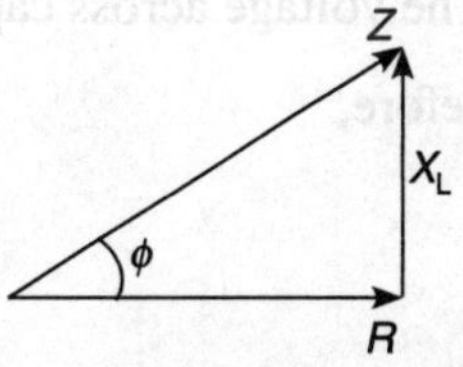

Fig. 4.15 Impedance triangle.

Power in an *RL* series circuit

The equations for the voltage and current are $v = V_m \sin \omega t$ and $i = I_m \sin (\omega t - \phi)$. So, the average power,

$$P = \frac{1}{2\pi}\int_0^{2\pi} V_m I_m \sin \omega t \sin (\omega t - \phi) d(\omega t)$$

$$= \frac{V_m I_m}{4\pi}\int_0^{2\pi} \cos \phi - \cos (2\omega t - \phi) d(\omega t)$$

$$\because [2 \sin A \sin B = \cos (A - B) - \cos (A + B)]$$

$$= \frac{V_m I_m}{4\pi}\left[\cos \phi \omega t - \left(\frac{\sin (2\omega t - \phi)}{2\omega}\right)\right]_0^{2\pi}$$

$$= \frac{V_m I_m}{4\pi}\left[(2\pi \cos \phi - 0) - \left(\frac{\sin (4\pi - \phi)}{2\omega} - \frac{\sin (0 - \phi)}{2\omega}\right)\right]$$

$$= \frac{V_m I_m}{4\pi}\left[(2\pi\cos\phi - 0) - \left(\frac{\sin(-\phi)}{2\omega} - \frac{\sin(-\phi)}{2\omega}\right)\right]$$

$$= \frac{V_m I_m}{4\pi}[2\pi\cos\phi]$$

$$= \frac{V_m}{\sqrt{2}}\frac{I_m}{\sqrt{2}}\cos\phi$$

$$= VI\cos\phi$$

where '$\cos\phi$' is called the power factor of the circuit. It varies between 0 and 1.

4.1.5 *RC* Series Circuit

Consider the *RC* circuit in Fig. 4.16. Let the applied voltage be *V* volts and the current flowing is *I* amperes. V_R and V_C are the voltage drops across resistance *R* and capacitance *C*, respectively.

The voltage across resistor *R*, $V_R = IR$
The voltage across capacitor *C*, $V_C = IX_C$

Therefore,

$$V = \sqrt{V_R^2 + V_C^2} = \sqrt{(IR)^2 + (IX_C)^2}$$

$$= I\sqrt{R^2 + X_C^2} = IZ$$

where $Z = \sqrt{R^2 + X_C^2}$ is called the impedance of the *RC* circuit. Using $Z = \sqrt{R^2 + X_C^2}$, the impedance triangle can be drawn and is shown in Fig. 4.17. From Fig. 4.18, ϕ is the angle by which the total current *I* is out of phase from voltage *V*.

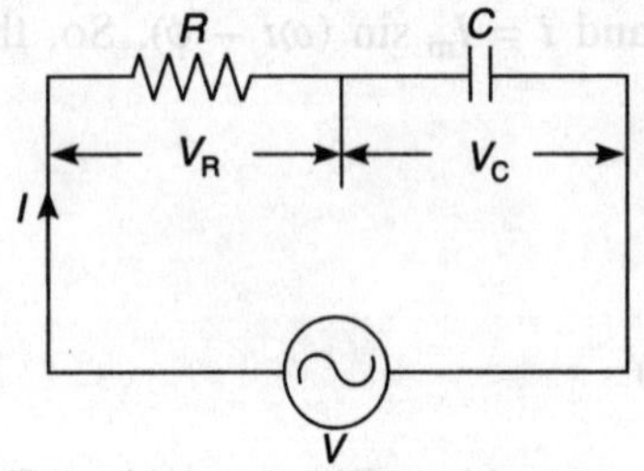

Fig. 4.16 *RC* series circuit.

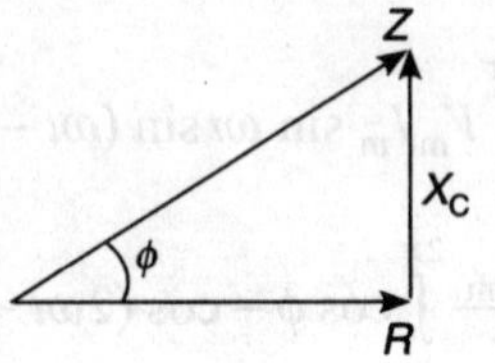

Fig. 4.17 Impedance triangle.

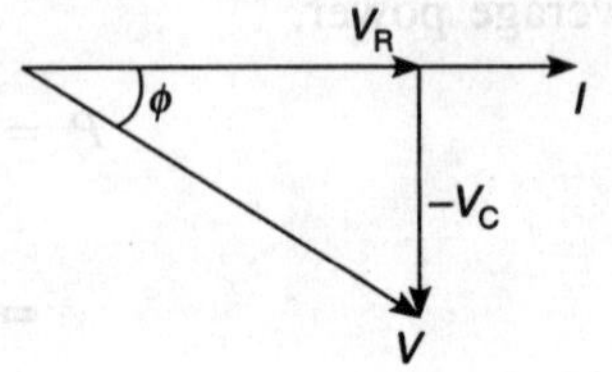

Fig. 4.18 Phasor diagram.

Power in an RC series circuit

The equations for the voltage and current would be

$$v = V_m \sin\omega t$$

$$i = I_m \sin(\omega t + \phi)$$

Hence, the average power,

$$P = \frac{1}{2\pi}\int_0^{2\pi} V_m I_m \sin \omega t \sin(\omega t + \phi)\, d(\omega t)$$

Integrating and simplifying,

$$P = VI \cos \phi$$

$\cos \phi$, power factor of the circuit varies between 0 and 1 for different values of R and C.

4.1.6 *RLC* Series Circuit

Consider the *RLC* series circuit in Fig. 4.19. If $X_L > X_C$ then the circuit behaves like an *RL* circuit and if $X_C > X_L$, the circuit is like an *RC* circuit. Fig. 4.20 shows the phasor diagrams for the two cases.

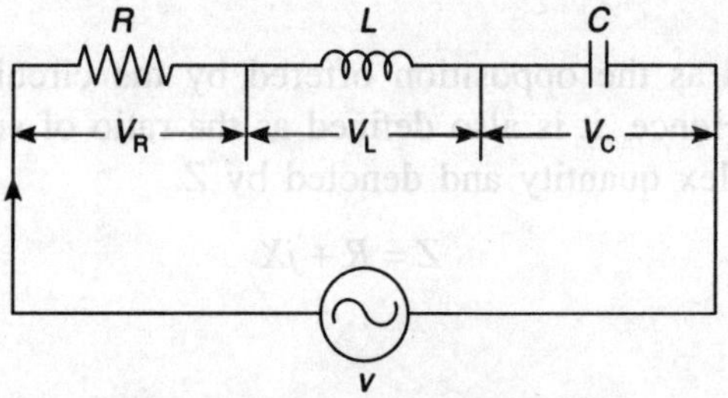

Fig. 4.19 *RLC* series circuit.

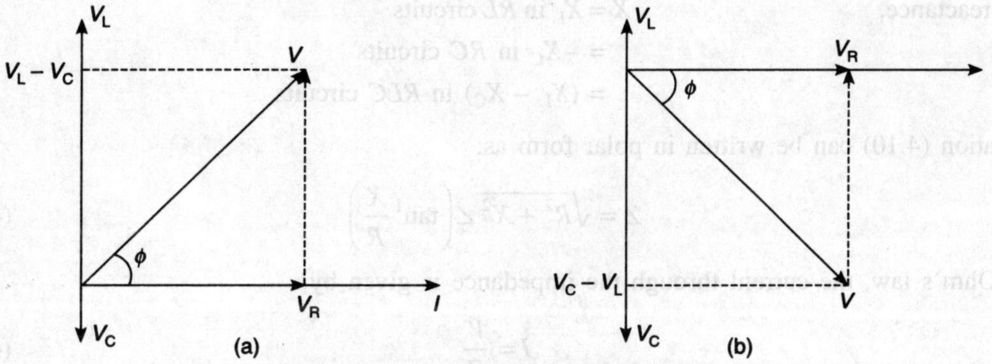

Fig. 4.20 Phasor diagrams of *RLC* circuit.

The equations are written as:

$$V_R = IR, \quad V_L = IX_L, \quad V_C = IX_C$$

$$V = \sqrt{I^2 R^2 + I^2 (X_L - X_C)^2}$$

$$V = I\sqrt{R^2 + (X_L - X_C)^2}$$

$$V = I\sqrt{R^2 + (X_L - X_C)^2} = IZ$$

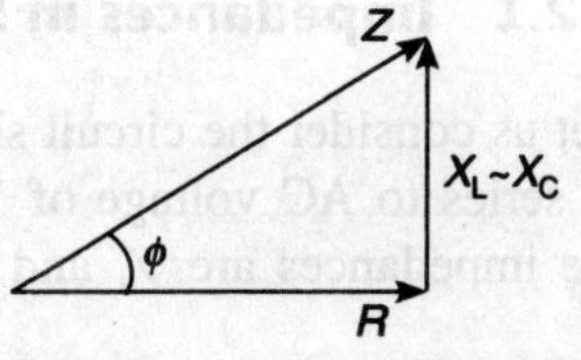

Fig. 4.21 Impedance triangle.

where $Z=\sqrt{R^2+(X_L-X_C)^2}$ is called the impedance of the *RLC* circuit. The impedance triangle for *RLC* circuit is shown in Fig. 4.21.

Power in an RLC series circuit

The equations for the voltage and current would be $v = V_m \sin \omega t$ and $i = I_m \sin (\omega t \pm \phi)$. So, the average power,

$$P=\frac{1}{2\pi}\int_0^{2\pi} V_m I_m \sin \omega t \sin(\omega t \pm \phi) d(\omega t)$$

Integrating and simplifying,

$$P = VI \cos \phi$$

4.2 IMPEDANCE

The impedance is defined as the opposition offered by the circuit to the flow of sinusoidal current. Its unit is ohms. Hence, it is also defined as the ratio of sinusoidal voltage to current. The impedance is a complex quantity and denoted by Z.

$$Z = R + jX \tag{4.10}$$

where

R = Resistance

X = Reactance

The reactance, $X = X_L$ in *RL* circuits

$= -X_C$ in *RC* circuits

$= (X_L - X_C)$ in *RLC* circuits

Equation (4.10) can be written in polar form as:

$$Z=\sqrt{R^2+X^2}\angle\left(\tan^{1}\frac{X}{R}\right) \tag{4.11}$$

By Ohm's law, the current through the impedance is given by

$$I=\frac{V}{Z} \tag{4.12}$$

4.2.1 Impedances in Series

Let us consider the circuit shown in Fig. 4.22 in which two impedances Z_1 and Z_2 are connected in series to AC voltage of V volts. The current through the circuit is I and the voltages across the impedances are V_1 and V_2.

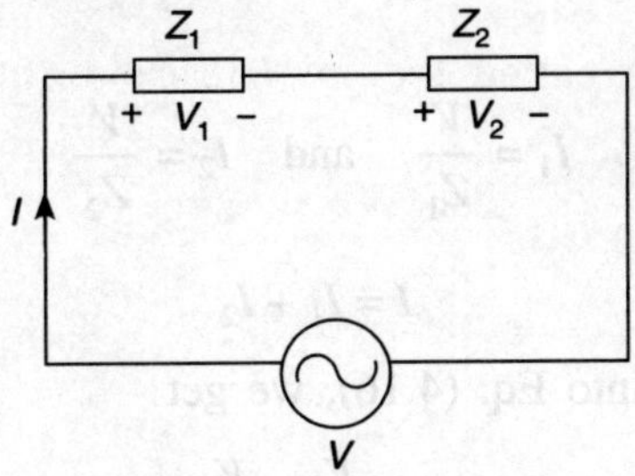

Fig. 4.22 Impedances in series.

From Eq. (4.12),

$$V_1 = IZ_1 \quad \text{and} \quad V_2 = IZ_2 \tag{4.13}$$

Using KVL,

$$V = V_1 + V_2 \tag{4.14}$$

Putting the values of V_1 and V_2 into Eq. (4.14), we get

$$V = IZ_1 + IZ_2$$

So,

$$V = I(Z_1 + Z_2)$$

or

$$\frac{V}{I} = Z_1 + Z_2$$

or

$$Z_{eq} = Z_1 + Z_2$$

Similarly, the equivalent impedance of n number of impedances connected in series is written as:

$$Z_{eq} = Z_1 + Z_2 + \cdots + Z_n$$

4.2.2 Impedances in Parallel

Let us consider the circuit shown in Fig. 4.23 in which two impedances Z_1 and Z_2 are connected in parallel to AC voltage of V volts. The current through the impedances are I_1 and I_2 and the voltage across them is V.

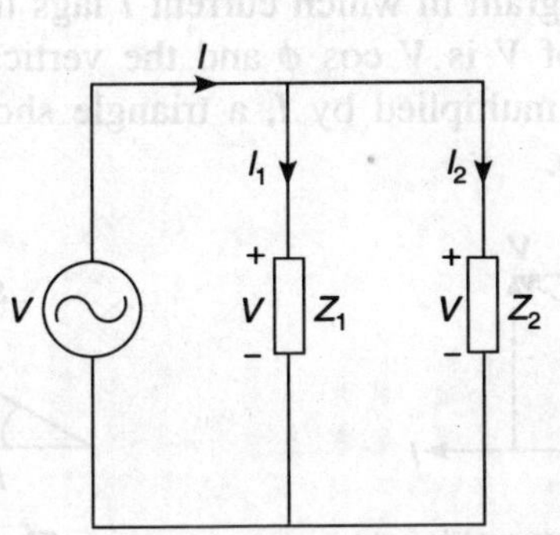

Fig. 4.23 Impedances in parallel.

From Eq. (4.12),

$$I_1 = \frac{V}{Z_1} \quad \text{and} \quad I_2 = \frac{V}{Z_2} \tag{4.15}$$

Using KCL,

$$I = I_1 + I_2 \tag{4.16}$$

Putting the values of I_1 and I_2 into Eq. (4.16), we get

$$I = \frac{V}{Z_1} + \frac{V}{Z_2}$$

Hence,

$$I = V\left(\frac{1}{Z_1} + \frac{1}{Z_2}\right)$$

or

$$\frac{V}{I} = \frac{1}{\left(\frac{1}{Z_1} + \frac{1}{Z_2}\right)}$$

or

$$Z_{eq} = \frac{1}{\left(\frac{1}{Z_1} + \frac{1}{Z_2}\right)}$$

or

$$\frac{1}{Z_{eq}} = \frac{1}{Z_1} + \frac{1}{Z_2} = \frac{Z_1 Z_2}{Z_1 + Z_2}$$

Similarly, the equivalent admittance of n number of impedances connected in parallel is written as:

$$\frac{1}{Z_{eq}} = \frac{1}{Z_1} + \frac{1}{Z_2} + \cdots + \frac{1}{Z_n}$$

4.3 POWER AND POWER FACTOR IN AC CIRCUITS

Figure 4.24 shows a phasor diagram in which current I lags the applied voltage V by an angle ϕ. The horizontal component of V is $V \cos \phi$ and the vertical component of V is $V \sin \phi$. If each of the voltage phasors is multiplied by I, a triangle shown in Fig. 4.25 is obtained and is known as the *power triangle*.

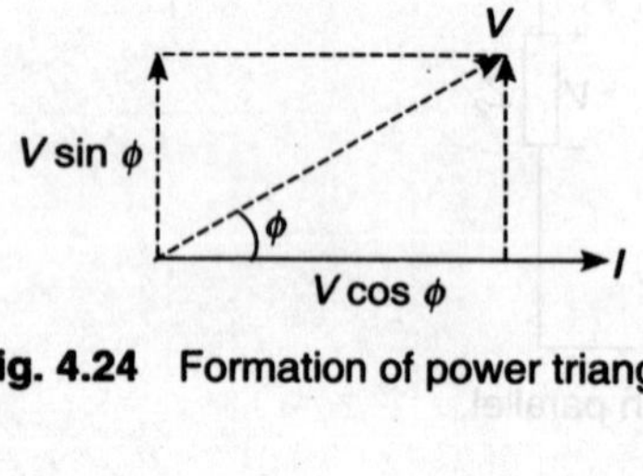

Fig. 4.24 Formation of power triangle.

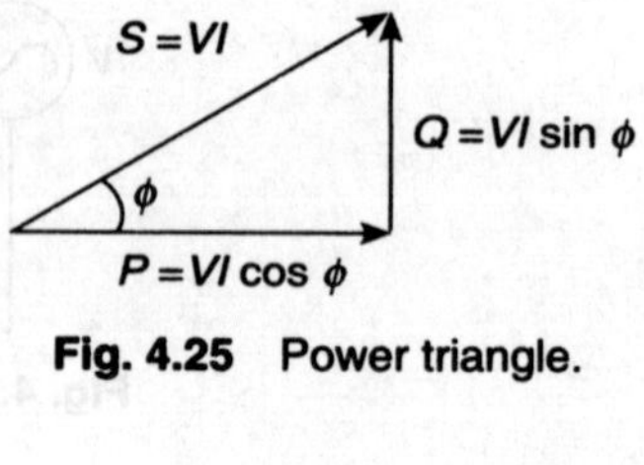

Fig. 4.25 Power triangle.

True or active power, $P = VI \cos \phi$ watts (W)

Reactive power, $Q = VI \sin \phi$ reactive voltamperes (var)

Apparent power, $S = VI$ voltamperes (VA)

Power factor is defined as the ratio of true power to apparent power.

$$\text{Powerfactor} = \frac{\text{True power, } P}{\text{Apparent power, } S} = \frac{VI \cos \phi}{VI} = \cos \phi$$

From impedance triangle (Fig. 4.21), power factor can also be written as the ratio of resistance to impedance.

$$\text{Power factor} = \cos \phi = \frac{R}{Z}$$

4.4 PARALLEL CIRCUITS

Consider two elements in parallel having two different values of resistances and inductances as shown in Fig. 4.26. As the two impedances Z_1 and Z_2 are in parallel, the voltage across each is the same. But the currents through them are different. The phasor diagram for voltage and current is shown in Fig. 4.27. I_1 and I_2 are respective currents through impedances Z_1 and Z_2. ϕ_1 and ϕ_2 are the phase differences of I_1 and I_2 exist with the applied voltage. The vector sum of I_1 and I_2 gives the total current taken from the supply and it lags behind the voltage by an angle, ϕ.

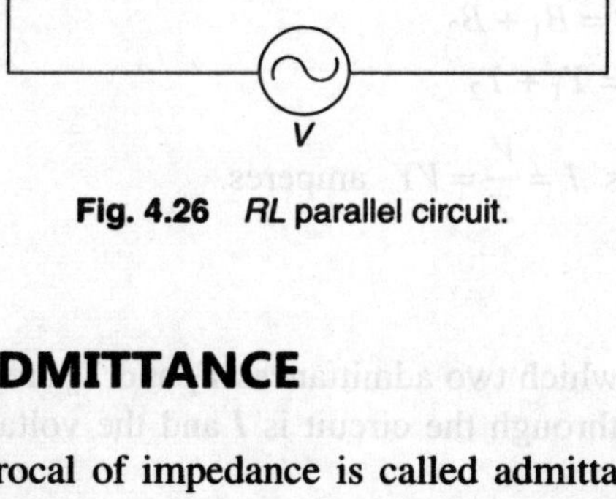

Fig. 4.26 *RL* parallel circuit.

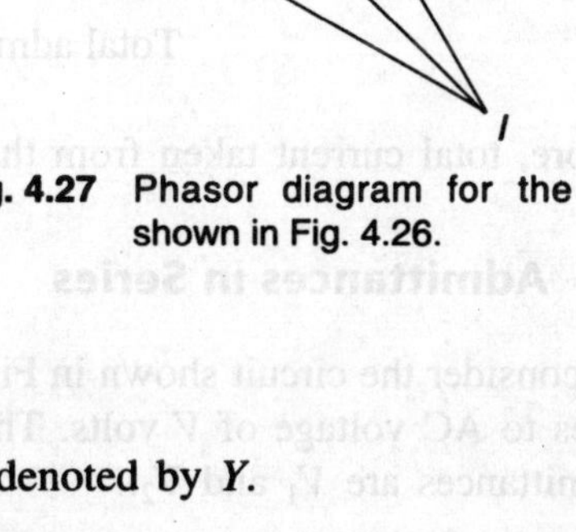

Fig. 4.27 Phasor diagram for the circuit shown in Fig. 4.26.

4.5 ADMITTANCE

The reciprocal of impedance is called admittance. It is denoted by Y.

$$\therefore \qquad \text{Admittance, } Y = \frac{1}{Z} = \frac{I}{V} \qquad (4.17)$$

Unit of admittance is Siemens (S) or mho (℧).

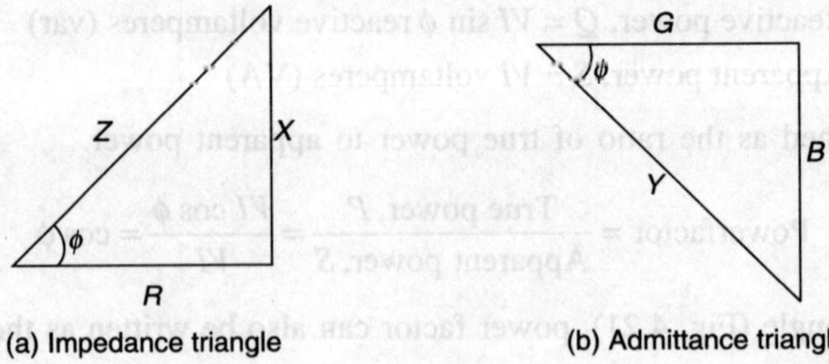

Fig. 4.28

Figures 4.28(a) and (b) shows the impedance triangle and the admittance triangle, respectively. Just as impedance has two components—resistance (R) and reactance (X), admittance is made of conductance (G) and susceptance (B). Obviously, $G = 1/R$ and $B = 1/X$.

For admittance Δ, conductance $G = Y \cos \phi$

$$= \frac{1}{Z}\frac{R}{Z} = \frac{R}{Z^2} \text{ mho (or) Siemens}$$

Similarly, susceptance $B = Y \sin \phi$

$$= \frac{1}{Z}\frac{X}{Z} = \frac{X}{Z^2} \text{ mho (or) Siemens}$$

Admittance, $$Y = \sqrt{\text{Conductance}^2 + \text{Susceptance}^2}$$

$$Y = \sqrt{G^2 + B^2}$$

If G_1 and G_2 are the conductances of the two reactors in parallel, and B_1 and B_2 are their susceptances, then

Total conductance, $G = G_1 + G_2$

Total susceptance, $B = B_1 + B_2$

Total admittance, $Y = Y_1 + Y_2$

Therefore, total current taken from the supply is $I = \frac{V}{Z} = VY$ amperes.

4.5.1 Admittances in Series

Let us consider the circuit shown in Fig. 4.29 in which two admittances Y_1 and Y_2 are connected in series to AC voltage of V volts. The current through the circuit is I and the voltages across the admittances are V_1 and V_2.

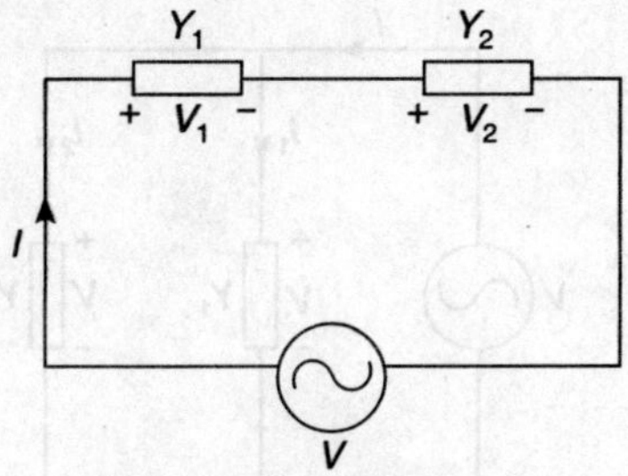

Fig. 4.29 Admittances in series.

From Eq. (4.17),

$$V_1 = \frac{I}{Y_1} \text{ and } V_2 = \frac{I}{Y_2} \tag{4.18}$$

Using KVL,

$$V = V_1 + V_2 \tag{4.19}$$

Putting the values of V_1 and V_2 into Eq. (4.19), we get

$$V = \frac{I}{Y_1} + \frac{I}{Y_2}$$

Hence,

$$V = I\left(\frac{1}{Y_1} + \frac{1}{Y_2}\right)$$

or

$$\frac{I}{V} = \frac{1}{\left(\frac{1}{Y_1} + \frac{1}{Y_2}\right)}$$

or

$$Y_{eq} = \frac{1}{\left(\frac{1}{Y_1} + \frac{1}{Y_2}\right)} = \frac{Y_1 Y_2}{Y_1 + Y_2}$$

or

$$\frac{1}{Y_{eq}} = \frac{1}{Y_1} + \frac{1}{Y_2}$$

Similarly, the equivalent admittance of n number of admittances connected in series is written as:

$$\frac{1}{Y_{eq}} = \frac{1}{Y_1} + \frac{1}{Y_2} + \dots + \frac{1}{Y_n}$$

4.5.2 Admittances in Parallel

Let us consider the circuit shown in Fig. 4.30 in which two admittances Y_1 and Y_2 are connected in parallel to AC voltage of V volts. The current through the admittances are I_1 and I_2 and the voltage across them is V.

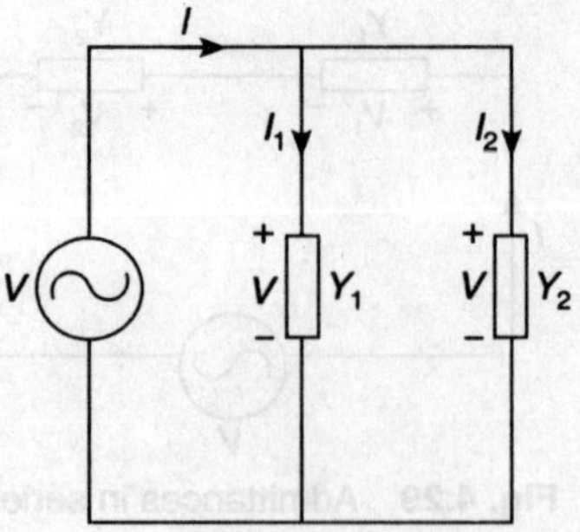

Fig. 4.30 Admittances in parallel.

From Eq. (4.17),

$$I_1 = VY_1 \text{ and } I_2 = VY_2 \tag{4.20}$$

Using KCL,

$$I = I_1 + I_2 \tag{4.21}$$

Putting the values of I_1 and I_2 into Eq. (4.21), we get

$$I = VY_1 + VY_2$$

So,

$$I = V(Y_1 + Y_2)$$

or

$$\frac{I}{V} = Y_1 + Y_2$$

or

$$Y_{eq} = Y_1 + Y_2$$

Similarly, the equivalent admittance of n number of admittances connected in parallel is written as:

$$Y_{eq} = Y_1 + Y_2 + \cdots + Y_n$$

EXAMPLE 4.1 A series RL circuit (Fig. 4.31) with resistance $R = 25\ \Omega$ and inductance $L = 0.02$ H is connected across a 250 V, 50 Hz single-phase AC supply. Calculate (i) the impedance, (ii) current, (iii) power factor and (iv) power.

Solution

(i) $X_L = 2\pi f L = 2\pi \times 50 \times 0.02 = 6.28\ \Omega$

Impedance,

$$Z = \sqrt{R^2 + X_L^2} = \sqrt{25^2 + 6.28^2} = 25.78\ \Omega$$

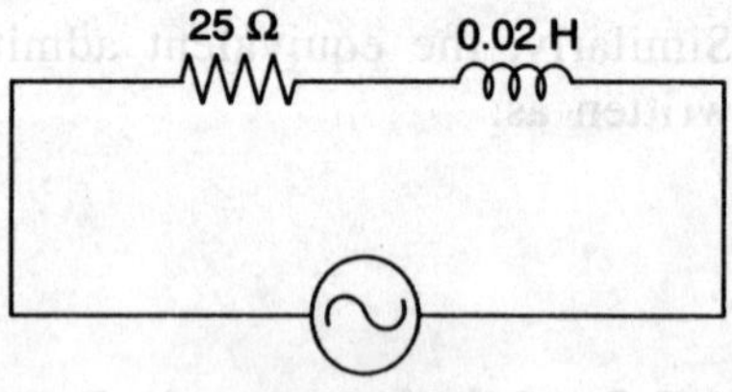

Fig. 4.31

(ii) Current, $I = \dfrac{V}{Z} = \dfrac{250}{25.78} = 9.7$ A

(iii) Power factor, $\cos\phi = \dfrac{R}{Z} = \dfrac{25}{25.78} = 0.97\text{(lag)}$

(iv) Power $= VI\cos\phi = 250 \times 9.7 \times 0.97 = 2352$ W

EXAMPLE 4.2 A load comprising of R and L consumes 20 kW at 0.8 power factor when connected to a 400 V, 50 Hz supply. Calculate (i) the current, (ii) power factor angle, (iii) the impedance, (iv) the resistance and (v) the reactance of the load. Write the equations for voltage and current.

Solution

(i) Power, $P = VI \cos\phi = 20 \text{ kW} = 20{,}000 \text{ W}$

$$\text{Current, } I = \frac{P}{V\cos\phi} = \frac{20{,}000}{400 \times 0.8} = 62.5 \text{ A}$$

(ii) Power factor angle, $\phi = \cos^{-1}(0.8) = 36.8°$ (lag)

(iii) Impedance, $Z = \dfrac{V}{I} = \dfrac{400}{62.5} = 6.4\ \Omega$

(iv) Resistance, $R = Z\cos\phi = 6.4 \times 0.8 = 5.12\ \Omega$

(v) Reactance, $X_L = Z\sin\phi = 6.4 \times 0.6 = 3.84\ \Omega$

Equation for voltage

$$v = V_m \sin \omega t = 400\sqrt{2} \sin(314t) \qquad (\omega = 2\pi f = 2\pi \times 50 = 314)$$

Equation for current

In RL circuit, current I lags behind the supply voltage V by ϕ degrees.

$$\therefore \quad i = I_m \sin(\omega t - \phi) = 62.5\sqrt{2} \sin(314t - 36.8°)$$

EXAMPLE 4.3 When a voltage of 100 V at 50 Hz is applied to a choking coil A (RL circuit), the current taken is 8 A and the power is 120 W [Fig. 4.32(a)]. When applied to coil B, the current is 10 A and the power is 500 W [Fig. 4.32(b)]. What current and power will be taken when 100 V is applied to the two coils connected in series?

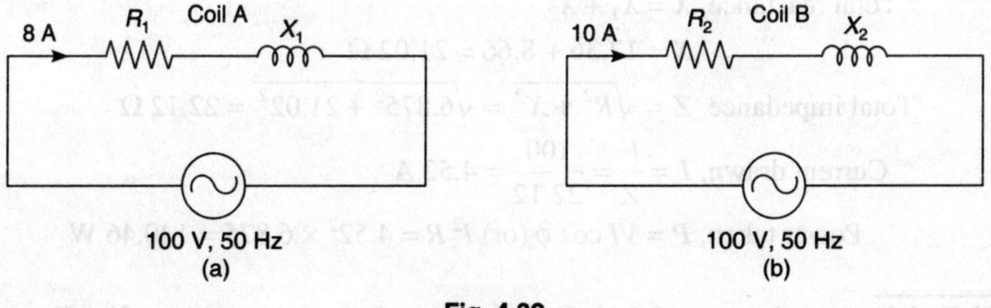

Fig. 4.32

Solution

For coil A

Impedance, $Z_1 = \frac{V}{I_1} = \frac{100}{8} = 12.5\ \Omega$

Power, $P_1 = I_1^2 R_1 = 120$

$= 8^2 R_1 = 120$

So, $R_1 = \frac{120}{8^2} = 1.875\ \Omega$

Reactance, $X_1 = \sqrt{Z_1^2 - R_1^2}$

$= \sqrt{12.5^2 - 1.875^2} = 12.36\ \Omega$

For coil B

Impedance, $Z_2 = \frac{V}{I_2} = \frac{100}{10} = 10\ \Omega$

Power, $P_2 = I_2^2 R_2 = 500$

$= 10^2 R_2 = 500$

So, $R_2 = \frac{500}{10^2} = 5\ \Omega$

Reactance, $X_2 = \sqrt{Z_2^2 - R_2^2}$

$= \sqrt{10^2 - 5^2} = 8.66\ \Omega$

Unknown values determined are:

$R_1 = 1.875\ \Omega;\quad X_1 = 12.36\ \Omega$

$R_2 = 5\ \Omega;\quad X_2 = 8.66\ \Omega$

When coils A and B are in series (Fig. 4.33)

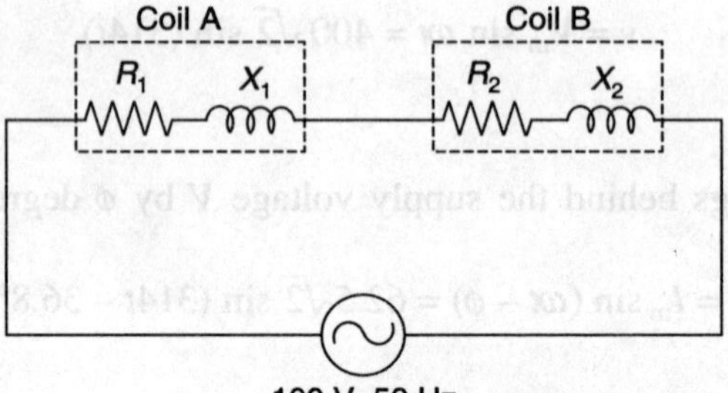

Fig. 4.33

Total resistance, $R = R_1 + R_2$

$R = 1.875 + 5 = 6.875\ \Omega$

Total reactance, $X = X_1 + X_2$

$X = 12.36 + 8.66 = 21.02\ \Omega$

Total impedance, $Z = \sqrt{R^2 + X^2} = \sqrt{6.875^2 + 21.02^2} = 22.12\ \Omega$

Current drawn, $I = \frac{V}{Z} = \frac{100}{22.12} = 4.52$ A

Power taken, $P = VI\cos\phi$ (or) $I^2R = 4.52^2 \times 6.875 = 140.46$ W

EXAMPLE 4.4 A resistance of 100 Ω is connected in series with a 50 μF capacitor to a supply at 200 V, 50 Hz. Find the (i) impedance, current, power factor and phase angle,

(ii) active power, reactive power and apparent power, and (iii) the voltage across resistor and across capacitor.

Solution $R = 100\ \Omega$; $C = 50\ \mu F$

$V = 200$ V; $f = 50$ Hz

$$\text{Capacitive reactance, } X_C = \frac{1}{\omega C} = \frac{1}{2\pi fC} = \frac{1}{2\pi \times 50 \times 50 \times 10^{-6}}$$

$$= 63.66\ \Omega$$

$$\text{Impedance, } Z = \sqrt{R^2 + X_C^2} = \sqrt{100^2 + 63.66^2}$$

$$= 118.54\ \Omega$$

$$\text{Current, } I = \frac{V}{Z} = \frac{200}{118.54} = 1.69 \text{ A}$$

$$\text{Power factor, } \cos\phi = \frac{R}{Z} = \frac{100}{118.54} = 0.844 \text{ (lead)}$$

$$\text{Phase angle, } \phi = \cos^{-1} 0.844 = 32.44°$$

$$\text{Aeactive power, } P = VI\cos\phi = 200 \times 1.69 \times 0.844 = 285.27 \text{ W}$$

$$\text{Reactive power, } Q = VI\sin\phi = 200 \times 1.69 \times 0.536 = 181.17 \text{ var}$$

$$\text{Apparent power, } S = VI = 200 \times 1.69 = 338 \text{ VA}$$

$$\text{Voltage across resistor} = IR = 1.69 \times 100$$

$$V_R = \mathbf{169\ V}$$

$$\text{Voltage across capacitor} = IX_C = 1.69 \times 63.66$$

$$V_C = \mathbf{107.59\ V}$$

EXAMPLE 4.5 In a circuit, the applied voltage of 150 V lags the current of 8 A by 40°.

(i) Is power factor lagging or leading?
(ii) What is the value of power factor?
(iii) Is the circuit inductive or capacitive?
(iv) What is the value of active power and reactive power?

Solution

(i) The 'applied voltage lags behind the current' means current leads the voltage. Therefore, power factor is leading.
(ii) Power factor = $\cos\phi = \cos 40° = 0.766$ (leading)
(iii) Power factor 'leading' means the circuit is capacitive.
(iv) Active power = $VI\cos\phi = 150 \times 8 \times 0.766 = 919.2$ W; reactive power = $VI\sin\phi$ $= 150 \times 8 \times 0.643 = 771.6$ var

EXAMPLE 4.6 Find the circuit constants of a two-element series circuit which consumes 700 W with 0.707 leading power factor. The applied voltage is $v = 141.4 \sin 314t$.

Solution Leading power factor means the given two-element series circuit is an *RC* circuit.

The supply voltage, $v = 141.4 \sin 314t$

Comparing it with $v = V_m \sin \omega t$, we get maximum value of supply voltage,

$$V_m = 141.4 \text{ V}$$

Angular frequency, $\omega = 314$

So, frequency,

$$f = \frac{\omega}{2\pi} = \frac{314}{2\pi} = 50$$

RMS value of supply voltage,

$$V = \frac{141.4}{\sqrt{2}} = 99.98 \text{ V}$$

$$\text{Power} = 700 \text{ W}; \qquad \cos \phi = 0.707 \text{ (lead)}$$

$$I = \frac{P}{V \cos \phi} = \frac{700}{99.98 \times 0.707} = 9.9 \text{ A}$$

$$\text{Impedance, } Z = \frac{V}{I} = \frac{99.98}{9.9} = 10.09 \, \Omega$$

$$R = Z \cos \phi = 10.09 \times 0.707 = \mathbf{7.13 \, \Omega}$$

$$X_C = Z \sin \phi = 10.09 \times 0.707 = 7.13 \, \Omega$$

$$X_C = \frac{1}{2\pi f C} = 7.13 \, \Omega$$

Hence,

$$C = \frac{1}{2\pi f X_C} = \frac{1}{2\pi \times 50 \times 7.13} = 4.466 \times 10^{-4} \text{ F}$$

$$C = \mathbf{446.6 \times 10^{-6} \text{ F}} \quad \text{or} \quad \mathbf{446.6 \, \mu F}$$

EXAMPLE 4.7 A coil of resistance 8 Ω and reactance 10 Ω is connected in parallel with another resistance of 10 Ω (Figure 4.34). If the voltage across the whole combination is 200 V AC, find the total current taken from the supply mains.

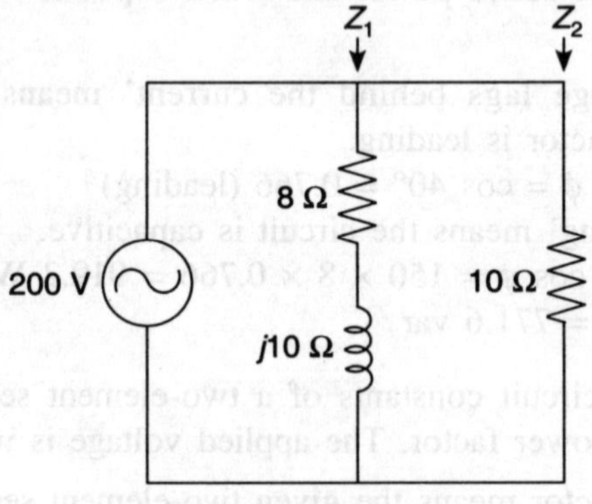

Fig. 4.34

Solution

$$\text{Impedance, } Z_1 = (8 + j10)\ \Omega = \sqrt{8^2 + 10^2} = 12.81\ \Omega$$

$$\text{Impedance, } Z_2 = 10\ \Omega$$

For any impedance,

$$\text{Conductance, } G_1 = \frac{R_1}{Z_1^2} = \frac{8}{12.81^2} = 0.049 \text{ mho}$$

$$\text{Conductance, } G_2 = \frac{R_2}{Z_2^2} = \frac{10}{10^2} = 0.1 \text{ mho}$$

Similarly,

$$\text{Susceptance, } B_1 = \frac{X_1}{Z_1^2} = \frac{10}{12.81^2} = 0.061 \text{ mho}$$

$$\text{Susceptance, } B_2 = \frac{X_2}{Z_2^2} = 0 \text{ mho}$$

$$\text{Total conductance, } G = G_1 + G_2$$
$$= 0.049 + 0.1 = 0.149 \text{ mho}$$

$$\text{Total susceptance, } B = B_1 + B_2$$
$$= 0.061 \text{ mho}$$

$$\text{Total admittance, } Y = \sqrt{(G_1^2 + G_2^2) + (B_1^2 + B_2^2)}$$
$$= \sqrt{0.149^2 + 0.061^2}$$
$$= 0.161 \text{ mho}$$

$$\text{Total current, } I = VY$$
$$= 200 \times 0.161$$
$$= \mathbf{32.2\ A}$$

EXAMPLE 4.8 A 50 μF capacitor is connected in parallel with a choke coil which has a resistance of 20 Ω and an inductance of 0.05 H across a 200 V, 50 Hz supply (Fig. 4.35). Calculate the total current, branch currents, power factor and active power of the circuit.

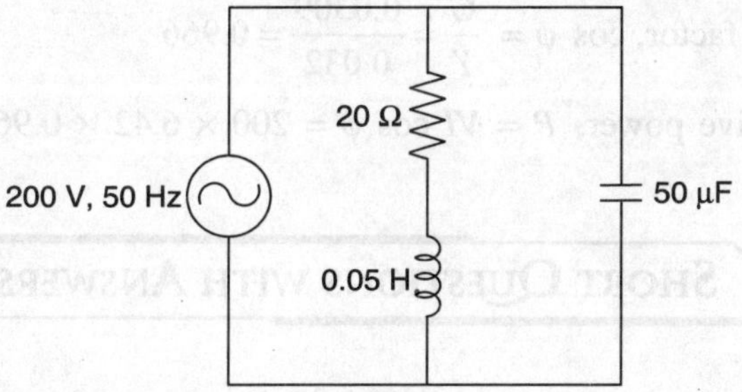

Fig. 4.35

Solution Inductive reactance, $X_L = 2\pi fL$

$$= 2\pi \times 50 \times 0.05$$
$$= 15.71\ \Omega$$

$$\therefore \qquad Z_1 = \sqrt{20^2 + 15.71^2} = 25.43\ \Omega$$

$$\text{Capacitive reactance, } X_C = \frac{1}{2\pi fC}$$

$$= \frac{1}{2\pi \times 50 \times 50 \times 10^{-6}} = 63.66\ \Omega$$

Hence,

$$Z_2 = 63.66\ \Omega$$

$$\text{Conductance, } G_1 = \frac{R_1}{Z_1^2} = \frac{20}{25.43^2} = 0.0309 \text{ mho}$$

$$\text{Susceptance, } B_1 = \frac{X_1}{Z_1^2} = \frac{15.71}{25.43^2} = 0.0243 \text{ mho}$$

$$\text{Conductance, } G_2 = 0$$

$$\text{Susceptance, } B_2 = \frac{X_2}{Z_2^2} = \frac{63.66}{63.66^2} = 0.0157 \text{ mho}$$

$$\text{Total conductance, } G = G_1 + G_2 = 0.0309 \text{ mho}$$

$$\text{Total susceptance, } B = B_1 \sim B_2 = 0.0243 - 0.0157 = 0.0086 \text{ mho}$$

$$\text{Total admittance, } Y = \sqrt{G^2 + B^2}$$

$$= \sqrt{0.0309^2 + 0.0086^2}$$

$$= 0.032 \text{ mho}$$

$$\text{Total current, } I = VY = 200 \times 0.032$$

$$= 6.4 \text{ A}$$

$$\text{Current through } Z_1,\ I_1 = \frac{V}{Z_1} = \frac{200}{25.43} = 7.86 \text{ A}$$

$$\text{Current through } Z_2,\ I_2 = \frac{V}{Z_2} = \frac{200}{63.66} = 3.14 \text{ A}$$

$$\text{Power factor, } \cos\phi = \frac{G}{Y} = \frac{0.0309}{0.032} = 0.966$$

$$\text{Active power, } P = VI\cos\phi = 200 \times 6.42 \times 0.964 = 1236.48 \text{ W}$$

SHORT QUESTIONS WITH ANSWERS

1. **What is the power consumed by a pure resistive circuit?**

   The power consumed by a pure resistive circuit is $P = VI$ watts, where V and I are the RMS values of voltage and current, respectively.

2. What is the power consumed by a pure inductive circuit?

The power consumed by a pure inductive circuit is zero.

3. What is the power consumed by a pure capacitive circuit?

The power consumed by a pure inductive circuit is zero.

4. Define inductive reactance.

The ratio of voltage to current is the opposition offered by the circuit element in a pure inductive circuit. For *RL* circuits, it is $V_m/I_m = \omega L$. The quantity ωL is called inductive reactance (X_L) and is measured in Ω.

5. Define capacitive reactance.

The ratio of voltage to current is the opposition offered by the circuit element in a pure capacitive circuit. For *RC* circuits, it is $V_m/I_m = 1/\omega C$. The quantity $1/\omega C$ is called capacitive reactance (X_C) and is measured in Ω.

6. Define impedance.

The ratio of voltage to current in an AC circuit is called impedance. The property of impedance is to oppose the flow of current. Mathematically, $Z = \sqrt{R^2 + X^2}$, where *Z*, *R* and *X* are called as impedance, resistance and reactance, respectively.

7. What is the impedance of an *RL* circuit?

The impedance of an *RL* circuit is given by $Z = \sqrt{R^2 + X_L^2}$

8. What is the impedance of an *RC* circuit?

The impedance of an *RC* circuit is given by $Z = \sqrt{R^2 - X_C^2}$

9. What is the impedance of an *RLC* circuit?

The impedance of an *RLC* circuit is given by $Z = \sqrt{R^2 + (X_L - X_C)^2}$

10. Draw the impedance triangle.

R = resistance of the circuit

X_L = inductive reactance of the circuit

X_C = capacitive reactance of the circuit

$Z = \sqrt{R^2 + (X_L - X_C)^2}$ is called the impedance of the *RLC* circuit.

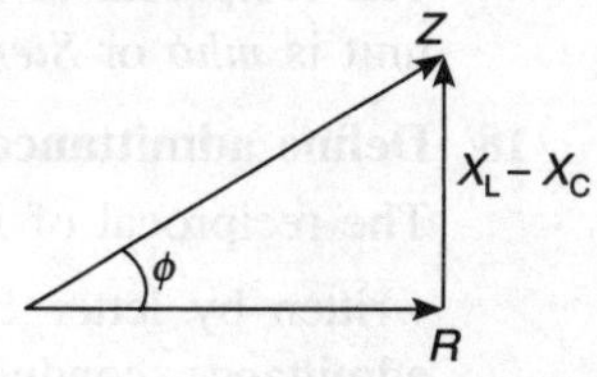

Fig. S4.1 Impedance triangle.

11. What is the power consumed by an *RL* circuit?

The power consumed by an *RL* circuit is $P = VI \cos \phi$ W.

12. What is the power consumed by an *RC* circuit?

The power consumed by an *RC* circuit is $P = VI \cos \phi$ W.

13. What is the power consumed by an *RLC* circuit?

The power consumed by an *RLC* circuit is $P = VI \cos \phi$ W.

14. Define power factor.

Power factor is defined as the ratio of true power to apparent power.

$$\text{Power factor} = \frac{\text{true power, } P}{\text{apparent power, } S} = \frac{VI\cos\phi}{VI} = \cos\phi$$

Power factor can also be written as the ratio of resistance to impedance.

$$\text{Power factor} = \cos\phi = \frac{R}{Z}$$

15. Draw the power triangle.

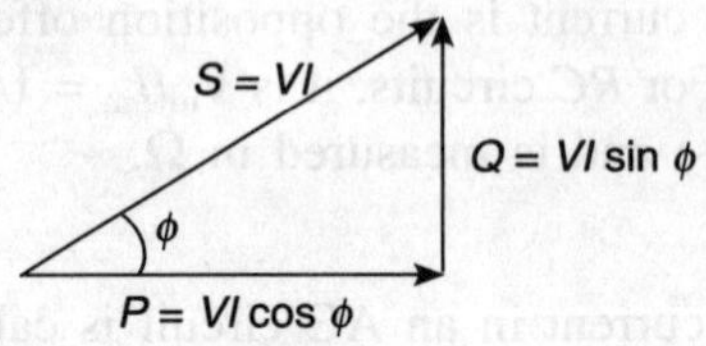

Fig. S4.2 Power triangle.

True or active power, $P = VI \cos\phi$ W

Reactive power, $Q = VI \sin\phi$ reactive voltamperes (var)

Apparent power, $S = VI$ voltamperes (VA)

16. Define conductance.

The reciprocal of resistance is called as conductance. It is written by letter G. Its unit is *mho* or *Siemens*.

17. Define susceptance.

The reciprocal of reactance is called as susceptance. It is written by the letter B. Its unit is *mho* or *Siemens*.

18. Define admittance.

The reciprocal of impedance is called as admittance. Its unit is *mho* or *Siemens*. It is written by letter Y. Mathematically, $Y = \sqrt{G^2 + B^2}$, where Y, G and B are called as admittance, conductance and susceptance, respectively.

19. Write the formula to find the equivalent impedance in a network when a number of impedances are connected in parallel.

If there are n number of impedances connected in parallel, then

$$\frac{1}{Z_{eq}} = \frac{1}{Z_1} + \frac{1}{Z_2} + \frac{1}{Z_3} + \cdots + \frac{1}{Z_n}$$

where Z_{eq} is the equivalent value of impedance.

20. Write the formula to find the equivalent impedance in a network when a number of impedances are connected in series.

If there are *n* number of impedances connected in series, then

$$Z_{eq} = Z_1 + Z_2 + Z_3 + \cdots + Z_n$$

where Z_{eq} is the equivalent value of impedance.

21. Write the formula to find the equivalent admittance in a network when a number of admittances are connected in parallel.

If there are *n* number of admittances connected in parallel, then

$$Y_{eq} = Y_1 + Y_2 + Y_3 + \cdots + Y_n$$

where Y_{eq} is the equivalent value of admittance.

22. Write the formula to find the equivalent admittance in a network when a number of admittances are connected in series.

If there are *n* number of admittances connected in series, then

$$\frac{1}{Y_{eq}} = \frac{1}{Y_1} + \frac{1}{Y_2} + \frac{1}{Y_3} + \cdots + \frac{1}{Y_n}$$

where Y_{eq} is the equivalent value of admittance.

UNSOLVED PROBLEMS

1. An alternating voltage (80 + *j*60) V is applied to a circuit and the current flowing is (–4 + *j*10) A. Find (a) the impedance of the circuit, (b) the power consumed and (c) the phase angle. **[Ans.** (a) 9.28 Ω; (b) 280 W; (c) 75° leading]

2. Calculate (a) the admittance *Y*, (b) the conductance *G* and (c) the susceptance *B* of a circuit consisting of a resistor of 10 Ω in series with an inductor of 0.1 H, when the frequency is 50 Hz.

[Ans. (a) Y = 0.0303 ℧; (b) G = 0.0092 ℧; (c) B = –0.029 ℧]

3. A voltage of 125 V at 60 Hz is applied across a non-inductive resistor connected in series with a capacitor. The current is 2.2 A. The power loss in the resistor is 96.8 W and the power in capacitor is negligible. Calculate the resistance and the capacitance.

[Ans. 20 Ω, 50 μF]

4. A current of 5 A flows through a non-inductive resistance in series with a coil when the circuit is excited with 250 V, 50 Hz supply. If the voltage across the resistor is 125 V and across the coil 200 V, calculate (a) the impedance, resistance and reactance of the coil, (b) the power absorbed by the coil and (c) the total power.

[Ans. (a) 40 Ω, 39.6 Ω, 5.5 Ω; (b) 137.5 W; (c) 762.5 W]

5. Coil A takes 2 A at power factor 0.8 lagging with an applied voltage of 10 V. Coil B takes 2 A with power factor of 0.7 lagging with an applied voltage of 5 V. What voltage will be required to produce a total current of 2 A with coils A and B in series? Find power factor in this case. **[Ans.** 14.96 V; 0.77 lagging]

6. A 200 V, 50 Hz supply is connected to a 20 Ω resistor in series with a coil. The reading of the voltmeter across the resistor is 120 V and across the coil 144 V. Calculate the power and the reactive VA in the coil and the power factor of the circuit.
[**Ans.** 121 W, 855 VAR, 0.7]

7. Find an expression for the current and calculate the power, when a voltage represented by $v = 283 \sin 100\pi t$ is applied to a coil having $R = 50\ \Omega$ and $L = 0.159$ H.
[**Ans.** $4 \sin (100\pi t - \pi/4)$, 400 W]

8. A voltage of 200 V is applied to a series circuit consisting of a resistor, an inductor and a capacitor. The respective voltages across these components are 170 V, 150 V and 100 V and the current is 4 A. Find the power factor of the inductor and of the circuit.
[**Ans.** 0.16, 0.97]

9. A 230 V, 50 Hz voltage is applied to a coil $L = 5$ H and $R = 2\ \Omega$ in series with capacitance C. What value must C have in order that the voltage across the coil shall be 250 V?
[**Ans.** 26 μF]

10. A coil of resistance 15 Ω and inductance 0.05 H is connected in parallel with a non-inductive resistor of 20 Ω. Find (a) the current in each branch, (b) the total current supplied and (c) the phase angle of the combination, when a voltage of 200 V at 50 Hz is applied.
[**Ans.** (a) 9.2 A, 10 A; (b) 17.6 A; (c) 22°]

11. An equipment consumes 2 kW when connected across a 110 V, 100 Hz supply and takes a lagging current of 25 A. If a capacitor is connected in parallel with equipment to make the power factor unity, find its capacitance.
[**Ans.** 248.29 μF]

12. Two parallel circuits consist respectively (i) a coil of 20 Ω and inductance 0.07 H, and (ii) a capacitance of 60 μF in series with a resistance of 50 Ω. Calculate the current in the mains and the power factor of the arrangement when connected across a 200 V, 50 Hz supply.
[**Ans.** 7.05 A; 0.907 lag]

CHAPTER 5

AC Network Analysis

5.1 INTRODUCTION

In Chapter 2, we have discussed mesh analysis and nodal analysis to solve circuits containing resistors only with DC sources. Circuits with AC sources can also be analysed through these techniques. The techniques discussed in Chapter 2 with DC circuits are slightly modified to deal with AC circuits. Here, phasors of voltage, current, impedances and admittances are used to form equations based on KVL and KCL. The following sections discuss how mesh analysis and nodal analysis are used to solve network containing resistor, inductor and capacitor.

5.2 MESH ANALYSIS FOR AC CIRCUITS

As in DC circuits, the mesh currents are selected to write Kirchhoff's voltage law equations and they are put in matrix form. The mesh equations expressed as matrix equations are of the form:

$$[Z]\,[I] = [V]$$

where $[Z]$ is the impedance matrix.

By inspection, the matrix equations are written for the circuit shown in Fig. 5.1 as:

$$\begin{bmatrix} Z_1 + Z_2 & -Z_2 & 0 \\ -Z_2 & Z_2 + Z_3 + Z_4 & -Z_4 \\ 0 & -Z_4 & Z_4 + Z_5 \end{bmatrix} \begin{bmatrix} I_1 \\ I_2 \\ I_3 \end{bmatrix} = \begin{bmatrix} V_A \\ 0 \\ -V_B \end{bmatrix}$$

The above equations can be solved for currents I_1, I_2 and I_3 using Cramer's rule. We shall illustrate mesh analysis with the following examples.

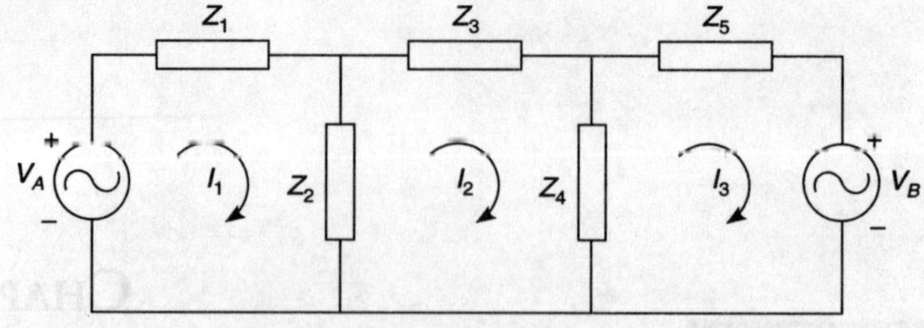

Fig. 5.1 Network with three meshes.

EXAMPLE 5.1 In the circuit shown in Fig. 5.2, find the currents in the various elements.

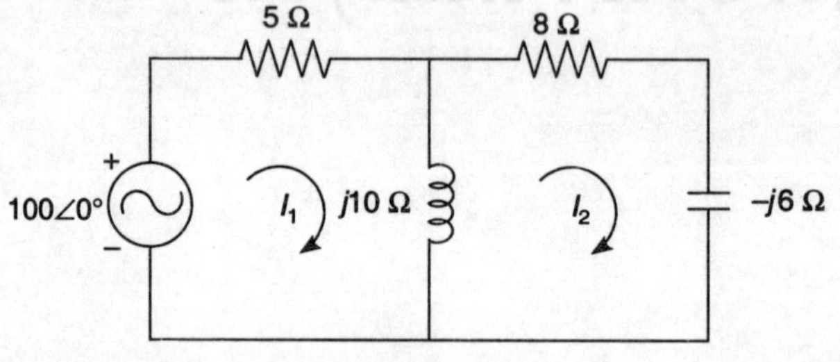

Fig. 5.2

Solution By inspection, the mesh equations are written as:

$$\begin{bmatrix} 5+j10 & -j10 \\ -j10 & 8+j4 \end{bmatrix}\begin{bmatrix} I_1 \\ I_2 \end{bmatrix} = \begin{bmatrix} 100\angle 0^\circ \\ 0 \end{bmatrix}$$

$$I_1 = \frac{\begin{vmatrix} 100\angle 0^\circ & -j10 \\ 0 & 8+j4 \end{vmatrix}}{\begin{vmatrix} 5+j10 & -j10 \\ -j10 & 8+j4 \end{vmatrix}} = \frac{800+j400}{(0+j100)-(-100+j0)} = \frac{800+j400}{100+j100} = (6-j2)\text{ A}$$

$$I_2 = \frac{\begin{vmatrix} 5+j10 & 100\angle 0^\circ \\ -j10 & 0 \end{vmatrix}}{\begin{vmatrix} 5+j10 & -j10 \\ -j10 & 8+j4 \end{vmatrix}} = \frac{0+j1000}{100+j100} = (5+j5)\text{ A}$$

Hence, current through 5 Ω resistance = **(6 – *j*2) A**
Current through 8 Ω resistance and (– *j*6 Ω) reactance = **(5+ *j*5) A**
Current through (*j*10 Ω) reactance = (6 – *j*2) – (5 + *j*5) = **(1 – *j*7) A**

EXAMPLE 5.2 Find the value of the voltage source which makes a voltage drop of 20 V across 5 Ω resistor.

Solution By inspection, the matrix equations are written for the circuit shown in Fig. 5.3 as:

$$\begin{bmatrix} 3+j5 & -3 & 0 \\ -3 & 3 & j2 \\ 0 & j2 & 5-j2 \end{bmatrix} \begin{bmatrix} I_1 \\ I_2 \\ I_3 \end{bmatrix} = \begin{bmatrix} V \\ 0 \\ 0 \end{bmatrix}$$

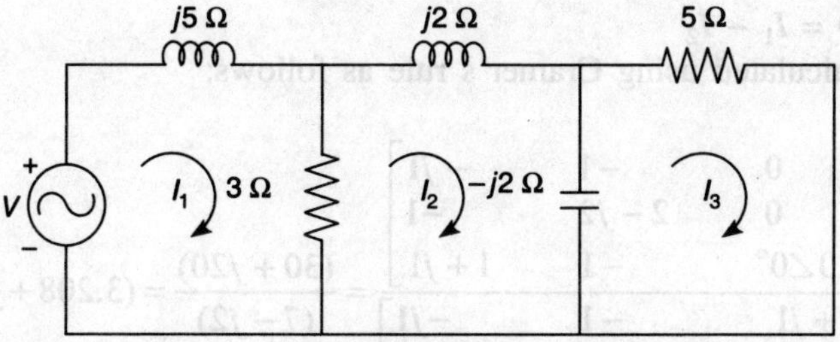

Fig. 5.3

Solving for I_3 using Cramer's rule

$$I_3 = \frac{\begin{bmatrix} 3+j5 & -3 & V \\ -3 & 3 & 0 \\ 0 & j2 & 0 \end{bmatrix}}{\begin{bmatrix} 3+j5 & -3 & 0 \\ -3 & 3 & j2 \\ 0 & j2 & 5-j2 \end{bmatrix}} = \frac{(-j6)V}{(42+j95)} \text{ A}$$

So, voltage across 5 Ω resistor $= \dfrac{(-j6)V}{(42+j95)} \times 5$

$$20 = \frac{(-j6)V}{(42+j95)} \times 5 = -63.33 + j28$$

Solving,

$$\mathbf{V = 69.24\angle 156.15^\circ\ V}$$

EXAMPLE 5.3 Find the current in the arm BD of the bridge network shown in Fig. 5.4.

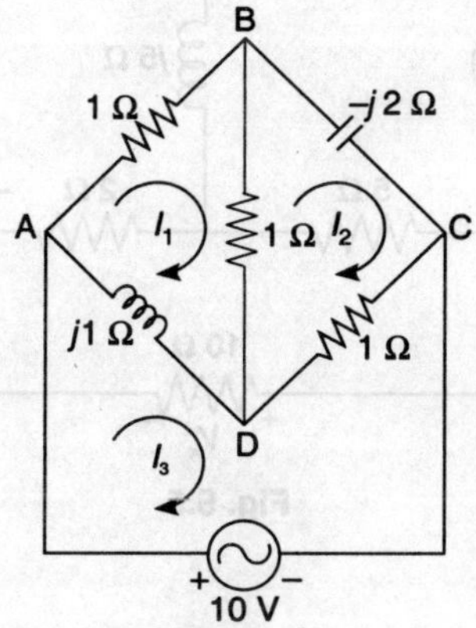

Fig. 5.4

Solution The loop equations are written by inspection as follows:

$$\begin{bmatrix} 2+j1 & -1 & -j1 \\ -1 & 2-j2 & -1 \\ -j1 & -1 & 1+j1 \end{bmatrix} \begin{bmatrix} I_1 \\ I_2 \\ I_3 \end{bmatrix} = \begin{bmatrix} 0 \\ 0 \\ 10\angle 0^\circ \end{bmatrix}$$

Current in arm BD = $I_1 - I_2$

I_1 and I_2 are calculated using Cramer's rule as follows:

$$I_1 = \frac{\begin{bmatrix} 0 & -1 & -j1 \\ 0 & 2-j2 & -1 \\ 10\angle 0^\circ & -1 & 1+j1 \end{bmatrix}}{\begin{bmatrix} 2+j1 & -1 & -j1 \\ -1 & 2-j2 & -1 \\ -j1 & -1 & 1+j1 \end{bmatrix}} = \frac{(30+j20)}{(7-j2)} = (3.208 + j3.773)\text{ A}$$

$$I_2 = \frac{\begin{bmatrix} 2+j1 & 0 & -j1 \\ -1 & 0 & -1 \\ -j1 & 10\angle 0^\circ & 1+j1 \end{bmatrix}}{\begin{bmatrix} 2+j1 & -1 & -j1 \\ -1 & 2-j2 & -1 \\ -j1 & -1 & 1+j1 \end{bmatrix}} = \frac{(20+j20)}{(7-j2)} = (1.887 + j3.396)\text{ A}$$

Current in arm BD = $I_1 - I_2 = \mathbf{(1.321 + j0.377)\ A}$ (or) $\mathbf{1.374\angle 15.93^\circ A}$

EXAMPLE 5.4 In the circuit shown in Fig. 5.5, find V_x.

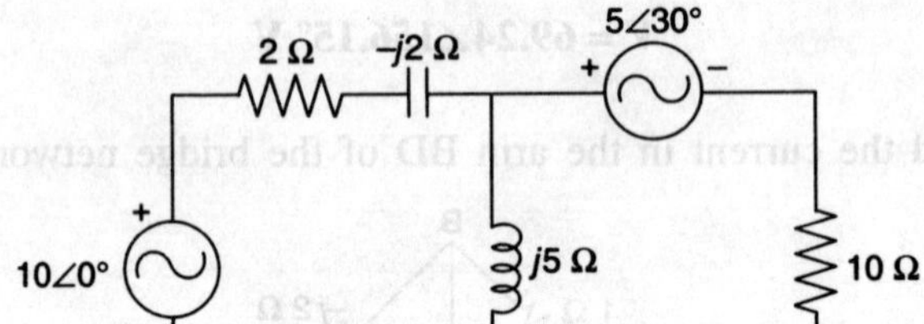

Fig. 5.5

Solution The mesh currents of the given network are assumed as shown in Fig. 5.6.

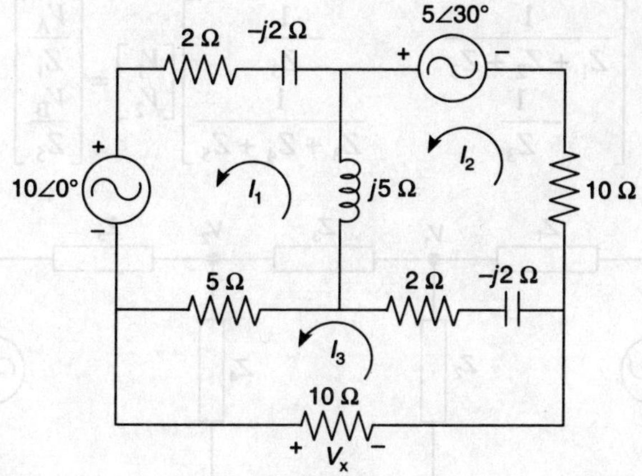

Fig. 5.6

By inspection, the matrix equations are written as

$$\begin{bmatrix} 7+j3 & -j5 & -5 \\ -j5 & 12+j3 & -2+j2 \\ -5 & -2+j2 & 17-j2 \end{bmatrix} \begin{bmatrix} I_1 \\ I_2 \\ I_3 \end{bmatrix} = \begin{bmatrix} -10\angle 0^\circ \\ 5\angle 30^\circ \\ 0 \end{bmatrix}$$

Solving for I_3 using Cramer's rule

$$I_3 = \frac{\begin{bmatrix} 7+j3 & -j5 & -10\angle 0^\circ \\ -j5 & 12+j3 & 5\angle 30^\circ \\ -5 & -2+j2 & 0 \end{bmatrix}}{\begin{bmatrix} 7+j3 & -j5 & -5 \\ -j5 & 12+j3 & -2+j2 \\ -5 & -2+j2 & 17-j2 \end{bmatrix}} = \frac{-655.9 - j126.39}{1390 + j650} = -0.4221 + j0.1065 \text{ A}$$

$$= 0.435\angle 165.8^\circ \text{ A}$$

Therefore, $V_X = 10 \times 0.435\angle 165.8^\circ = \mathbf{4.35\angle 165.8^\circ\ V}$

5.3 NODAL ANALYSIS FOR AC CIRCUITS

The nodal voltage method described in Chapter 2 with DC circuits is extended to networks containing complex impedances driven by sinusoidal sources. As in DC circuits, one of the nodes is taken as reference node and the voltages of the other nodes are determined by writing KCL at those nodes. The equations formed are then put in matrix form. The nodal equations expressed as matrix equations are of the form:

$$[Y]\,[V] = [I]$$

where $[Y]$ is the admittance matrix.

By inspection, the matrix equations are written for the circuit shown in Fig. 5.7 as:

$$\begin{bmatrix} \dfrac{1}{Z_1+Z_2+Z_3} & -\dfrac{1}{Z_3} \\ -\dfrac{1}{Z_3} & \dfrac{1}{Z_3+Z_4+Z_5} \end{bmatrix} \begin{bmatrix} V_1 \\ V_2 \end{bmatrix} = \begin{bmatrix} \dfrac{V_A}{Z_1} \\ \dfrac{V_B}{Z_5} \end{bmatrix}$$

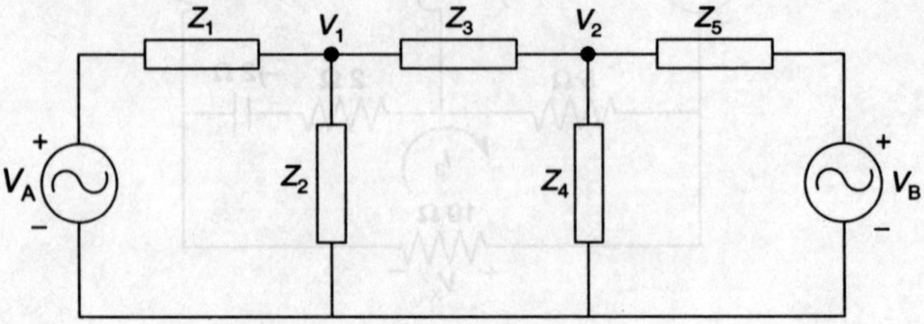

Fig. 5.7 Network with three meshes.

The above equations can be solved for voltages V_1 and V_2 using Cramer's rule. Nodal analysis is better understood with the following example.

EXAMPLE 5.5 Using nodal voltage method, find the power consumed by each resistor of the circuit shown in Fig. 5.8.

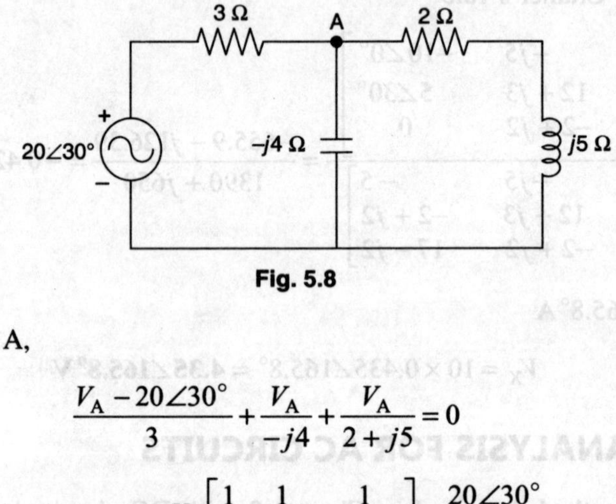

Fig. 5.8

Writing KCL at node A,

$$\frac{V_A - 20\angle 30^\circ}{3} + \frac{V_A}{-j4} + \frac{V_A}{2+j5} = 0$$

$$V_A\left[\frac{1}{3} - \frac{1}{j4} + \frac{1}{2+j5}\right] = \frac{20\angle 30^\circ}{3}$$

$$V_A[0.33 + j0.25 + 0.069 - j0.17] = 6.67\angle 30^\circ$$

Hence, $$V_A = \frac{6.67\angle 30^\circ}{0.399 + j0.08} = \frac{6.67\angle 30^\circ}{0.407\angle 11.34^\circ} = 16.39\angle 18.66^\circ$$

$$\text{Current in the 2 }\Omega\text{ resistor} = \frac{V_A}{2+j5} = \frac{16.39\angle 18.66^\circ}{5.385\angle 68.2^\circ} = 3.04\angle -49.54^\circ$$

Therefore, the power consumed by 2 Ω resistor = $3.04^2 \times 2$ = **18.48 W**

$$\text{Current in the 3 }\Omega\text{ resistor} = \frac{V_A - 20\angle 30^\circ}{3} = \frac{16.39\angle 18.64^\circ - 20\angle 30^\circ}{3}$$

$$= 5.46\angle 18.64^\circ - 6.67\angle 30^\circ$$

$$= (5.17 + j1.75) - (5.78 + j3.34) = (-0.61 - j1.59)$$

$$= 1.7\angle -111^\circ \text{ A}$$

Hence, the power consumed by 3 Ω resistor = $1.7^2 \times 3$ = **8.67 W.**

EXAMPLE 5.6 Using nodal voltage method, determine the current through various elements in the circuit shown in Fig. 5.9.

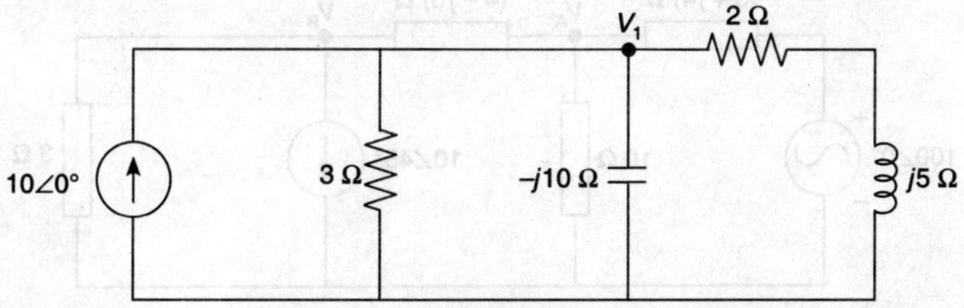

Fig. 5.9

Applying KCL,

$$\left(\frac{1}{3} + \frac{1}{-j10} + \frac{1}{2+j5}\right) V_1 = 10\angle 0^\circ$$

$$(0.33 + j0.1 + 0.0689 - j0.1724) V_1 = 10\angle 0^\circ$$

$$(0.3989 - j0.0724)\, V_1 = 10\angle 0^\circ$$

So, $$V_1 = \frac{10\angle 0^\circ}{0.3989 - j0.0724} = 24.27 + j4.404 = 24.67\angle 10.28^\circ \text{ V}$$

$$\text{Current through 3 }\Omega\text{ resistor} = \frac{24.67\angle 10.28^\circ}{3} = \mathbf{8.2233\angle 10.28^\circ A}$$

$$\text{Current through } -j10\ \Omega\text{ reactance} = \frac{24.67\angle 10.28^\circ}{-j10} = \mathbf{2.467\angle 100.28^\circ A}$$

$$\text{Current through } (2 + j5)\ \Omega\text{ impedance} = \frac{24.67\angle 10.28^\circ}{2+j5} = \mathbf{4.58\angle -57.9^\circ A}$$

EXAMPLE 5.7 Using nodal voltage method, find the power consumed by 10 Ω resistor in the circuit shown in Fig. 5.10.

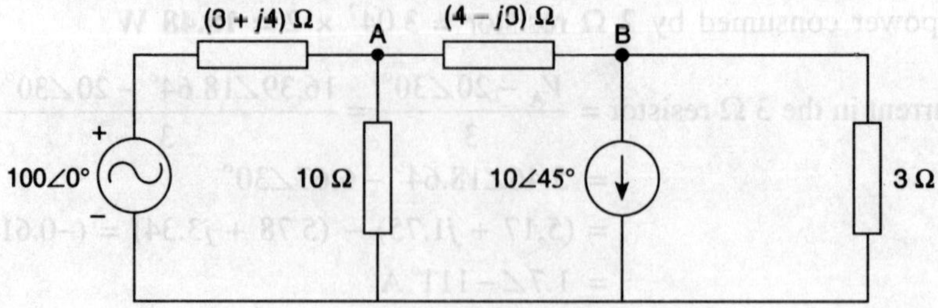

Fig. 5.10

Solution Let us assume the voltages at nodes A and B as V_A and V_B, respectively.

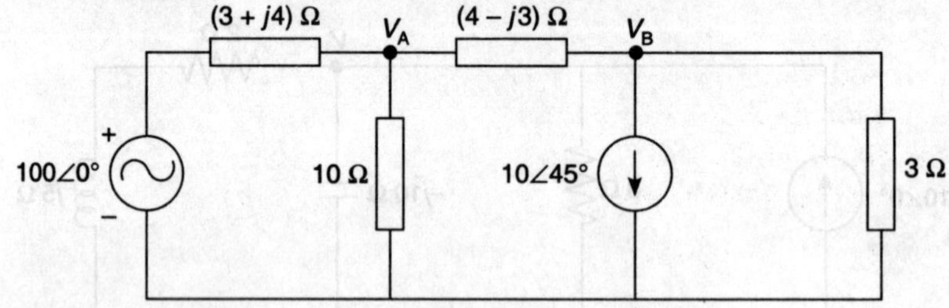

Fig. 5.11

The nodal equations are written as follows:

At node A

$$\frac{V_A - 100\angle 0^\circ}{3 + j4} + \frac{V_A}{10} + \frac{V_A - V_B}{(4 - j3)} = 0$$

$$V_A\left(\frac{1}{3 + j4} + \frac{1}{10} + \frac{1}{4 - j3}\right) + V_B\left(-\frac{1}{4 - j3}\right) = \frac{100\angle 0^\circ}{3 + j4} \quad \text{(i)}$$

At node B

$$\frac{V_B - V_A}{(4 - j3)} + \frac{V_B}{3} + 10\angle 45^\circ = 0$$

$$\frac{V_B - V_A}{(4 - j3)} + \frac{V_B}{3} + 7.07 + j7.07 = 0$$

$$V_A\left(-\frac{1}{4 - j3}\right) + V_B\left(\frac{1}{4 - j3} + \frac{1}{3}\right) = -7.07 - j7.07 \quad \text{(ii)}$$

Putting Eqs. (i) and (ii) in matrix form, we get

$$\begin{bmatrix} \frac{1}{3+j4}+\frac{1}{10}+\frac{1}{4-j3} & -\frac{1}{4-j3} \\ -\frac{1}{4-j3} & \frac{1}{4-j3}+\frac{1}{3} \end{bmatrix} \begin{bmatrix} V_A \\ V_B \end{bmatrix} = \begin{bmatrix} \frac{100\angle 0^\circ}{(3+j4)} \\ -7.07-j7.07 \end{bmatrix}$$

$$\begin{bmatrix} 0.38-j0.04 & -0.16-j0.12 \\ -0.16-j0.12 & 0.49+j0.12 \end{bmatrix} \begin{bmatrix} V_A \\ V_B \end{bmatrix} = \begin{bmatrix} 12-j16 \\ -7.07-j7.07 \end{bmatrix}$$

Using Cramer's rule, V_A is calculated as follows:

$$\frac{\begin{bmatrix} 12-j16 & -0.16-j0.12 \\ -7.07-j7.07 & 0.49+j0.12 \end{bmatrix}}{\begin{bmatrix} 0.38-j0.04 & -0.16-j0.12 \\ -0.16-j0.12 & 0.49+j0.12 \end{bmatrix}} = \frac{7.5172-j8.3796}{0.1798-j0.0124} = 44.81-j43.51 \text{ V}$$

$$V_A = (44.81-j43.51) = 62.46\angle -44.16^\circ \text{ V}$$

Hence, the power consumed by 10 Ω resistor $= \frac{62.46^2}{10} = 390.125$ W

EXAMPLE 5.8 In the network shown in Fig. 5.12, obtain the voltage, V_2.

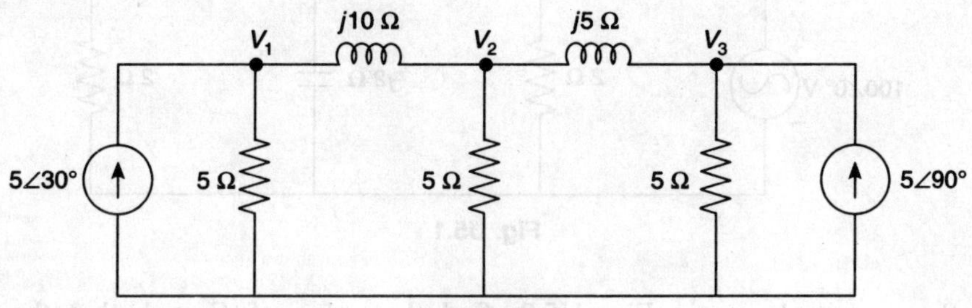

Fig. 5.12

The nodal equations are written by inspection as:

$$\begin{bmatrix} \left(\frac{1}{5}+\frac{1}{j10}\right) & -\frac{1}{j10} & 0 \\ -\frac{1}{j10} & \left(\frac{1}{5}+\frac{1}{j10}+\frac{1}{j5}\right) & -\frac{1}{j5} \\ 0 & -\frac{1}{j5} & \left(\frac{1}{5}+\frac{1}{j5}\right) \end{bmatrix} \begin{bmatrix} V_1 \\ V_2 \\ V_3 \end{bmatrix} = \begin{bmatrix} 5\angle 30^\circ \\ 0 \\ 5\angle 90^\circ \end{bmatrix}$$

$$\begin{bmatrix} (0.2-j0.1) & -j0.1 & 0 \\ -j0.1 & (0.2-j0.3) & j0.2 \\ 0 & j0.2 & (0.2-j0.2) \end{bmatrix} \begin{bmatrix} V_1 \\ V_2 \\ V_3 \end{bmatrix} = \begin{bmatrix} (4.33+j2.5) \\ 0 \\ j5 \end{bmatrix}$$

$$V_2 = \frac{\begin{bmatrix} (0.2-j0.1) & (4.33+j2.5) & 0 \\ -j0.1 & 0 & j0.2 \\ 0 & j5 & (0.2-j0.2) \end{bmatrix}}{\begin{bmatrix} (0.2-j0.1) & -j0.1 & 0 \\ -j0.1 & (0.2-j0.3) & j0.2 \\ 0 & j0.2 & (0.2-j0.2) \end{bmatrix}} = \frac{0.2366+j0.0366}{-0.004-j0.024}$$

Therefore, $V_2 = (-3.0824 + j9.3446) = \mathbf{9.8398\angle 108.26^\circ\ V.}$

UNSOLVED PROBLEMS

1. Find the current supplied by the voltage source of the circuit in Fig. U5.1 by mesh analysis. **[Ans.** 12.4124∠–37° A]

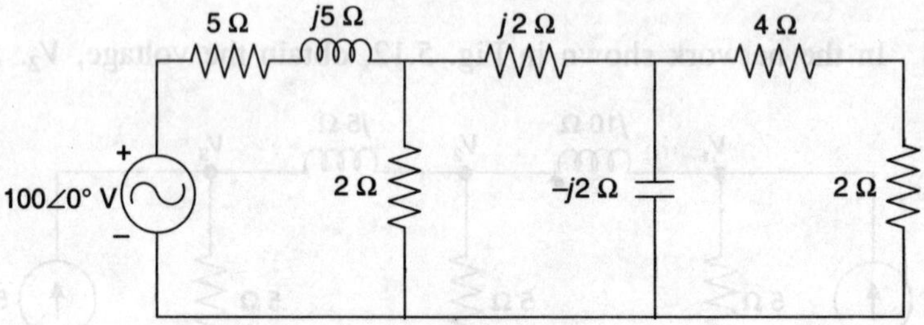

Fig. U5.1

2. In the circuit shown in Fig. U5.2, find the value of E such that the current in $(1 + j1)\ \Omega$ is zero. **[Ans.** 96∠36.87° V]

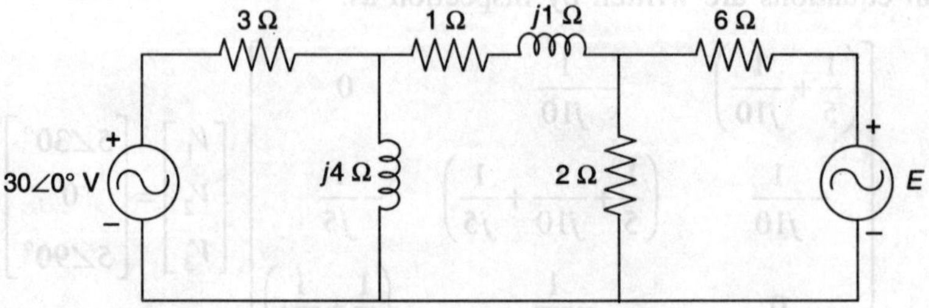

Fig. U5.2

3. Find the voltage V_{BD} in the network in Fig. U5.3. **[Ans.** 0 V]

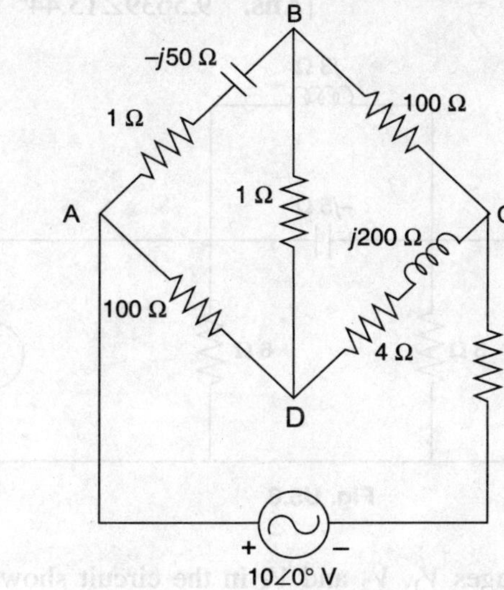

Fig. U5.3

4. Use mesh current method to find current flowing through $j3\ \Omega$ reactance (Fig. U5.4). **[Ans.** 3∠–90° A]

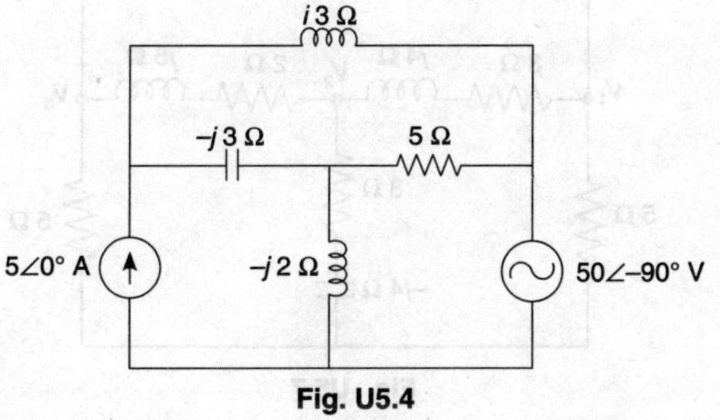

Fig. U5.4

5. Determine the power consumed by 10 Ω resistor in the circuit shown in Fig. U5.5 by nodal analysis method. **[Ans.** 389.5 W]

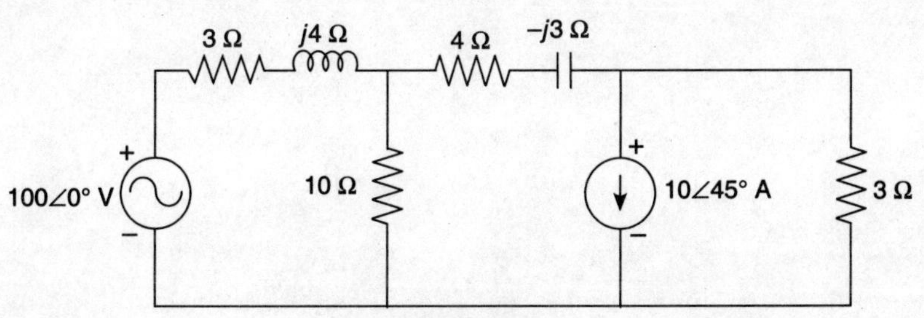

Fig. U5.5

6. Find the voltages across the two current sources in the circuit shown in Fig. U5.6 by nodal analysis method. **[Ans.** $9.5639\angle 13.44°$ V; $48.5754\angle 31.69°$ V]

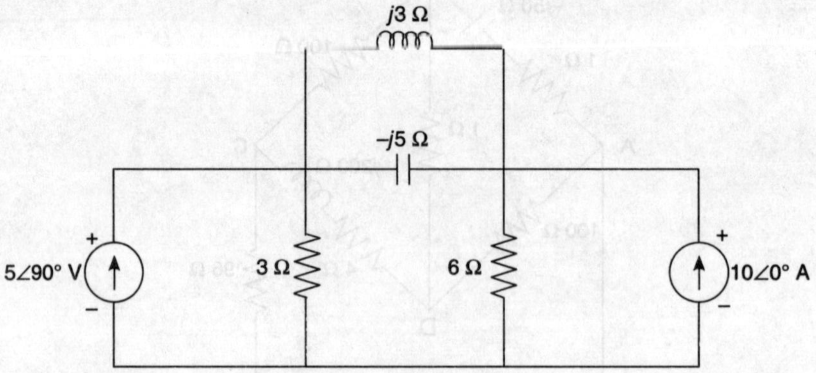

Fig. U5.6

7. Determine the node voltages V_1, V_2 and V_3 in the circuit shown in Fig. U5.7.
[Ans. $V_1 = 15.2583\angle 27.4°$ V; $V_2 = 2.3292\angle 63.4°$ V; $V_3 = 44.8764\angle -111.8°$ V]

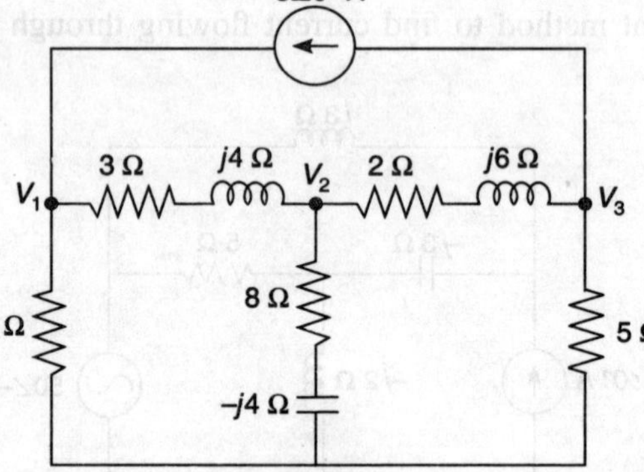

Fig. U5.7

CHAPTER 6

Network Theorems

6.1 INTRODUCTION

In the previous chapters, we solved network problems using Ohm's law and Kirchhoff's laws. In certain networks, application of these laws does not bring the solution easily. To overcome this difficulty, some network theorems have been developed which are very useful in analysing both simple and complex networks. These theorems find their application not only in DC and AC circuits but also in linear and some nonlinear electric circuits. With these theorems, it is possible to either simplify the network or to obtain the solution quickly. In this chapter, the following theorems are discussed.

1. Superposition theorem
2. Thevenin's theorem
3. Norton's theorem
4. Reciprocity theorem
5. Maximum power transfer theorem
6. Millman's theorem
7. Substitution theorem
8. Compensation theorem
9. Tellegen's theorem

6.2 SUPERPOSITION THEOREM

Superposition theorem is stated as:

In a linear, lumped, bilateral network which is energised by two or more sources, the response (current/voltage) in any element is equal to the algebraic sum of the separate responses in the element when each source acts separately. While one source is active, other sources are replaced by their internal resistances.

If the internal resistances are not known, voltage source is replaced by short-circuit and current source is replaced by open-circuit. This theorem is not useful for nonlinear networks.

Steps for solving electrical networks using superpostion theorem

1. Firstly, all internal impedances are represented as external impedances. For a voltage source, internal impedance is represented as external impedance in series with ideal voltage source. For a current source, it is represented as shunt resistance in parallel with ideal current source.
2. Response (voltage or current) due to each source is determined by considering only one source at a time while other sources are made inactive.
3. Total response is found by adding all individual responses.

Superposition theorem is better understood by the following examples.

EXAMPLE 6.1 By superposition theorem, find the current flows through 6 Ω resistor in the circuit shown in Fig. 6.1.

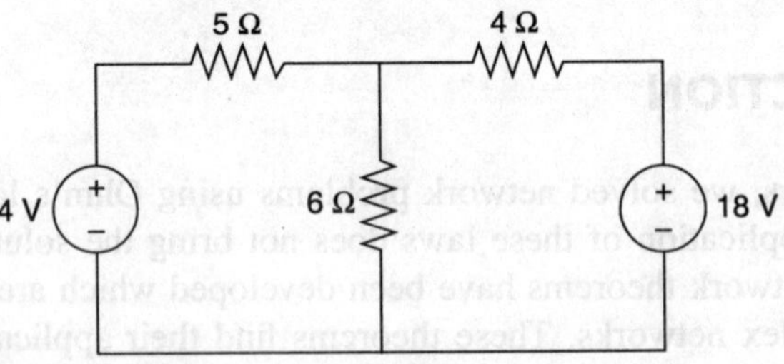

Fig. 6.1

Solution The network has two voltage sources of 24 V and 18 V. First 18 V source is made inactive by short-circuiting it as shown in Fig. 6.2.

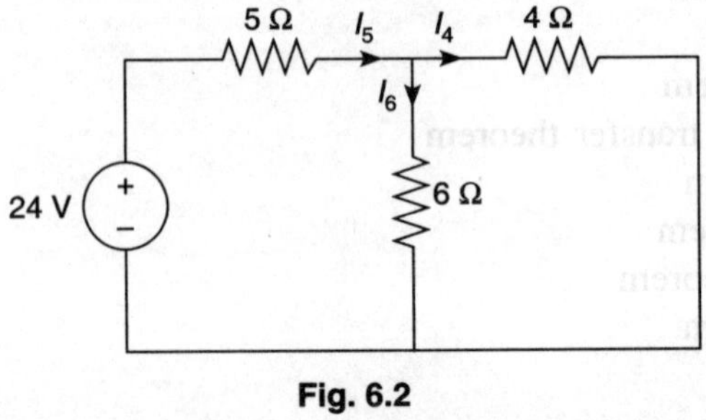

Fig. 6.2

The equivalent resistance across the 24 V source is

$$(4\|6) + 5 = \frac{(4 \times 6)}{(4+6)} + 5 = 2.4 + 5 = 7.4\ \Omega$$

So, the current supplied by the 24 V source alone (I_5) = $\dfrac{24}{7.4} = 3.24$ A

Using current division technique, the current through 6 Ω, $I_6 = 3.24 \times \frac{4}{4+6} = 1.3\text{ A}$

Now, the 24 V source is short-circuited and 18 V source is made operative as shown in Fig. 6.3.

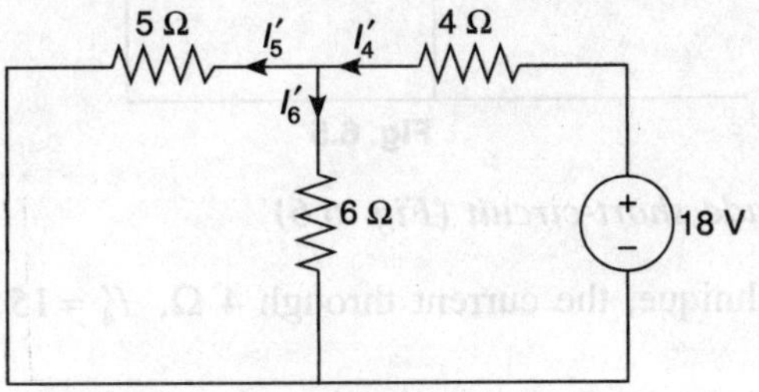

Fig. 6.3

The equivalent resistance across the 18 V source is

$$(5\|6) + 4 = \frac{(5\times 6)}{(5+6)} + 4 = 2.73 + 4 = 6.73\ \Omega$$

Hence the current supplied by the 18 V source alone (I_4') = $\frac{18}{6.73} = 2.67\text{ A}$

Using current division technique, the current through 6 Ω, $I_6' = 2.67 \times \frac{5}{5+6} = 1.21\text{ A}$

Therefore, the current through 6 Ω resistor, by superposition principle is

$$I_6 + I_6' = 1.3 + 1.21 = \mathbf{2.51\ A}$$

EXAMPLE 6.2 By using superposition theorem, find the current flows through 4 Ω resistor in the network shown in Fig. 6.4.

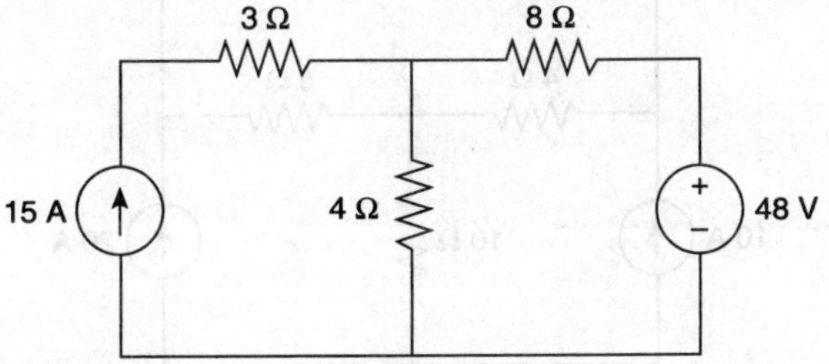

Fig. 6.4

Solution The network has two sources of 15 A and 48 V.
15 A current source is made open-circuit (Fig. 6.5)

Current through 4 Ω, $I_4 = \frac{48}{8+4} = 4\text{ A}$

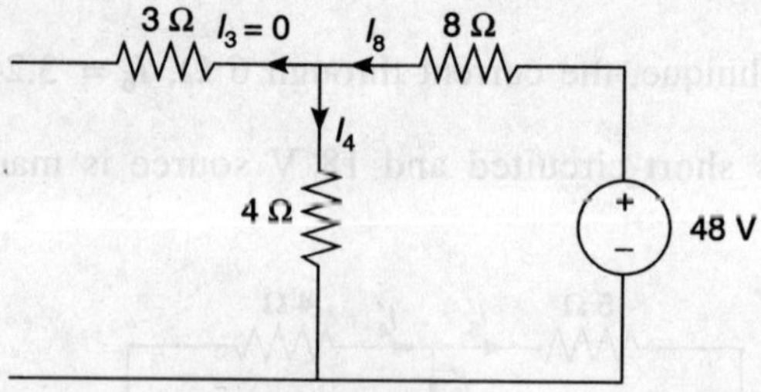

Fig. 6.5

48 V voltage source is made short-circuit (Fig. 6.6)

Using current division technique, the current through 4 Ω, $I_4' = 15 \times \frac{8}{8+4} = 10$ A

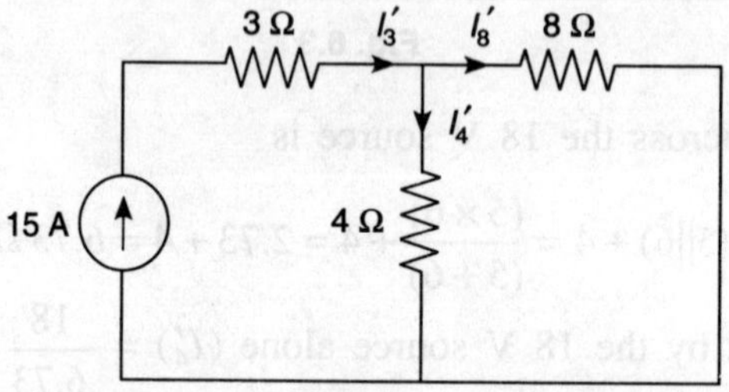

Fig. 6.6

Therefore, the current through 4 Ω resistor, by superposition principle is

$$I_4 + I_4' = 4 + 10 = \mathbf{14\ A}$$

EXAMPLE 6.3 By superposition theorem, find the current flows through 10 Ω resistor in the circuit shown in Fig. 6.7.

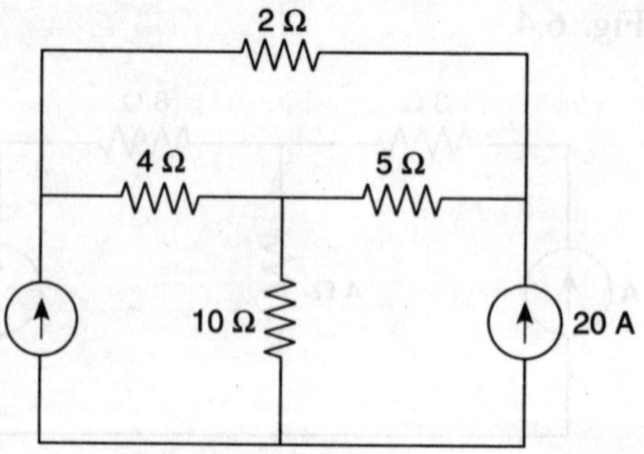

Fig. 6.7

Solution

20 A current source is made open (Fig. 6.8)

The current through 10 Ω resistor due to 10 A current source, I_{10} = 10 A

10 A current source is made open (Fig. 6.9)

The current through 10 Ω resistor due to 20 A current source, I_{10}' = 20 A

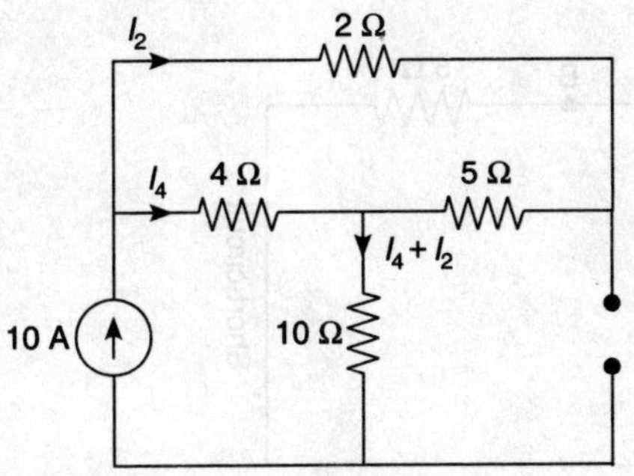

Fig. 6.8

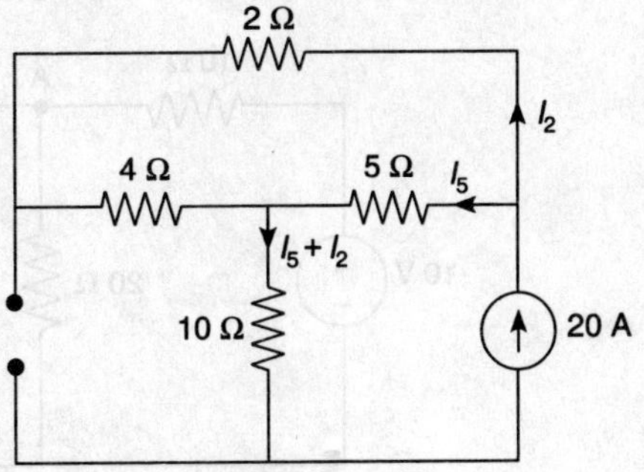

Fig. 6.9

By principle of superposition, the current through 10 Ω resistor is

$$I_{10} + I'_{10} = 10 + 20 = \mathbf{30\ A}$$

EXAMPLE 6.4 Find the voltage across 2 Ω resistor using superposition theorem in the circuit shown in Fig. 6.10.

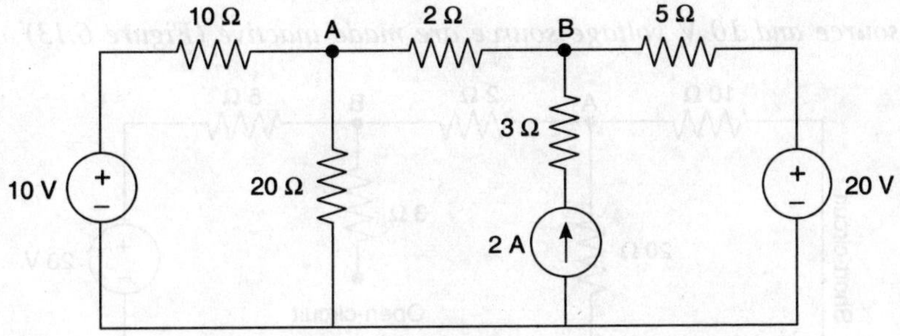

Fig. 6.10

Solution To find the voltage across 2 Ω resistor, the current through it is to be determined first. *2 A current source and 20 V voltage source are made inactive (Fig. 6.11)*

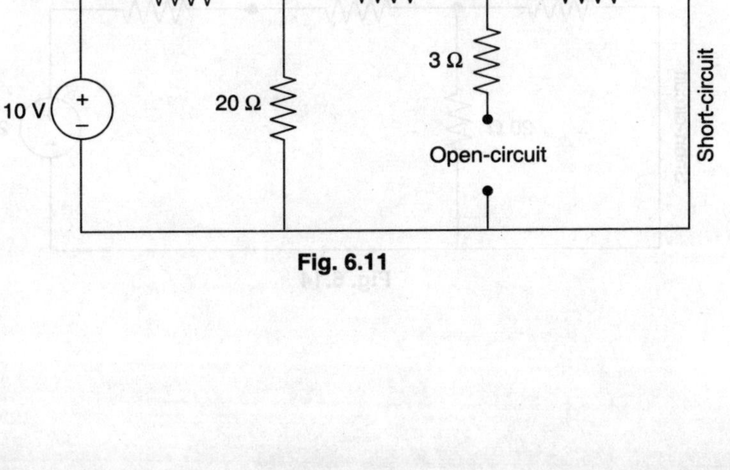

Fig. 6.11

As no current flows through 3 Ω, the equivalent circuit becomes as shown in Fig. 6.12.

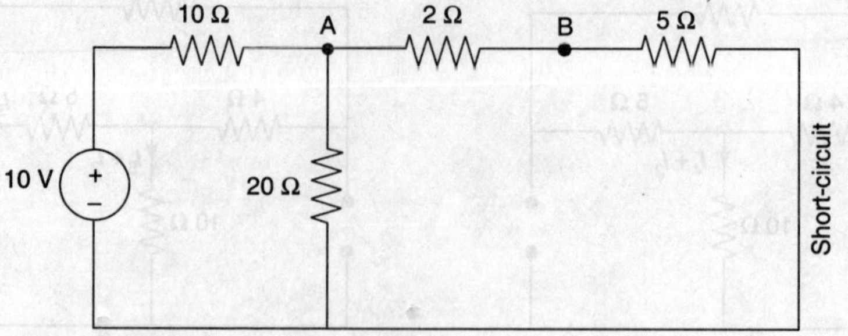

Fig. 6.12

From the circuit shown in Fig. 6.12, the current supplied by the 10 V source

$$\frac{10}{10+\dfrac{7\times 20}{7+20}} = \frac{10}{15.19} = 0.66 \text{ A}$$

Using current division, current through 2 Ω = $0.66 \times \dfrac{20}{20+2+5} = 0.488$ A (flows from A to B)

2 A current source and 10 V voltage source are made inactive (Figure 6.13)

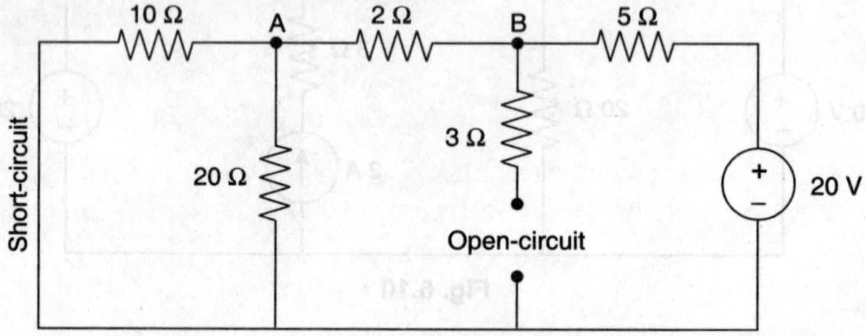

Fig. 6.13

As no current flows through 3 Ω, the equivalent circuit becomes as shown in Fig. 6.14.

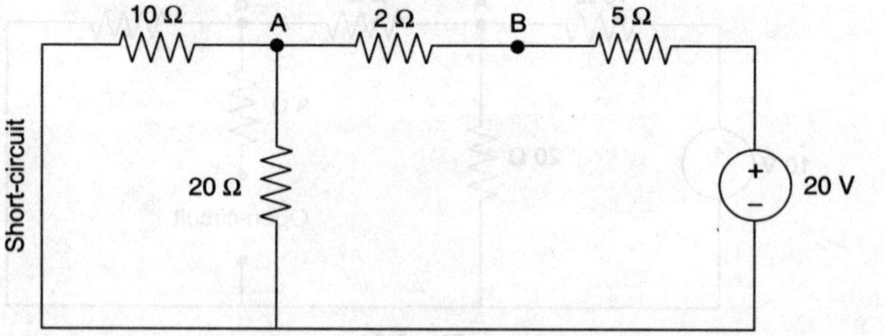

Fig. 6.14

The current supplied by the 20 V source $= \dfrac{20}{7+\left(\dfrac{10\times 20}{10+20}\right)} = \dfrac{20}{13.67} = 1.46$ A

Current through 2 Ω = 1.46 A (flows from B to A)

10 V voltage source and 20 V voltage source are made inactive (Fig. 6.15)

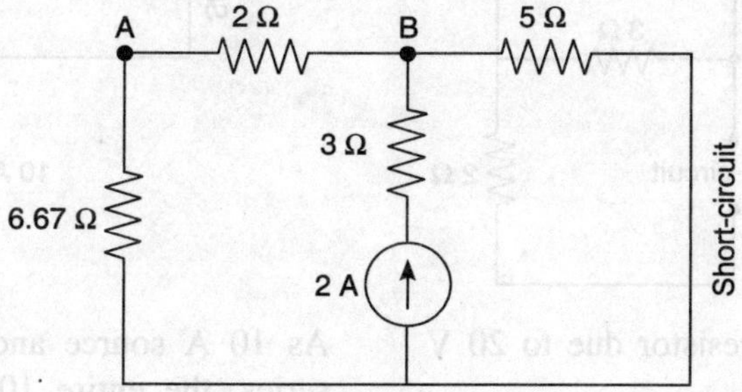

Fig. 6.15

Using current division rule, the current through 2 Ω $= 2\times\dfrac{5}{6.67+2+5} = 0.73$ A (flows from B to A). Now,

Current through 2 Ω resistance due to 10 V source only = 0.488 A (flows from A to B)
Current through 2 Ω resistance due to 20 V source only = 1.46 A (flows from B to A)
Current through 2 Ω resistance due to 2 A source only = 0.73 A (flows from B to A)

Then, current through 2 Ω due to all the sources = 1.702 A (flows from B to A).
Hence, the voltage across 2 Ω resistor = 2 × 1.702 = **3.404 V** (with terminal 'B' positive and 'A' negative).

EXAMPLE 6.5 Find the current through 2 Ω resistor using superposition theorem in the network shown in Fig. 6.16.

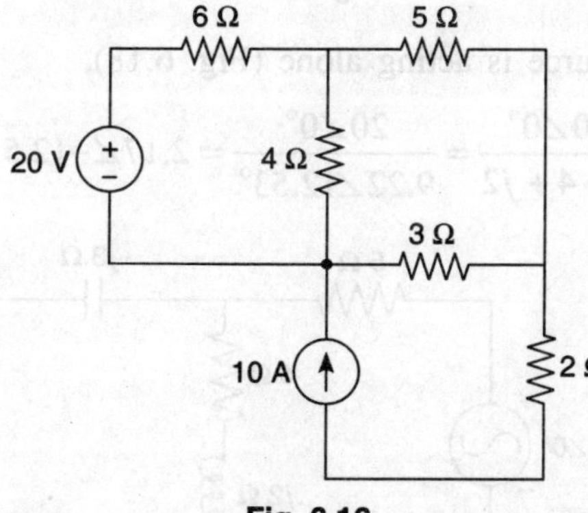

Fig. 6.16

Solution

10 A source is open-circuited

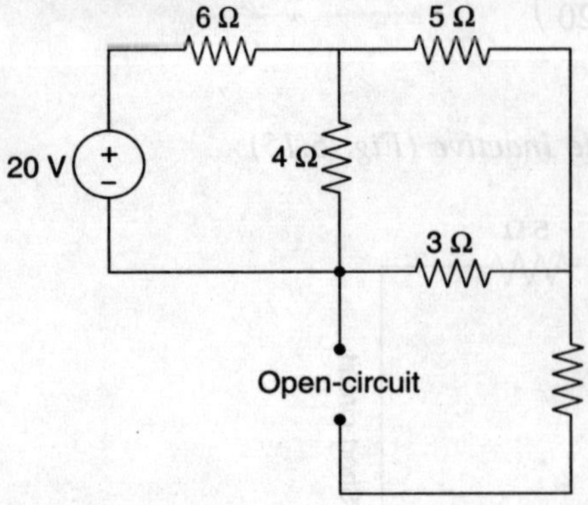

Current through 2 Ω resistor due to 20 V source is zero.

$$I_{2\Omega} = 0$$

20 V source is short-circuited

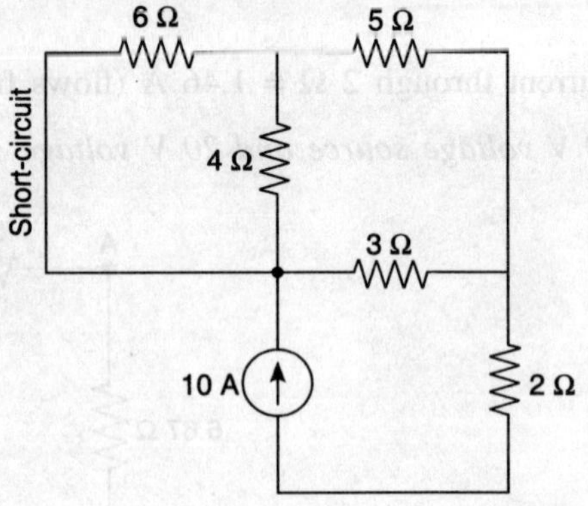

As 10 A source and 2 Ω resistance are in series, the entire 10 A flows through 2 Ω resistance.

$$I'_{2\Omega} = 10$$

Hence, the current through 2 Ω resistor = $I_{2\Omega} + I'_{2\Omega}$ = 0 + 10 A = **10 A.**

EXAMPLE 6.6 For the circuit shown in Fig. 6.17, find the current in (4 + *j*2) Ω by superposition principle.

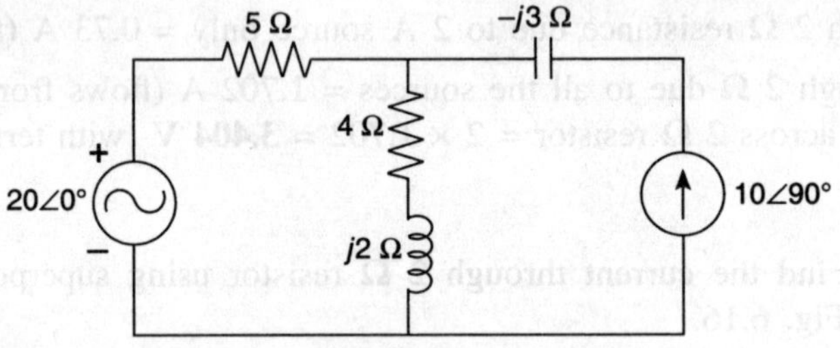

Fig. 6.17

Solution When 20∠0° V source is acting alone (Fig. 6.18),

Current, $$I' = \frac{20\angle 0^\circ}{5 + 4 + j2} = \frac{20\angle 0^\circ}{9.22\angle 12.53^\circ} = 2.17\angle{-12.53^\circ}\text{ A}$$

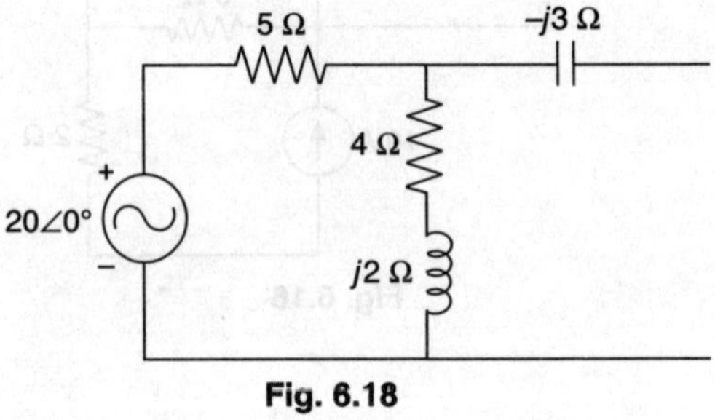

Fig. 6.18

When 10∠90° A source is acting alone (Fig. 6.19), by current division,

$$I'' = 10\angle 90^\circ \times \frac{5}{5+4+j2} = \frac{50\angle 90^\circ}{9.22\angle 12.53^\circ}$$

$$= 5.42\angle 77.47^\circ \text{ A}$$

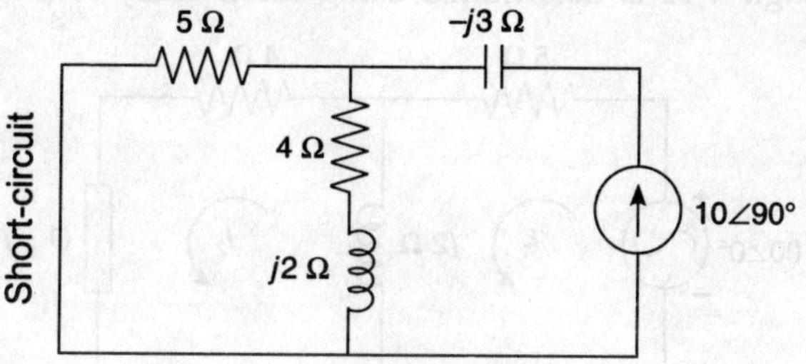

Fig. 6.19

By superposition principle, total current through (4 + j2) Ω is

$$I' + I'' = 2.17\angle -12.53^\circ + 5.42\angle 77.47^\circ$$

$$= (2.12 - j0.47) + (1.17 + j5.29)$$

$$= (3.29 + j4.82) \text{ A}$$

$$= \mathbf{5.836\angle 55.68^\circ \text{ A}}$$

EXAMPLE 6.7 For the circuit shown in Fig. 6.20, find the current in 4 Ω resistance by superposition principle.

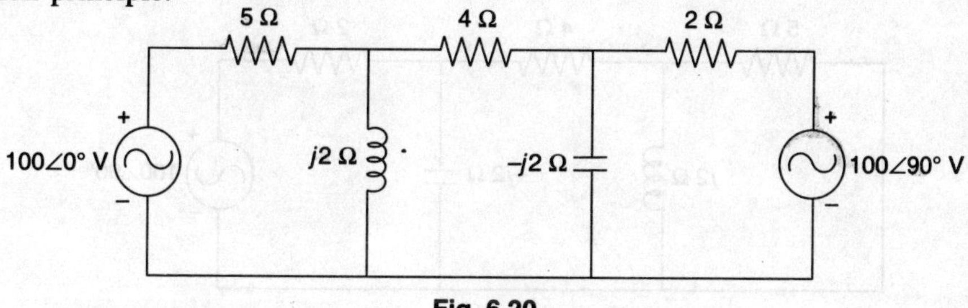

Fig. 6.20

Solution 100∠0° source is made active and 100∠90° V source is shorted as shown in Fig. 6.21.

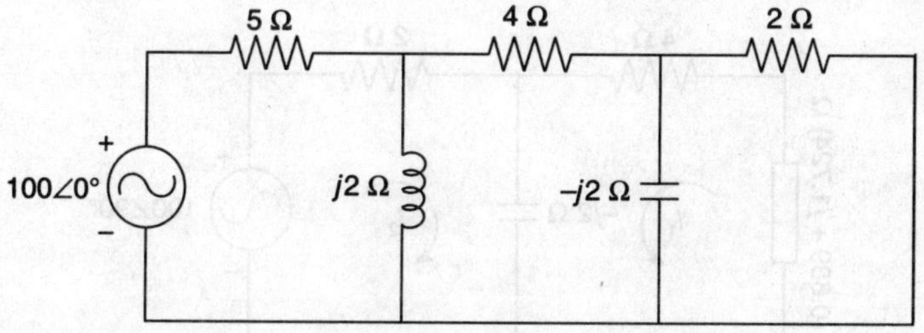

Fig. 6.21

Now the circuit is simplified by combining 2 Ω and –*j*2 Ω as:

$$(2\|-j2)=\frac{2\times(-j2)}{(2-j2)}=(1-j1)\ \Omega$$

Then, the current through 4 Ω is determined using mesh analysis (Fig. 6.22).

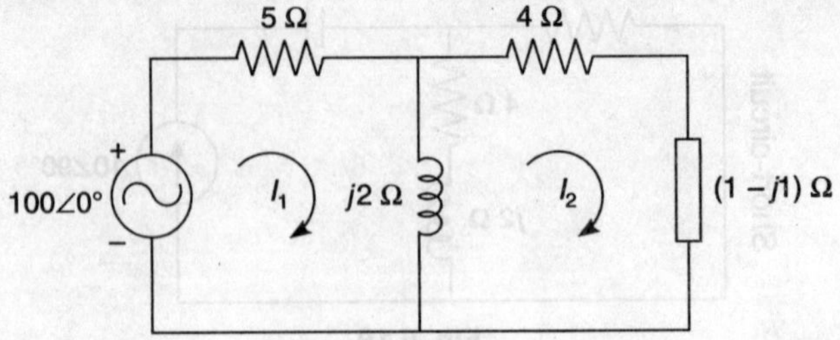

Fig. 6.22

$$I' = I_2 = \frac{\begin{vmatrix} 5+j2 & 100\angle 0^\circ \\ -j2 & 0 \end{vmatrix}}{\begin{vmatrix} 5+j2 & -j2 \\ -j2 & 5+j1 \end{vmatrix}} = \frac{j200}{(23+j15)-(-4+j0)} = \frac{j200}{27+j15}$$

$$= 3.145 + j5.66 = 6.475\angle 60.94^\circ\ \text{A}$$

Now, 100∠90° V is made active and 100∠0° V is shorted as shown in Fig. 6.23.

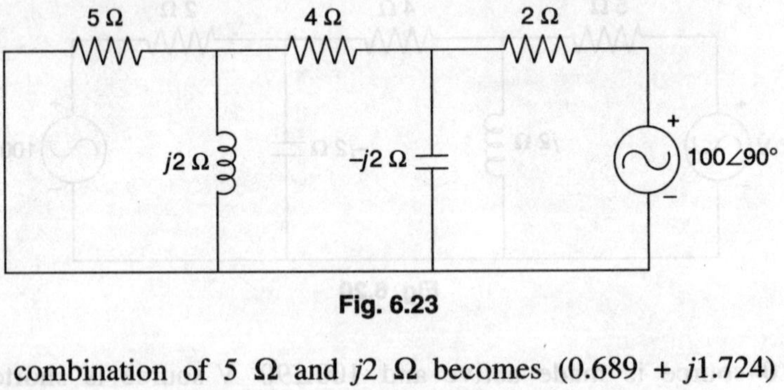

Fig. 6.23

The parallel combination of 5 Ω and *j*2 Ω becomes (0.689 + *j*1.724) Ω as shown in Fig. 6.24 and the current through 4 Ω is determined using mesh analysis.

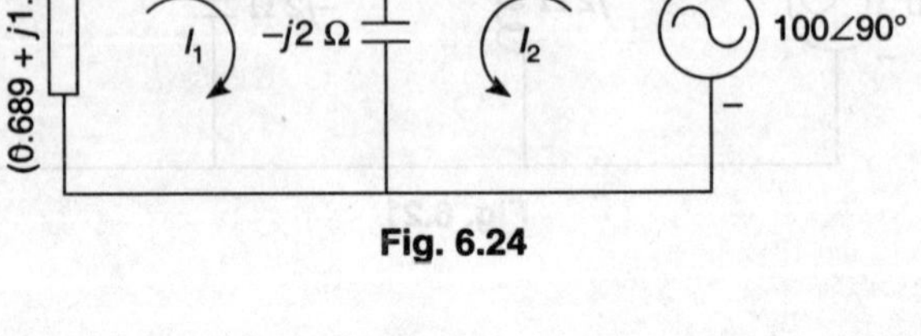

Fig. 6.24

$$I'' = I_1 = \frac{\begin{vmatrix} 0 & -j2 \\ 100\angle 90^\circ & 2-j2 \end{vmatrix}}{\begin{vmatrix} 4.689 - j0.276 & -j2 \\ -j2 & 2-j2 \end{vmatrix}} = \frac{-200}{(8.826 - j9.93) - (-4 + j0)} = \frac{-200}{12.826 - j9.93}$$

$$= -9.7495 - j7.5481 = 12.3299\angle -142.25^\circ \text{ A}$$

Therefore, by superposition principle, total current through 4 Ω is

$$I' + I'' = (3.145 + j5.66) + (-9.7495 - j7.5481)$$

$$= (-6.6045 - j1.8881) \text{ A} = \mathbf{6.8691\angle -164.04^\circ \text{ A}.}$$

6.3 THEVENIN'S THEOREM

Thevenin's theorem is stated as:

A linear, lumped, bilateral network with two terminals can be reduced to an equivalent circuit consisting of a voltage source in series with a resistance (or impedance).

The voltage source is called Thevenin's voltage (V_{th}) and it is equal to the open-circuit voltage across the two terminals of the network, and resistance (R_{th}) is equal to the equivalent resistance measured between the terminals with all the energy sources replaced by their internal resistances. The Thevenin's equivalent circuits for DC and AC networks are shown respectively in Figs. 6.25(a) and (b).

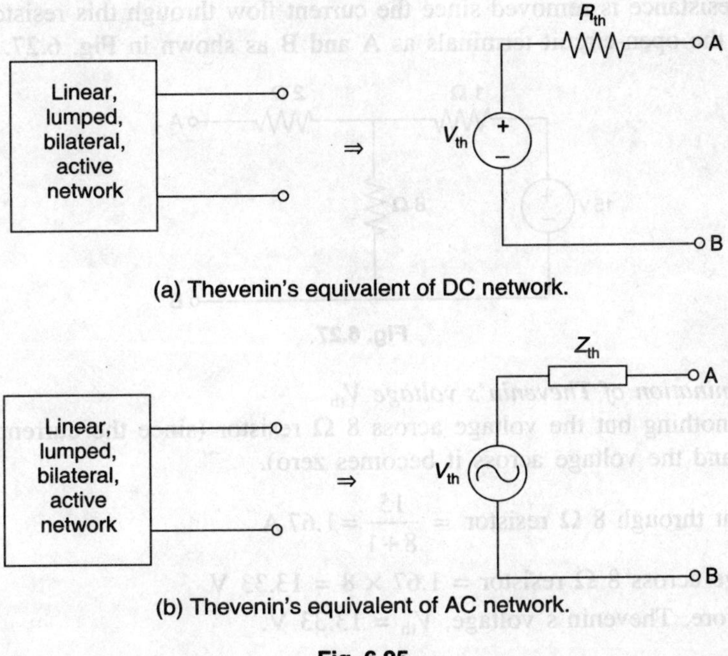

(a) Thevenin's equivalent of DC network.

(b) Thevenin's equivalent of AC network.

Fig. 6.25

Steps for solving electrical networks using Thevenin's theorem

1. Disconnect load resistance R_L (in the case of AC networks impedance, Z_L) through which current is needed and label the two terminals A and B.
2. Calculate the open circuit voltage across A–B, called Thevenin voltage V_{th}, using methods like nodal analysis/ohm's law/voltage division.
3. Calculate Thevenin resistance R_{th} (Thevenin impedance Z_{th}, in the case of AC networks) by network reduction method by looking back the circuit through A–B.
4. Draw Thevenin's equivalent circuit and reconnect R_L in series with R_{th}.

Thevenin's theorem is better understood by the following examples.

EXAMPLE 6.8 Find the current through 6 Ω resistor in the circuit shown in Fig. 6.26.

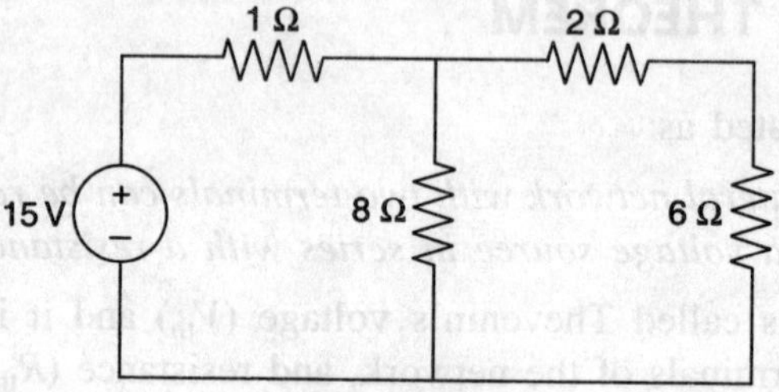

Fig. 6.26

Solution

Step 1 *Removal of resistor in question*

6 Ω resistance is removed since the current flow through this resistor is in question. Mark the open-circuit terminals as A and B as shown in Fig. 6.27.

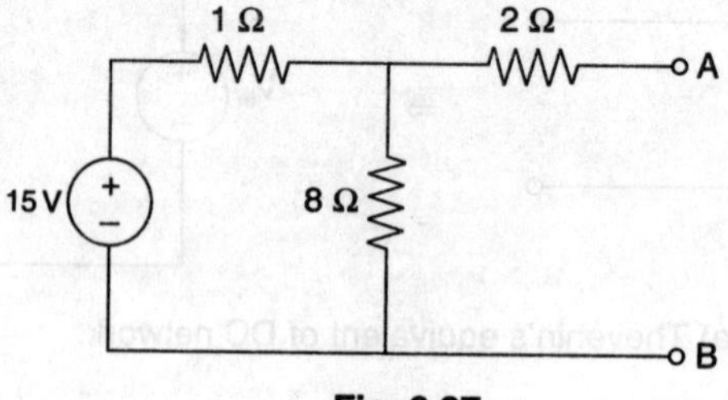

Fig. 6.27

Step 2 *Determination of Thevenin's voltage V_{th}*

V_{th} is nothing but the voltage across 8 Ω resistor (since the current through 2 Ω is zero, and the voltage across it becomes zero).

$$\text{Current through } 8\ \Omega \text{ resistor} = \frac{15}{8+1} = 1.67\text{ A}$$

Voltage across 8 Ω resistor = 1.67 × 8 = 13.33 V

Therefore, Thevenin's voltage, V_{th} = 13.33 V.

Step 3 *Determination of Thevenin's resistance R_{th}*
R_{th} is calculated by short-circuiting the 15 V voltage source and looking back the network from terminals A and B as shown in Fig. 6.28.

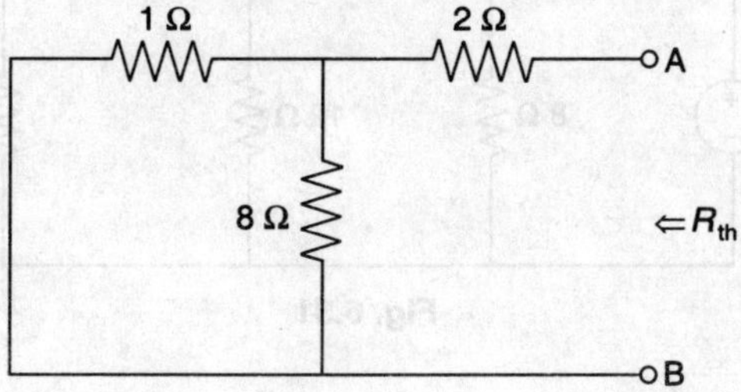

Fig. 6.28

From Fig. 6.28, $R_{th} = R_{AB} = 2 + \dfrac{1\times 8}{1+8} = 2.89\ \Omega.$

Step 4 *Construction of Thevenin's equivalent circuit*
Thevenin's equivalent circuit is shown in Fig. 6.29.

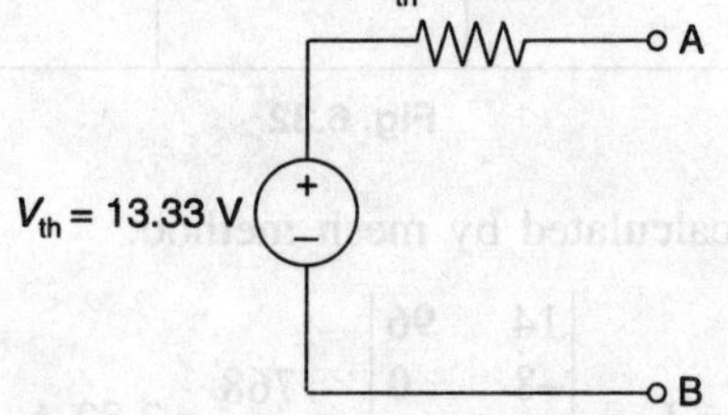

Fig. 6.29

Step 5 *Reconnection of removed 6 Ω resistor in Thevenin's equivalent ciruit and obtaining the result.*

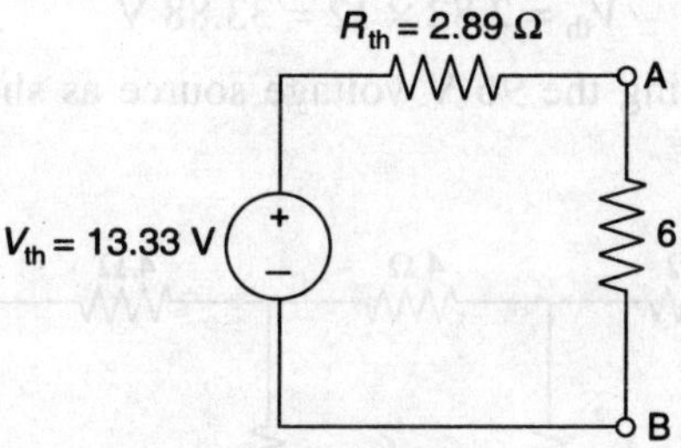

Fig. 6.30

As shown in Fig. 6.30, current through 6 Ω resistor $= \dfrac{13.33}{2.89+6}$ = **1.5 A.**

EXAMPLE 6.9 For the circuit shown in Fig. 6.31, determine the load current using Thevenin's theorem.

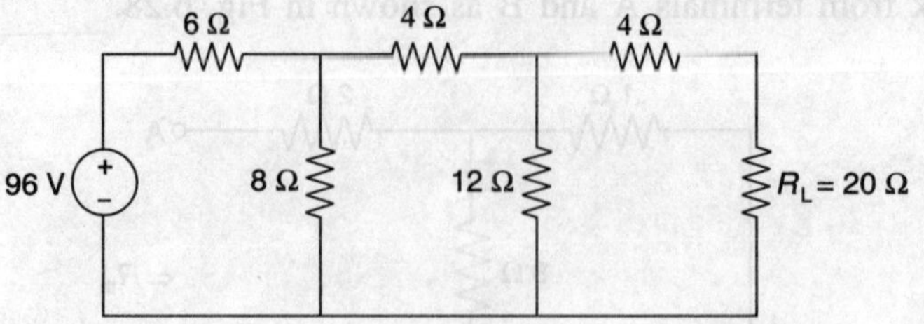

Fig. 6.31

Solution The circuit is redrawn by opening the load R_L. V_{th} is nothing but the voltage across 12 Ω resistor (Fig. 6.32).

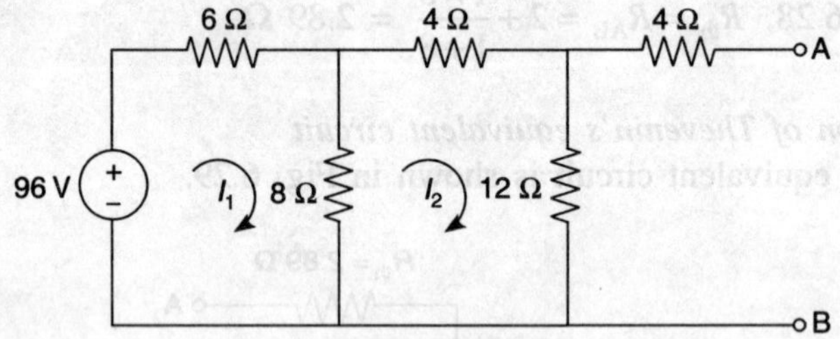

Fig. 6.32

The current through 12 Ω is calculated by mesh method.

$$I_2 = \frac{\begin{vmatrix} 14 & 96 \\ -8 & 0 \end{vmatrix}}{\begin{vmatrix} 14 & -8 \\ -8 & 24 \end{vmatrix}} = \frac{768}{272} = 2.82 \text{ A}$$

Therefore, Thevenin's voltage,

$$V_{th} = 2.82 \times 12 = 33.88 \text{ V}$$

R_{th} is calculated by short-circuiting the 96 V voltage source as shown in Fig. 6.33 and looking back from terminals A and B.

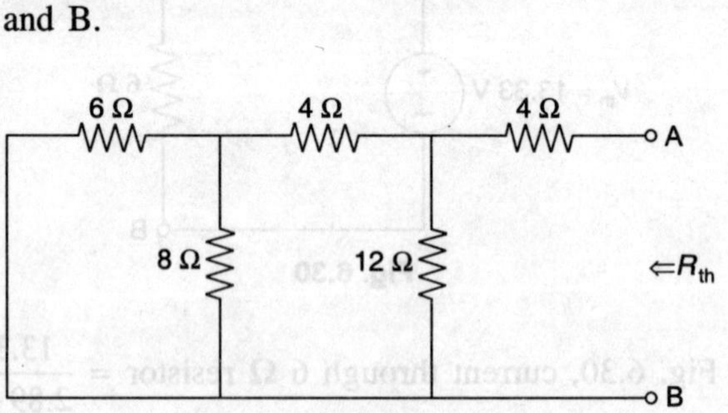

Fig. 6.33

Equivalent resistance for the parallel combination of 6 Ω and 8 Ω = $\dfrac{6 \times 8}{6+8} = 3.43\ \Omega$

3.43 Ω is in series with 4 Ω, which gives 7.43 Ω.

Equivalent resistance for the parallel combination of 7.43 Ω and 12 Ω = $\dfrac{7.43 \times 12}{7.43+12} = 4.59\ \Omega$

4.59 Ω is in series with 4 Ω, which gives 8.59 Ω. Therefore,

$$R_{th} = R_{AB} = 8.59\ \Omega$$

Thevenin's equivalent circuit

As shown in Fig. 6.34, current through 20 Ω resistor

$$= \frac{33.88}{8.59+20} = \mathbf{1.185\ A}$$

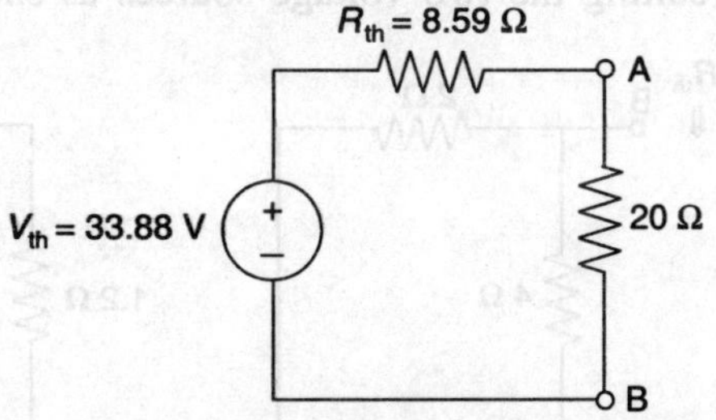

Fig. 6.34

EXAMPLE 6.10 Using Thevenin's theorem, find the current through 10 Ω resistor in the circuit shown in Fig. 6.35.

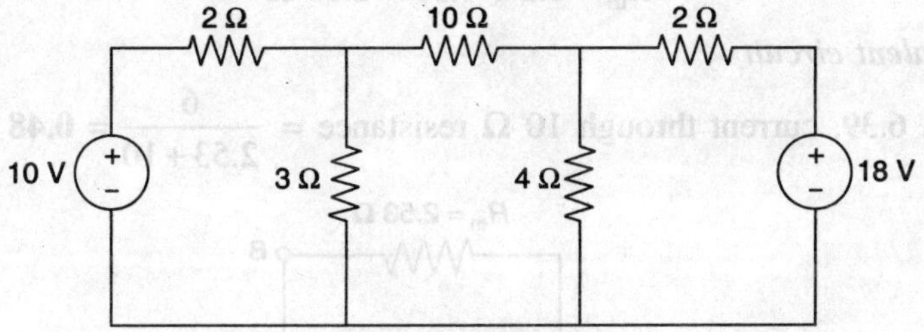

Fig. 6.35

Solution The circuit is redrawn by opening the 10 Ω resistance. V_{th} is the voltage between terminals A and B.

To find V_{th}

From Fig. 6.36,

$$I_1 = \frac{10}{2+3} = 2\ \text{A and } I_2 = \frac{18}{2+4} = 3\ \text{A}$$

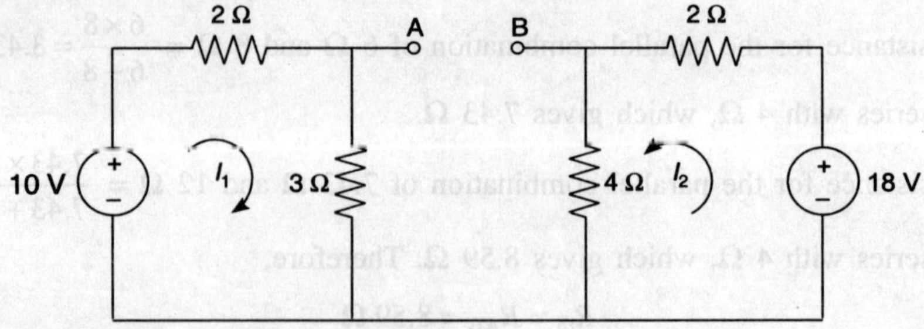

Fig. 6.36

$$V_A = 3 \times I_1 = 3 \times 2 = 6\ \text{V} \quad \text{and} \quad V_B = 4 \times I_2 = 4 \times 3 = 12\ \text{V}$$

Therefore, $V_{AB} = V_{th} = 12 - 6 = 6$ V (terminal B positive)

To find R_{th}

R_{th} is calculated by short-circuiting the two voltage sources as shown in Fig. 6.37.

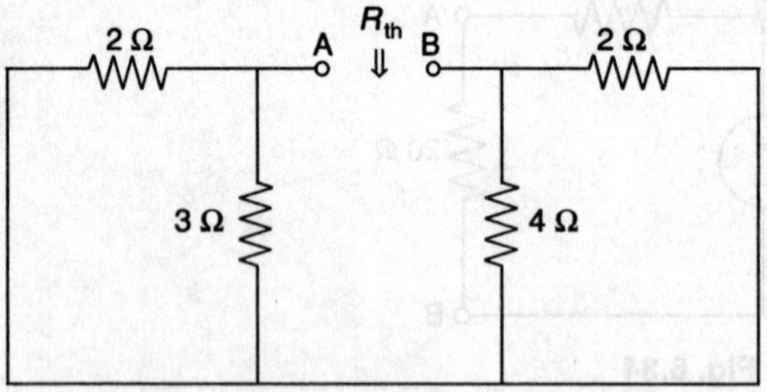

Fig. 6.37

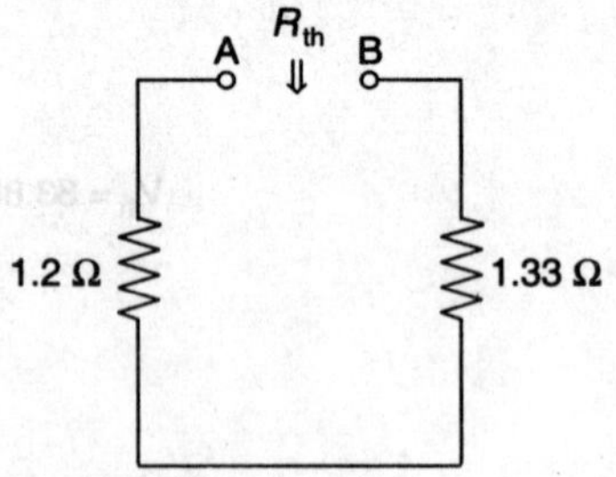

Fig. 6.38

From Fig. 6.38,

$$R_{th} = 1.2 + 1.33 = 2.53\ \Omega$$

Thevenin's equivalent circuit

As shown in Fig. 6.39, current through 10 Ω resistance $= \dfrac{6}{2.53 + 10} = \mathbf{0.48}$ **A**.

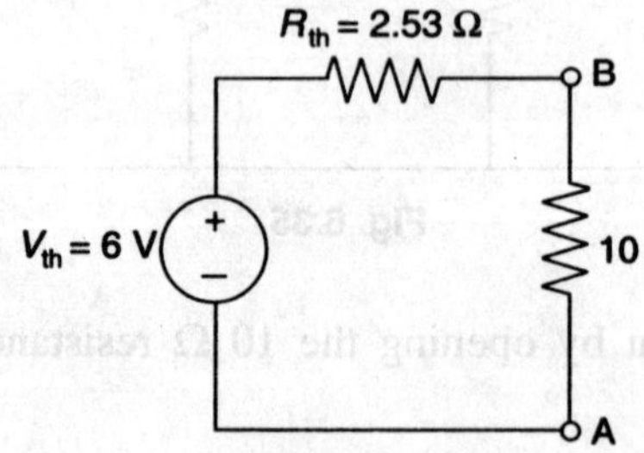

Fig. 6.39

EXAMPLE 6.11 Using Thevenin's theorem, find the current through 5 Ω resistor in the circuit shown in Fig. 6.40.

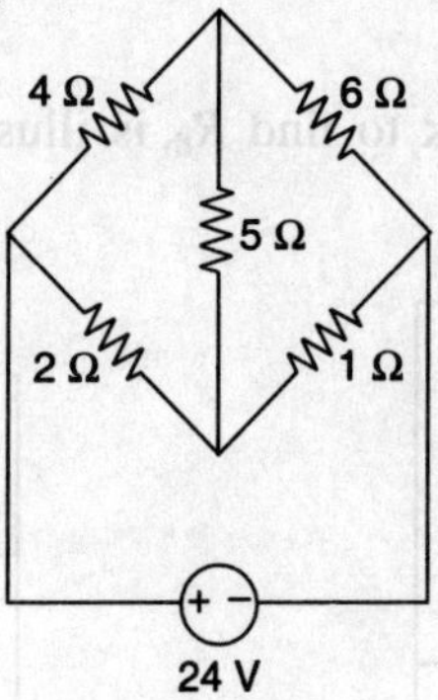

Fig. 6.40

Solution

To find V_{th}

The circuit is drawn by opening the 5 Ω resistance (Fig. 6.41). V_{th} is the voltage between terminals A and B. It can also been redrawn as shown in Fig. 6.42.

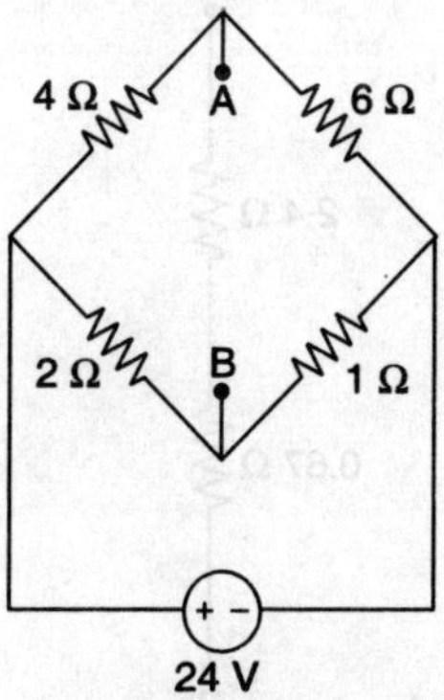

Fig. 6.41

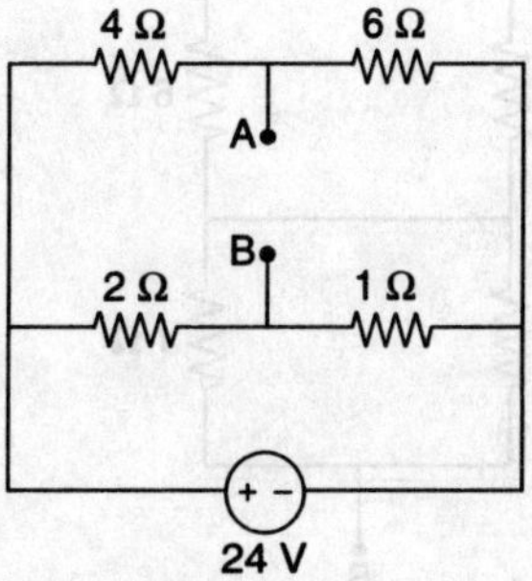

Fig. 6.42

Current through 4 Ω and 6 Ω branch = $\dfrac{24}{4+6} = 2.4$ A

Current through 2 Ω and 1 Ω branch = $\dfrac{24}{2+1} = 8$ A

From Fig. 6.43, V_{th} = 9.6 − 16 = 6.4 V (terminal B negative).

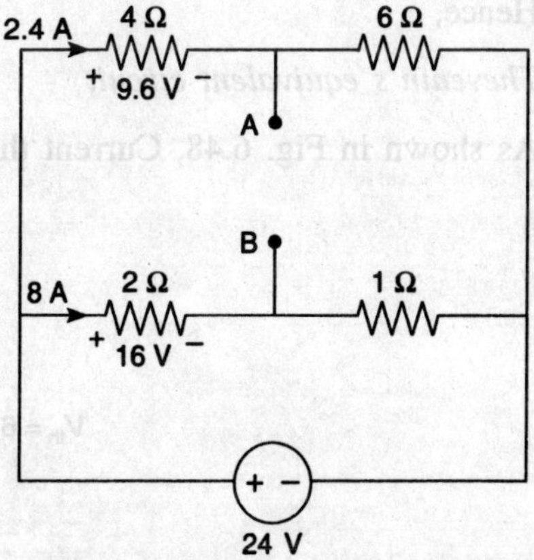

Fig. 6.43

To find R_{th}

The step-by-step reduction of network to find R_{th} is illustrated in Figs. 6.44 through 6.47.

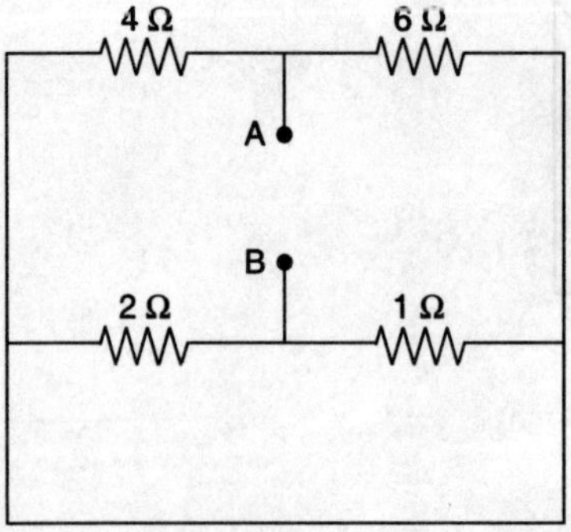

Fig. 6.44

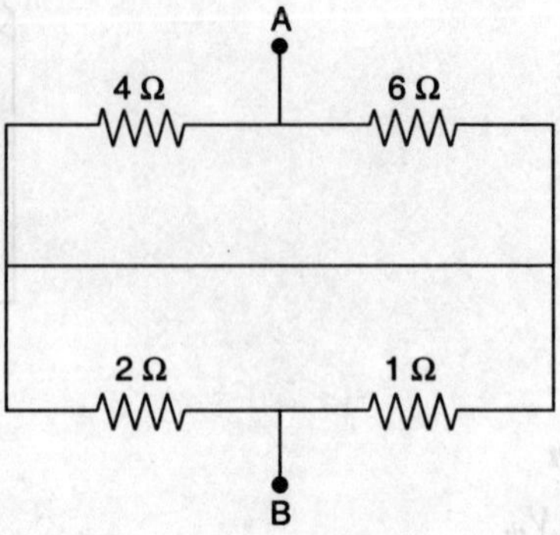

Fig. 6.45

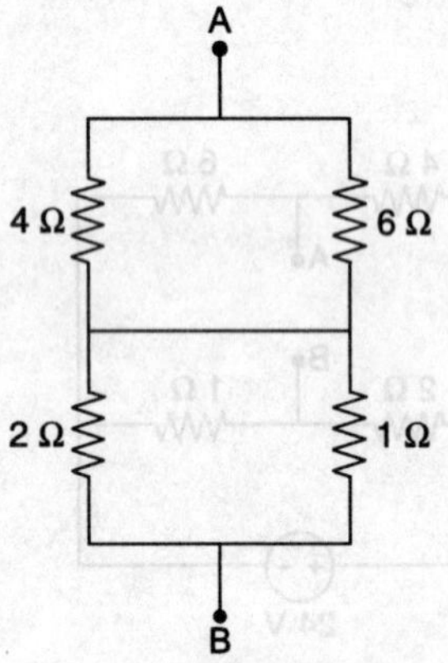

Fig. 6.46

Fig. 6.47

Hence, $R_{th} = R_{AB} = 3.07\ \Omega$

Thevenin's equivalent circuit

As shown in Fig. 6.48, Current through 5 Ω resistance $= \dfrac{6.4}{3.07 + 5} = \mathbf{0.79\ A}$.

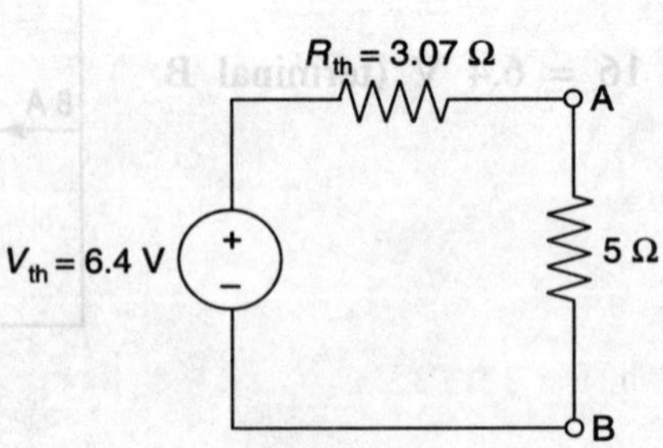

Fig. 6.48

EXAMPLE 6.12 In the network shown in Fig. 6.49, find the current through 3 Ω resistor. Use Thevenin's theorem.

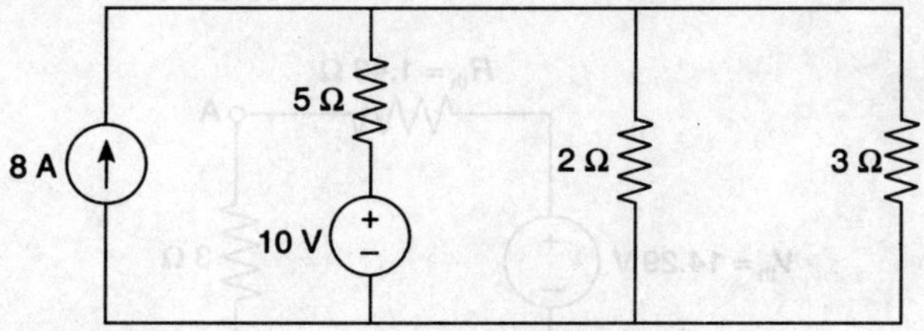

Fig. 6.49

Solution

To find V_{th}

The 3 Ω resistor is first removed as shown in Fig. 6.50.

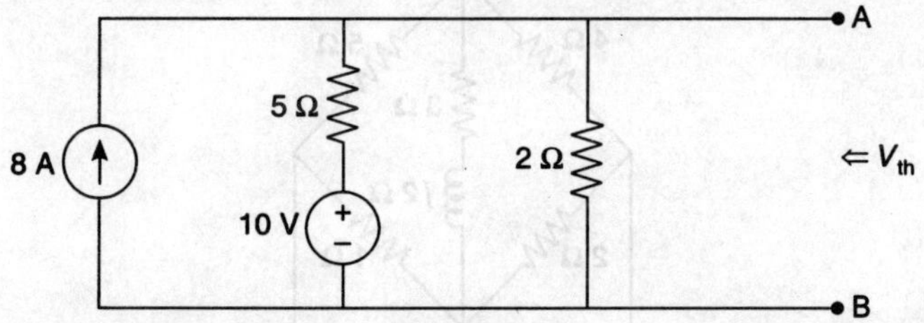

Fig. 6.50

The voltage at A w.r.t B is found by nodal analysis.

$$-8+\frac{V_A-10}{5}+\frac{V_A}{2}=0$$

∴ $$V_A = V_{th}\ 14.29\ \text{V}$$

To find R_{th}

The two sources are made dead sources as shown in Fig. 6.51.

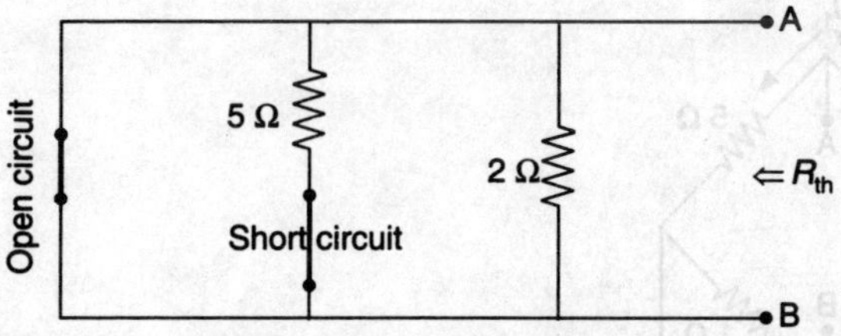

Fig. 6.51

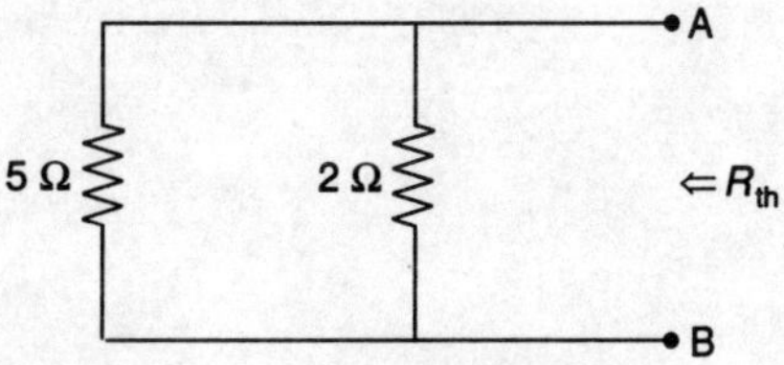

Fig. 6.52

From Fig. 6.52, $R_{AB} = R_{th} = (5\|2) = 1.43\ \Omega$

Thevenin's equivalent circuit

As shown in Fig. 6.53, current through 3 Ω resistance = $\dfrac{14.29}{1.43+3}$ = **3.23** A.

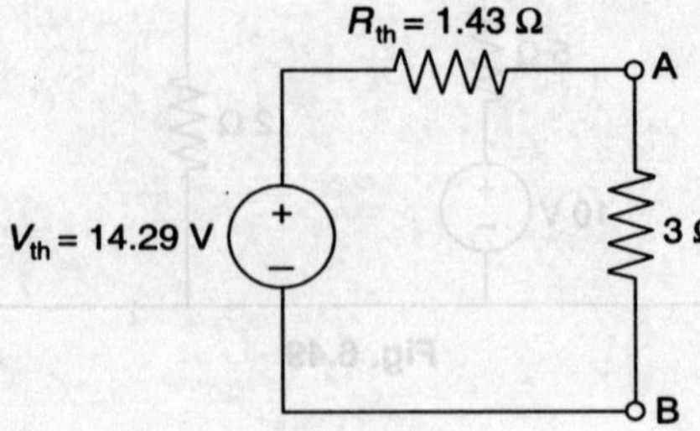

Fig. 6.53

EXAMPLE 6.13 Using Thevenin's theorem, find the current through 3 Ω resistor in Fig. 6.54.

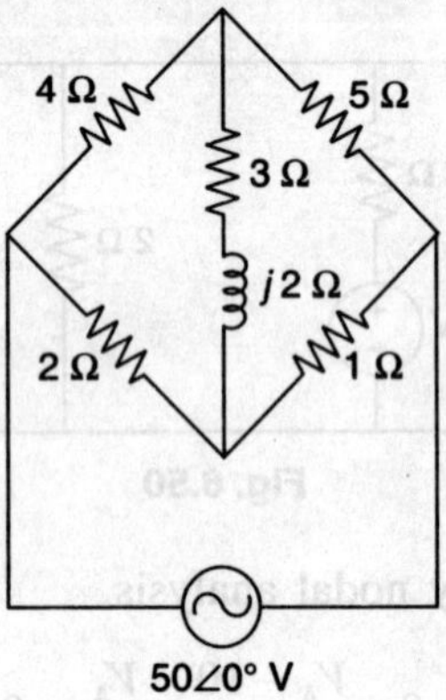

Fig. 6.54

Solution

To find V_{th}

The circuit is drawn by opening the 3 Ω resistance as shown in Fig. 6.55. V_{th} is the voltage between terminals A and B.

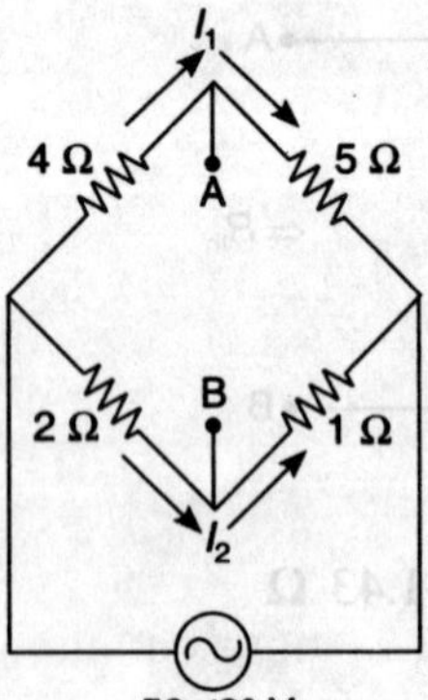

Fig. 6.55

Current through 4 Ω and 5 Ω branch (I_1) = $\dfrac{50\angle 0^\circ}{4+5} = 5.56\angle 0^\circ$ A

Current through 2 Ω and 1 Ω branch (I_2) = $\dfrac{50\angle 0^\circ}{2+1} = 16.67\angle 0^\circ$ A

$$V_{th} = -4I_1 + 2I_2$$
$$= (-4 \times 5.56\angle 0^\circ) + (2 \times 16.67\angle 0^\circ)$$
$$= 11.1\angle 0^\circ \text{ V}$$

To find Z_{th}

The voltage source is short-circuited and the circuit is redrawn as shown in Fig. 6.56.

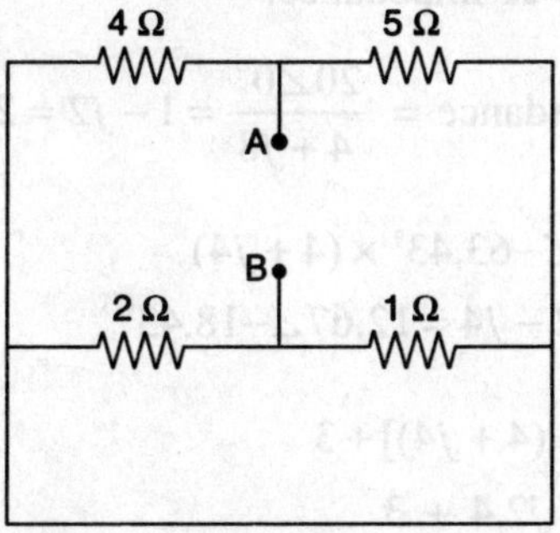

Fig. 6.56

From Fig. 6.56, $Z_{AB} = Z_{th} = \left(\dfrac{4\times 5}{4+5}\right) + \left(\dfrac{2\times 1}{2+1}\right) = 2.89\ \Omega.$

Thevenin's equivalent circuit

As shown in Fig. 6.57, current through (3 + *j*2) Ω resistance = $\dfrac{11.1\angle 0^\circ}{5.89 + j2} = \mathbf{1.785\angle -18.76^\circ}$ **A**.

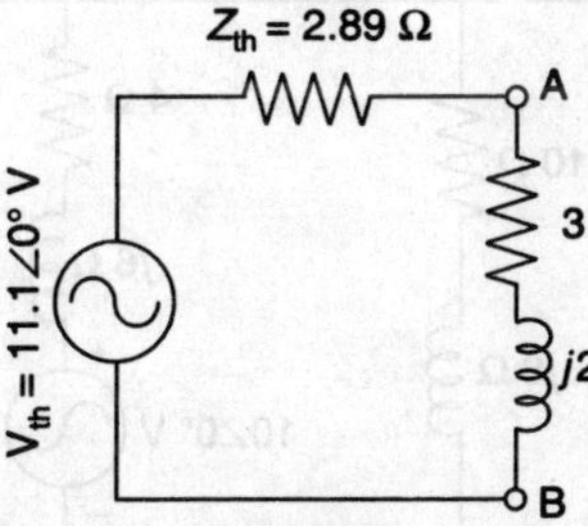

Fig. 6.57

EXAMPLE 6.14 Find the Thevenin's equivalent circuit across terminals A–B in the circuit shown in Fig. 6.58.

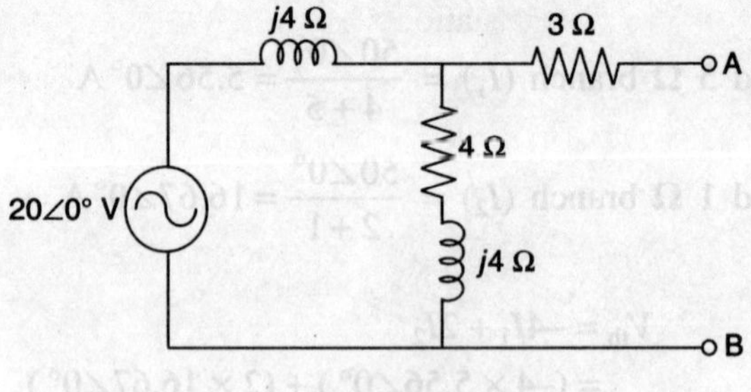

Fig. 6.58

Solution

To find V_{th}

V_{th} is the voltage across (4 + j4) Ω impedance.

$$\text{Current through } (4 + j4)\ \Omega \text{ impedance} = \frac{20\angle 0^\circ}{4 + j8} = 1 - j2 = 2.24\angle -63.43^\circ$$

Therefore,

$$V_{th} = 2.24\angle -63.43^\circ \times (4 + j4)$$
$$= 12.02 - j4 = 12.67\angle -18.43^\circ$$

To find Z_{th}

$$Z_{th} = [(j4) \| (4 + j4)] + 3$$
$$= 0.8 + j2.4 + 3$$
$$= (3.8 + j2.4)\ \Omega$$

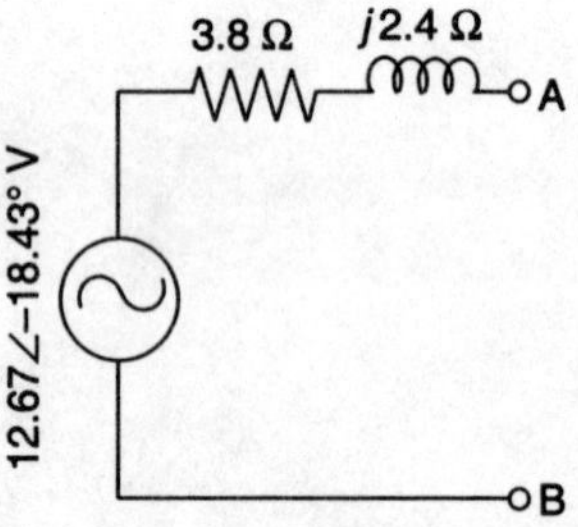

Fig. 6.59

Using the values of V_{th} and Z_{th}, Thevenin's equivalent circuit is drawn and shown in Fig. 6.59.

EXAMPLE 6.15 Determine the current through Z_L in the circuit shown in Fig. 6.60 using Thevenin's theorem.

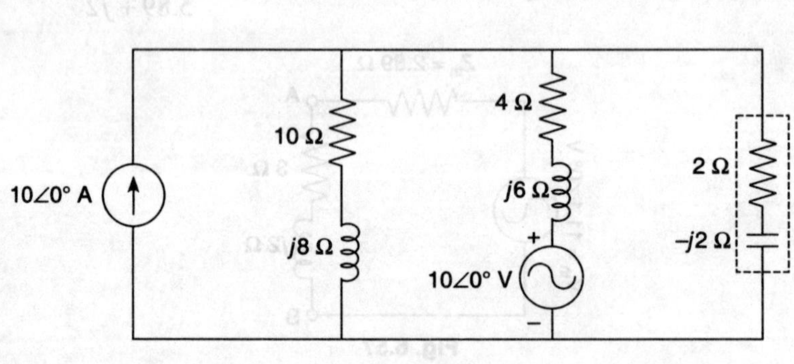

Fig. 6.60

Solution The load impedance Z_L is removed as shown in Fig. 6.61.

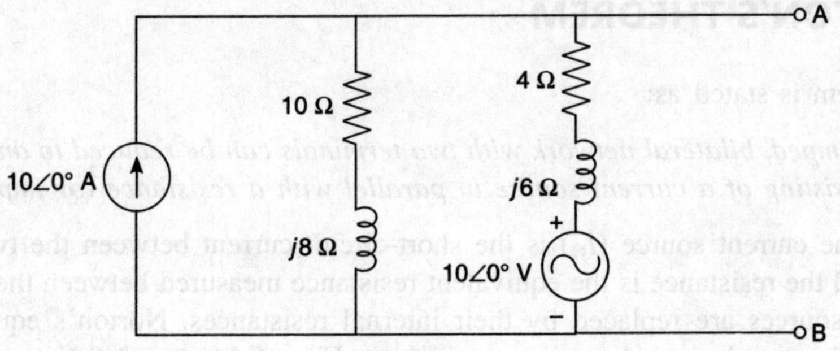

Fig. 6.61

To find V_{th}

Taking node B as reference node and KCL is written at node A as follows:

$$-10+\frac{V_A}{10+j8}+\frac{V_A-10}{4+j6}=0$$

$$V_A\left(\frac{1}{10+j8}+\frac{1}{4+j6}\right)-\left(\frac{10}{4+j6}\right)=10$$

Solving,

$$V_A = V_{th} = 50.56\angle 43.8^\circ \text{ V}$$

To find Z_{th}

The circuit to find Z_{th} is shown in Fig. 6.62.

$$Z_{th} = (10 + j8)\|(4 + j6)$$

$$= (3 + j3.5714)\ \Omega = 4.664\angle 49.97^\circ\ \Omega$$

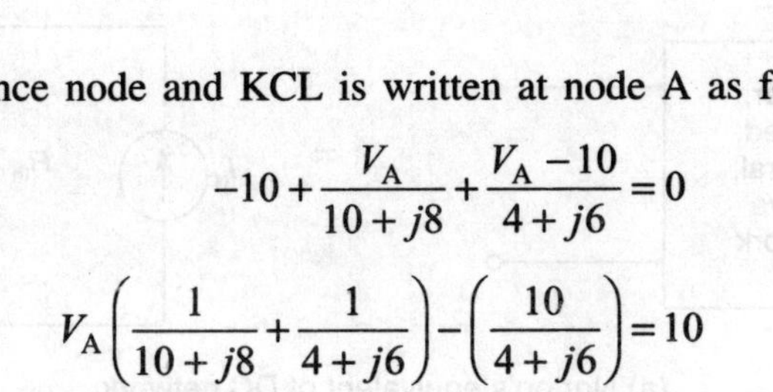

Fig. 6.62

Thevenin's equivalent circuit

As shown in Fig. 6.63, current through the load impedance,

$$Z_L = \frac{50.56\angle 43.8^\circ}{(3+j3.5714)+(2-j2)}$$

$$= \mathbf{8.6442 + j4.2822 = 9.6468\angle 26.35^\circ\ A.}$$

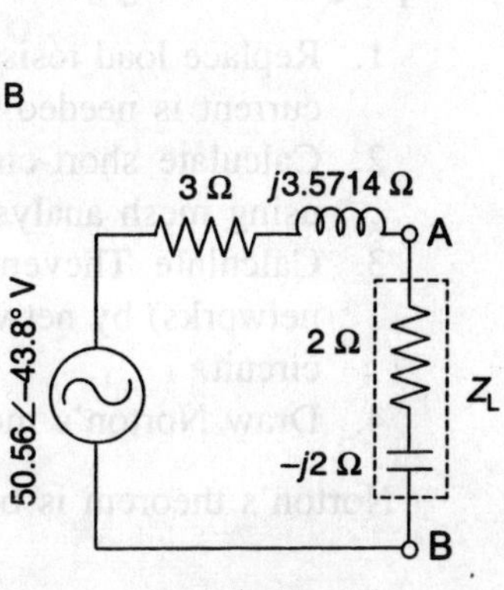

Fig. 6.63

6.4 NORTON'S THEOREM

Norton's theorem is stated as:

A linear, lumped, bilateral network with two terminals can be reduced to an equivalent circuit consisting of a current source in parallel with a resistance (or impedance).

The value of the current source (I_N) is the short-circuit current between the two terminals of the network and the resistance is the equivalent resistance measured between the terminals with all the energy sources are replaced by their internal resistances. Norton's equivalent circuits for DC and AC networks are shown respectively in Figs. 6.64(a) and (b).

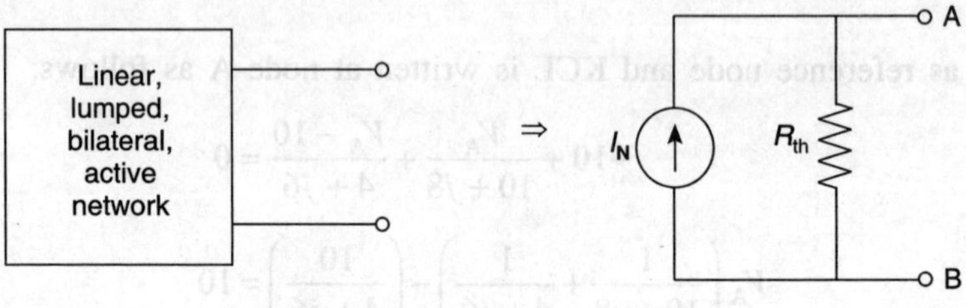

(a) Norton's equivalent of DC network.

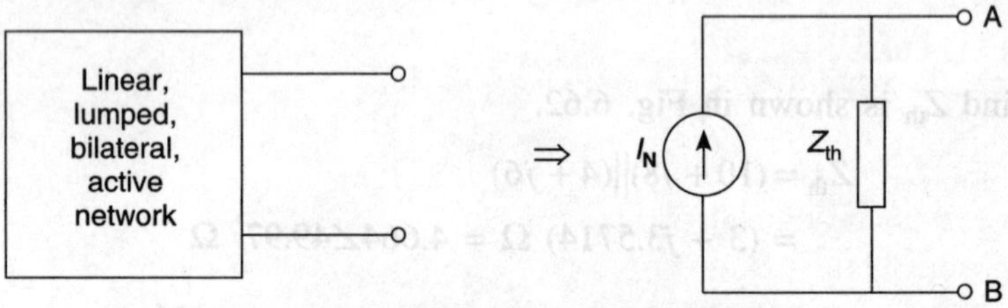

(b) Norton's equivalent of AC network.

Fig. 6.64

Steps for solving electrical networks using Norton's theorem

1. Replace load resistance R_L (in the case of AC networks impedance, Z_L) through which current is needed by a short-circuit.
2. Calculate short-circuit current, called Norton's current (I_N) through the short-circuit using mesh analysis/ohm's law/current division, etc.
3. Calculate Thevenin's resistance R_{th} (Thevenin impedance Z_{th}, in the case of AC networks) by network reduction method by looking back the circuit through the short-circuit.
4. Draw Norton's model and reconnect R_L in parallel with R_{th}.

Norton's theorem is better understood by the following examples.

EXAMPLE 6.16 Find the current through 6 Ω resistor in the network shown in Fig. 6.65.

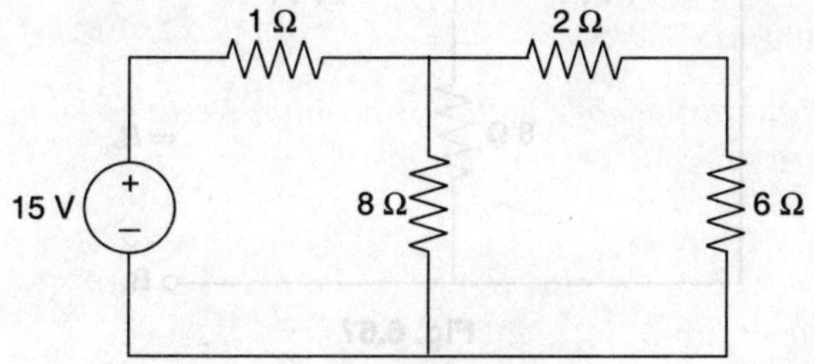

Fig. 6.65

Solution

Step 1 *Removal of resistor in question*

6 Ω resistance is removed and it is replaced by short-circuit as shown in Fig. 6.66. The current through the short-circuit is called Norton current (I_N).

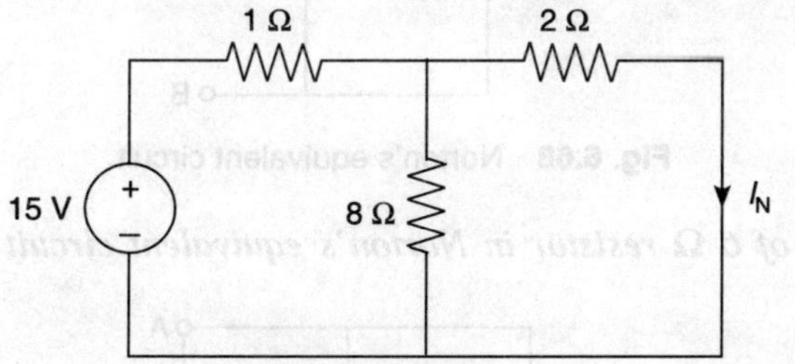

Fig. 6.66

Step 2 *Determination of Norton's current I_N*

I_N is nothing but the current flowing through the short-circuit. As seen in Fig. 6.66, 8 Ω and 2 Ω are in parallel and this parallel equivalent resistance is in series with 1 Ω. Therefore, the effective resistance of the network is

$$1 + (8 \| 2) = 1 + \frac{8 \times 2}{8 + 2} = 2.6\ \Omega$$

Hence, the current supplied by 15 V source $= \dfrac{15}{2.6} = 5.77$ A

Current through short-circuit (I_N) is nothing but the current through 2 Ω resistor.

Using current division, I_N is determined as $5.77 \times \dfrac{8}{8+2} = 4.62$ A

Step 3 *Determination of Thevenin's resistance R_{th}*

R_{th} is calculated by short-circuiting the 15 V voltage source and looking back the network from the terminals A and B as shown in Fig. 6.67.

$$R_{th} = R_{AB} = 2 + \frac{1 \times 8}{1 + 8} = 2.89\ \Omega$$

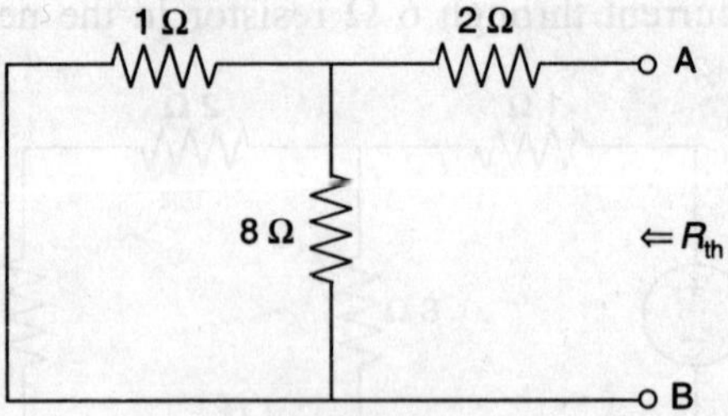

Fig. 6.67

Step 4 *Construction of Norton's equivalent circuit (Fig. 6.68)*

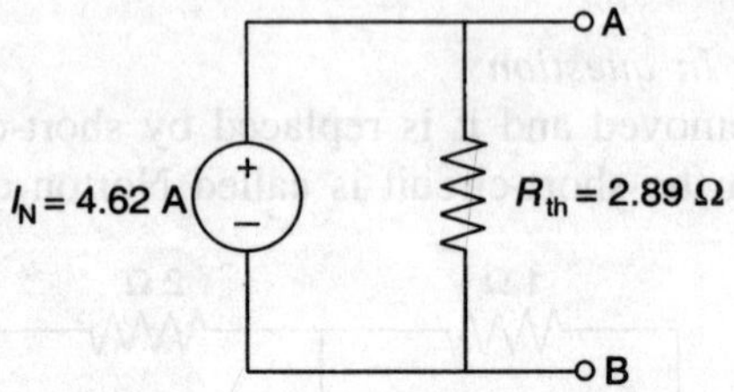

Fig. 6.68 Norton's equivalent circuit.

Step 5 *Reconnection of 6 Ω resistor in Norton's equivalent circuit (Fig. 6.69)*

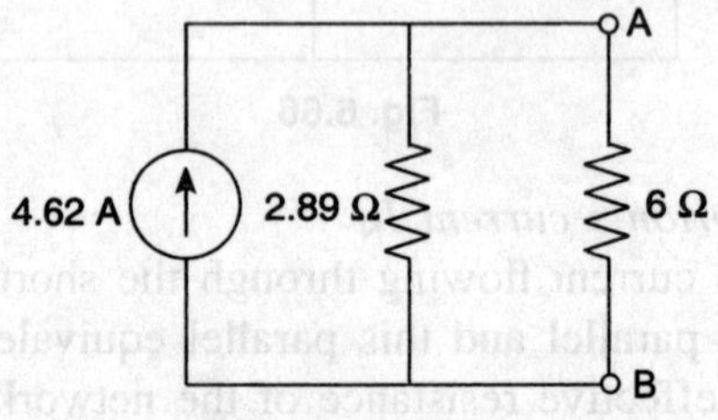

Fig. 6.69

Therefore, the current through 6 Ω resistor $= 4.62 \times \dfrac{2.89}{2.89 + 6} = \mathbf{1.5\ A.}$

EXAMPLE 6.17 For the circuit shown in Fig. 6.70, determine the load current using Norton's theorem.

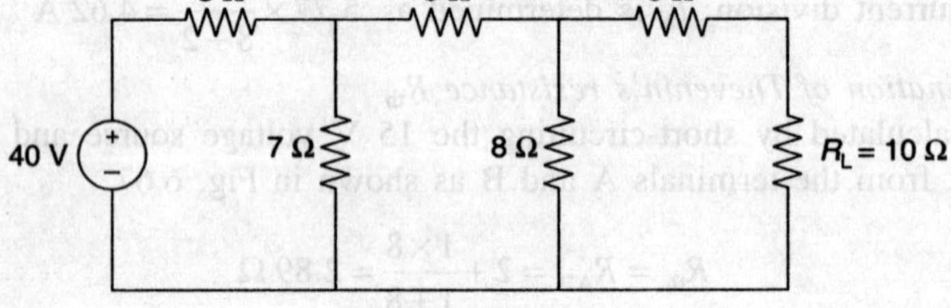

Fig. 6.70

Solution

To find Norton's current (I_N)

The circuit is redrawn by short-circuiting the load resistance 10 Ω as shown in Fig. 6.71.

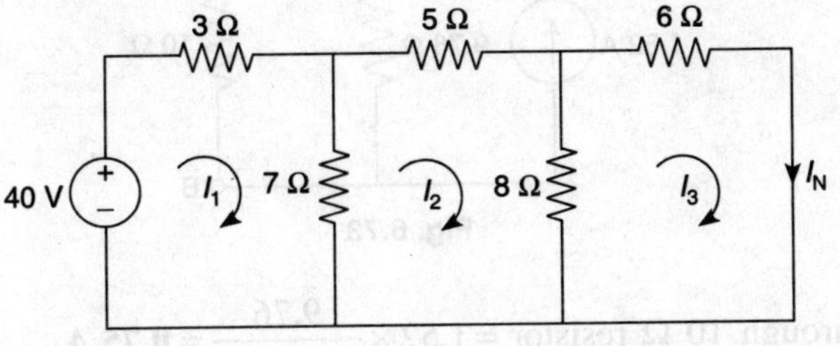

Fig. 6.71

The short-circuit current I_N is nothing but I_3. It is calculated by mesh analysis method.

$$I_3 = \frac{\begin{vmatrix} 10 & -7 & 40 \\ -7 & 20 & 0 \\ 0 & -8 & 0 \end{vmatrix}}{\begin{vmatrix} 10 & -7 & 0 \\ -7 & 20 & -8 \\ 0 & -8 & 14 \end{vmatrix}} = \frac{2240}{1474} = 1.52 \text{ A}$$

So, Norton's current, $I_N = 1.52$ A.

To find R_{th}

R_{th} is calculated by short-circuiting the 40 V voltage source as shown in Fig. 6.72 and looking back terminals A and B.

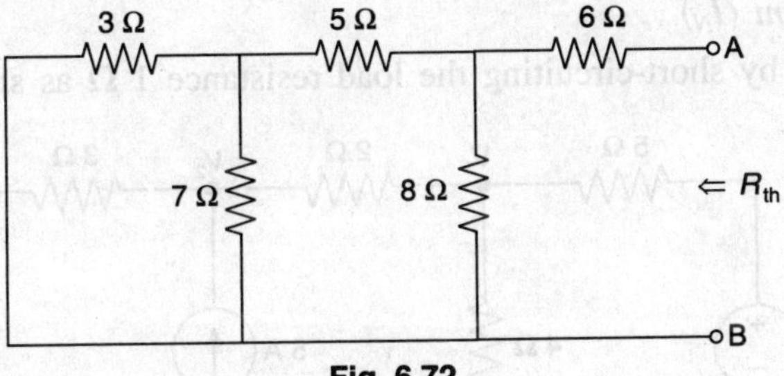

Fig. 6.72

Equivalent resistance for the parallel combination of 3 Ω and 7 Ω = $\dfrac{3 \times 7}{3+7} = 2.1\ \Omega$

2.1 Ω is in series with 5 Ω, which gives 7.1 Ω.

Equivalent resistance for the parallel combination of 7.1 Ω and 8 Ω = $\dfrac{7.1 \times 8}{7.1+8} = 3.76\ \Omega$

3.76 Ω is in series with 6 Ω, which gives 9.76 Ω.

So, $$R_{th} = R_{AB} = 9.76\ \Omega$$

Norton's equivalent circuit (Fig. 6.73)

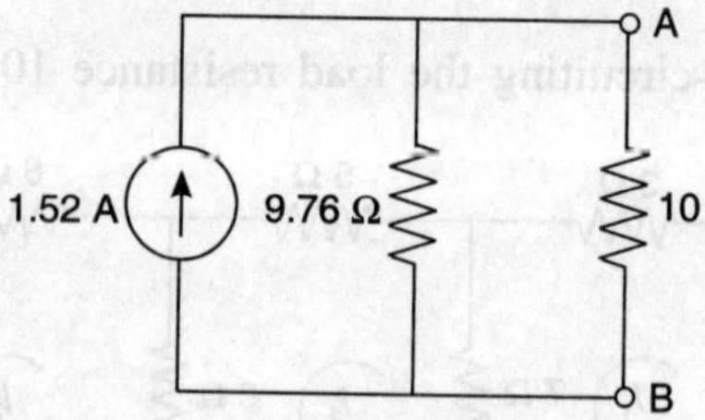

Fig. 6.73

Hence, the current through 10 Ω resistor $= 1.52 \times \dfrac{9.76}{9.76 + 10} = \mathbf{0.75\ A.}$

EXAMPLE 6.18 Using Norton's theorem, find the current through 1 Ω resistor in the circuit shown in Fig. 6.74.

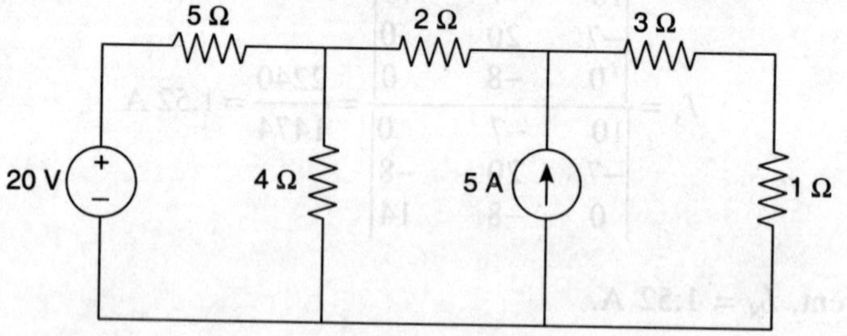

Fig. 6.74

Solution

To find Norton's current (I_N)

The circuit is redrawn by short-circuiting the load resistance 1 Ω as shown in Fig. 6.75.

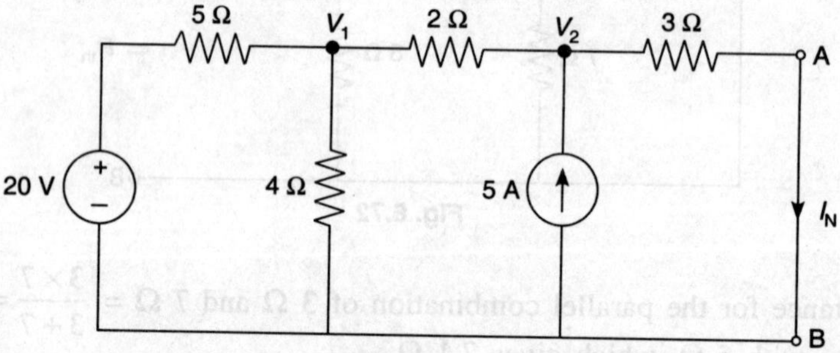

Fig. 6.75

Using nodal analysis, voltage V_2 is determined as follows:

$$\frac{V_1 - 20}{5} + \frac{V_1}{4} + \frac{V_1 - V_2}{2} = 0 \qquad \text{(i)}$$

$$\frac{V_2 - V_1}{2} + \frac{V_2}{3} = 5 \quad \text{(ii)}$$

Solving Eqs. (i) and (ii), we get

$$V_2 = 12.53 \text{ V}$$

Hence,
$$I_N = \frac{V_2}{3} = \frac{12.53}{3} = 4.18 \text{ A}$$

To find R_{th}

The two sources are made dead as shown in Fig. 6.76.

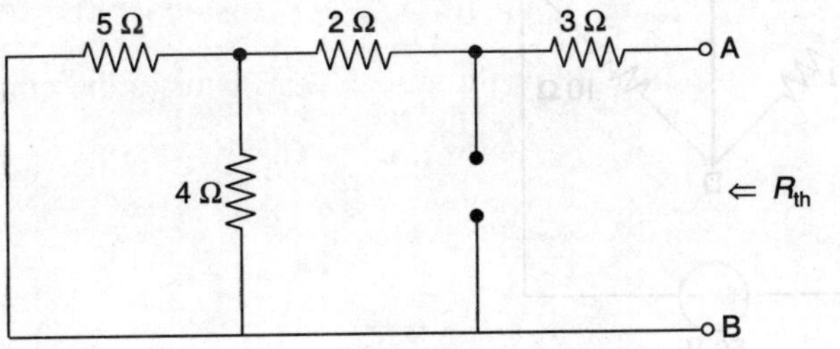

Fig. 6.76

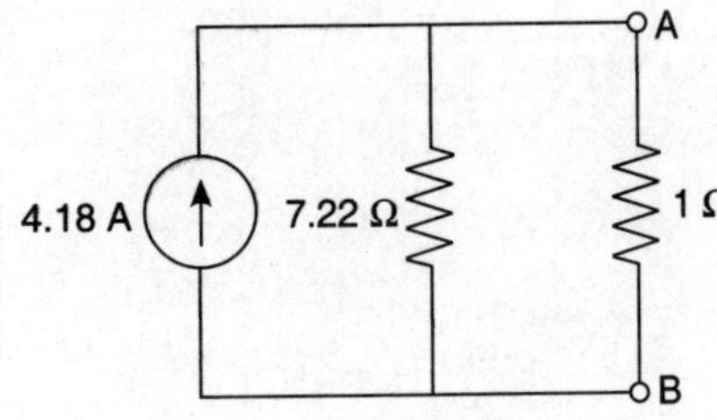

Fig. 6.77

Therefore,
$$R_{th} = (5\|4) + 2 + 3 = \left(\frac{5\times 4}{5+4}\right) + 5 = 7.22\ \Omega$$

From the Norton's equivalent circuit shown in Fig. 6.77, current through 1 Ω resistor $= 4.18 \times \frac{7.22}{7.22+1} = \mathbf{3.67\ A}$.

EXAMPLE 6.19 Using Norton's theorem, find the current through the galvanometer in the circuit shown in Fig. 6.78.

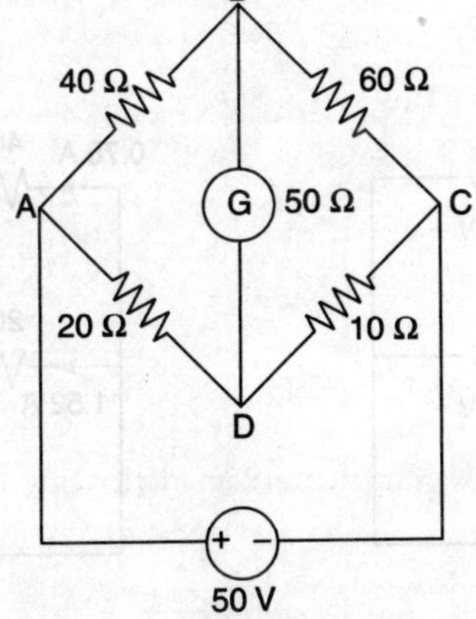

Fig. 6.78

Solution

To find I_N

50 Ω resistance is shorted as shown in Fig. 6.79. To find Norton's current, I_N, voltage drops across each resistance is first determined. The circuit is simplified and the voltage across each resistance is found as illustrated in Figs. 6.80 through 6.83.

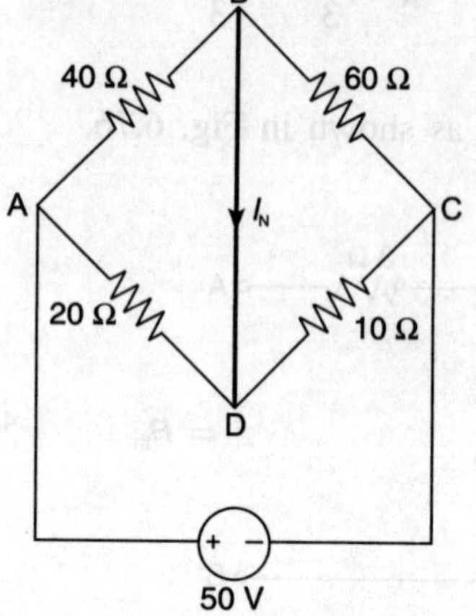

Fig. 6.79

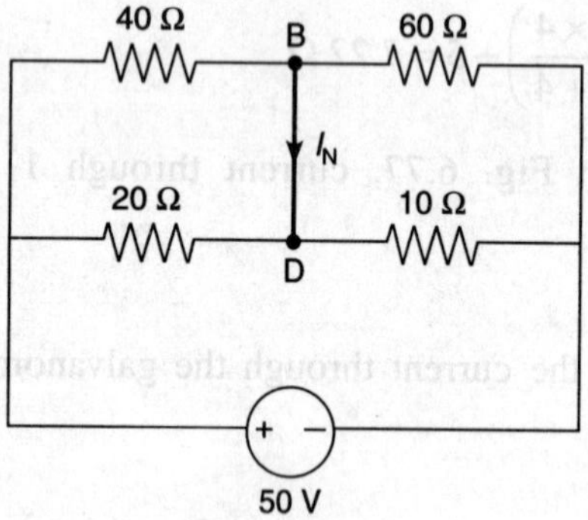

Fig. 6.80

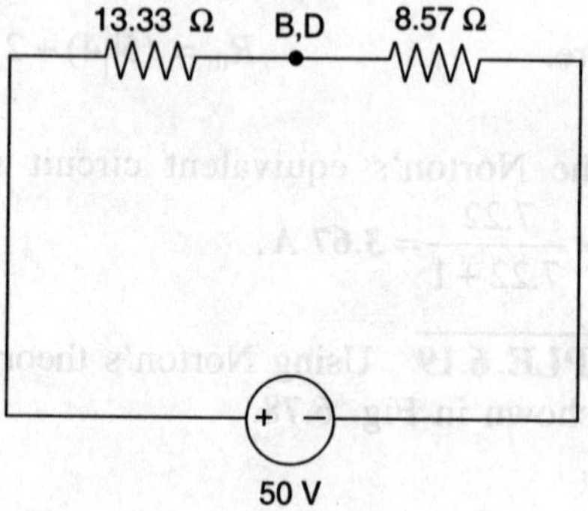

Fig. 6.81

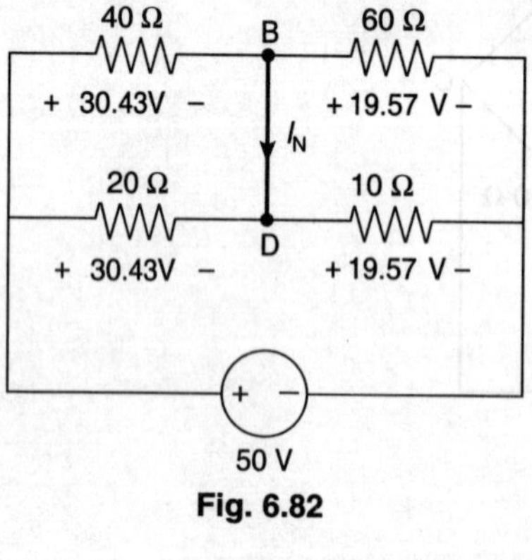

Fig. 6.82

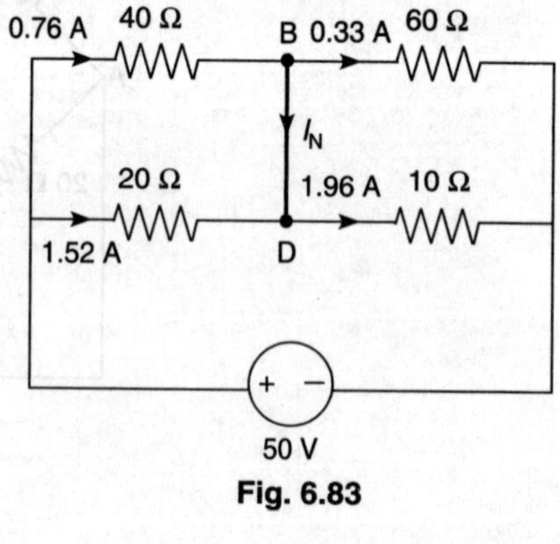

Fig. 6.83

Current through 40 Ω = $\dfrac{30.43}{40} = 0.76$ A, Current through 20 Ω = $\dfrac{30.43}{20} = 1.52$ A

Current through 60 Ω = $\dfrac{19.57}{60} = 0.33$ A, Current through 10 Ω = $\dfrac{19.57}{10} = 1.96$ A

As seen Fig. 6.83, Norton's current, $I_N = 0.76 - 0.33 = 0.43$ A.

To find R_{th}

The step-by-step reduction of network to find R_{th} is given in Fig. 6.84(a)–(d).

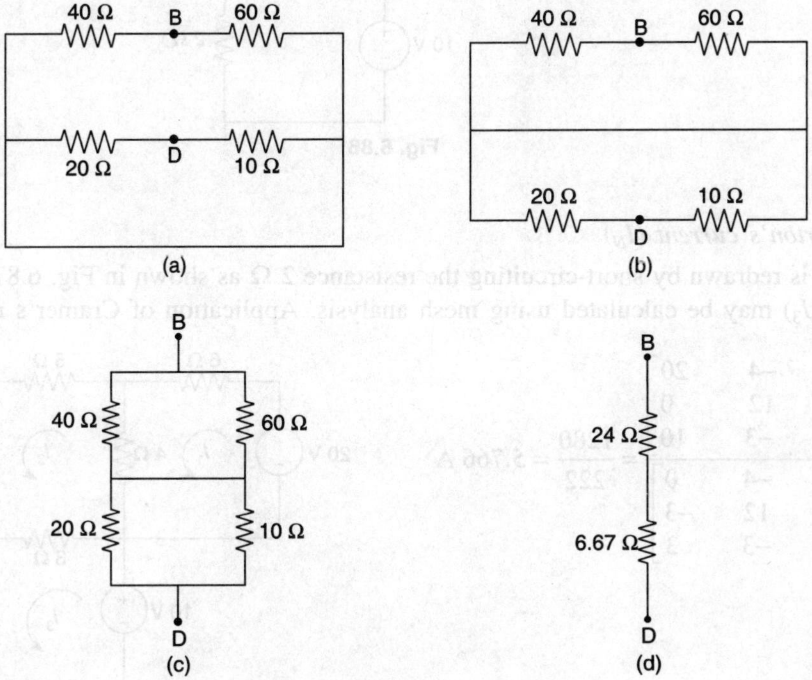

Fig. 6.84

From Fig. 6.84(d), $R_{th} = R_{BD} = 30.67$ Ω.

Norton's equivalent circuit (Fig. 6.85)

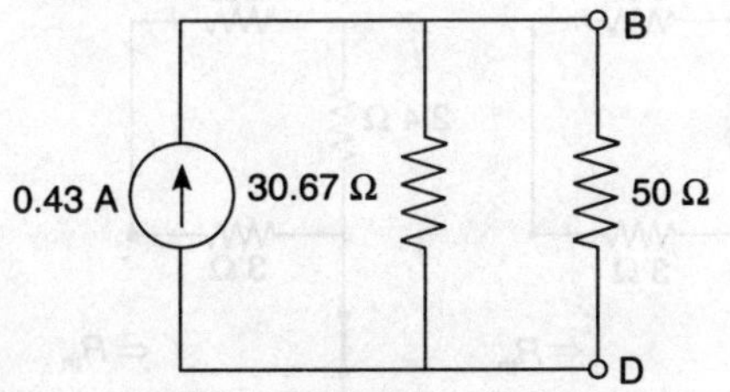

Fig. 6.85

Hence, the current through 50 Ω resistance $= 0.43 \times \dfrac{30.67}{30.67 + 50}$ = **0.1635 A.**

EXAMPLE 6.20 In the network shown in Fig. 6.86, find the current through 2 Ω resistor. Use Norton's and Thevenin's theorems.

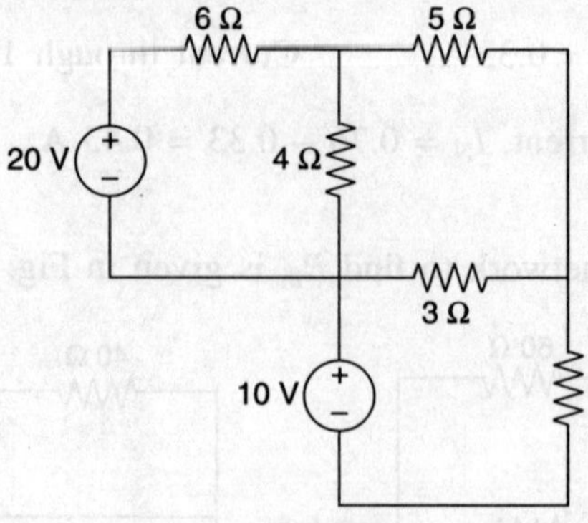

Fig. 6.86

Solution

To find Norton's current (I_N)

The circuit is redrawn by short-circuiting the resistance 2 Ω as shown in Fig. 6.87. The Norton current I_N (I_3) may be calculated using mesh analysis. Application of Cramer's rule gives

$$I_N = \frac{\begin{bmatrix} 10 & -4 & 20 \\ -4 & 12 & 0 \\ 0 & -3 & 10 \end{bmatrix}}{\begin{bmatrix} 10 & -4 & 0 \\ -4 & 12 & -3 \\ 0 & -3 & 3 \end{bmatrix}} = \frac{1280}{222} = 5.766 \text{ A}$$

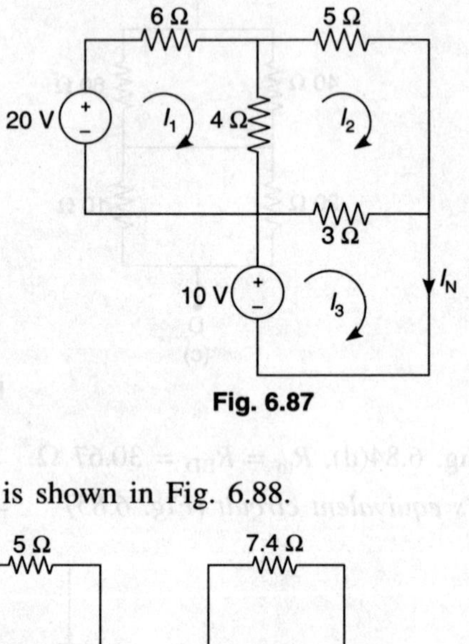

Fig. 6.87

To find R_{th}

The step-by-step reduction of network to find R_{th} is shown in Fig. 6.88.

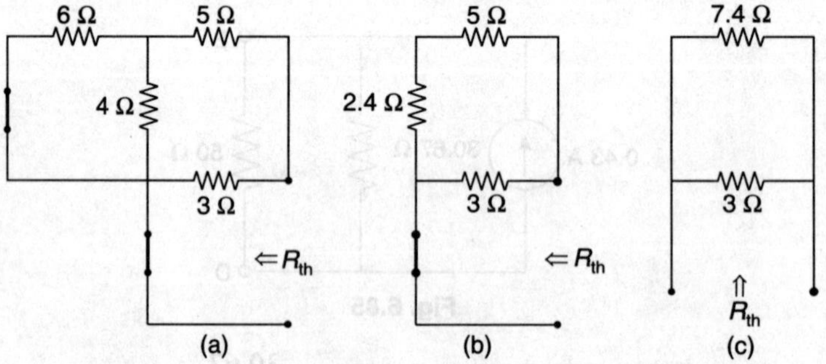

Fig. 6.88 Step-by-step reduction of network to compute R_{th}.

From Fig. 6.88(c), R_{th} is found as $\frac{7.4 \times 3}{7.4+3} = 2.14\ \Omega$.

Norton's equivalent circuit

As shown in Fig. 6.89, current through 2 Ω resistor $= 5.76 \times \frac{2.14}{2.14+2} = \mathbf{2.97\ A}$.

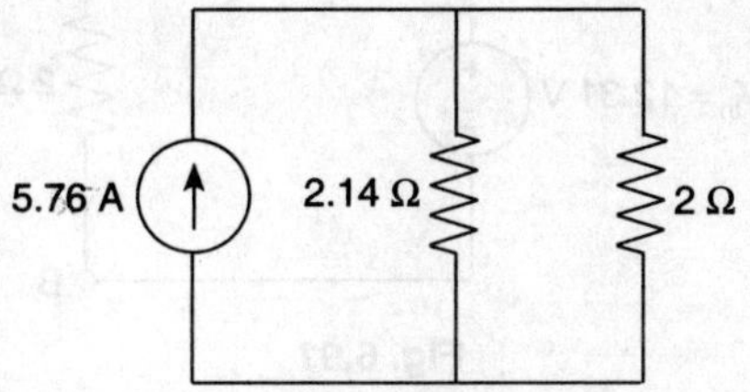

Fig. 6.89

To find V_{th}

The circuit is redrawn by opening the 2 Ω resistance. V_{th} is the voltage between terminals A and B. To find V_{th}, the voltage across 3 Ω is determined first. To do so, the current through 3 Ω is found using mesh analysis (Fig. 6.90).

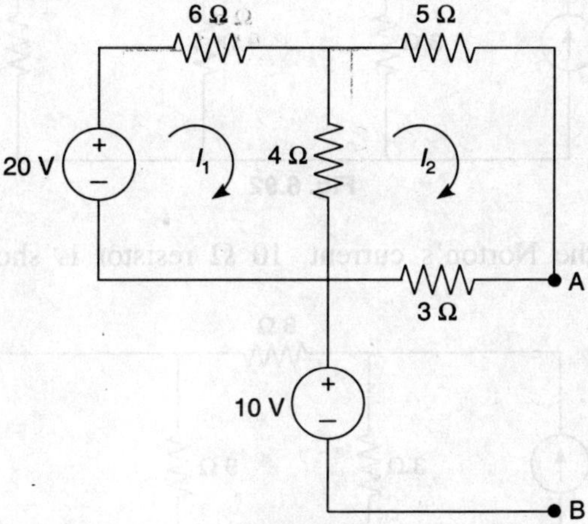

Fig. 6.90

Application of Cramer's rule gives

$$I_2 = \frac{\begin{bmatrix} 10 & 20 \\ -4 & 0 \end{bmatrix}}{\begin{bmatrix} 10 & -4 \\ -4 & 12 \end{bmatrix}} = \frac{80}{104} = 0.77\ \text{A}$$

Voltage across 3 Ω resistance = 3 × 0.77 = 2.31 V

Therefore, $V_{th} = 10 + 2.31 = 12.31\ \text{V}$

Thevenin's equivalent circuit

As shown in Fig. 6.91, current through 2 Ω resistance = $\frac{12.31}{2.14+2}$ = **2.97 A.**

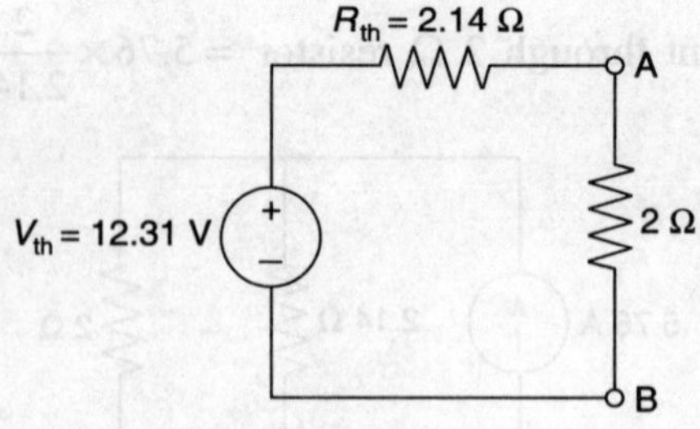

Fig. 6.91

EXAMPLE 6.21 Determine the current in the 10 Ω resistor of the circuit shown in Fig. 6.92 by Norton's theorem.

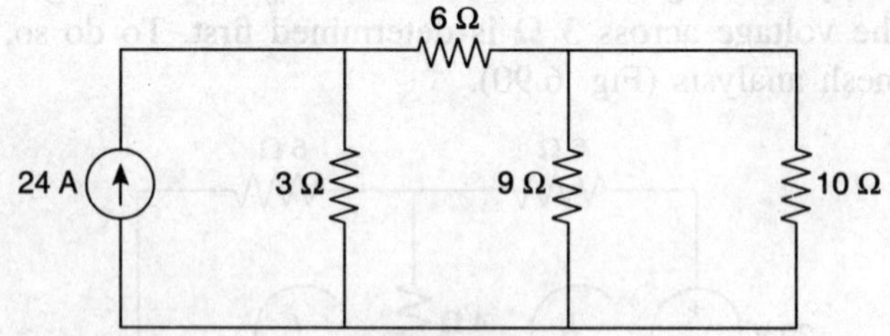

Fig. 6.92

Solution To determine the Norton's current, 10 Ω resistor is short-circuited as shown in Fig. 6.93.

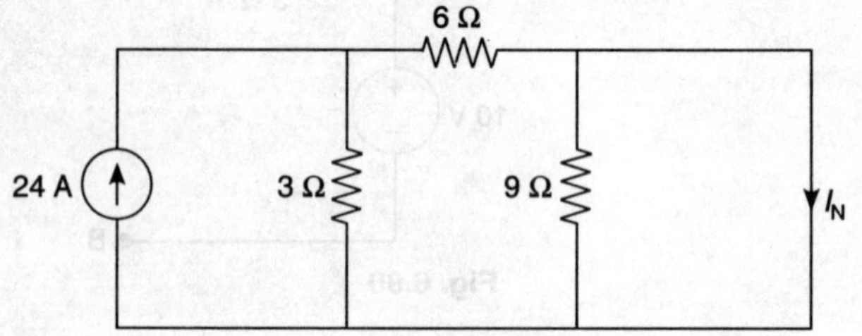

Fig. 6.93

To find I_N

By current division rule,

$$I_N = 24 \times \frac{3}{3+6} = 8 \text{ A}$$

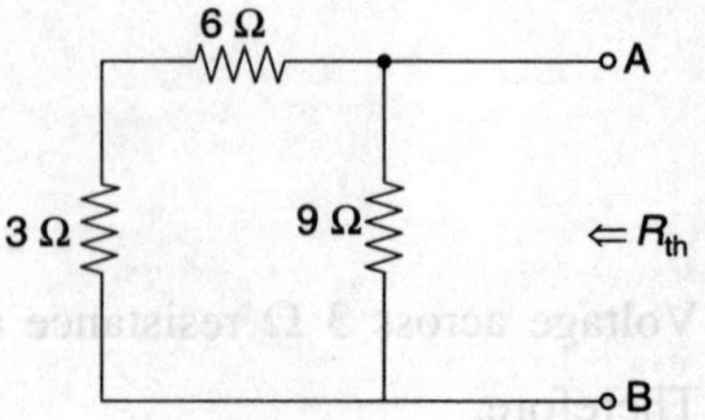

Fig. 6.94

To find R_{th}

From Fig. 6.94,

$$R_{\text{th}} = (3+6)\|9 = \left(\frac{9\times 9}{9+9}\right) = 4.5\ \Omega$$

Norton's equivalent circuit

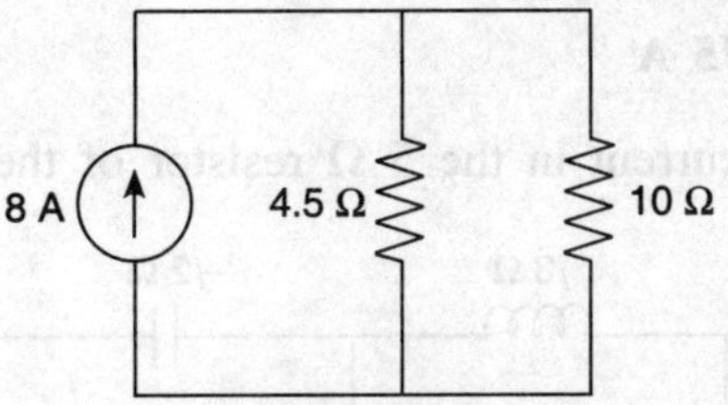

Fig. 6.95

From Fig. 6.95, current through 10 Ω resistance $= 8\times\dfrac{4.5}{4.5+10}$ = **2.48 A.**

EXAMPLE 6.22 Determine the current in the 50 Ω resistor of the circuit shown in Fig. 6.96 by Norton's theorem.

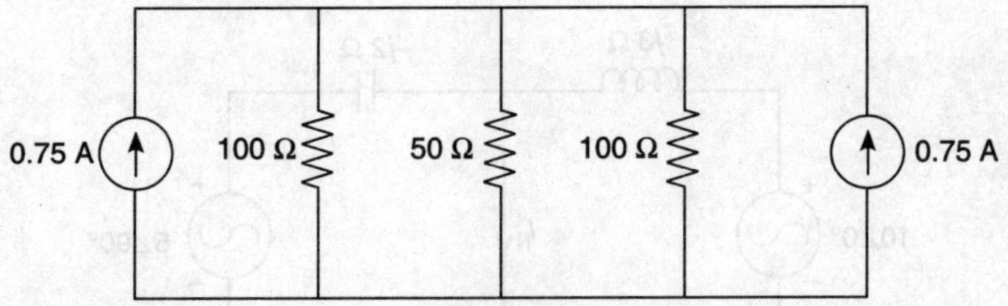

Fig. 6.96

Solution If the 50 Ω resistor is short-circuited, the entire current supplied by the two sources would flow through the short-circuit only. Thus,

$$I_N = 0.75 + 0.75 = 1.5\text{ A}$$

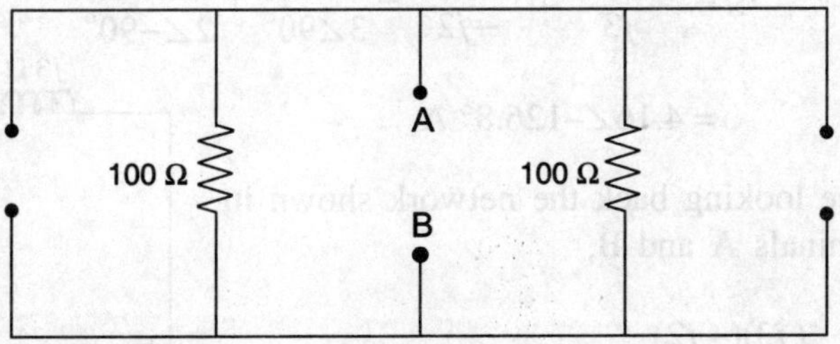

Fig. 6.97

From Fig. 6.97, $R_{th} = \dfrac{100 \times 100}{100 + 100} = 50\ \Omega$

Norton's equivalent circuit

As shown in Fig. 6.98, current through 50 Ω resistance

$$= 1.5 \times \frac{50}{50 + 50}$$

$$= \mathbf{0.75\ A}$$

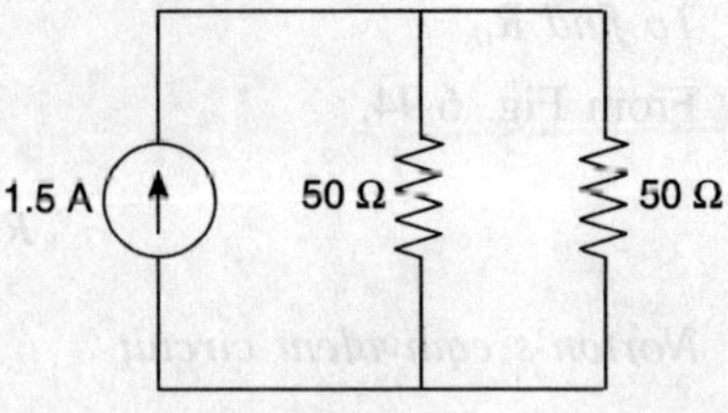

Fig. 6.98

EXAMPLE 6.23 Find the current in the 5 Ω resistor of the circuit shown in Fig. 6.99 by Norton's theorem.

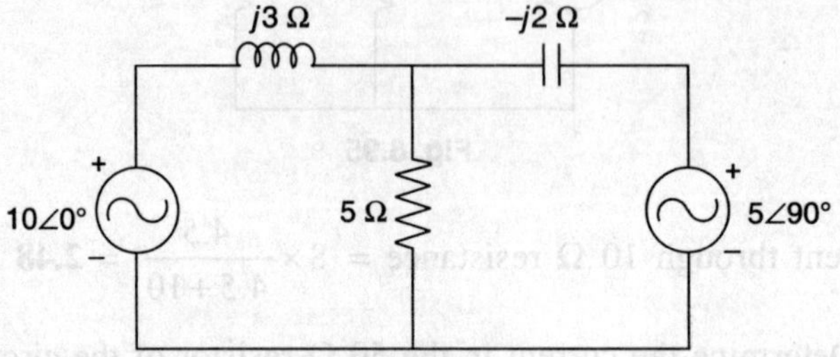

Fig. 6.99

Solution

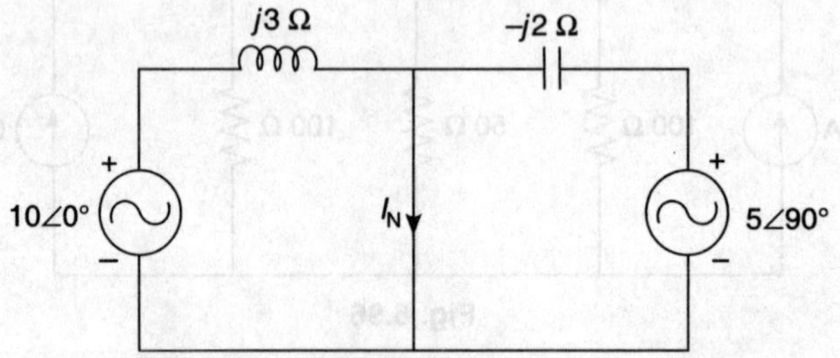

Fig. 6.100

5 Ω resistor is short-circuited (Fig. 6.100) and the Norton's current is calculated as:

$$I_N = \frac{10\angle 0^\circ}{j3} + \frac{5\angle 90^\circ}{-j2} = \frac{10\angle 0^\circ}{3\angle 90^\circ} + \frac{5\angle 90^\circ}{2\angle -90^\circ}$$

$$= 4.16\angle -126.8^\circ \text{ A}$$

Thevenin's impedance looking back the network shown in Fig. 6.101, from terminals A and B,

$$Z_{th} = \frac{(j3)(-j2)}{(j3) - (j2)} = -j6\ \Omega$$

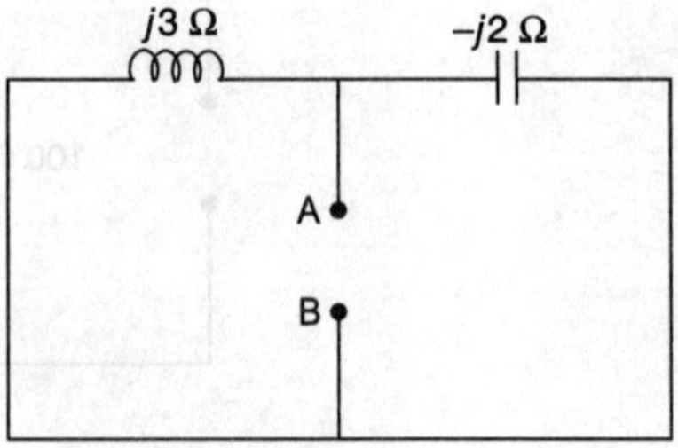

Fig. 6.101

Norton's equivalent circuit

From Fig. 6.102, current through 50 Ω resistance

$$= 4.16\angle -126.8 \times \frac{(-j6)}{(5-j6)}$$

$$= (-3.103 - j0.738)\ \text{A} = 3.19\angle -166.62°\ \text{A}$$

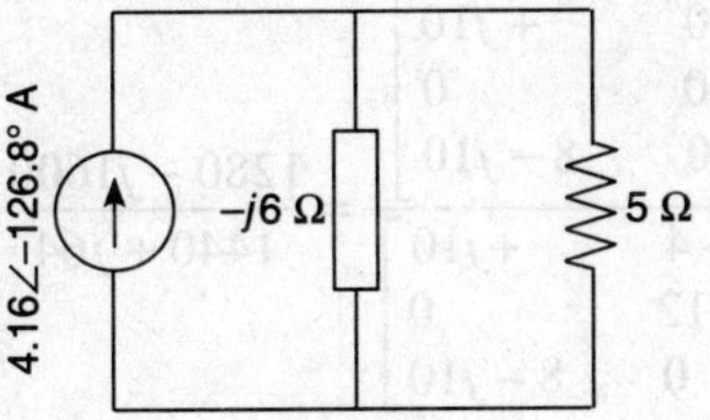

Fig. 6.102

EXAMPLE 6.24 Find the current through (2 + *j*5) Ω impedance in the network shown in Fig. 6.103 using Norton's theorem.

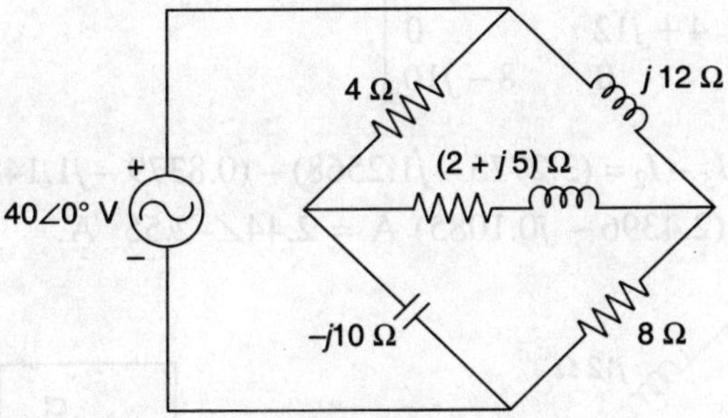

Fig. 6.103

Solution To find Norton's current, (2 + *j*5) Ω impedance is shorted and the directions of mesh currents are assumed as shown in Fig. 6.104.

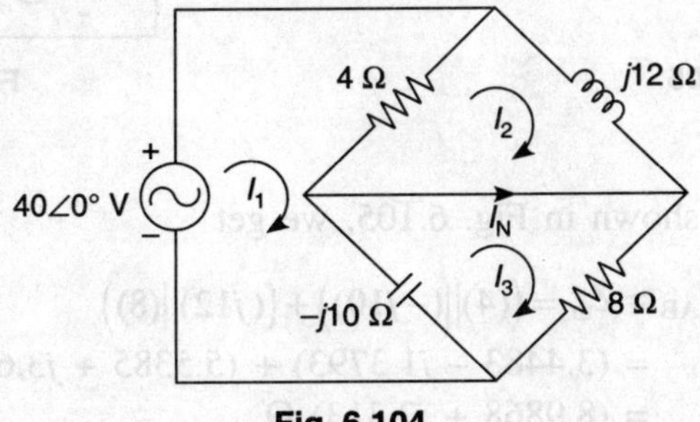

Fig. 6.104

To find I_N

With reference to Fig. 6.104, the mesh equations are written as:

$$\begin{bmatrix} 4-j10 & -4 & +j10 \\ -4 & 4+j12 & 0 \\ +j10 & 0 & 8-j10 \end{bmatrix} \begin{bmatrix} I_1 \\ I_2 \\ I_3 \end{bmatrix} = \begin{bmatrix} 40\angle 0^\circ \\ 0 \\ 0 \end{bmatrix}$$

I_2 and I_3 are calculated using Cramer's rule as follows:

$$I_2 = \frac{\begin{bmatrix} 4-j10 & 40\angle 0^\circ & +j10 \\ -4 & 0 & 0 \\ +j10 & 0 & 8-j10 \end{bmatrix}}{\begin{bmatrix} 4-j10 & -4 & +j10 \\ -4 & 4+j12 & 0 \\ +j10 & 0 & 8-j10 \end{bmatrix}} = \frac{1280-j1600}{1440+j64} = (0.8379-j1.1483)\text{ A}$$

$$I_3 = \frac{\begin{bmatrix} 4-j10 & -4 & 40\angle 0^\circ \\ -4 & 4+j12 & 0 \\ +j10 & 0 & 0 \end{bmatrix}}{\begin{bmatrix} 4-j10 & -4 & +j10 \\ -4 & 4+j12 & 0 \\ +j10 & 0 & 8-j10 \end{bmatrix}} = \frac{4800-j1600}{1440+j64} = (3.2775-j1.2568)\text{ A}$$

Therefore,

$$I_N = I_3 - I_2 = (3.2775 - j1.2568) - (0.8379 - j1.1483)\text{ A}$$
$$= (2.4396 - j0.1085)\text{ A} = 2.44\angle -2.55^\circ\text{ A}.$$

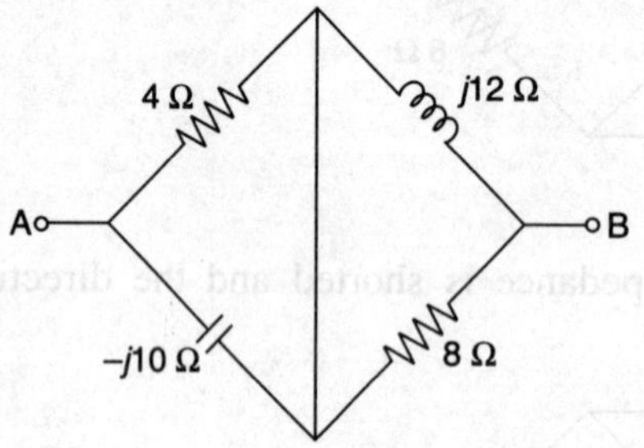

Fig. 6.105

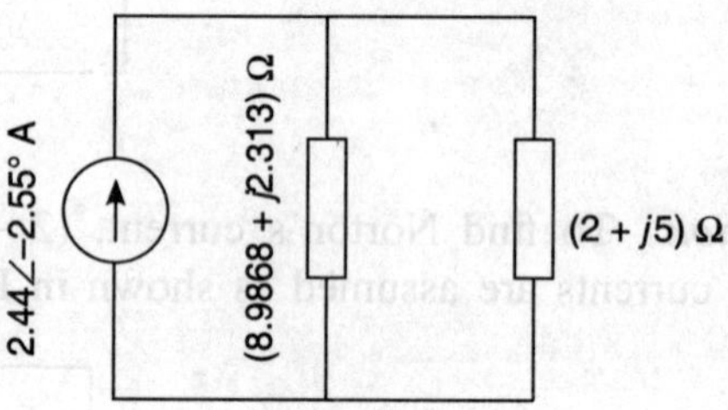

Fig. 6.106

To find Z_{th}

The circuit is redrawn as shown in Fig. 6.105, we get

$$Z_{AB} = Z_{th} = \big((4)\|(-j10)\big) + \big((j12)\|(8)\big)$$
$$= (3.4483 - j1.3793) + (5.5385 + j3.6923)$$
$$= (8.9868 + j2.313)\ \Omega$$
$$= 9.2797\angle 14.4^\circ\ \Omega$$

Norton's equivalent circuit

From Fig. 6.106, current through (2 + *j*5) Ω resistance

$$= 2.44\angle -2.55 \times \frac{(8.9868 + j2.313)}{(8.9868 + j2.313) + (2 + j5)}$$

$$= \mathbf{1.72\angle -21.8^\circ \ A}$$

6.5 RECIPROCITY THEOREM

Reciprocity theorem is stated as:

> *In a linear, lumped, bilateral, single source network, the ratio of excitation to response is constant when the position of excitation and response are interchanged.*

Reciprocity theorem is well understood by the following examples.

EXAMPLE 6.25 Verify the reciprocity theorem for the network shown in Fig. 6.107.

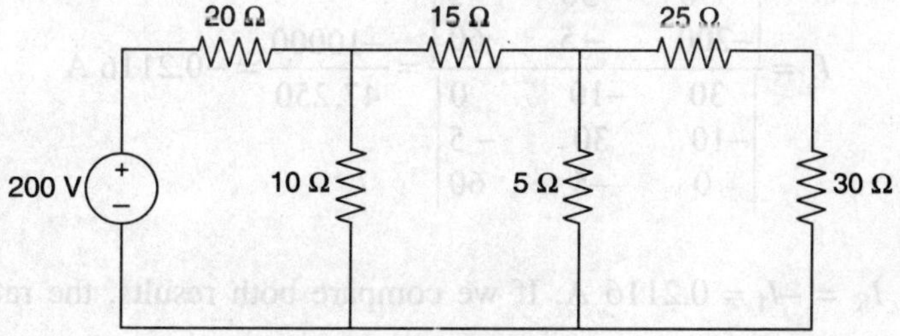

Fig. 6.107

Solution One of the responses, the current through 30 Ω (I_X), is taken for verification. Using mesh analysis, $I_X(I_3)$ is found for which the circuit shown in Fig. 6.108 is used.

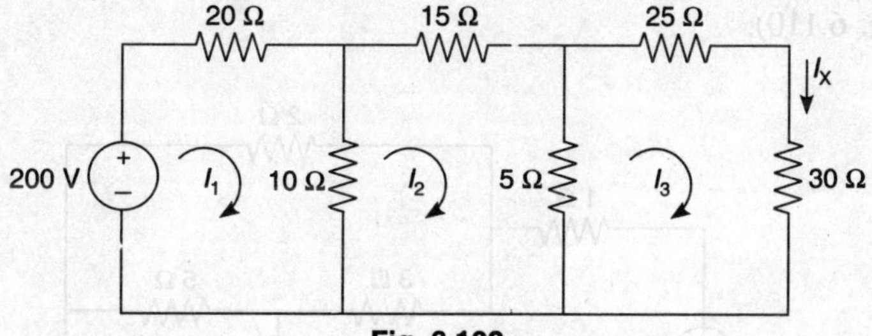

Fig. 6.108

$$I_3 = \frac{\begin{vmatrix} 30 & -10 & 200 \\ -10 & 30 & 0 \\ 0 & -5 & 0 \end{vmatrix}}{\begin{vmatrix} 30 & -10 & 0 \\ -10 & 30 & -5 \\ 0 & -5 & 60 \end{vmatrix}} = \frac{10{,}000}{47{,}250} = 0.2116 \text{ A}$$

To verify reciprocity theorem, the excitation 200 V source and the response I_X are interchanged as shown in Fig. 6.109. Now, $I_X = -I_1$ is found using mesh analysis.

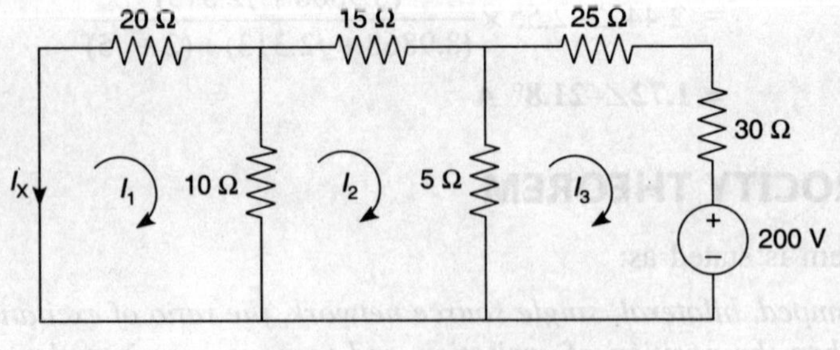

Fig. 6.109

$$I_1 = \frac{\begin{vmatrix} 0 & -10 & 0 \\ 0 & 30 & -5 \\ -200 & -5 & 60 \end{vmatrix}}{\begin{vmatrix} 30 & -10 & 0 \\ -10 & 30 & -5 \\ 0 & -5 & 60 \end{vmatrix}} = \frac{-10000}{47,250} = -0.2116 \text{ A}$$

The response, $I_X = -I_1 = 0.2116$ A. If we compare both results, the ratio of excitation to response $\left(\frac{200}{0.2116} = 945.18\right)$ is found same. Thus, reciprocity theorem is verified.

EXAMPLE 6.26 Prove reciprocity theorem by interchanging the position of 20 V source and current I (Fig. 6.110).

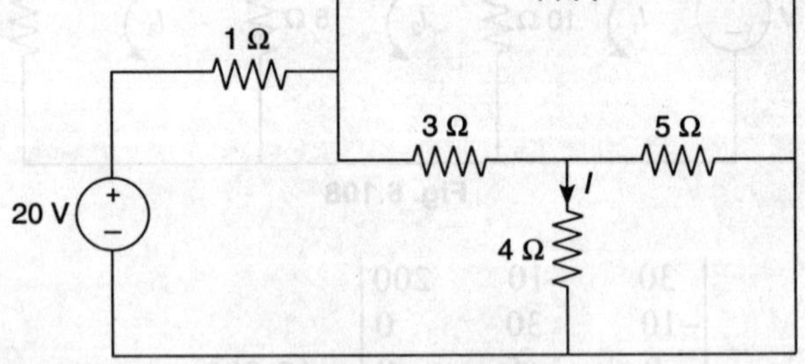

Fig. 6.110

Solution The response (current I through 4 Ω resistor shown in Fig. 6.111) is found using mesh analysis.

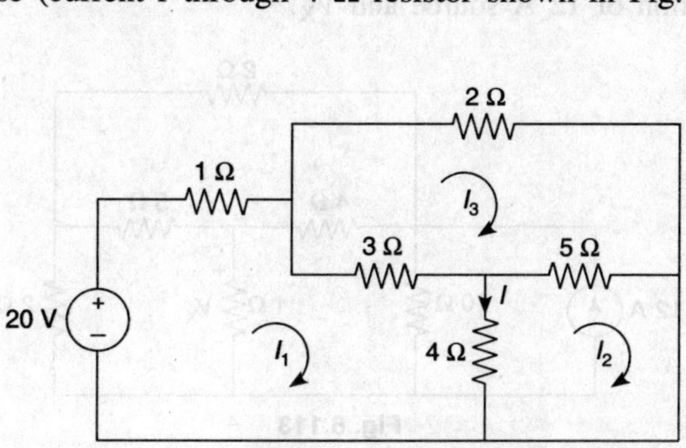

Fig. 6.111

By applying Cramer's rule,

$$I_1 = \frac{\begin{vmatrix} 20 & -4 & 3 \\ 0 & 9 & -5 \\ 0 & -5 & 10 \end{vmatrix}}{\begin{vmatrix} 8 & -4 & -3 \\ -4 & 9 & -5 \\ -3 & -5 & 10 \end{vmatrix}} = \frac{1300}{159} = 8.18\text{ A} \quad \text{and} \quad I_2 = \frac{\begin{vmatrix} 8 & 20 & -3 \\ -4 & 0 & -5 \\ -3 & 0 & 10 \end{vmatrix}}{\begin{vmatrix} 8 & -4 & -3 \\ -4 & 9 & -5 \\ -3 & -5 & 10 \end{vmatrix}} = \frac{1100}{159} = 6.92\text{ A}$$

Hence, the current through 4 Ω resistance,

$$I = I_1 - I_2 = 8.18 - 6.92 = 1.26\text{ A}$$

To verify reciprocity theorem, the excitation 20 V source and response I are interchanged as shown in Fig. 6.112. Now, $I = -I_1$ is found using mesh analysis.

$$I_1 = \frac{\begin{vmatrix} -20 & -4 & -3 \\ 20 & 9 & -5 \\ 0 & -5 & 10 \end{vmatrix}}{\begin{vmatrix} 8 & -4 & -3 \\ -4 & 9 & -5 \\ -3 & -5 & 10 \end{vmatrix}} = \frac{-200}{159} = -1.26\text{ A}$$

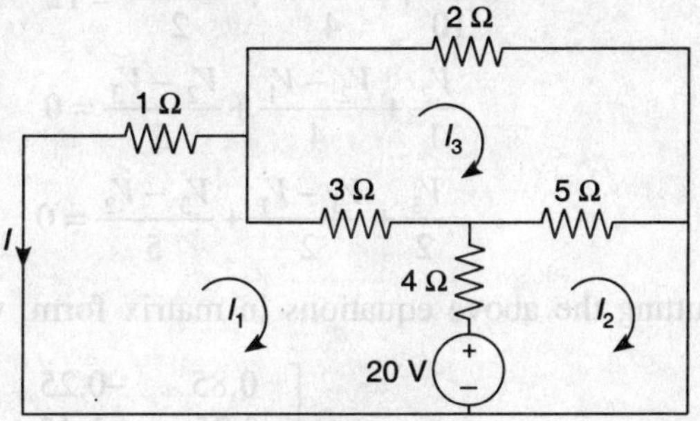

Fig. 6.112

The response, $I = -I_1 = 1.26$ A. If we compare both results, the ratio of excitation to response $\left(\frac{20}{1.26} = 15.873\right)$ is found same. Thus, reciprocity theorem is verified.

EXAMPLE 6.27 Prove reciprocity theorem in the network shown in Fig. 6.113 by interchanging the position of 12 A source and V_X.

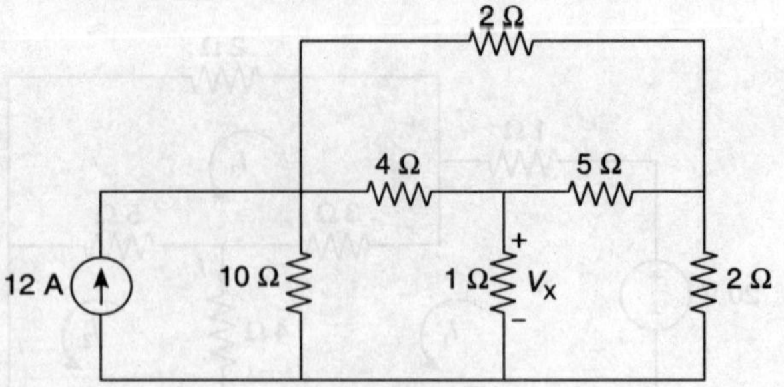

Fig. 6.113

Solution Using nodal analysis, response V_X is calculated. The nodal equations are formed as follows:

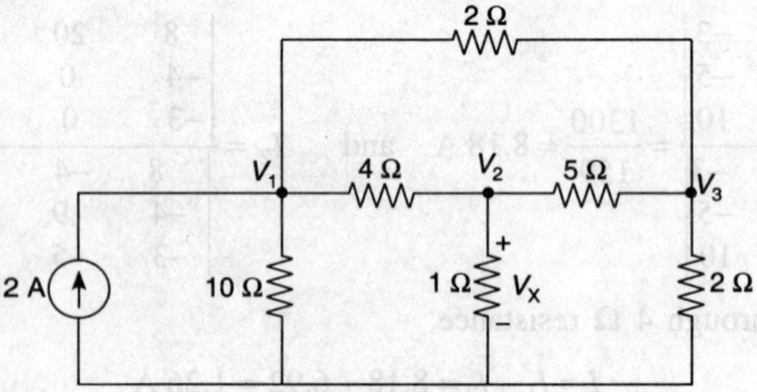

Fig. 6.114

$$\frac{V_1}{10} + \frac{V_1 - V_2}{4} + \frac{V_1 - V_3}{2} = 12 \quad \text{or} \quad 0.85V_1 - 0.25V_2 - 0.5V_3 = 12 \qquad \text{(i)}$$

$$\frac{V_2}{1} + \frac{V_2 - V_1}{4} + \frac{V_2 - V_3}{5} = 0 \quad \text{or} \quad -0.25V_1 + 1.45V_2 - 0.2V_3 = 0 \qquad \text{(ii)}$$

$$\frac{V_3}{2} + \frac{V_3 - V_1}{2} + \frac{V_3 - V_2}{5} = 0 \quad \text{or} \quad -0.5V_1 - 0.2V_2 + 1.2V_3 = 0 \qquad \text{(iii)}$$

Putting the above equations in matrix form, we get

$$\begin{bmatrix} 0.85 & -0.25 & -0.5 \\ -0.25 & 1.45 & -0.2 \\ -0.5 & -0.2 & 1.2 \end{bmatrix} \begin{bmatrix} V_1 \\ V_2 \\ V_3 \end{bmatrix} = \begin{bmatrix} 12 \\ 0 \\ 0 \end{bmatrix}$$

$$V_X = V_2 = \frac{\begin{bmatrix} 0.85 & 12 & -0.5 \\ -0.25 & 0 & -0.2 \\ -0.5 & 0 & 1.2 \end{bmatrix}}{\begin{bmatrix} 0.85 & -0.25 & -0.5 \\ -0.25 & 1.45 & -0.2 \\ -0.5 & -0.2 & 1.2 \end{bmatrix}} = \frac{4.8}{0.9575} = 5.0131 \text{ V}$$

To verify reciprocity theorem, the excitation 12 A source and response V_X are interchanged as shown in Fig. 6.115. Now $V_X = V_1$ is found using nodal analysis.

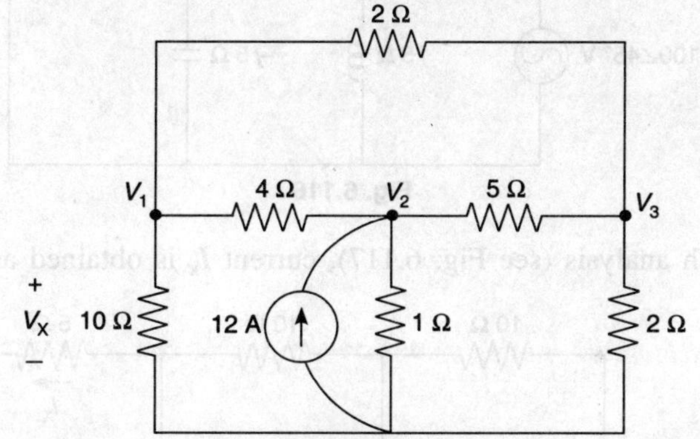

Fig. 6.115

The nodal equations are formed as follows:

$$\frac{V_1}{10}+\frac{V_1-V_2}{4}+\frac{V_1-V_3}{2}=0 \quad \text{or} \quad 0.85V_1-0.25V_2-0.5V_3=0 \tag{iv}$$

$$\frac{V_2}{1}+\frac{V_2-V_1}{4}+\frac{V_2-V_3}{5}=12 \quad \text{or} \quad -0.25V_1+1.45V_2-0.2V_3=12 \tag{v}$$

$$\frac{V_3}{2}+\frac{V_3-V_1}{2}+\frac{V_3-V_2}{5}=0 \quad \text{or} \quad -0.5V_1-0.2V_2+1.2V_3=0 \tag{vi}$$

Putting the above in matrix form, we get

$$\begin{bmatrix} 0.85 & -0.25 & -0.5 \\ -0.25 & 1.45 & -0.2 \\ -0.5 & -0.2 & 1.2 \end{bmatrix}\begin{bmatrix} V_1 \\ V_2 \\ V_3 \end{bmatrix} = \begin{bmatrix} 0 \\ 12 \\ 0 \end{bmatrix}$$

$$V_X = V_1 = \frac{\begin{vmatrix} 0 & -0.25 & -0.5 \\ 12 & 1.45 & -0.2 \\ 0 & -0.2 & 1.2 \end{vmatrix}}{\begin{vmatrix} 0.85 & -0.25 & -0.5 \\ -0.25 & 1.45 & -0.2 \\ -0.5 & -0.2 & 1.2 \end{vmatrix}} = \frac{4.8}{0.9575} = 5.0131 \text{ V}$$

If we compare both results, the ratio of excitation to response is found same. Thus, reciprocity theorem is verified.

EXAMPLE 6.28 In the network shown in Fig. 6.116, find I_X and verify reciprocity theorem.

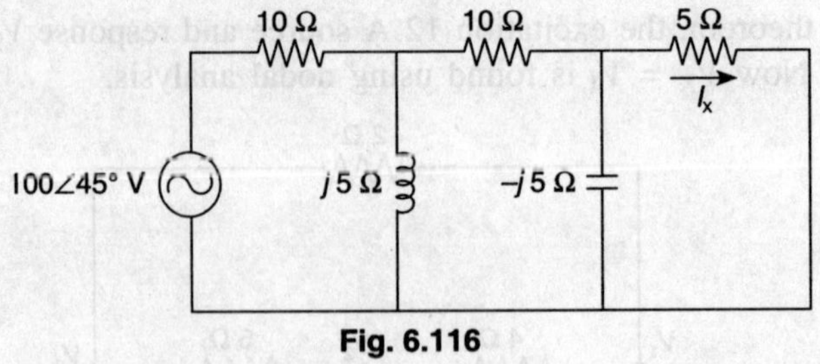

Fig. 6.116

Solution Using mesh analysis (see Fig. 6.117), current I_x is obtained as:

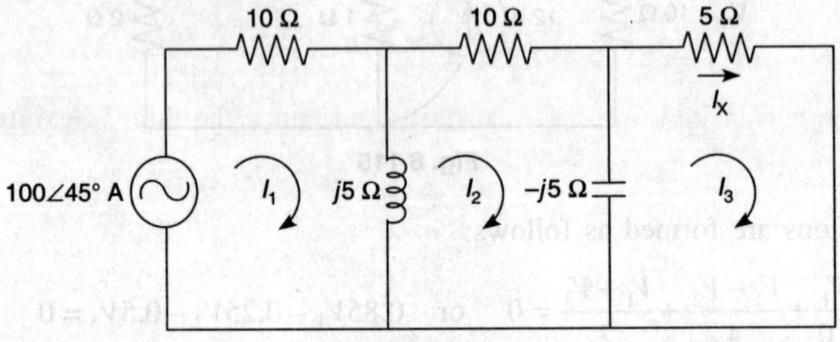

Fig. 6.117

$$I_X = I_3 = \frac{\begin{bmatrix} 10+j5 & -j5 & 100\angle 45^\circ \\ -j5 & 10 & 0 \\ 0 & j5 & 0 \end{bmatrix}}{\begin{bmatrix} 10+j5 & -j5 & 0 \\ -j5 & 10 & j5 \\ 0 & j5 & 5-j5 \end{bmatrix}} = \frac{100\angle 45^\circ(-j5)(j5)}{(10+j5)(50-j50+25)+j5(-j25-25)}$$

$$= \frac{1767.77 + j1767.77}{1125 - j250}$$

$$= 1.1646 + j1.83 = 2.169\angle 57.53^\circ \text{ A}$$

To verify reciprocity theorem, the excitation 100∠45° V source and the response I_X are interchanged as shown in Fig. 6.118.

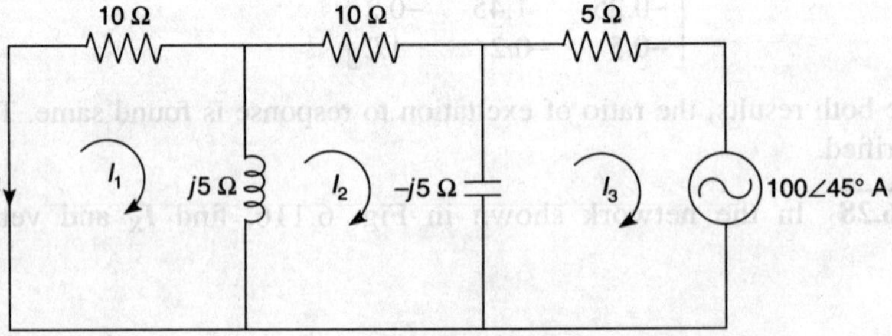

Fig. 6.118

Again using mesh analysis (see Fig. 6.118), current I_x is obtained as:

$$I_X = -I_1 = \frac{\begin{bmatrix} 0 & -j5 & 0 \\ 0 & 10 & j5 \\ 100\angle 45^\circ & j5 & 5-j5 \end{bmatrix}}{\begin{bmatrix} 10+j5 & -j5 & 0 \\ -j5 & 10 & j5 \\ 0 & j5 & 5-j5 \end{bmatrix}} = \frac{-(j5)(100\angle 45^\circ)(j5)}{(10+j5)(50-j50+25)+j5(-j25-25)}$$

$$= \frac{1767.77 + j1767.77}{1125 - j250}$$

$$= 1.1646 + j1.83 = 2.169\angle 57.53^\circ \text{ A}$$

In both cases, the current through 5 Ω resistance and hence the ratio of excitation to response is found same. Thus, reciprocity theorem is verified.

EXAMPLE 6.29 In the network shown in Fig. 6.119, compute V_X and verify reciprocity theorem.

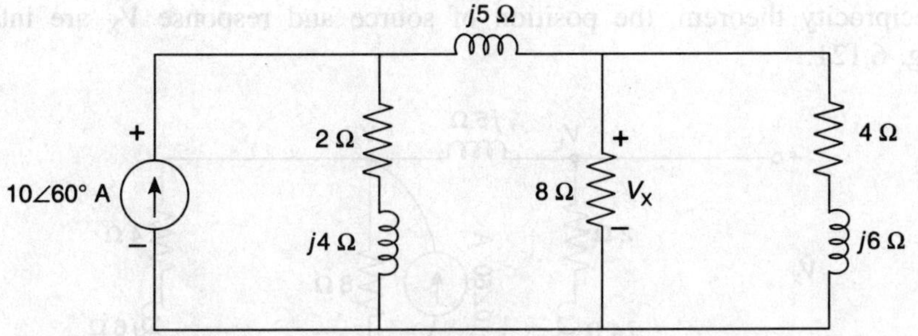

Fig. 6.119

Solution Referring to Fig. 6.120, V_X or V_2 is found using nodal analysis.

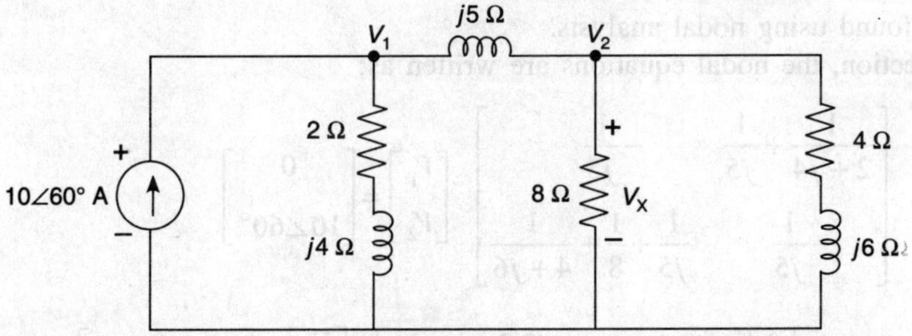

Fig. 6.120

By inspection, the nodal equations are written as:

$$\begin{bmatrix} \dfrac{1}{2+j4}+\dfrac{1}{j5} & -\dfrac{1}{j5} \\ -\dfrac{1}{j5} & \dfrac{1}{j5}+\dfrac{1}{8}+\dfrac{1}{4+j6} \end{bmatrix} \begin{bmatrix} V_1 \\ V_2 \end{bmatrix} = \begin{bmatrix} 10\angle 60^\circ \\ 0 \end{bmatrix}$$

$$= \begin{bmatrix} 0.1-j0.4 & j0.2 \\ j0.2 & 0.2019-j0.3154 \end{bmatrix} \begin{bmatrix} V_1 \\ V_2 \end{bmatrix} = \begin{bmatrix} 5+j8.6603 \\ 0 \end{bmatrix}$$

$$V_2 = \frac{\begin{bmatrix} 0.1-j0.4 & 5+j8.6603 \\ j0.2 & 0 \end{bmatrix}}{\begin{bmatrix} 0.1-j0.4 & j0.2 \\ j0.2 & 0.2019-j0.3154 \end{bmatrix}} = \frac{1.7321-j1}{-0.066-j0.1123}$$

$$= -0.1189 + j15.3539 = \mathbf{15.35\angle 90.44^\circ\ V}$$

To verify reciprocity theorem, the position of source and response V_X are interchanged as shown in Fig. 6.121.

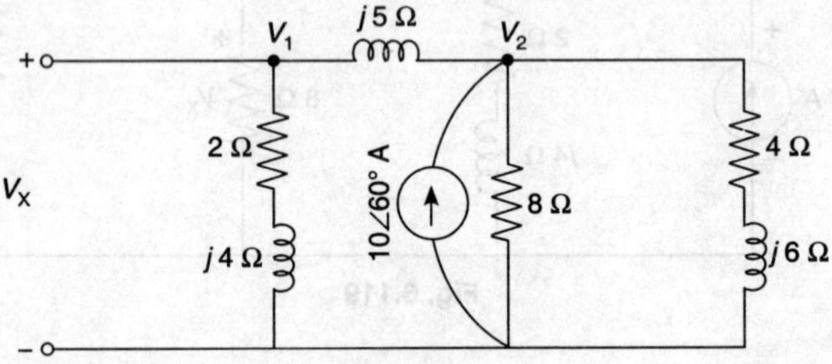

Fig. 6.121

V_X or V_1 is found using nodal analysis.

By inspection, the nodal equations are written as:

$$\begin{bmatrix} \dfrac{1}{2+j4}+\dfrac{1}{j5} & -\dfrac{1}{j5} \\ -\dfrac{1}{j5} & \dfrac{1}{j5}+\dfrac{1}{8}+\dfrac{1}{4+j6} \end{bmatrix} \begin{bmatrix} V_1 \\ V_2 \end{bmatrix} = \begin{bmatrix} 0 \\ 10\angle 60^\circ \end{bmatrix}$$

$$= \begin{bmatrix} 0.1-j0.4 & j0.2 \\ j0.2 & 0.2019-j0.3154 \end{bmatrix} \begin{bmatrix} V_1 \\ V_2 \end{bmatrix} = \begin{bmatrix} 0 \\ 5+j8.6603 \end{bmatrix}$$

$$V_1 = \frac{\begin{bmatrix} 0 & j0.2 \\ 5 + j8.6603 & 0.2019 - j0.3154 \end{bmatrix}}{\begin{bmatrix} 0.1 - j0.4 & j0.2 \\ j0.2 & 0.2019 - j0.3154 \end{bmatrix}} = \frac{0 - (-1.7321 + j1)}{-0.066 - j0.1123}$$

$$= -0.1189 + j15.3539 = \mathbf{15.35\angle 90.44^\circ \ V}$$

It is observed that, the response remains same in both cases and hence the ratio of excitation to response is found same. Thus, reciprocity theorem is verified.

6.6 MAXIMUM POWER TRANSFER THEOREM

Maximum power transfer theorem is used to find the maximum power that is transferred from a source to load. It is stated as:

Maximum power is transferred from source to load, when the load resistance is equal to the source resistance.

Case (i) DC source with internal resistance connected to a variable resistive load

Let us consider a DC source of V volts with its internal resistance, R_s ohms. A load resistance R_L is connected to it as shown in Fig. 6.122.

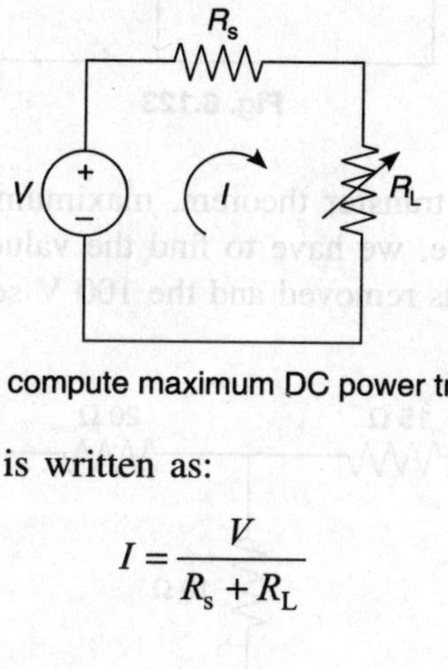

Fig. 6.122 Circuit to compute maximum DC power transferred to R load.

With Ohm's law, the current is written as:

$$I = \frac{V}{R_s + R_L} \tag{i}$$

Then, the power delivered to load,

$$P = I^2 R_L \tag{ii}$$

Using Eq. (i),

$$P = \left(\frac{V}{R_s + R_L}\right)^2 R_L = \frac{V^2 R_L}{(R_s + R_L)^2}$$

Maximum power is obtained by differentiating P with respect to R_L and equating it to zero, i.e.,

$$\frac{dP}{dR_L} = \frac{V^2(R_s + R_L)^2 - 2V^2 R_L(R_s + R_L)}{(R_s + R_L)^4} = 0$$

$$= V^2 (R_s + R_L)^2 - 2V^2R_L (R_s + R_L) = 0$$

or

$$V^2 (R_s + R_L)^2 = 2V^2R_L (R_s + R_L)$$

$$R_s + R_L = 2R_L$$

so,

$$R_s = R_L \qquad \text{(iii)}$$

Equation (iii) states that the maximum power is transferred from source to load when load resistance is equal to source resistance.

EXAMPLE 6.30 Find the value of adjustable resistor R in the circuit shown in Fig. 6.123, so that maximum power is transferred to it.

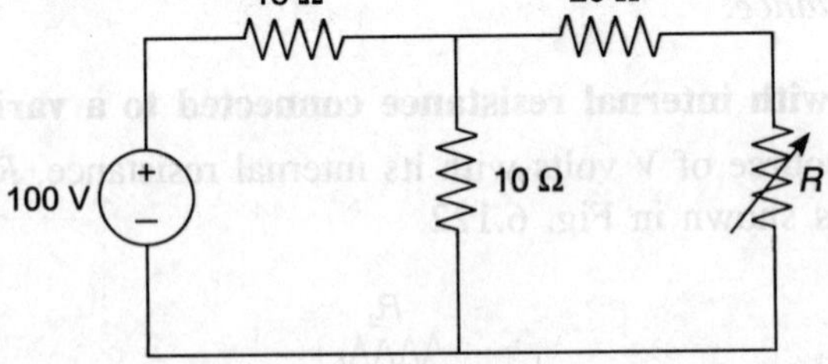

Fig. 6.123

Solution By maximum power transfer theorem, maximum power is transferred to R, if the value of R is equal to R_{th}. Hence, we have to find the value of Thevenin's resistance R_{th}. To find R_{th}, adjustable resistance R is removed and the 100 V source is replaced by a short-circuit as shown in Fig. 6.124.

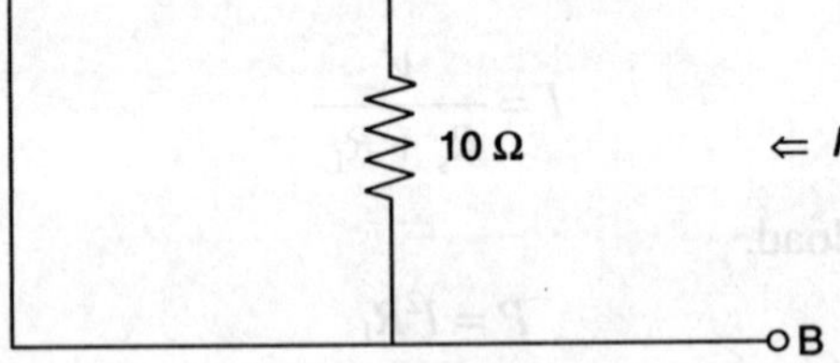

Fig. 6.124

$$R_{th} = R_{AB} = 20 + \frac{10 \times 15}{10 + 15} = 26\ \Omega$$

Hence, the value of R for maximum power transfer is **26 Ω**.

EXAMPLE 6.31 Determine the maximum power delivered to R (Fig. 6.125).

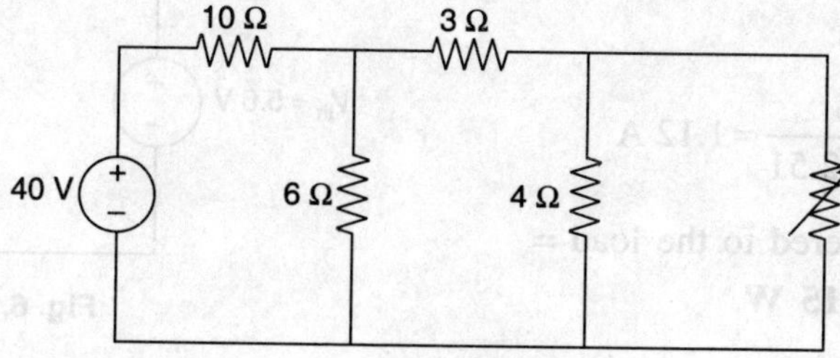

Fig. 6.125

Solution To determine the maximum power delivered to R, let us find the Thevenin's equivalent circuit across AB as shown in Fig. 6.126. By mesh analysis, current through 4 Ω is determined as:

$$I_{4\Omega} = \frac{\begin{bmatrix} 16 & 40 \\ -6 & 0 \end{bmatrix}}{\begin{bmatrix} 16 & -6 \\ -6 & 13 \end{bmatrix}} = \frac{240}{172} = 1.4 \text{ A}$$

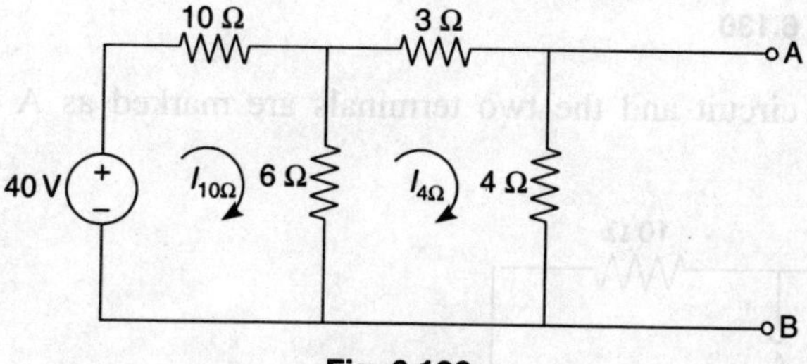

Fig. 6.126

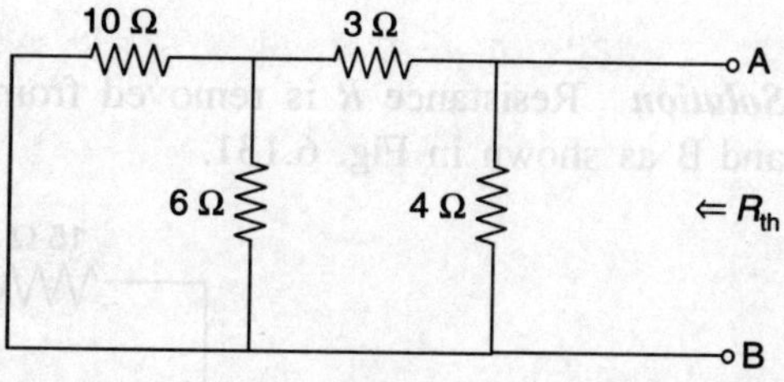

Fig. 6.127

$$V_{th} = V_{AB} = \text{voltage across 4 }\Omega\text{ resistance}$$
$$= 4 \times 1.4 = 5.6 \text{ V}$$

From Fig. 6.127,

$$R_{th} = \left[(10 \| 6) + 3 \right] \| 4 = 2.51\ \Omega$$

Thevenin's equivalent circuit is shown in Fig. 6.128.

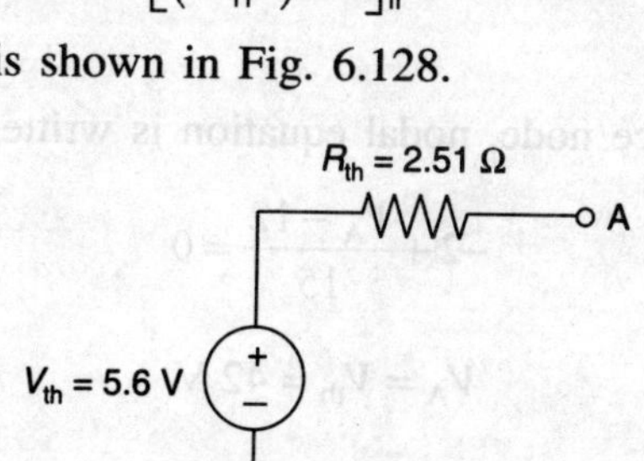

Fig. 6.128

Therefore, the value of R for maximum power transfer is 2.51 Ω and the current through R (Fig. 6.129) is found as:

$$I_L = \frac{5.6}{2.51+2.51} = 1.12 \text{ A}$$

Hence, the power delivered to the load = $I_L^2 R = 1.12^2 \times 2.51 = \mathbf{3.15\ W}$.

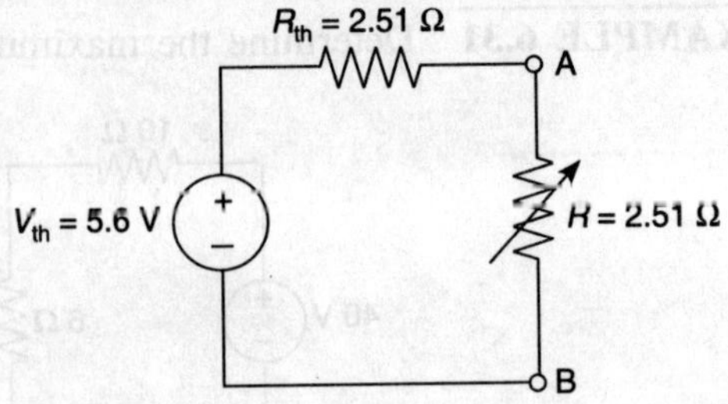

Fig. 6.129

EXAMPLE 6.32 In the circuit shown in Fig. 6.130, determine the maximum power delivered to R.

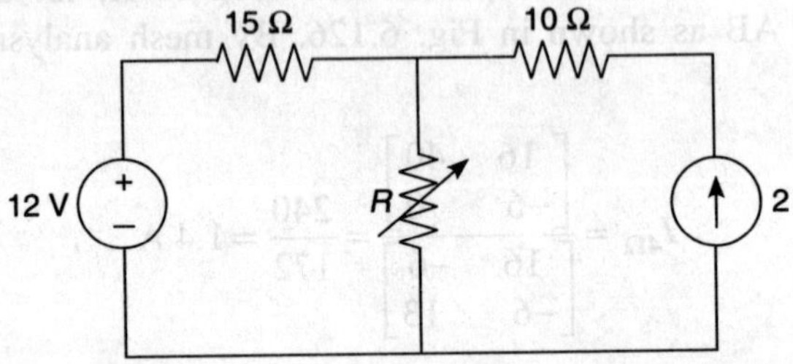

Fig. 6.130

Solution Resistance R is removed from the circuit and the two terminals are marked as A and B as shown in Fig. 6.131.

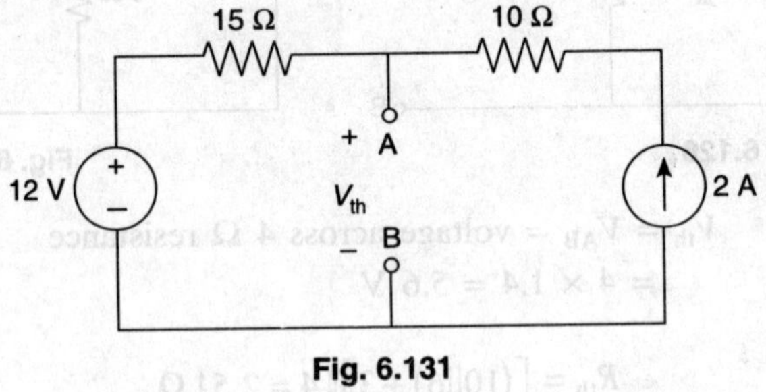

Fig. 6.131

To find V_{th}

Considering node B as reference node, nodal equation is written at node A as:

$$-2 + \frac{V_A - 12}{15} = 0$$

Therefore, $$V_A = V_{th} = 42 \text{ V}$$

To find R_{th}

The voltage source is replaced by a short-circuit and, the current source is replaced by an open-circuit as shown in Fig. 6.132. As seen from Fig. 6.131, $R_{AB} = R_{th} = 15\ \Omega$.

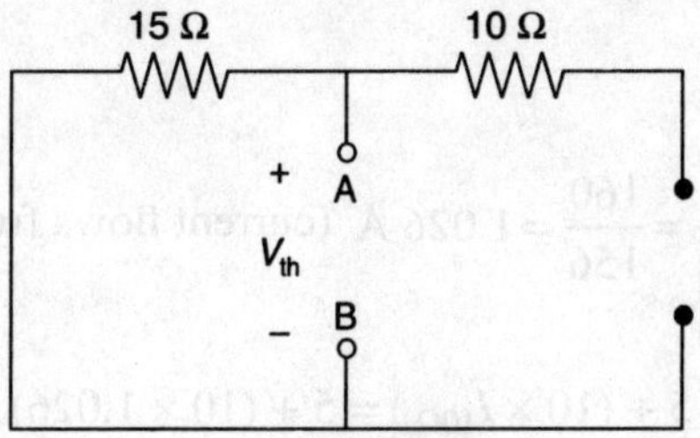

Fig. 6.132

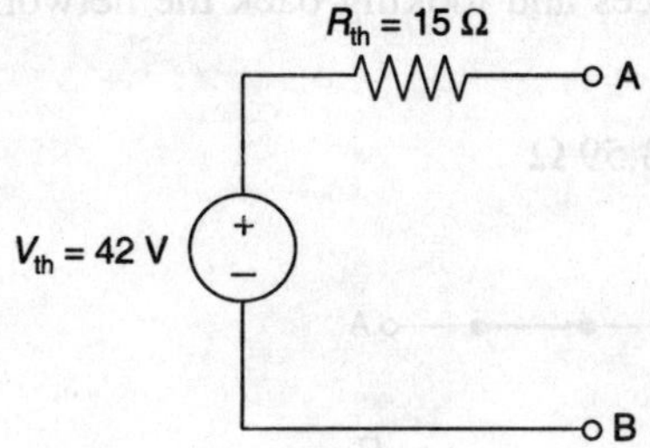

Fig. 6.133

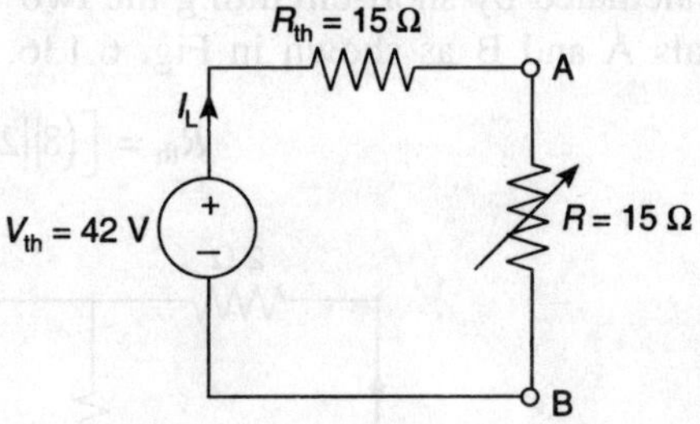

Fig. 6.134

Thevenin's equivalent circuit is shown in Fig. 6.133. The value of R for maximum power transfer is 15 Ω and the current through R (Fig. 6.134), I_L is found as:

$$I_L = \frac{42}{15+15} = 1.4 \text{ A}$$

Hence, the power delivered to the load = $I_L^2 R$ = $1.4^2 \times 15$ = **29.4 W**.

EXAMPLE 6.33 Determine the value of load resistance that may be connected across terminals A and B, in the circuit shown in Fig. 6.135, so that maximum power is transferred to it. Also find the maximum power.

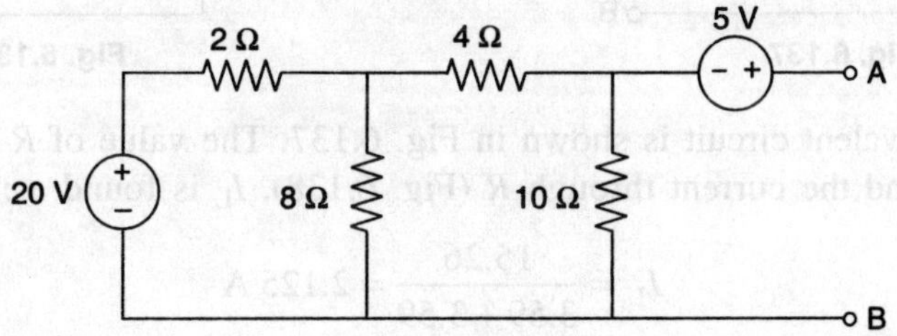

Fig. 6.135

Solution

To find V_{th}

To find Thevenin's voltage, V_{th} across terminals A and B, voltage across 10 Ω resistor is determined. First, the current through 10 Ω resistor is found using mesh analysis as follows.

Current through 10 Ω resistor,

$$I_{10\Omega} = \frac{\begin{bmatrix} 10 & 20 \\ -8 & 0 \end{bmatrix}}{\begin{bmatrix} 10 & -8 \\ -8 & 22 \end{bmatrix}} = \frac{160}{156} = 1.026 \text{ A (current flows from top to bottom)}$$

$$\therefore \quad V_{th} = V_{AB} = 5 + (10 \times I_{10\Omega}) = 5 + (10 \times 1.026) = 15.26 \text{ V}$$

To find R_{th}

R_{th} is calculated by short-circuiting the two voltage sources and looking back the network from terminals A and B as shown in Fig. 6.136.

$$R_{th} = \left[(8\|2) + 4\right]\|10 = 3.59\ \Omega$$

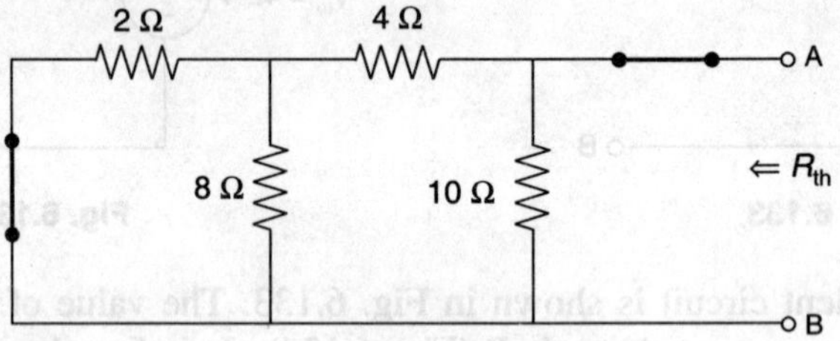

Fig. 6.136

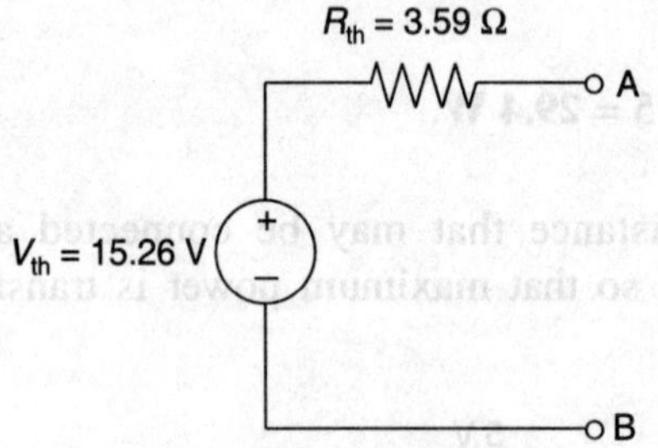

Fig. 6.137

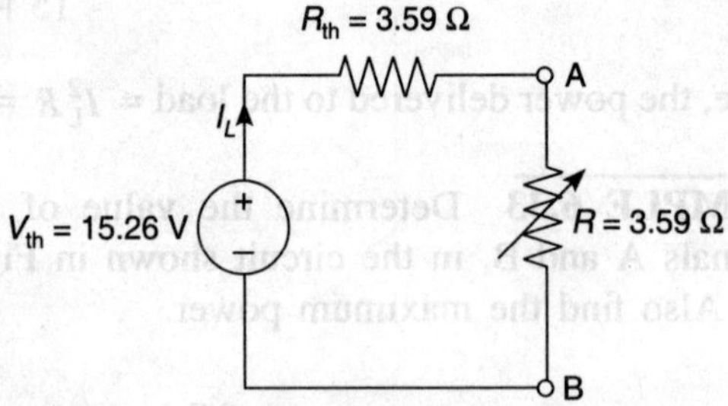

Fig. 6.138

Thevenin's equivalent circuit is shown in Fig. 6.137. The value of R for maximum power transfer is **3.59 Ω** and the current through R (Fig. 6.138), I_L is found as:

$$I_L = \frac{15.26}{3.59 + 3.59} = 2.125 \text{ A}$$

Hence, the power delivered to the load $= I_L^2 R = 2.125^2 \times 3.59 =$ **16.21 W**.

EXAMPLE 6.34 In the circuit shown in Fig. 6.139, determine the value of maximum power transferred to R.

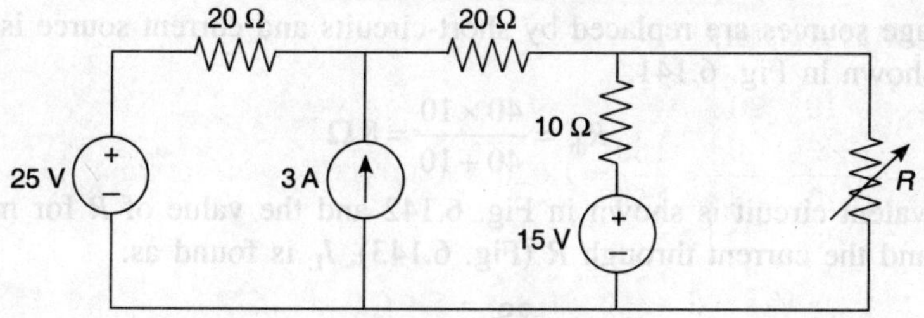

Fig. 6.139

Solution Resistance R is removed from the circuit and the two terminals are marked as A and B as shown in Fig. 6.140.

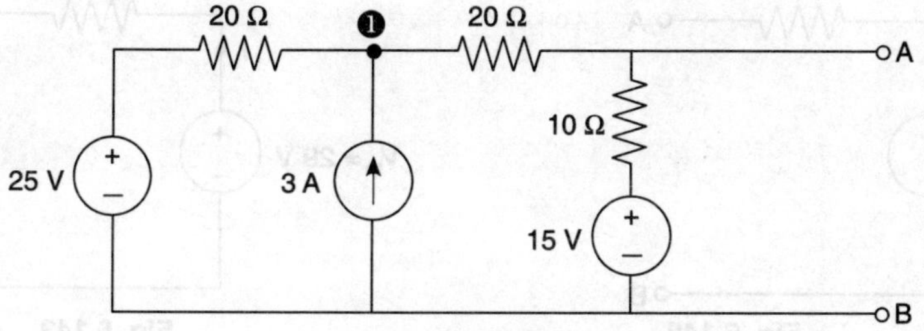

Fig. 6.140

To find V_{th}

To find Thevenin's voltage, V_{th} across terminals A and B, voltage across 10 Ω resistor is determined. Using nodal analysis, the voltage at node 1 is written as:

$$\frac{V_1 - 25}{20} + \frac{V_1 - 15}{30} - 3 = 0$$

$$5V_1 - 285 = 0$$

$$V_1 = 57 \text{ V}$$

Current through 10 Ω resistor $= \dfrac{57 - 15}{30} = 1.4$ A

Then the voltage across 10 Ω resistor = 1.4 × 10 = 14 V.

So, the voltage across terminals A and B (V_{th}) = 15 + 14 = 29 V.

To find R_{th}

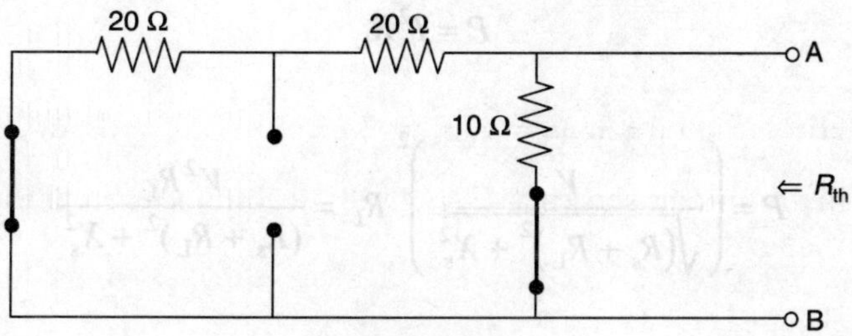

Fig. 6.141

To find R_{th}, voltage sources are replaced by short-circuits and current source is replaced by an open-circuit as shown in Fig. 6.141.

$$R_{th} = \frac{40 \times 10}{40 + 10} = 8\ \Omega$$

Thevenin's equivalent circuit is shown in Fig. 6.142 and the value of R for maximum power transfer is 8 Ω and the current through R (Fig. 6.143), I_L is found as:

$$I_L = \frac{29}{8+8} = 1.8125\ \text{A}$$

Hence, the power delivered to the load = $I_L^2 R = 1.8125^2 \times 8 = \mathbf{26.28\ W}$.

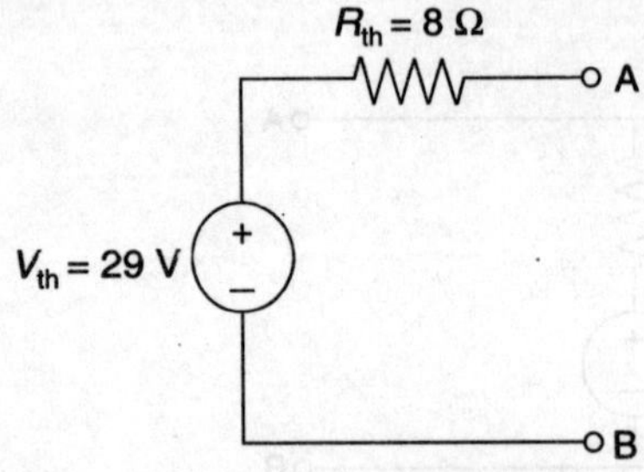

Fig. 6.142

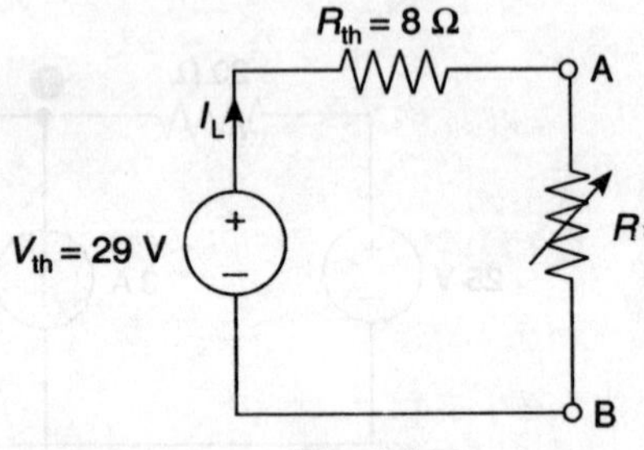

Fig. 6.143

Case (ii) AC source with internal impedance connected to a variable resistive load

In this case, an AC source of V volts with internal impedance of Z_s ohms is connected with a variable resistive load as shown in Fig. 6.144.

The equation for current is written as:

$$I = \frac{V}{(R_s + jX_s) + R_L}$$

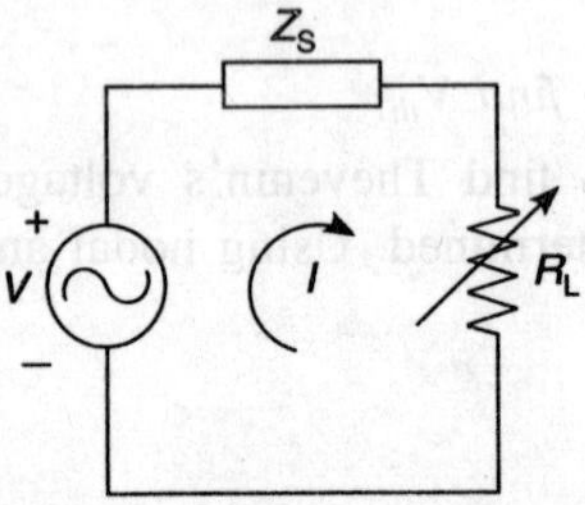

Fig. 6.144 Circuit to compute maximum AC power transferred to variable R load.

Magnitude of current,

$$I = \frac{V}{\sqrt{(R_s + R_L)^2 + X_s^2}} \quad \text{(i)}$$

Then, the power delivered to the load,

$$P = I^2 R_L \quad \text{(ii)}$$

Using Eq. (i),

$$P = \left(\frac{V}{\sqrt{(R_s + R_L)^2 + X_s^2}} \right)^2 R_L = \frac{V^2 R_L}{(R_s + R_L)^2 + X_s^2}$$

Maximum power is obtained by differentiating P with respect to R_L and equating it to zero,

$$\frac{dP}{dR_L} = \frac{V^2\left[(R_s + R_L)^2 + X_s^2\right] - V^2 R_L\left[2(R_s + R_L)\right]}{\left[(R_s + R_L)^2 + X_s^2\right]^2} = 0$$

$$(R_s + R_L)^2 + X_s^2 = 2R_sR_L + 2R_L^2$$

$$R_s^2 + R_L^2 + 2R_sR_L + X_s^2 = 2R_sR_L + 2R_L^2$$

$$R_s^2 + X_s^2 = R_L^2$$

Hence,

$$R_L = \sqrt{R_s^2 + X_s^2} \tag{iii}$$

From Eq. (iii), for a variable resistive load, the maximum power is transferred only when its value is equal to the magnitude of source impedance.

EXAMPLE 6.35 In the circuit shown in Fig. 6.145, determine the value of R, so that maximum power is transferred to it.

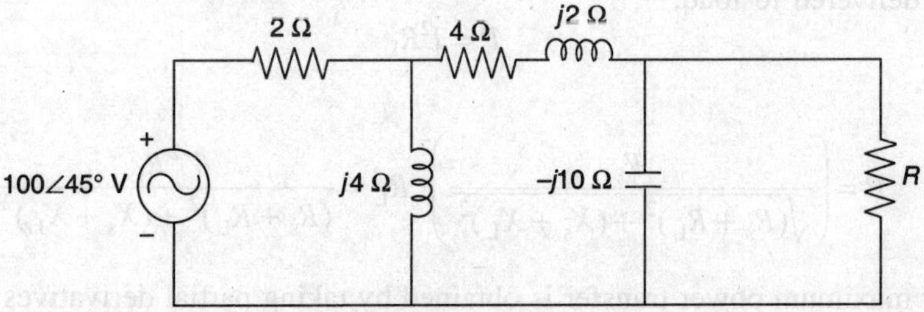

Fig. 6.145

Solution If the load contains resistance only, the maximum power is transferred to it, when its value is equal to magnitude of Z_s. The step-by-step procedure of finding Z_s is shown in Fig. 6.146.

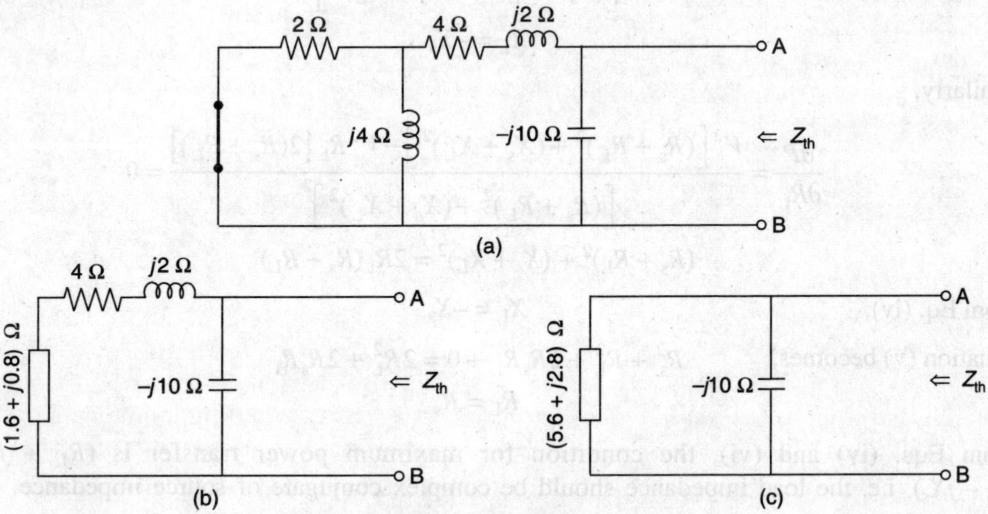

Fig. 6.146 Step-by-step procedure of computing Z_s.

$$Z_{th} = \frac{(5.6 + j2.8) \times (-j10)}{(5.6 + j2.8) + (-j10)} = (6.7308 - j1.3462) = 6.8641\angle{-11.3^\circ}\ \Omega$$

By maximum power transfer theorem, the value of R = **6.8641 Ω.**

Case (iii) AC source with internal impedance connected to a variable reactive load

In this case, an AC source of V volts with internal impedance of Z_s ohms is connected with a variable reactive load as shown in Fig. 6.147.

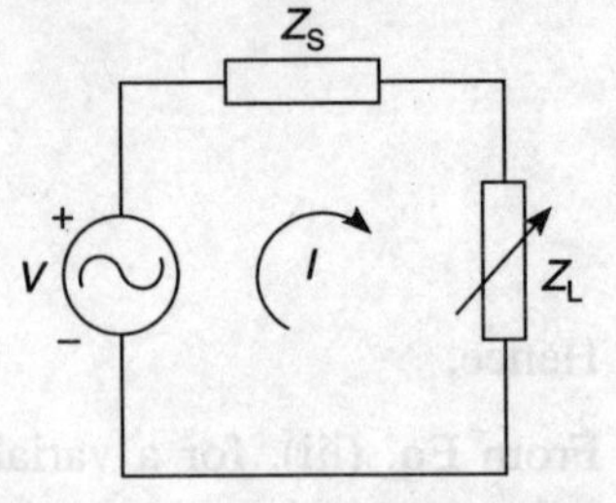

Fig. 6.147 Circuit to compute maximum AC power transferred to Z load.

The equation for current is written as:

$$I = \frac{V}{(R_s + jX_s) + (R_L + jX_L)}$$

Magnitude of current,

$$I = \frac{V}{\sqrt{(R_s + R_L)^2 + (X_s + X_L)^2}} \qquad \text{(i)}$$

Then, power delivered to load,

$$P = I^2 R_L \qquad \text{(ii)}$$

Using Eq. (i),

$$P = \left(\frac{V}{\sqrt{(R_s + R_L)^2 + (X_s + X_L)^2}} \right)^2 R_L = \frac{V^2 R_L}{(R_s + R_L)^2 + (X_s + X_L)^2} \qquad \text{(iii)}$$

Condition for maximum power transfer is obtained by taking partial derivatives $\frac{\partial P}{\partial X_L}$ and $\frac{\partial P}{\partial R_L}$ and equating them to zero.

$$\frac{\partial P}{\partial X_L} = \frac{0 \times \left[(R_s + R_L)^2 + (X_s + X_L)^2 \right] - V^2 R_L \left[0 + 2(X_s + X_L) \right]}{\left[(R_s + R_L)^2 + (X_s + X_L)^2 \right]^2} = 0$$

or

$$X_L = -X_s \qquad \text{(iv)}$$

Similarly,

$$\frac{\partial P}{\partial R_L} = \frac{V^2 \left[(R_s + R_L)^2 + (X_s + X_L)^2 \right] - V^2 R_L \left[2(R_s + R_L) \right]}{\left[(R_s + R_L)^2 + (X_s + X_L)^2 \right]^2} = 0$$

or

$$(R_s + R_L)^2 + (X_s + X_L)^2 = 2R_L(R_s + R_L) \qquad \text{(v)}$$

From Eq. (iv),

$$X_L = -X_s$$

Equation (v) becomes,

$$R_s^2 + R_L^2 + 2R_s R_L + 0 = 2R_L^2 + 2R_s R_L$$

or

$$R_L = R_s \qquad \text{(vi)}$$

From Eqs. (iv) and (vi), the condition for maximum power transfer is $(R_L + jX_L) = (R_s - jX_s)$, i.e. the load impedance should be complex conjugate of source impedance.

EXAMPLE 6.36 In the circuit shown in Fig. 6.148, determine the value of Z so that maximum power is transferred to it. Also determine the power.

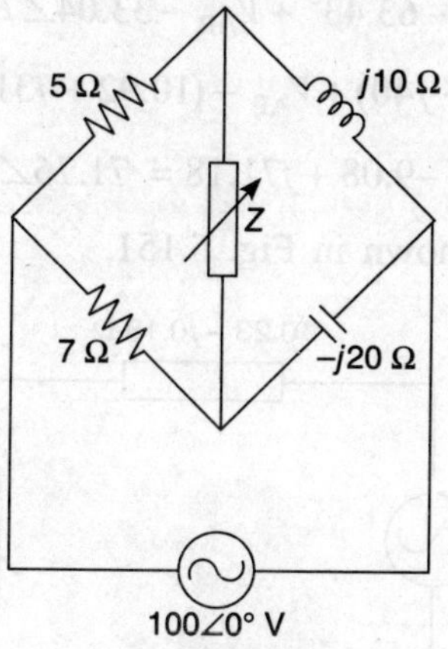

Fig. 6.148

Solution

To find Z

From Fig. 6.149,

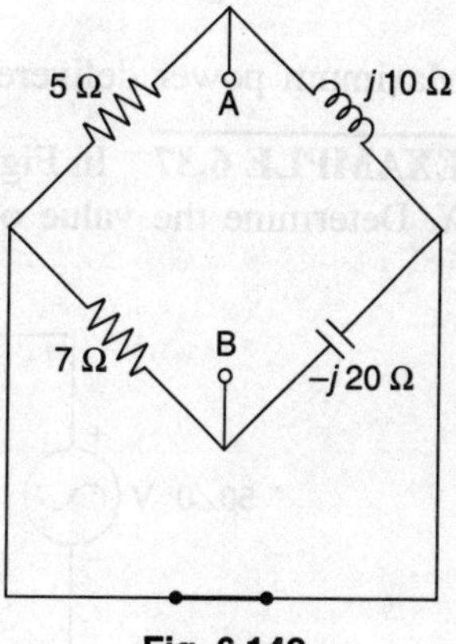

Fig. 6.149

$$Z_{AB} = \frac{5(j10)}{5+j10} + \frac{7(-j20)}{7-j20}$$

$$= \frac{50\angle 90^\circ}{11.18\angle 63.43^\circ} + \frac{140\angle -90^\circ}{21.19\angle -70.7^\circ}$$

$$= 10.23 - j0.18$$

To get maximum power, the load should be complex conjugate of the source impedance, i.e.

$$Z = 10.23 + j0.18$$

To find V_{th}

The circuit shown in Fig. 6.150 is used to calculate V_{th}.

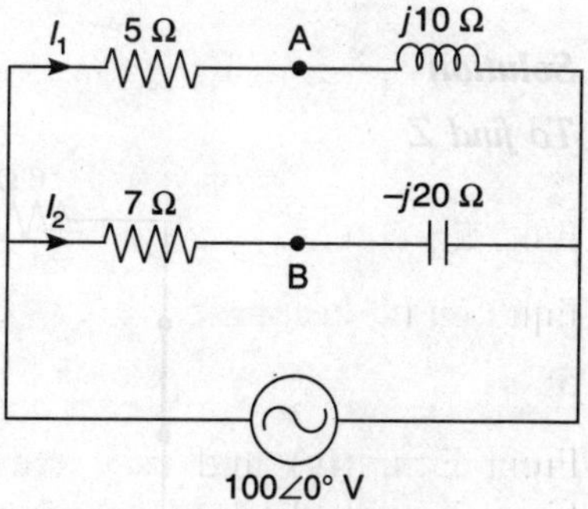

Fig. 6.150

$$I_1 = \frac{100\angle 0^\circ}{5+j10} = \frac{100\angle 0^\circ}{11.18\angle 63.43^\circ} = 8.944\angle -63.43^\circ$$

$$I_2 = \frac{100\angle 0^\circ}{7-j20} = \frac{100\angle 0^\circ}{21.19\angle -70.7^\circ} = 4.72\angle 70.7^\circ$$

Voltage drop across 5 Ω resistance

$$= 5 \times 8.944\angle -63.43^\circ = 44.72\angle -63.43^\circ \text{ V}$$

Voltage drop across 7 Ω resistance

$$= 7 \times 4.72\angle 70.7^\circ = 33.04\angle 70.7^\circ \text{ V.}$$

$$V_{5\Omega} + V_{AB} - V_{7\Omega} = 0$$

$$44.72\angle{-63.43^\circ} + V_{AB} - 33.04\angle 70.7^\circ = 0$$

$$(20 - j40) + V_{AB} - (10.92 + j31.18) = 0$$

$$\therefore \quad V_{AB} = -9.08 + j71.18 = 71.76\angle 97.27^\circ$$

Thevenin's equivalent circuit is shown in Fig. 6.151.

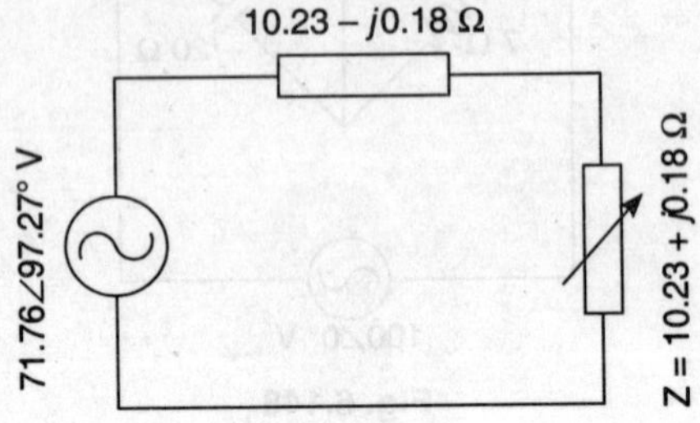

Fig. 6.151

$$\text{Current through the load} = \frac{71.76\angle 97.27^\circ}{20.46} = 3.51\angle 97.27^\circ$$

Maximum power delivered to the load = $(3.51)^2 \times 10.23$ = **126.03 W.**

EXAMPLE 6.37 In Fig. 6.152, load Z_L has a fixed resistance of 5 Ω and a variable reactance X. Determine the value of X to receive maximum power.

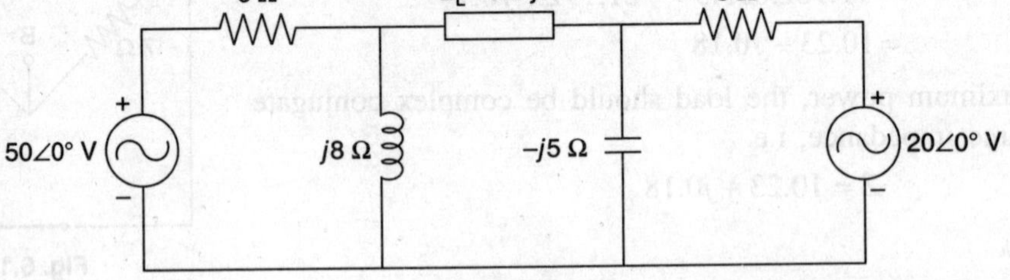

Fig. 6.152

Solution

To find Z

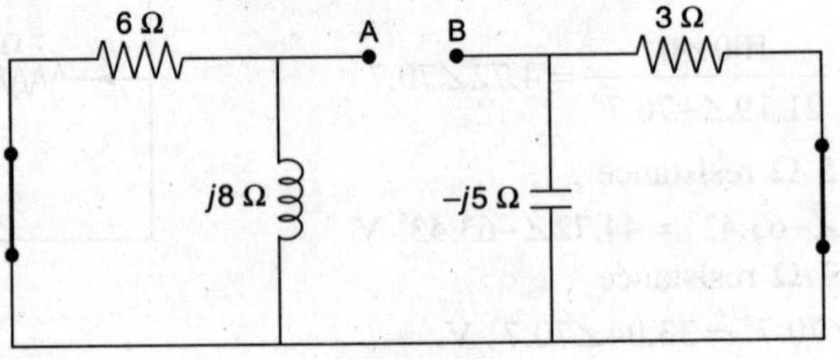

Fig. 6.153

From Fig. 6.153,

$$Z_{AB} = \frac{6(j8)}{6+j8} + \frac{3(-j5)}{3-j5}$$

$$= \frac{48\angle 90^\circ}{10\angle 53.13^\circ} + \frac{15\angle -90^\circ}{5.83\angle -59.04^\circ}$$

$$= 4.8\angle 36.87^\circ + 2.57\angle -30.96^\circ$$

$$= 3.84 + j2.88 + 2.2 - j1.32$$

$$= 6.04 + j1.56$$

To get maximum power, the load should be the complex conjugate of the source impedance, i.e.

$$6.04 - j1.56$$

then, $\mathbf{X = -j1.56\ \Omega}$

6.7 MILLMAN'S THEOREM

Millman's theorem is used to combine a number of voltage sources connected in parallel into a single equivalent voltage source.

Millman's theorem is stated as:

If 'n' number of voltage sources with their internal impedances are connected in parallel then they can be combined into a single equivalent voltage source with an internal impedance.

Consider n number of parallel connected voltage sources with their internal impedances as shown in Fig. 6.154. By applying Millman's theorem, the voltage sources which are in parallel can be converted to a single equivalent voltage source as shown in Fig. 6.155.

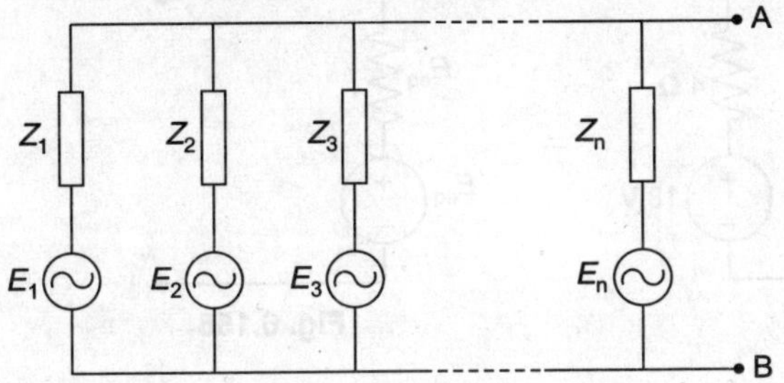

Fig. 6.154 *n*-Voltage sources connected in parallel.

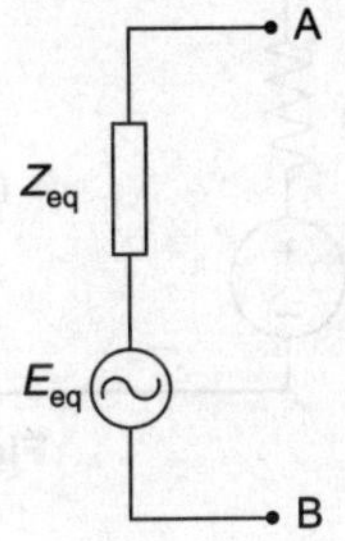

Fig. 6.155 Equivalent single voltage source.

Let,

$E_1, E_2, E_3, \ldots, E_n$ are e.m.f. of voltage sources in parallel.

$Z_1, Z_2, Z_3, \ldots, Z_n$ are internal impedance of voltage sources.

$Y_1, Y_2, Y_3, \ldots, Y_n$ are respective admittances of internal impedances $Z_1, Z_2, Z_3, \ldots, Z_n$.

E_{eq} = e.m.f. of equivalent voltage source.

Z_{eq} = internal impedance of equivalent voltage source.

Then, by Millman's theorem

$$E_{eq} = \frac{E_1Y_1 + E_2Y_2 + E_3Y_3 + \cdots + E_nY_n}{Y_1 + Y_2 + Y_3 + \cdots + Y_n}$$

and

$$Z_{eq} = \frac{1}{Y_1 + Y_2 + Y_3 + \cdots + Y_n}$$

Note: For DC circuits, the admittance (*Y*) and impedance (*Z*) are replaced by conductance (*G*) and resistance (*R*), respectively.

EXAMPLE 6.38 Calculate the current flowing through 6 Ω resistor in the circuit shown in Fig. 6.156, using Millman's theorem.

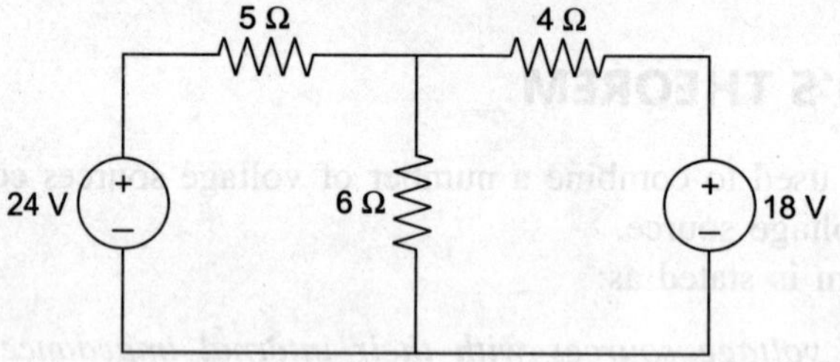

Fig. 6.156

Solution The given network is redrawn as shown in Fig. 6.157. According to Millman's theorem, the two voltage sources can be replaced by a single equivalent voltage source in series with equivalent resistance as shown in Fig. 6.158.

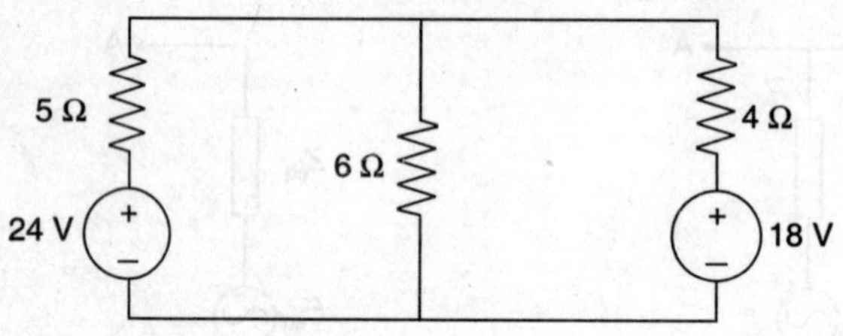

Fig. 6.157

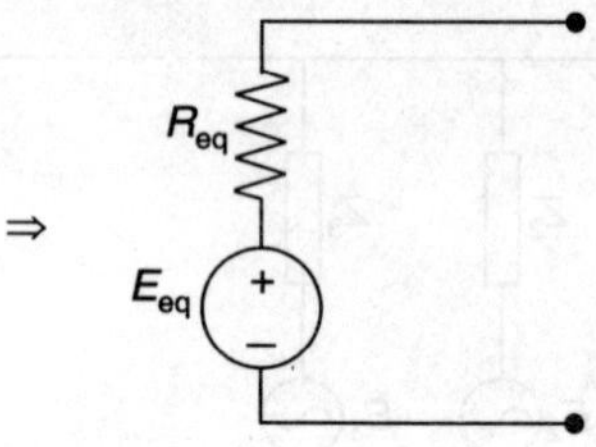

Fig. 6.158

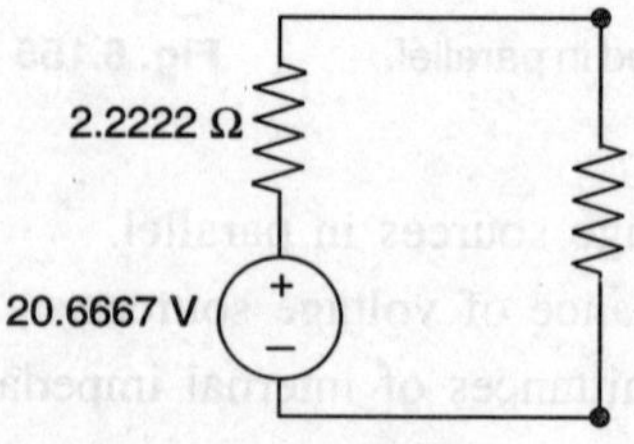

Fig. 6.159

$$E_{eq} = \frac{24\left(\frac{1}{5}\right) + 18\left(\frac{1}{4}\right)}{\left(\frac{1}{5}\right) + \left(\frac{1}{4}\right)} = 20.6667 \text{ V}$$

and $$R_{eq} = \frac{1}{\left(\frac{1}{5}\right) + \left(\frac{1}{4}\right)} = 2.2222\ \Omega$$

From Fig. 6.159, current through 6 Ω resistance = $\frac{20.6667}{2.2222 + 6}$ = **2.5135 A.**

EXAMPLE 6.39 Calculate the current flowing through 3 Ω resistor in the network shown in Fig. 6.160, using Millman's theorem.

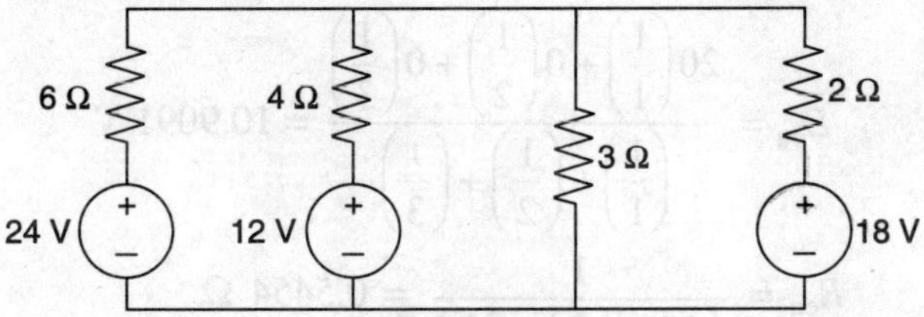

Fig. 6.160

Solution The given network is redrawn as shown in Fig. 6.161. Millman's equivalent circuit is shown in Fig. 6.162.

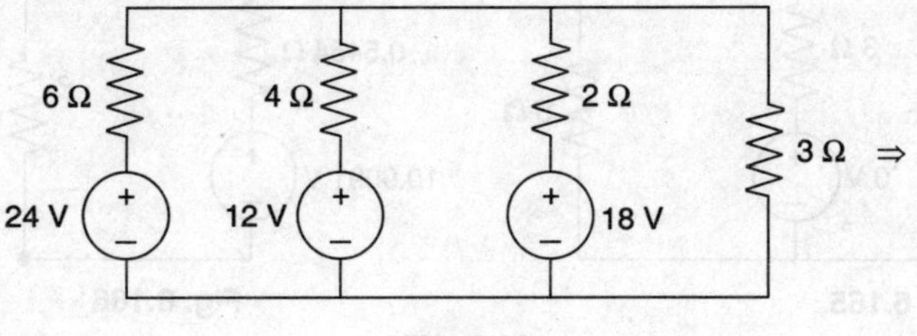

Fig. 6.161

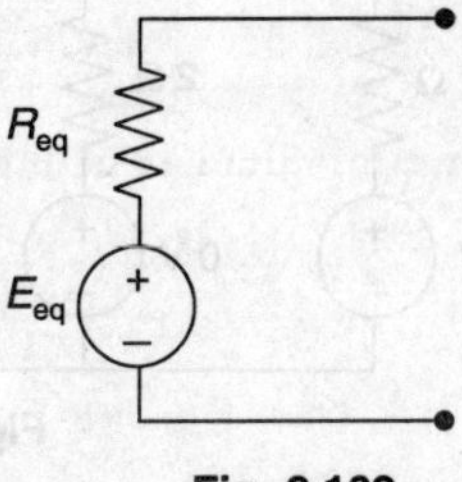

Fig. 6.162

$$E_{eq} = \frac{24\left(\frac{1}{6}\right) + 12\left(\frac{1}{4}\right) + 18\left(\frac{1}{2}\right)}{\left(\frac{1}{6}\right) + \left(\frac{1}{4}\right) + \left(\frac{1}{2}\right)} = 17.4545 \text{ V}$$

and $$R_{eq} = \frac{1}{\left(\frac{1}{6}\right) + \left(\frac{1}{4}\right) + \left(\frac{1}{2}\right)} = 1.0909\ \Omega$$

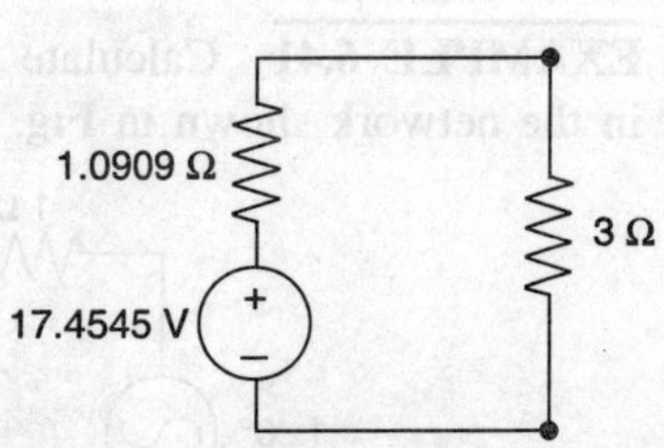

Fig. 6.163

From Fig. 6.163, current through 3 Ω resistance = $\frac{17.4545}{1.0909 + 3}$ = **4.2667 A.**

EXAMPLE 6.40 Using Millman's theorem, calculate the current flowing through 5 Ω resistor in the circuit shown in Fig. 6.164.

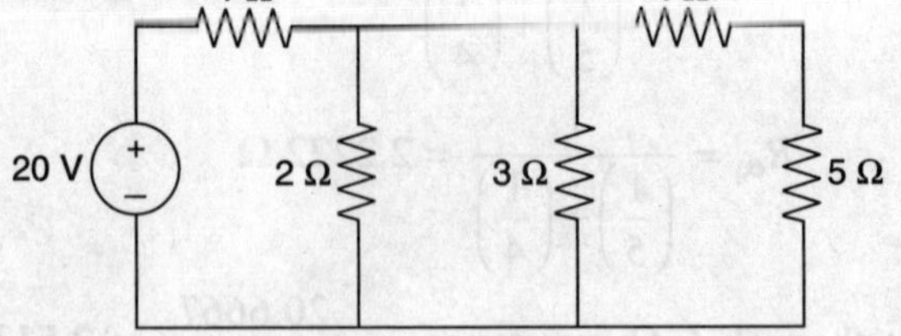

Fig. 6.164

The given network is redrawn as shown in Fig. 6.165, wherein the branches with 2 Ω and 3 Ω resistances are assumed to have voltage sources of zero value. By Millman's theorem,

$$E_{eq} = \frac{20\left(\frac{1}{1}\right)+0\left(\frac{1}{2}\right)+0\left(\frac{1}{3}\right)}{\left(\frac{1}{1}\right)+\left(\frac{1}{2}\right)+\left(\frac{1}{3}\right)} = 10.9091 \text{ V}$$

and $$R_{eq} = \frac{1}{\left(\frac{1}{1}\right)+\left(\frac{1}{2}\right)+\left(\frac{1}{3}\right)} = 0.5454\ \Omega$$

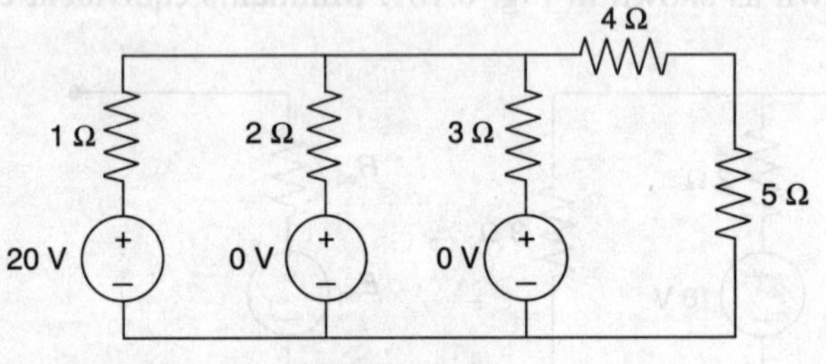

Fig. 6.165

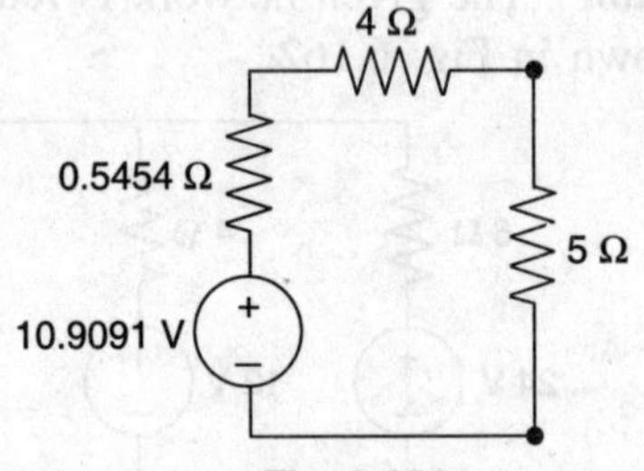

Fig. 6.166

From Fig. 6.166, current through 5 Ω resistance $= \dfrac{10.9091}{0.5454+4+5}$ = **1.1429 A.**

EXAMPLE 6.41 Calculate the current flowing through Z_L by applying Millman's theorem in the network shown in Fig. 6.167.

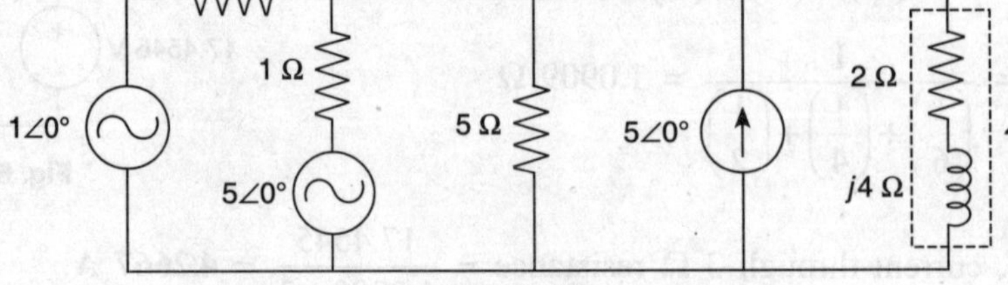

Fig. 6.167

Solution Since Millman's theorem is applicable only if the circuit is having voltage sources, the current source is converted to a voltage source as shown in Fig. 6.168.

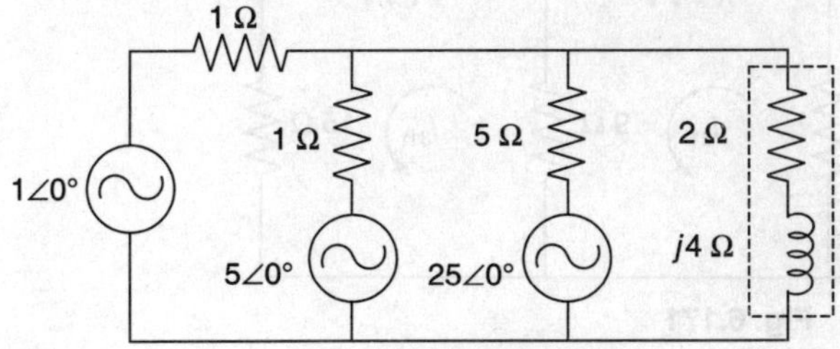

Fig. 6.168

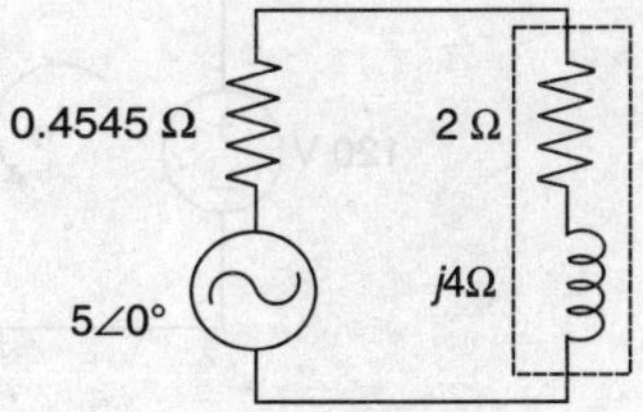

Fig. 6.169

$$E_{eq} = \frac{1\angle 0^\circ\left(\frac{1}{1}\right) + 5\angle 0^\circ\left(\frac{1}{1}\right) + 25\angle 0^\circ\left(\frac{1}{5}\right)}{\left(\frac{1}{1}\right)+\left(\frac{1}{1}\right)+\left(\frac{1}{5}\right)} = 5\angle 0^\circ \text{ V}$$

$$Z_{eq} = \frac{1}{\left(\frac{1}{1}\right)+\left(\frac{1}{1}\right)+\left(\frac{1}{5}\right)} = 0.4545\ \Omega$$

From Fig. 6.169, current through $Z_L = \dfrac{5\angle 0^\circ}{0.4545 + 2 + j4} = \dfrac{5\angle 0^\circ}{4.693\angle 58.47^\circ}$

$$= \mathbf{1.0654\angle{-58.47}^\circ\ A}$$

6.8 SUBSTITUTION THEOREM

Substitution theorem is stated as,

A branch of a network carrying a current of I amperes can be replaced by a voltage source of value IZ volts, where Z is its branch impedance, without affecting the other parts of the network.

According to substitution theorem, any branch of a network can be replaced by a different branch as long as its voltage and current remain unchanged. When the original branch is passive, either a voltage source or a current source is generally used for the substituted branch.

EXAMPLE 6.42 In the network shown in Fig. 6.170, substitute the branch of 9 Ω resistance by an equivalent branch and verify substitution theorem.

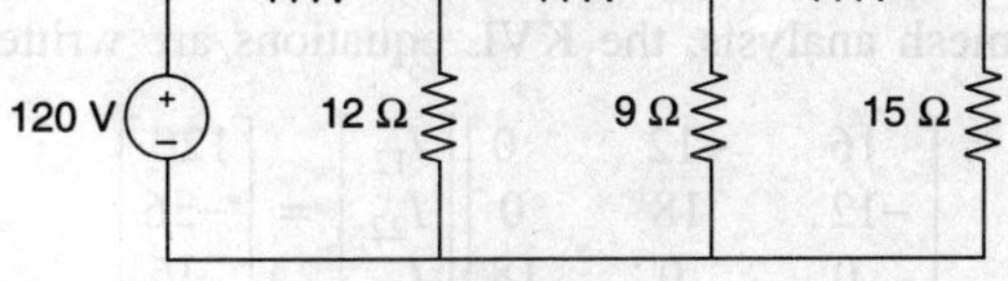

Fig. 6.170

Solution

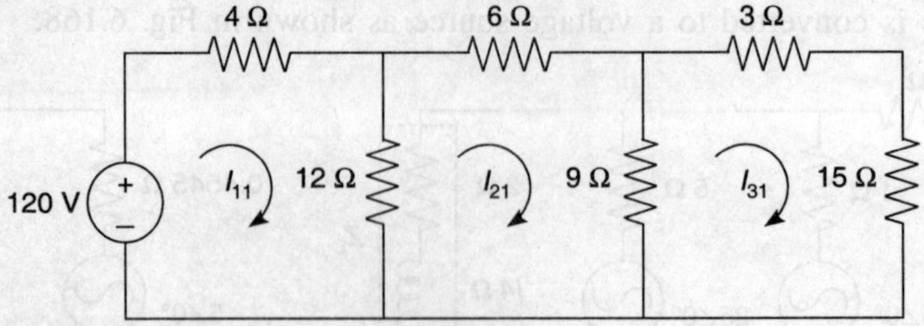

Fig. 6.171

Applying mesh analysis to the circuit shown in Fig. 6.171, the KVL equations are written as:

$$\begin{bmatrix} 16 & -12 & 0 \\ -12 & 27 & -9 \\ 0 & -9 & 27 \end{bmatrix} \begin{bmatrix} I_{11} \\ I_{21} \\ I_{31} \end{bmatrix} = \begin{bmatrix} 120 \\ 0 \\ 0 \end{bmatrix}$$

Using Cramer's rule, the mesh currents are calculated as:

$$I_{11} = \frac{\Delta I_{11}}{\Delta} = \frac{77760}{6480} = 12 \text{ A}$$

$$I_{21} = \frac{\Delta I_{21}}{\Delta} = \frac{38880}{6480} = 6 \text{ A}$$

$$I_{31} = \frac{\Delta I_{31}}{\Delta} = \frac{12960}{6480} = 2 \text{ A}$$

From which the current through 9 Ω resistance = $I_{21} - I_{31}$ = 6 – 2 = 4 A.
Hence, the voltage drop across 9 Ω resistance = 4 × 9 = 36 V.

Let us substitute the 9 Ω resistive branch by a voltage source of 36 V as shown in Fig. 6.172.

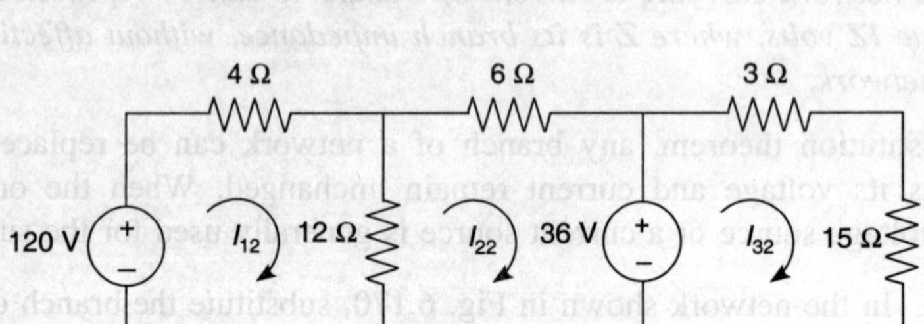

Fig. 6.172

Once again by applying mesh analysis, the KVL equations are written as:

$$\begin{bmatrix} 16 & -12 & 0 \\ -12 & 18 & 0 \\ 0 & 0 & 18 \end{bmatrix} \begin{bmatrix} I_{12} \\ I_{22} \\ I_{32} \end{bmatrix} = \begin{bmatrix} 120 \\ -36 \\ 36 \end{bmatrix}$$

Using Cramer's rule, the mesh currents are calculated as:

$$I_{12} = \frac{\Delta I_{12}}{\Delta} = \frac{31104}{2592} = 12 \text{ A}$$

$$I_{22} = \frac{\Delta I_{22}}{\Delta} = \frac{15552}{2592} = 6 \text{ A}$$

$$I_{32} = \frac{\Delta I_{32}}{\Delta} = \frac{5184}{2592} = 2 \text{ A}$$

It is observed that the two sets of mesh currents (I_{11}, I_{21}, I_{31}) and (I_{12}, I_{22}, I_{32}) are equal. Thus, it is concluded that the substitution of 36 V voltage source in the place of 9 Ω resistive branch does not alter the values of currents in other parts of the circuit. Hence, the substitution theorem is proved.

EXAMPLE 6.43 In the network shown in Fig. 6.173, substitute the branch with 2 Ω resistance by an equivalent branch and verify the substitution theorem.

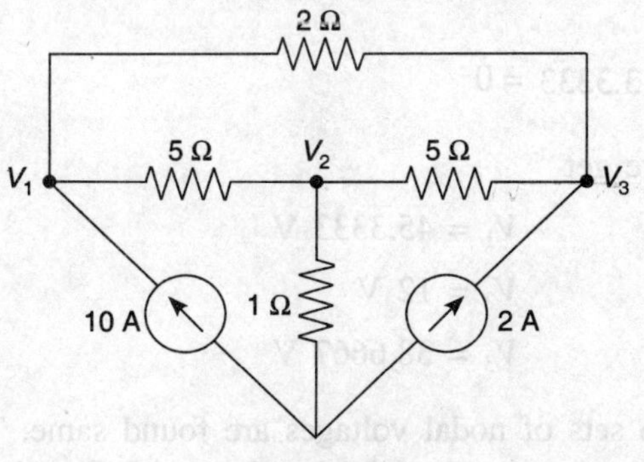

Fig. 6.173

Solution Applying KCL

at node 1, $$\frac{V_1 - V_2}{5} + \frac{V_1 - V_3}{2} - 10 = 0$$

at node 2, $$\frac{V_2 - V_1}{5} + \frac{V_2 - V_3}{5} + V_2 = 0$$

at node 3, $$\frac{V_3 - V_2}{5} + \frac{V_3 - V_1}{2} - 2 = 0$$

Solving these equations, we get

$$V_1 = 45.3333 \text{ V}$$

$$V_2 = 12 \text{ V}$$

$$V_3 = 38.6667 \text{ V}$$

From which, the current through 2 Ω resistance is obtained as:

$$\frac{V_1 - V_3}{2} = \frac{45.3333 - 38.6667}{2} = 3.3333 \text{ A}$$

Let us substitute the 2 Ω resistive branch by a current source of 3.3333 A as shown in Fig. 6.174.

Once again by applying KCL

at node 1,

$$\frac{V_1 - V_2}{5} + 3.3333 - 10 = 0$$

at node 2,

$$\frac{V_2 - V_1}{5} + \frac{V_2 - V_3}{5} + V_2 = 0$$

at node 3,

$$\frac{V_3 - V_2}{5} - 2 - 3.3333 = 0$$

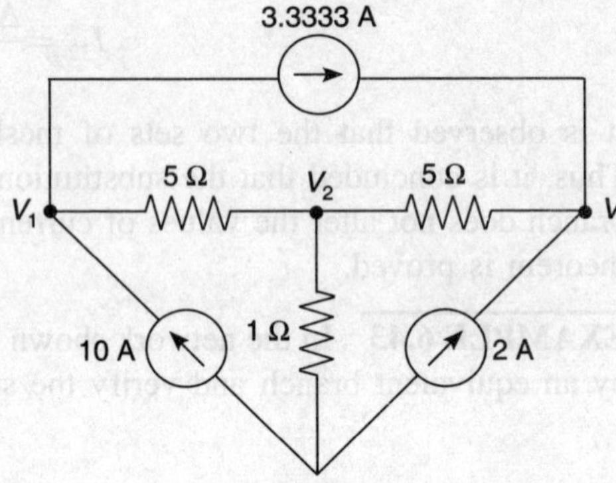

Fig. 6.174

Solving these equations, we get

$$V_1 = 45.3333 \text{ V}$$

$$V_2 = 12 \text{ V}$$

$$V_3 = 38.6667 \text{ V}$$

It is observed that the two sets of nodal voltages are found same. Thus, it is concluded that the substitution of 3.3333 A current source in the place of 2 Ω resistive branch does not alter the values of voltages in other parts of the circuit. Hence, the substitution theorem is proved.

EXAMPLE 6.44 In the network shown in Fig. 6.175, substitute the branch of (3 + *j*5) Ω impedance by a voltage source and verify the substitution theorem.

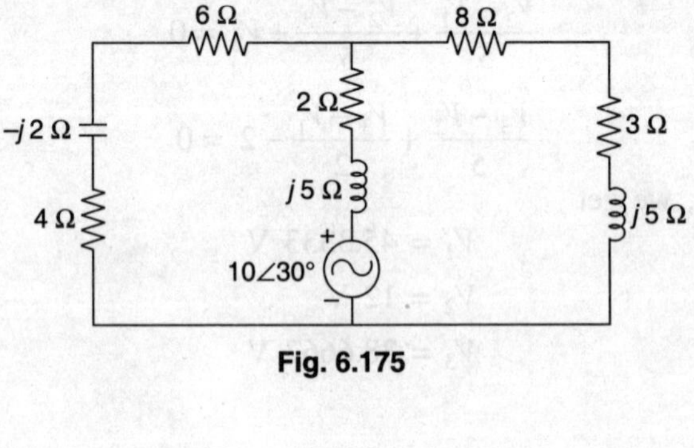

Fig. 6.175

Solution Let us assume mesh currents I_1 and I_2 as shown in Fig. 6.176. The mesh equations are written by inspection as:

$$\begin{bmatrix} 12+j3 & 2+j5 \\ 2+j5 & 13+j10 \end{bmatrix} \begin{bmatrix} I_1 \\ I_2 \end{bmatrix} = \begin{bmatrix} 10\angle 30^\circ \\ 10\angle 30^\circ \end{bmatrix}$$

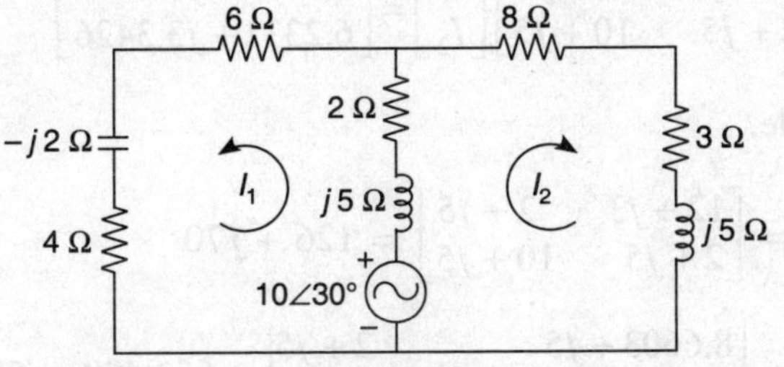

Fig. 6.176

Applying Cramer's rule,

$$\Delta = \begin{vmatrix} 12+j3 & 2+j5 \\ 2+j5 & 13+j10 \end{vmatrix} = 147 + j139$$

$$\Delta_1 = \begin{vmatrix} 10\angle 30^\circ & 2+j5 \\ 10\angle 30^\circ & 13+j10 \end{vmatrix} = 70.2633 + j98.3015$$

$$\Delta_2 = \begin{vmatrix} 12+j3 & 10\angle 30^\circ \\ 2+j5 & 10\angle 30^\circ \end{vmatrix} = 96.603 + j32.6794$$

$$I_1 = \frac{\Delta_1}{\Delta} = \frac{70.2633 + j98.3015}{147 + j139} = 0.5862 + j0.1144 \text{ A}$$

$$I_2 = \frac{\Delta_2}{\Delta} = \frac{96.603 + j32.6794}{147 + j139} = 0.4579 - j0.2107 \text{ A}$$

Therefore, the voltage across (3 + $j5$) Ω impedance

$$= (0.4579 - j0.2107) \times (3 + j5)\ \Omega$$

$$= 2.4272 + j1.6574 \text{ V} = 2.9391\angle 34.33^\circ \text{ V}$$

To verify substitution theorem, let us substitute the (3 + $j5$) Ω impedance by a voltage source of 2.9391∠34.33° V as shown in Fig. 6.177.

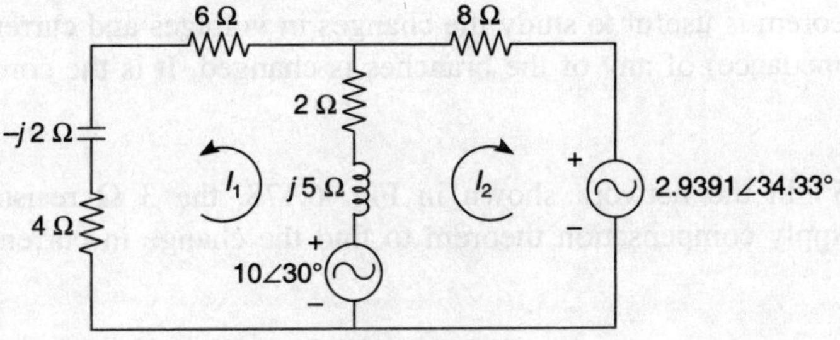

Fig. 6.177

The mesh equations are written by inspection as:

$$\begin{bmatrix} 12+j3 & 2+j5 \\ 2+j5 & 10+j5 \end{bmatrix} \begin{bmatrix} I_1 \\ I_2 \end{bmatrix} = \begin{bmatrix} 10\angle 30^\circ \\ 10\angle 30^\circ - 2.9391\angle 34.33^\circ \end{bmatrix}$$

$$\begin{bmatrix} 12+j3 & 2+j5 \\ 2+j5 & 10+j5 \end{bmatrix} \begin{bmatrix} I_1 \\ I_2 \end{bmatrix} = \begin{bmatrix} 8.6603+j5 \\ 6.2331+j3.3426 \end{bmatrix}$$

Applying Cramer's rule,

$$\Delta = \begin{bmatrix} 12+j3 & 2+j5 \\ 2+j5 & 10+j5 \end{bmatrix} = 126 + j70$$

$$\Delta_1 = \begin{vmatrix} 8.6603+j5 & 2+j5 \\ 6.2331+j3.3426 & 10+j5 \end{vmatrix} = 65.8498 + j55.4508$$

$$\Delta_2 = \begin{vmatrix} 12+j3 & 8.6603+j5 \\ 2+j5 & 6.2331+j3.3426 \end{vmatrix} = 72.4488 + j5.509$$

$$I_1 = \frac{\Delta_1}{\Delta} = \frac{65.8498+j55.4508}{126+j70} = 0.5862 + j0.1144 \text{ A}$$

$$I_2 = \frac{\Delta_2}{\Delta} = \frac{72.4488+j5.509}{126+j70} = 0.4579 - j0.2107 \text{ A}$$

On comparing the two sets of mesh currents, it is found that both are same. Hence, substitution theorem is verified.

6.9 COMPENSATION THEOREM

Compensation theorem is stated as:

> *When a resistance R (impedance Z) of a branch carrying current I is increased to R + dR (Z + dZ), the changes that occur could be determined by connecting a series voltage called compensation voltage V_C opposing the current flowing in that branch of resistance R (impedance Z). The magnitude of V_C is $I \times dR$ ($I \times dZ$). The polarity of V_C is such that it opposes the original current that was flowing in the circuit.*

Compensation theorem is useful to study the changes in voltages and currents of a circuit when the resistance (impedance) of any of the branches is changed. It is the corollary of substitution theorem.

EXAMPLE 6.45 In the network shown in Fig. 6.178, the 3 Ω resistance is replaced by 6 Ω resistance. Apply compensation theorem to find the change in current in 5 Ω resistance.

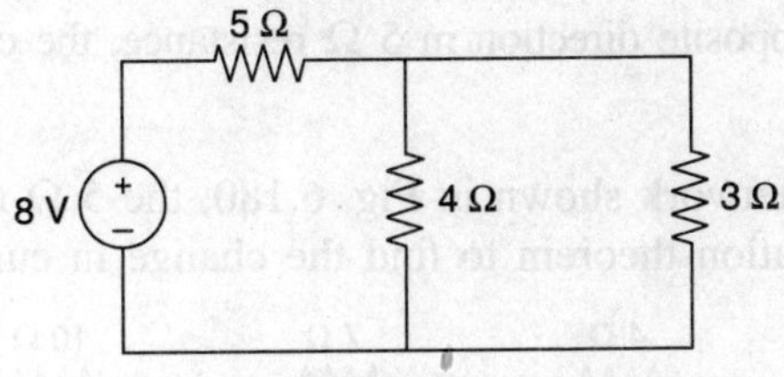

Fig. 6.178

Solution Let us first find the current through 3 Ω resistance.

Total resistance connected to 8 V source = $5+\left(\frac{3\times 4}{3+4}\right) = 6.7143\ \Omega$.

Current supplied by the source (current through 5 Ω) = $\frac{8}{6.7143} = 1.1915$ A.

By current division rule, current through 3 Ω resistance = $1.1915\times\frac{4}{7} = 0.6808$ A.

Compensation voltage,

$$V_C = I \times dR$$

where

I = current through 3 Ω resistance, i.e. 0.6808 A

dR = change in resistance, i.e. 6 – 3 = 3 Ω

So, the compensation voltage, $V_C = 0.6808 \times 3 = 2.0424$ V.

To apply compensation theorem, the original circuit is modified as shown in Fig. 6.179. Here, the compensation voltage is introduced in the branch where replacement is done while the 8 V source is short-circuited.

Note: The polarity of V_C is such that it opposes the original current.

Referring to Fig. 6.179,

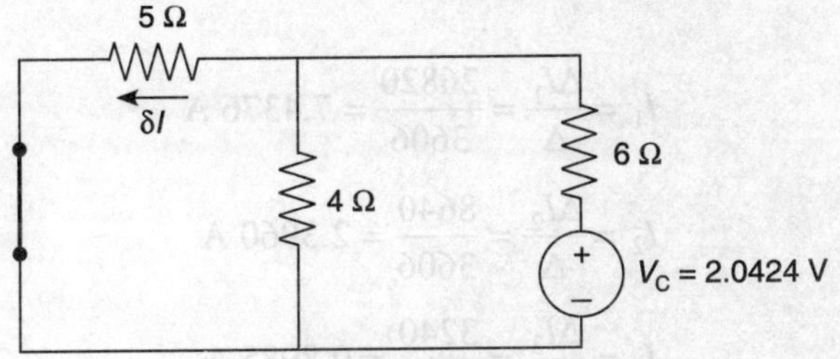

Fig. 6.179

Total resistance connected to 2.0424 V compensation voltage = $6+\left(\frac{5\times 4}{5+4}\right) = 8.2222\ \Omega$.

Current supplied by the compensation voltage = $\frac{2.0424}{8.2222} = 0.2484$ A.

Change in current δI is nothing but the current through 5 Ω resistance due to V_C.

By current division rule, current through 5 Ω resistance = $0.2484\times\frac{4}{9} = 0.1104$ A.

Since δI flows in the opposite direction in 5 Ω resistance, the original current (i.e. 1.1915 A) is reduced by 0.1104 A.

EXAMPLE 6.46 In the network shown in Fig. 6.180, the 5 Ω resistance is replaced by 8 Ω resistance. Apply compensation theorem to find the change in current through 4 Ω resistance.

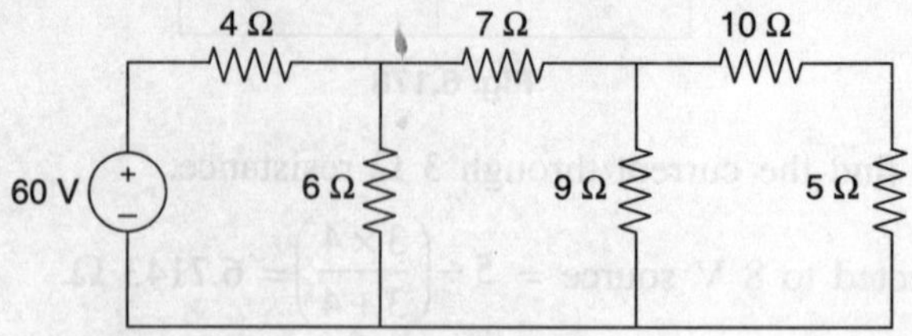

Fig. 6.180

Solution The mesh currents are assumed as shown in Fig. 6.181 and by applying mesh analysis method, the KVL equations are written by inspection as:

$$\begin{bmatrix} 10 & -6 & 0 \\ -6 & 22 & -9 \\ 0 & -9 & 24 \end{bmatrix} \begin{bmatrix} I_1 \\ I_2 \\ I_3 \end{bmatrix} = \begin{bmatrix} 60 \\ 0 \\ 0 \end{bmatrix}$$

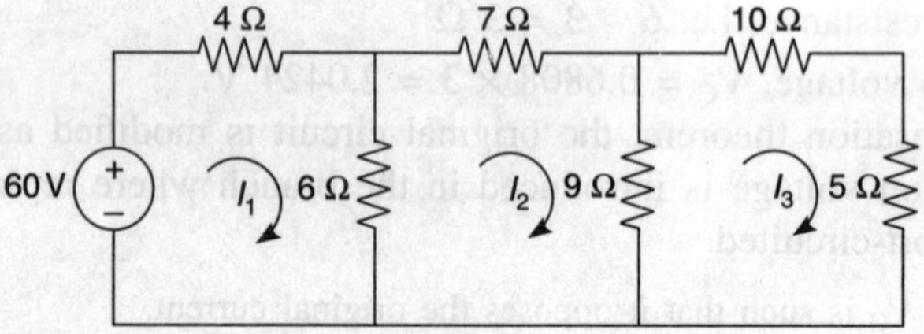

Fig. 6.181

Using Cramer's rule,

$$I_1 = \frac{\Delta I_1}{\Delta} = \frac{26820}{3606} = 7.4376 \text{ A}$$

$$I_2 = \frac{\Delta I_2}{\Delta} = \frac{8640}{3606} = 2.3960 \text{ A}$$

$$I_3 = \frac{\Delta I_3}{\Delta} = \frac{3240}{3606} = 0.8985 \text{ A}$$

Compensation voltage,

$$V_C = I \times dR$$

where

I = current through 5 Ω resistance, i.e. 0.8985 A

dR = change in resistance, i.e. 8 – 5 = 3 Ω

Therefore, the compensation voltage, $V_C = 0.8985 \times 3 = 2.6955$ V.

To apply compensation theorem, the original circuit is modified as shown in Fig. 6.182. By applying mesh analysis method, the KVL equations are written by inspection as:

$$\begin{bmatrix} 10 & -6 & 0 \\ -6 & 22 & -9 \\ 0 & -9 & 27 \end{bmatrix} \begin{bmatrix} \delta I_1 \\ \delta I_2 \\ \delta I_3 \end{bmatrix} = \begin{bmatrix} 0 \\ 0 \\ -2.6955 \end{bmatrix}$$

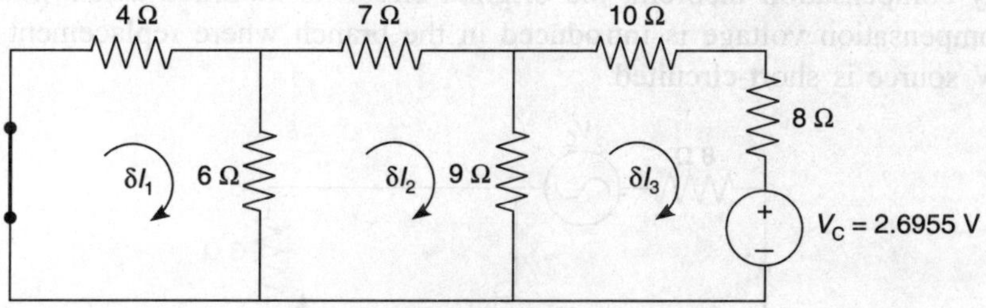

Fig. 6.182

Using Cramer's rule, change in current in 4 Ω resistance,

$$\delta I_1 = \frac{\Delta I_1}{\Delta} = \frac{-145.557}{4158} = -0.035 \text{ A}$$

Since δI flows in the opposite direction in 4 Ω resistance, the original current (i.e. 0.8985 A) is reduced by 0.035 A due to the replacement of 5 Ω resistance by 8 Ω resistance.

EXAMPLE 6.47 In the network shown in Fig. 6.183, the 5 Ω resistance is replaced by 8 Ω resistance. Apply compensation theorem to find the change in current through (3 + *j*4) Ω impedance.

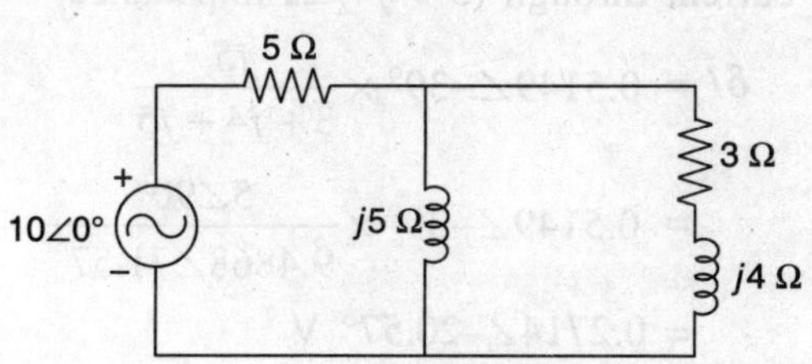

Fig. 6.183

Solution Let us first find the current through 5 Ω resistance.

Total impedance connected to 10 V source $= 5 + \dfrac{(3+j4)(j5)}{3+j4+j5} = (5.8333 + j2.5)\ \Omega.$

Current supplied by the source (current through 5 Ω) $= \dfrac{10\angle 0^\circ}{5.8333 + j2.5}$

$$= \frac{10\angle 0^\circ}{6.3464\angle 23.2^\circ} = 1.5757\angle{-23.2^\circ} \text{ A.}$$

Compensation voltage,

$$V_C = I \times dR.$$

where

I = current through 5 Ω resistance, i.e. $1.5757\angle{-23.2°}$ A

dR = change in resistance, i.e. 8 – 5 = 3 Ω

Therefore, the compensation voltage, $V_C = 1.5757\angle{-23.2°} \times 3 = 4.7271\angle{-23.2°}$ V.

To apply compensation theorem, the original circuit is modified as shown in Fig. 6.184. Here, the compensation voltage is introduced in the branch where replacement is done while the $10\angle 0°$ V source is short-circuited.

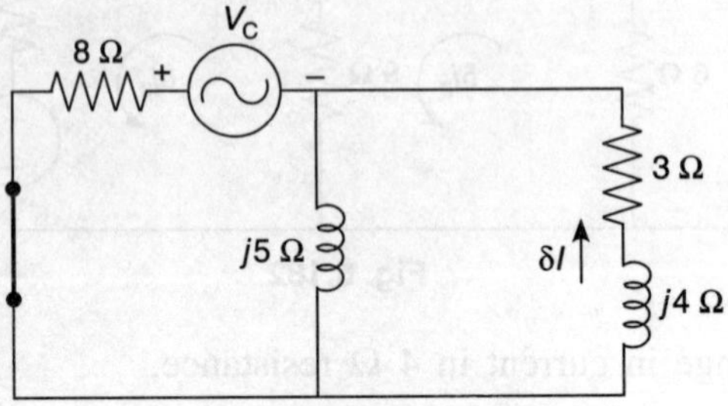

Fig. 6.184

$$\text{Total impedance connected to } V_C = 8 + \frac{(3+j4)(j5)}{3+j4+j5} = (8.8333 + j2.5)\ \Omega$$

$$\text{Current supplied by } V_C = \frac{4.7271\angle{-23.2°}}{8.8333 + j2.5} = \frac{4.7271\angle{-23.2°}}{9.1803\angle 15.8°} = 0.5149\angle{-39°}\ \text{V}$$

Change in current δI is nothing but the current through (3 + $j4$) Ω impedance due to V_C. By current division rule, current through (3 + $j4$) Ω impedance

$$\delta I = 0.5149\angle{-39°} \times \frac{j5}{3+j4+j5}$$

$$= 0.5149\angle{-39°} \times \frac{5\angle 90°}{9.4868\angle 71.57°}$$

$$= 0.2714\angle{-20.57°}\ \text{V}$$

Note: The change in current δI is directed upwards.

6.10 TELLEGEN'S THEOREM

Tellegen's theorem is stated as:

In a lumped network, the algebraic sum of the instantaneous powers in all the branches at any instant is zero. All the branch currents and the voltages in that network must obey Kirchhoff's laws. In other words, it can be stated as, in a given network, the algebraic sum of the powers delivered by all the sources is equal to the algebraic sum of the powers absorbed by all the elements.

Mathematically this theorem is expressed as:

$$\sum_{k=1}^{b} v_k i_k = 0$$

Tellegen's theorem is applicable to all networks consisting of lumped elements which may be linear or nonlinear, passive or active, time invariant or time varying.

EXAMPLE 6.48 Verify Tellegen's theorem for the network shown in Fig. 6.185.

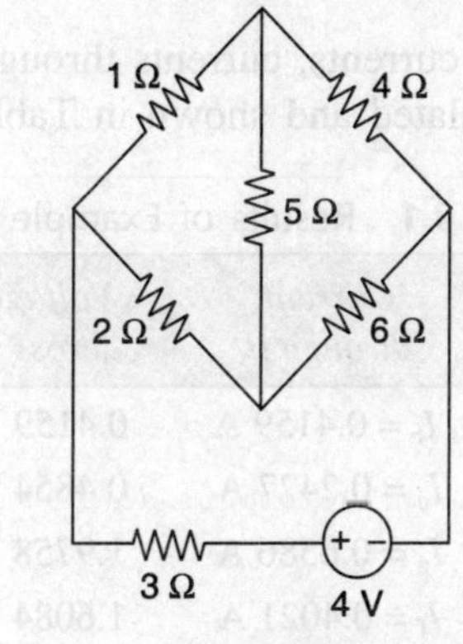

Fig. 6.185

Solution The mesh currents and various branch currents are marked as shown in Fig. 6.186.

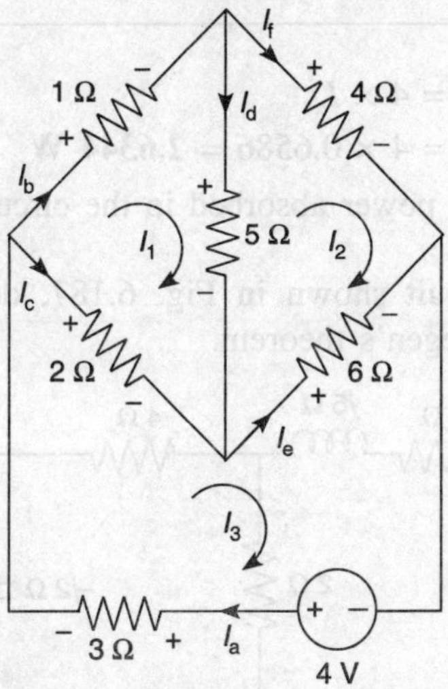

Fig. 6.186

By applying mesh analysis, the KVL equations are written by inspection as:

$$\begin{bmatrix} 8 & -5 & -2 \\ -5 & 15 & -6 \\ -2 & -6 & 11 \end{bmatrix} \begin{bmatrix} I_1 \\ I_2 \\ I_3 \end{bmatrix} = \begin{bmatrix} 0 \\ 0 \\ 4 \end{bmatrix}$$

Using Cramer's rule,

$$I_1 = \frac{\Delta I_1}{\Delta} = \frac{240}{577} = 0.4159 \text{ A}$$

$$I_2 = \frac{\Delta I_2}{\Delta} = \frac{232}{577} = 0.4021 \text{ A}$$

$$I_3 = \frac{\Delta I_3}{\Delta} = \frac{380}{577} = 0.6586 \text{ A}$$

From the knowledge of three mesh currents, currents through various elements, voltage across them and power absorbed are calculated and shown in Table 6.1.

Table 6.1 Results of Example 6.48

Element	*Current through it*	*Voltage across it*	*Power absorbed by it*
Resistance, 1 Ω	$I_b = 0.4159$ A	0.4159 V	0.1730 W
Resistance, 2 Ω	$I_c = 0.2427$ A	0.4854 V	0.1178 W
Resistance, 3 Ω	$I_a = 0.6586$ A	1.9758 V	1.3012 W
Resistance, 4 Ω	$I_f = 0.4021$ A	1.6084 V	0.6467 W
Resistance, 5 Ω	$I_d = 0.0138$ A	0.0690 V	0.0009 W
Resistance, 6 Ω	$I_e = 0.2565$ A	1.5390 V	0.3948 W
Sum of power absorbed by all elements			**2.6344 W**

Power delivered by the source $= 4 \times I_a$

$= 4 \times 0.6586 =$ **2.6344 W**

As power delivered is equal to power absorbed in the circuit, Tellegen's theorem is verified.

EXAMPLE 6.49 In the circuit shown in Fig. 6.187, determine the branch currents and voltages and hence verify Tellegen's theorem.

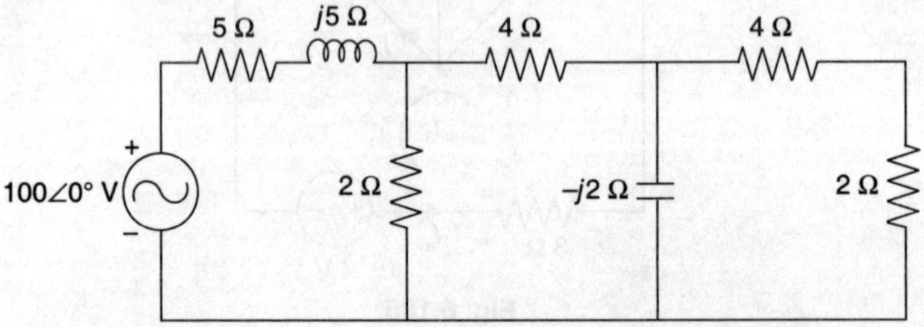

Fig. 6.187

Solution The mesh currents and various branch currents are marked as shown in Fig. 6.188. By applying mesh analysis method, the KVL equations are written by inspection as:

$$\begin{bmatrix} 7+j5 & -2 & 0 \\ -2 & 6-j2 & j2 \\ 0 & j2 & 6-j2 \end{bmatrix} \begin{bmatrix} I_1 \\ I_2 \\ I_3 \end{bmatrix} = \begin{bmatrix} 100\angle 0^\circ \\ 0 \\ 0 \end{bmatrix}$$

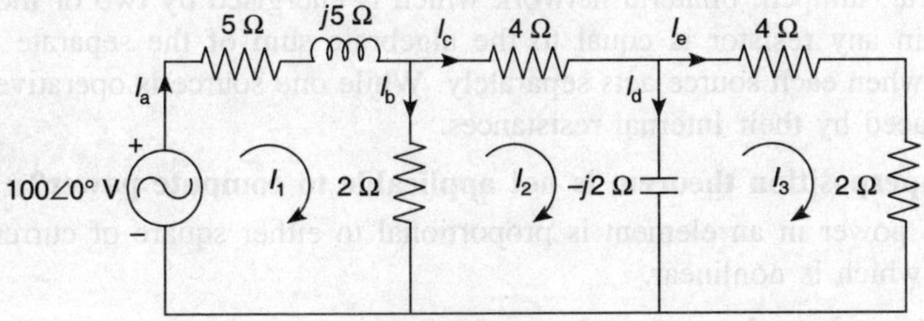

Fig. 6.188

Using Cramer's rule,

$$I_1 = \frac{\Delta I_1}{\Delta} = \frac{3600 - j2400}{348 + j20}$$
$$= 9.9157 - j7.4664 = 12.412\angle -37^\circ \text{ A}$$

$$I_2 = \frac{\Delta I_2}{\Delta} = \frac{1200 - j400}{348 + j20}$$
$$= 3.3711 - j1.3432 = 3.629\angle -21.7^\circ \text{ A}$$

$$I_3 = \frac{\Delta I_3}{\Delta} = \frac{-j400}{348 + j20}$$
$$= -0.0658 - j1.1456 = 1.148\angle -93.3^\circ \text{ A}$$

From the knowledge of three mesh currents, currents through various elements, voltage across them and power absorbed are calculated and shown in Table 6.2.

Table 6.2 Results of Example 6.49

Element	*Current through it*	*Voltage across it*	*Complex power (VA)*
Impedance (5 + j5) Ω	$I_a = I_1 = 12.412\angle -37^\circ$ A	87.7661∠8° V	1089.36∠–29°
Resistance, 2 Ω	$I_b = I_1 - I_2 = 8.962\angle -43.1^\circ$ A	17.924∠–43.1° V	160.63∠–86.2°
Resistance, 4 Ω	$I_c = I_2 = 3.629\angle -21.7^\circ$ A	14.516∠–21.7° V	52.68∠–43.4°
Reactance, –j2 Ω	$I_d = I_2 - I_3 = 3.443\angle -3.3^\circ$ A	6.886∠–93.3° V	23.71∠–96.6°
Resistance, 4 + 2 = 6 Ω	$I_e = I_3 = 1.148\angle -93.3^\circ$ A	6.888∠–93.3° V	7.91∠–186.6°
Sum of complex power absorbed by all elements			**1241.24∠–37°**

Complex power delivered by the source $= 100\angle 0^\circ \times I_a$
$= 100\angle 0^\circ \times 12.412\angle -37^\circ =$ **1241.2∠–37° VA**

As both delivered and absorbed complex powers are same in the circuit, Tellegen's theorem is verified.

Short Questions with Answers

1. State superposition theorem.

In a linear, lumped, bilateral network which is energised by two or more sources, the current in any resistor is equal to the algebraic sum of the separate currents in the resistor when each source acts separately. While one source is operative, other sources are replaced by their internal resistances.

2. Why superposition theorem is not applicable to compute power?

Because power in an element is proportional to either square of current or square of voltage which is nonlinear.

3. State Thevenin's theorem.

A linear, lumped, bilateral network with two terminals can be reduced to an equivalent circuit consisting of a voltage source in series with a resistance (or impedance).

4. State Norton's theorem.

A linear, lumped, bilateral network with two terminals can be reduced to an equivalent circuit consisting of a current source in parallel with a resistance (or impedance).

5. State reciprocity theorem.

In a linear, lumped, bilateral, single source network, the ratio of excitation to response is constant when the position of excitation and response are interchanged.

6. State maximum power transfer theorem.

Maximum power is transferred from source to load, when the load resistance is equal to source resistance.

7. What is the condition for maximum power transfer in DC circuits?

In DC circuits maximum power is transferred from source to load when load resistance is equal to source resistance, i.e.

$$R_s = R_L.$$

8. What is the condition for maximum power transfer in AC circuits?

- For a variable resistive load, the maximum power is transferred only when its value is equal to the magnitude of source impedance, i.e.

$$R_L = Z_s.$$

- For a variable reactive load, the maximum power is transferred only when the load impedance is equal to the complex conjugate of source impedance, i.e.

$$(R_L + jX_L) = (R_s - jX_s)$$

9. Give some applications of maximum power transfer theorem.

- In communication circuits, where the power is relatively small.
- In power amplifiers for impedance matching.

10. State Millman's theorem.

If '*n*' number of voltage sources with their internal impedances are connected in parallel, they can be combined into a single equivalent voltage source with an internal impedance.

11. What is the use of Millman's theorem?

Millman's theorem is used to combine a number of voltage sources connected in parallel into a single equivalent voltage source.

12. Write the expression for Millman's equivalent source of *n* number of parallel connected sources.

$$E_{eq} = \frac{E_1Y_1 + E_2Y_2 + E_3Y_3 + \cdots + E_nY_n}{Y_1 + Y_2 + Y_3 + \cdots + Y_n}$$

and
$$Z_{eq} = \frac{1}{Y_1 + Y_2 + Y_3 + \cdots + Y_n}$$

where

$E_1, E_2, E_3, \ldots, E_n$ are e.m.f. of voltage sources in parallel.

$Z_1, Z_2, Z_3, \ldots, Z_n$ are internal impedance of voltage sources.

$Y_1, Y_2, Y_3, \ldots, Y_n$ are respective admittances of internal impedances $Z_1, Z_2, Z_3, \ldots, Z_n$.

E_{eq} = e.m.f. of equivalent voltage source.

Z_{eq} = internal impedance of equivalent voltage source.

13. State substitution theorem.

A branch of a network carrying a current of *I* amperes can be replaced by a voltage source of value *IZ* volts, where *Z* is its branch impedance, without affecting the other parts of the network.

14. State compensation theorem.

When resistance *R* (impedance *Z*) of a branch carrying current *I* is increased to *R* + *dR* (*Z* + *dZ*), the changes that occur could be determined by connecting a series voltage called compensation voltage V_C opposing the current flowing in that branch of resistance *R* (impedance *Z*). The magnitude of V_C is $I \times dR$ ($I \times dZ$). The polarity of V_C is such that it opposes the original current that was flowing in the circuit.

15. State Tellegen's theorem.

In a lumped network, the algebraic sum of the instantaneous powers in all the branches at any instant is zero. All the branch currents and the voltages in that network must obey Kirchhoff's laws. In other words, it can be stated as, in a given network, the algebraic sum of the powers delivered by all the sources is equal to the algebraic sum of the powers absorbed by all the elements.

UNSOLVED PROBLEMS

1. **Find the current in the 5 Ω resistor shown in Fig. U6.1, using superposition method.**
[Ans. 12.7272 A]

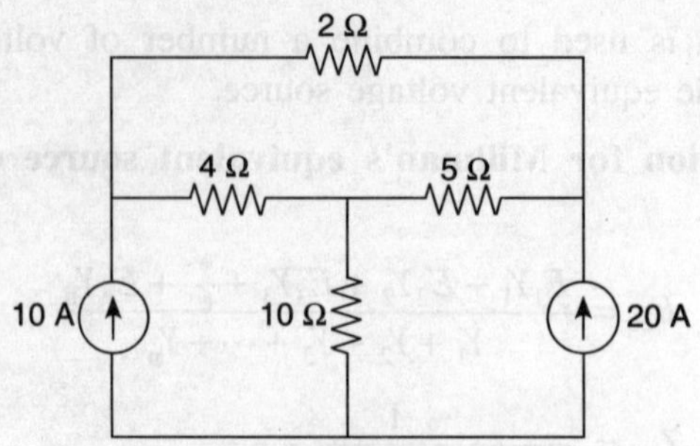

Fig. U6.1

2. **Find the voltage across 6 Ω resistor in the circuit shown in Fig. U6.2, using superposition theorem.** **[Ans.** 3 V]

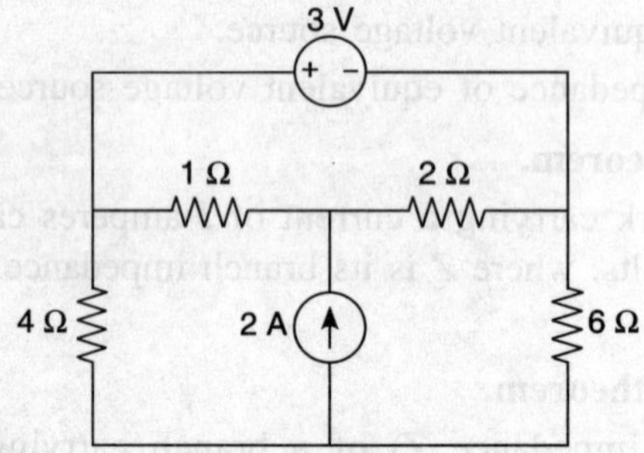

Fig. U6.2

3. **Compute the current I_L in the circuit shown in Fig. U6.3, using superposition theorem.**
[Ans. 11.2335 A]

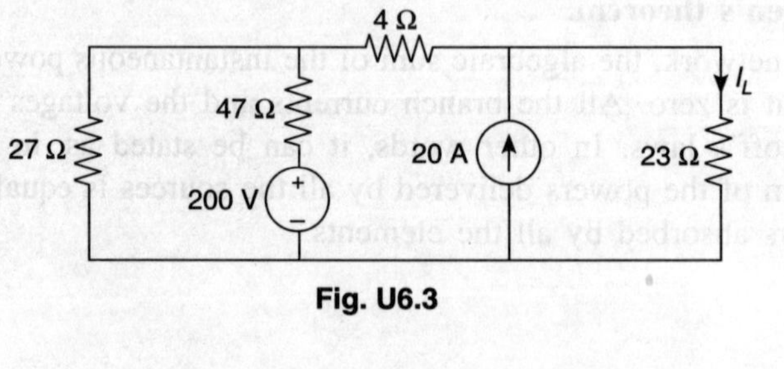

Fig. U6.3

4. Using superposition theorem, obtain the current delivered by voltage source of 1 V in the network shown in Fig. U6.4. **[Ans.** 0.0723 A]

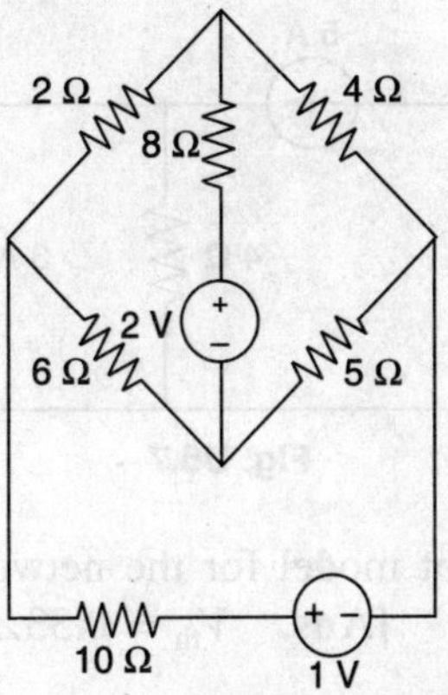

Fig. U6.4

5. Find the current through (3 + *j*4) Ω impedance by superposition theorem in the circuit shown in Fig. U6.5. **[Ans.** 8.31∠85.3° A]

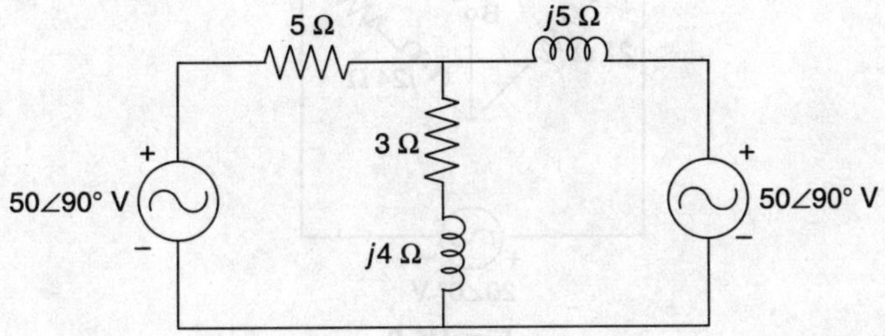

Fig. U6.5

6. Find the current flowing through the capacitor in the circuit shown in Fig. U6.6, using superposition theorem. **[Ans.** 6.41∠119.12° A]

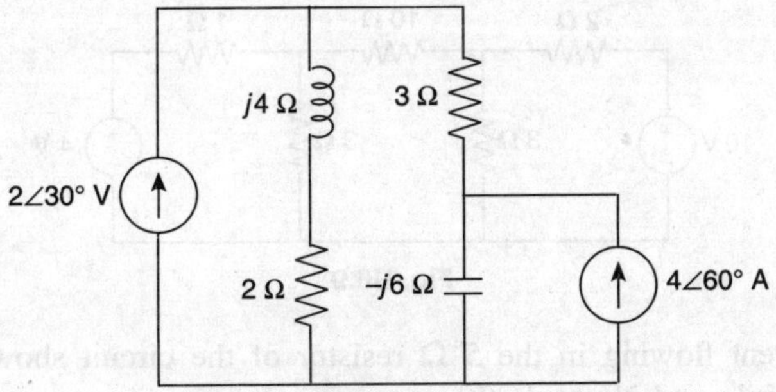

Fig. U6.6

7. Find the current through 6 Ω resistor in the circuit shown in Fig. U6.7, using by Thevenin's and Norton's theorems. [**Ans.** 3.2 A; V_{th} = 32 V; R_{th} = 4 Ω; I_N = 8 A]

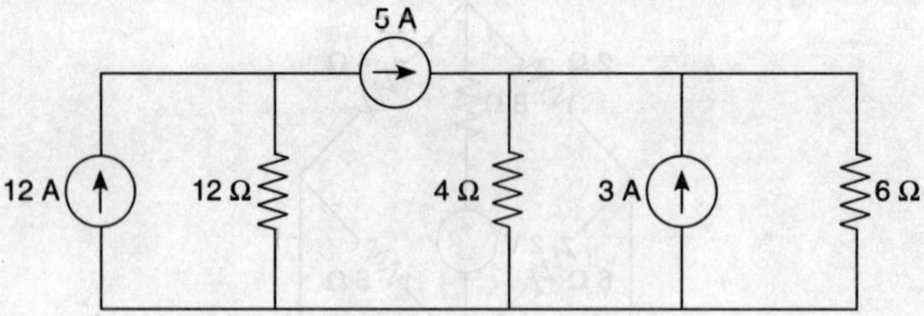

Fig. U6.7

8. Obtain Thevenin's equivalent model for the network shown in Fig. U6.8.
[**Ans.** V_{th} = 0.33∠171.12° V; Z_{th} = 47.35∠26.81° Ω]

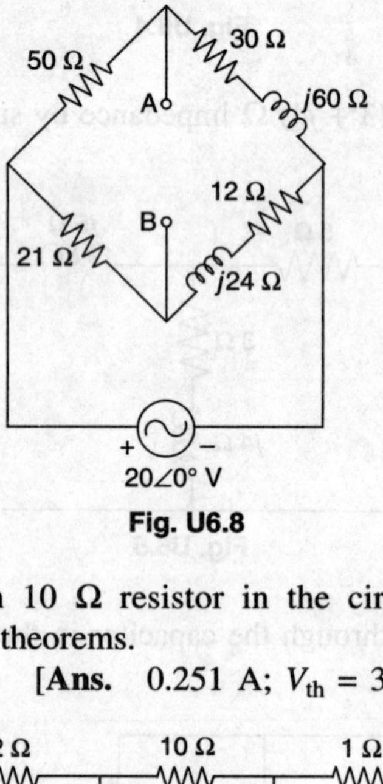

Fig. U6.8

9. Find the current through 10 Ω resistor in the circuit shown in Fig. U6.9, using Thevenin's and Norton's theorems.
[**Ans.** 0.251 A; V_{th} = 3 V; R_{th} = 1.95 Ω; I_N = 1.538 A]

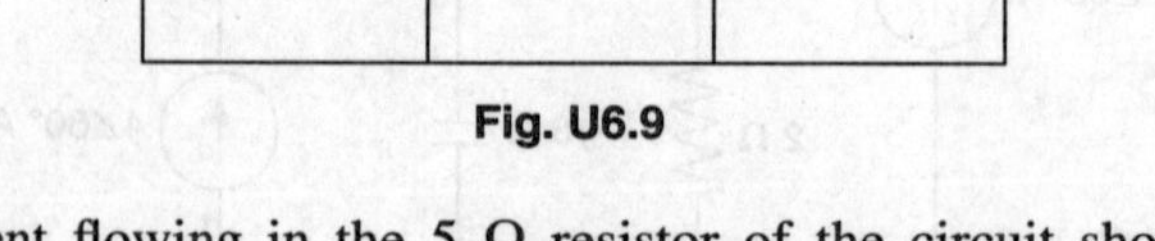

Fig. U6.9

10. Find the current flowing in the 5 Ω resistor of the circuit shown in Fig. U6.10, by using Thevenin's and Norton's theorems.
[**Ans.** 0.8876∠82.3° A; V_{th} = 10.93∠79.5° V;
Z_{th} = 7.32∠–4.78° Ω; I_N = 1.49∠84.28° A]

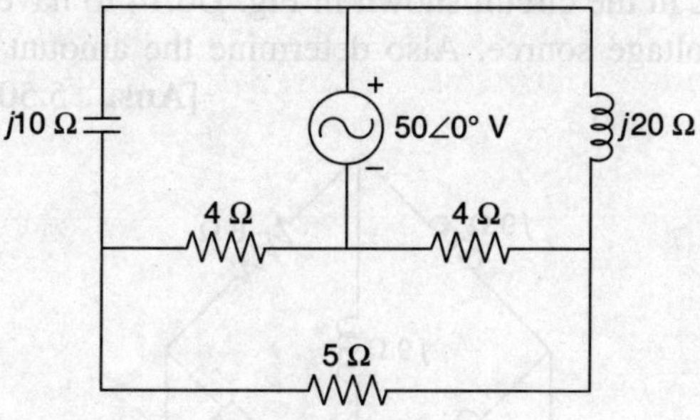

Fig. U6.10

11. Determine the value of resistance *R* in the network shown in Fig. U6.11, for maximum power transfer to it and the maximum power. **[Ans.** 2.7273 Ω; 2.7273 W]

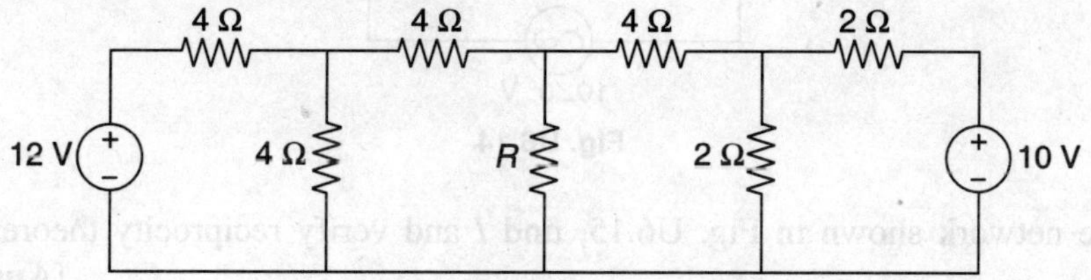

Fig. U6.11

12. Determine the value of *R* in the circuit shown in Fig. U6.12, for maximum power transfer to it. **[Ans.** 14.5 Ω]

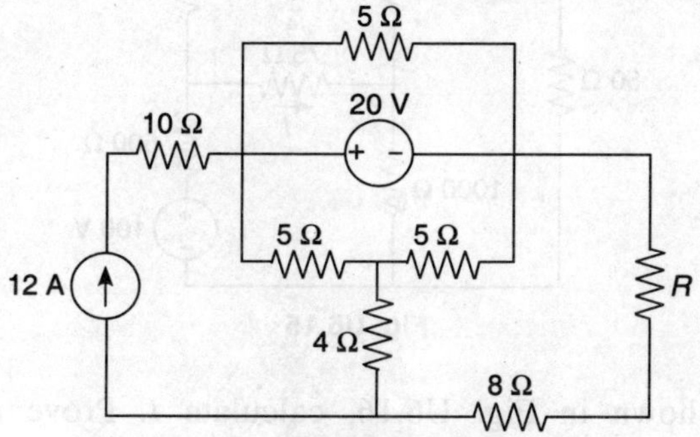

Fig. U6.12

13. Determine the value of *X* in the circuit shown in Fig. U6.13, for maximum power transfer to it. **[Ans.** 0.1583 Ω]

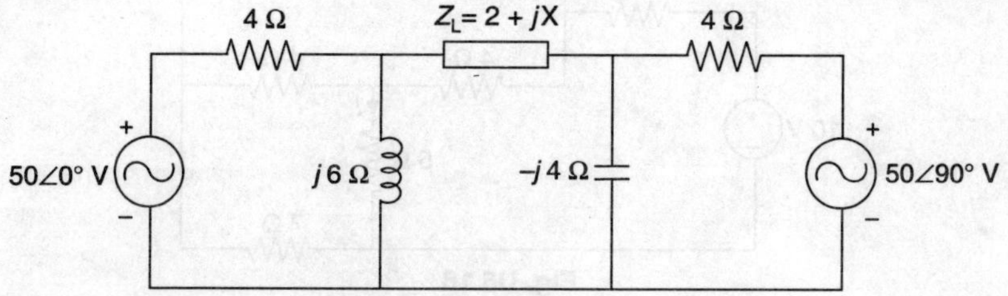

Fig. U6.13

14. Find the value of Z in the circuit shown in Fig. U6.14 to have maximum power transfer from the $10\angle 0°$ voltage source. Also determine the amount of maximum power.

[**Ans.** $5.5073\angle -82.49°\ \Omega$; 25.32 W]

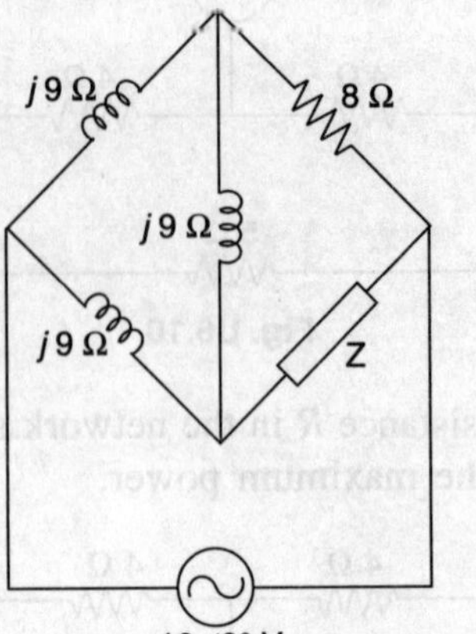

Fig. U6.14

15. In the network shown in Fig. U6.15, find I and verify reciprocity theorem.

[**Ans.** 0.043 A]

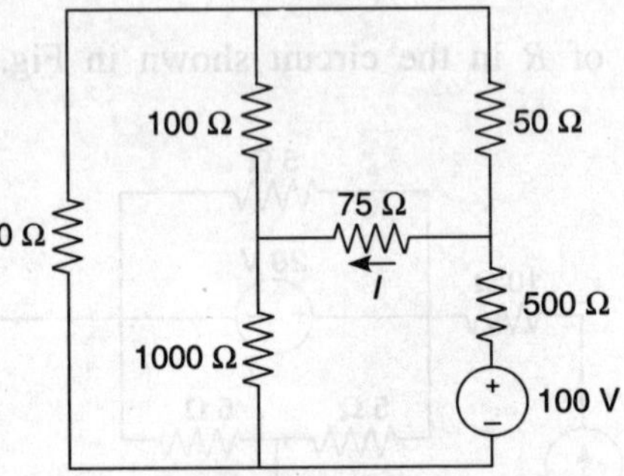

Fig. U6.15

16. In the circuit shown in Fig. U6.16, calculate I. Prove reciprocity theorem by interchanging the positions of 10 V and I.

[**Ans.** 0.8326 A]

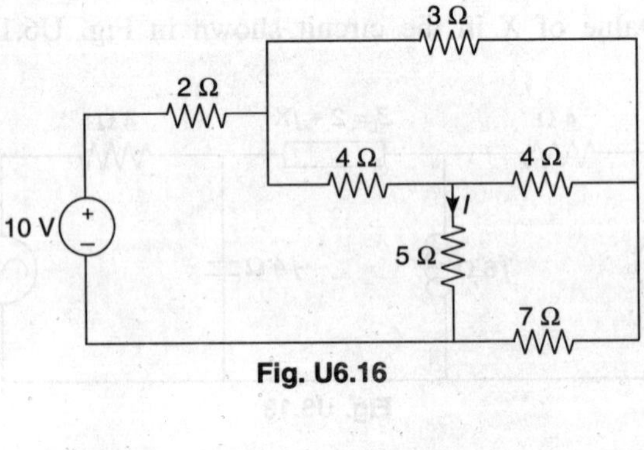

Fig. U6.16

17. In the circuit shown in Fig. U6.17, compute V_x. Verify reciprocity theorem by interchanging the positions of the source and response. [**Ans.** $15.3559\angle 90.4°$ V]

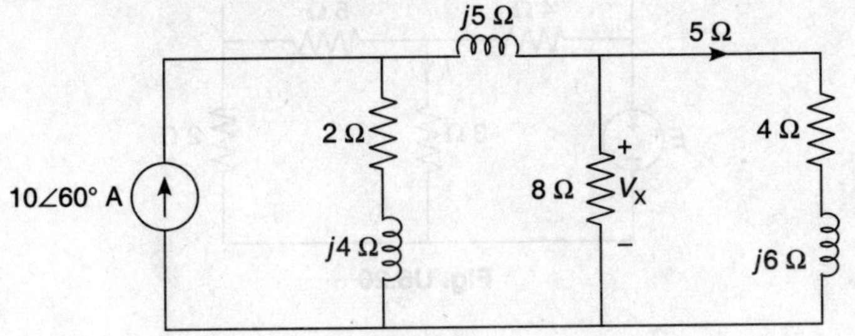

Fig. U6.17

18. In the circuit shown in Fig. U6.18, substitute the 8 Ω resistance branch by a voltage source and prove substitution theorem. [**Ans.** 20.8848 V]

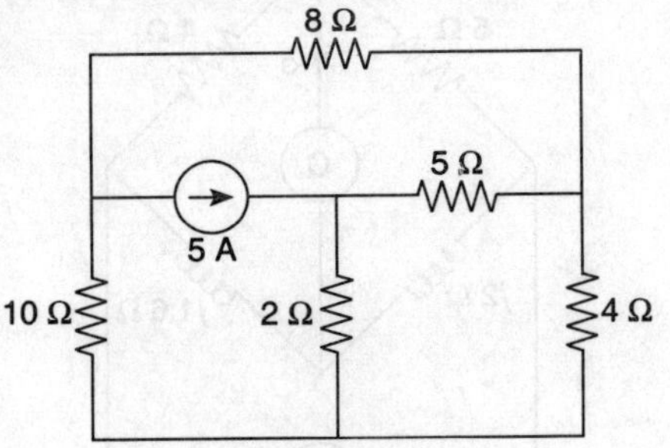

Fig. U6.18

19. In the circuit shown in Fig. U6.19, substitute the $(1 - j1)$ Ω branch by an equivalent branch and verify substitution theorem. [**Ans.** $10.5473\angle 95°$ A]

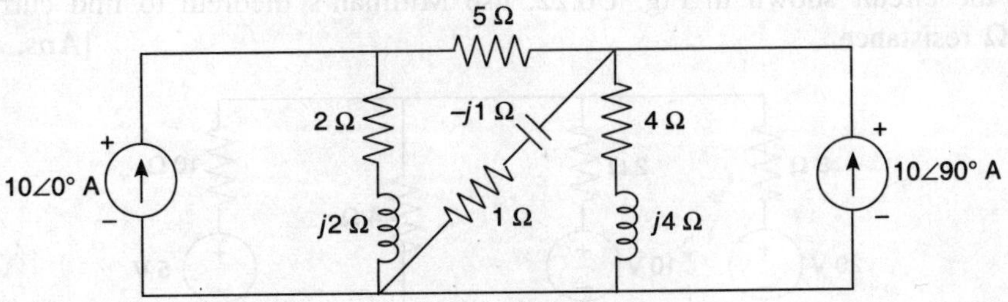

Fig. U6.19

20. In the circuit shown in Fig. U6.20, if the 8 Ω resistance is changed to 10 Ω resistance then find the change in current I_L using compensation theorem.
[**Ans.** $V_C = 10$ V; $\delta I_L = -0.1374$ A]

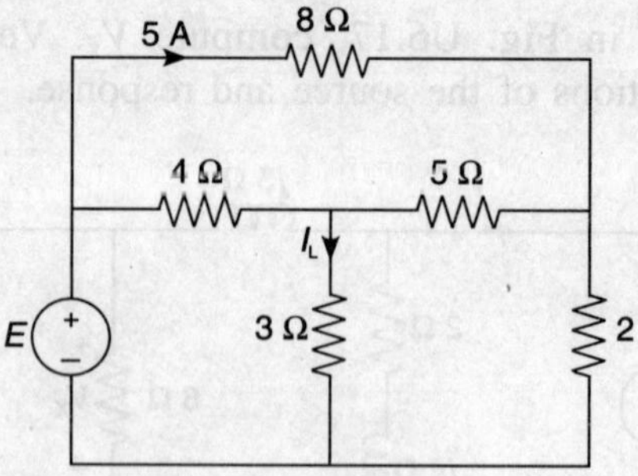

Fig. U6.20

21. The bridge circuit shown in Fig. U6.21 is balanced and so current through galvanometer is zero. If the $j1.6\ \Omega$ reactance is changed to $j1.2\ \Omega$, then determine the current through galvanometer using compensation theorem. **[Ans.** $0.1481\angle{-18.6^\circ}$ A]

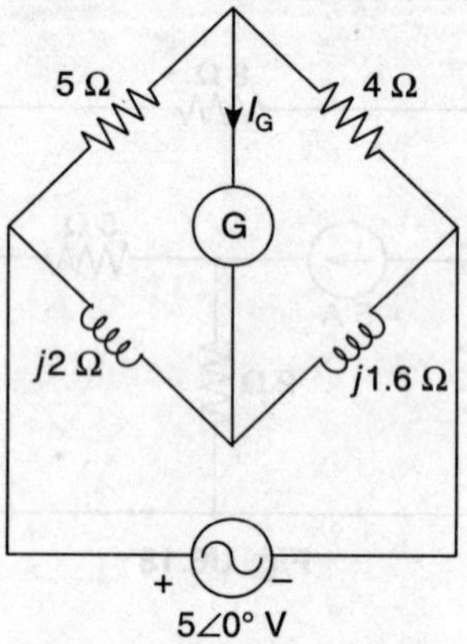

Fig. U6.21

22. In the circuit shown in Fig. U6.22, use Millman's theorem to find current through 4 Ω resistance. **[Ans.** 2.0513 A]

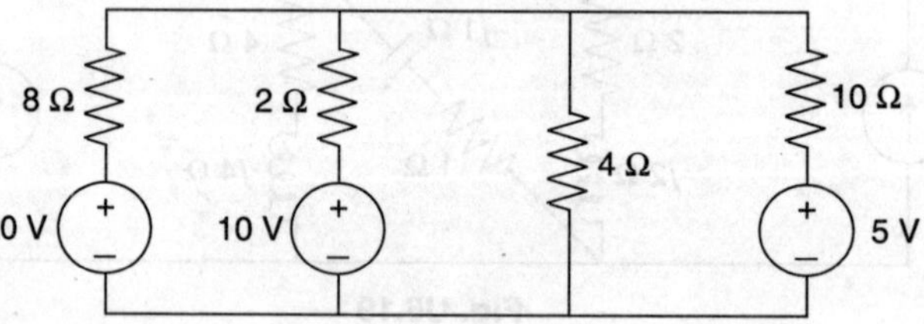

Fig. U6.22

23. In the circuit shown in Fig. U6.23, find the current through 2 Ω by using Millman's theorem. **[Ans.** $0.7433\angle{-138^\circ}$ A]

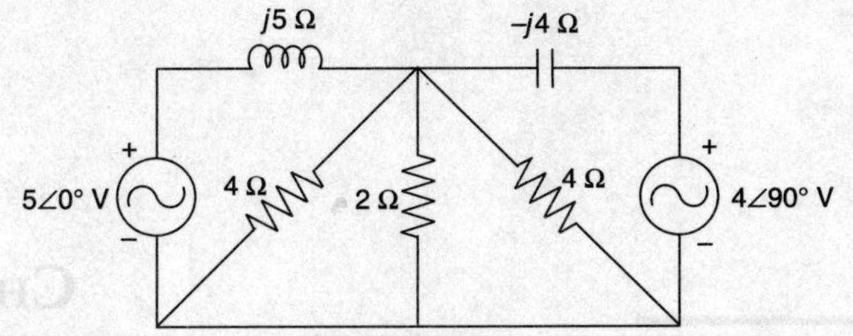

Fig. U6.23

24. In the circuit shown in Fig. U6.24, determine the branch currents and voltages and hence verify Tellegen's theorem.

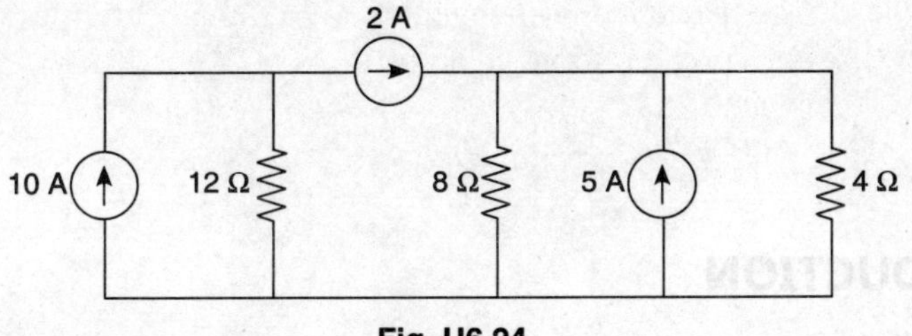

Fig. U6.24

25. In the circuit shown in Fig. U6.25, determine the branch currents and voltages and hence verify Tellegen's theorem.

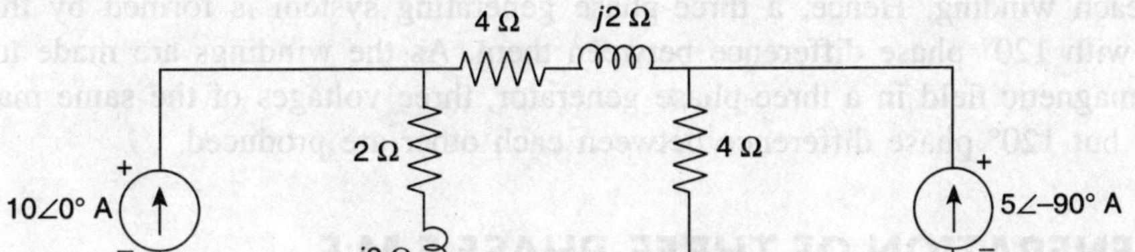

Fig. U6.25

CHAPTER 7

Three-Phase Circuits

7.1 INTRODUCTION

The AC voltages and currents discussed in the previous chapters are single-phase quantities because they are derived from a single source or single armature winding. In general, generation, transmission and distribution of electrical energy are in three phase. Three-phase system can be viewed as the combination of three single-phase systems with a phase difference of 120° between each winding. Hence, a three-phase generating system is formed by three separate windings with 120° phase difference between them. As the windings are made to rotate in a common magnetic field in a three-phase generator, three voltages of the same magnitude and frequency but 120° phase difference between each other are produced.

7.2 GENERATION OF THREE-PHASE E.M.F.

(i) Figure 7.1(a) shows a single-phase AC generator with stationary armature and rotating field. The two-pole magnetic field is assumed to be rotating in anticlockwise direction of angular velocity ω. Then, the e.m.f. induced in the winding is given by

$$e_{A_1A_2} = V_m \sin \omega t$$

(ii) Figure 7.1(b) shows a three-phase AC generator having three identical windings A, B and C displaced electrically with 120° from each other. The two-pole field is assumed to be rotating in anticlockwise direction of angular velocity ω. As shown in Fig. 7.1(b), the order of coils in the rotating field is A, B, C. It is to be noted that the e.m.f. in coil B is lagging coil A by 120° while leading coil C by 120°. Hence, the three e.m.f.s induced across each winding are

$$e_{A_1A_2} = V_m \sin \omega t$$

$$e_{B_1B_2} = V_m \sin(\omega t - 120°)$$

$$e_{C_1C_2} = V_m \sin(\omega t - 240°)$$

The order in which the alternating quantities reach their maximum values is called *phase sequence*. In the waveform seen in Fig. 7.1(b), the sequence of voltages generated is $e_{A_1A_2}$, $e_{B_1B_2}$ and $e_{C_1C_2}$. This direction of rotation of field system is called as positive phase sequence. On the other hand, if it is rotated in clock-wise direction, then the order will be $e_{A_1A_2}$, $e_{C_1C_2}$ and $e_{B_1B_2}$. It is called negative phase sequence. Generally, the letters R, Y and B are used to name the phasors to mean the colours red, yellow and blue, respectively. Hence, RYB is known as positive phase sequence and RBY is negative phase sequence.

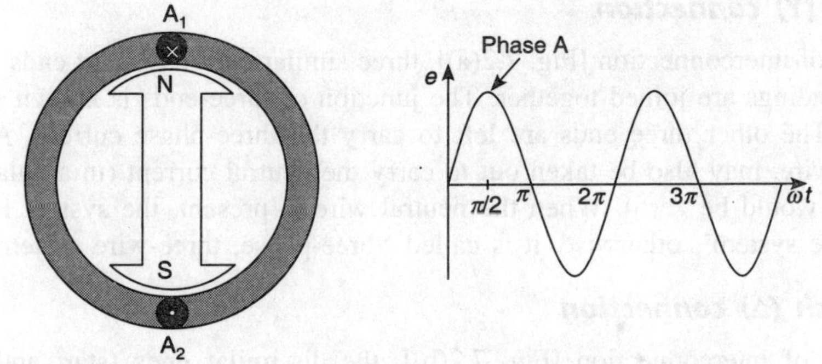

(a) Generation of single-phase e.m.f.

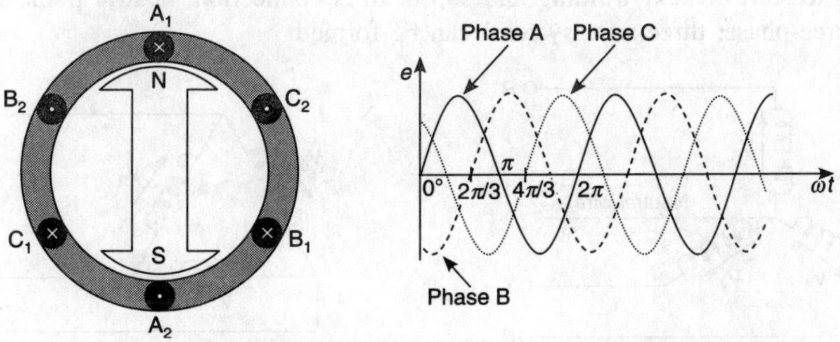

(b) Generation of three-phase e.m.f.

Fig. 7.1

7.3 ADVANTAGES OF THREE-PHASE SYSTEM

The three-phase system has the following advantages over a single-phase system.

(i) Generation, transmission and distribution of three-phase power is more economical than that of single-phase one because of less conductive material required.

(ii) Three-phase induction motors are self starting unlike single-phase induction motors.

(iii) For the same size, three-phase machine gives more output than single-phase machine.
(iv) Three-phase machine provides better power factor and efficiency.
(v) Converting AC to DC using rectifiers is smoother in three-phase system than single phase system.
(vi) In three-phase systems, better voltage regulation is obtained.

7.4 INTERCONNECTION OF THREE PHASES

In a three-phase system, there are three separate windings or phases and their interconnection is done on either of the two methods described below.

Star or Wye (Y) connection

In this method of interconnection [Fig. 7.2(a)], three similar ends say, start ends (or finish ends) of the three windings are joined together. The junction of three ends is known as *star point* or *neutral point*. The other three ends are left to carry the three-phase current. A separate line, called neutral wire, may also be taken out to carry the neutral current (in a balanced load, this neutral current would be zero). When the neutral wire is present, the system is called 'three-phase, four-wire system', otherwise, it is called 'three-phase, three-wire system'.

Delta or Mesh (Δ) connection

In this method of interconnection [Fig. 7.2(b)], the dissimilar ends (start and finish) of the three-phase windings are joined together to form a closed loop. That is, finish of one winding is connected to start of next winding and so on. In Δ connection, neutral point does not exist and only 'three-phase, three-wire system' can be formed.

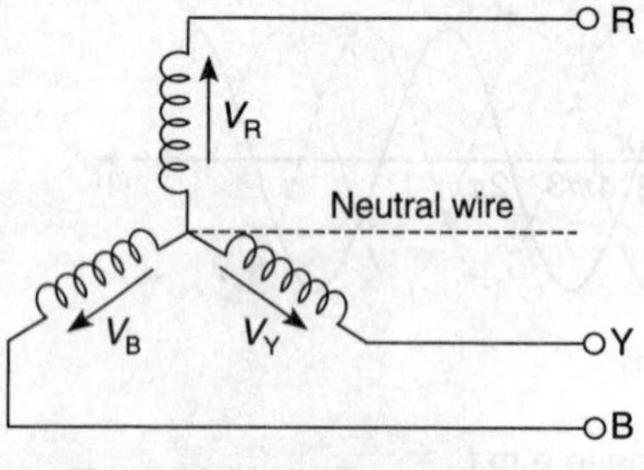

Fig. 7.2(a) Star connection.

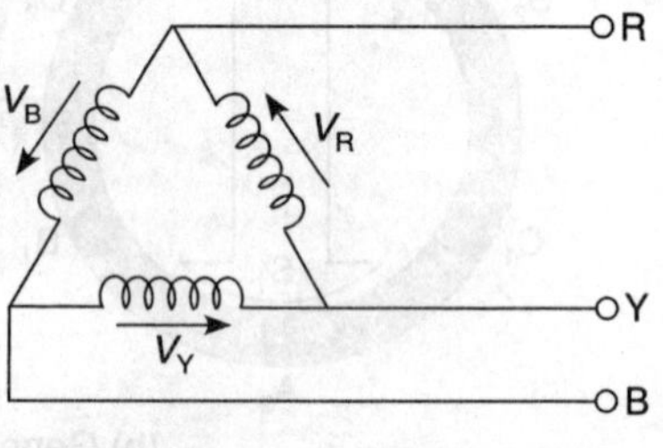

Fig. 7.2(b) Delta connection.

7.4.1 Balanced Y-Connection

Referring to Fig. 7.3, let the RMS values of the voltages measured across the three windings, known as phase voltages, be V_R, V_Y and V_B. Consequently, the line-to-line voltages are,

Voltage between lines R and Y, $V_{RY} = V_R - V_Y$
Voltage between lines Y and B, $V_{YB} = V_Y - V_B$
Voltage between lines B and R, $V_{BR} = V_B - V_R$

As the line voltage V_{RY} is equal to the phasor difference of V_R and V_Y, it is obtained by subtracting V_Y from V_R. To do this, phasor V_Y is reversed as seen in Fig. 7.4, such that $(V_R - V_Y)$ becomes $[V_R + (-V_Y)]$. From the phasor diagram, it is observed that the two phasors V_R and $-V_Y$ are equal in magnitude and are displaced with 60°.

$$V_{RY} = 2V_R \cos (60°/2)$$
$$= 2V_R \cos 30°$$
$$= \sqrt{3}V_R$$

Thus, in a balanced three-phase star connection, line voltage = $\sqrt{3}$ × phase voltage. It is also noted from Fig. 7.4, the phase voltage V_R lags behind the line voltage V_{RY} by 30°.

i.e. $$V_R = \frac{V_{RY}}{\sqrt{3}} \angle(0° - 30°)$$

Similarly, for a phase sequence of RYB,

$$V_Y = \frac{V_{YB}}{\sqrt{3}} \angle(-120° - 30°)$$

and $$V_B = \frac{V_{BR}}{\sqrt{3}} \angle(-240° - 30°)$$

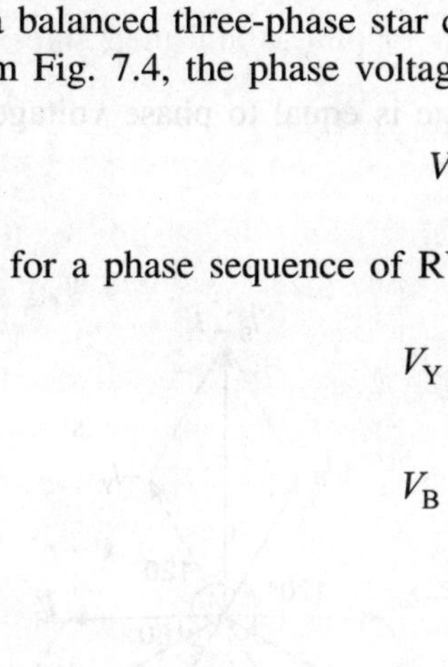

Fig. 7.3 Phase and line voltages in Y connection.

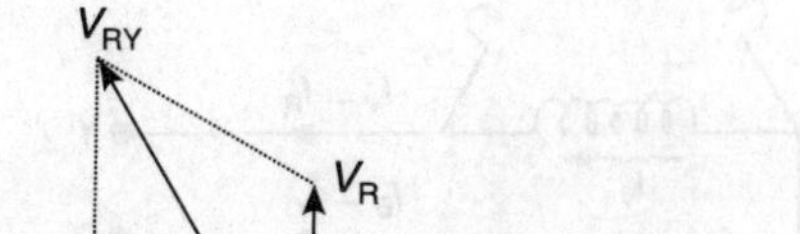

Fig. 7.4 Phasor diagram.

In star connection, the line current is same as the phase current, since each line conductor is connected in series with a separate winding (phase).

Hence, line current, $$I_L = I_{ph}$$

Total power, $$P = 3 \times \text{power in each phase}$$
$$= 3 \times V_{ph} I_{ph} \cos \phi$$
$$= 3V_{ph} I_{ph} \cos \phi \quad (7.1)$$

For Y connection, $$V_{ph} = \frac{V_L}{\sqrt{3}}$$

and $$I_L = I_{ph}$$

$$P = 3 \times \frac{V_L}{\sqrt{3}} \times I_L \times \cos\phi$$

$$P = \sqrt{3}\, V_L I_L \cos\phi \tag{7.2}$$

where ϕ is the phase difference between a phase voltage and the corresponding phase current.

7.4.2 Balanced Δ Connection

From Fig. 7.5, it is clear that in Δ connection, line voltage is equal to phase voltage, as each winding is connected between any two lines. Therefore,

$$V_L = V_{ph}$$

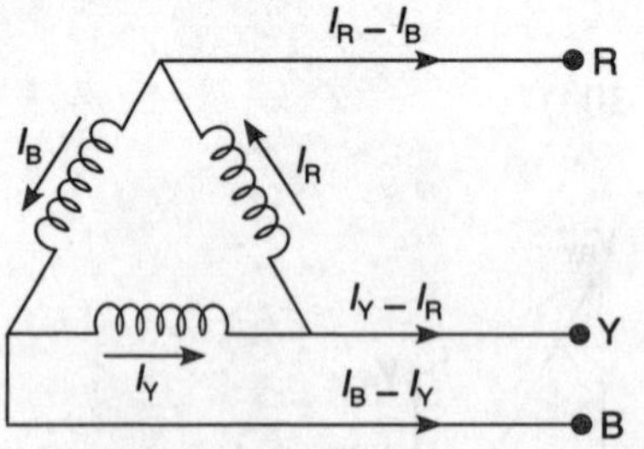

Fig. 7.5 Phase and line currents in Δ connection.

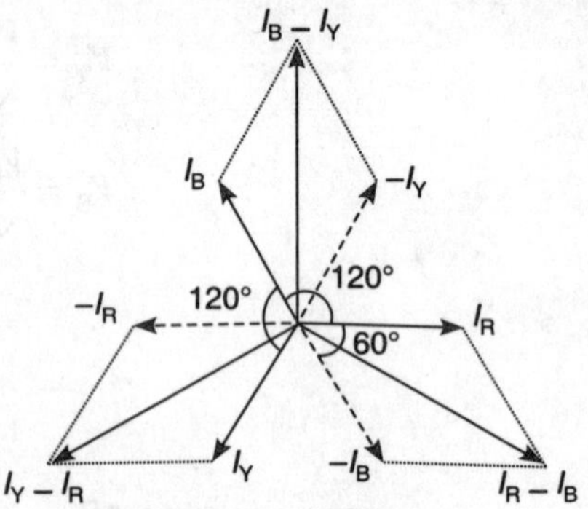

Fig. 7.6 Phasor diagram.

As shown in Fig. 7.5, in Δ connection, current in any line is equal to the phasor difference of the currents in the two phases connected to that line. Thus,

Current in line R = $I_R - I_B$
Current in line Y = $I_Y - I_R$
Current in line B = $I_B - I_Y$

As the current in line R is the phasor difference between I_R and I_B, it is obtained by subtracting I_B from I_R. To do this, phasor I_B is reversed as seen in Fig. 7.6 such that $[I_R - I_B]$ becomes $[I_R + (-I_B)]$. From the phasor diagram, it is observed that the two phasors I_R and $-I_B$ are equal in magnitude and are displaced with 60°.

So, the current in line R,

$$I_R - I_B = 2I_{ph} \cos(60°/2)$$

$$= 2I_{ph} \cos 30° = \sqrt{3} I_{ph}$$

Similarly,

$$\text{Current in line Y, } I_Y - I_R = \sqrt{3} I_{ph}$$

$$\text{Current in line B, } I_B - I_Y = \sqrt{3} I_{ph}$$

Thus, in a balanced three-phase delta connection, line current = $\sqrt{3}$ × phase current. It is also noted from Fig. 7.6, the line current $I_R - I_B$ lags behind the phase current I_R by 30°.

Total power,
$$P = 3 \times \text{power in each phase}$$
$$= 3 \times V_{ph} I_{ph} \cos \phi \quad (7.3)$$

For Δ connection,
$$V_{ph} = V_L$$

and
$$I_{ph} = \frac{I_L}{\sqrt{3}}$$

$$P = 3 \times V_L \times \frac{I_L}{\sqrt{3}} \times \cos \phi$$

$$P = \sqrt{3}\, V_L I_L \cos\phi \quad (7.4)$$

where cos ϕ is the power factor of each phase.

EXAMPLE 7.1 Three coils, each having a resistance of 10 Ω and an inductive reactance of 15 Ω, are connected in star to a 440 V, three-phase, 50 Hz supply, as shown in Fig. 7.7. Calculate (i) the line current, (ii) power factor and (iii) power supplied.

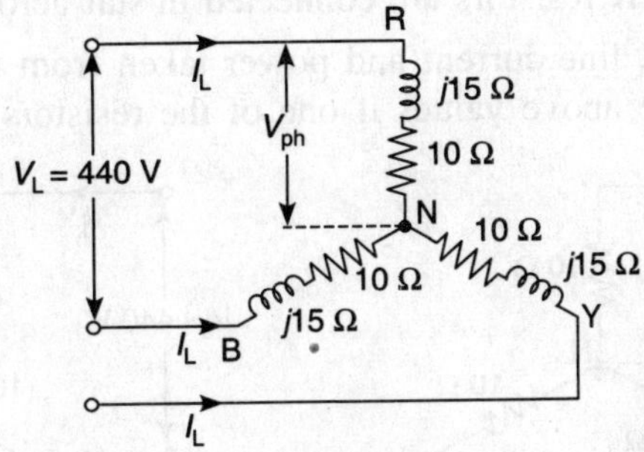

Fig. 7.7

Solution
$$V_{ph} = \frac{V_L}{\sqrt{3}} = \frac{440}{\sqrt{3}} = 254.03 \text{ V}$$

$$Z_{ph} = \sqrt{10^2 + 15^2} = 18.03\ \Omega$$

$$I_{ph} = \frac{V_{ph}}{Z_{ph}} = \frac{254.03}{18.03} = 14.09 \text{ A}$$

(i) Line current, $I_L = I_{ph}$ = **14.09 A.**

(ii) Power factor, $\cos \phi = \frac{R_{ph}}{Z_{ph}} = \frac{10}{18.03}$ = **0.5546 (lag)**

(iii) Power, $P = \sqrt{3}\, V_L I_L \cos\phi = \sqrt{3} \times 440 \times 14.09 \times 0.5546$ = **5955.3 W**

EXAMPLE 7.2 A balanced star-connected load of impedance (3 + $j4$) ohms per phase is connected to a three-phase, 230 V, 50 Hz supply. Find the line current and power absorbed by each phase.

Solution

$$Z_{ph} = \sqrt{3^2 + 4^2} = 5\ \Omega$$

$$V_{ph} = \frac{V_L}{\sqrt{3}} = \frac{230}{\sqrt{3}} = 132.8\ \text{V}$$

$$\cos\phi = \frac{R_{ph}}{Z_{ph}} = \frac{3}{5} = 0.6\ \text{(lag)}$$

$$I_{ph} = \frac{V_{ph}}{Z_{ph}} = \frac{132.8}{5} = 26.56\ \text{A}$$

Therefore, line current, $I_L = I_{ph}$ = **26.56 A**

Power absorbed per phase = $V_{ph}\, I_{ph} \cos\phi$

= 132.8 × 26.56 × 0.6 = **2116.3 W.**

EXAMPLE 7.3 Three 10 Ω resistors are connected in star across 440 V, three-phase supply.

(i) Find phase current, line current and power taken from the mains.

(ii) What would be the above values if one of the resistors were disconnected?

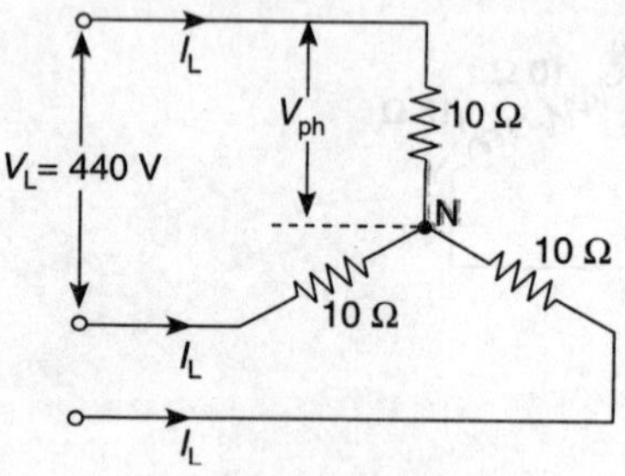

Fig. 7.8

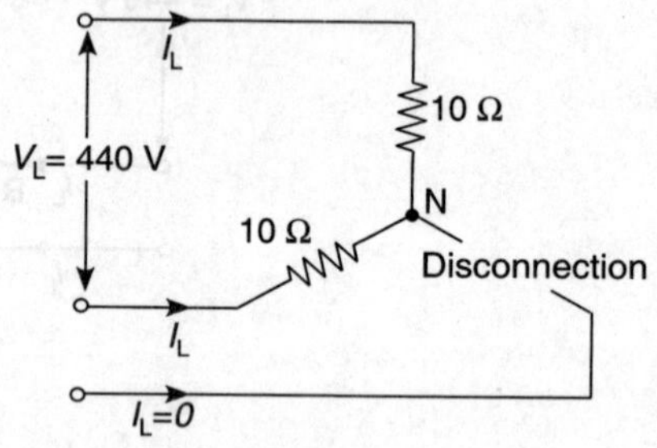

Fig. 7.9

Solution

$$V_{ph} = 440/\sqrt{3} = 254.03\ \text{V}$$

$$Z_{ph} = R_{ph} = 10\ \Omega$$

$$\cos\phi = 1$$

(i) *When the three resistors are star-connected (Fig. 7.8)*

$$I_{ph} = \frac{V_{ph}}{Z_{ph}} = \frac{254.03}{10} = \mathbf{25.4\ A}$$

$$I_L = I_{ph} = \mathbf{25.4\ A}$$

Power taken, $P = \sqrt{3}\, V_L I_L \cos\phi = \sqrt{3} \times 440 \times 25.4 \times 1$ = **19357.4 W**

(ii) *When one of the resistors is disconnected (Fig. 7.9)*
When one of the resistors is disconnected, the remaining two resistors behave as if they were connected in series across the line voltage. In fact, the circuit behaves as a single-phase circuit.

Therefore, $$I_{ph} = I_L = \frac{440}{10+10} = 22\text{ A}$$

Power taken, $$V_L I_L \cos\phi = 440 \times 22 \times 1 = \mathbf{9680\ W}.$$

EXAMPLE 7.4 Three 20 Ω non-inductive resistances are connected in delta across 400 V, threee-phase lines. Calculate the power taken from the mains. If one of the resistances is disconnected, what would be the power taken from the mains?

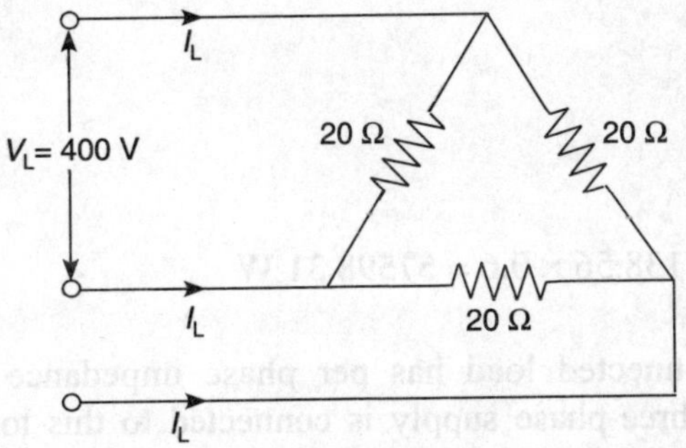

Fig. 7.10

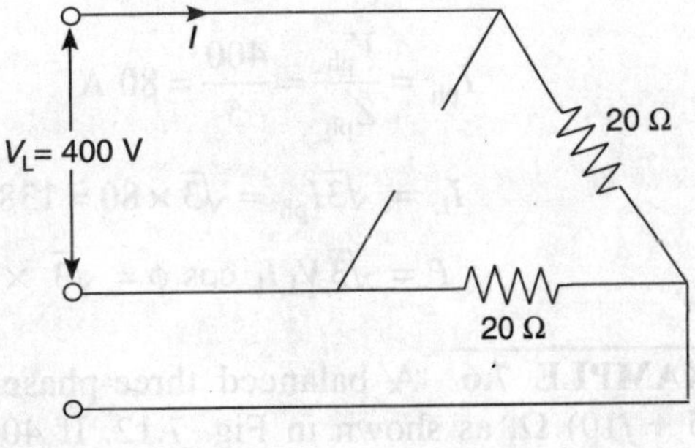

Fig. 7.11

Solution *When the three resistors are delta-connected (Fig. 7.10)*

$$V_{ph} = V_L = 400\text{ V}$$

$$I_{ph} = \frac{V_{ph}}{Z_{ph}} = \frac{400}{20} = 20\text{ A}$$

$$I_L = \sqrt{3}I_{ph} = \sqrt{3} \times 20 = 34.64\text{ A}$$

$$P = \sqrt{3}\,V_L I_L \cos\phi = \sqrt{3} \times 400 \times 34.64 \times 1 = \mathbf{24000\ W}$$

When one resistor is removed
Figure 7.11 shows the circuit with one of the resistors is removed.

Current in each resistor, $I = 400/20 = 20$ A

Power consumed in the two resistors = $2I^2R = 2 \times 20^2 \times 20 =$ **16000 W**

EXAMPLE 7.5 A balanced three-phase load consists of three coils, each with resistance 3 Ω and inductive reactance of 4 Ω. Determine the line current and power absorbed when the coils are (i) star-connected and (ii) delta-connected across 400 V, three-phase supply.

Solution $$Z_{ph} = \sqrt{3^2 + 4^2} = 5\ \Omega$$

$$\cos\phi = \frac{R_{ph}}{Z_{ph}} = \frac{3}{5} = 0.6\text{ (lag)}$$

Star connection

$$V_{ph} = \frac{V_L}{\sqrt{3}} = \frac{400}{\sqrt{3}} = 230.94 \text{ V}$$

$$I_{ph} = \frac{V_{ph}}{Z_{ph}} = \frac{230.94}{5} = 46.19 \text{ A}$$

$$I_L = I_{ph} = 46.19 \text{ A}$$

$$P = \sqrt{3} V_L I_L \cos\phi = \sqrt{3} \times 400 \times 46.19 \times 0.6 = \mathbf{19200.82 \text{ W}}$$

Delta connection

$$V_{ph} = V_L = 400 \text{ V}$$

$$I_{ph} = \frac{V_{ph}}{Z_{ph}} = \frac{400}{5} = 80 \text{ A}$$

$$I_L = \sqrt{3} I_{ph} = \sqrt{3} \times 80 = 138.56 \text{ A}$$

$$P = \sqrt{3} V_L I_L \cos\phi = \sqrt{3} \times 400 \times 138.56 \times 0.6 = \mathbf{57598.31 \text{ W}}$$

EXAMPLE 7.6 A balanced three-phase, Δ-connected load has per phase impedance of (15 + *j*10) Ω, as shown in Fig. 7.12. If 400 V, three-phase supply is connected to this load, find (i) the phase currents (ii) line currents and (iii) power supplied by the load. The phase sequence is RYB.

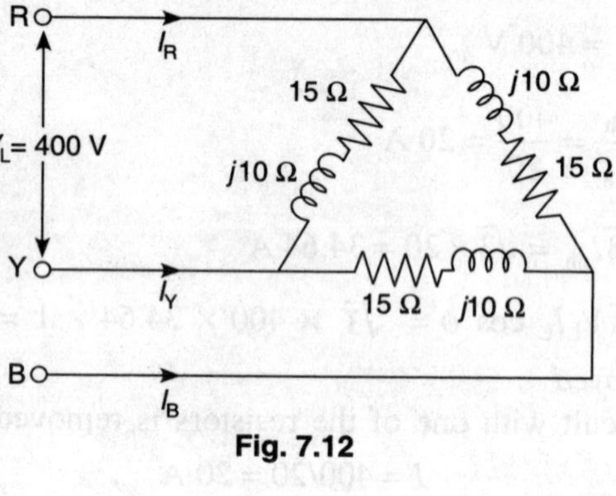

Fig. 7.12

Solution

$$V_{RY} = 400\angle 0^\circ$$

$$V_{YB} = 400\angle -120^\circ$$

$$V_{BR} = 400\angle -240^\circ$$

(i) The phase currents are

$$I_{RY} = \frac{V_{RY}}{Z_{RY}} = \frac{400\angle 0^\circ}{15 + j10} = 18.46 - j12.31 = 22.19\angle -33.69^\circ \text{ A}$$

Similarly,

$$I_{YB} = 22.19\angle{-153.69°} \text{ A}$$
$$I_{BR} = 22.19\angle{-273.69°} \text{ A}$$

(ii) Line currents are

$$I_R = \sqrt{3}\angle{-30°} \times I_{RY}$$ (In Δ-connected loads, the line current lags behind the phase current by 30°)

$$= \sqrt{3}\angle{-30°} \times 22.19\angle{-33.69°}$$
$$= 38.43\angle{-63.69°} \text{ A}$$
$$I_Y = 38.43\angle{-183.69°} \text{ A}$$
$$I_B = 38.43\angle{-303.69°} \text{ A}$$

(iii) Power supplied, $P = \sqrt{3}\, V_L I_L \cos\phi$

$$= \sqrt{3} \times 400 \times 38.43 \times \cos(33.69°) = \mathbf{22152\ W}$$

EXAMPLE 7.7 A balanced, three-phase, Y-connected load has per phase impedance of (25 + *j*40) Ω, as shown in Fig. 7.13. If 400 V, three-phase supply is connected to load, find (i) the phase currents (ii) line currents and (iii) power supplied by the load. The phase sequence is RYB.

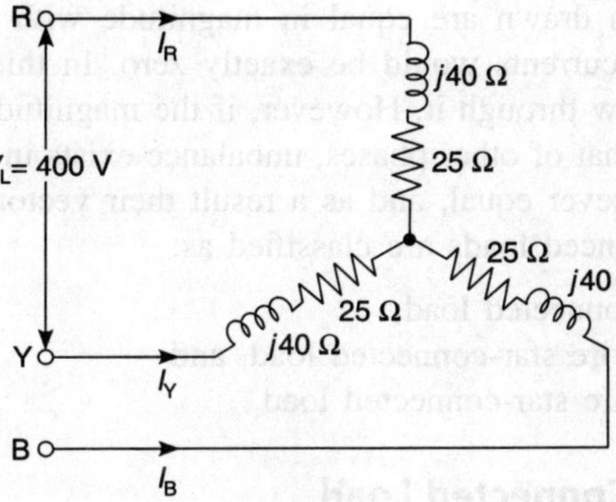

Fig. 7.13

Solution For the given phase sequence of RYB, the phase voltages are as follows:

$$V_{RN} = \frac{400\angle 0°}{\sqrt{3}} = 230.94\angle{-30°}$$ (In Y-connected loads, the phase voltage lags behind the line voltage by 30°)

$$V_{YN} = \frac{400\angle{-120°}}{\sqrt{3}} = 230.94\angle{-150°}$$

$$V_{BN} = \frac{400\angle{-240°}}{\sqrt{3}} = 230.94\angle{-270°}$$

(i) The phase currents are

$$I_{RN} = \frac{V_{RN}}{Z_{RY}} = \frac{230.94\angle -30^\circ}{25 + j40} = \frac{230.94\angle -30^\circ}{47.17\angle 58^\circ} = 4.9\angle -88^\circ \text{ A}$$

Similarly,

$$I_{YN} = 4.9\angle -208^\circ \text{ A}$$

$$I_{BN} = 4.9\angle 8^\circ \text{ A}$$

(ii) In Y-connected loads, the line currents and phase currents are same. Therefore, the line currents are

$$I_R = I_{RN} = 4.9\angle -88^\circ \text{ A}$$

$$I_Y = I_{YN} = 4.9\angle -208^\circ \text{ A}$$

$$I_B = I_{BN} = 4.9\angle 8^\circ \text{ A}$$

(iii) Power supplied, $P = \sqrt{3}\, V_L I_L \cos\phi$

$$= \sqrt{3} \times 400 \times 4.9 \times \cos(58^\circ) = \mathbf{1798.98\ W}$$

7.5 UNBALANCED LOADS

In a balanced system, all the three impedances of the load are identical in ohmic value. Therefore, the line currents drawn are equal in magnitude with 120° phase shift each. The vector sum of three-phase currents would be exactly zero. In this scenario, if a neutral wire exists, no current would flow through it. However, if the magnitude of impedances of any one of the phases differs from that of other phases, unbalance exists in the system. In this case, the phase or line currents are never equal, and as a result their vector sum is not equal to zero.

The three phase unbalanced loads are classified as:

1. three-phase delta-connected load,
2. three-phase three-wire star-connected load, and
3. three-phase four-wire star-connected load.

7.5.1 Unbalanced Δ-Connected Load

A three-phase unbalanced load connected to a balanced three-phase supply system is shown in Fig. 7.14. In three-phase unbalanced delta-connected loads, the phase voltages and the line voltages are equal. Hence, the phase currents could easily be obtained using Ohm's law. Subsequently, the line currents could be computed by applying Kirchhoff's current law.

Let Z_{RY} is the impedance connected between R and Y lines, then the phase current

$$I_{RY} = \frac{V_{RY}}{Z_{RY}} = |I_{RY}|\angle\alpha$$

Similarly,

$$I_{YB} = \frac{V_{YB}}{Z_{YB}} = |I_{YB}|\angle\beta$$

$$I_{BR} = \frac{V_{BR}}{Z_{BR}} = |I_{BR}|\angle\gamma$$

where $|I_{RY}|$, $|I_{YB}|$ and $|I_{BR}|$ are magnitude of phase currents and α, β and γ are their respective phase angles.

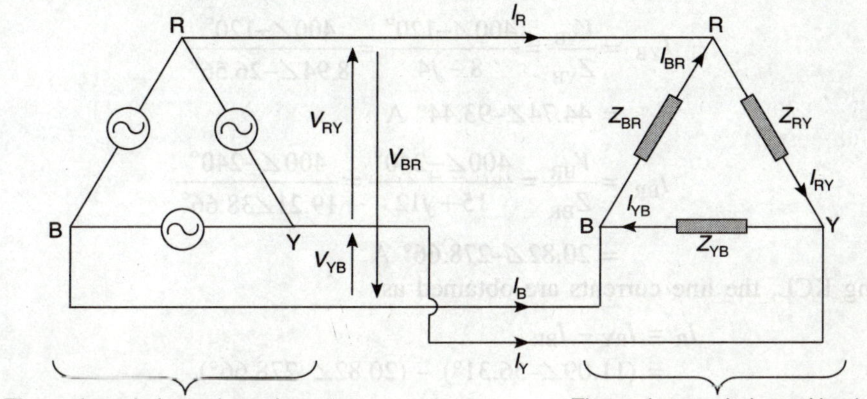

Fig. 7.14 Three-pase unbalanced Δ-connected load.

Using KCL at nodes R, Y and B, the line currents are calculated as:

$$I_R = I_{RY} - I_{BR}$$

$$I_Y = I_{YB} - I_{RY}$$

and

$$I_B = I_{BR} - I_{YB}$$

EXAMPLE 7.8 Determine the line currents for the unbalanced delta-connected load shown in Fig. 7.15. Assume the phase sequence be RYB. Take line voltage as 400 V.

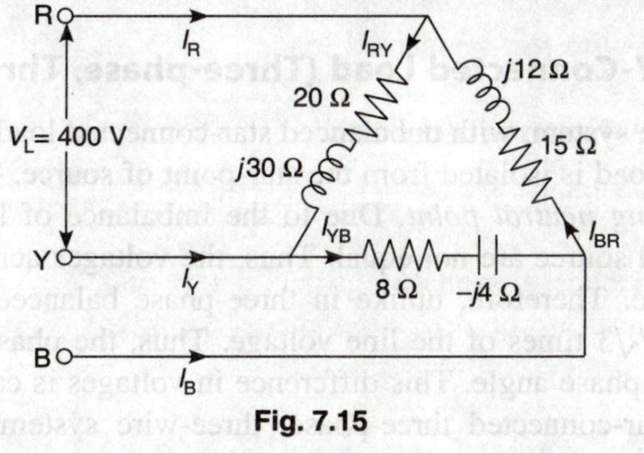

Fig. 7.15

Solution

$$V_{RY} = 400\angle 0°$$

$$V_{YB} = 400\angle -120°$$

$$V_{BR} = 400\angle -240°$$

The phase currents I_{RY}, I_{YB} and I_{BR} are calculated as:

$$I_{RY} = \frac{V_{RY}}{Z_{RY}} = \frac{400\angle 0^\circ}{20 + j30} = \frac{400\angle 0^\circ}{36.06\angle 56.31^\circ}$$

$$= 11.09\angle -56.31^\circ \text{ A}$$

$$I_{YB} = \frac{V_{YB}}{Z_{YB}} = \frac{400\angle -120^\circ}{8 - j4} = \frac{400\angle -120^\circ}{8.94\angle -26.56^\circ}$$

$$= 44.74\angle -93.44^\circ \text{ A}$$

$$I_{BR} = \frac{V_{BR}}{Z_{BR}} = \frac{400\angle -240^\circ}{15 + j12} = \frac{400\angle -240^\circ}{19.21\angle 38.66^\circ}$$

$$= 20.82\angle -278.66^\circ \text{ A}$$

By applying KCL, the line currents are obtained as:

$$I_R = I_{RY} - I_{BR}$$
$$= (11.09\angle -56.31^\circ) - (20.82\angle -278.66^\circ)$$
$$= 3.0167 - j29.81$$
$$= \mathbf{29.96\angle -84.22^\circ \text{ A}}$$

$$I_Y = I_{YB} - I_{RY}$$
$$= (44.74\angle -93.44^\circ) - (11.09\angle -56.31^\circ)$$
$$= -8.8362 - j35.43$$
$$= \mathbf{36.52\angle -104^\circ \text{ A}}$$

$$I_B = I_{BR} - I_{YB}$$
$$= (20.82\angle -278.66^\circ) - (44.74\angle -93.44^\circ)$$
$$= 5.8194 + j65.24$$
$$= \mathbf{65.5\angle 84.9^\circ \text{ A}}$$

7.5.2 Unbalanced Y-Connected Load (Three-phase, Three-wire System)

In three-phase, three-wire system, with unbalanced star-connected loads, there is no neutral wire. Hence, the star-point of load is isolated from the star-point of source. This isolated neutral point of a load is called *floating neutral point*. Due to the imbalance of load, the potentials of the star point of the load and source are not equal. Thus, the voltages across the different branches of the load are not same. Therefore, unlike in three-phase balanced star-connected load, the phase voltages are not $1/\sqrt{3}$ times of the line voltage. Thus, the phase voltages differ not only in magnitude but also in phase angle. This difference in voltages is called *neutral shift voltage*.

Such unbalanced star-connected three-phase, three-wire system can be analysed by the following methods:

1. Star–delta conversion
2. Mesh analysis
3. Millman's theorem

EXAMPLE 7.9 Three impedances of (5 + *j*3) Ω, (4 + *j*6) Ω and (2 + *j*5) Ω are connected in star across a three-phase, 440 V supply as shown in Fig. 7.16. Find the three line currents. Assume ABC as phase sequence.

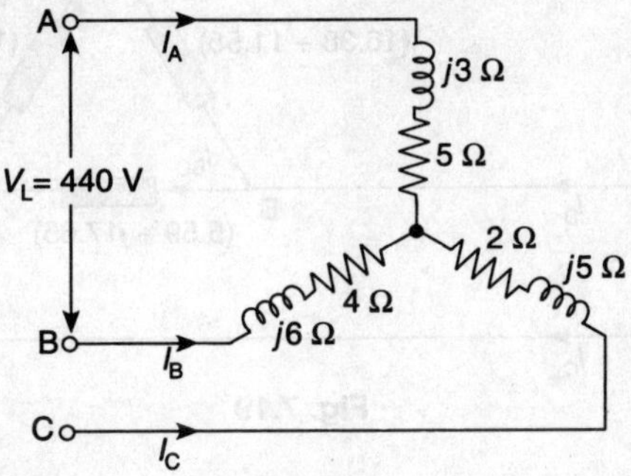

Fig. 7.16

Solution

Method I (Star–Delta conversion)

Figures 7.17 and 7.18 illustrate the conversion of star network into an equivalent delta.

$$Z_{AB} = \frac{Z_A Z_B + Z_B Z_C + Z_A Z_C}{Z_C}$$

$$= \frac{(5+j3)(4+j6)+(4+j6)(2+j5)+(5+j3)(2+j5)}{(2+j5)}$$

$$= 16.38 + j11.55$$

$$Z_{BC} = \frac{Z_A Z_B + Z_B Z_C + Z_A Z_C}{Z_A}$$

$$= \frac{(5+j3)(4+j6)+(4+j6)(2+j5)+(5+j3)(2+j5)}{(5+j3)}$$

$$= 5.59 + j17.65$$

$$Z_{AC} = \frac{Z_A Z_B + Z_B Z_C + Z_A Z_C}{Z_B}$$

$$= \frac{(5+j3)(4+j6)+(4+j6)(2+j5)+(5+j3)(2+j5)}{(4+j6)}$$

$$= 10.19 + j10.96$$

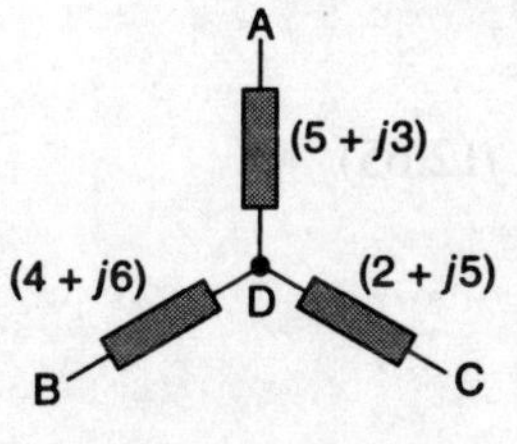

Fig. 7.17

⇒

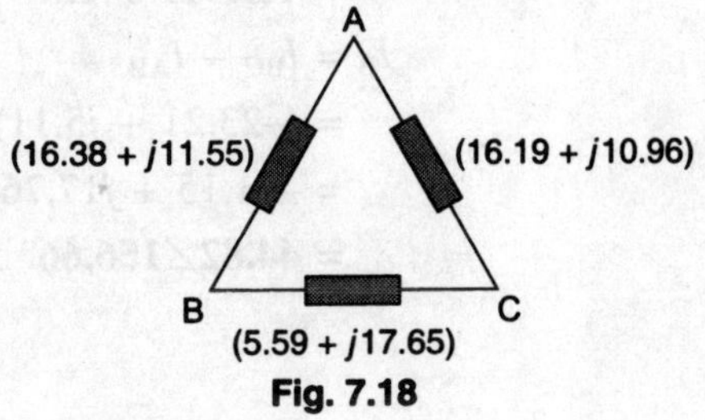

Fig. 7.18

The star connection is replaced with its equivalent delta as shown in Fig. 7.19.

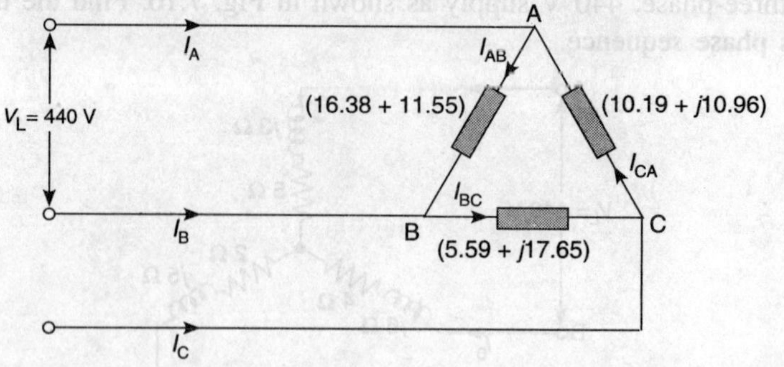

Fig. 7.19

Since the line voltages and phase voltages are same in delta connection, the phase currents are calculated as follows:

$$I_{AB} = \frac{V_{AB}}{Z_{AB}}$$

$$= \frac{440\angle 0°}{16.38 + j11.55} = 17.94 - j12.65$$

$$I_{BC} = \frac{V_{BC}}{Z_{BC}}$$

$$= \frac{440\angle -120°}{5.59 + j17.65} = -23.21 + j5.11$$

$$I_{AC} = \frac{V_{AC}}{Z_{AC}}$$

$$= \frac{440\angle -240°}{10.19 + j10.96} = 8.64 + j28.1$$

Applying KCL at nodes A, B and C, the line currents are calculated as follows:

$$I_A = I_{AB} - I_{AC}$$
$$= (17.94 - j12.65) - (8.64 + j28.1)$$
$$= 9.3 - j40.75$$
$$= \mathbf{41.79\angle -77.14°\ A}$$

$$I_B = I_{BC} - I_{AB}$$
$$= (-23.21 + j5.11) - (17.94 - j12.65)$$
$$= -41.15 + j17.76$$
$$= \mathbf{44.82\angle 156.66°\ A}$$

$$I_C = I_{AC} - I_{BC}$$
$$= (8.64 + j28.1) - (-23.21 + j5.11)$$
$$= 31.85 + j22.99$$
$$= \mathbf{39.28\angle 35.82^\circ\ A}$$

Method II (mesh analysis)

In this method, the line voltages are assumed to be individual voltage sources and are connected as shown in Fig. 7.20. Two mesh currents I_1 and I_2 are also assumed in the shown directions. Now,

$$I_A = I_1$$
$$I_B = I_2 - I_1$$
$$I_C = -I_2$$

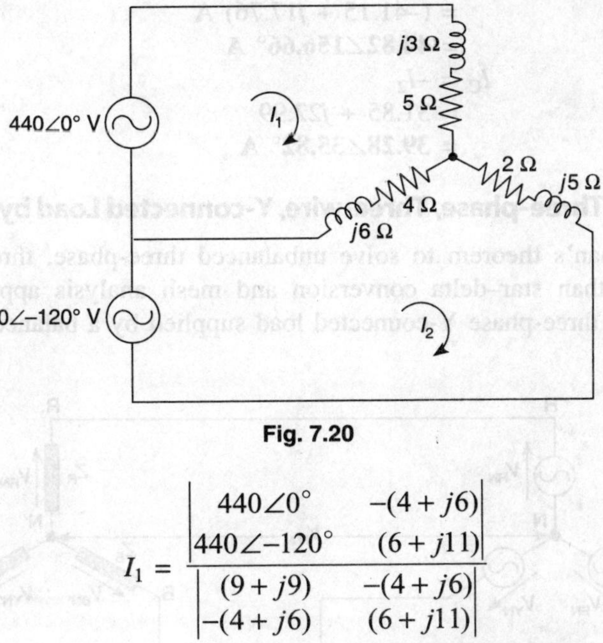

Fig. 7.20

$$I_1 = \frac{\begin{vmatrix} 440\angle 0^\circ & -(4+j6) \\ 440\angle -120^\circ & (6+j11) \end{vmatrix}}{\begin{vmatrix} (9+j9) & -(4+j6) \\ -(4+j6) & (6+j11) \end{vmatrix}}$$

$$= \frac{(2640+j4840)-(-1406.31+j2844.2)}{(-45+j153)-(-20+j48)}$$

$$= \frac{(4046.31+j1995.8)}{(-25+j105)}$$

$$= 9.3 - j40.75$$
$$= 41.79\angle -77.14^\circ\ A$$

$$I_2 = \frac{\begin{vmatrix} (9+j9) & 440\angle 0^\circ \\ -(4+j6) & 440\angle -120^\circ \end{vmatrix}}{\begin{vmatrix} (9+j9) & -(4+j6) \\ -(4+j6) & (6+j11) \end{vmatrix}}$$

$$= \frac{(1449.46 - j5409.46) - (-1760 - j2640)}{(-45 + j153) - (-20 + j48)}$$

$$= \frac{(3209.46 - j2769.46)}{(-25 + j105)}$$

$$= -31.85 - j22.99$$

$$= 39.28\angle -144.18° \text{ A}$$

$$I_A = I_1$$

$$= (9.3 - j40.75) \text{ A}$$

$$= \mathbf{41.79\angle -77.14° \text{ A}}$$

$$I_B = I_2 - I_1$$

$$= (-31.85 - j22.99) - (9.3 - j40.75)$$

$$= (-41.15 + j17.76) \text{ A}$$

$$= \mathbf{44.82\angle 156.66° \text{ A}}$$

$$I_C = -I_2$$

$$= 31.85 + j22.99$$

$$= \mathbf{39.28\angle 35.82° \text{ A}}$$

7.5.3 Solution of Three-phase, Three-wire, Y-connected Load by Millman's Method

Application of Millman's theorem to solve unbalanced three-phase, three-wire, Y-connected load is quite easier than star–delta conversion and mesh analysis approaches. Figure 7.21 shows an unbalanced three-phase Y-connected load supplied by a balanced three-phase supply without neutral wire.

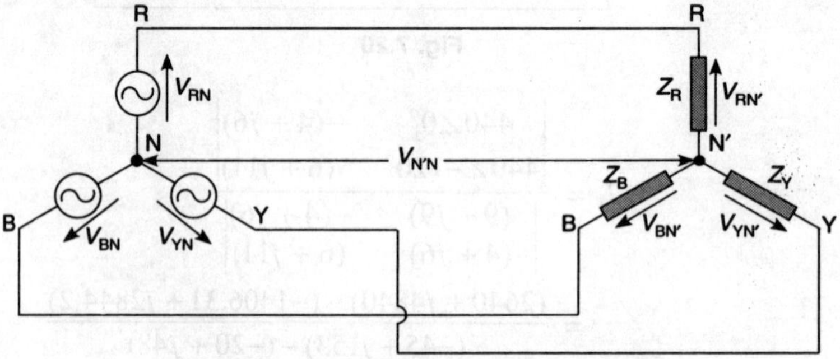

Fig. 7.21 Three-phase, three-wire system.

Let,

V_{RN} = phase voltage of generator with respect to R phase
V_{YN} = phase voltage of generator with respect to Y phase
V_{BN} = phase voltage of generator with respect to B phase
$V_{RN'}$ = phase voltage of unbalanced load with respect to R phase
$V_{YN'}$ = phase voltage of unbalanced load with respect to Y phase
$V_{BN'}$ = phase voltage of unbalanced load with respect to B phase
$V_{N'N}$ = neutral shift voltage due to imbalance in load

Figure 7.21 has been redrawn and its equivalent single-source models are shown in Figs. 7.22 and 7.23.

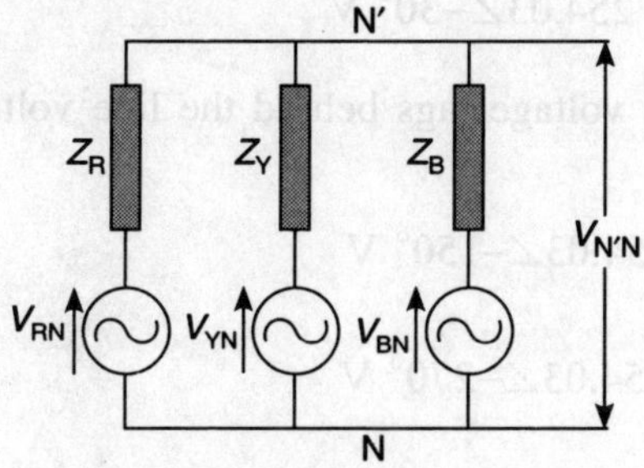

Fig. 7.22

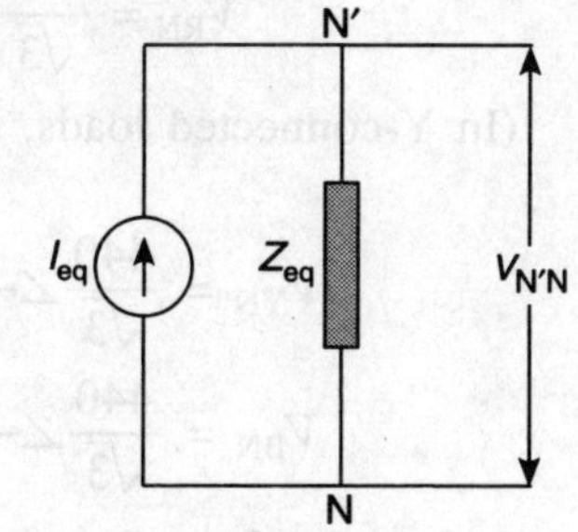

Fig. 7.23

From Fig. 7.23,

$$I_{eq} = \frac{V_{RN}}{Z_R} + \frac{V_{YN}}{Z_Y} + \frac{V_{BN}}{Z_B}$$

or
$$I_{eq} = V_{RN}Y_R + V_{YN}Y_Y + V_{BN}Y_B \tag{7.5}$$

$$\frac{1}{Z_{eq}} = \frac{1}{Z_R} + \frac{1}{Z_Y} + \frac{1}{Z_B}$$

or
$$Y_{eq} = Y_R + Y_Y + Y_B \tag{7.6}$$

So,
$$V_{N'N} = I_{eq} \times Z_{eq} \tag{7.7}$$

$$= \frac{I_{eq}}{Y_{eq}}$$

Putting Eqs. (7.5) and (7.6) in Eq. (7.7), we get

$$V_{N'N} = \frac{V_{RN}Y_R + V_{YN}Y_Y + V_{BN}Y_B}{Y_R + Y_Y + Y_B}$$

The phase voltages of the load are then obtained as:

$$V_{RN'} = V_{RN} - V_{N'N}$$
$$V_{YN'} = V_{YN} - V_{N'N}$$
$$V_{BN'} = V_{BN} - V_{N'N}$$

EXAMPLE 7.10 Determine the phase voltages for the unbalanced star-connected load shown in Fig. 7.24, using Millman's theorem. Assume the phase sequence be RYB. Take line voltage as 440 V.

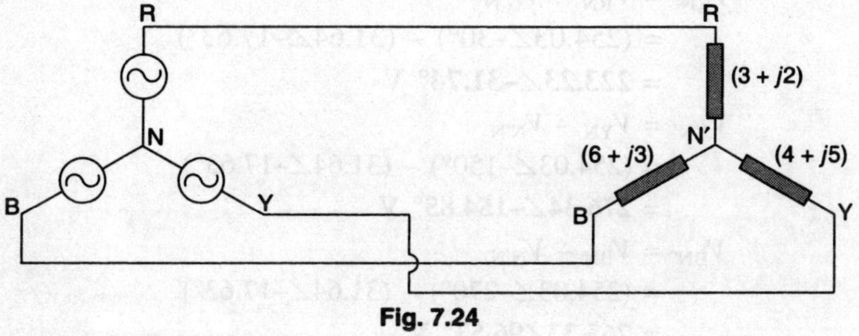

Fig. 7.24

Solution The phase voltages of the load are

$$V_{RN} = \frac{440}{\sqrt{3}}\angle{-30^\circ} = 254.03\angle{-30^\circ}\text{ V}$$

(In Y-connected loads, the phase voltage lags behind the line voltage by 30°)

Similarly,

$$V_{YN} = \frac{440}{\sqrt{3}}\angle{-150^\circ} = 254.03\angle{-150^\circ}\text{ V}$$

$$V_{BN} = \frac{440}{\sqrt{3}}\angle{-270^\circ} = 254.03\angle{-270^\circ}\text{ V}$$

The admittances of the load are found as follows:

$$Y_R = \frac{1}{3+j2} = 0.28\angle{-33.69^\circ}$$

$$Y_Y = \frac{1}{4+j5} = 0.16\angle{-51.34^\circ}$$

$$Y_B = \frac{1}{6+j3} = 0.15\angle{-26.57^\circ}$$

By applying Millman's theorem, the neutral shift voltage is obtained as:

$$V_{N'N} = \frac{V_{RN}Y_R + V_{YN}Y_Y + V_{BN}Y_B}{Y_R + Y_Y + Y_B}$$

$$= \frac{(254.03\angle{-30^\circ})(0.28\angle{-33.69^\circ}) + (254.03\angle{-150^\circ})(0.16\angle{-51.34^\circ}) + (254.03\angle{-270^\circ})(0.15\angle{-26.57^\circ})}{(0.28\angle{-33.69^\circ}) + (0.16\angle{-51.34^\circ}) + (0.15\angle{-26.57^\circ})}$$

$$= \frac{(71.13\angle{-63.69^\circ}) + (40.64\angle{-201.34^\circ}) + (38.1\angle{-296.57^\circ})}{0.58\angle{-36.64^\circ}}$$

$$= \frac{18.35\angle{-54.27^\circ}}{0.58\angle{-36.64^\circ}}$$

$$\mathbf{V_{N'N} = 31.64\angle{-17.63^\circ}\text{ V}}$$

By applying KVL, the phase voltages of the unbalanced load are calculated as follows:

$$V_{RN'} = V_{RN} - V_{N'N}$$
$$= (254.03\angle{-30^\circ}) - (31.64\angle{-17.63^\circ})$$
$$= \mathbf{223.23\angle{-31.74^\circ}\text{ V}}$$

$$V_{YN'} = V_{YN} - V_{N'N}$$
$$= (254.03\angle{-150^\circ}) - (31.64\angle{-17.63^\circ})$$
$$= \mathbf{276.34\angle{-154.85^\circ}\text{ V}}$$

$$V_{BN'} = V_{BN} - V_{N'N}$$
$$= (254.03\angle{-270^\circ}) - (31.64\angle{-17.63^\circ})$$
$$= \mathbf{265.33\angle{96.53^\circ}\text{ V}}$$

7.5.4 Unbalanced Y-Connected Load (Three-phase, Four-wire System)

In a three-phase, four-wire system, the star point of load and source are connected by a fourth wire as shown in Fig. 7.25. As a result, the phase voltages of load and source are forced to be same. Since the supply voltage is assumed to be a balanced one, the phase voltages of the load should also be balanced. The existence of fourth wire does not allow voltage unbalance even if the load is unbalanced. Therefore, in three-phase, four-wire, star-connected unbalanced load, the relation between line voltages and phase voltages are same as that of balanced load. If a phase sequence of RYB exists, then

$$V_R = \frac{V_{RY}}{\sqrt{3}} \angle(0° - 30°)$$

$$V_Y = \frac{V_{YB}}{\sqrt{3}} \angle(-120° - 30°)$$

and

$$V_B = \frac{V_{BR}}{\sqrt{3}} \angle(-240° - 30°)$$

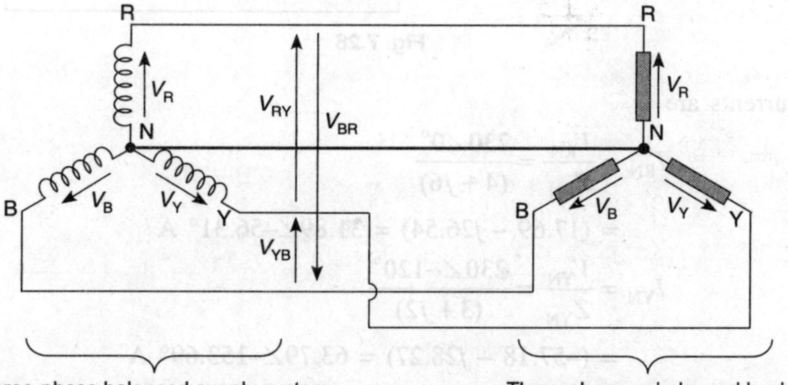

Fig. 7.25 Three-phase, four-wire system.

Therefore, the phase currents are

$$I_R = \frac{V_R}{Z_R} \angle -30°$$

$$I_Y = \frac{V_Y}{Z_Y} \angle -150°$$

and

$$I_B = \frac{V_B}{Z_B} \angle -270°$$

Also the current through neutral wire is the vector sum of the currents in the three lines.

EXAMPLE 7.11 A three-phase, four-wire system having a 230 V line-to-neutral connected to an unbalanced load having phase impedances of (4 + *j*6) Ω, (3 + *j*2) Ω and (5 + *j*7) Ω, as shown in Fig. 7.26. Calculate the current in neutral wire when phase sequence is (i) RYB and (ii) RBY.

Solution

(i) *Phase sequence* RYB

$$V_{RN} = 230\angle 0^\circ$$

$$V_{YN} = 230\angle -120^\circ$$

$$V_{BN} = 230\angle -240^\circ$$

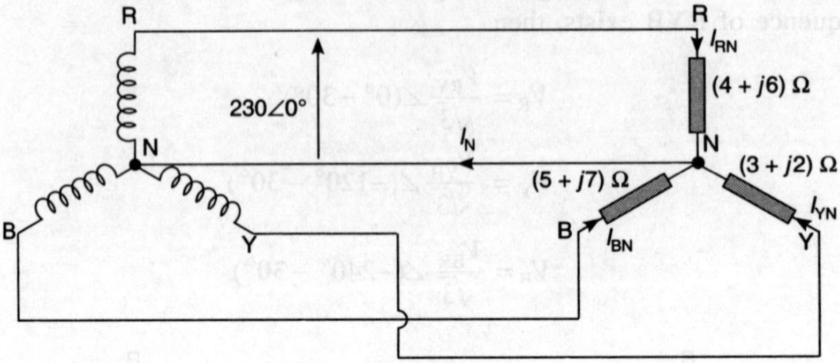

Fig. 7.26

The phase currents are

$$I_{RN} = \frac{V_{RN}}{Z_{RN}} = \frac{230\angle 0^\circ}{(4 + j6)}$$

$$= (17.69 - j26.54) = 31.89\angle -56.31^\circ \text{ A}$$

$$I_{YN} = \frac{V_{YN}}{Z_{YN}} = \frac{230\angle -120^\circ}{(3 + j2)}$$

$$= (-57.18 - j28.27) = 63.79\angle -153.69^\circ \text{ A}$$

$$I_{BN} = \frac{V_{BN}}{Z_{BN}} = \frac{230\angle -240^\circ}{(5 + j7)}$$

$$= (11.07 + j24.34) = 26.74\angle 65.54^\circ \text{ A}$$

Hence, the neutral current, $I_N = -(I_{RN} + I_{YN} + I_{BN})$

$$= -[(17.69 - j26.54) + (-57.18 - j28.27) + (11.07 + j24.34)]$$

$$= (28.42 + j30.47)$$

$$= \mathbf{41.67\angle 47 \text{ A}}$$

(ii) *Phase sequence* RBY

$$V_{RN} = 230\angle 0^\circ$$

$$V_{YN} = 230\angle -240^\circ$$

$$V_{BN} = 230\angle -120^\circ$$

The phase currents are

$$I_{RN} = \frac{V_{RN}}{Z_{RN}} = \frac{230\angle 0^\circ}{(4 + j6)}$$

$$= (17.69 - j26.54) = 31.89\angle -56.31^\circ \text{ A}$$

$$I_{YN} = \frac{V_{YN}}{Z_{YN}} = \frac{230\angle -240^\circ}{(3 + j2)}$$

$$= (4.11 + j63.66) = 63.79\angle 86.31^\circ \text{ A}$$

$$I_{BN} = \frac{V_{BN}}{Z_{BN}} = \frac{230\angle -120^\circ}{(5 + j7)}$$

$$= (-26.61 - j2.58) = 26.74\angle -174.46^\circ \text{ A}$$

Hence, the neutral current, $I_N = -(I_{RN} + I_{YN} + I_{BN})$

$$= -[(17.69 - j26.54) + (4.11 + j63.66) + (-26.61 - j2.58)]$$

$$= (4.81 - j34.54)$$

$$= \mathbf{34.87\angle -82^\circ \text{ A}}$$

7.6 MEASUREMENT OF POWER AND POWER FACTOR

To measure power in AC circuits, wattmeters are used. A wattmeter has a low-resistance current coil which is inserted in series with the line carrying the current and a high-resistance pressure coil (or potential coil) across the two points of the load. It shows a reading which is proportional to the product of the current through its current coil, the voltage across the pressure coil and cosine of the angle between the current and the voltage.

Power in a three-phase circuit may be measured by any one of the following methods:

1. three-wattmeter method,
2. two-wattmeter method, and
3. single-wattmeter method.

While the first two methods can be used to measure both balanced and unbalanced power, the single-wattmeter method is suitable to measure power in balanced load only.

7.6.1 Three-wattmeter Method

Methods of connecting three wattmeters to measure three-phase power of delta- and star-connected loads are shown in Figs. 7.27(a) and (b), respectively. The current coil of each wattmeter would carry the phase current and each pressure coil would measure the phase voltage. Hence, each wattmeter measures the power of the phase in which it is connected. The algebraic sum of the three readings gives the total power supplied to the load.

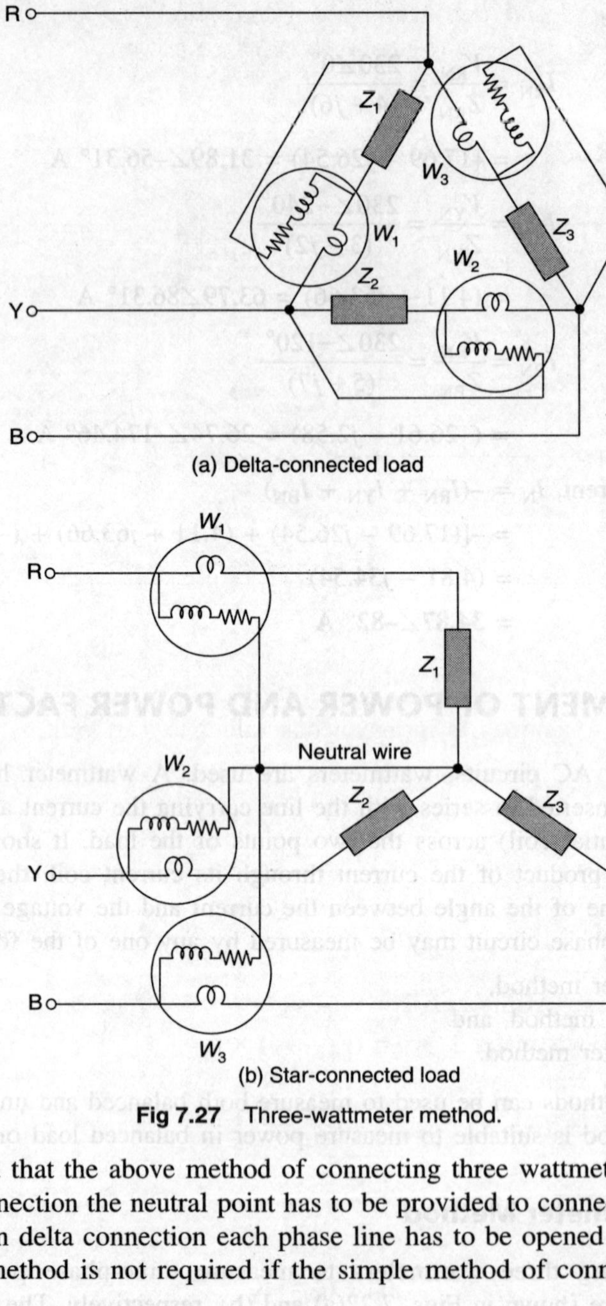

(a) Delta-connected load

(b) Star-connected load

Fig 7.27 Three-wattmeter method.

It is understood that the above method of connecting three wattmeters is difficult. This is because, in star connection the neutral point has to be provided to connect the ends of the three pressure coils and in delta connection each phase line has to be opened for connecting current coil. This tedious method is not required if the simple method of connecting two wattmeters is known.

7.6.2 Two-wattmeter Method

Three-phase power can be measured with just two wattmeters connected as shown in Fig. 7.28. The principle involved applies to both balanced and unbalanced loads. Also, it is applicable for star- and delta-connected loads. Let us take a star-connected load and prove that the algebraic sum of the two wattmeter readings results in three-phase power.

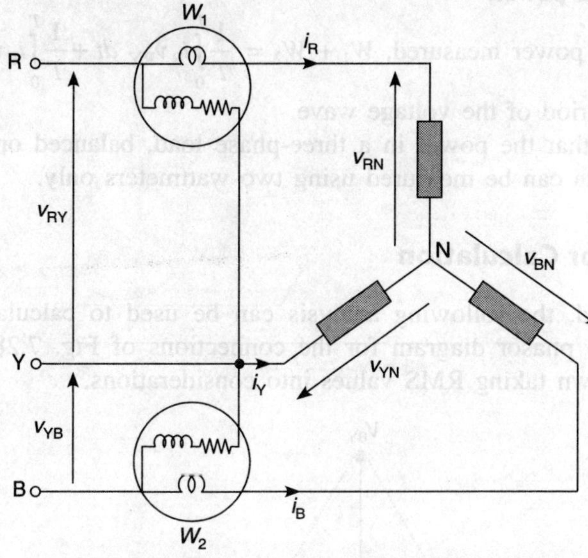

Fig. 7.28 Two-wattmeter method.

Let us compute all the quantities as instantaneous values to start with.

Current through current coil 1 = i_R

Voltage across pressure coil 1 = v_{RY}

$= v_{RN} - v_{YN}$

or $= v_R - v_Y$

Similarly,

Current through current coil 2 = i_B

Voltage across pressure coil 2 = $v_B - v_Y$

Power read by wattmeter 1, $w_1 = (v_R - v_Y) \times i_R$

Power read by wattmeter 2, $w_2 = (v_B - v_Y) \times i_B$

Total instantaneous power, $w_1 + w_2 = (v_R - v_Y) \times i_R + (v_B - v_Y) \times i_B$

$= v_R i_R - v_Y i_R + v_B i_B - v_Y i_B$

$= v_R i_R + v_B i_B - v_Y(i_R + i_B)$

By Kirchhoff's current law, $i_R + i_Y + i_B = 0$

So, $i_R + i_B = -i_Y$

Substituting this into the above power equation, we get

$$w_1 + w_2 = v_R i_R + v_B i_B - v_Y(-i_Y)$$

$$= v_R i_R + v_B i_B + v_Y i_Y$$

= sum of three-phase power

Hence, it is quite sufficient to employ only two wattmeters to read three-phase power. It is to be noted that the moving systems of wattmeter do not measure the instantaneous power but indicate the average power.

Therefore, the average power measured, $W_1 + W_2 = \frac{1}{T}\int_0^T i_R v_{RY}\, dt + \frac{1}{T}\int_0^T i_B v_{BY}\, dt$

where T is the time period of the voltage wave.

Hence, it is clear that the power in a three-phase load, balanced or unbalanced, type of connection, star or delta can be measured using two wattmeters only.

7.6.3 Power Factor Calculation

If the load is balanced, the following analysis can be used to calculate the power factor. Figure 7.29 shows the phasor diagram for the connections of Fig. 7.28 for balanced loads. Phasors have been drawn taking RMS values into considerations.

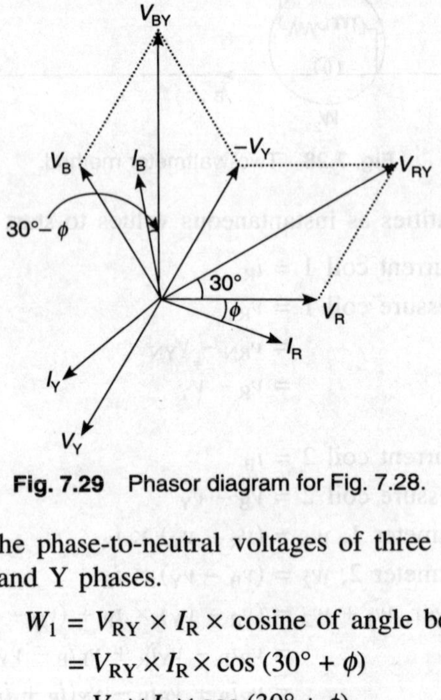

Fig. 7.29 Phasor diagram for Fig. 7.28.

Let V_R, V_Y and V_B be the phase-to-neutral voltages of three phases and I_R and I_B be the line or phase currents of R and Y phases.

Power read by wattmeter 1, $W_1 = V_{RY} \times I_R \times$ cosine of angle between V_{RY} and I_R

$= V_{RY} \times I_R \times \cos(30° + \phi)$ [from Fig. 7.29]

$$= V_L \times I_L \times \cos(30° + \phi) \quad (7.8)$$

Power read by wattmeter 2, $W_2 = V_{BY} \times I_B \times$ cosine of angle between V_{BY} and I_B

$= V_{BY} \times I_B \times \cos(30° - \phi)$ [from Fig. 7.29]

$$= V_L \times I_L \times \cos(30° - \phi) \quad (7.9)$$

Adding Eqs. (7.8) and (7.9), we get

$$W_1 + W_2 = V_L\, I_L\, [\cos(30° + \phi) + \cos(30° - \phi)]$$
$$= V_L\, I_L\, [2\cos 30° \cos \phi]$$

$$= V_L I_L\left(2\frac{\sqrt{3}}{2}\cos\phi\right)$$

$$W_1 + W_2 = \sqrt{3}\, V_L\, I_L \cos\phi \qquad (7.10)$$

Substracting Eq. (7.9) from Eq. (7.8), we get

$$W_2 - W_1 = V_L\, I_L\, [\cos(30° - \phi) - \cos(30 + \phi)]$$

$$= V_L\, I_L\, [2 \sin 30° \sin\phi]$$

$$= V_L I_L\left(2\frac{1}{2}\sin\phi\right)$$

$$W_2 - W_1 = V_L\, I_L \sin\phi \qquad (7.11)$$

From Eqs. (7.10) and (7.11), we get

$$\frac{W_2 - W_1}{W_1 + W_2} = \frac{V_L I_L \sin\phi}{\sqrt{3} V_L I_L \cos\phi}$$

$$\sqrt{3}\frac{W_2 - W_1}{W_1 + W_2} = \frac{V_L I_L \sin\phi}{V_L I_L \cos\phi}$$

$$\sqrt{3}\frac{W_2 - W_1}{W_1 + W_2} = \tan\phi$$

$$\therefore \qquad \phi = \tan^{-1}\left(\sqrt{3}\frac{W_2 - W_1}{W_1 + W_2}\right)$$

Hence, using the above formula we can find the value of ϕ and also load power factor.

7.6.4 Effect of Power Factor on Wattmeter Readings

In previous section, it is seen that the readings of the two wattmeters depend on the load power factor angle ϕ such that

$$W_1 = V_L I_L \cos(30° + \phi)$$

$$W_2 = V_L I_L \cos(30° - \phi)$$

Case (i) When power factor is unity, i.e. $\phi = 0$

$$W_1 = V_L I_L \cos(30° + 0) = V_L I_L \cos 30° = \frac{\sqrt{3}}{2} V_L I_L = \text{half of total power}$$

$$W_2 = V_L I_L \cos(30° - 0) = V_L I_L \cos 30° = \frac{\sqrt{3}}{2} V_L I_L = \text{half of total power}$$

Therefore, both wattmeters read equal and positive readings.

Case (ii) When power factor is 0.5, i.e. $\phi = 60°$

$$W_1 = V_L I_L \cos(30° + 60°) = V_L I_L \cos 90° = 0$$

$$W_2 = V_L I_L \cos(30° - 60°) = V_L I_L \cos(-30°) = \frac{\sqrt{3}}{2} V_L I_L = \text{half of total power}$$

Hence, total power is measured by wattmeter W_2 alone.

Case (iii) When power factor is 0 to 0.5, i.e. $90° > \phi > 60°$

For this case, W_2 indicates positive value and W_1 reads down scale reading. To obtain positive reading on W_1, connection of either current coil or pressure coil is reversed. The reading for W_1 after reversing the coil connection is recorded with negative sign.

Therefore,
$$\text{total power} = W_2 + (-W_1)$$
$$= W_2 - W_1$$

Thus, if the load power factor is less than 0.5, one of the wattmeters will give negative reading.

Case (iv) When power factor is zero, i.e. $\phi = 90°$

$$W_1 = V_L I_L \cos(30° + 90°) = V_L I_L \cos 120° = -\frac{1}{2} V_L I_L$$

$$W_2 = V_L I_L \cos(30° - 90°) = V_L I_L \cos(-60°) = \frac{1}{2} V_L I_L$$

Thus, the two wattmeters read equal and opposite. Hence,

$$W_1 + W_2 = 0$$

Therefore,

(i) when the wattmeter readings are equal, the power factor is unity.
(ii) when one of the wattmeter readings is zero, the power factor is 0.5.
(iii) when one of the wattmeter readings is negative, the power factor is less than 0.5.
(iv) when both the wattmeter readings are positive, the power factor will be greater than 0.5.

7.6.5 Reactive Power with Two-wattmeter Method

The reactive power in a three-phase balanced load can be determined from the readings of the two-wattmeter method. From Eq. (7.11),

$$W_2 - W_1 = V_L I_L \sin\phi$$
$$\sqrt{3}(W_2 - W_1) = \sqrt{3} V_L I_L \sin\phi$$
$$= \text{reactive power}$$

Hence,
$$\text{reactive power} = \sqrt{3}(W_2 - W_1)$$

7.6.6 One-wattmeter Method

This method can be used when the load is balanced. In this method, the current coil of the wattmeter is connected in any one line and one of the two ends of the pressure coil is connected in this line. The other end of the pressure coil is alternatively connected between the other two lines as shown in Fig. 7.30. The algebraic sum of the two readings gives the total power drawn by the balanced three-phase load. The two wattmeter readings are obtained first by connecting terminal 1 and 2, and changed to 1 and 3.

Then, the two readings are

$$= (V_R - V_Y) \times I_R + (V_R - V_B) \times I_R$$
$$= V_R I_R - V_Y I_R + V_R I_R - V_B I_R$$
$$= 2V_R I_R - I_R(V_B + V_Y)$$
$$= 3V_R I_R$$

as $$V_R + V_Y + V_B = 0$$

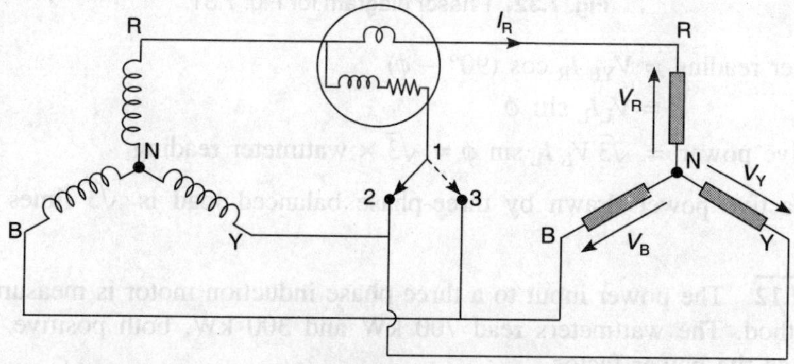

Fig. 7.30 Real power measurement by one wattmeter method.

7.6.7 Reactive Power with One-wattmeter Method

A single wattmeter may be used to measure the reactive power, if the load is a balanced one. Figure 7.31 gives the connection for reactive power measurement. Here, current coil of the wattmeter is connected in one line and the pressure coil is connected across other two lines.

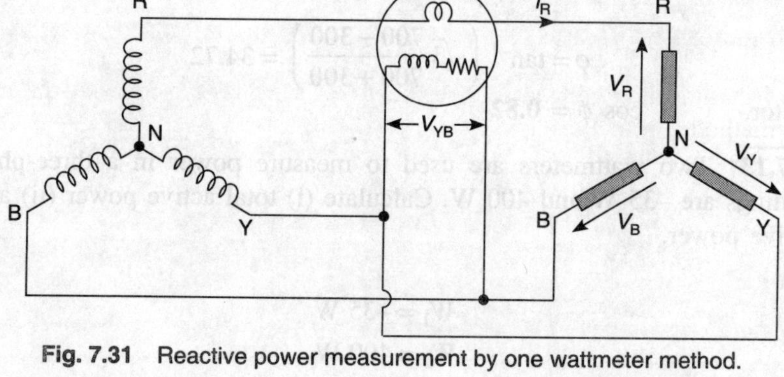

Fig. 7.31 Reactive power measurement by one wattmeter method.

Current through current coil = I_R

Voltage across pressure coil = V_{YB}

and wattmeter reading = $I_R \times V_{YB} \times$ cosine of angle between I_R and V_{YB}

From the phasor diagram (Fig. 7.32), the angle between I_R and V_{YB} is seen as 90° – ϕ.

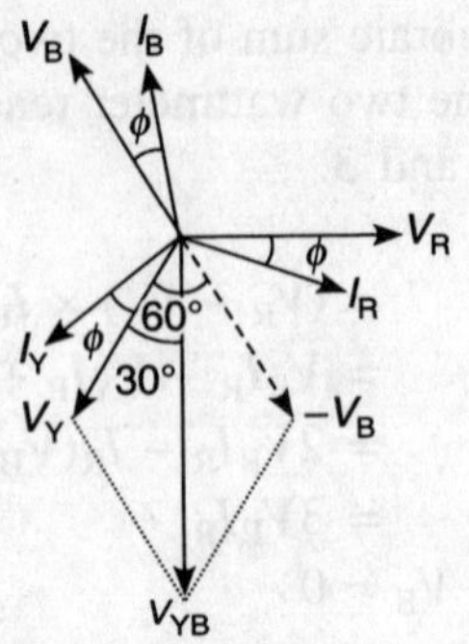

Fig. 7.32 Phasor diagram for Fig. 7.31.

Thus, wattmeter reading = $V_{YB}\, I_R \cos (90° - \phi)$

$= V_L I_L \sin \phi$

But reactive power = $\sqrt{3}\, V_L\, I_L \sin \phi = \sqrt{3} \times$ wattmeter reading

Hence, the reactive power drawn by three-phase balanced load is $\sqrt{3}$ times the wattmeter reading.

EXAMPLE 7.12 The power input to a three-phase induction motor is measured by the two-wattmeter method. The wattmeters read 700 kW and 300 kW, both positive. Find the total power input and the power factor.

Solution Let

$$W_1 = 300 \text{ kW}$$

$$W_2 = 700 \text{ kW}$$

Total three-phase power input = $W_1 + W_2$ = 300 + 700 = **1000 kW**

$$\tan \phi = \sqrt{3}\,\frac{W_2 - W_1}{W_1 + W_2}$$

$$\phi = \tan^{-1}\left(\sqrt{3}\,\frac{700 - 300}{700 + 300}\right) = 34.72$$

So, power factor, $\cos \phi =$ **0.82**

EXAMPLE 7.13 Two wattmeters are used to measure power in a three-phase load. The wattmeter readings are –35 W and 400 W. Calculate (i) total active power (ii) apparent power and (iii) reactive power.

Solution Let

$$W_1 = -35 \text{ W}$$

$$W_2 = 400 \text{ W}$$

(i) Total active power $= W_1 + W_2$

$= -35 + 400$

$= \mathbf{365\ W}$

$$\text{Power factor, } \cos\phi = \cos\left[\tan^{-1}\left(\sqrt{3}\,\frac{W_2 - W_1}{W_1 + W_2}\right)\right]$$

$$= \cos\left[\tan^{-1}\left(\sqrt{3}\,\frac{400-(-35)}{400-35}\right)\right]$$

$$= \cos 64.15° = 0.436$$

$$\text{Active power} = \sqrt{3}\,V_L I_L \cos\phi = 365$$

(ii) Apparent power
$$= \frac{\text{active power}}{\text{power factor}} = \frac{365}{0.436}$$

$= \mathbf{837.16\ VA}$

(iii) Reactive power $= \sqrt{3}\,V_L I_L \sin\phi = \text{apparent power} \times \sin\phi$

$= 837.16 \times \sin 64.15°$

$= \mathbf{753.4\ var}$

EXAMPLE 7.14 A 500 V, three-phase motor has an output of 3.73 kW and operates at a power factor of 0.85, with an efficiency of 90%. Calculate the reading on each of the two wattmeters connected to measure the input.

Solution

Power output = 3.73 kW

Efficiency = 90%

$$\text{Power input} = \frac{\text{power output}}{\text{efficiency}} = \frac{3.73}{0.9}$$

$= 4.144$ kW

If two wattmeters are connected to measure total power, then,

$$W_1 + W_2 = 4.144 \text{ kW} \qquad \text{(i)}$$

Power factor, $\cos\phi = 0.85$ (given)

Hence,
$$\tan\phi = \tan(\cos^{-1}\phi) = 0.619$$

$$0.619 = \sqrt{3}\,\frac{W_2 - W_1}{W_1 + W_2} = \sqrt{3}\,\frac{W_2 - W_1}{4.144}$$

$$W_2 - W_1 = \frac{4.144 \times 0.619}{\sqrt{3}}$$

$$W_2 - W_1 = 1.48 \text{ kW} \qquad \text{(ii)}$$

Solving Eqs. (i) and (ii), we get

$$\mathbf{W_1 = 1.332\ kW}$$

$$\mathbf{W_2 = 2.812\ kW}$$

SHORT QUESTIONS WITH ANSWERS

1. What is balanced voltage?

In three-phase system, if the magnitude of phase voltages and the phase angle difference between consecutive phasors are equal, then the voltages are called balanced voltages.

2. What is phase sequence?

The order in which the alternating quantities reach their maximum values is called phase sequence.

3. What are the advantages of three-phase system?

The three-phase system has the following advantages over a single-phase system.

- Generation, transmission and distribution of three-phase power is more economical than single-phase power because of less conductive material required.
- Three-phase induction motors are self starting unlike single-phase induction motors.
- For the same size, three-phase machine gives more output than single-phase machine.
- Three-phase machine provides better power factor and efficiency.
- Converting AC to DC using rectifiers is smoother in three-phase system than in single-phase system.
- In three-phase systems, better voltage regulation is obtained.

4. Write the relation between the line and phase value of voltage and current in a balanced star connection.

In a balanced three-phase Y-connection,

$$\text{line voltage} = \sqrt{3} \times \text{phase voltage}$$
$$\text{line current} = \text{phase current}$$

5. Write the relation between the line and phase value of voltage and current in a balanced delta connection.

In a balanced three-phase Δ-connection,

$$\text{line voltage} = \text{phase voltage}$$
$$\text{line current} = \sqrt{3} \times \text{phase current}$$

6. What is floating neutral point?

The isolated neutral (star) point of a load is called floating neutral point.

7. Name any two methods used to analyse unbalanced star-connected loads having isolated neutral point.

1. Star–delta conversion
2. Mesh analysis

8. What is neutral shift voltage?

In three-wire star-connected system, the neutral points of source and load are not tied together because of the nonexistence of fourth wire. In this circumstance, if the load is unbalanced there will be some difference in neutral voltages of both source and load and this difference in voltages is called neutral shift voltage.

9. Write the expression for power factor in two-wattmeter method of power measurement.

$$\cos\phi = \cos\left[\tan^{-1}\left(\sqrt{3}\,\frac{W_2 - W_1}{W_1 + W_2}\right)\right]$$

10. What is the effect of power factor on wattmeter readings?

1. When power factor is unity, the two wattmeters read equal and positive readings.
2. When power factor is 0.5, the total power is measured by only one wattmeter.
3. When power factor is less than 0.5, one of the wattmeters will give negative reading.
4. When both wattmeter readings are positive, the power factor will be greater than 0.5.

Unsolved Problems

1. A balanced star-connected load of impedance (15 + *j*20) Ω per phase is connected to a three-phase, 440 V, 50 Hz supply. Find the line currents and power absorbed by the load. Assume RYB sequence.

[Ans. $I_R = 10.1614\angle{-83.1°}$ A; $I_Y = 10.1614\angle{-203.1°}$A; $I_B = 10.1614\angle{-323.1°}$ A; $I_{RN} = 10.1614\angle{-83.1°}$ A; $I_{YN} = 10.1614\angle{-203.1°}$ A; $I_{BN} = 10.1614\angle{-323.1°}$ A; 4.6497 kW]

2. A balanced star-connected load of impedance (4 + *j*8) Ω per phase is connected to a three-phase, 440 V, 60 Hz supply. Find the line currents, active and reactive power of the load.

[Ans. $I_R = 25.8198\angle{-93.4°}$ A; $I_Y = 25.8198\angle{-213.4°}$ A; $I_B = 25.8198\angle{-333.4°}$ A; $P = 8.0097$ kW; $Q = 15.9951$ kVAR]

3. A three-phase delta-connected RYB system with an effective voltage of 400 V has a balanced load with impedances (3 + *j*4) Ω. Calculate the (i) phase currents (ii) line currents and (iii) power in each phase.

[Ans. (i) $I_R = 138.55\angle{-83.09°}$ A, $I_Y = 138.55\angle{-203.09°}$ A, $I_B = 138.55\angle{-323.09°}$ A; (ii) $I_{RN} = 80\angle{-53.1°}$ A, $I_{YN} = 80\angle{-173.1°}$ A, $I_{BN} = 80\angle{-293.1°}$ A; (iii) 57.6 kW]

4. A three-phase balanced delta-connected load with line voltage of 200 V has line currents as $I_R = 10\angle 90°$ A; $I_Y = 10\angle{-150°}$ A; $I_B = 10\angle{-30°}$ A. (i) What is the phase sequence? (ii) What are the impedances? **[Ans.** (i) RBY; (ii) 34.64 Ω]

5. A three-phase delta-connected load has $Z_{RY} = (100 + j50)\ \Omega$, $Z_{YB} = (20 - j75)\ \Omega$, $Z_{BR} = (70.7 + j70.7)\ \Omega$ and is connected to balanced three-phase, 400 V supply. Determine the line currents and the power consumed by the load. Assume RYB phase sequence.

[**Ans.** $I_R = 5.8788\angle 51.6°$ A, $I_Y = 2.0851\angle 42.5°$ A, $I_B = 7.9445\angle -130.8°$ A, 2.9412 kW]

6. An unbalanced four-wire star-connected load is connected to a balanced supply of 380 V. Estimate the line currents, neutral current and the power consumed by the load. Take $Z_R = (2 + j4)\ \Omega$; $Z_Y = (3 + j2)\ \Omega$; $Z_B = (6 + j10)\ \Omega$.

[**Ans.** $I_R = 49.0582\angle -93.4°$ A, $I_Y = 60.8479\angle -183.7°$ A, $I_B = 18.8128\angle 31°$ A; $I_N = 59.2178\angle -143.3°$ A; $P = 18.0513$ kW]

7. An unbalanced four-wire star-connected load has balanced supply voltage of 400 V. The load impedances are $Z_R = (4 + j8)\ \Omega$; $Z_Y = (3 + j4)\ \Omega$; $Z_B = (15 + j10)\ \Omega$. Calculate the line currents, neutral current and the total power.

[**Ans.** $I_R = 25.8198\angle -93.4°$ A, $I_Y = 46.188\angle -203.1°$ A, $I_B = 12.8102\angle 56.3°$ A; $I_N = 37.0304\angle 175.3°$ A; $P = 11.5356$ kW]

8. An unbalanced three-wire star-connected load is connected to a symmetrical supply of 415 V. If the load impedances are $Z_R = 7\angle 30°\ \Omega$; $Z_Y = 9\angle 60°\ \Omega$; $Z_B = 8\angle 45°\ \Omega$, calculate the line currents, phase voltages and displacement neutral voltage.

[**Ans.** $I_R = 34.9247\angle -70.2°$ A, $I_Y = 30.8275\angle 154.6°$ A, $I_B = 25.3461\angle 50.8°$ A; $V_R = 244.4729\angle -40.2°$ V, $V_Y = 277.4475\angle 214.6°$ V, $V_B = 202.7688\angle 95.8°$ V; $V_{NN'} = 43.3042\angle 61.3°$ V]

9. An unbalanced delta-connected load is connected to a three-phase 415 V supply mains. Determine the phase and line currents for RYB sequence by taking V_{BR} as reference phasor. The impedances of three branches are $(70 + j45)\ \Omega$, $(10 - j20)\ \Omega$ and $(30 + j10)\ \Omega$.

[**Ans.** $I_{RY} = 4.987\angle -152.7°$ A, $I_{YB} = 18.5593\angle -176.6°$ A, $I_{BR} = 13.1234\angle -18.4°$ A; $I_R = 16.9856\angle 173.7°$ A, $I_Y = 14.145\angle 175.2°$ A, $I_B = 31.1281\angle -5.6°$ A]

10. Two wattmeters are connected to measure power in a three-phase load. Both wattmeters show equal readings and is 8 kW, when the power factor is unity. What will be the reading of wattmeters if the power factor falls to (i) 0.7 lagging (ii) 0.5 lagging (iii) 0.4 lagging and the total three-phase power remains unaltered?

[**Ans.** (i) $P_1 = 3.2834$ kW, $P_2 = 12.7166$ kW; (ii) $P_1 = 0$, $P_2 = 16$ kW; (iii) $P_1 = -2.572$ kW, $P_2 = 18.572$ kW]

11. A 415 V, three-phase motor has an output of 5.595 kW and operates at a power factor of 0.82 with an efficiency of 85%. Calculate the readings of the two wattmeters connected to measure the input power. [**Ans.** $P_1 = 1.9656$ kW; $P_2 = 4.6168$ kW]

12. A three-phase load connected to a 415 V, four-phase, three-wire system draws a current of $12\angle -40°$ A in R phase and $16\angle -200°$ A in the B phase. Determine the readings of the two wattmeters connected to measure the power if current coil of one wattmeter is connected to R phase and that of other to B phase.

[**Ans.** $P_1 = 3.8149$ kW; $P_2 = -1.153$ kW]

CHAPTER 8

Magnetic Circuits

8.1 INTRODUCTION

Flux lines are imaginary lines and these imaginary lines form a closed path. A *magnetic circuit* is made up of closed path which permits the flow of magnetic flux. It is analogous to electric circuit which permits the flow of electric current. The knowledge of magnetic circuit is important in the design and analysis of electrical machines such as generators, motors and transformers.

8.2 MAGNETIC CIRCUIT

Consider a coil of N turns wound on an iron core as shown in Fig. 8.1. When current I amperes is passed through the coil, magnetic flux ϕ is set up in the core. The flux follows the closed path shown in dotted lines and hence is a magnetic circuit.

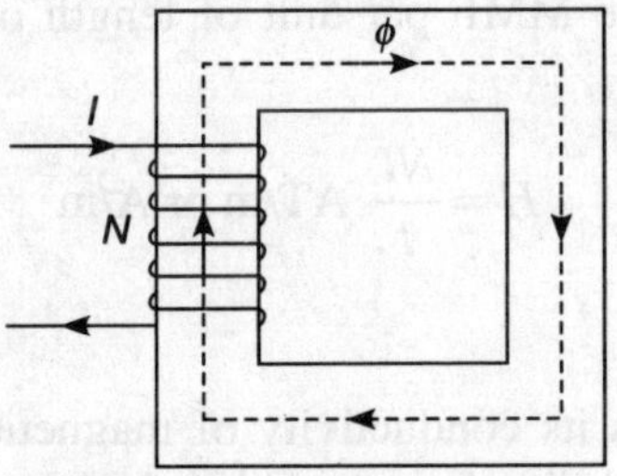

Fig. 8.1 Magnetic circuit.

The amount of flux set up in the core depends upon the number of turns (N) and the magnitude of current (I) flows. The product NI is called *magnetomotive force* (MMF) and it is analogous to *electromotive force* (e.m.f.) which sends electric current in electric circuit.

The opposition to the passage of flux in magnetic circuit is called *reluctance*. It depends upon the length of magnetic circuit, the cross-sectional area of the circuit and the nature of materials used. Materials having high permeability offer low reluctance to the flow of magnetic flux.

8.3 TERMINOLOGY

The fundamental magnetic quantities that form the basis for the study of magnetic circuits are introduced in this section.

Magnetic flux

Magnetic flux is the magnetic lines of force produced by a magnetic source. The symbol for magnetic flux is ϕ (Greek letter 'phi'), and its unit is *weber* (Wb). One Wb flux is considered to be composed of 10^8 flux lines.

Magnetic flux density

Magnetic flux density is the flux per unit cross-sectional area. It is denoted by B and its unit is Wb/m^2, or tesla (T), i.e.

$$B = \frac{\phi}{A} \text{ Wb/m}^2 \text{ or T}$$

Magnetomotive force

Magnetomotive force (MMF) is the cause of existence of magnetic flux in a magnetic circuit. It is equal to the product of current and number of turns of the coil, i.e.

$$\text{MMF} = NI \text{ ampere-turns (or) AT}$$

Magnetomotive force in a magnetic circuit is analogous to e.m.f. in an electric circuit. Sometimes, its unit is expressed as 'ampere' only as 'turns' have no dimensions.

Magnetic field strength (magnetising force)

Magnetic field strength (H) is the MMF per unit of length of magnetic circuit, the unit being ampere-turns/metre, i.e.

$$H = \frac{NI}{l} \text{ AT/m or A/m}$$

Permeability

Permeability of a material means its conductivity of magnetic flux. Obviously, good magnetic materials would have high permeability. Every medium is supposed to possess two permeabilities: (1) absolute permeability (μ_0) and (2) relative permeability (μ_r).

To measure relative permeability, vacuum or free space is chosen as the reference medium. It is allotted an absolute permeability of $\mu_0 = 4\pi \times 10^{-7}$ henry/metre. The relative permeability of a medium is defined as:

$$\mu_r = \frac{\text{absolute permeabilty of the material}}{\text{absolute permeabilty of free space}}$$

Hence for vacuum or free space the relative permeability is unity and for any other medium

$$\mu_r = \frac{\mu}{\mu_0}$$

where

μ = absolute permeability of the material

μ_0 = absolute permeability of free space

μ_r = relative permeability of the material

Reluctance

The opposition that the magnetic circuit offers to flux is called reluctance (S) and its unit is AT/Wb.

$$S = \frac{l}{a\mu} \text{ AT/Wb}$$

or

$$S = \frac{l}{a\mu_0\mu_r} \text{ AT/Wb}$$

It is analogous to resistance in an electric circuit.

Permeance

It is the magnetic flux per ampere of total MMF in the path of magnetic field. It is the reciprocal of reluctance. Its unit is Wb/AT.

$$\text{Permeance} = \frac{1}{\text{reluctance}} = \frac{a\mu_0\mu_r}{l}$$

It is analogous to conductance in an electric circuit.

Table 8.1 Analogous Electrical and Magnetic Circuit Quantities

Electrical	*Magnetic*	*Magnetic Units*
E.M.F., e	Magnetomotive force, F	Ampere-turns
Current, I	Magnetic flux, ϕ	Webers (Wb)
Resistance, R	Reluctance, S	Ampere-turns/Wb
Conductance, G	Permeance, ρ	Wb/Ampere-turns
Current density, J	Magnetic flux density, B	Wb/m^2 (or) tesla, T

EXAMPLE 8.1 A magnetic pole face has a rectangular section having dimensions 300 mm × 100 mm. If the total flux emerging from the pole is 150 μWb, calculate the flux density.

Solution Flux, $\phi = 150\ \mu\text{Wb} = 150 \times 10^{-6}$ Wb

Cross sectional area, $a = 300 \times 100 = 30{,}000\ \text{mm}^2$

Flux density, $B = \dfrac{\phi}{a} = \dfrac{150 \times 10^{-6}}{30,000 \times 10^{-6}}$

$= \mathbf{0.005\ T}$

EXAMPLE 8.2 A magnetising force of 4000 A/m is applied to a circular magnetic circuit of mean diameter 30 cm by passing a current through a coil wound on the circuit. If the coil is uniformly wound around the circuit and has 750 turns, find the current in the coil.

Solution Magnetising force, $H = 4000$ A/m

Length, $l = \pi D = \pi \times 30 \times 10^{-2}$ m

Magnetising force, $H = \dfrac{NI}{l}$ and current, $I = \dfrac{Hl}{N}$

Therefore, current, $I = \dfrac{4000 \times \pi \times 30 \times 10^{-2}}{750} = \mathbf{5.0265\ A}$

EXAMPLE 8.3 A torroidal air cored coil with 2000 turns has a mean radius of 25 cm, the diameter of each turn being 5 cm. If the current in the coil is 10 A, find (i) the MMF, (ii) reluctance, (iii) flux and (iv) flux density.

Solution

(i) MMF = Ampere turns

$= 10 \times 2000$

$= \mathbf{20000\ AT}$

(ii) Reluctance, $S = \dfrac{l}{a\mu_0\mu_r}$ AT/Wb

Length of coil, $l = 2\pi r$

$= 2\pi \times 25$

$= 157.08$ cm $= 1.5708$ m

Cross-sectional area, $a = \dfrac{\pi d^2}{4} = \dfrac{\pi \times 5^2}{4}$

$= 19.63\ \text{cm}^2 = 0.001963\ \text{m}^2$

Absolute permeability, $\mu_0 = 4\pi \times 10^{-7}$

Relative permeability, $\mu_r = 1$ (for air)

Hence, reluctance, $S = \dfrac{1.5708}{0.001963 \times 4\pi \times 10^{-7} \times 1}$

$= \mathbf{63.678 \times 10^7\ AT/Wb}$

(iii) Flux $= \dfrac{\text{MMF}}{\text{reluctance}} = \dfrac{20000}{63.678 \times 10^7}$

$= \mathbf{3.14 \times 10^{-5}\ Wb}$

(iv) Flux density $= \dfrac{\text{flux}}{\text{area of cross-section}} = \dfrac{3.14 \times 10^{-5}}{0.001963}$

$= \mathbf{0.016\ T}$

EXAMPLE 8.4 Flux produced in the air gap of two electro magnetic poles is 5×10^{-2} Wb. If the cross-sectional area of the air gap is 0.2 sq.m, find (i) the flux density (ii) magnetic field intensity, (iii) reluctance and (iv) permeance of the gap. Find also the MMF dropped in the air gap given the length of the air gap to be 1 cm.

Solution

(i) Flux density, $B = \dfrac{\phi}{a} = \dfrac{5\times10^{-2}}{0.2} = \mathbf{0.25\ T}$

(ii) Magnetic flux intensity, $H = \dfrac{B}{\mu} = \dfrac{0.25}{4\pi\times10^{-7}\times1} = \mathbf{1.9894 \times 10^5\ AT/m}$

(iii) Reluctance of the air gap $= \dfrac{l}{a\mu} = \dfrac{0.01}{0.2\times4\pi\times10^{-7}\times1} = \mathbf{39788.74\ AT/Wb}$

(iv) Permeance $= \dfrac{1}{\text{Reluctance}} = \dfrac{1}{39788.74} = \mathbf{2.5133 \times 10^{-5}\ Wb/AT}$

(v) MMF in the air gap $= \phi \times$ reluctance $= 5 \times 10^{-2} \times 39788.74 =$ **1989 AT**

8.4 COMPOSITE MAGNETIC CIRCUIT

Figure 8.2 shows a composite magnetic circuit in which there are different materials with different dimensions. The mean lengths of the materials are l_1, l_2, l_3 and l_g, with cross-sectional areas of a_1, a_2, a_3 and a_g, respectively. Length l_g with cross-sectional area a_g is known as air gap length. Since the relative permeability of each material differs from that of other, their own reluctance is also different. Therefore, the total reluctance is equal to the sum of reluctances of individual parts.

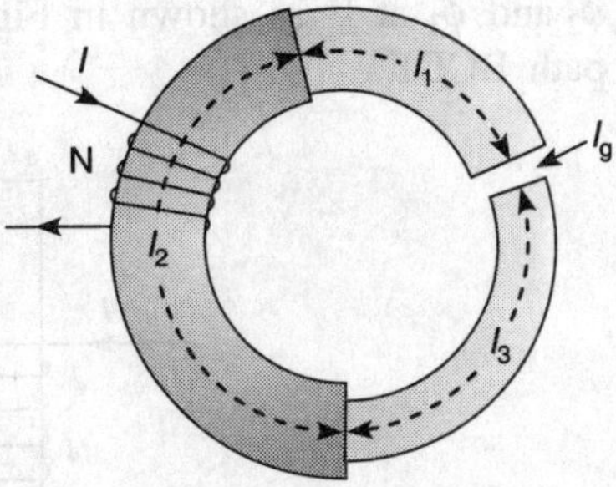

Fig. 8.2 Composite circuit.

i.e. $$\text{total reluctance} = \frac{l_1}{a_1\mu_0\mu_{r1}} + \frac{l_2}{a_2\mu_0\mu_{r2}} + \frac{l_3}{a_3\mu_0\mu_{r3}} + \frac{l_g}{a_g\mu_0} \qquad [\text{for air, } \mu_r = 1]$$

$$\text{Total MMF} = \text{flux} \times \text{total reluctance}$$

$$= \phi\left[\frac{l_1}{a_1\mu_0\mu_{r1}} + \frac{l_2}{a_2\mu_0\mu_{r2}} + \frac{l_3}{a_3\mu_0\mu_{r3}} + \frac{l_g}{a_g\mu_0}\right]$$

$$= \left(\frac{\phi}{a_1\mu_0\mu_{r1}}l_1\right) + \left(\frac{\phi}{a_2\mu_0\mu_{r2}}l_2\right) + \left(\frac{\phi}{a_3\mu_0\mu_{r3}}l_3\right) + \left(\frac{\phi}{a_g\mu_0}l_g\right)$$

$$= \left(\frac{B_1}{\mu_0\mu_{r1}}l_1\right) + \left(\frac{B_2}{\mu_0\mu_{r2}}l_2\right) + \left(\frac{B_3}{\mu_0\mu_{r3}}l_3\right) + \left(\frac{B_4}{\mu_0}l_g\right)$$

$$= H_1l_1 + H_2l_2 + H_3l_3 + H_gl_g$$

= Sum of ampere-turns required for individual parts of the magnetic circuit

8.4.1 Determination of Ampere-Turns

Flux created in a magnetic circuit, $\phi = \dfrac{\text{MMF}}{\text{Reluctance}}$

$$= \frac{\text{AT}}{l/\mu a} = \frac{\text{AT}}{l/\mu_0\mu_r a}$$

or $\quad$ AT required $= \dfrac{\phi}{a} \times \dfrac{l}{\mu_0\mu_r}$

$$= \frac{B}{\mu_0\mu_r} \times l = Hl$$

8.5 PARALLEL MAGNETIC CIRCUIT

In series magnetic circuit, same flux passes through all of its parts. If the magnetic circuit has more than one path for flux, then it is called parallel magnetic circuit. For example, consider a parallel magnetic circuit shown in Fig. 8.3. It has a coil of N turns wound on it and a current of I amperes is made to pass through the coil. The total flux ϕ_1 set up by the coil divides as ϕ_2 and ϕ_3 at B as shown in Fig. 8.3. Flux ϕ_2 passes along path BE, whereas ϕ_3 passes along path BCDE.

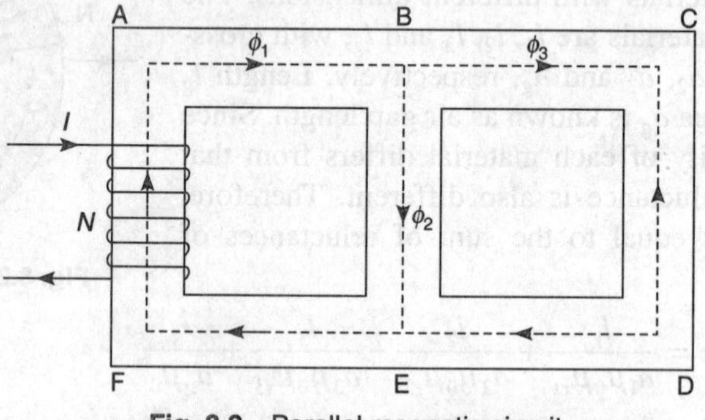

Fig. 8.3 Parallel magnetic circuit.

Magnetic paths BE and BCDE are in parallel and form a magnetic circuit. The ampere-turns required for this parallel circuit is equal to ampere-turns required for any one of the paths.

Hence, $\quad$ total MMF required = MMF to create ϕ_1 + MMF to create ϕ_2

= MMF to create ϕ_1 + MMF to create ϕ_3

If reluctances S_1, S_2 and S_3 are known, the total MMF required,

$$NI = \phi_1 S_1 + \phi_2 S_2 = \phi_1 S_1 + \phi_3 S_3$$

8.6 LEAKAGE FLUX AND FRINGING

The flux which does not follow the desired path of magnetic material is called *leakage flux*. Generally, it is not possible to confine the entire flux to the iron path only, unlike an electric current which can be confined to a wire by providing proper insulation. In any magnetic circuit, it is impossible to avoid magnetic leakage since there is no magnetic insulator available in practice. For example, air is a good insulator of electricity, though, it conducts magnetic flux. The flux in the air gap is known as *useful flux* because it can be utilised for practical purposes.

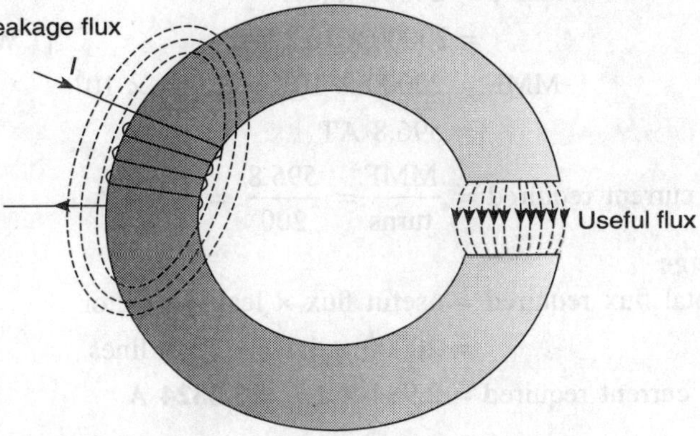

Fig. 8.4 Illustration of leakage flux.

Figure 8.4 shows an iron ring wound with a coil and having a narrow air gap. The total flux produced by the coil does not pass through the air gap as some of it leaks through the air surrounding the ring. These flux lines are called *leakage flux*.

The ratio of total flux (flux in the iron path) to the useful flux (flux in the air gap) is known as *leakage factor* (λ).

i.e.

$$\text{leakage factor, } \lambda = \frac{\text{total flux}}{\text{useful flux}}$$

or

$$\text{leakage factor, } \lambda = \frac{\text{flux in the iron path}}{\text{flux in the air gap}}$$

It may be noted that the useful flux passing across the air gap tends to bulge outwards as shown in Fig. 8.4, thereby increasing the effective area of the gap and reducing the flux density in the gap. This effect is known as *fringing*.

EXAMPLE 8.5 An iron rod of 1 cm radius is bent to a ring of mean diameter 30 cm and wound with 200 turns of wire. Assume the relative permeability of iron to be 800. Calculate the current required to produce a useful flux of 20000 lines if (i) leakage is neglected and (ii) leakage factor is 1.1.

Solution

(i) *Neglecting leakage*

$$\text{Reluctance of iron path, } S = \frac{l}{a\mu}$$

$$= \frac{\pi(0.3)}{\pi(0.01)^2 \times 4\pi \times 10^{-7} \times 800}$$

$$= 2.984 \times 10^6 \text{ AT/Wb}$$

$$\text{MMF} = \phi \times S$$

$$\text{Flux, } \phi = 20000 \text{ lines}$$

$$= 20000 \times 10^{-8} \text{ Wb} \qquad [1 \text{ Wb} = 1 \times 10^8 \text{ flux lines}]$$

$$\text{MMF} = 20000 \times 10^{-8} \times 2.984 \times 10^6$$

$$= 596.8 \text{ AT}$$

Hence, $\text{current required} = \frac{\text{MMF}}{\text{turns}} = \frac{596.8}{200} = \mathbf{2.984 \text{ A}}$

(ii) *With leakage*

Total flux required = useful flux × leakage factor

$= 20000 \times 1.1 = 22000$ lines

Hence, current required = $2.984 \times 1.1 =$ **3.2824 A**

EXAMPLE 8.6 An iron ring of cross-sectional area of 8 cm^2 is wound with a wire of 200 turns and has a saw-cut of 5 mm as shown in Fig. 8.5. Calculate the magnetising current required to produce a flux of 0.1 mWb, if mean length of magnetic path is 20 cm and relative permeability of iron is 470.

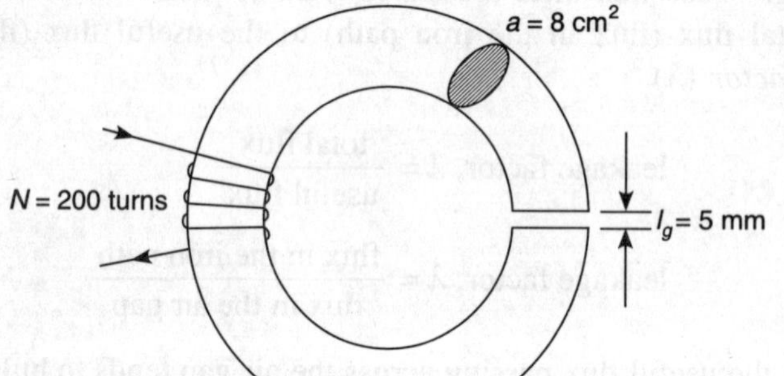

Fig. 8.5

Solution Flux density $= \frac{\phi}{a} = \frac{0.1 \times 10^{-3}}{8 \times 10^{-4}} = 0.125$ T

Total AT required = AT required for iron + AT required for air gap

$= (H_i \times l_i) + (H_g \times l_g)$

$$= \left(\frac{B}{\mu_0 \mu_r} \times l_i\right) + \left(\frac{B}{\mu_0} \times l_g\right)$$

$$= \left(\frac{0.125}{4\pi \times 10^{-7} \times 470} \times 0.2\right) + \left(\frac{0.125}{4\pi \times 10^{-7}} \times 0.005\right)$$

$$= 42.33 + 497.36 = 539.7$$

Magnetising current, $I = 539.7/200 =$ **2.7 A**

EXAMPLE 8.7 A circular iron ring has a mean circumference of 1.5 m and a cross-sectional area of 0.01 m^2. A saw-cut of 4 mm wide is made in the ring. Calculate the magnetising current required to produce a flux of 0.8 mWb in the air gap if the ring is wound with a coil of 175 turns. Assume relative permeability of iron as 400 and leakage factor 1.25.

Solution

$$\text{Magnetising current} = \frac{\text{total AT}}{\text{number of turns}}$$

$$= \frac{\text{AT for air gap + AT for iron part}}{\text{number of turns}}$$

AT for air gap

$$B_g = \frac{\phi_g}{a} = \frac{0.8 \times 10^{-3}}{0.01} = 0.08 \text{ Wb/m}^2$$

$$H_g = \frac{B_g}{\mu_0} = \frac{0.08}{4\pi \times 10^{-7}} = 63662 \text{ AT/m}$$

$$AT_g = H_g \times l_g = 63662 \times 0.004 = 254.65 \text{ AT}$$

AT for iron path

$$\text{Leakage factor } (\lambda) = \frac{\text{flux in the iron path } (\phi_i)}{\text{flux in the air gap } (\phi_g)}$$

$$\phi_i = \phi_g \times \lambda = 0.8 \times 10^{-3} \times 1.25 = 1 \text{ mWb}$$

$$B_i = \frac{\phi_i}{a} = \frac{1 \times 10^{-3}}{0.01} = 0.1 \text{ Wb/m}^2$$

$$H_i = \frac{B_i}{\mu_0 \mu_r} = \frac{0.1}{4\pi \times 10^{-7} \times 400} = 198.94 \text{ AT/m}$$

$$AT_i = H_i \times l_i = 198.94 \times 1.5 = 298.41 \text{ AT}$$

Hence, total AT required $= 254.65 + 298.41 = 553.06$

and magnetising current, $I = 553.06/175 =$ **3.16 A**

EXAMPLE 8.8 A steel ring 30 cm mean diameter and of circular section 4 cm in diameter has an air gap 3 mm long. It is wound uniformly with 500 turns of wire carrying current of 5 A. Find (i) total MMF, (ii) total reluctance and (iii) flux. Neglect magnetic leakage. The iron path takes 40% of the total MMF.

Solution

(i) Total MMF = $NI = 500 \times 5 =$ **2500 AT**

MMF for iron part = 40% of total MMF

$= 0.4 \times 2500 = 1000$ AT

So, MMF for air gap = 2500 – 1000 = 1500 AT

Reluctance, $$S = \frac{\text{MMF}}{\text{flux}}$$

Reluctance of iron, $$S_i = \frac{\text{MMF for iron}}{\text{flux}}$$

and reluctance of air gap, $$S_g = \frac{\text{MMF for air gap}}{\text{flux}}$$

Since flux being constant,

$$\frac{S_i}{S_g} = \frac{\text{MMF for iron}}{\text{MMF for air gap}} = \frac{1000}{1500}$$

$$= 0.67$$

$$S_g = \frac{l_g}{a\mu_0} = \frac{3 \times 10^{-3}}{\pi(2 \times 10^{-2})^2 \times 4\pi \times 10^{-7}}$$

$$= 1899772 \text{ AT/Wb}$$

$$S_i = 0.67 \times 1899772 = 1272847 \text{ AT/Wb}$$

(ii) Total reluctance, $S = S_g + S_i = 1899772 + 1272847 =$ **3172619 AT/Wb**

(iii) Flux $= \dfrac{\text{total MMF}}{\text{total reluctance}} = \dfrac{2500}{3172619} =$ **0.7879 mWb**

EXAMPLE 8.9 The magnetic circuit frame shown in Fig. 8.6 is built up of iron of square cross-section 5 cm side. Each air gap is 2 mm wide. The relative permeability of part A and part B may be taken as 1000 and 1200, respectively. Find reluctances of parts A and B and reluctance of two air gaps.

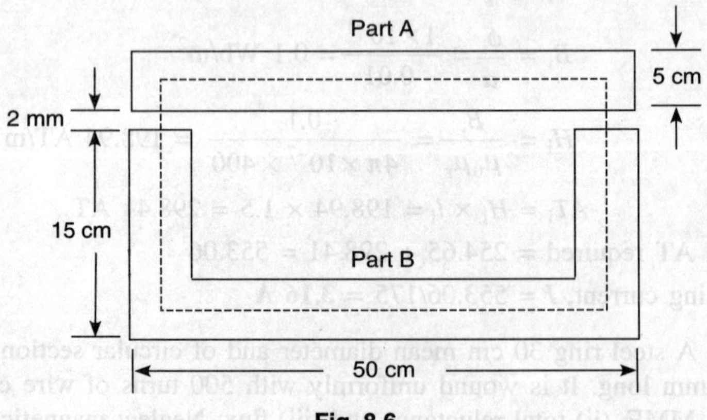

Fig. 8.6

Solution The dotted line shows the mean path of flux.

Part A

$$\text{Mean length, } l_A = 50 - 2.5 - 2.5 + 2.5 + 2.5 = 50 \text{ cm}$$

$$\text{Cross-sectional area, } a = 5 \times 5 = 25 \text{ cm}^2 = 25 \times 10^{-4} \text{ m}^2$$

So, $$\text{reluctance, } S = \frac{l_A}{a\mu_0\mu_r} = \frac{0.5}{25\times10^{-4}\times4\pi\times10^{-7}\times1000}$$

$$= \mathbf{159155 \text{ AT/Wb}}$$

Part B

$$\text{Mean length, } l_B = 50 - 2.5 - 2.5 + 12.5 + 12.5 = 70 \text{ cm}$$

$$\text{Cross-sectional area, } a = 5 \times 5 = 25 \text{ cm}^2 = 25 \times 10^{-4} \text{ m}^2$$

Hence, $$\text{reluctance, } S = \frac{l_B}{a\mu_0\mu_r} = \frac{0.7}{25\times10^{-4}\times4\pi\times10^{-7}\times1200}$$

$$= \mathbf{185681 \text{ AT/Wb}}$$

For air gap

$$\text{Air gap length, } l_g = 2 + 2 = 4 \text{ mm} = 0.004 \text{ m}$$

Hence, $$\text{reluctance, } S = \frac{l_g}{a\mu_0} = \frac{0.004}{25\times10^{-4}\times4\pi\times10^{-7}}$$

$$= \mathbf{1273240 \text{ AT/Wb}}$$

EXAMPLE 8.10 A cast steel magnetic structure has the dimensions as shown in Fig. 8.7. The cross-section is 4 cm^2 everywhere. Find the current that 600 turn coil on left limb should carry so that a flux of 2 mWb is produced in right limb. Take relative permeability as 600.

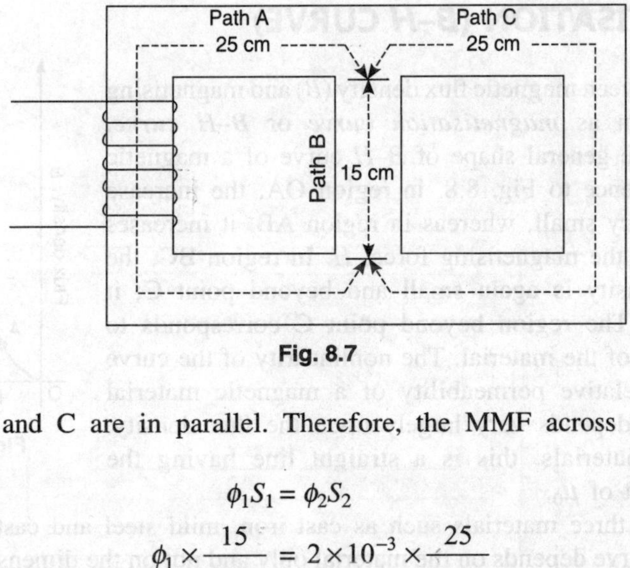

Fig. 8.7

Solution Paths B and C are in parallel. Therefore, the MMF across these two paths are equal, i.e.

$$\phi_1 S_1 = \phi_2 S_2$$

Therefore, $$\phi_1 \times \frac{15}{\mu_0\mu_r} = 2\times10^{-3}\times\frac{25}{\mu_0\mu_r}$$

and $$\phi_1 = 3.33 \text{ mWb}$$

So, total flux required $\phi = \phi_1 + \phi_2 = 3.33 + 2 = 5.33$ mWb

Total AT required = AT required for path A + AT required for path B or C

$$\text{Flux density in path A} = \frac{\phi}{a} = \frac{5.33\times10^{-3}}{4\times10^{-4}}$$

$$= 13.325 \text{ Wb/m}^2$$

$$\text{AT required} = \text{H} \times l = \frac{B}{\mu_0\mu_r} \times l$$

$$= \frac{13.325}{4\pi\times10^{-7}\times600} \times 0.25$$

$$= 4418.2 \text{ AT}$$

$$\text{Flux density in path C} = \frac{\phi}{a} = \frac{2\times10^{-3}}{4\times10^{-4}}$$

$$= 5 \text{ Wb/m}^2$$

$$\text{AT required} = \text{H} \times l = \frac{B}{\mu_0\mu_r} \times l = \frac{5}{4\pi\times10^{-7}\times600} \times 0.25$$

$$= 1657.8 \text{ AT}$$

$$\text{Total AT required} = 4418.2 + 1657.8 = 6076$$

Hence, current $= \dfrac{6076}{600} = \mathbf{10.13\ A}$

8.7 MAGNETISATION (*B–H* CURVE)

The curve drawn between magnetic flux density (*B*) and magnetising force (*H*) is known as *magnetisation curve* or *B–H curve.* Figure 8.8 shows the general shape of *B–H* curve of a magnetic material. With reference to Fig. 8.8, in region OA, the increase in flux density is very small, whereas in region AB, it increases almost linearly with the magnetising force, *H*. In region BC, the increase in flux density is again small and beyond point C; it is almost constant. The region beyond point C corresponds to *magnetic saturation* of the material. The nonlinearity of the curve indicates that the relative permeability of a magnetic material is not constant but depends very largely upon the flux density. For nonmagnetic materials, this is a straight line having the approximate gradient of μ_0.

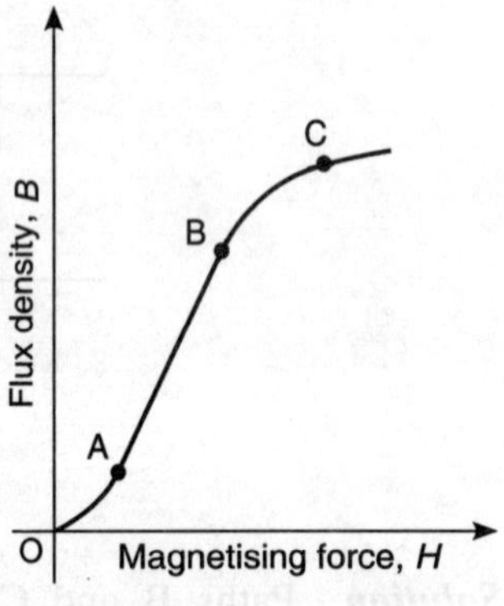

Fig. 8.8 *B–H* curve.

B–H curves for three materials such as cast iron, mild steel and cast steel are shown in Fig. 8.9. The *B–H* curve depends on the material only and not on the dimensions of it. In making calculations, values of *H* and μ_r corresponding to working flux density, *B* are to be taken. The use of *B–H* curves permits the calculations of magnetic circuits with fair degree of accuracy.

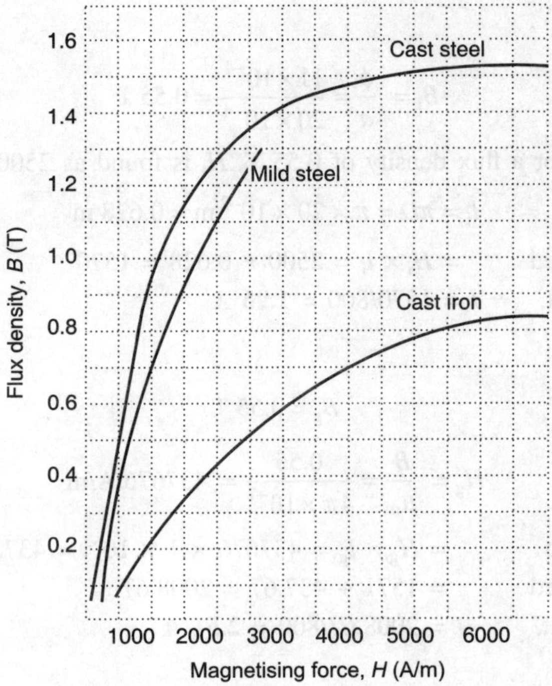

Fig. 8.9 *B–H* curves for different materials.

EXAMPLE 8.11 A cast iron ring of mean diameter 20 cm having a circular cross-section of 20 cm^2 is uniformly wound with 800 turns. Determine the magnetising current required to establish a flux of 11×10^{-4} Wb with no air gap and with a radial air gap of 1 mm.

The magnetisation curve for cast iron is given below.

B(T)	0.2	0.3	0.4	0.5	0.6	0.7	0.8
H(A/m)	400	900	1500	2100	3000	4200	7000

Solution The *B-H* curve is plotted from the data given as shown in Fig. 8.10.

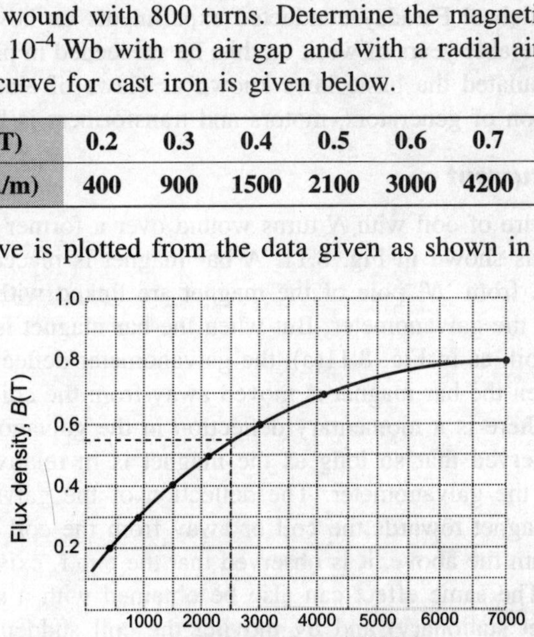

Fig. 8.10

With no air gap

Flux density in iron, $B_i = \dfrac{\phi}{a} = \dfrac{11 \times 10^{-4}}{20 \times 10^{-4}} = 0.55$ T

From the *B-H* curve, for a flux density of 0.55 T, H_i is found as 2500 A/m.

Length of iron, $l_i = \pi D = \pi \times 20 \times 10^{-2}$ m $= 0.628$ m

Hence, total AT required $= H_i \times l_i = 2500 \times 0.628 = 1570$

So, magnetising current, $I = 1570/800 =$ **1.96 A**

With air gap of 1 mm

Flux density in iron, $B_g = 0.55$ T

Magnetising force, $H_g = \dfrac{B}{\mu_0} = \dfrac{0.55}{4\pi \times 10^{-7}} = 437676$ A/m

AT required for air gap $= H_g \times l_g = 437676 \times 1 \times 10^{-3} = 437.67$

Hence, total AT required $= 1571 + 437.67 = 2008.67$

and magnetising current, $I = 2008.67/800 =$ **2.51 A**

8.8 ELECTROMAGNETIC INDUCTION

Oersted, Danish Physics professor, discovers the phenomenon that *electric current induces magnetic field*. Then, Michael Faraday conducted experiments on how to get electricity from magnetism. After considerable years of work on this, he succeeded in producing electricity from magnetism and he formulated the basic laws known as "laws of electromagnetic induction". The principle of operation of generators, motors and transformers is based upon this theory.

Induced e.m.f and current

Consider an insulated wire of coil with *N* turns wound over a former which is connected to a sensitive galvanometer as shown in Fig. 8.11. A bar magnet is placed very close to the coil. Now some lines of flux from '*N*' pole of the magnet are linked with the coil. At this stage, there is no deflection in the galvanometer. But when the bar magnet is suddenly moved closer to the coil or into the coil, as in Fig. 8.11(a), the galvanometer deflects momentarily and then comes to rest. Also, when the bar magnet is moved away from the coil as in Fig. 8.11(b), then it is found that, again, there is a momentary deflection in the galvanometer.

From this, it is observed that so long as the magnet is in relative motion with the coil, there is a deflection in the galvanometer. The deflection of the galvanometer during sudden movement of the bar magnet towards the coil or away from the coil indicates the production of e.m.f. in the coil. From the above, it is observed that the e.m.f. exists so long as the change in lines of flux exists. The same effect can also be obtained with a stationary magnetic field (keeping the bar magnet stationary) and by moving the coil suddenly towards or away the magnetic field as shown in Fig. 8.12.

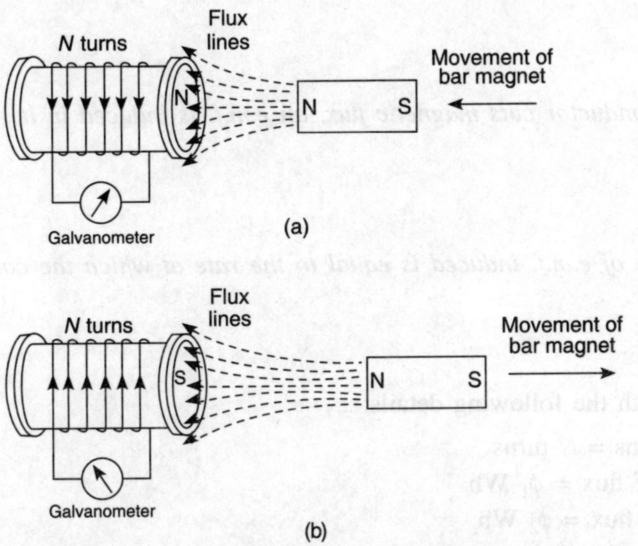

Fig. 8.11 Generation of e.m.f. due to changing magnetic field.

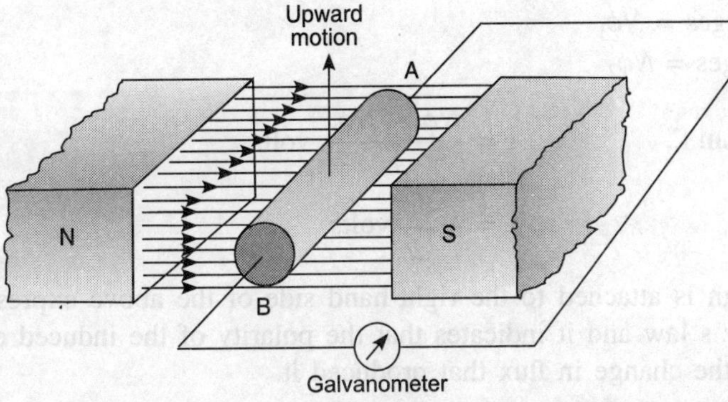

Fig. 8.12 Generation of e.m.f. due to stationary magnetic field.

In Fig. 8.12, AB is a conductor placed within a magnetic field. A galvanometer is connected between the two ends of the conductor. If the conductor is moved in upward direction or in downward direction, so as to cut the magnetic lines of flux, a momentary deflection is produced in the galvanometer. It means that an e.m.f. is induced in the conductor and its magnitude depends on how fast the conductor AB is moved with respect to the magnetic field.

From this experiment, it is learnt that whenever a conductor cuts magnetic flux, an e.m.f. is induced in it. It is also found that if the conductor is moved parallel to the direction of the flux so that it does not cut it, then no e.m.f. is induced in it.

8.9 FARADAY'S LAWS OF ELECTROMAGNETIC INDUCTION

Faraday summed up the above findings into two laws known as Faraday's laws of electromagnetic induction.

First Law

It states that,

Whenever a conductor cuts magnetic flux, an e.m.f. is induced in it.

Second Law

It states that,

The magnitude of e.m.f. induced is equal to the rate at which the conductor cuts the magnetic flux.

Explanation:

Consider a coil with the following details:

Number of turns = N turns
Initial value of flux = ϕ_1 Wb
Final value of flux = ϕ_2 Wb
Time taken for the change in flux from ϕ_1 to ϕ_2 = t seconds

Flux linkages means the product of flux and number of turns. Therefore,

Initial flux linkages = $N\phi_1$
Final flux linkages = $N\phi_2$

Therefore, induced e.m.f.,
$$e = \frac{N\phi_2 - N\phi_1}{t} \text{ volts}$$

In differential form,
$$e = N\frac{d\phi}{dt} \text{ volts}$$

Usually, a *minus* sign is attached to the right-hand side of the above expression. This minus sign is to obey Lenz's law and it indicates that the polarity of the induced e.m.f is in such a direction to oppose the change in flux that produced it.

Hence, induced e.m.f.,
$$e = -N\frac{d\phi}{dt} \text{ volts}$$

EXAMPLE 8.12 A coil of 50 turns is placed in a magnetic field of 0.2 Wb. This coil is moved from this magnetic field to a magnetic field of 0.1 Wb, in 0.2 seconds. Find the average e.m.f. induced in the coil.

Solution

Number of turns = 50 turns
Initial value of flux = 0.2 Wb
Final value of flux = 0.1 Wb
Time taken for the change in flux, dt = 0.2 seconds

Hence, induced e.m.f.,
$$e = N\frac{d\phi}{dt} = 50 \times \frac{0.2 - 0.1}{0.2} = \mathbf{25\ V}$$

EXAMPLE 8.13 A coil of 200 turns is linked by a flux of 10 mWb. If this flux is reversed in a time of 2 ms, calculate the average e.m.f. induced in the coil.

Solution

Number of turns = 200 turns

Change in flux, $d\phi$ = 10 – (–10) = 20 mWb

Time taken for the change in flux, dt = 2 ms = 2×10^{-3} s

Hence, induced e.m.f., $$e = N\frac{d\phi}{dt} = 200 \times \frac{20 \times 10^{-3}}{2 \times 10^{-3}} = \mathbf{2000\ V}$$

8.10 DIRECTION OF INDUCED E.M.F. AND CURRENT

The direction of induced e.m.f. is determined by applying either Lenz's law or Fleming's right-hand rule.

8.10.1 Lenz's Law

In 1985, Lenz, a German scientist, formulated this law to find the direction of current supplied by the induced e.m.f. It states that,

The induced current flows in such a direction so as to oppose the very cause producing it.

With reference to Fig. 8.11(a), it is found that when N-pole of the magnet is brought nearer to the coil, the current due to the induced e.m.f. flows in the anticlockwise direction in the coil as seen from the magnet side. As a result, the face of the coil becomes an N-pole and so tends to oppose the onward approach of the N-pole magnet. The mechanical energy spent in overcoming this repulsion is converted into electrical energy which appears in the coil. On the other hand, if the magnet is withdrawn as in Fig. 8.11(b), the induced current flows in the clockwise direction thus making the face of the coil an S-pole. Therefore, the N-pole of the magnet has to be withdrawn against this force of the S-pole of the coil. Again, the mechanical energy required to overcome this force of attraction is converted into electric energy.

Lenz's law is all about conservation of energy. If we assume that the magnetic field associated with the induced current were in the same direction as the change in magnetic field that created it, these two magnetic fields would combine to give a net magnetic field which would in turn induce a current with twice the magnitude. This process would continue creating infinite current from just moving a magnet, a violation of the law of conservation of energy.

8.10.2 Fleming's Right-hand Rule

Fleming's right hand rule is suitable to find the direction of induced e.m.f. and hence current when the conductor moves at right angles to a stationary magnetic field. It may be stated as:

If the thumb, forefinger and middle finger of right hand are held mutually perpendicular to each other, forefinger pointing into the direction of the field and thumb in the direction of motion of conductor then the middle finger will point in the direction of the induced current.

This is illustrated in Fig. 8.13.

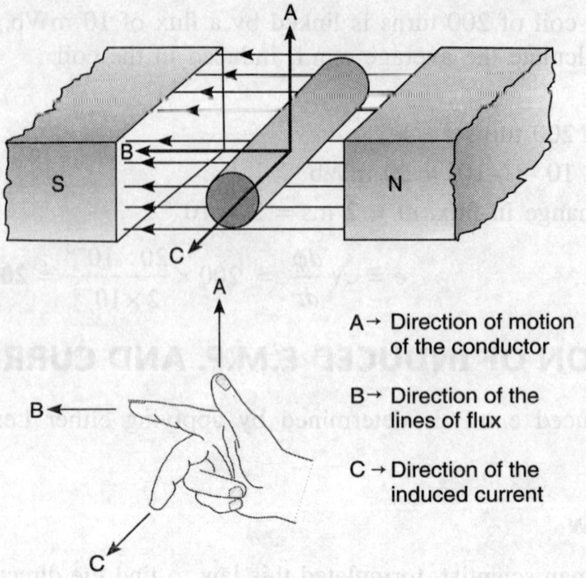

Fig. 8.13 Fleming's right hand rule.

8.11 INDUCED E.M.F.

According to Faraday's laws of electromagnetic induction, whenever the flux linked with a conductor (or coil) changes, an e.m.f. is induced in it. The flux linkage can be made either by moving the conductor inside a stationary magnet or by changing the magnetic field while keeping the conductor stationary. In the former case, the e.m.f. induced is called *dynamically induced e.m.f.* and in the latter case, it is called as *statically induced e.m.f.*

8.11.1 Dynamically Induced e.m.f.

When the conductor is moving and the field is kept stationary, the e.m.f. induced is called *dynamically induced e.m.f.* Consider a single conductor of length l metres moving at right angles to a uniform magnetic field of B Wb/m^2 with a velocity of v m/s. Suppose the conductor moves through a small distance dx in dt seconds. Then area swept by the conductor is 'ldx'. Hence,

$$\text{flux cut, } d\phi = \text{flux density} \times \text{area swept} = B \times ldx \text{ Wb}$$

According to Faraday's laws of electromagnetic induction, e.m.f. e induced in the conductor is given by

$$= N\frac{d\phi}{dt} = 1\frac{Bldx}{dt} \qquad (N, \text{ number of conductors} = 1)$$

$$= Blv \text{ volts} \qquad \left(v, \text{velocity} = \frac{dx}{dt}\right)$$

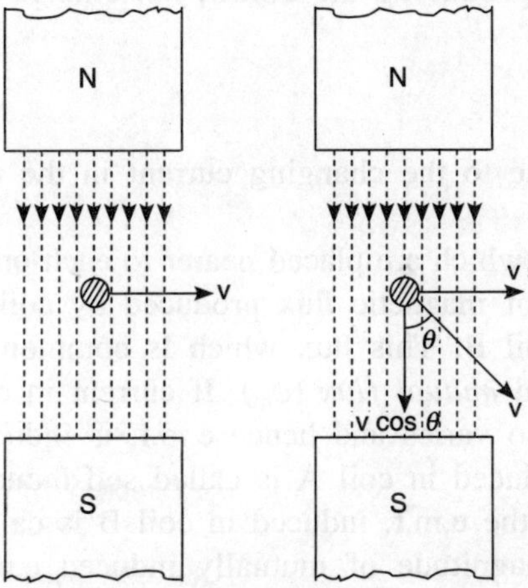

Fig. 8.14 Illustration of generation of dynamically induced e.m.f.

As shown in Fig. 8.14, if the conductor moves at an angle θ to the magnetic field, then the velocity at which the conductor moves across the field is $v \sin \theta$. Hence,

$$e = Blv \sin \theta$$

The direction of induced e.m.f. can be determined by Fleming's right-hand rule.

EXAMPLE 8.14 A conductor 200 mm long moves at a uniform speed of 10 m/s at right angles to a uniform magnetic field of flux density 3.5 T. (i) Determine the induced e.m.f. in the conductor? (ii) What will be the e.m.f. induced if the conductor moves at an angle 30° to the field?

Solution

(i) $e = Blv = 3.5 \times 0.2 \times 10 = 7$ V

(ii) $e = Blv \sin \theta = 3.5 \times 0.2 \times 10 \times \sin 30° =$ **3.5 V**.

8.11.2 Statically Induced E.M.F.

When the conductor is stationary and the field is moving, the e.m.f. induced in the conductor is called statically induced e.m.f. It is further divided into *self-induced e.m.f.* and *mutually induced e.m.f.*

Self-induced e.m.f.

The e.m.f. induced in the coil due to the change in its own flux linked with it is called *self-induced e.m.f.*

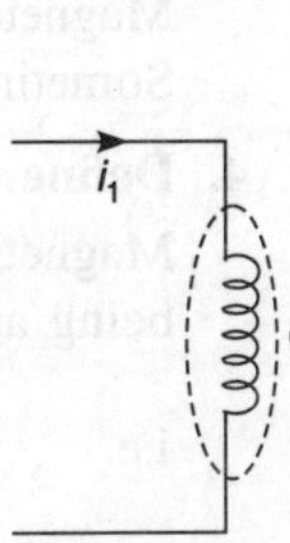

Fig. 8.15

When a coil is carrying current (Fig. 8.15), a magnetic field is established around the coil. If the current in the coil changes, then flux linking the coil also changes. Hence by Faraday's laws, an e.m.f. will be induced in the coil. This is known as self-induced e.m.f. The magnitude of the self-induced e.m.f. $= Nd\phi/dt$. The direction of induced e.m.f. (by Lenz's law) is always such that it opposes the cause producing the e.m.f., i.e. change of current (and hence the field) in the coil.

Mutually induced e.m.f.

The e.m.f. induced in a coil due to the changing current in the neighbouring coil is called *mutually induced e.m.f.*

Consider two coils A and B which are placed nearer to each other as shown in Fig. 8.16. A part of magnetic flux produced by coil A passes through or links with coil B. This flux which is common to both the coils A and B is called *mutual flux* (ϕ_m). If current in coil A is varied, the mutual flux also varies and hence e.m.f. is induced in both the coils. The e.m.f. induced in coil A is called self-induced e.m.f. as already discussed and the e.m.f. induced in coil B is called mutually induced e.m.f. The magnitude of mutually induced e.m.f. $e = N_B d\phi_{AB}/dt$, where N_B is the number of turns of coil B and ϕ_{AB} is the flux linked in coil B due to flux set up in coil A.

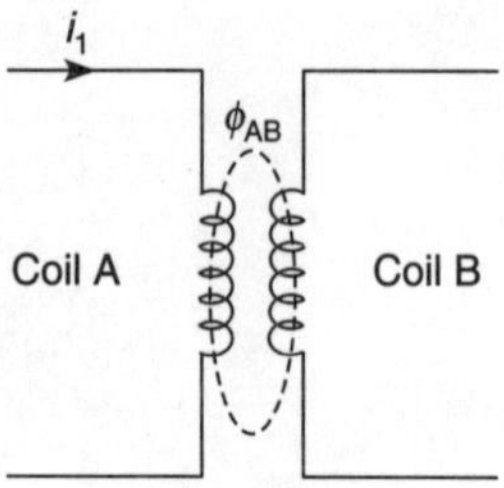

Fig. 8.16

SHORT QUESTIONS WITH ANSWERS

1. **Define magnetic flux.**

   Magnetic flux is the magnetic lines of force produced by a magnetic source. The symbol for magnetic flux is ϕ (Greek letter 'phi'), and its unit is *weber* (Wb).

2. **Define magnetic flux density.**

   Magnetic flux density is the flux per unit cross-sectional area. It is denoted by B and its unit is Wb/m², or tesla (T), i.e.

$$B = \frac{\phi}{A} \text{ Wb/m}^2 \text{ or T}$$

3. **Define magnetomotive force.**

   Magnetomotive force (MMF) is the cause of existence of magnetic flux in a magnetic circuit. It is equal to the product of current and number of turns of the coil, i.e.

$$\text{MMF} = NI \text{ ampere-turns (or) AT}$$

   Magnetomotive force in a magnetic circuit is analogous to e.m.f. in an electric circuit. Sometimes, its unit is expressed as 'ampere' only as 'turns' have no dimensions.

4. **Define magnetic field strength.**

   Magnetic field strength (H) is the MMF per unit of length of magnetic circuit, the unit being ampere-turns/metre.

   i.e.
$$H = \frac{NI}{l} \text{ AT/m or A/m}$$

5. Define permeability.

Permeability of a material means its conductivity of magnetic flux. Obviously, good magnetic materials would have high permeability.

6. What are the two types of permeabilities? How they are measured?

The two types of permeabilities are:

(i) absolute permeability (μ_0) and (ii) relative permeability (μ_r).

To measure relative permeability, vacuum or free space is chosen as the reference medium. It is allotted an absolute permeability of $\mu_0 = 4\pi \times 10^{-7}$ henry/metre. The relative permeability of a medium is defined as:

$$\mu_r = \frac{\text{absolute permeabilty of the material}}{\text{absolute permeabilty of free space}}$$

Hence, for vacuum or free space, the relative permeability is unity and for any other medium

$$\mu_r = \frac{\mu}{\mu_0}$$

where

μ = absolute permeability of the material

μ_0 = absolute permeability of free space

μ_r = relative permeability of the material

7. Define reluctance.

The opposition that the magnetic circuit offers to flux is called reluctance (S) and its unit is AT/Wb.

$$S = \frac{l}{a\mu} \text{ AT/Wb}$$

or

$$S = \frac{l}{a\mu_0\mu_r} \text{ AT/Wb}$$

It is analogous to resistance in an electric circuit.

8. Define permeance.

It is the magnetic flux per ampere of total MMF in the path of magnetic field. It is the reciprocal of reluctance. Its unit is Wb/AT.

$$\text{Permeance} = \frac{1}{\text{reluctance}} = \frac{a\mu_0\mu_r}{l}$$

It is analogous to conductance in an electric circuit.

9. What quantity in magnetic circuit is analogous to resistance in electric circuit?

Reluctance (Ampere-turns/Wb).

10. What quantity in electric circuit is analogous to magnetic flux?

Current (A).

11. What quantity in magnetic circuit is analogous to electromotive force in electric circuit?

Magnetomotive force (Ampere-turns).

12. What quantity in magnetic circuit is analogous to conductance in electric circuit?

Permeance (Wb/Ampere-turns).

13. What quantity in electric circuit is analogous to magnetic flux density?

Current density (A/m^2).

14. What is composite magnetic circuit?

A series magnetic circuit which has several parts of different dimensions and materials is called composite magnetic circuit.

15. What is parallel magnetic circuit?

If a magnetic circuit has more than one path for flux then it is called parallel magnetic circuit.

16. What is leakage flux?

The flux does not follow the desired path of magnetic medium is called leakage flux.

17. What is leakage factor?

The ratio of total flux (flux in the iron path) to the useful flux (flux in the air gap) is known as *leakage factor*.

i.e. $$\text{leakage factor} = \frac{\text{total flux}}{\text{useful flux}}$$

or $$\text{leakage factor} = \frac{\text{flux in the iron path}}{\text{flux in the air gap}}$$

18. Define fringing effect.

In a flux carrying magnetic circuit (say iron ring) with an air gap, the magnetic flux bulges outwards when it passes through the air gap. This phenomenon is called fringing effect.

19. What is meant by magnetisation curve?

The curve drawn between magnetic flux density (B) and magnetising force (H) is known as magnetisation curve or B–H curve.

20. State Faraday's laws of Electromagnetic Induction.

Faraday's laws of electromagnetic induction are:

First Law

It states that,

Whenever a conductor cuts magnetic flux, an e.m.f. is induced in it.

Second Law

It states that,

The magnitude of e.m.f. induced is equal to the rate at which the conductor cuts the magnetic flux.

21. State Lenz's law.

Lenz's law states that,

The induced current flows in such a direction so as to oppose the very cause producing it.

22. State Fleming's right-hand rule.

Fleming's right-hand rule states that,

If the thumb, fore-finger and middle finger of right-hand are held mutually perpendicular to each other, fore-finger pointing into the direction of the field and thumb in the direction of motion of conductor, then the middle finger will point in the direction of the induced current.

23. What is induced e.m.f and what are its types?

According to Faraday's laws of electromagnetic induction, whenever the flux linked with a conductor (or coil) changes, an e.m.f. is induced in it. The flux linkage can be made either by moving the conductor inside a stationary magnet or by changing the magnetic field while keeping the conductor stationary. In the former case, the e.m.f. induced is called dynamically induced e.m.f. and in the latter case, it is called as statically induced e.m.f.

UNSOLVED PROBLEMS

1. An air cored toroidal coil has 200 turns. The length of magnetic circuit is 20 cm and the cross-sectional area is 5 cm^2. Find the current required to establish a flux of 1×10^{-6} Wb. [**Ans.** 1.59 A]

2. A mild steel ring has a mean diameter of 20 cm and cross-sectional area of 400 mm^2. Calculate (i) the reluctance of the ring and (ii) the MMF to produce a flux of 500 μWb. Assume the relative permeability of mild steel to be 1000.

[**Ans.** (i) 1.25×10^6 AT/Wb; (ii) 625 AT]

3. An air cored toroidal coil has 3000 turns and carries a current of 0.1 A, the length of magnetic circuit is 150 mm and the cross-sectional area of the coil is 400 mm^2. Calculate (i) the magnetic field strength, (ii) flux density and (iii) total flux within the coil. [**Ans.** (i) 2000 AT/m; (ii) 2.51 Wb/m^2; (iii) 1.004 μWb]

4. A cast steel ring of 200 mm mean diameter and 300 mm^2 cross-section has a winding of 200 turns. If the relative permeability of steel is 1000, what current will be required to produce a flux of 3.75×10^{-4} Wb. [**Ans.** 3.17 A]

5. An iron ring has a cross-sectional area of 10^{-3} m^2 and an air gap of 2 mm wide. The mean circumference of the ring is 1 m. If the relative permeability of the material is 500 and the coil has 200 turns, find the current required to set up a flux of 0.001 Wb. Neglect leakage. [**Ans.** 15.9 A]

6. A coil of 200 turns is wound uniformly over a wooden ring having a mean circumference of 600 mm and a uniform cross-sectional area of 500 mm^2. If the current through the coil is 4 A, calculate (i) magnetic field strength (ii) flux density and (iii) total flux.
[**Ans.** (i) 1333 A/m; (ii) 0.001675 Wb/m^2; (iii) 0.8375 μWb]

7. An iron ring has a mean circumference of 0.4 m and an area of 10^{-4} m^2. It is wound uniformly with 500 turns. When the current in the winding is 0.06 A, the flux in the ring is 6×10^{-6} Wb. Find (i) the flux density, (ii) field strength and (iii) relative permeability. [**Ans.** (i) 0.06 Wb/m^2; (ii) 75 AT/m; (ii) 636.6]

8. An iron ring of mean length 50 cm has an air gap of 1 mm and a winding of 200 turns. If the relative permeability of iron is 300, when a current 1 A flows through the coil, find the flux density. [**Ans.** 94.2×10^{-3} Wb/m^2]

9. An electromagnet has 100 turns. The length of the core is 8 cm and the current in the coil is 0.2 A. If $B = 0.13$ T, find (i) the magnetic field strength, (ii) permeability and (iii) relative permeability. [**Ans.** (i) 250 AT/m; (ii) 5.2×10^{-4} H/m; (iii) 413.8]

10. A coil of 500 turns and resistance 20 Ω is wound uniformly on an iron ring of mean circumference 50 cm and cross-sectional area of 4 cm^2. It is connected to a 24 V DC supply. The relative permeability is 800. Calculate the values of (i) the m.m.f of the coil, (ii) the magnetising force, (iii) the total flux in the iron and (iv) the reluctance of the ring.
[**Ans.** (i) 600 AT; (ii) 1200 AT/m; (iii) 0.483 mWb; (iv) 1.24×10^6 AT/Wb]

11. A magnetic ring has a mean circumference of 1.5 m and is of 0.01 m^2 in cross-section and is wound with 175 turns. A saw-cut of 4 mm width is made in the ring. Calculate the magnetising current required to produce a flux of 0.8 mWb in the air gap. Assume leakage factor as 1.25 and permeability of iron as 400. [**Ans.** 3.16 A]

12. A cast steel magnet has an air gap of length 2 mm and an iron path of 30 cm. Find the number of ampere-turns necessary to produce a flux density of 1.2 Wb/m^2. The relative permeability of cast steel is 900. Neglect leakage and fringing.
[**Ans.** 2228.3 AT]

13. A magnetic circuit consists of an iron ring of mean circumference 80 cm with cross-sectional area of 12 cm^2 throughout. A current of 1 A in the magnetising coil of 200 turns produces a total flux of 1.2 mWb in the iron. Calculate (i) the flux density in the iron (ii) the absolute and relative permeability of iron and (iii) the reluctance of the circuit. [**Ans.** (i) 1 Wb/m^2; (ii) 4×10^{-3}, 3183; (iii) 166.67×10^3 AT/Wb]

14. A cast steel structure made from a bar of section 2 cm × 2 cm is shown in Fig. U8.1. Determine the current that the 600 turn magnetising coil on the left limb should carry so that a flux of 2 mWb is produced in the right limb. [**Ans.** 20.26 A]

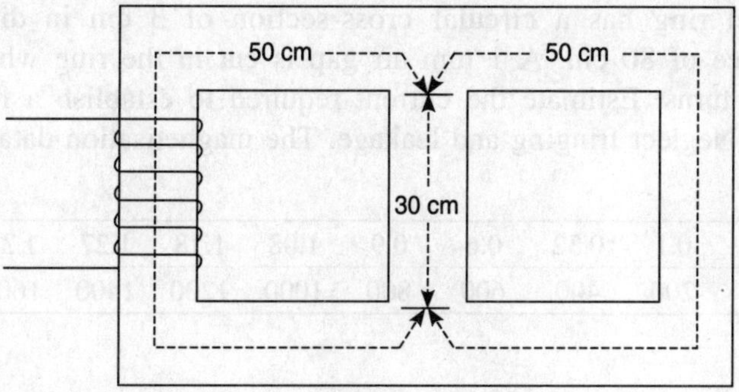

Fig. U8.1

15. Find the number of ampere-turns required to produce a flux of 0.55 mWb in a ring with mean circumference of 70 cm and area of cross-section of 5 cm^2. The flux density to relative permeability test for the iron gave the following results:

B (T)	0.8	1.0	1.2	1.3
H (A/m)	2300	2000	1600	1100

If a saw-cut 1.5 mm wide is made in the above ring, how many extra ampere-turns would be required? **[Ans.** 1312.299 AT]

16. A soft iron specimen has the following *B–H* curve data.

B (T)	0.126	0.252	0.378	0.428	0.441
H (A/m)	1000	2000	3000	4000	5000

(a) Find the values of permeability and relative permeability.
(b) At what value of *H* saturation does start?

[Ans. (a) 1.26×10^{-4} H/m, 100.27; (b) 4000 AT/m]

17. An iron ring has a mean length of 70 cm and cross-sectional area of 2 cm^2. It is wound with 1500 turns. If a current of 2 A flows through the coil, find the flux density in the ring. Assume the relative permeability of iron to be 300. Also calculate current in the coil, if a 2 mm air gap is created in the ring, to have the same flux density in the ring. **[Ans.** 1.6156 Wb/m^2; 3.7141 A]

18. A cast steel electromagnet has an air gap of length 2 mm and an iron path of length 30 cm. Find the number of ampere-turns required to produce flux density of 0.8 Wb/m^2 in the gap. Neglect leakage and fringing. The magnetisation curve is given below.

B (Wb/m^2)	0.05	0.2	0.6	0.8	1.0	1.3
H (A/m)	100	300	600	730	900	1500

[Ans. 1274 AT]

19. A cast steel ring has a circular cross-section of 3 cm in diameter and a mean circumference of 80 cm. A 1 mm air gap is cut in the ring which is wound with a coil of 600 turns. Estimate the current required to establish a flux of 0.75 mWb in the air gap. Neglect fringing and leakage. The magnetisation data of the material is as follows:

B (T)	0.1	0.32	0.6	0.9	1.08	1.18	1.27	1.32	1.36
H (A/m)	200	400	600	800	1000	1200	1400	1600	1800

[**Ans.** 2.72 A]

20. The magnetic circuit shown in Fig. U8.2 is made of silicon sheet steel and has a square cross-section. It is required to set up a flux of 25×10^{-4} Wb. Find MMF and exciting current if the coil has 100 turns. [**Ans.** 268.85 AT; 2.6885 A]

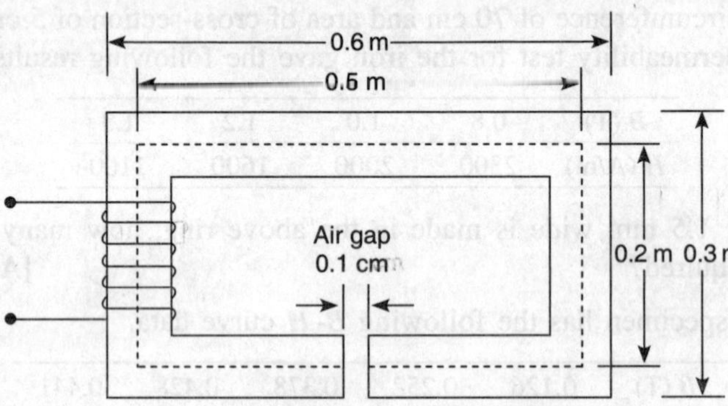

Fig. U8.2

21. A mild steel ring has a mean circumference of 1000 mm and a uniform area of cross-section of 600 mm^2. Calculate the m.m.f required to produce a flux of 500 nWb, assuming permeability of mild steel as 1200. An air gap of 2 mm in length is now cut in the ring. Determine the flux produced, if the m.m.f remains constant.

[**Ans.** 147.1397 nWb]

22. A cast steel electromagnet has an air gap of length 1 mm and an iron path of length 40 cm. Find the number of ampere-turns necessary to produce a flux density of 1.5 Wb/m^2 in the gap. Neglect leakage and fringing. Take μ_r = 800 for cast steel.

[**Ans.** 1790.4931 AT]

23. A coil of 1500 turns carrying a current of 5 A produces a flux of 2.5 mWb. If the current is reversed in 0.2 second, find the average value of e.m.f. induced in the coil. Also find the self-inductance of the coil. [**Ans.** 37.5 V; 0.75 H]

24. A flux of 4×10^{-4} Wb is created by current of 10 A flowing through a 100 turns coil. Calculate the inductance of the coil corresponding to a complete reversal of the current in 0.01 second. Also find the magnitude of e.m.f. induced in the coil.

[**Ans.** 4 mH; 8 V]

CHAPTER 9

Coupled Circuits

9.1 INTRODUCTION

Coupled circuits refer to the circuits which are coupled magnetically, not conductively. If the magnetic flux produced by an element in one circuit links the elements of another circuit, then both circuits are said to be magnetically coupled circuits. In such circuits (e.g. transformer), transfer of energy takes place from one circuit to another through magnetic flux. This chapter discusses some magnetic coupling between two or more circuits and its effects.

9.2 SELF-INDUCTANCE

If a current passing through a coil in a circuit changes, the flux set up by the current also does change and hence an e.m.f. is induced in the circuit. This e.m.f. is proportional to the rate of change of current.

i.e.
$$e \propto \frac{di}{dt}$$

or
$$e = L\frac{di}{dt} \tag{9.1}$$

where L is called self inductance of the coil.

According to Faraday's law of electromagnetic induction, e.m.f. induced is proportional to the rate of flux linkages.

i.e.
$$e \propto \frac{d\phi}{dt}$$

or
$$e = N\frac{d\phi}{dt} \tag{9.2}$$

where N is number of turns of the coil.

From Eqs. (9.1) and (9.2), we can write:

$$L\frac{di}{dt} = N\frac{d\phi}{dt}$$

or

$$L = N\frac{d\phi}{di}$$

If the permeability is constant, then

$$L = N\frac{\phi}{I} \tag{9.3}$$

9.3 MUTUAL INDUCTANCE

Consider two coils A and B, having turns N_1 and N_2, respectively, which are placed very close to each other as shown in Fig. 9.1. Coil A is supplied with current i_1 and as a result, flux (ϕ_1) sets up in it. Since coil B is kept nearer to coil A, a part of this flux (ϕ_{12}), called mutual flux, links coil B. If current in coil A changes, the mutual flux also changes and hence e.m.f. is induced in coil B. This e.m.f. is called mutually induced e.m.f. This effect also happens in the opposite way, i.e. a change of current in coil B results in mutually induced e.m.f. in coil A.

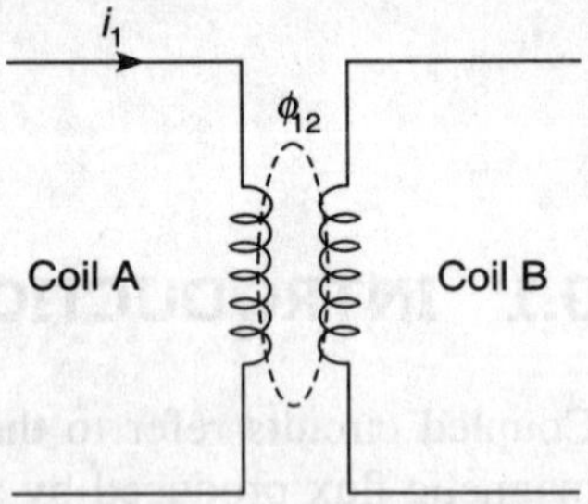

Fig. 9.1 Coupled circuit.

Hence, mutually induced e.m.f. in coil B ∝ rate of change of current in coil A

or

$$e_2 \propto \frac{di_1}{dt}$$

or

$$e_2 = M\frac{di_1}{dt} \tag{9.4}$$

where M is called mutual inductance between the two coils.

Also, e_2 is proportional to the rate of flux linkage, i.e.

$$e_2 \propto \frac{d\phi_{12}}{dt}$$

or

$$e_2 = N_2\frac{d\phi_{12}}{dt} \tag{9.5}$$

Equating Eqs. (9.4) and (9.5), we get

$$M = N_2\frac{d\phi_{12}}{di_1}$$

If the permeability is constant, then

$$M = N_2\frac{\phi_{12}}{I_1}$$

The mutual inductance is bilateral and if I_2 changes, a mutually induced e.m.f. e_1 will be induced and is proportional to the rate of change of current in coil B. Then,

$$e_1 = N_1 \frac{d\phi_{21}}{dt}$$

and

$$M = N_1 \frac{\phi_{21}}{I_2}$$

where ϕ_{21} is part of flux linked in coil A due to the current flows in coil B.

9.4 COEFFICIENT OF COUPLING

The fraction of the total flux produced by one coil linking another coil is called *coefficient of coupling*. Thus,

$$k = \frac{\phi_{12}}{\phi_1} = \frac{\phi_{21}}{\phi_2} \tag{9.6}$$

Since $\phi_{12} < \phi_1$ and $\phi_{21} < \phi_2$, the value of k is always less than or equal to 1. We know that

$$M = \frac{N_2\phi_{12}}{i_1} \tag{9.7}$$

Also

$$M = \frac{N_1\phi_{21}}{i_2} \tag{9.8}$$

Multiplying Eqs. (9.7) and (9.8), we get

$$M^2 = \left(\frac{N_2\phi_{12}}{i_1}\right)\left(\frac{N_1\phi_{21}}{i_2}\right)$$

$$= \left(\frac{N_2k\phi_1}{i_1}\right)\left(\frac{N_1k\phi_2}{i_2}\right) \qquad \text{[From Eq. 9.6]}$$

$$= k^2\left(\frac{N_1\phi_1}{i_1}\right)\left(\frac{N_2\phi_2}{i_2}\right)$$

$$= k^2 L_1L_2$$

or

$$M = k\sqrt{L_1L_2}$$

EXAMPLE 9.1 Two coupled coils with self-inductances 0.9 H and 0.4 H have a coupling coefficient of 0.3. Find the mutual inductance and turns ratio. What will be the maximum possible value of mutual inductance?

Solution Mutual inductance,

$$M = k\sqrt{L_1L_2} \tag{i}$$

Putting $L_1 = 0.9$ H and $L_2 = 0.4$ H in Eq. (i), we get

$$M = 0.3\sqrt{0.9 \times 0.4}$$

$$= \mathbf{0.18\ H}$$

From Eq. (9.3), $L_1 = \dfrac{N_1\phi_1}{I_1}$ and $L_2 = \dfrac{N_2\phi_2}{I_2}$

$$\therefore \qquad \frac{L_1}{L_2} = \frac{\left(\dfrac{N_1\phi_1}{I_1}\right)}{\left(\dfrac{N_2\phi_2}{I_2}\right)} = \frac{N_1\phi_1}{N_2\phi_2} \times \frac{I_2}{I_1} \qquad \text{(ii)}$$

As the flux is same in coupled coils

$$\frac{L_1}{L_2} = \frac{N_1}{N_2} \times \frac{I_2}{I_1} \qquad \text{(iii)}$$

Also, $\dfrac{N_1}{N_2} = \dfrac{I_2}{I_1}$

Hence, Eq. (iii) can be written as:

$$\frac{L_1}{L_2} = \frac{N_1}{N_2} \times \frac{N_1}{N_2} = \left(\frac{N_1}{N_2}\right)^2$$

Turns ratio, $\dfrac{N_1}{N_2} = \sqrt{\dfrac{L_1}{L_2}} = \sqrt{\dfrac{0.9}{0.4}} = \mathbf{1.5}$

The maximum possible value of M is obtained if $k = 1$. Therefore, the maximum possible value of $M = K\sqrt{L_1L_2} = \sqrt{0.9 \times 0.4} = \mathbf{0.6\ H}$

EXAMPLE 9.2 The coefficient of coupling between two coils is 0.6. There are 500 turns in coil 1. The total flux of coil 1 is 0.5 mWb, when the current in this coil is 4 A. When the current is changed from 4 A to zero amperes in 4 milliseconds, the voltage induced in coil 2 is 50 V. Calculate the self-inductance and mutual inductance of the coils and the number of turns in coil 2.

Solution Self-inductance of coil 1, $L_1 = \dfrac{N_1\phi_1}{I_1} = \dfrac{500 \times 0.5 \times 10^{-3}}{4} = \mathbf{62.5\ mH}$

Voltage across coil 2 due to current in coil 1, $e_2 = M\dfrac{di_1}{dt}$

$$50 = M\frac{di_1}{dt} = M\frac{4}{4 \times 10^{-3}}$$

So, mutual inductance, $M = \mathbf{50 \times 10^{-3}\ H}$

Since $$M = K\sqrt{L_1 L_2}, \quad L_2 = \frac{M^2}{k^2 L_1}$$

Hence, $$L_2 = \frac{(50\times10^{-3})^2}{0.6^2\times62.5\times10^{-3}} = \mathbf{0.1111\ H}$$

Turns ratio, $$\frac{N_1}{N_2} = \sqrt{\frac{L_1}{L_2}}$$

$\therefore$ $$N_2 = N_1\times\sqrt{\frac{L_2}{L_1}} = 500\times\sqrt{\frac{0.1111}{62.5\times10^{-3}}} = \mathbf{667\ turns}$$

EXAMPLE 9.3 Two coils A and B with turns of 1200 and 1500 respectively lie in a parallel plane so that 40% of flux produced by coil A links with coil B. A current of 4 A in coil A produces a flux 0.05 mWb while the same current in coil B produces a flux of 0.08 mWb. Determine the coefficient of coupling between the coils.

Solution Self-inductance of coil A, $L_1 = \frac{N_A\phi_A}{I_A} = \frac{1200\times0.05\times10^{-3}}{4}$

$$= 0.015\ \text{H}$$

Self-inductance of coil B, $L_2 = \frac{N_B\phi_B}{I_B} = \frac{1500\times0.08\times10^{-3}}{4} = 0.03\ \text{H}$

Given that 40% flux produced by coil A links with coil B, i.e.

$$\frac{\phi_{AB}}{\phi_A} = 40\%$$

or $$\phi_{AB} = 0.4\phi_A$$

Hence, mutual inductance, $$M = \frac{N_B\phi_{AB}}{I_A} = \frac{1500\times0.4\times0.05\times10^{-3}}{4} = 7.5\ \text{mH}$$

So, coefficient of coupling, $$k = \frac{M}{\sqrt{L_1L_2}} = \frac{7.5\times10^{-3}}{\sqrt{0.015\times0.03}} = \mathbf{0.3536}$$

9.5 DOT CONVENTION

The e.m.f. induced due to mutual inductance may aid or oppose the e.m.f. induced due to self-inductance of a coil in a coupled circuit. Its polarity (sign) depends on the relative directions of currents and winding sense. Generally, the manufacturers of coupled coils mark one end of each coupled coil by a dot, which represents the polarity of mutual induced e.m.f. Dot rule may be stated as:

In coupled coils, a current enters at a dotted end of first coil induces a positive e.m.f. at the dotted end of the second coil.

Figure 9.2 clearly explains the sign of mutually induced e.m.f.

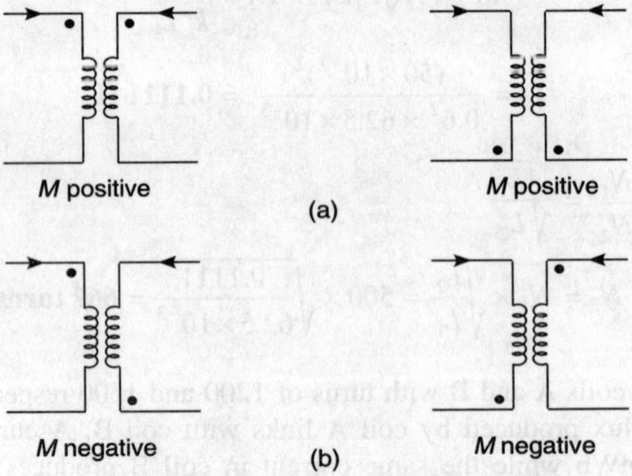

Fig. 9.2 Dot convention.

9.6 ANALYSIS OF COUPLED CIRCUITS

Consider the coupled circuit shown in Fig. 9.3. Each circuit contains a voltage source and the directions of currents are assumed as shown in Fig. 9.3. As both currents enter the coils at dotted ends, the mutual inductance M is taken as positive. The loop equations are written as:

$$R_1 i_1 + L_1 \frac{di_1}{dt} + M \frac{di_2}{dt} = e_1$$

and
$$R_2 i_2 + L_2 \frac{di_2}{dt} + M \frac{di_1}{dt} = e_2$$

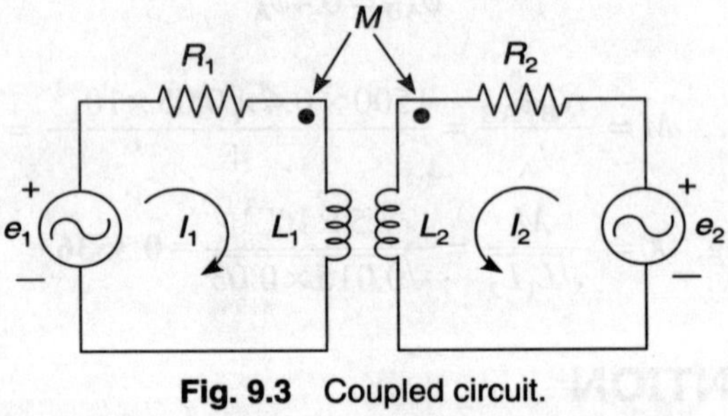

Fig. 9.3 Coupled circuit.

For sinusoidal steady-state analysis, the above equations become

$$(R_1 + j\omega L_1)I_1 + j\omega M I_2 = E_1$$
$$j\omega M I_1 + (R_2 + j\omega L_2)I_2 = E_2$$

In matrix form:

$$\begin{bmatrix} R_1 + j\omega L_1 & j\omega M \\ j\omega M & R_2 + j\omega L_2 \end{bmatrix} \begin{bmatrix} I_1 \\ I_2 \end{bmatrix} = \begin{bmatrix} E_1 \\ E_2 \end{bmatrix}$$

The above two equations may be rewritten as:

$$[R_1 + j\omega(L_1 - M + M)]I_1 + j\omega MI_2 = E_1 \tag{9.9}$$

and

$$j\omega MI_1 + [R_2 + j\omega (L_2 - M + M)]I_2 = E_2 \tag{9.10}$$

Figure 9.3 shows that there is no electrical coupling between the coils, but the equations show that there is coupling between the two loops. Thus, the coupling is said to be mutual coupling. Using Eqs. (9.9) and (9.10), Fig. 9.3 may now be redrawn as Fig. 9.4. It is called conductively coupled circuit because of the common coupling element M.

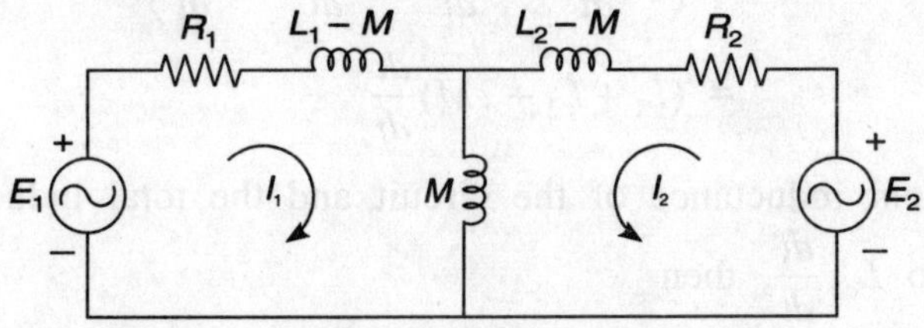

Fig. 9.4 Equivalent circuit for the coupled circuit shown in Fig. 9.3.

9.7 INDUCTIVE COUPLING IN SERIES

When two inductors having self-inductances L_1 and L_2 are coupled in series, mutual inductance M exists between them. Two kinds of series connection are possible; series aiding and series opposing.

9.7.1 Series Aiding

If the mutual e.m.f. and the self-induced e.m.f. in series connection have same polarity, then the connection is called *series aiding*.

In this connection, two coils are connected in series so that their voltages become additive in nature as shown in Fig. 9.5. Here, current i enters both the coils at dotted end and so the self and mutual e.m.f. will have same polarity.

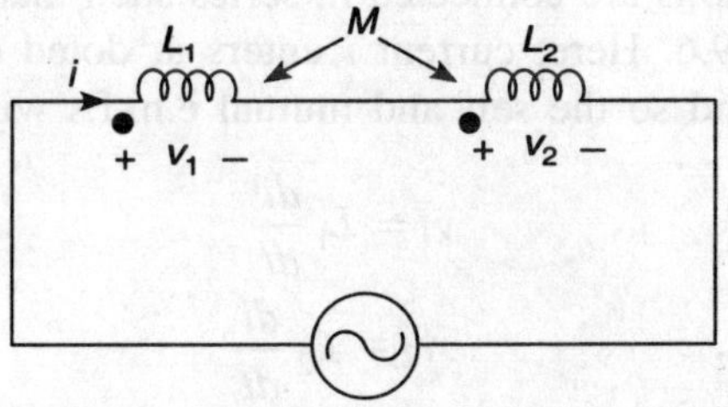

Fig. 9.5 Connection for series aiding.

Self-induced voltage in coil 1, $v_1 = L_1 \dfrac{di}{dt}$

Self induced voltage in coil 2, $v_2 = L_2 \dfrac{di}{dt}$

Mutually induced voltage in coil 1 due to change in current in coil 2, $v_1' = M\dfrac{di}{dt}$

Mutually induced voltage in coil 2 due to change in current in coil 1, $v_2' = M\dfrac{di}{dt}$

Therefore, total induced voltage = $v_1 + v_2 + v_1' + v_2'$

$$= \left(L_1\frac{di}{dt} + L_2\frac{di}{dt} + M\frac{di}{dt} + M\frac{di}{dt}\right)$$

$$= (L_1 + L_2 + 2M)\frac{di}{dt}$$

If L_{eq} is equivalent inductance of the circuit and the total induced voltage in single inductance is equal to $L_{eq}\dfrac{di}{dt}$, then

$$L_{eq} = L_1 + L_2 + 2M$$

9.7.2 Series Opposing

When the mutual and self-induced e.m.f.s have opposite polarity, the connection is called *series opposing.*

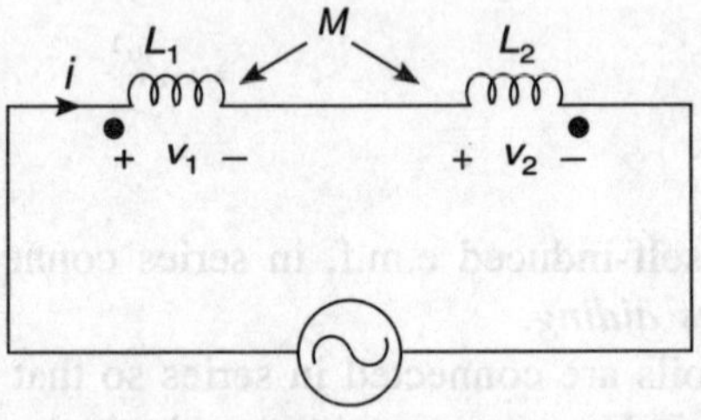

Fig. 9.6 Connection for series opposing.

In this connection, two coils are connected in series such that their voltages are of opposite polarities as shown in Fig. 9.6. Here, current i enters at doted end at one coil and leaves at dotted end in another coil and so the self and mutual e.m.f.s will have opposite polarity.

Self-induced voltage in coil 1, $v_1 = L_1\dfrac{di}{dt}$

Self induced voltage in coil 2, $v_2 = L_2\dfrac{di}{dt}$

Mutually induced voltage in coil 1 due to change in current in coil 2, $v_1' = -M\dfrac{di}{dt}$

Mutually induced voltage in coil 2 due to change in current in coil 1, $v_2' = -M\dfrac{di}{dt}$

Therefore, total induced voltage $= v_1 + v_2 + v_1' + v_2'$

$$= \left(L_1 \frac{di}{dt} + L_2 \frac{di}{dt} - M\frac{di}{dt} - M\frac{di}{dt}\right)$$

$$= (L_1 + L_2 - 2M)\frac{di}{dt}$$

If L_{eq} is equivalent inductance of the circuit and the total induced voltage in single inductance is equal to $L_{eq}\frac{di}{dt}$, then

$$L_{eq} = L_1 + L_2 - 2M$$

9.8 INDUCTIVE COUPLING IN PARALLEL

When two inductors having self-inductances L_1 and L_2 are coupled in parallel, again two kinds of connection are possible: parallel aiding and parallel opposing.

9.8.1 Parallel Aiding

Consider the coupled coils connected in parallel as shown in Fig. 9.7.

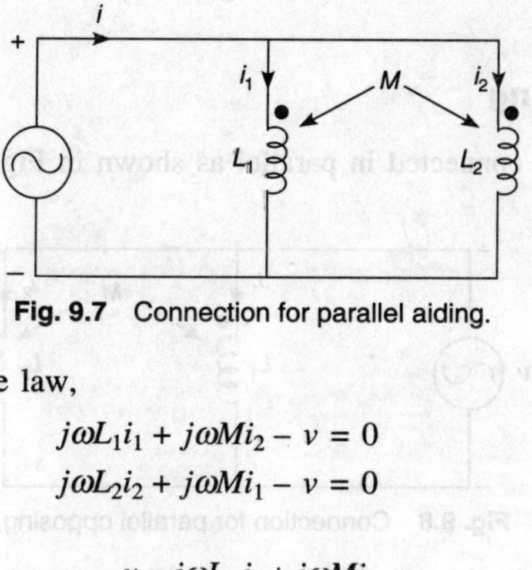

Fig. 9.7 Connection for parallel aiding.

Applying Kirchhoff's voltage law,

$$j\omega L_1 i_1 + j\omega M i_2 - v = 0$$

$$j\omega L_2 i_2 + j\omega M i_1 - v = 0$$

Rearranging,

$$v = j\omega L_1 i_1 + j\omega M i_2 \tag{9.11}$$

$$v = j\omega L_2 i_2 + j\omega M i_1 \tag{9.12}$$

Equating Eqs. (9.11) and (9.12), we get

$$j\omega L_1 i_1 + j\omega M i_2 = j\omega L_2 i_2 + j\omega M i_1 \tag{9.13}$$

From Fig. 9.7,

$$i_2 = i - i_1$$

Putting the value of i_2 in Eq. (9.13), we get

$$j\omega L_1 i_1 + j\omega M\,(i - i_1) = j\omega L_2\,(i - i_1) + j\omega M i_1$$
$$j\omega\, i_1\,(L_1 + L_2 - 2M) = j\omega i\,(L_2 - M)$$

Hence,
$$i_1 = \left(\frac{L_2 - M}{L_1 + L_2 - 2M}\right) i$$

Similarly,
$$i_2 = \left(\frac{L_1 - M}{L_1 + L_2 - 2M}\right) i$$

Substituting the values of i_1 and i_2 into Eq. (9.11), we get

$$v = j\omega\left(\frac{L_1 L_2 - M^2}{L_1 + L_2 - 2M}\right) i \qquad (9.14)$$

If L_{eq} is equivalent inductance of parallel combination then,

$$v = j\omega\, L_{eq}\, i \qquad (9.15)$$

Comparing Eqs. (9.14) and (9.15) we have

$$L_{eq} = \left(\frac{L_1 L_2 - M^2}{L_1 + L_2 - 2M}\right)$$

9.8.2 Parallel Opposing

Consider the coupled coils connected in parallel as shown in Fig. 9.8.

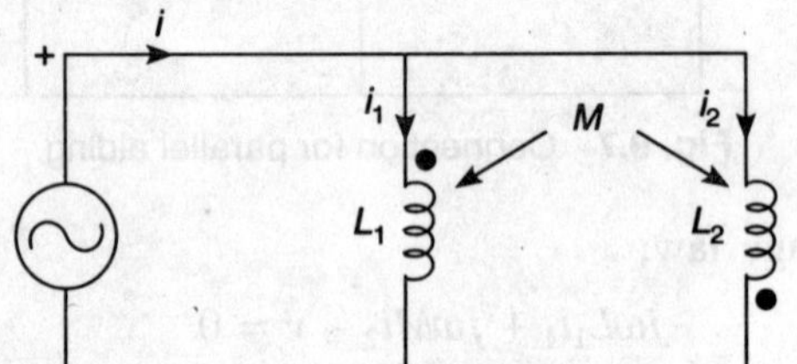

Fig. 9.8 Connection for parallel opposing.

Applying Kirchhoff's voltage law,

$$j\omega L_1 i_1 - j\omega M i_2 - v = 0$$
$$j\omega L_2 i_2 - j\omega M i_1 - v = 0$$

Rearranging,

$$v = j\omega L_1 i_1 - j\omega M i_2 \qquad (9.16)$$
$$v = j\omega L_2 i_2 - j\omega M i_1 \qquad (9.17)$$

Equating Eqs. (9.16) and (9.17), we get

$$j\omega L_1 i_1 - j\omega M i_2 = j\omega L_2 i_2 - j\omega M i_1 \quad (9.18)$$

From Fig. 9.8,

$$i_2 = i - i_1$$

Putting the value of i_2 in Eq. (9.18), we get

$$j\omega L_1 i_1 - j\omega M (i - i_1) = j\omega L_2 (i - i_1) - j\omega M i_1$$

$$j\omega i_1 (L_1 + L_2 + 2M) = j\omega i (L_2 + M)$$

Hence,

$$i_1 = \left(\frac{L_2 + M}{L_1 + L_2 + 2M}\right) i$$

Similarly,

$$i_2 = \left(\frac{L_1 + M}{L_1 + L_2 + 2M}\right) i$$

Substituting the values of i_1 and i_2 into Eq. (9.16), we get

$$v = j\omega \left(\frac{L_1 L_2 - M^2}{L_1 + L_2 + 2M}\right) i \quad (9.19)$$

If L_{eq} is equivalent inductance of parallel combination, then

$$v = j\omega L_{eq} i \quad (9.20)$$

Comparing Eqs. (9.19) and (9.20), we have

$$L_{eq} = \left(\frac{L_1 L_2 - M^2}{L_1 + L_2 + 2M}\right)$$

EXAMPLE 9.4 Find the equivalent inductance between terminals A and B of the circuit shown in Fig. 9.9.

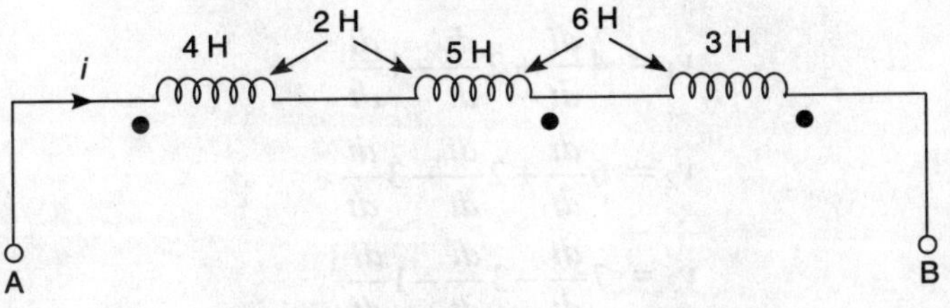

Fig. 9.9

Solution Assume a voltage of v volts is applied between terminals A and B. Applying Kirchhoff's voltage law,

$$-v + 4\frac{di}{dt} - 2\frac{di}{dt} + 5\frac{di}{dt} - 2\frac{di}{dt} + 3\frac{di}{dt} + 6\frac{di}{dt} + 6\frac{di}{dt} = 0$$

$$v = 20\frac{di}{dt} = L_{eq}\frac{di}{dt}$$

Therefore, equivalent inductance of the circuit, L_{eq} = **20 H**

EXAMPLE 9.5 Find the equivalent inductance between terminals A and B of the circuit shown in Fig. 9.10.

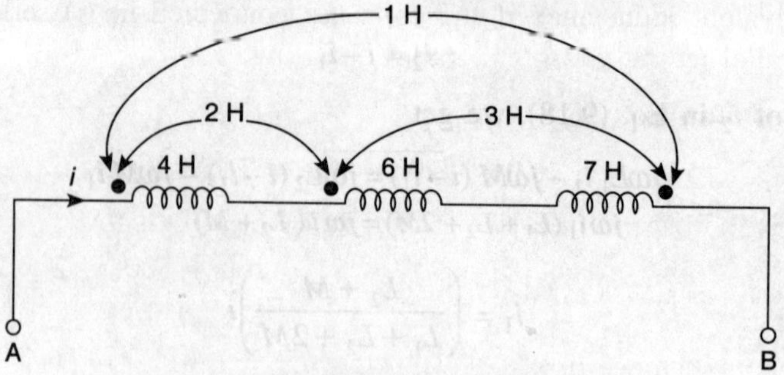

Fig. 9.10

Solution Let v be the voltage applied between A and B. The voltage induced across 4 H, 6 H and 7 H inductances are respectively v_1, v_2, and v_3 as shown in Fig. 9.11.

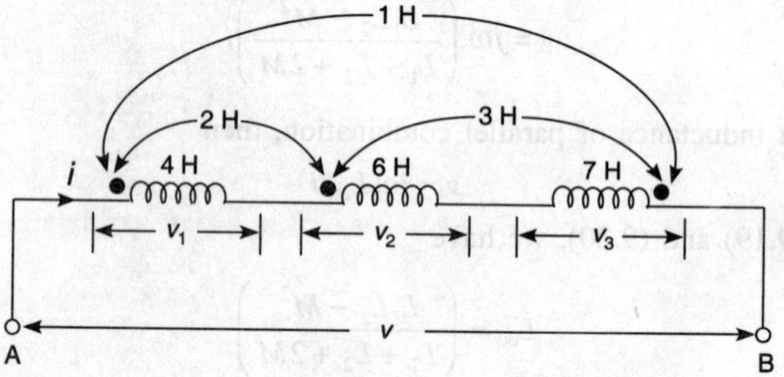

Fig. 9.11

From Fig. 9.11,

$$v_1 = 4\frac{di}{dt} + 2\frac{di}{dt} - 1\frac{di}{dt}$$

$$v_2 = 6\frac{di}{dt} + 2\frac{di}{dt} - 3\frac{di}{dt}$$

$$v_3 = 7\frac{di}{dt} - 3\frac{di}{dt} - 1\frac{di}{dt}$$

Applying Kirchhoff's voltage law,

$$v = v_1 + v_2 + v_3$$

$$v = \frac{di}{dt}(4 + 2 - 1 + 6 + 2 - 3 + 7 - 3 - 1)$$

$$v = 13\frac{di}{dt} = L_{eq}\frac{di}{dt}$$

Hence, equivalent inductance of the circuit, L_{eq} **= 13 H**

EXAMPLE 9.6 A coil having an inductance of 200 mH is magnetically coupled to another coil having an inductance of 500 mH. The coefficient of coupling between the coils is 0.65. Calculate the equivalent inductance if the coils are connected in (i) series aiding, (ii) series opposing, (iii) parallel aiding and (iv) parallel opposing.

Solution Mutual inductance, $M = k\sqrt{L_1 L_2}$

$$= 0.65\sqrt{200 \times 500} = 205.55 \text{ mH}$$

(i) Equivalent inductance in series aiding, $L_{eq} = L_1 + L_2 + 2M$

$$= 200 + 500 + (2 \times 205.55)$$

$$= \mathbf{1111.1\ mH}$$

(ii) Equivalent inductance in series opposing, $L_{eq} = L_1 + L_2 - 2M$

$$= 200 + 500 - (2 \times 205.55)$$

$$= \mathbf{288.9\ mH}$$

(iii) Equivalent inductance in parallel aiding, $L_{eq} = \dfrac{L_1 L_2 - M^2}{L_1 + L_2 - 2M}$

$$= \frac{(200 \times 500) - 205.55^2}{200 + 500 - (2 \times 205.55)}$$

$$= \mathbf{199.89\ mH}$$

(iv) Equivalent inductance in parallel opposing, $L_{eq} = \dfrac{L_1 L_2 - M^2}{L_1 + L_2 + 2M}$

$$= \frac{200 \times 500 - 205.55^2}{200 + 500 + (2 \times 205.55)}$$

$$= \mathbf{51.97\ mH}$$

EXAMPLE 9.7 Two coils offer an equivalent inductance of 0.8 H when connected in series aiding and 0.4 H in series opposing. Determine the self-inductance of the coils. Assume the coefficient of coupling be 0.55.

Solution

Equivalent inductance in series aiding, $L_{eq} = L_1 + L_2 + 2M = 0.8$ (i)

Equivalent inductance in series opposing, $L_{eq} = L_1 + L_2 - 2M = 0.4$ (ii)

Solving Eqs. (i) and (ii),

Mutual inductance, $$M = \frac{0.4}{4} = 0.1 \text{ H}$$

$$M = k\sqrt{L_1 L_2}$$

$$M^2 = k^2 L_1 L_2$$

$$\therefore \quad L_2 = \frac{M^2}{k^2 L_1} = \frac{0.1^2}{0.55^2 L_1} = \frac{0.033}{L_1} \quad \text{(iii)}$$

Substituting the value of $L_2 = \dfrac{0.033}{L_1}$ into Eq. (i), we get

$$L_1 + \frac{0.033}{L_1} + 2M = 0.8$$

$$L_1^2 + 0.033 + 2ML_1 = 0.8L_1$$

$$L_1^2 + 0.033 + 2(0.1)L_1 = 0.8L_1$$

$$L_1^2 - 0.6L_1 + 0.033 = 0 \tag{iv}$$

Solving Eq. (iv), we get

$$L_1 = 0.54 \text{ or } 0.06$$

Substituting the value of $L_1 = 0.54$ into Eq. (iii), L_2 becomes 0.06.

Hence, self-inductance of coil 1, $\quad L_1 = \mathbf{0.54\ H}$

and self-inductance of coil 2, $\quad L_2 = \mathbf{0.06\ H}$

EXAMPLE 9.8 Determine the equivalent inductance across terminals A and B of the circuit with coupled coils shown in Fig. 9.12.

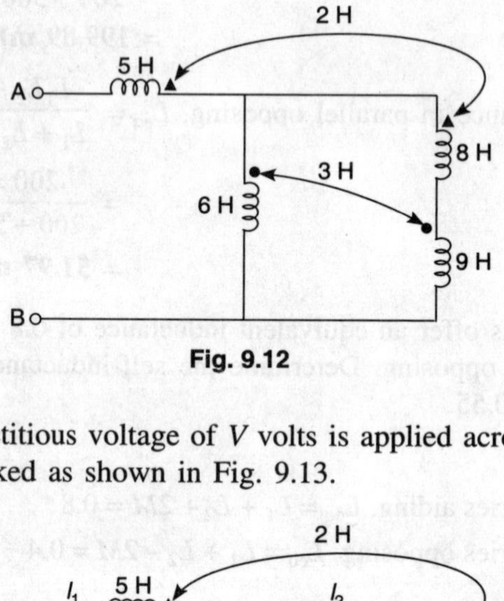

Fig. 9.12

Solution Assume that a fictitious voltage of V volts is applied across terminals A and B and the branch currents are marked as shown in Fig. 9.13.

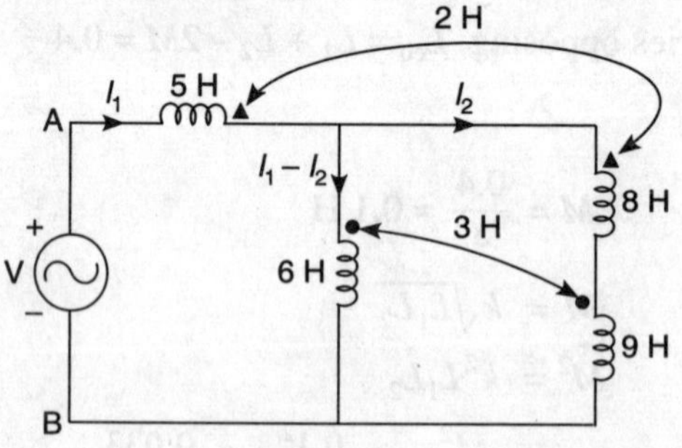

Fig. 9.13

There are two mutual inductances, 2 H and 3 H. The current leaves 5 H inductance in dotted terminal whereas it enters 8 H at dotted terminal. So, the mutual inductance 2 H is taken as negative. In 6 H and 9 H inductances, the respective currents enter at dotted terminals and hence 3 H is taken as positive.

Applying KVL to the first loop,

$$5j\omega I_1 - 2j\omega I_2 + 6j\omega(I_1 - I_2) + 3j\omega I_2 = V$$

$$11j\omega I_1 - 5j\omega I_2 = V \quad \text{(i)}$$

Applying KVL to the second loop,

$$5j\omega I_1 - 2j\omega I_2 + 8j\omega I_2 - 2j\omega I_1 + 9j\omega I_2 + 3j\omega(I_1 - I_2) = V$$

$$6j\omega I_1 + 12j\omega I_2 = V \quad \text{(ii)}$$

Putting Eqs. (i) and (ii) in matrix form,

$$\begin{bmatrix} j11\omega & -j5\omega \\ j6\omega & j12\omega \end{bmatrix}\begin{bmatrix} I_1 \\ I_2 \end{bmatrix} = \begin{bmatrix} V \\ V \end{bmatrix}$$

Using Cramer's rule, I_1 is calculated as:

$$I_1 = \frac{\begin{vmatrix} V & -j5\omega \\ V & j12\omega \end{vmatrix}}{\begin{vmatrix} j11\omega & -j5\omega \\ j6\omega & j12\omega \end{vmatrix}}$$

$$= \frac{V(j17\omega)}{-132\omega^2 - 30\omega^2} = \frac{V(j17\omega)}{-162\omega^2} = \frac{V(j17\omega)}{j^2 162\omega^2}$$

$$\frac{V}{I_1} = \frac{j162\omega}{17} = j\omega L_{eq}$$

Hence, equivalent inductance across terminals A and B, $L_{eq} = \dfrac{162}{17}$ = **9.53 H**

EXAMPLE 9.9 Find the mutual reactance X_m in the coupled coils shown in Fig. 9.14, if the average power in 6 Ω resistance is 80 W.

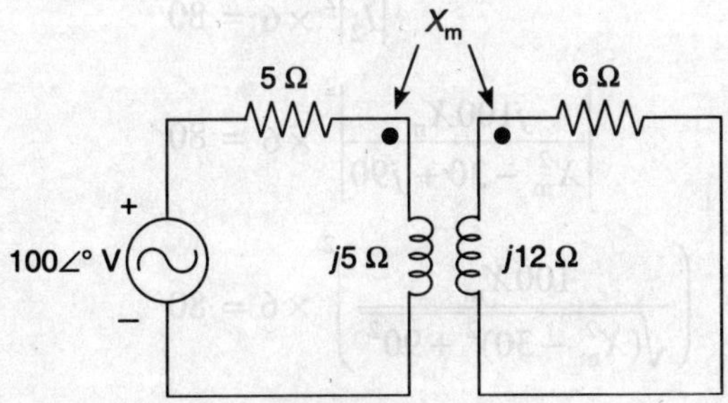

Fig. 9.14

Solution Two mesh currents I_1 and I_2 are assumed in the direction shown in Fig. 9.15. The electrical equivalent circuit of the coupled coils is shown in Fig. 9.16.

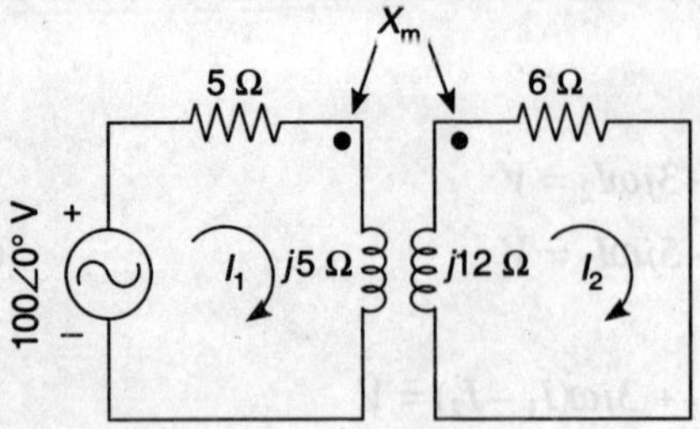

Fig. 9.15

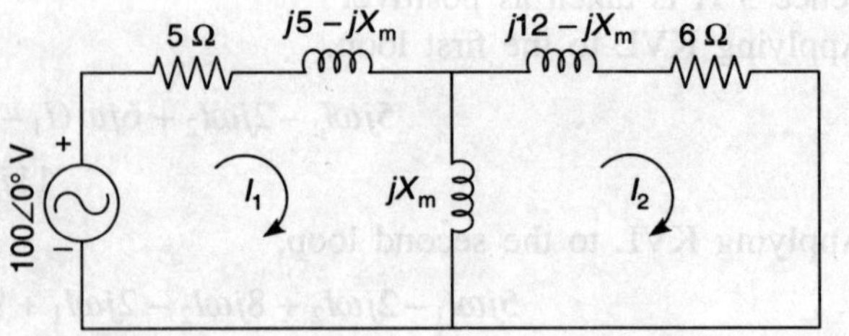

Fig. 9.16

The mesh equations for the circuit shown in Fig. 9.16 is written as:

$$\begin{bmatrix} 5 + j5 - jX_m + jX_m & -jX_m \\ -jX_m & jX_m + j12 - jX_m + 6 \end{bmatrix} \begin{bmatrix} I_1 \\ I_2 \end{bmatrix} = \begin{bmatrix} 100\angle 0^\circ \\ 0 \end{bmatrix}$$

$$\begin{bmatrix} 5 + j5 & -jX_m \\ -jX_m & 6 + j12 \end{bmatrix} \begin{bmatrix} I_1 \\ I_2 \end{bmatrix} = \begin{bmatrix} 100\angle 0^\circ \\ 0 \end{bmatrix}$$

$$I_2 = \frac{\begin{vmatrix} 5 + j5 & 100\angle 0^\circ \\ -jX_m & 0 \end{vmatrix}}{\begin{vmatrix} 5 + j5 & -jX_m \\ -jX_m & 6 + j12 \end{vmatrix}}$$

$$= \frac{j100X_m}{(5 + j5)(6 + j12) - (jX_m)^2}$$

$$= \frac{j100X_m}{X_m^2 - 30 + j90}$$

Power in 6 Ω resistance is given as 80 W.

i.e.

$$|I_2|^2 \times 6 = 80$$

$$\left| \frac{j100X_m}{X_m^2 - 30 + j90} \right|^2 \times 6 = 80$$

$$\left(\frac{100X_m}{\sqrt{(X_m^2 - 30)^2 + 90^2}} \right)^2 \times 6 = 80$$

$$\left(\frac{100^2 X_m^2}{X_m^4 + 30^2 - 60X_m^2 + 90^2} \right) = \frac{80}{6}$$

$$\left(\frac{100^2 X_m^2}{X_m^4 - 60X_m^2 + 9000}\right) = \frac{80}{6}$$

$$100^2 X_m^2 = \frac{80}{6} X_m^4 - 800X_m^2 + 120000$$

$$X_m^4 - 810X_m^2 + 9000 = 0$$

By substituting $x = X_m^2$ into the above equation, it becomes

$$x^2 - 810x + 9000 = 0$$

Solving, $x = 11.27$ and 798.73

or **$X_m = 3.3571\ \Omega$ and $28.2618\ \Omega$**

EXAMPLE 9.10 Find the voltage across the 10 Ω resistor for the network shown in Fig. 9.17.

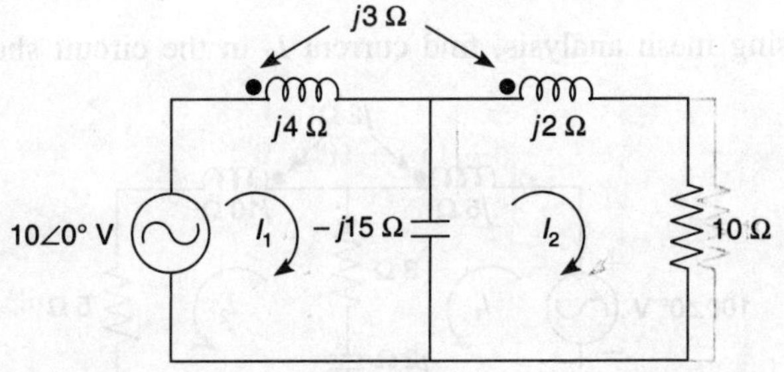

Fig. 9.17

Solution Current I_1 and I_2 are entering the dot, hence the polarity of mutually induced voltage and the self-induced voltage are same.
Applying KVL to the first mesh,

$$j4I_1 - j15\,(I_1 - I_2) + j3I_2 = 10\angle 0°$$

$$-j11I_1 + j18I_2 = 10\angle 0° \quad \text{(i)}$$

Applying KVL to the second mesh,

$$j2I_2 + 10I_2 - j15\,(I_2 - I_1) + j3I_1 = 0$$

$$j18I_1 + (10 - j13)\,I_2 = 0 \quad \text{(ii)}$$

Putting Eqs. (i) and (ii) in matrix form,

$$\begin{bmatrix} -j11 & j18 \\ j18 & 10 - j13 \end{bmatrix}\begin{bmatrix} I_1 \\ I_2 \end{bmatrix} = \begin{bmatrix} 10\angle 0° \\ 0 \end{bmatrix}$$

Using Cramer's rule, I_2 is calculated as:

$$I_2 = \frac{\begin{vmatrix} -j11 & 10\angle 0^\circ \\ j18 & 0 \end{vmatrix}}{\begin{vmatrix} -j11 & j18 \\ j18 & 10-j13 \end{vmatrix}}$$

$$= \frac{-j180}{181-j110}$$

$$= \frac{180\angle -90^\circ}{211.8\angle -31.29^\circ}$$

$$= 0.85\angle -58.71^\circ$$

The voltage across 10 Ω resistor $= I_2 \times 10$

$$= 0.85\angle -58.71^\circ \times 10$$

$$= \mathbf{8.5\angle -58.71^\circ\ V}$$

EXAMPLE 9.11 Using mesh analysis, find current I_2 in the circuit shown in Fig. 9.18.

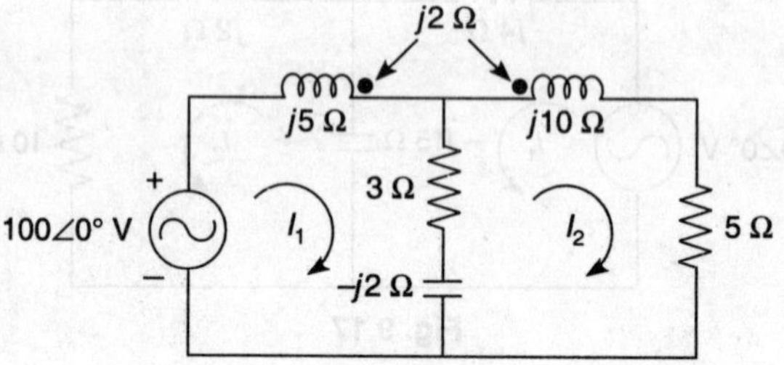

Fig. 9.18

Solution Current I_1 is leaving the dot and I_2 is entering the dot, hence the polarities of mutually induced voltage and self-induced voltage are opposite.

Applying KVL to the first mesh,

$$j5I_1 - j2I_2 + 3(I_1 - I_2) - j2\,(I_1 - I_2) = 100\angle 0^\circ$$

$$(3 + j3)I_1 - 3I_2 = 100\angle 0^\circ \qquad \text{(i)}$$

Applying KVL to the second mesh,

$$j10I_2 - j2I_1 + 5I_2 - j2\,(I_2 - I_1) + 3(I_2 - I_1) = 0$$

$$-3I_1 + (8 + j8)I_2 = 0 \qquad \text{(ii)}$$

Putting Eqs. (i) and (ii) in matrix form,

$$\begin{bmatrix} 3+j3 & -3 \\ -3 & 8+j8 \end{bmatrix}\begin{bmatrix} I_1 \\ I_2 \end{bmatrix} = \begin{bmatrix} 100\angle 0^\circ \\ 0 \end{bmatrix}$$

Applying Cramer's rule, I_2 is calculated as:

$$I_2 = \frac{\begin{vmatrix} 3+j3 & 100\angle 0^\circ \\ -3 & 0 \end{vmatrix}}{\begin{vmatrix} 3+j3 & -3 \\ -3 & 8+j8 \end{vmatrix}}$$

$$= \frac{300\angle 0^\circ}{(0+j48)-9}$$

$$I_2 = \mathbf{6.143\angle{-100.62^\circ}\ A}$$

EXAMPLE 9.12 For the circuit shown in Fig. 9.19, find the ratio of output voltage, V_2 to input voltage V_1.

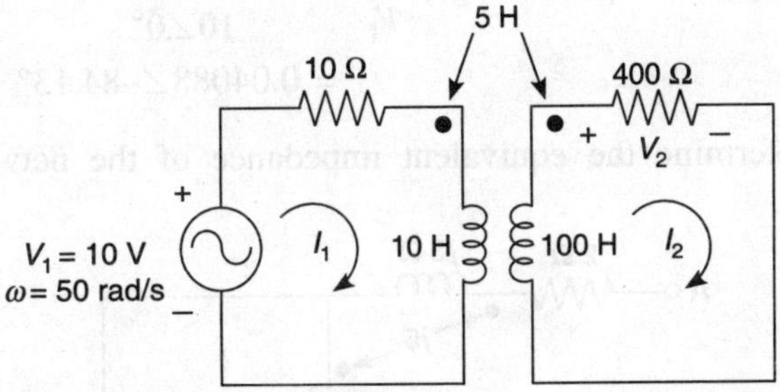

Fig. 9.19

Solution Current I_1 is entering the dot and I_2 is leaving the dot, hence the polarities of mutually induced voltage and self-induced voltage are opposite.
Applying KVL to the first mesh,

$$10I_1 + (j50 \times 10)\, I_1 - (j50 \times 5)I_2 = 10\angle 0^\circ$$

$$10I_1 + j500I_1 - j250I_2 = 10\angle 0^\circ \quad \text{(i)}$$

Applying KVL to the second mesh,

$$400I_2 + (j50 \times 100)\, I_2 - (j50 \times 5)I_1 = 0$$

$$400I_2 + j5000I_2 - j250I_1 = 0 \quad \text{(ii)}$$

Putting Eqs. (i) and (ii) in matrix form,

$$\begin{bmatrix} 10+j500 & -j250 \\ -j250 & 400+j5000 \end{bmatrix}\begin{bmatrix} I_1 \\ I_2 \end{bmatrix} = \begin{bmatrix} 10\angle 0^\circ \\ 0 \end{bmatrix}$$

Using Cramer's law, I_2 is calculated as:

$$I_2 = \frac{\begin{vmatrix} 10+j500 & 10\angle 0^\circ \\ -j250 & 0 \end{vmatrix}}{\begin{vmatrix} 10+j500 & -j250 \\ -j250 & 400+j5000 \end{vmatrix}}$$

$$= \frac{j2500}{-24,96,000 + j250,000 + 62,500}$$

$$= \frac{j2500}{-24,33,500 + j250,000} = \frac{2500\angle 90^\circ}{24,46,308\angle 174.13^\circ}$$

$$= 1.022 \times 10^{-3}\angle -84.13^\circ \text{ A}$$

Therefore, output voltage, $V_2 = I_2 \times 400$

$$= 1.022 \times 10^{-3}\angle -84.13^\circ \times 400$$

$$= 0.4088\angle -84.13^\circ \text{ V}$$

and ratio of output voltage to input voltage, $\frac{V_2}{V_1} = \frac{0.4088\angle -84.13^\circ}{10\angle 0^\circ}$

$$= 0.04088\angle -84.13^\circ \text{ A}$$

EXAMPLE 9.13 Determine the equivalent impedance of the network with coupled coils shown in Fig. 9.20.

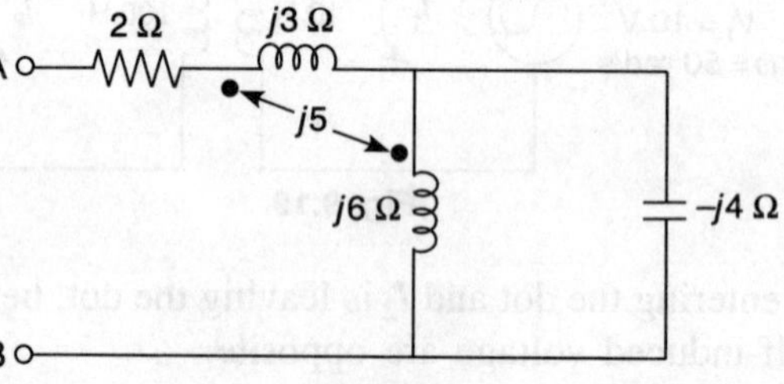

Fig. 9.20

Solution Assume that a voltage of V volts is applied across terminals A and B and the branch currents are marked as shown in Fig. 9.21.

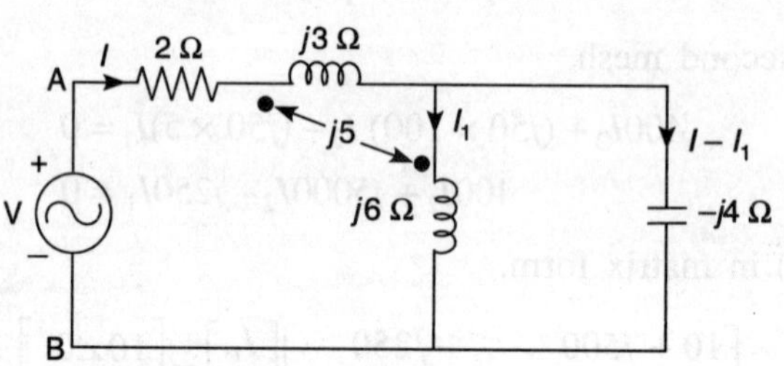

Fig. 9.21

Applying KVL to the first loop,

$$2I + j3I + j5I + j6I_1 + j5I_1 = V$$

$$(2 + j8)I + j11I_1 = V \qquad \text{(i)}$$

Applying KVL to the second loop,

$$2I + j3I + j5I_1 - j4(I - I_1) = V$$

$$(2 - j1)I + j9I_1 = V \qquad \text{(ii)}$$

Putting Eqs. (i) and (ii) in matrix form,

$$\begin{bmatrix} 2+j8 & j11 \\ 2-j1 & j9 \end{bmatrix}\begin{bmatrix} I \\ I_1 \end{bmatrix} = \begin{bmatrix} V \\ V \end{bmatrix}$$

Using Cramer's rule, source current I is calculated as:

$$I = \frac{\begin{vmatrix} V & j11 \\ V & j9 \end{vmatrix}}{\begin{vmatrix} 2+j8 & j11 \\ 2-j1 & j9 \end{vmatrix}}$$

$$= \frac{V(j9 - j11)}{(-72 + j18) - (11 + j22)}$$

$$= \frac{V(-j2)}{(-83 - j4)}$$

$$\frac{V}{I} = \frac{83 + j4}{j2} = \mathbf{(2 + j41.5)\ \Omega}$$

EXAMPLE 9.14 Obtain the dotted equivalent circuit for the coupled circuit shown in Fig. 9.22 and use it to find the voltage V across the 10 Ω capacitive reactance.

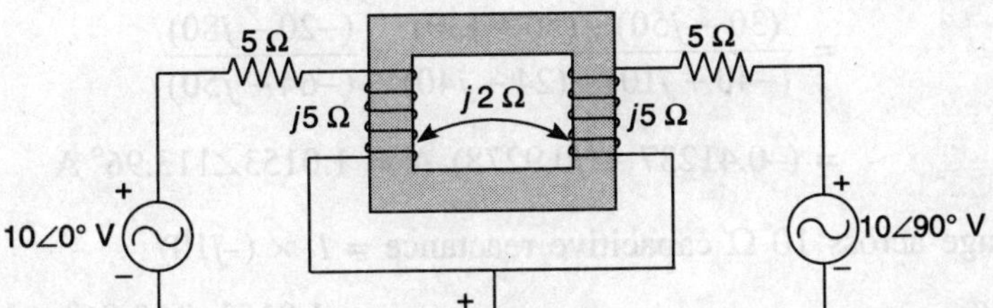

Fig. 9.22

Solution The dotted equivalent circuit considering the winding sense is shown in Fig. 9.23. However, the current directions shown are assumed to conveniently apply Kirchhoff's laws.

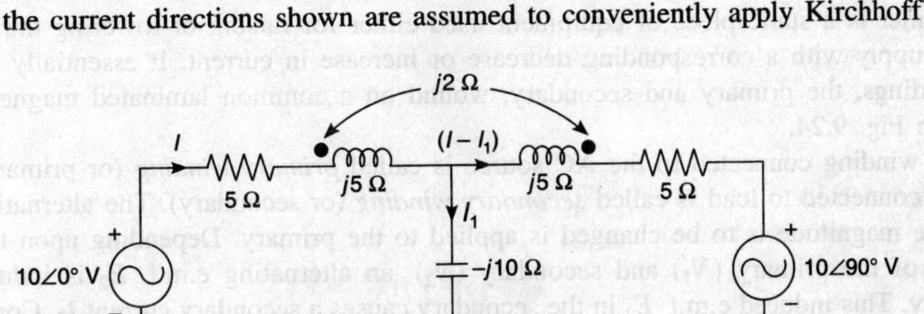

Fig. 9.23

Applying KVL to the first mesh,

$$5I + j5I - j2(I - I_1) - j10I_1 = 10\angle 0^\circ$$

$$(5 + j3)I - j8I_1 = 10 + j0 \qquad \text{(i)}$$

Applying KVL to the second mesh,

$$j5(I - I_1) - j2I + 5(I - I_1) + 10\angle 90^\circ + j10I_1 = 0$$

$$j5I - j5I_1 - j2I + 5I - 5I_1 + (0 + j10) + j10I_1 = 0$$

$$(5 + j3)I + (-5 + j5)I_1 = -j10 \qquad \text{(ii)}$$

Putting Eqs. (i) and (ii) in matrix form,

$$\begin{bmatrix} 5+j3 & -j8 \\ 5+j3 & -5+j5 \end{bmatrix} \begin{bmatrix} I \\ I_1 \end{bmatrix} = \begin{bmatrix} 10 \\ -j10 \end{bmatrix}$$

Using Cramer's rule, the current I_1 is calculated as:

$$I_1 = \frac{\begin{vmatrix} 5+j3 & 10 \\ 5+j3 & -j10 \end{vmatrix}}{\begin{vmatrix} 5+j3 & -j8 \\ 5+j3 & -5+j5 \end{vmatrix}}$$

$$= \frac{(30 - j50) - (50 + j30)}{(-40 + j10) - (24 - j40)} = \frac{(-20 - j80)}{(-64 + j50)}$$

$$= (-0.41237 + j0.9278)\text{ A} = 1.0153\angle 113.96^\circ \text{ A}$$

Therefore, voltage across 10 Ω capacitive reactance $= I_1 \times (-j10)$

$$= 1.0153\angle 113.96^\circ \times 10\angle -90^\circ$$

$$= \mathbf{10.153\angle 23.96^\circ\ V}$$

9.9 TRANSFORMER

Transformer is a static piece of equipment used either for raising or lowering the voltage of an AC supply with a corresponding decrease or increase in current. It essentially consists of two windings, the primary and secondary, wound on a common laminated magnetic core as shown in Fig. 9.24.

The winding connected to the AC source is called *primary winding* (or primary) and the winding connected to load is called *secondary winding* (or secondary). The alternating voltage V_1 whose magnitude is to be changed is applied to the primary. Depending upon the number of turns of the primary (N_1) and secondary (N_2), an alternating e.m.f. E_2 is induced in the secondary. This induced e.m.f. E_2 in the secondary causes a secondary current I_2. Consequently, terminal voltage V_2 will appear across the load. If $V_2 > V_1$, it is called as *step-up transformer*. On the other hand, if $V_2 < V_1$, it is called *step-down transformer*.

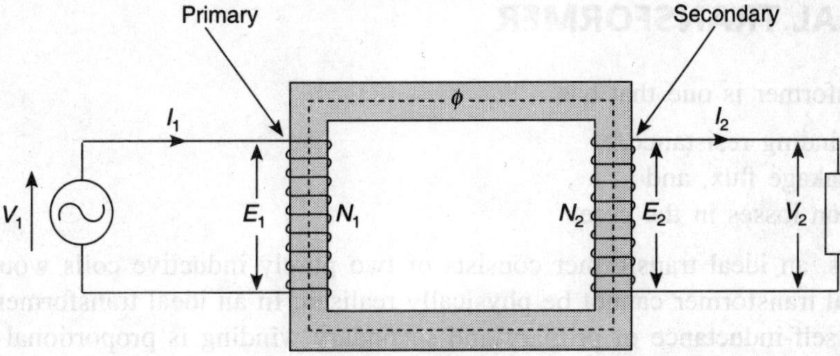

Fig. 9.24

When an alternating voltage V_1 is applied to the primary, an alternating flux ϕ is set up in the core. This alternating flux links both the windings and induces e.m.f.s E_1 and E_2 in them according to Faraday's laws of electromagnetic induction. The e.m.f. E_1 is termed as primary e.m.f. and e.m.f. E_2 is termed as secondary e.m.f.

Clearly,
$$E_1 = -N_1 \frac{d\phi}{dt}$$

and
$$E_2 = -N_2 \frac{d\phi}{dt}$$

Hence,
$$\frac{E_2}{E_1} = \frac{N_2}{N_1}$$

Note that magnitudes of E_1 and E_2 depend upon the number of turns on the secondary and primary, respectively. If $N_2 > N_1$, then $E_2 > E_1$ (or $V_2 > V_1$) and we get a step-up transformer. On the other hand, if $N_2 < N_1$, then $E_2 < E_1$ (or $V_2 < V_1$) and we get a step-down transformer. If load is connected across the secondary winding, the secondary e.m.f. E_2 causes a current I_2 to flow through the load. Thus, a transformer transfers AC power from one circuit to another with a change in voltage level. The following points may be noted carefully:

(i) The transformer action is based on the laws of electromagnetic induction.
(ii) There is no electrical connection between the primary and the secondary. The AC power is transferred from primary to secondary through magnetic flux.
(iii) There is no change in frequency, i.e. output power has the same frequency as the input power.
(iv) The losses that occur in a transformer are:
  (a) core losses—eddy current and hysteresis losses
  (b) copper losses—in the resistance of the windings

In practice, these losses are very small so that output power is nearly equal to the input primary power. In other words, a transformer has very high efficiency.

9.10 IDEAL TRANSFORMER

An ideal transformer is one that has

(i) no winding resistance,
(ii) no leakage flux, and
(iii) no iron losses in the core.

In other words, an ideal transformer consists of two purely inductive coils wound on a loss-free core. Ideal transformer cannot be physically realised. In an ideal transformer, $E_1 = V_1$ and $E_2 = V_2$. The self-inductance of primary and secondary winding is proportional to the square of number of turns of the coil. The turns ratio is represented by a. The relationship between the number of turns and self-inductances is given by

$$\frac{L_2}{L_1} = \left(\frac{N_2}{N_1}\right)^2 = a^2$$

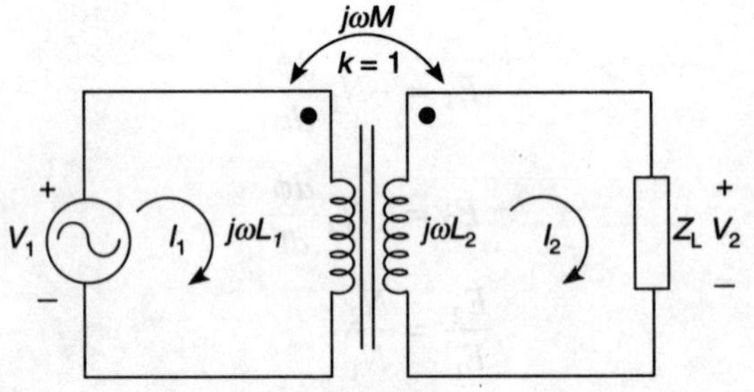

Fig. 9.25

Figure 9.25 shows an ideal transformer to which the secondary is connected to a load impedance Z_L. The coefficient of coupling is $k = 1$, and the turns ratio N_2 to N_1 is a:1.

The input impedance Z_{in} of the transformer can be determined by writing mesh equations as:

$$(j\omega L_1)I_1 - (j\omega M)I_2 = V_1 \quad (9.21)$$

$$(Z_L + j\omega L_2)I_2 - (j\omega M)I_1 = 0 \quad (9.22)$$

From Eq. (9.22),

$$I_2 = \frac{j\omega M}{(Z_L + j\omega L_2)} I_1$$

Substituting the value of I_2 into Eq. (9.21), we get

$$V_1 = (j\omega L_1)I_1 - (j\omega M)\frac{j\omega M}{(Z_L + j\omega L_2)} I_1$$

$$V_1 = (j\omega L_1)I_1 + \frac{\omega^2 M^2}{(Z_L + j\omega L_2)} I_1$$

$$V_1 = \left[j\omega L_1 + \frac{\omega^2 M^2}{(Z_L + j\omega L_2)} \right] I_1$$

$$Z_{in} = \frac{V_1}{I_1} = j\omega L_1 + \frac{\omega^2 M^2}{(Z_L + j\omega L_2)}$$

As the coefficient of coupling is assumed to be equal to unity,

$$M = \sqrt{L_1 L_2} \quad \text{or} \quad M^2 = L_1 L_2$$

Substituting the value of M^2 into the equation for Z_{in},

$$Z_{in} = j\omega L_1 + \frac{\omega^2 L_1 L_2}{(Z_L + j\omega L_2)}$$

$$Z_{in} = \frac{j\omega L_1 (Z_L + j\omega L_2) + \omega^2 L_1 L_2}{(Z_L + j\omega L_2)}$$

$$Z_{in} = \frac{j\omega L_1 Z_L - \omega^2 L_1 L_2 + \omega^2 L_1 L_2}{(Z_L + j\omega L_2)}$$

$$Z_{in} = \frac{j\omega L_1 Z_L}{(Z_L + j\omega L_2)}$$

Since $\dfrac{L_2}{L_1} = a^2$ and $L_2 = a^2 L_1$,

$$Z_{in} = \frac{j\omega L_1 Z_L}{Z_L + j\omega a^2 L_1}$$

As L_1 is assumed to be infinitely large, Z_L is neglected.

Therefore,
$$Z_{in} = \frac{j\omega L_1 Z_L}{j\omega a^2 L_1} = \frac{Z_L}{a^2} = \left(\frac{N_1}{N_2}\right)^2 Z_L$$

Thus, an ideal transformer transforms the magnitude of the impedance only and not the phase angle.

9.11 TUNED CIRCUITS

The eddy current and hysteresis losses are directly proportional to frequency and they increase with increase in frequency. Hence, air cored transformers are preferred over iron cored transformers at radio frequencies. Tuned circuits are mainly used to transfer energy from a weak source to a load or used for maximum power transfer from one circuit to another. This is possible only when both the coils work at resonance condition. There are two types of tuned circuits, namely

(i) single-tuned circuits
(ii) double-tuned circuits

Generally, capacitors are added to primary and secondary of tuned circuits to bring them to resonance condition. If capacitor is added only in secondary, this coil is referred to as *single-tuned circuit*, and if capacitors are added both to primary and secondary coils, it is called as *double-tuned circuit*. In single-tuned circuit, it is possible to resonate by adding capacitor to secondary alone, if the primary inductive reactance is low.

9.11.1 Single-tuned Circuits

Figure 9.26 shows a single-tuned circuit which is connected to a voltage source of V_i volts. R_1 is the winding resistance of primary coil L_1. It may include the source resistance as well. Similarly, R_2 is the secondary coil resistance and may include load resistance. Usually these circuits are employed for coupling an amplifier and radio receiver circuits. In such applications, maximum secondary voltage V_o is desired rather than maximum power transfer.

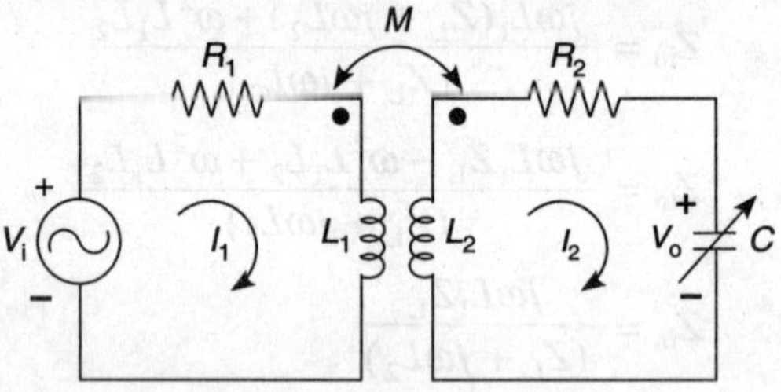

Fig. 9.26 Single-tuned circuit.

With reference to Fig. 9.26, the matrix equation by mesh analysis method is written as:

$$\begin{bmatrix} R_1 + j\omega L_1 & -j\omega M \\ -j\omega M & R_2 + j\omega L_2 - j\dfrac{1}{\omega C} \end{bmatrix} \begin{bmatrix} I_1 \\ I_2 \end{bmatrix} = \begin{bmatrix} V_i \\ 0 \end{bmatrix}$$

Using Cramer's rule,

$$I_2 = \frac{\begin{vmatrix} R_1 + j\omega L_1 & V_i \\ -j\omega M & 0 \end{vmatrix}}{\begin{vmatrix} R_1 + j\omega L_1 & -j\omega M \\ -j\omega M & R_2 + j\omega L_2 - j\dfrac{1}{\omega C} \end{vmatrix}}$$

$$I_2 = \frac{j\omega M V_i}{(R_1 + j\omega L_1)\left(R_2 + j\omega L_2 - j\dfrac{1}{\omega C}\right) - (j\omega M)^2} \tag{9.23}$$

The single-tuned circuits are used only when the primary inductive reactance is very low. Hence, in the primary circuit,

$$R_1 >> j\omega L_1$$

Therefore, $$R_1 + j\omega L_1 \approx R_1$$

Therefore, Eq. (9.23) can be written as:

$$I_2 = \frac{j\omega MV_i}{(R_1)\left(R_2 + j\omega L_2 - j\dfrac{1}{\omega C}\right) - (j\omega M)^2}$$

or

$$I_2 = \frac{j\omega MV_i}{R_1R_2 + \omega^2M^2 + jR_1\left(\omega L_2 - \dfrac{1}{\omega C}\right)} \quad (9.24)$$

Maximum secondary voltage is obtained when I_2 is large. To have maximum I_2, the reactive term in the denominator of Eq. (9.24) is made zero. This is achieved by adjusting the value of C so that resonance condition is reached. Hence,

$$I_2 = \frac{j\omega MV_i}{R_1R_2 + \omega^2M^2} \quad (9.25)$$

Then, the maximum secondary voltage is given by

$$V_o = I_2 \times \left(-j\frac{1}{\omega C}\right)$$

$$= \frac{j\omega MV_i}{R_1R_2 + \omega^2M^2} \times \left(-j\frac{1}{\omega C}\right)$$

$$V_o = \frac{MV_i}{C(R_1R_2 + \omega^2M^2)} \quad (9.26)$$

From Eqs. (9.25) and (9.26), it is clear that the values of secondary current and secondary voltage are functions of mutual inductance. Hence, it is possible to vary the secondary current and output voltage by varying mutual inductance which in turn can be done by varying the coefficient of coupling k.

By differentiating Eq. (9.26) with respect to M and equating it to zero, condition for maximum secondary voltage $V_{o(max)}$ is obtained. It is obtained at resonance frequency, ω_r only.

$$\frac{dV_o}{dM} = \frac{V_i}{C}\left(\frac{(R_1R_2 + \omega_r^2M^2) - M(2\omega_r^2M)}{(R_1R_2 + \omega_r^2M^2)^2}\right) = 0$$

$$R_1R_2 + \omega_r^2M^2 = 2\omega_r^2M^2$$

$$R_1R_2 = \omega_r^2M^2$$

Therefore, $$M = \frac{\sqrt{R_1R_2}}{\omega_r}$$

The value of secondary voltage is then computed as:

$$V_o = \frac{MV_i}{C(R_1R_2 + \omega_r^2 M^2)}$$

or

$$V_{o(max)} = \frac{\frac{\sqrt{R_1R_2}}{\omega_r}V_i}{C\left(R_1R_2 + \omega_r^2 \frac{R_1R_2}{\omega_r^2}\right)}$$

$$= \frac{V_i\sqrt{R_1R_2}}{2\omega_r CR_1R_2}$$

$$V_{o(max)} = \frac{V_i}{2\omega_r C\sqrt{R_1R_2}}.$$

The maximum amplification is given by

$$A_m = \frac{V_o}{V_i} = \frac{1}{2\omega_r C\sqrt{R_1R_2}} \text{ and } I_2 = j\frac{V_i}{2\sqrt{R_1R_2}}$$

The variation of secondary voltage with angular frequency ω for various values of coefficient of coupling is shown in Fig. 9.27.

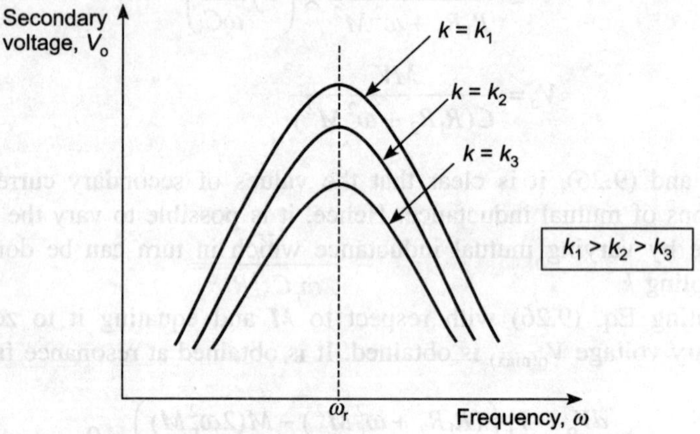

Fig. 9.27 Variation of output voltage with frequency in single-tuned circuit.

EXAMPLE 9.15 Consider the single-tuned circuit shown in Fig. 9.28 and determine (i) the resonance frequency, (ii) the output voltage at $f = 25$ kHz and (iii) the maximum output voltage. Assume $k = 0.9$.

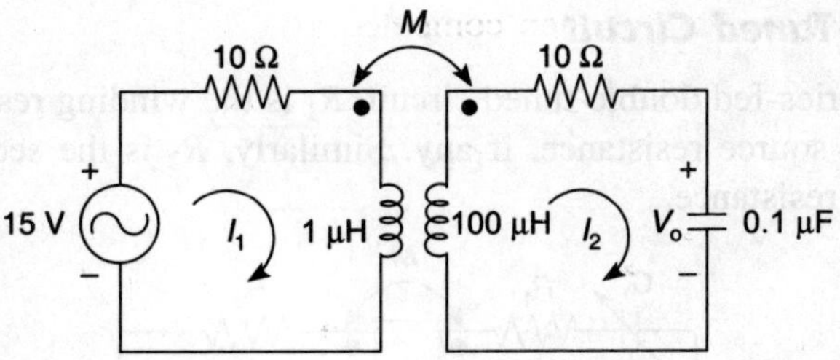

Fig. 9.28

Solution Mutual inductance,

$$M = k\sqrt{L_1 L_2}$$

$$= 0.9\sqrt{1\times10^{-6}\times100\times10^{-6}}$$

$$= 9\ \mu\text{H}$$

At resonance,

$$\omega L_2 = \frac{1}{\omega C}$$

(i) Resonance frequency,

$$\omega_r = \frac{1}{\sqrt{L_2 C}}$$

$$= \frac{1}{\sqrt{100\times10^{-6}\times0.1\times10^{-6}}}$$

$$= \mathbf{316227.766\ rad/s}$$

(ii) Output voltage,

$$V_o = \frac{MV_i}{C(R_1R_2+\omega^2M^2)}$$

$$= \frac{9\times10^{-6}\times15}{0.1\times10^{-6}\left[(10\times10)+(2\pi\times25\times10^3)^2(9\times10^{-6})^2\right]}$$

$$= \mathbf{13.24\ V}$$

(iii) Maximum output voltage,

$$V_{o(max)} = \frac{V_i}{2\omega_r C\sqrt{R_1R_2}}$$

$$= \frac{15}{2\times316227.766\times0.1\times10^{-6}\times\sqrt{100}}$$

$$= \mathbf{23.72\ V}$$

9.11.2 Double-tuned Circuits

In double-tuned circuits, the capacitance is connected in both primary and secondary coils. Further, it may be connected either in series or in parallel with the primary coil. If it is connected in series with the primary coil, it is called *series-fed double-tuned circuit*, and if it is connected in parallel with the primary coil, the tuned circuit is called *parallel-fed double-tuned circuit*.

Series-Fed Double-Tuned Circuit

Figure 9.29 shows a series-fed double-tuned circuit. R_1 is the winding resistance of primary coil L_1. It may include the source resistance, if any. Similarly, R_2 is the secondary coil resistance and may include load resistance.

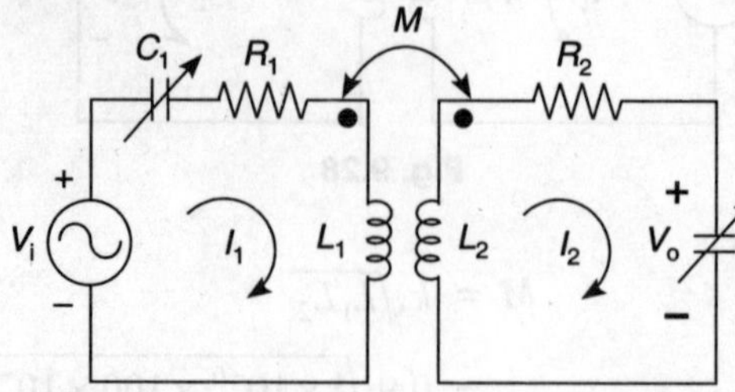

Fig. 9.29 Series-fed double-tuned circuit.

With reference to Fig. 9.29, the matrix equation by mesh analysis method is written as:

$$\begin{bmatrix} R_1 + j\omega L_1 - j\dfrac{1}{\omega C_1} & -j\omega M \\ -j\omega M & R_2 + j\omega L_2 - j\dfrac{1}{\omega C_2} \end{bmatrix} \begin{bmatrix} I_1 \\ I_2 \end{bmatrix} = \begin{bmatrix} V_i \\ 0 \end{bmatrix}$$

Using Cramer's rule,

$$I_2 = \frac{\begin{vmatrix} R_1 + j\omega L_1 - j\dfrac{1}{\omega C_1} & V_i \\ -j\omega M & 0 \end{vmatrix}}{\begin{vmatrix} R_1 + j\omega L_1 - j\dfrac{1}{\omega C_1} & -j\omega M \\ -j\omega M & R_2 + j\omega L_2 - j\dfrac{1}{\omega C_2} \end{vmatrix}}$$

$$= \frac{j\omega M V_i}{\left(R_1 + j\omega L_1 - j\dfrac{1}{\omega C_1}\right)\left(R_2 + j\omega L_2 - j\dfrac{1}{\omega C_2}\right) - (j\omega M)^2}$$

Output voltage, V_o

$$= I_2\left(-j\frac{1}{\omega C_2}\right) \tag{9.27}$$

$$= \frac{j\omega M V_i\left(-j\dfrac{1}{\omega C_2}\right)}{\left(R_1 + j\omega L_1 - j\dfrac{1}{\omega C_1}\right)\left(R_2 + j\omega L_2 - j\dfrac{1}{\omega C_2}\right) - (j\omega M)^2}$$

$$= \frac{M V_i / C_2}{\left(R_1 + j\omega L_1 - j\dfrac{1}{\omega C_1}\right)\left(R_2 + j\omega L_2 - j\dfrac{1}{\omega C_2}\right) + \omega^2 M^2} \tag{9.28}$$

In double-tuned circuits, both primary and secondary are tuned to resonance frequency ω_r. Hence, at resonance,

$$\omega_r L_1 - \frac{1}{\omega_r C_1} = 0 \quad \text{and} \quad \omega_r L_2 - \frac{1}{\omega_r C_2} = 0$$

Using Eqs. (9.27) and (9.28), secondary current at resonance is

$$I_2 = \frac{j\omega_r M V_i}{R_1 R_2 + \omega_r^2 M^2} \tag{9.29}$$

and the output voltage at resonance is

$$V_o = \frac{\dfrac{MV_i}{C_2}}{R_1 R_2 + \omega_r^2 M^2} \tag{9.30}$$

Equations (9.29) and (9.30) are found to be similar to that of Eqs. (9.25) and (9.26) of single-tuned circuit, respectively, obtained earlier. Hence, the condition for maximum output voltage will be same as that of single-tuned circuit. Therefore, by an analysis similar to that of single-tuned circuit, we can obtain the following equations.

Condition for maximum output voltage is

$$M = \frac{\sqrt{R_1 R_2}}{\omega_r}$$

Then,

$$V_{o(max)} = \frac{V_i}{2\omega_r C \sqrt{R_1 R_2}}$$

The maximum amplification is given by

$$A_m = \frac{V_o}{V_i} = \frac{1}{2\omega_r C \sqrt{R_1 R_2}} \quad \text{and} \quad I_2 = j\frac{V_i}{2\sqrt{R_1 R_2}}$$

The variation of secondary voltage with angular frequency ω for various values of coefficient of coupling is shown in Fig. 9.30.

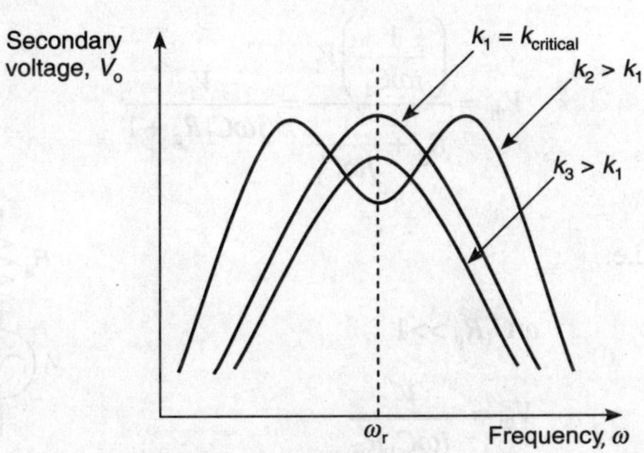

Fig. 9.30 Variation of output voltage with frequency in double-tuned circuit.

Parallel-Fed Double-Tuned Circuit

Figure 9.31 shows a parallel-fed double-tuned circuit in which R_g is the source resistance and R_1 is the primary winding resistance. R_2 is the secondary coil resistance and it may include load resistance as well.

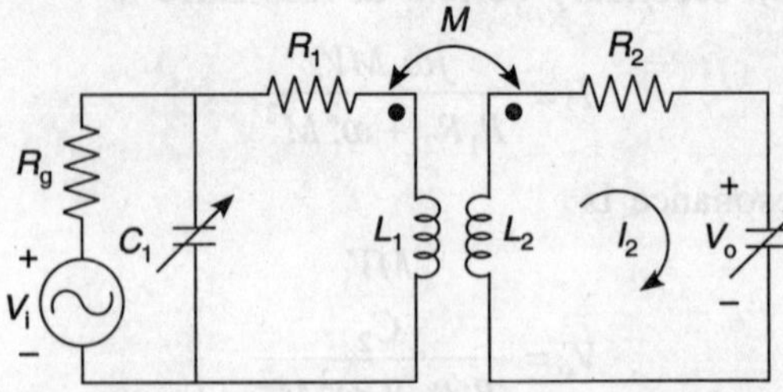

Fig. 9.31 Parallel-led double-tuned circuit.

The parallel-fed double-tuned circuit can be converted into an equivalent series-fed double-tuned circuit by finding the Thevenin's equivalent circuit across terminals A and B as shown in Fig. 9.32.

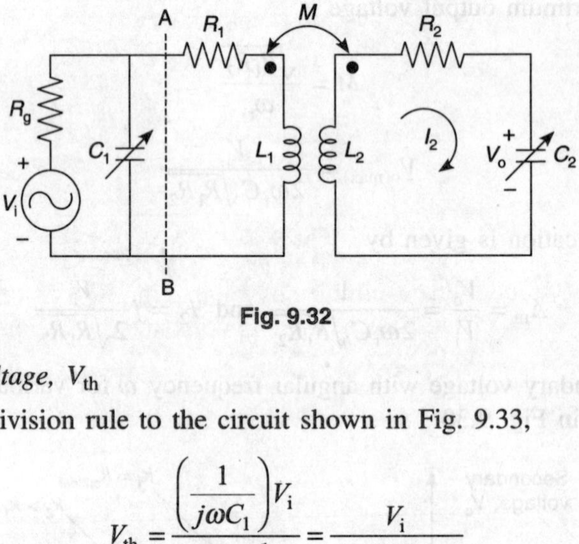

Fig. 9.32

To find Thevenin's voltage, V_{th}

By applying voltage division rule to the circuit shown in Fig. 9.33,

$$V_{th} = \frac{\left(\dfrac{1}{j\omega C_1}\right)V_i}{R_g + \dfrac{1}{j\omega C_1}} = \frac{V_i}{j\omega C_1 R_g + 1}$$

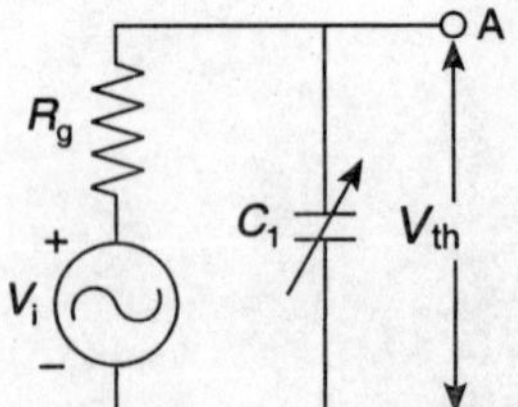

Fig. 9.33

Assume $R_g >> \dfrac{1}{\omega C_1}$, i.e.

$$\omega C_1 R_g >> 1$$

Hence,

$$V_{th} = \frac{V_i}{j\omega C_1 R_g}$$

To find Thevenin's impedance, Z_{th}

With reference to Fig. 9.34,

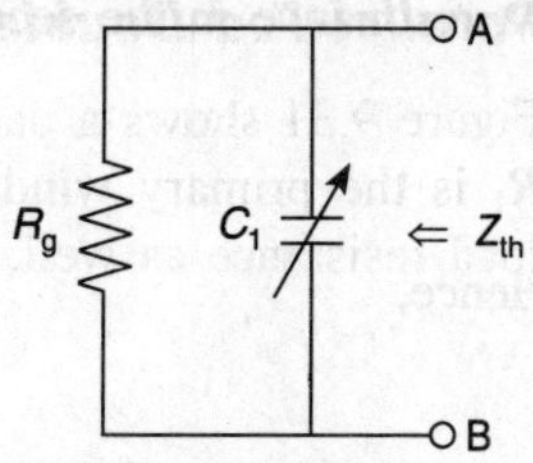

Fig. 9.34

$$Z_{th} = \frac{R_g\left(\frac{1}{j\omega C_1}\right)}{R_g + \frac{1}{j\omega C_1}} = \frac{R_g}{j\omega C_1 R_g + 1}$$

Again by assuming $R_g >> \frac{1}{\omega C_1}$, i.e.

$$\omega C_1 R_g >> 1$$

Hence,
$$Z_{th} = \frac{R_g}{j\omega C_1 R_g} = \frac{1}{j\omega C_1}$$

Using the Thevenin's equivalent circuit, the parallel-fed double-tuned circuit in Fig. 9.32 is drawn as shown in Fig. 9.35. The circuit in Fig. 9.35 is similar to that of series-fed double-tuned circuit. Therefore, the analysis will be similar to that of series-fed double-tuned circuit.

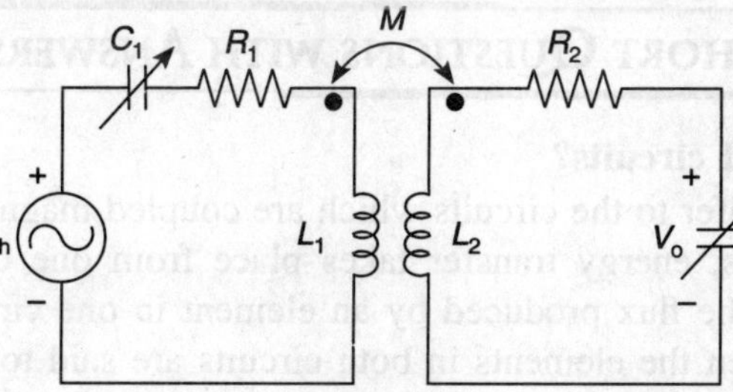

Fig. 9.35

EXAMPLE 9.16 The double-tuned circuit shown in Fig. 9.36 is tuned to a frequency of 1000 rad/s. Calculate the self-inductance of the two coils and the optimum value of the mutual inductance and coefficient of coupling.

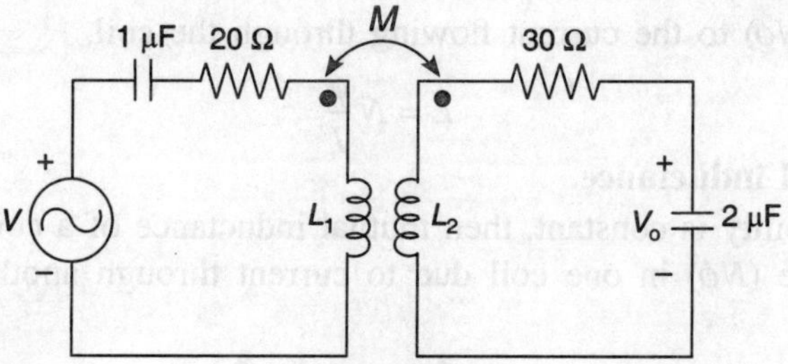

Fig. 9.36

Solution At resonance in the primary side,

$$\omega_1 L_1 = \frac{1}{\omega_r C_1}$$

Hence,
$$L_1 = \frac{1}{\omega_r^2 C_1} = \frac{1}{1000^2 \times 1 \times 10^{-6}} = 1\text{ H}$$

At resonance in the secondary side,

$$\omega_r L_2 = \frac{1}{\omega_r C_2}$$

Hence,
$$L_2 = \frac{1}{\omega_r^2 C_2} = \frac{1}{1000^2 \times 2 \times 10^{-6}} = 0.5 \text{ H}$$

Optimum value of mutual inductance,
$$M_C = \frac{\sqrt{R_1 R_2}}{\omega_r}$$

$$= \frac{\sqrt{20 \times 30}}{1000} = \mathbf{0.0245\ H}$$

Optimum value of coefficient of coupling,
$$k_C = \frac{M_C}{\sqrt{L_1 L_2}}$$

$$= \frac{0.0245}{\sqrt{1 \times 0.5}} = \mathbf{0.0346\ H}$$

SHORT QUESTIONS WITH ANSWERS

1. **What are coupled circuits?**
   Coupled circuits refer to the circuits which are coupled magnetically, not conductively. In coupled circuits, energy transfer takes place from one circuit to another through magnetic flux. If the flux produced by an element in one circuit links the elements of another circuit, then the elements in both circuits are said to be magnetically coupled.
2. **What are coupled coils?**
   When two or more coils are linked by magnetic flux, then the coils are said to be coupled coils.
3. **Define self-inductance.**
   If the permeability is constant, then self-inductance of a coil is defined as the ratio of flux linkage ($N\phi$) to the current flowing through the coil.

   $$L = N\frac{\phi}{I}$$

4. **Define mutual inductance.**
   If the permeability is constant, then mutual inductance of a coil is defined as the ratio of flux linkage ($N\phi$) in one coil due to current through another coil. Hence, mutual inductance,

   $$M = N_1\frac{\phi_{21}}{I_2} \quad \text{or} \quad N_2\frac{\phi_{12}}{I_1}$$

5. **Define coefficient of coupling.**
   The fraction of the total flux produced by one coil linking a second coil is called coefficient of coupling. Thus, hence, coefficient of coupling,

   $$k = \frac{\phi_{12}}{\phi_1} = \frac{\phi_{21}}{\phi_2}$$

where

ϕ_{12} = flux produced by coil 1 linking coil 2.

ϕ_{21} = flux produced by coil 2 linking coil 1.

ϕ_1 = flux produced by coil 1.

ϕ_2 = flux produced by coil 2.

6. Give the relation between self and mutual inductances.

The mutual inductance (M) existing between two coils of self-inductances L_1 and L_2 is given by

$$M = k\sqrt{L_1 L_2}$$

where k is the coefficient of coupling.

7. What is dot convention? Why is it required?

The e.m.f. induced due to mutual inductance may aid or oppose the e.m.f. induced due to self-inductance of a coil in a coupled circuit. Its polarity (sign) depends on the relative directions of currents and winding sense. Generally, the manufacturers of coupled coils mark one end of each coupled coil by a dot, which represents the polarity of mutual induced e.m.f.

8. State dot rule for coupled circuits.

Dot rule may be stated as:

In coupled circuits, current enters at a dotted end of first coil induces a positive e.m.f. at the dotted end of the second coil.

9. Write the expression for equivalent inductance of two coupled coils connected in series.

In series aiding, $L_{eq} = L_1 + L_2 + 2M$

In series opposing, $L_{eq} = L_1 + L_2 - 2M$

10. Write the expression for equivalent inductance of two coupled coils connected in parallel.

In parallel aiding, $L_{eq} = \left(\dfrac{L_1 L_2 - M^2}{L_1 + L_2 - 2M}\right)$

In parallel opposing, $L_{eq} = \left(\dfrac{L_1 L_2 - M^2}{L_1 + L_2 + 2M}\right)$

11. What is tuned coupled circuit?

If capacitors are added to primary and secondary of the coupled circuits to bring them to resonance condition to achieve maximum power transfer, then they are called tuned coupled circuits.

12. Where the tuned coupled circuits are used?

The tuned coupled circuits are mainly used to transfer energy from a weak source to a load or employed for maximum power transfer from one circuit to another. This is possible only when both the coils work at resonance condition.

13. What is single-tuned and double-tuned circuit?

In a coupled circuit, if a capacitor is added only in secondary, this circuit is called as single-tuned circuit, and if capacitors are added both in primary and secondary coils, it is called as double-tuned circuit.

UNSOLVED PROBLEMS

1. A primary coil having an inductance of 100 μH is connected in series with a secondary coil having an inductance of 240 μH and the total inductance of the combination is measured as 146 μH. Determine the coefficient of coupling. **[Ans.** 0.6261]
2. Two coils connected in series have an equivalent inductance of 0.8 H when connected in aiding and an equivalent inductance of 0.4 H when connected in opposing. Determine the mutual inductance. Calculate the self-inductance of the coils, by taking $k = 0.55$.
[Ans. $M = 0.1$ H; $L_1 = 0.5386$ H; $L_2 = 0.0615$ H]
3. Two magnetically coupled coils are connected in series and their total effective inductance is found to be 4.4 mH. When one coil is reversed in connection, the combined inductance drops to 1.6 mH. Here, all the flux due to the first coil links the second coil but only 40% of the flux due to the second coil links with the first coil. Find the self-inductance of each coil and the mutual inductance between the coils.
[Ans. $M = 0.7$ mH; $L_1 = 2.5124$ mH; $L_2 = 0.4876$ mH]
4. Two coupled coils with self-inductances 0.9 H and 0.4 H have a coupling coefficient of 0.3. Find the mutual inductance and turns ratio. What will be the maximum possible value of mutual inductance?
[Ans. $M = 0.18$ H; turns ratio = 1.5; maximum $M = 0.6$ H]
5. Two coupled coils have a coefficient of coupling of 0.65. If the self-inductances of the coils are 0.02 H and 0.08 H, calculate the equivalent inductance if the two coils are connected in (i) series aiding, (ii) series opposing, (iii) parallel aiding and (iv) parallel opposing. **[Ans.** (i) 0.152 H; (ii) 0.048 H; (iii) 0.01925 H; (iv) 6.0789×10^{-3}]
6. Determine the equivalent inductance of the series-connected coupled coils shown in Fig. U9.1. **[Ans.** 9 H]

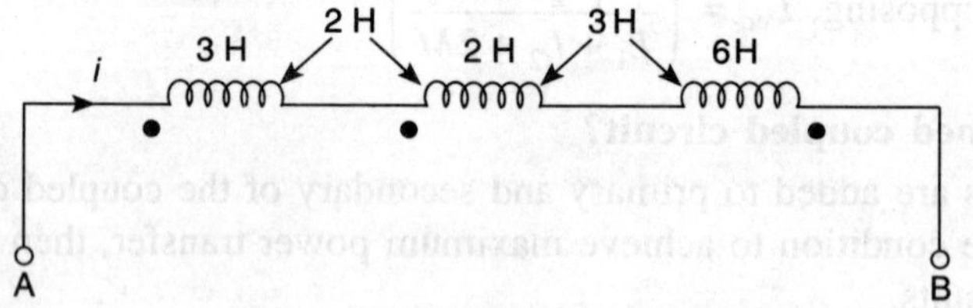

Fig. U9.1

7. Determine the equivalent inductance across terminals A and B of the circuit with coupled coils shown in Fig. U9.2. **[Ans.** 8 H]

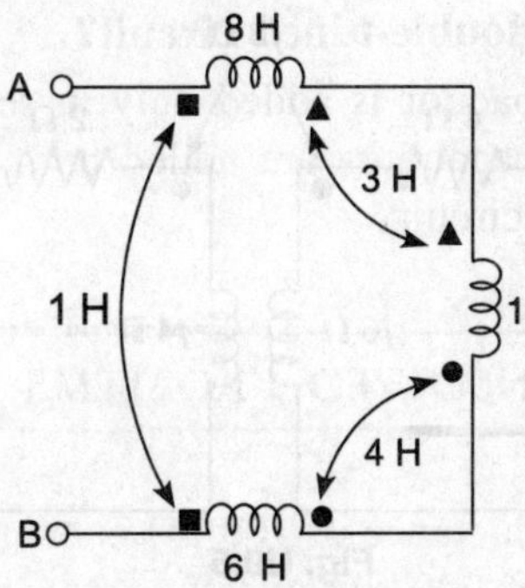

Fig. U9.2

8. Determine the equivalent inductance across terminals A and B of the circuit with coupled coils shown in Fig. U9.3. **[Ans.** 1.3077 H]

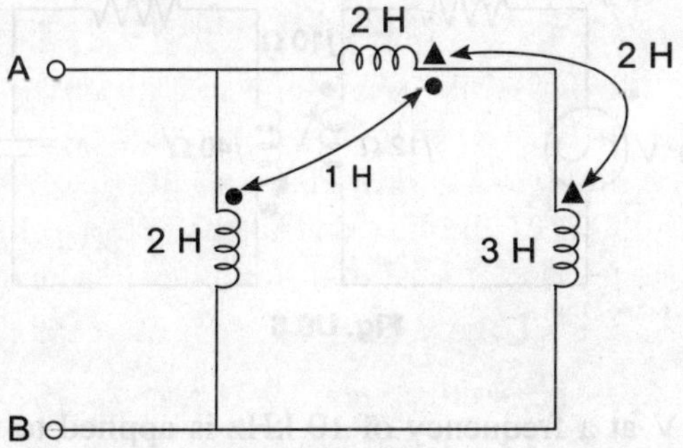

Fig. U9.3

9. Determine the equivalent impedance of the parallel-connected impedance with magnetic coupling shown in Fig. U9.4. **[Ans.** $(1 + j3.5)\ \Omega$]

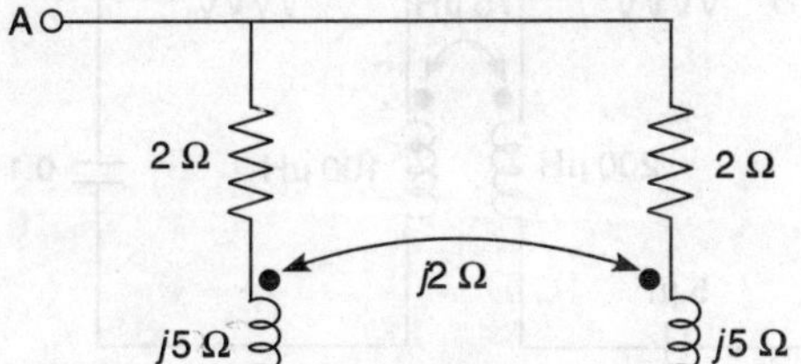

Fig. U9.4

10. For the circuit shown in Fig. U9.5, determine V_X such that the current supplied by $5\angle 0°$ voltage source is zero. **[Ans.** $7.45\angle -26.56°$ V]

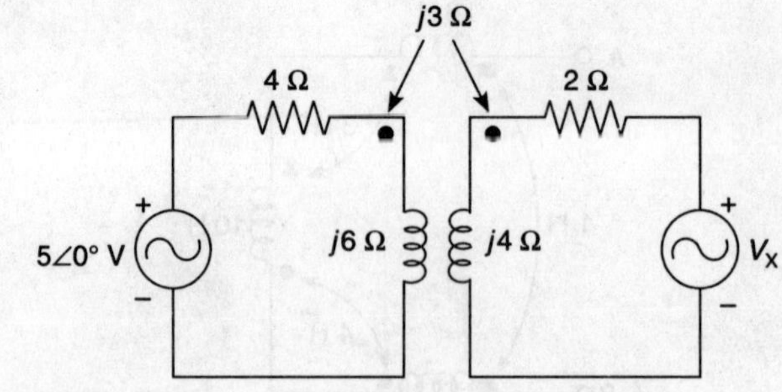

Fig. U9.5

11. Calculate the primary and secondary currents in the coupled circuit shown in Fig. U9.6.

[**Ans.** 24.3462∠–31.3° A; 11.3024∠170.5° A]

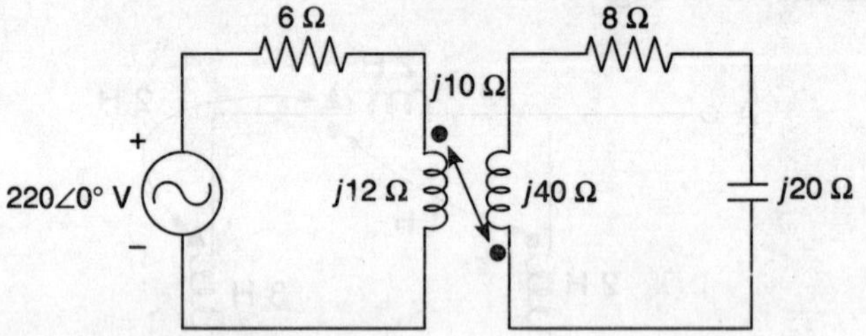

Fig. U9.6

12. A voltage of 115 V at a frequency of 10 kHz is applied to the primary of the coupled circuit shown in Fig. U9.7. Determine the total impedance referred to primary and the currents in both primary and secondary windings.

[**Ans.** 15.3286∠38.4° Ω; I_p = 7.5023∠–38.4° A; I_s = 0.231∠138.6° A]

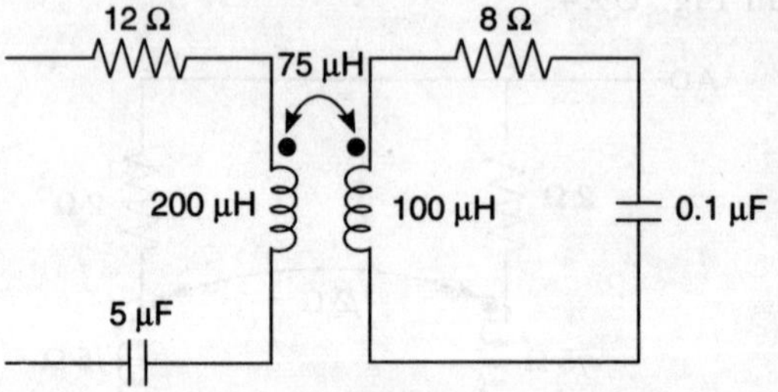

Fig. U9.7

CHAPTER 10

Resonance

10.1 SERIES RESONANCE

In communication circuits, resonance plays a vital role. For instance, the ability of a radio receiver to receive signals of one particular frequency and eliminate signals of other frequency is based on the principle of resonance. In general, resonance is a phenomenon in which applied voltage and the resulting current in an AC circuit are in phase. In other words, the circuit behaves as a purely resistive circuit and the power factor is unity. This condition is achieved by varying the frequency of the source so that inductive reactance X_L and capacitive reactance X_C become equal, thus canceling each other. Therefore, the total reactance of the circuit becomes zero and the circuit behaves as purely resistive circuit. The circuit is now said to be in *resonance* and the frequency at which it occurs is called *resonant frequency*.

Consider the *RLC* series circuit excited by a sinusoidal source of variable frequency as shown in Fig. 10.1.

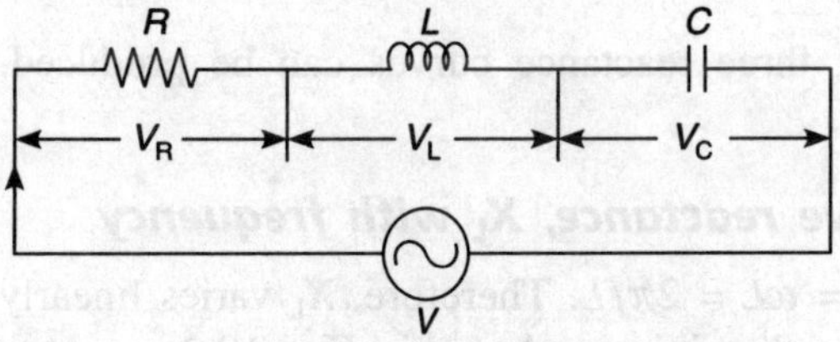

Fig. 10.1 *RLC* series circuit.

The total impedance of the circuit is given by

$$Z = R + j\left(\omega L - \frac{1}{\omega C}\right)$$

At resonance,

$$\omega L = \frac{1}{\omega C}$$

$$2\pi f_r L = \frac{1}{2\pi f_r C}$$

Hence, resonant frequency, $$f_r = \frac{1}{2\pi\sqrt{LC}}$$

In a series *RLC* circuit, resonance is achieved either by varying the frequency keeping the circuit elements constant or by varying *L* or *C* for a constant frequency.

10.1.1 Effects of Series Resonance

(i) Since at resonance $X_L = X_C$, the circuit will behave like a pure resistive circuit. Therefore, the impedance of the circuit is low, and is equal to the resistance of the circuit, i.e.

$$Z_r = R$$

(ii) Since the impedance is minimum, current is maximum, i.e.

$$I_r = \frac{V}{Z_r} = \frac{V}{R}$$

(iii) Since the current is maximum at resonance, power taken by the circuit is also maximum, i.e. $P_r = I_r^2 R = \dfrac{V^2}{R}$

(iv) Since at series resonance the current is large, the voltage drops across *L* and *C* are also very large. In fact, these drops are much greater than the applied voltage. However, voltage drops across *L–C* combination as a whole will be zero because these drops are equal in magnitude but 180° out of phase with each other.

10.1.2 Reactance Curves

In a series *RLC* circuit, three reactance curves can be produced. They are drawn between reactance and frequency.

Variation of inductive reactance, X_L with frequency

Inductive reactance, $X_L = \omega L = 2\pi f L$. Therefore, X_L varies linearly with frequency and it is a straight line passing through origin as shown in Fig. 10.2.

Variation of capacitive reactance, X_C with frequency

Capacitive reactance, $X_C = 1/\omega C = 1/2\pi f C$. Therefore, X_C varies indirectly with frequency and it is a rectangular hyperbola as shown in Fig. 10.3.

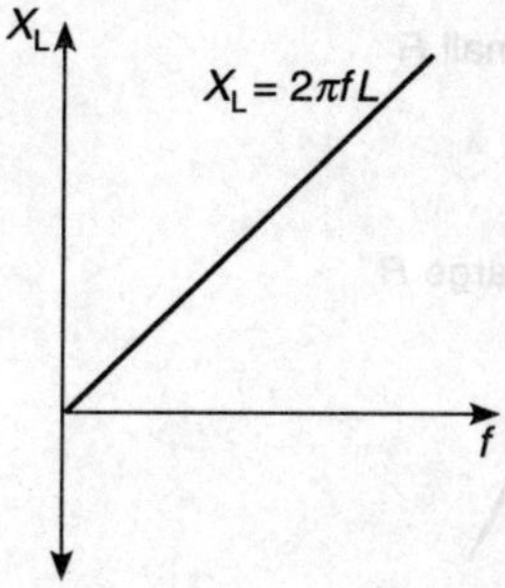

Fig. 10.2 Inductive reactance versus frequency.

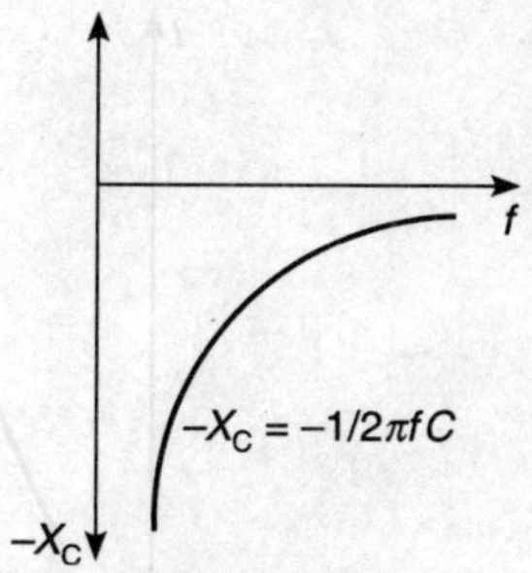

Fig. 10.3 Capacitive reactance versus frequency.

Variation of total reactance, X with frequency

Total reactance, $X = X_L - X_C$. For frequencies lesser than the value of resonant frequency (f_r), X is capacitive and for frequencies greater than the resonant frequency, X is inductive. At $f = f_r$, $X = 0$. The total reactance curve is shown in Fig. 10.4.

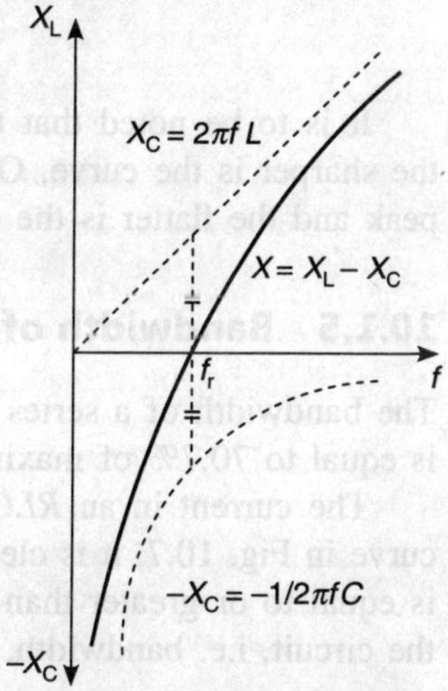

Fig. 10.4 Total reactance versus frequency.

10.1.3 Variation of Impedance with Frequency

The impedance of a series *RLC* circuit is

$$Z = R + j\left(\omega L - \frac{1}{\omega C}\right)$$

From the above equation, it is evident that at resonance, since $\omega L = 1/\omega C$, $|Z| = R$. At all other frequencies, $|Z| > R$. At lesser frequencies than resonant frequency, the impedance is capacitive (and hence the power factor is leading) and at frequencies above resonant frequency, the impedance is inductive (and hence power factor is lagging). The impedance as a function of frequency is shown in Fig. 10.5.

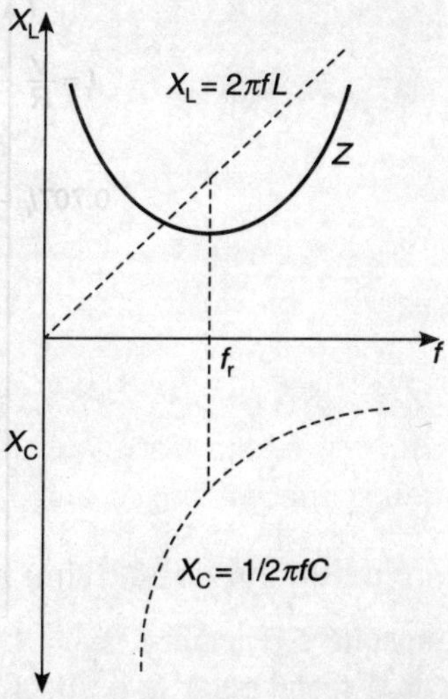

Fig. 10.5 Variation of impedance with frequency.

10.1.4 Resonance Curve

The curve between current and frequency is known as resonance curve. Figure 10.6 shows the resonance curve of a typical *RLC* series circuit. Note that the current reaches its maximum value at the resonant frequency (f_r), falling off rapidly on either side at that point. It is because if the frequency is below f_r, $X_C > X_L$ and the net reactance is no longer zero. If the frequency is above f_r, $X_L > X_C$ and the net reactance is again not zero. In both cases, the circuit impedance will be more than the impedance Z_r (= R) at resonance. The result is that the magnitude of circuit current decreases rapidly as the frequency changes from the resonant frequency.

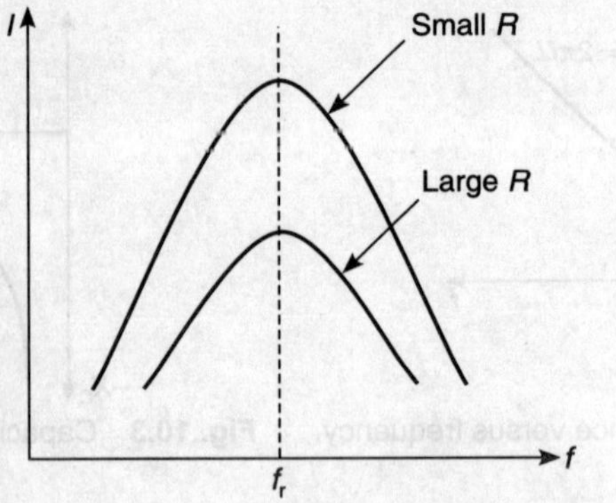

Fig. 10.6 Resonance curve

It is to be noted that the smaller the resistance, the greater the current at resonance and the sharper is the curve. On the other hand, the greater the resistance, the lower the resonant peak and the flatter is the curve.

10.1.5 Bandwidth of Series *RLC* Circuit

The bandwidth of a series circuit is defined as the range of frequencies for which the current is equal to 70.7% of maximum current or 70.7% of current at resonant frequency.

The current in an *RLC* series circuit changes with frequency. Referring to the resonance curve in Fig. 10.7, it is clear that for any frequency lying between f_1 and f_2, the circuit current is equal to or greater than 70.7% of maximum current. Therefore, $f_2 - f_1$ is the bandwidth of the circuit, i.e. bandwidth,

$$\beta = f_2 - f_1$$

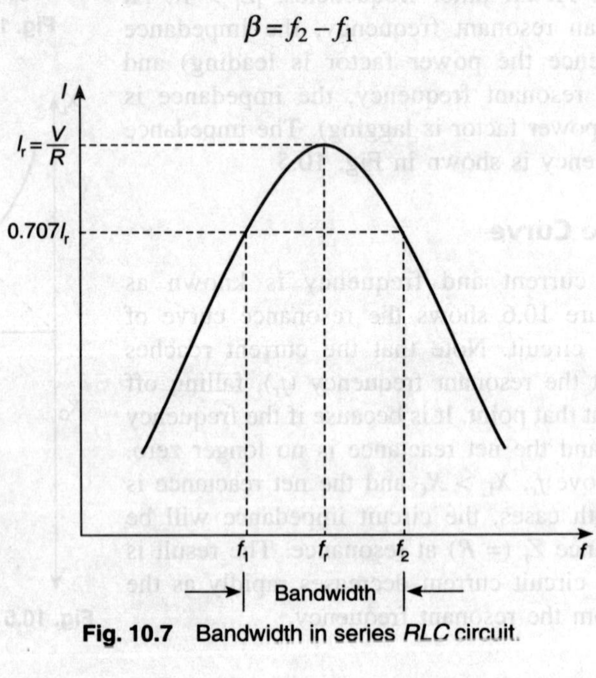

Fig. 10.7 Bandwidth in series *RLC* circuit.

Note that f_1 and f_2 are the limiting frequencies at which current is exactly equal to 70.7% of the maximum value. Frequency f_1 is called the *lower cut-off frequency* and frequency f_2 is called the *upper cut-off frequency*. They are also called as *lower half-power frequency* and *upper half-power frequency*, respectively, because at these frequencies, the power is half the value at resonance.

$$\text{Power at resonance} = I_r^2 R$$

$$\text{Power at cut-off frequencies} = (0.707 I_r)^2 R$$

$$= \frac{1}{2} I_r^2 R \qquad \text{(half of the power at resonance)}$$

Current in a series *RLC* circuit is given by

$$I = \frac{V}{Z}$$

$$= \frac{V}{\sqrt{R^2 + (X_L - X_C)^2}} = \frac{V}{\sqrt{R^2 + \left(\omega L - \dfrac{1}{\omega C}\right)^2}} \qquad (10.1)$$

At resonance,

$$I_r = \frac{V}{R} \qquad \text{(i.e. maximum value)}$$

$$\frac{V}{R} = \sqrt{2} I$$

$$\frac{V}{R} = \sqrt{2} \frac{V}{\sqrt{R^2 + \left(\omega L - \dfrac{1}{\omega C}\right)^2}} \qquad \text{[from Eq. (10.1)]}$$

$$\sqrt{R^2 + \left(\omega L - \frac{1}{\omega C}\right)^2} = \sqrt{2} R$$

$$R^2 + \left(\omega L - \frac{1}{\omega C}\right)^2 = 2R^2$$

$$\left(\omega L - \frac{1}{\omega C}\right)^2 = R^2$$

$$\omega L - \frac{1}{\omega C} = \pm R \qquad (10.2)$$

On multiplying Eq. (10.2) by ωC, we get

$$\omega^2 LC - 1 = \pm \omega CR$$

or

$$\omega^2 - \frac{1}{LC} = \pm \frac{R}{L} \omega$$

or $$\omega^2 \mp \frac{R}{L}\omega - \frac{1}{LC} = 0$$

or $$\omega^2 - \frac{R}{L}\omega - \frac{1}{LC} = 0$$

and $$\omega^2 + \frac{R}{L}\omega - \frac{1}{LC} = 0$$

Solving the above two quadratic equation, we get the roots of

$$\omega_1 = \frac{R}{2L} \pm \sqrt{\left(\frac{R}{2L}\right)^2 + \frac{1}{LC}} \quad \text{and} \quad \omega_2 = -\frac{R}{2L} \pm \sqrt{\left(\frac{R}{2L}\right)^2 + \frac{1}{LC}}$$

The cut-off frequencies are given by the positive roots of the two quadratic equations.

Hence, upper cut-off frequency, $f_2 = \frac{1}{2\pi}\left[\frac{R}{2L} \pm \sqrt{\left(\frac{R}{2L}\right)^2 + \frac{1}{LC}}\right]$

and lower cut-off frequency, $f_1 = \frac{1}{2\pi}\left[-\frac{R}{2L} \pm \sqrt{\left(\frac{R}{2L}\right)^2 + \frac{1}{LC}}\right]$

Bandwidth, $\beta = f_2 - f_1$

$$= \frac{1}{2\pi}\left[\frac{R}{2L} \pm \sqrt{\left(\frac{R}{2L}\right)^2 + \frac{1}{LC}}\right] - \frac{1}{2\pi}\left[-\frac{R}{2L} \pm \sqrt{\left(\frac{R}{2L}\right)^2 + \frac{1}{LC}}\right]$$

$$= \frac{R}{2\pi L}$$

i.e. bandwidth, $\beta = \frac{R}{2\pi L}$

Also, the bandwidth is written as $2 \times \left(\frac{R}{4\pi L}\right)$ and $\frac{R}{4\pi L}$ is marked on either side of resonant frequency as shown in Fig. 10.8. Now we can write:

Lower half-power frequency, $f_1 = f_r - \frac{R}{4\pi L}$

Upper half-power frequency, $f_2 = f_r + \frac{R}{4\pi L}$

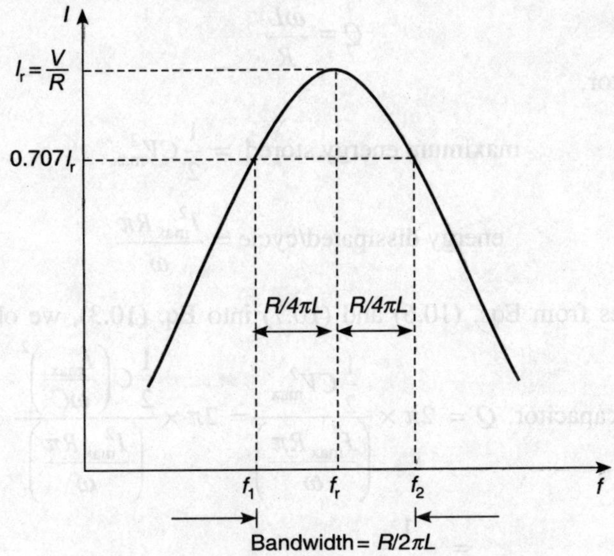

Fig. 10.8 Bandwidth and half-power frequencies.

10.1.6 Quality Factor of Series *RLC* Circuit

A *quality factor* or *figure of merit*, Q, can be assigned to a component or to a complete circuit. It is defined as:

$$Q = 2\pi\left(\frac{\text{maximum energy stored}}{\text{energy dissipated per cycle}}\right) \tag{10.3}$$

In an inductor,

$$\text{maximum energy stored} = \frac{1}{2}LI_{\text{max}}^2 \tag{10.4}$$

$$\text{energy dissipated/cycle} = I_{\text{rms}}^2 R \times t$$

$$= I_{\text{rms}}^2 R \times \frac{1}{f} \tag{10.5}$$

$$= \left(\frac{I_{\text{max}}}{\sqrt{2}}\right)^2 R\left(\frac{2\pi}{\omega}\right) = \frac{I_{\text{max}}^2 R\pi}{\omega}$$

Substituting the values from Eqs. (10.4) and (10.5) into Eq. (10.3), we obtain

quality factor of the coil, $$Q = 2\pi \times \frac{\left(\frac{1}{2}LI_{\text{max}}^2\right)}{\left(\frac{I_{\text{max}}^2 R\pi}{\omega}\right)}$$

or
$$Q = \frac{\omega L}{R}$$

Similarly in a capacitor,

$$\text{maximum energy stored } = \frac{1}{2} C V_{\text{max}}^2 \tag{10.6}$$

$$\text{energy dissipated/cycle} = \frac{I_{\text{max}}^2 R \pi}{\omega} \tag{10.7}$$

Substituting the values from Eqs. (10.6) and (10.7) into Eq. (10.3), we obtain

$$\text{quality factor of the capacitor, } Q = 2\pi \times \frac{\frac{1}{2} C V_{\text{max}}^2}{\left(\frac{I_{\text{max}}^2 R \pi}{\omega} \right)} = 2\pi \times \frac{\frac{1}{2} C \left(\frac{I_{\text{max}}}{\omega C} \right)^2}{\left(\frac{I_{\text{max}}^2 R \pi}{\omega} \right)}$$

$$= \frac{1}{\omega C R}$$

Also, the quality factor, $Q = \dfrac{\omega_r L}{R} = \dfrac{1}{\omega_r C R} = \dfrac{1}{R}\sqrt{\dfrac{L}{C}}$

Relation between Q-factor and Bandwidth

We know that,

$$Q\text{-factor} = \frac{\omega_r L}{R} = \frac{2\pi f_r L}{R}$$

$$= \frac{2\pi L}{R} f_r$$

$$= \frac{f_r}{\beta} \qquad \left(\because \beta = \frac{R}{2\pi L} \right)$$

It is to be noted here that the smaller the *Q*-factor, the greater the bandwidth.

10.1.7 Voltage Magnification in Series Resonance

Let us assume a voltage of *V* volts is applied to the *RLC* series circuit. Then,

$$\text{voltage across inductor, } V_L = I X_L = \frac{V}{R} \omega_r L$$

$$\text{voltage across capacitor, } V_C = I X_C = \frac{V}{R} \frac{1}{\omega_r C}$$

Since
$$Q = \frac{\omega_r L}{R} = \frac{1}{\omega_r C R}$$

where ω_r is the frequency at resonance.

Therefore, $V_L = VQ$

and $V_C = VQ$

The ratio of voltage across either L or C to the applied voltage at resonance is called *magnification.*

Therefore, magnification, $Q = \dfrac{V_L}{V}$ or $\dfrac{V_C}{V}$

EXAMPLE 10.1 A coil of resistance 100 Ω and inductance 100 μH is connected in series with a 100 pF capacitor. The circuit is connected to a 20 V variable frequency supply. Calculate (i) the resonant frequency, (ii) current at resonance, (iii) Q-factor of the circuit and (iv) voltage across L and C at resonant frequency.

Solution

(i) Resonant frequency, $f_r = \dfrac{1}{2\pi\sqrt{LC}} = \dfrac{1}{2\pi\sqrt{100\times10^{-6}\times100\times10^{-12}}}$

$= \mathbf{1591549\ Hz}$

(ii) Current at resonance, $I_r = \dfrac{V}{R} = \dfrac{20}{100} = \mathbf{0.2\ A}$

(iii) Q-factor $= \dfrac{1}{R}\sqrt{\dfrac{L}{C}} = \dfrac{1}{100}\sqrt{\dfrac{100\times10^{-6}}{100\times10^{-12}}} = \mathbf{10}$

(iv) At resonance, $X_L = 2\pi f_r L = 2\pi \times 1591549 \times 100 \times 10^{-6} = 1000\ \Omega$

Hence, $V_L = I_r X_L = 0.2 \times 1000 = \mathbf{200\ V}$

Similarly,

At resonance, $X_C = \dfrac{1}{2\pi f_r C} = \dfrac{1}{2\pi\times1591549\times100\times10^{-12}} = 1000\ \Omega$

So, $V_C = I_r X_C = 0.2 \times 1000 = \mathbf{200\ V}$

It is to be noted that at resonance, voltages across inductor and capacitor are equal since their respective reactance values are equal. Also the voltages are Q times the applied voltage value.

EXAMPLE 10.2 A series RLC circuit with $R = 10\ \Omega$, $L = 1$ mH and $C = 1000$ pF capacitor is connected across a sinusoidal source of 20 V with variable frequency. Calculate (i) the resonant frequency, (ii) Q-factor of the circuit at resonant frequency and (iii) the cut-off frequencies.

Solution

(i) Resonant frequency, $f_r = \dfrac{1}{2\pi\sqrt{LC}} = \dfrac{1}{2\pi\sqrt{1\times10^{-3}\times1000\times10^{-12}}}$

$= \mathbf{159155\ Hz}$

(ii) Q-factor $= \dfrac{1}{R}\sqrt{\dfrac{L}{C}} = \dfrac{1}{10}\sqrt{\dfrac{1\times10^{-3}}{1000\times10^{-12}}}$

$= \mathbf{100}$

(iii) Lower cut-off angular frequency,

$$\omega_1 = -\frac{R}{2L} + \sqrt{\left(\frac{R}{2L}\right)^2 + \frac{1}{LC}}$$

$$= -\frac{10}{2(1\times10^{-3})} + \sqrt{\left(\frac{10}{2(1\times10^{-3})}\right)^2 + \frac{1}{(1\times10^{-3})(1000\times10^{-12})}}$$

$$= -5000 + \sqrt{5000^2 + 10^{12}} = 995012.5 \text{ rad/s.}$$

Hence, lower cut-off frequency, $f_1 = \dfrac{\omega_1}{2\pi}$

$$= \frac{995012.5}{2\pi} = \mathbf{158361\ Hz}$$

Upper cut-off angular frequency,

$$\omega_2 = \frac{R}{2L} + \sqrt{\left(\frac{R}{2L}\right)^2 + \frac{1}{LC}}$$

$$= \frac{10}{2(1\times10^{-3})} + \sqrt{\left(\frac{10}{2(1\times10^{-3})}\right)^2 + \frac{1}{(1\times10^{-3})(1000\times10^{-12})}}$$

$$= 5000 + \sqrt{5000^2 + 10^{12}} = 1005012.5 \text{ rad/s}$$

Therefore, upper cut-off frequency, $f_2 = \dfrac{\omega_2}{2\pi}$

$$= \frac{1005012.5}{2\pi} = \mathbf{159952\ Hz}$$

Alternatively,

Lower cut-off frequency, $f_1 = \left(f_r - \dfrac{R}{4\pi L}\right) = \left(159155 - \dfrac{10}{4\pi \times 1\times10^{-3}}\right)$

$= \mathbf{158360\ Hz}$

Upper cut-off frequency, $f_2 = \left(f_r + \dfrac{R}{4\pi L}\right) = \left(159155 + \dfrac{10}{4\pi \times 1\times10^{-3}}\right)$

$= \mathbf{159952\ Hz}$

EXAMPLE 10.3 Compute the quality factor of an *RLC* series circuit, with $R = 20\ \Omega$, $L = 50$ mH and $C = 1\ \mu$F, using (i) $Q = \omega_r L/R$, (ii) $Q = 1/\omega_r CR$ and (iii) $Q = \omega_r/\beta$.

Solution

Resonant frequency, $\omega_r = \dfrac{1}{\sqrt{LC}} = \dfrac{1}{\sqrt{50\times10^{-3}\times1\times10^{-6}}} = 4472.136$ rad/s

Lower cut-off angular frequency,

$$\omega_1 = -\frac{R}{2L} + \sqrt{\left(\frac{R}{2L}\right)^2 + \frac{1}{LC}}$$

$$= -\frac{20}{2(50\times10^{-3})} + \sqrt{\left(\frac{20}{2(50\times10^{-3})}\right)^2 + \frac{1}{(50\times10^{-3})(1\times10^{-6})}}$$

$$= 4276.6 \text{ rad/s}$$

Upper cut-off angular frequency, ω_2

$$\omega_2 = \frac{R}{2L} + \sqrt{\left(\frac{R}{2L}\right)^2 + \frac{1}{LC}}$$

$$= \frac{20}{2(50\times10^{-3})} + \sqrt{\left(\frac{20}{2(50\times10^{-3})}\right)^2 + \frac{1}{(50\times10^{-3})(1\times10^{-6})}}$$

$$= 4676.6 \text{ rad/s}$$

Bandwidth, $\beta = \omega_2 - \omega_1 = 4676.6 - 4276.6 = 400$ rad/s

(i) Quality factor, $Q = \dfrac{\omega_r L}{R} = \dfrac{4472.136\times50\times10^{-3}}{20} = \mathbf{11.18}$

(ii) Quality factor, $Q = \dfrac{1}{\omega_r CR} = \dfrac{1}{4472.136\times1\times10^{-6}\times20} = \mathbf{11.18}$

(iii) Quality factor, $Q = \dfrac{\omega_r}{\beta} = \dfrac{4472.136}{400} = \mathbf{11.18}$

EXAMPLE 10.4 A series *RLC* circuit resonates at a frequency of 1400 Hz and consumes 60 W power for 55 V at resonant frequency. The bandwidth is 850 Hz. Determine the circuit parameters.

Solution

Power, $P = \dfrac{V^2}{R}$

Hence, resistance, $R = \dfrac{V^2}{P} = \dfrac{55^2}{60} = \mathbf{50.42\ \Omega}$

Bandwidth, $\beta = \dfrac{R}{2\pi L}$

Hence, inductance, $L = \dfrac{R}{2\pi\times\beta} = \dfrac{50.42}{2\pi\times850} = \mathbf{9.44\ mH}$

Resonant frequency, $f_r = \dfrac{1}{2\pi\sqrt{LC}}$

$$f_r^2 = \frac{1}{4\pi^2 LC}$$

So, capacitance,
$$C = \frac{1}{4\pi^2 f_r^2 L}$$
$$= \frac{1}{4\pi^2 \times 1400^2 \times 9.44 \times 10^{-3}} = \mathbf{1.37\ \mu F}$$

EXAMPLE 10.5 A series *RLC* circuit has an impedance of 40 Ω at a frequency of 200 rad/s. When the circuit is excited by a 10 V source of variable frequency, the circuit resonates at 250 rad/s. The current at resonance is 0.5 A and the quality factor at resonance is 10. Determine the circuit parameters.

Solution

Current at resonance,
$$I_r = \frac{V}{R}$$

So, resistance
$$R = \frac{V}{I_r} = \frac{10}{0.5} = \mathbf{20\ \Omega}$$

Q-factor at resonance
$$= \frac{\omega_r L}{R}$$

Hence, inductance
$$L = \frac{QR}{\omega_r} = \frac{10 \times 20}{250} = \mathbf{0.8\ H}$$

Impedance,
$$Z = \sqrt{R^2 + (X_L - X_C)^2}$$
$$(X_L - X_C) = \sqrt{Z^2 - R^2}$$
$$X_C = X_L - \sqrt{Z^2 - R^2}$$
$$\frac{1}{\omega C} = \omega L - \sqrt{Z^2 - R^2}$$

It is given as $Z = 40\ \Omega$ at a frequency of 200 rad/s
$$\frac{1}{200C} = 200(0.8) - \sqrt{40^2 - 20^2}$$

Therefore, capacitance
$$C = \mathbf{39.885\ \mu F}$$

10.2 PARALLEL RESONANCE

A parallel circuit containing reactive elements (*L* and *C*) is said to be in resonance when $X_C = X_L$. The frequency at which resonance occurs is called *resonant frequency*. At resonant frequency, the power factor is unity. Consider the practical parallel *RLC* circuit shown in Fig. 10.9. This circuit is also called as *tank circuit* and is widely used in communication networks.

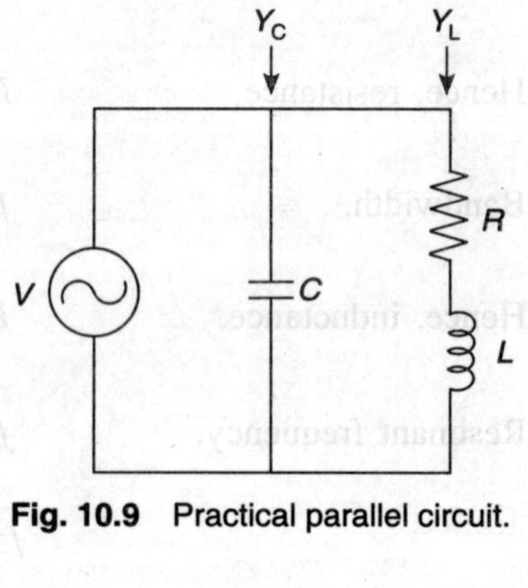

Fig. 10.9 Practical parallel circuit.

The admittance of the circuit, $Y = Y_C + Y_L$

$$= \frac{1}{Z_C} + \frac{1}{Z_L}$$

$$= \frac{1}{-jX_C} + \frac{1}{R + jX_L}$$

$$= \frac{j}{X_C} + \left(\frac{1}{R + jX_L} \times \frac{R - jX_L}{R - jX_L}\right)$$

$$= \frac{j}{X_C} + \left(\frac{R - jX_L}{R^2 + X_L^2}\right)$$

$$= \frac{j}{X_C} + \frac{R}{R^2 + X_L^2} - \frac{jX_L}{R^2 + X_L^2}$$

$$= \frac{R}{R^2 + X_L^2} + j\left(\frac{1}{X_C} - \frac{X_L}{R^2 + X_L^2}\right) \quad (10.8)$$

At resonance, the reactive part of the above equation must be zero. Therefore,

$$\frac{1}{X_C} - \frac{X_L}{R^2 + X_L^2} = 0$$

$$\frac{1}{X_C} = \frac{X_L}{R^2 + X_L^2}$$

$$R^2 + X_L^2 = X_L\, X_C$$

$$R^2 + X_L^2 = \omega_r L \times \frac{1}{\omega_r C}$$

$$R^2 + X_L^2 = \frac{L}{C} \quad (10.9)$$

$$X_L^2 = \frac{L}{C} - R^2$$

$$X_L = \sqrt{\frac{L}{C} - R^2}$$

or

$$\omega_r L = \sqrt{\frac{L}{C} - R^2}$$

$$\omega_r = \frac{1}{L}\sqrt{\frac{L}{C} - R^2}$$

$$= \sqrt{\frac{1}{LC} - \frac{R^2}{L^2}}$$

Hence,

$$f_r = \frac{1}{2\pi}\sqrt{\frac{1}{LC} - \frac{R^2}{L^2}}$$

If the coil resistance is small as compared to the inductive reactance, then

$$f_r = \frac{1}{2\pi\sqrt{LC}} \qquad \text{(same as that of series resonance)}$$

From Eq. (10.8), the admittance of the circuit at resonance is

$$Y_r = \frac{R}{R^2 + X_L^2}$$

Hence,

$$Z_r = \frac{R^2 + X_L^2}{R}$$

$$= \frac{(L/C)}{R} \qquad \text{[from Eq. (10.9)]}$$

or

$$Z_r = \frac{L}{CR}$$

The impedance offered by the parallel circuit at resonance is called as *dynamic impedance* of the circuit.

Note that $Z_r (= L/CR)$ is a pure resistance because there is no frequency term present. Further, the value of impedance at parallel resonance is at its maximum (because the L/C ratio is very large at resonance) and hence the resulting current is low.

A parallel resonant circuit is also called a rejecter circuit, since the current at resonance is low. In other words, a tank circuit rejects the current at resonance.

Hence, the current at resonance $= I_r = \dfrac{V}{Z_r} = \dfrac{VCR}{L}$.

10.2.1 Susceptance Curves

Consider a parallel *RLC* circuit shown in Fig. 10.9. The admittance of the circuit is given by

$$Y = G + j(B_C - B_L)$$

where

$$G = \frac{1}{R}; \quad B_C = \omega C; \quad B_L = \frac{1}{\omega L}$$

In parallel *RLC* circuit, susceptance curves are drawn between susceptance and frequency as follows.

Variation of capacitive susceptance B_C with frequency

Capacitive susceptance, $B_C = \omega C = 2\pi f C$. Therefore, B_C varies linearly with frequency and it is a straight line passing through origin as shown in Fig. 10.10.

Variation of inductive susceptance B_L with frequency

Inductive susceptance,

$$B_L = \frac{1}{\omega L} = \frac{1}{2\pi f L}$$

Therefore, B_L varies indirectly with frequency and it is a rectangular hyperbola as shown in Fig. 10.11.

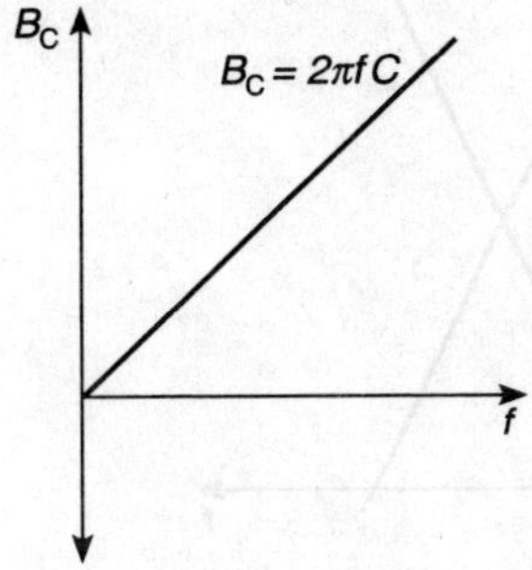

Fig. 10.10 Capacitve susceptance versus frequency.

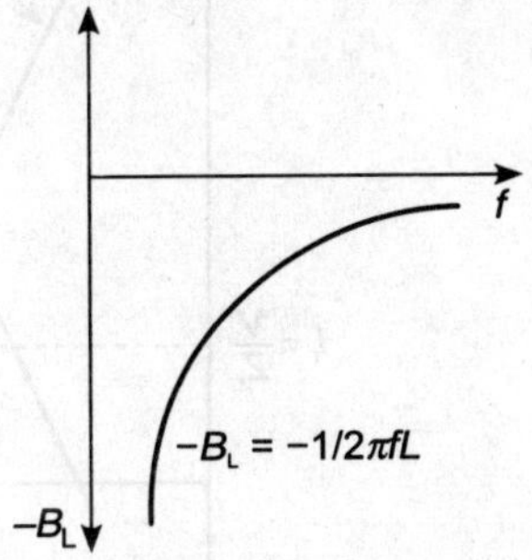

Fig. 10.11 Inductive susceptance versus frequency.

Variation of total susceptance B with frequency

Total susceptance, $B = B_C - B_L$. For frequencies lesser than the value of resonant frequency (f_r), B is inductive and for frequencies greater than the resonant frequency, B is capacitive. At $f = f_r$, $B = 0$. The total susceptance curve is shown in Fig. 10.12.

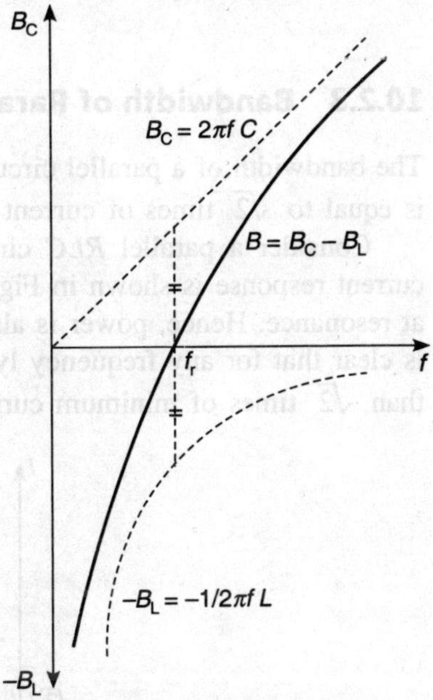

Fig. 10.12 Total susceptance versus frequency.

10.2.2 Variation of Impedance and Current with Frequency

As supply frequency changes, the value of circuit impedance also varies. The impedance versus frequency and current versus frequency are plotted in Fig. 10.13. It is to be noted that the value of the impedance is maximum (and hence the resulting current is minimum) at resonance. As the frequency changes from resonance, the circuit impedance decreases (current increases) very rapidly. This is because, at frequencies below resonant frequency, the capacitive reactance will be higher and thus more current will flow through the coil. This causes the line current to lag behind the applied voltage and the circuit appears inductive. For frequencies above resonance, inductive reactance is higher and more current will flow through the capacitor. Consequently, the line current leads the applied voltage and circuit appears capacitive. In either case, the circuit impedance is far less than its value at resonance.

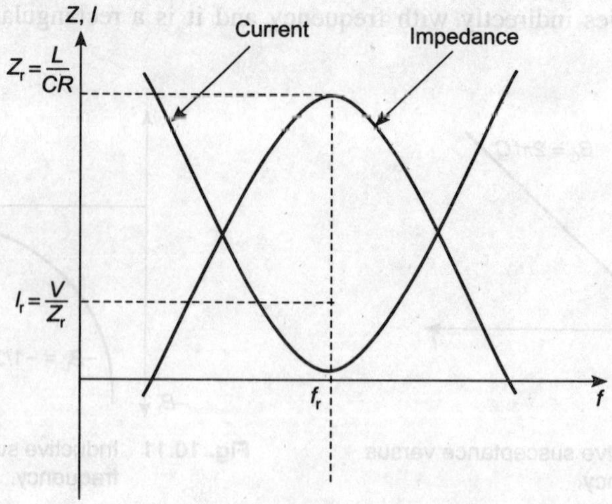

Fig. 10.13 Variation of circuit impedance and current with frequency.

10.2.3 Bandwidth of Parallel *RLC* Circuit

The bandwidth of a parallel circuit is defined as the range of frequencies for which the current is equal to $\sqrt{2}$ times of current at resonant frequency.

Consider a parallel *RLC* circuit excited by a voltage source as shown in Fig. 10.9. Its current response is shown in Fig. 10.14. The current changes with frequency and is minimum at resonance. Hence, power is also minimum at resonant frequency. Referring to Fig. 10.14, it is clear that for any frequency lying between f_1 and f_2, the circuit current is equal to or lesser than $\sqrt{2}$ times of minimum current.

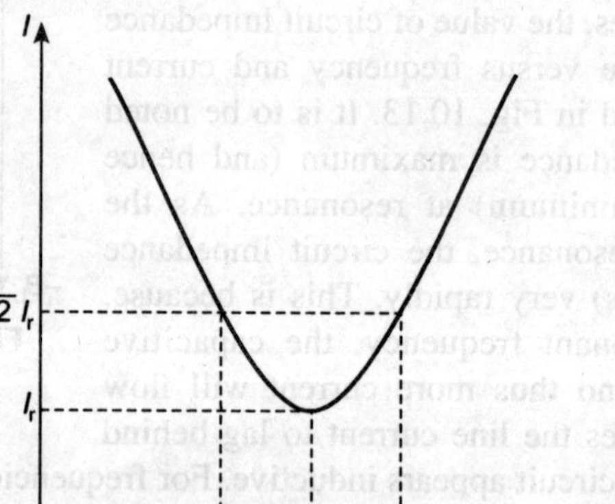

Fig. 10.14 Bandwidth in parallel RLC circuit.

Referring to Fig. 10.9,

Admittance, $$Y = G + jB$$

$$= \frac{1}{R} + j\omega C + \frac{1}{j\omega L}$$

$$= \frac{1}{R} + j\left(\omega C - \frac{1}{\omega L}\right)$$

$$= \sqrt{\left(\frac{1}{R}\right)^2 + \left(\omega C - \frac{1}{\omega L}\right)^2} \tag{10.10}$$

Power $$= I^2 R$$

$$= (VY)^2 R = V^2 Y^2 R$$

$$= V^2 R\left[\left(\frac{1}{R}\right)^2 + \left(\omega C - \frac{1}{\omega L}\right)^2\right] \quad \text{[from Eq. (10.10)]}$$

At cut-off frequencies, the power is twice that of power at resonance value. Therefore,

$$V^2 R\left[\left(\frac{1}{R}\right)^2 + \left(\omega C - \frac{1}{\omega L}\right)^2\right] = \frac{2V^2}{R}$$

$$\left(\frac{1}{R}\right)^2 + \left(\omega C - \frac{1}{\omega L}\right)^2 = \frac{2}{R^2}$$

$$\left(\omega C - \frac{1}{\omega L}\right)^2 = \frac{2}{R^2} - \frac{1}{R^2}$$

$$\left(\omega C - \frac{1}{\omega L}\right)^2 = \frac{1}{R^2}$$

$$\omega C - \frac{1}{\omega L} = \pm\frac{1}{R} \tag{10.11}$$

On multiplying Eq. (10.11) by ωL, we get

$$\omega^2 LC - 1 = \pm\frac{\omega L}{R} \tag{10.12}$$

or $$\omega^2 - \frac{1}{LC} = \pm\frac{\omega}{RC}$$

or $$\omega^2 \mp \frac{\omega}{RC} - \frac{1}{LC} = 0$$

or $$\omega^2 - \frac{\omega}{RC} - \frac{1}{LC} = 0 \quad \text{and} \quad \omega^2 + \frac{\omega}{RC} - \frac{1}{LC} = 0$$

Solving the above two quadratic equation, we get the roots of

$$\omega_1 = \frac{1}{2RC} \pm \sqrt{\left(\frac{1}{2RC}\right)^2 + \frac{1}{LC}} \quad \text{and} \quad \omega_2 = -\frac{1}{2RC} \pm \sqrt{\left(\frac{1}{2RC}\right)^2 + \frac{1}{LC}}$$

The cut-off frequencies are given by the positive roots of the two quadratic equations.

So, upper cut-off frequency, $$f_2 = \frac{1}{2\pi}\left[\frac{1}{2RC} \pm \sqrt{\left(\frac{1}{2RC}\right)^2 + \frac{1}{LC}}\right]$$

and lower cut-off frequency, $$f_1 = \frac{1}{2\pi}\left[-\frac{1}{2RC} \pm \sqrt{\left(\frac{1}{2RC}\right)^2 + \frac{1}{LC}}\right]$$

Bandwidth, $\beta = f_2 - f_1$

$$= \frac{1}{2\pi}\left[\frac{1}{2RC} \pm \sqrt{\left(\frac{1}{2RC}\right)^2 + \frac{1}{LC}}\right] - \frac{1}{2\pi}\left[-\frac{1}{2RC} \pm \sqrt{\left(\frac{1}{2RC}\right)^2 + \frac{1}{LC}}\right]$$

$$\beta = \frac{1}{2\pi RC}$$

10.2.4 Quality Factor of Parallel *RLC* Circuit

In parallel resonance, the current circulating between the two branches is many times greater than the current taken from the supply. As seen from Fig. 10.15, the circulating current between capacitor and coil branches is I_C.

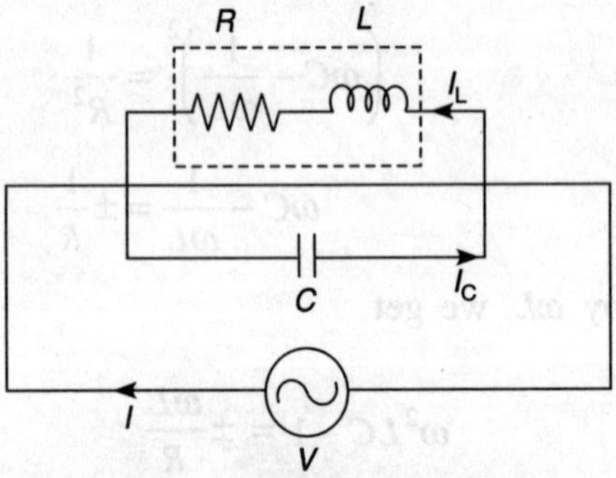

Fig. 10.15 Practical parallel circuit.

Therefore, the quality factor of the parallel circuit is defined as the ratio of the circulating current to the supply current.

i.e. $$Q = \frac{I_C}{I}$$

Now circulating current, $$I_C = \frac{V}{X_C} = \frac{V}{\left(\frac{1}{\omega C}\right)} = \omega CV$$

and supply current, $$I = \frac{V}{\left(\frac{L}{CR}\right)} = \frac{VCR}{L}$$

Hence, quality factor $$Q = \frac{I_C}{I} = \omega CV \times \frac{L}{VCR}$$

$$= \frac{\omega L}{R} = \frac{2\pi fL}{R} \quad \text{(same as that of series resonance)}$$

If the coil resistance is small as compared to the inductive reactance, then

$$f_r = \frac{1}{2\pi\sqrt{LC}}$$

Putting this value above,

$$Q = \frac{2\pi fL}{R} = \frac{2\pi L}{R} \times \frac{1}{2\pi\sqrt{LC}}$$

$$= \frac{L}{R} \times \frac{1}{\sqrt{LC}} = \frac{1}{R}\sqrt{\frac{L}{C}}$$

It should be noted that in series circuits, Q-factor gives the voltage magnification, whereas in parallel circuits, it gives the current magnification. Also,

$$Q = 2\pi\left(\frac{\text{maximum energy stored}}{\text{energy dissipated per cycle}}\right)$$

EXAMPLE 10.6 A coil of resistance 60 Ω and inductance 0.2 H is connected in parallel with a 20 μF capacitor across a 20 V variable frequency supply. Calculate (i) the resonant frequency, (ii) the dynamic resistance, (iii) current at resonance and (iv) Q-factor of the circuit at resonance.

Solution

(i) Resonant frequency, $$f_r = \frac{1}{2\pi}\sqrt{\frac{1}{LC} - \frac{R^2}{L^2}}$$

$$= \frac{1}{2\pi}\sqrt{\frac{1}{(0.2)(20 \times 10^{-6})} - \frac{60^2}{0.2^2}}$$

$$= \mathbf{63.66\ Hz}$$

(ii) Dynamic resistance, $$Z_r = \frac{L}{CR} = \frac{0.2}{20 \times 10^{-6} \times 60} = \mathbf{166.67\ \Omega}$$

(iii) Current at resonance, $I_r = \dfrac{V}{Z_r} = \dfrac{20}{166.67} = \mathbf{0.12\ A}$

(iv) Q-factor $= \dfrac{2\pi f_r L}{R} = \dfrac{2\pi \times 63.66 \times 0.2}{60} = \mathbf{1.3}$

EXAMPLE 10.7 A tuned circuit (Fig. 10.16) consists of a coil having an inductance of 200 μH and a resistance of 20 Ω is in parallel with a variable capacitor is connected in series with a resistor of 8000 Ω across a 60 V supply having a frequency of 1 MHz. Calculate:

(i) the value of C to give resonance;
(ii) the dynamic impedance and the Q-factor of the tuned circuit;
(iii) the current in each branch.

Solution

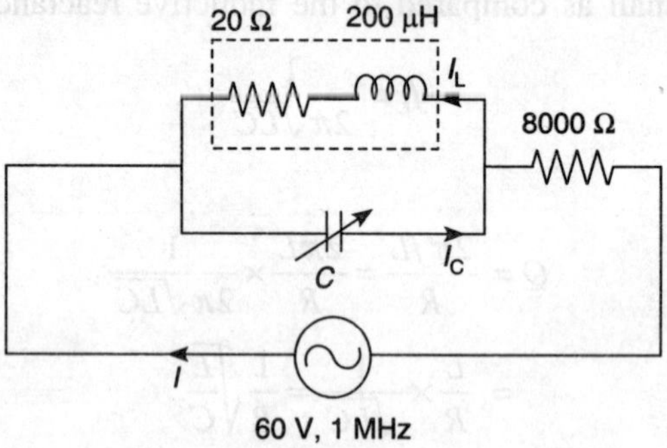

Fig. 10.16

(i) Resonant frequency, $f_r = \dfrac{1}{2\pi}\sqrt{\dfrac{1}{LC} - \dfrac{R^2}{L^2}}$

Rearranging,

Capacitor, $C = \dfrac{1}{L\left(4\pi^2 f_r^2 + \dfrac{R^2}{L^2}\right)}$

$$= \frac{1}{200 \times 10^{-6}\left(4\pi^2 \times 10^{12} + \dfrac{20^2}{(200 \times 10^{-6})^2}\right)}$$

$= 126.62 \times 10^{-12} = \mathbf{126.62\ pF}$

(ii) Dynamic impedance, $Z_r = \dfrac{L}{CR}$

$$= \frac{200 \times 10^{-6}}{126.62 \times 10^{-12} \times 20} = \mathbf{78976\ \Omega}$$

and Q-factor $= \dfrac{2\pi fL}{R} = \dfrac{2\pi \times 10^6 \times 200 \times 10^{-6}}{20} = \mathbf{62.83}$

(iii) Total equivalent resistance of the circuit (Fig. 10.17) is

$$78976 + 8000 = 86976\ \Omega$$

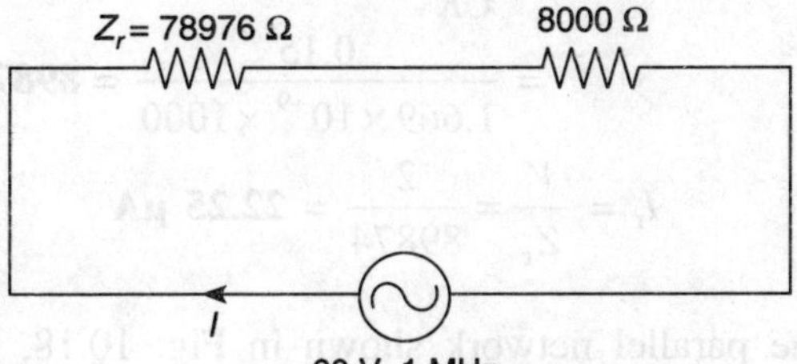

Fig. 10.17

So, current taken from supply $= \dfrac{60}{86976} = 0.689$ mA

p.d. across tuned circuit is

$$0.689 \times 10^{-3} \times 78976 = 54.41\text{ V}$$

Current through inductive branch of tuned circuit is

$$\frac{54.41}{\sqrt{(20)^2 + (2\pi \times 10^6 \times 200 \times 10^{-6})^2}} = \frac{54.41}{1256.9} = \mathbf{43.3\ mA}$$

and current through capacitor is

$$2\pi \times 10^6 \times 126.62 \times 10^{-12} \times 54.41 = \mathbf{43.3\ mA}$$

The results show that the current in each of the parallel circuit is about 62.83 (i.e. *Q*-factor) times the current taken from the supply.

EXAMPLE 10.8 A coil of 1 kΩ resistance and 0.15 H inductance is connected in parallel with a variable capacitor across a 2 V, 10 kHz, AC supply. Calculate:

(i) the value of *C* when the supply current is minimum;
(ii) the effective impedance of the circuit;
(iii) the supply current.

Solution The supply current is minimum when the circuit is at resonance.

(i) Resonant frequency, $f_r = \dfrac{1}{2\pi}\sqrt{\dfrac{1}{LC} - \dfrac{R^2}{L^2}}$

Rearranging,

Capacitor, $C = \dfrac{1}{L\left(4\pi^2 f_r^2 + \dfrac{R^2}{L^2}\right)}$

$$= \frac{1}{0.15\left(4\pi^2 \times 10^2 \times 10^6 + \dfrac{1000^2}{0.15^2}\right)}$$

$$= 1.669 \times 10^{-9} = \mathbf{1.669\ nF}$$

(ii) Effective impedance (dynamic impedance),

$$Z_r = \frac{L}{CR}$$

$$= \frac{0.15}{1.669 \times 10^{-9} \times 1000} = \mathbf{89874\ \Omega}$$

(iii) Supply current, $I_r = \frac{V}{Z_r} = \frac{2}{89874} = \mathbf{22.25\ \mu A}$

EXAMPLE 10.9 For the parallel network shown in Fig. 10.18, determine the value of R for resonance.

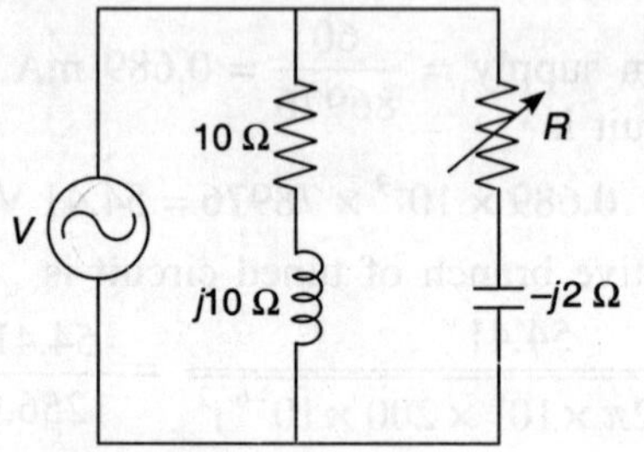

Fig. 10.18

Solution The total admittance,

$$Y = Y_1 + Y_2$$

$$= \frac{1}{10 + j10} + \frac{1}{R - j2}$$

$$= \left(\frac{1}{10 + j10} \times \frac{10 - j10}{10 - j10}\right) + \left(\frac{1}{R - j2} \times \frac{R + j2}{R + j2}\right)$$

$$= \left(\frac{10 - j10}{200}\right) + \left(\frac{R + j2}{R^2 + 4}\right)$$

$$= \left(\frac{10}{200} - \frac{j10}{200}\right) + \left(\frac{R}{R^2 + 4} + \frac{j2}{R^2 + 4}\right)$$

$$= \left(\frac{10}{200} + \frac{R}{R^2 + 4}\right) + j\left(\frac{2}{R^2 + 4} - \frac{10}{200}\right)$$

At resonance, the imaginary part of admittance is zero. Hence,

$$\frac{2}{R^2 + 4} - \frac{10}{200} = 0$$

Solving, $R = \mathbf{6\ \Omega}$

EXAMPLE 10.10 For the parallel network shown in Fig. 10.19, determine the value of C for resonance for the network to resonate at 2000 rad/s.

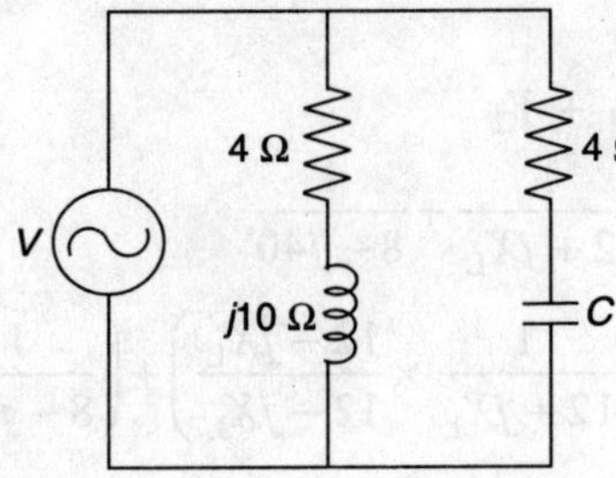

Fig. 10.19

Solution The total admittance,

$$Y = Y_1 + Y_2$$

$$= \frac{1}{4+j10} + \frac{1}{4-jX_C}$$

$$= \left(\frac{1}{4+j10} \times \frac{4-j10}{4-j10}\right) + \left(\frac{1}{4-jX_C} \times \frac{4+jX_C}{4+jX_C}\right)$$

$$= \left(\frac{4-j10}{116}\right) + \left(\frac{4+jX_C}{16+X_C^2}\right)$$

$$= \left(\frac{4}{116} - \frac{j10}{116}\right) + \left(\frac{4}{16+X_C^2} + \frac{jX_C}{16+X_C^2}\right)$$

$$= \left(\frac{4}{116} + \frac{4}{16+X_C^2}\right) + j\left(\frac{X_C}{16+X_C^2} - \frac{10}{116}\right)$$

At resonance, the imaginary part of admittance is zero. Therefore,

$$\frac{X_C}{16+X_C^2} - \frac{10}{116} = 0$$

Simplifying,

$$X_C^2 - 11.6X_C + 16 = 0$$

Solving, we get

$$X_C = 10 \text{ or } 1.6$$

Hence, capacitance,

$$C = \mathbf{50\ \mu F} \text{ or } \mathbf{312.5\ \mu F}$$

EXAMPLE 10.11 Find the value of L so that the circuit shown in Fig. 10.20 resonates. Assume the supply frequency is 50 Hz.

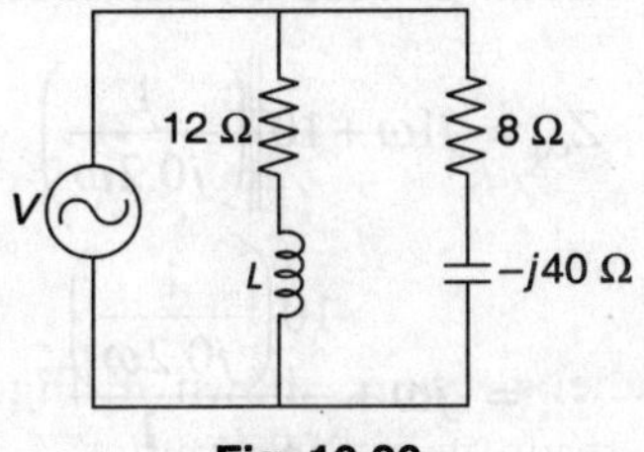

Fig. 10.20

Solution The total admittance,

$$Y = Y_1 + Y_2$$

$$= \frac{1}{12 + jX_L} + \frac{1}{8 - j40}$$

$$= \left(\frac{1}{12 + jX_L} \times \frac{12 - jX_L}{12 - jX_L}\right) + \left(\frac{1}{8 - j40} \times \frac{8 + j40}{8 + j40}\right)$$

$$= \left(\frac{12 - jX_L}{144 + X_L^2}\right) + \left(\frac{8 + j40}{64 + 1600}\right)$$

$$= \left(\frac{12}{144 + X_L^2} - \frac{jX_L}{144 + X_L^2}\right) + \left(\frac{8}{64 + 1600} + \frac{j40}{64 + 1600}\right)$$

$$= \left(\frac{12}{144 + X_L^2} + \frac{8}{1664}\right) + j\left(\frac{40}{1664} - \frac{X_L}{144 + X_L^2}\right)$$

At resonance, the imaginary part of admittance is zero. Hence,

$$\frac{40}{1664} - \frac{X_L}{144 + X_L^2} = 0$$

Simplifying, $X_L^2 - 41.6X_L + 144 = 0$

Solving, we get $X_L = 37.79$ or 3.81

So, inductance, $L =$ **0.12 H** or **0.012 H**

EXAMPLE 10.12 Compute the resonant frequency of the series-parallel circuit shown in Fig. 10.21.

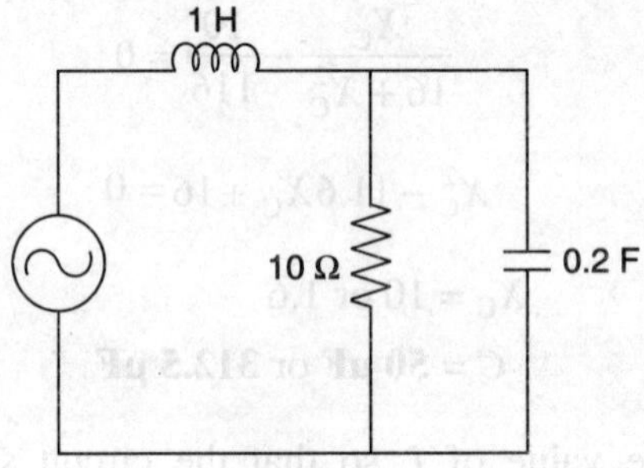

Fig. 10.21

Solution The impedance of the circuit as seen by the source is

$$Z_{eq} = j1\omega + 10 \,\Big\|\, \left(\frac{1}{j0.2\omega}\right)$$

$$= j\omega + \frac{10\left(\dfrac{1}{j0.2\omega}\right)}{10 + \dfrac{1}{j0.2\omega}}$$

$$= j\omega + \frac{\left(\dfrac{10}{j0.2\omega}\right)}{\left(\dfrac{j2\omega+1}{j0.2\omega}\right)} = j\omega + \frac{10}{1+j2\omega}$$

$$= \frac{j\omega(j2\omega+1)+10}{1+j2\omega} = \frac{(j\omega - 2\omega^2 + 10)}{(1+j2\omega)} \times \frac{(1-j2\omega)}{(1-j2\omega)}$$

$$= \frac{10 + j(4\omega^3 - 19\omega)}{1+4\omega^2}$$

At resonance, the imaginary part of Z_{eq} is zero, or

$$4\omega^3 - 19\omega = 0$$

Solving, resonant frequency,

$$\boldsymbol{\omega = 2.179 \text{ rad/s}}$$

SHORT QUESTIONS WITH ANSWERS

1. What is resonance?

The resonance is a phenomenon in which applied voltage and the resulting current in an *RLC* circuit are in phase. In other words, the circuit behaves as a purely resistive circuit and the power factor is unity.

2. What is resonant frequency?

The frequency at which resonance condition occurs is called resonant frequency.

3. Write the expressions for resonant frequency and current at resonance of an *RLC* series circuit.

Resonant frequency, $f_r = \dfrac{1}{2\pi\sqrt{LC}}$

Current at resonant frequency, $I_r = \dfrac{V}{Z_r} = \dfrac{V}{R}$

4. Draw the frequency response of *RLC* series circuit.

The variation of current with frequency is called frequency response, which is shown in Fig. S10.1, for *RLC* series circuit.

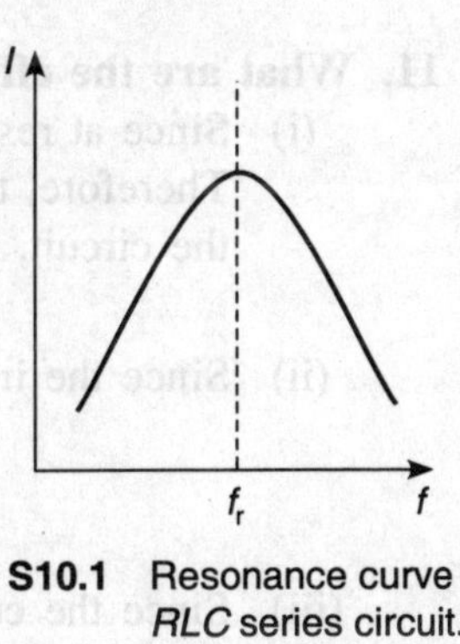

Fig. S10.1 Resonance curve for *RLC* series circuit.

5. Define quality factor.

A quality factor or figure of merit, Q, can be assigned to a component or to a complete circuit. It is defined as:

$$Q = 2\pi\left(\frac{\text{maximum energy stored}}{\text{energy dissipated per cycle}}\right)$$

6. Write the expression for quality factor of series *RLC* circuit.

$$\text{Quality factor at resonance, } Q = \frac{\omega_r L}{R} = \frac{1}{\omega_r CR} = \frac{1}{R}\sqrt{\frac{L}{C}}$$

7. Define bandwidth of series *RLC* circuit.

The bandwidth of a series circuit is defined as the range of frequencies for which the current is greater than or equal to 70.7% of maximum current or 70.7% of current at resonant frequency.

8. What are half-power frequencies?

In *RLC* circuits, the frequencies at which the power is half the maximum/minimum power are called half-power frequencies.

9. Write the expression for half-power frequencies of *RLC* series circuit.

$$\text{Lower cut-off frequency, } f_1 = \frac{1}{2\pi}\left[-\frac{R}{2L} \pm \sqrt{\left(\frac{R}{2L}\right)^2 + \frac{1}{LC}}\right]$$

$$\text{Upper cut-off frequency, } f_2 = \frac{1}{2\pi}\left[\frac{R}{2L} \pm \sqrt{\left(\frac{R}{2L}\right)^2 + \frac{1}{LC}}\right]$$

10. Write the expressions for bandwidth of *RLC* series circuit.

$$\text{Bandwidth,} \quad \beta = \frac{R}{2\pi L} \quad \text{or} \quad \frac{f_r}{Q_r}$$

11. What are the effects of series resonance?

(i) Since at resonance $X_L = X_C$, the circuit will behave like a pure resistive circuit. Therefore, the impedance of the circuit is low, and is equal to the resistance of the circuit, i.e.

$$Z_r = R$$

(ii) Since the impedance is minimum, current is maximum, i.e.

$$I_r = \frac{V}{Z_r} = \frac{V}{R}$$

(iii) Since the current is maximum at resonance, power taken by the circuit is also maximum, i.e. $P_r = I_r^2 R = \dfrac{V^2}{R}$

(iv) Since at series resonance the current is large, the voltage drops across L and C are also very large. In fact, these drops are much greater than the applied voltage. However, voltage drop across L–C combination as a whole will be zero because these drops are equal in magnitude but 180° out of phase with each other.

12. Draw the frequency response of *RLC* parallel circuit.

The variation of current with frequency is called frequency response, which is shown in Fig. S10.2.

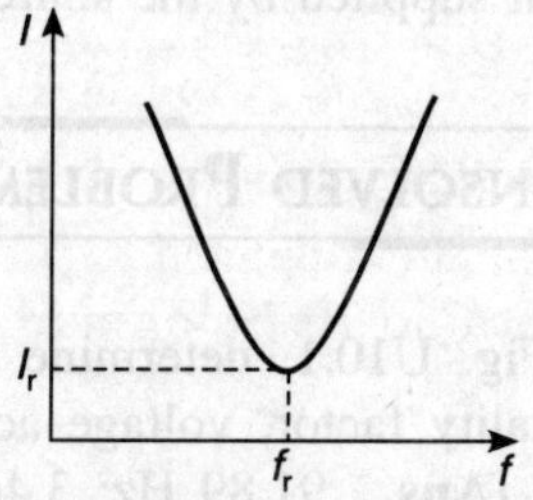

Fig. S10.2 Resonance curve for *RLC* parallel circuit.

13. Write the expression for quality factor of parallel *RLC* circuit.

Quality factor at resonance, $Q = \dfrac{\omega_r L}{R} = \dfrac{1}{\omega_r CR} = \dfrac{1}{R}\sqrt{\dfrac{L}{C}}$

14. Write the expressions for bandwidth of *RLC* parallel circuit.

Bandwidth, $\beta = \dfrac{1}{2\pi RC}$ or $\dfrac{f_r}{Q_r}$

15. Write the expression for half-power frequencies of *RLC* series circuit.

Lower cut-off frequency, $f_1 = \dfrac{1}{2\pi}\left[-\dfrac{1}{2RC} \pm \sqrt{\left(\dfrac{1}{2RC}\right)^2 + \dfrac{1}{LC}}\right]$

Upper cut-off frequency, $f_2 = \dfrac{1}{2\pi}\left[\dfrac{1}{2RC} \pm \sqrt{\left(\dfrac{1}{2RC}\right)^2 + \dfrac{1}{LC}}\right]$

16. What is dynamic resistance?

The impedance offered by the parallel circuit at resonance is called as *dynamic impedance* of the circuit.

$$Z_r = \frac{L}{CR}$$

17. What are the effects of parallel resonance?

(i) At resonance $X_L = X_C$, and the impedance of the circuit is high which is equal to L/CR.

(ii) At resonance, admittance is minimum and equal to conductance; therefore, current is minimum, i.e.

$$I_r = \frac{V}{Z_r} = \frac{VCR}{L}$$

(iii) The circuit behaves as inductive circuit for frequencies below resonant frequency and vice versa.

(iv) At resonance, the magnitude of current through inductance and capacitance will be Q times the current supplied by the source, but they are in phase opposition.

UNSOLVED PROBLEMS

1. For the circuit shown in Fig. U10.1, determine the frequency at which the circuit resonates. Also find the quality factor, voltage across inductance and voltage across capacitance at resonance. **[Ans.** 91.89 Hz; 3.4641; 69.28∠90° V; 69.28∠–90° V]

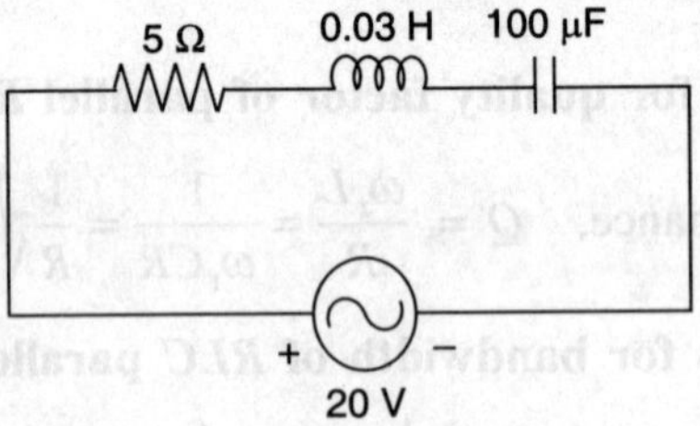

Fig. U10.1

2. An *RLC* series circuit consists of $R = 50\ \Omega$, $L = 0.16$ H and $C = 4\ \mu$F. Calculate the resonant frequency, quality factor, bandwidth and half-power frequencies.

[Ans. 198.9437 Hz; 4; 49.7359 Hz; 175.6239 Hz; 225.3599 Hz]

3. A series *RLC* circuit has an impedance of 40 Ω at a frequency of 200 rad/s. When the circuit is excited by a 10 V source of variable frequency, the circuit resonates at a frequency of 250 rad/s. The current at resonance is 0.5 A and the quality factor at resonance is 10. Determine the circuit parameters.

[Ans. $R = 20\ \Omega$; $L = 0.8$ H; $C = 40\ \mu$F]

4. A series *RLC* circuit is to be designed to produce a magnification of 10 at 100 rad/s. The 100 V source connected to *RLC* series circuit can supply a maximum current of 10 A. The half-power frequency impedance of the circuit should not be more than 14.14 Ω. Find the values of *R*, *L* and *C*. **[Ans.** $R = 10\ \Omega$; $L = 1$ H; $C = 100\ \mu$F]

5. The elements of an *RLC* series circuit are $R = 10\ \Omega$, $L = 0.04$ H and $C = 22\ \mu$F. When the circuit is excited by a variable frequency source of 100 V, (i) find the voltage across inductance and capacitance at resonance, (ii) determine the frequencies at which the voltages across L and C are maximum and (iii) find the maximum voltages across L and C.

[Ans. (i) $426.4\angle 90°$ V; $426.4\angle -90°$ V (ii) 1080.97 rad/s; 1051.24 rad/s (iii) $429.3656\angle 83.2°$ V; $429.3656\angle -83.2°$ V]

6. The parameters of an *RLC* parallel circuit excited by a current source are $R = 40\ \Omega$, $L = 2$ mH and $C = 3\ \mu$F. Determine the resonant frequency, quality factor, bandwidth and cut-off frequencies.

[Ans. 2.0547 kHz; 1.5492; 1.3263 kHz; 2.8222 kHz; 1.4959 kHz]

7. A coil of inductance 31.8 mH and resistance 10 Ω is connected in parallel with a capacitor across a 250 V, 50 Hz supply. Determine the value of capacitance if no reactive component is taken from the supply. **[Ans.** 159.15 μF]

8. In the *RLC* network shown in Fig. U10.2, determine the value of R_C for resonance. Find also the dynamic resistance. **[Ans.** 1.7857 Ω; 32.4676 Ω]

9. Determine the value of R_L for resonance in the network shown in Fig. U10.3.

[Ans. 24.4949 Ω]

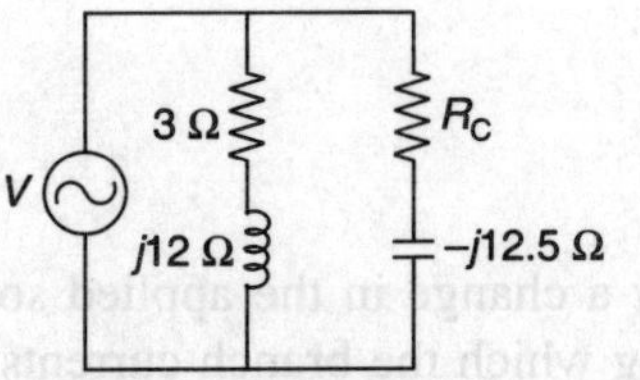

Fig. U10.2

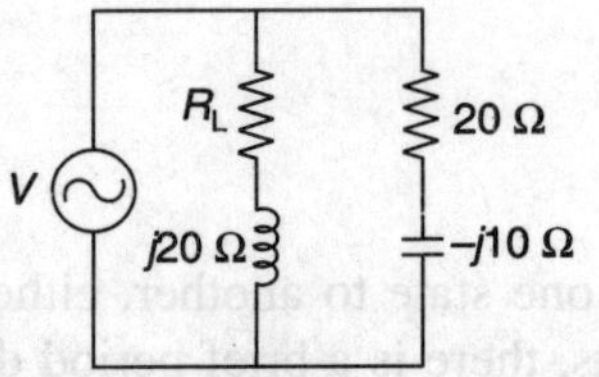

Fig. U10.3

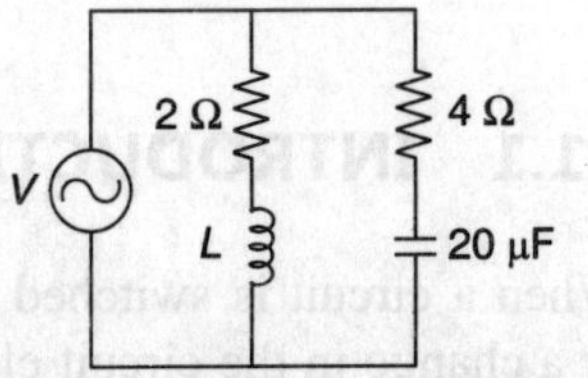

Fig. U10.4

10. In the *RLC* network shown in Fig. U10.4, determine the value of L for the network to resonate at 2000 rad/s. **[Ans.** 3.3693 mH (or) 74.2 μH]

CHAPTER 11

DC Transients

11.1 INTRODUCTION

When a circuit is switched from one state to another, either by a change in the applied source or a change in the circuit elements, there is a brief period during which the branch currents and voltage drops change from their previous values to new values. This brief period is called the *transient*. The value of current/voltage during the transient period is called *transient response*. After the transient period is passed, and if the circuit conditions are not altered, the currents and voltages settle to permanent values and the circuit now reaches a state called *steady state*. Transients occur in circuits which have energy storage elements such as inductance and/or capacitance and not in purely resistive circuits. In purely resistive circuits, the transition occurs in zero time.

11.2 FREE RESPONSE AND FORCED RESPONSE

The application of Kirchhoff's laws to circuit containing energy storage elements results in differential equation with constant coefficients. The solution of this equation represents the response of the circuit. In a circuit, which is independent of sources and if the response is entirely due to the internal energy stored in the elements, it is called *free response*, since the response is not forced by or driven by any external source. On the other hand, if the response is due to the application of external energy source, then it is called *forced response*.

In general, the solution of differential equation, i.e. the response, consists of two parts and they are complementary function and particular integral. The complementary function is the transient part of the solution and is also called *free response* or *transient response*. The particular integral depends on both the source and network elements and so it is called *forced response*.

11.3 INITIAL CONDITIONS

The free response depends on the initial conditions of the circuit elements. Therefore, it is necessary to know the initial conditions to evaluate the arbitrary constants of the differential equations. The number of initial conditions required is equal to the order of the differential equation.

In all transient studies, the switching instant is taken as time origin, i.e. $t = 0$. Also, two more time instants are considered such as $t = 0^-$ and $t = 0^+$. Time $t = 0^-$ is the instant just before switching takes place. And time $t = 0^+$ is the instant immediately after switching. These two moments are considered because current and/or voltage in certain elements may change instantaneously but in certain elements cannot change instantaneously.

11.3.1 Initial Conditions for Resistor

By Ohm's law, the relation between the applied voltage and the current through a resistor is given by

$$v = iR \text{ volts}$$

Above equation is linear and it shows that voltage across the resistor changes instantaneously if the current through it changes instantaneously and vice versa. The change happens instantaneously because there is no energy stored in resistor.

Also, a resistance allows sudden change in current and voltage. Therefore,

$$i_R(0^+) \neq i_R(0^-)$$

and

$$v_R(0^+) \neq v_R(0^-)$$

11.3.2 Initial Conditions for Inductor

In an inductor, the voltage developed across it due to a current flow through it is given by

$$v_L = \frac{di_L}{dt} \text{ volts}$$

If a DC current is made to pass through an inductor, the part di_L/dt in the above equation becomes zero. Hence, the voltage across the inductor is zero and it is equivalent to a short-circuit.

Also, an inductor does not allow sudden change in current but it allows sudden change in voltage. Hence

$$i_L(0^+) = i_L(0^-)$$

and

$$v_R(0^+) \neq v_R(0^-)$$

11.3.3 Initial Conditions for Capacitor

In a capacitor, the relation between the current through it and the voltage across it is given by

$$i_C = \frac{dv_C}{dt} \text{ amperes}$$

If a DC voltage is applied across the capacitor, the part dv_C/dt in the above equation becomes zero. Hence, the current through the capacitor is zero and it is equivalent to an open-circuit.

Also, a capacitor does not allow sudden change in voltage but it allows sudden change in current. Therefore,

$$i_C(0^+) \neq i_C(0^-)$$

and

$$v_C(0^+) = v_C(0^-)$$

Table 11.1 shows the equivalent circuit for three elements at $t = 0^+$ and $t = \infty$.

Table 11.1 Equivalent circuit of three basic electrical elements

Element	*Symbol*	*Behaviour immediately after excitation is given (i.e. at t = 0⁺)*	*Behaviour at steady state (i.e. at t = ∞)*
Resistor			
Inductor with no initial current		open-circuit 	short-circuit 
Capacitor with no initial voltage	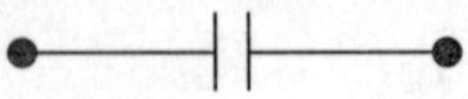	short-circuit 	open-circuit 
Inductor with initial current, I_0	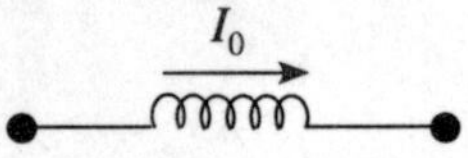	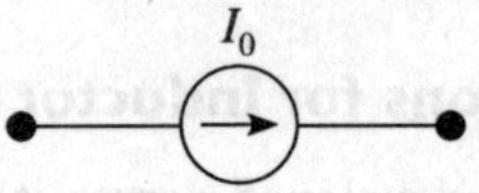	short-circuit 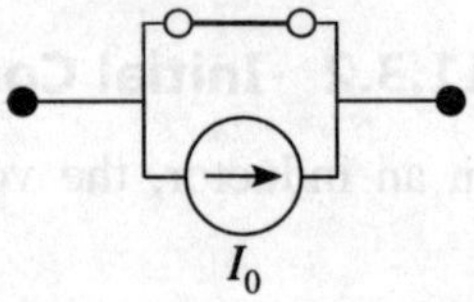
Capacitor with initial voltage, V_0			open-circuit 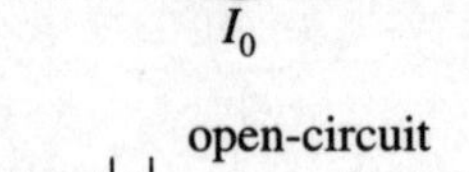

The following section describes the response of *RL* circuit, *RC* circuit and *RLC* circuit when they are subjected to DC excitation and the solution is attempted by both classical and Laplace transformation methods.

11.4 *RL* TRANSIENTS

11.4.1 Current Rise in *RL* Series Circuit (by Classical Method)

Let switch *S* be closed at $t = 0$ so that the series *RL* circuit is excited by the DC voltage *V* as shown in Fig. 11.1(a). Assume that the current through the inductor before closing the switch is zero. At $t = 0^+$, inductor acts as an open-circuit and hence $i(0^-) = i(0^+) = 0$. The equivalent circuit at $t = 0^+$ is shown in Fig. 11.1(b).

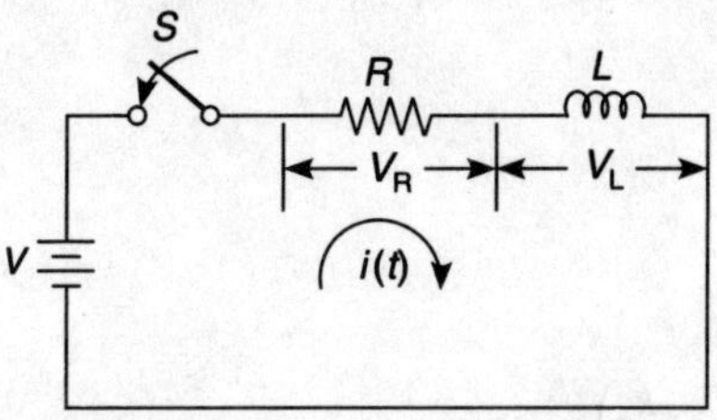

Fig. 11.1(a) *RL* series circuit.

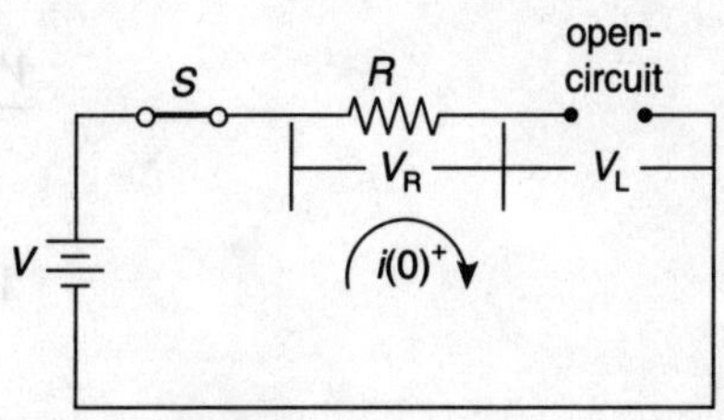

Fig. 11.1(b) Equivalent circuit of Fig. 11.1(a) at $t = 0^+$.

Applying KVL to the circuit in Fig. 11.1(a), after t seconds of closing switch S, we get

$$Ri + L\frac{di}{dt} = V \tag{11.1}$$

$$L\frac{di}{dt} = V - Ri$$

$$\frac{di}{dt} = \frac{V - Ri}{L}$$

Separating the variables,

$$\frac{di}{V - Ri} = \frac{dt}{L}$$

Multiplying by '$-R$' on both sides, we get

$$\left(\frac{-R}{V - Ri}\right)di = \left(\frac{-R}{L}\right)dt$$

Integrating both sides, we get

$$\int\left(\frac{-R}{V - Ri}\right)di = \int\left(\frac{-R}{L}\right)dt$$

$$\ln(V - Ri) = \frac{-R}{L}t + \mathrm{K} \tag{11.2}$$

where K is a constant of integration and is evaluated by applying initial conditions.

The initial condition is that at $t = 0$, $i = 0$.
Substituting this into Eq. (11.1), we get

$$\ln V = \mathrm{K}$$

Therefore, Eq. (11.2) becomes

$$\ln(V - Ri) = \frac{-R}{L}t + \ln V$$

$$\ln(V - Ri) - \ln V = \frac{-R}{L}t$$

$$\ln\left(\frac{V-Ri}{V}\right)=\frac{-R}{L}t$$

$$\frac{V-Ri}{V}=e^{\frac{-R}{L}t}$$

$$1-\frac{Ri}{V}=e^{\frac{-R}{L}t}$$

$$i=\frac{V}{R}\left(1-e^{\frac{-R}{L}t}\right) \tag{11.3}$$

In Eq. (11.3), terms $\frac{V}{R}$ and $\frac{V}{R}e^{\frac{-R}{L}t}$ are respectively the steady state part and the transient part of the solution of current. Equation (11.3) can also be written as

$$i(t)=\frac{V}{R}\left(1-e^{-\frac{t}{\tau}}\right) \tag{11.4}$$

where $\tau = L/R$ is known as *time constant* of the circuit. It is defined as the time required for the current to reach 0.632 times of its final value.

Let us examine the values of current $i(t)$ for different values of time t.

At $t = 0^+$, $\quad i(0^+) = \frac{V}{R}(1-e^0)=\frac{V}{R}(1-1) = 0$

At $t = 1\tau$, $\quad i(\tau) = \frac{V}{R}\left(1-e^{-\frac{\tau}{\tau}}\right)=\frac{V}{R}(1-e^{-1})=0.632\frac{V}{R}$

At $t = \infty$, $\quad i(\infty) = \frac{V}{R}(1-e^{-\infty})=\frac{V}{R}(1-0)=\frac{V}{R}$

The variation of current $i(t)$ with respect to time is plotted in Fig. 11.2.

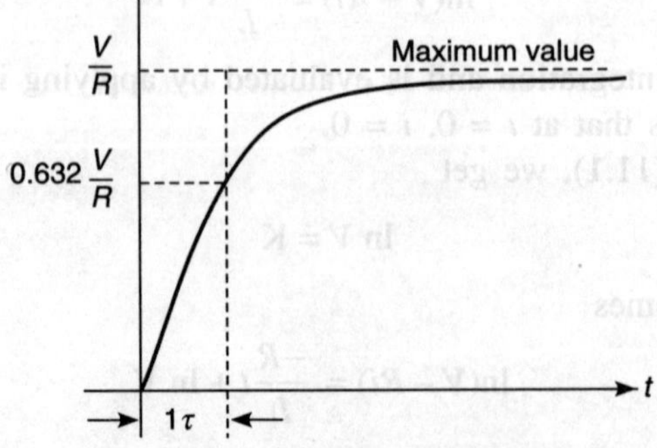

Fig. 11.2 Current rise in *RL* circuit.

Referring to Fig. 11.2, initially the current rises fast and at later stage the rate of rise reduces and the curve becomes almost flat. The current reaches its maximum value when t becomes infinity.

The voltage across resistor v_R is given by

$$v_R = iR$$

and the voltage across inductor v_L is given by

$$v_L = L\frac{di}{dt}$$

Substituting the value of i from Eq. (11.3) into the above equations, we get

$$v_R = V\left(1 - e^{\frac{-R}{L}t}\right) \quad (11.5)$$

$$v_L = L\frac{d}{dt}\left[\frac{V}{R}\left(1 - e^{\frac{-R}{L}t}\right)\right]$$

$$= L\frac{V}{R}\left[0 - \left(\frac{-R}{L}\right)\left(e^{\frac{-R}{L}t}\right)\right]$$

$$v_L = Ve^{\frac{-R}{L}t} \text{ volts} \quad (11.6)$$

The variation of v_R and v_L with respect to time is shown in Fig. 11.3.

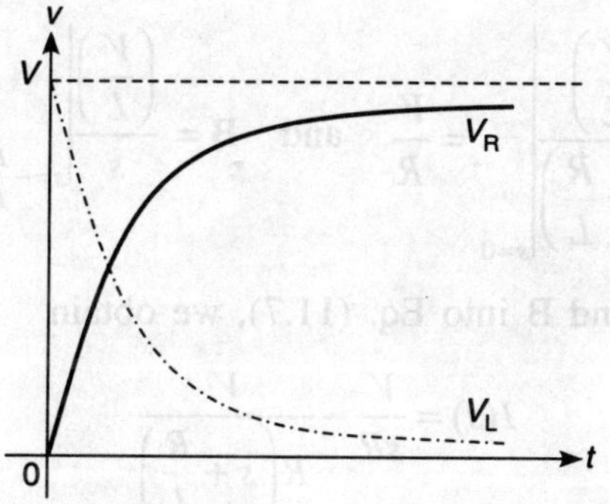

Fig. 11.3 Variation of v_R and v_L with respect to time t.

11.4.2 Current Rise in *RL* Series Circuit (by Laplace Transformation Method)

In view of Eq. (11.1),

$$Ri + L\frac{di}{dt} = V$$

Taking Laplace transformation on both sides, we get

$$RI(s) + L[sI(s) - i(0)] = \frac{V}{s}$$

As the initial current $i0^+$ is assumed zero,

$$RI(s) + L[sI(s)] = \frac{V}{s}$$

$$I(s)[R + sL] = \frac{V}{s}$$

Hence, $$I(s) = \frac{V}{s(sL + R)}$$

$$= \frac{V}{sL\left(s + \frac{R}{L}\right)} = \frac{\left(\frac{V}{L}\right)}{s\left(s + \frac{R}{L}\right)}$$

Applying partial fraction technique,

$$I(s) = \frac{\left(\frac{V}{L}\right)}{s\left(s + \frac{R}{L}\right)} = \frac{\text{A}}{s} + \frac{\text{B}}{\left(s + \frac{R}{L}\right)} \tag{11.7}$$

$$\frac{V}{L} = \text{A}\left(s + \frac{R}{L}\right) + \text{B}s$$

Solving for constants A and B,

$$\text{A} = \left.\frac{\left(\frac{V}{L}\right)}{\left(s + \frac{R}{L}\right)}\right|_{s=0} = \frac{V}{R} \quad \text{and} \quad \text{B} = \left.\frac{\left(\frac{V}{L}\right)}{s}\right|_{s=-\frac{R}{L}} = -\frac{V}{R}$$

Substituting the values of A and B into Eq. (11.7), we obtain

$$I(s) = \frac{V}{sR} - \frac{V}{R\left(s + \frac{R}{L}\right)}$$

$$I(s) = \frac{V}{R}\left(\frac{1}{s} - \frac{1}{\left(s + \frac{R}{L}\right)}\right)$$

Taking inverse Laplace transformation on both sides, we get

$$i(t) = \frac{V}{R}\left(1 - e^{-\frac{R}{L}t}\right)$$

as obtained earlier in Eq. (11.2). The above equation can also be written as:

$$i(t) = \frac{V}{R}\left(1 - e^{-\frac{t}{\tau}}\right) \tag{11.8}$$

where $\tau = L/R$ is known as *time constant* of the circuit as explained in Eq. (11.4).

11.4.3 Current Decay in *RL* Series Circuit (by Classical Method)

Assume that switch S in the circuit shown in Fig. 11.4 is kept at position 1 for sufficiently longer period so that the current reaches its steady-state value which is given by $I = V/R$. After this instant the switch is thrown to position 2. Let us take this time instant as $t' = 0$.

Applying KVL at this instant (i.e. moving the switch from position 1 to position 2),

$$Ri + L\frac{di}{dt'} = 0 \tag{11.9}$$

$$\frac{di}{dt'} = -\frac{Ri}{L}$$

Separating the variables,

$$\frac{di}{i} = \frac{-R}{L}dt'$$

Integrating both sides, we get

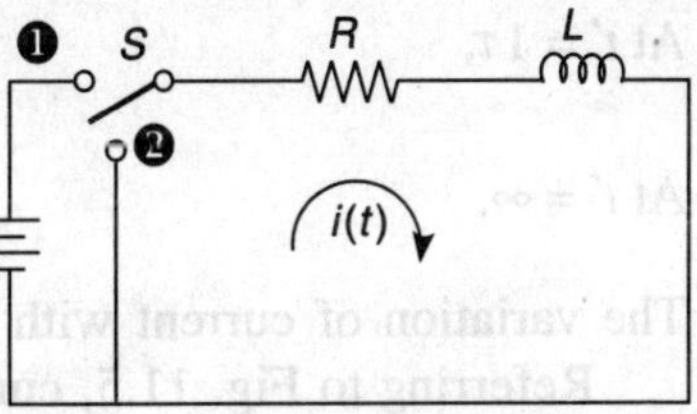

Fig. 11.4 *RL* series circuit.

$$\int\frac{di}{i} = \int\frac{-R}{L}dt'$$

$$\ln i = \frac{-R}{L}t' + K' \tag{11.10}$$

The constant of integration K′ is evaluated by applying initial conditions.

The initial condition is that at $t' = 0$, $i = I = \frac{V}{R}$.

Substituting $t' = 0$ and $i = I$ into Eq. (11.10), we obtain

$$\ln I = 0 + K' \tag{11.11}$$

Substituting the value of K′ obtained in Eq. (11.11) into (11.10), we get

$$\ln i = \frac{-R}{L}t' + \ln I$$

$$\ln i - \ln I = \frac{-R}{L}t'$$

$$\ln\left(\frac{i}{I}\right) = \frac{-R}{L}t'$$

$$\left(\frac{i}{I}\right) = e^{\frac{-R}{L}t'}$$

$$i = Ie^{\frac{-R}{L}t'}$$

$$i = \frac{V}{R}e^{\frac{-R}{L}t'} \qquad (11.12)$$

or

$$i(t') = \frac{V}{R}e^{\frac{-t'}{\tau}}$$

where τ is equal to L/R.

Let us examine the values of current $i(t')$ for different values of time t'.

At $t' = 0^+$, $$i(0^+) = \frac{V}{R}e^0 = \frac{V}{R}$$

At $t' = 1\tau$, $$i(\tau) = \frac{V}{R}e^{\frac{-\tau}{\tau}} = \frac{V}{R}e^{-1} = 0.368\frac{V}{R}$$

At $t' = \infty$, $$i(\infty) = \frac{V}{R}e^{-\infty} = 0$$

The variation of current with respect to time t' is drawn in Fig. 11.5.

Referring to Fig. 11.5, current i decays exponentially from V/R to zero as t' increases from zero to infinity.

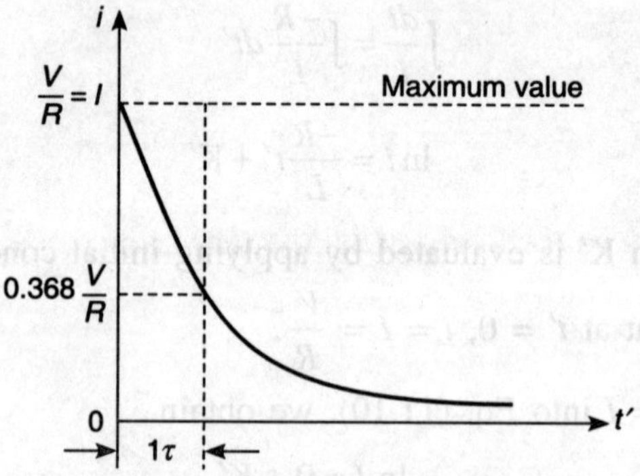

Fig. 11.5 Decaying current in *RL* circuit.

The voltage across resistor v_R is given by,

$$v_R = iR$$

Substituting the value of i from Eq. (11.12),

$$v_R = Ve^{\frac{-R}{L}t'} \qquad (11.13)$$

and the voltage across inductor v_L is given by

$$v_L = L\frac{di}{dt'}$$

Again, substituting the value of i from Eq. (11.12) into the above, we get

$$v_L = L\frac{d}{dt'}\left(\frac{V}{R}e^{\frac{-R}{L}t'}\right)$$

$$= L\frac{V}{R}\left(\frac{-R}{L}\right)\left(e^{\frac{-R}{L}t'}\right)$$

$$= -Ve^{\frac{-R}{L}t'} \text{ volts} \tag{11.14}$$

The variation of v_R and v_L against time t' is shown in Fig. 11.6.

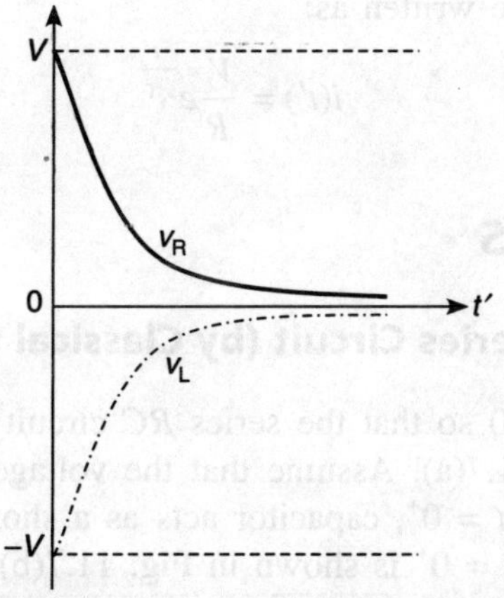

Fig. 11.6 Variation of v_R and v_L with respect to time t'.

11.4.4 Current Decay in *RL* Series Circuit (by Laplace Transformation Method)

In view of Eq. (11.9),

$$Ri + L\frac{di}{dt'} = 0$$

Taking Laplace transformation on both sides, we get

$$RI(s) + L[sI(s) - i(0)] = 0 \tag{11.15}$$

Here, $$i(0) = I = \frac{V}{R}$$

Therefore, $$RI(s) + L\left[sI(s) - \frac{V}{R}\right] = 0$$

$$RI(s) + sLI(s) - L\frac{V}{R} = 0$$

$$I(s)[R + sL] = L\frac{V}{R}$$

Hence,
$$I(s) = \frac{L(V/R)}{(sL+R)}$$
$$= \frac{(V/R)}{\left(s+\frac{R}{L}\right)}$$

Taking inverse Laplace transformation on both sides, we get

$$i(t') = \frac{V}{R}e^{-\frac{R}{L}t'} = Ie^{-\frac{R}{L}t'}$$

as obtained earlier in Eq. (11.12).

The above equation can also be written as:

$$i(t') = \frac{V}{R}e^{-\frac{t'}{\tau}} \tag{11.16}$$

11.5 *RC* TRANSIENTS

11.5.1 Charging in *RC* Series Circuit (by Classical Method)

Let switch S be closed at $t = 0$ so that the series RC circuit is excited by the DC voltage V in the circuit shown in Fig. 11.7(a). Assume that the voltage across the capacitor, v_C before closing the switch is zero. At $t = 0^+$, capacitor acts as a shortcircuit and hence $v(0^-) = v(0^+) = 0$. The equivalent circuit at $t = 0^+$ is shown in Fig. 11.7(b).

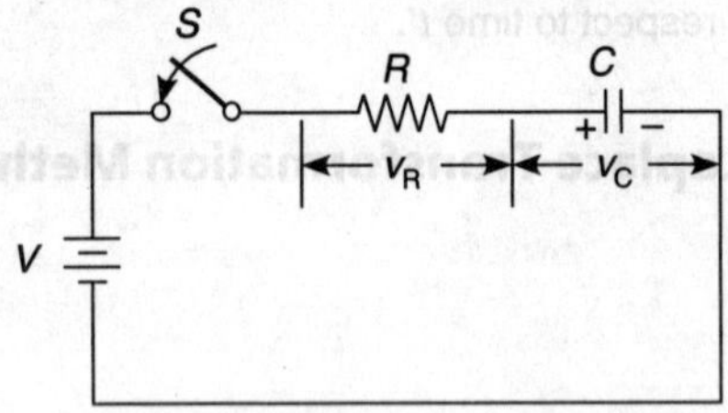

Fig. 11.7(a) *RC* series circuit.

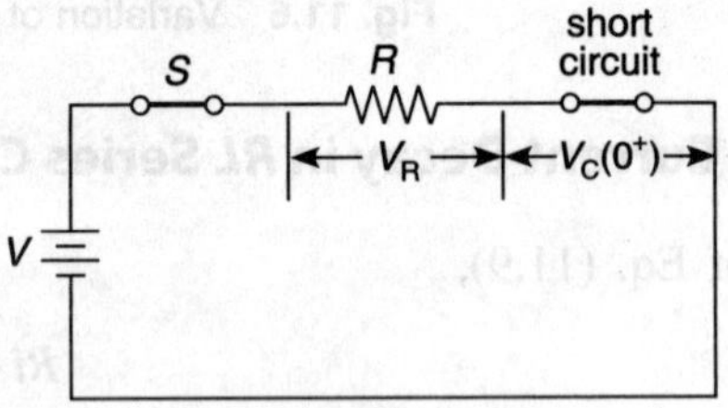

Fig. 11.7(b) Equivalent circuit of Fig. 11.7(a) at $t = 0^+$.

Applying KVL to the circuit in Fig. 11.7(b),

$$v_R + v_C = V$$
$$Ri + v_C = V \tag{11.17}$$

Since $i = C\frac{dv_C}{dt}$, Eq. (11.17) can be written as:

$$RC\frac{dv_C}{dt} + v_C = V$$

$$RC\frac{dv_C}{dt} = V - v_C$$

Separating the variables,

$$\frac{dv_C}{V - v_C} = \frac{dt}{RC}$$

Multiplying by "–1" on both sides, we get

$$\left(\frac{-1}{V - v_C}\right) dv_C = \left(\frac{-1}{RC}\right) dt$$

Integrating both sides, we get

$$\int \left(\frac{-1}{V - v_C}\right) dv_C = \int \left(\frac{-1}{RC}\right) dt$$

$$\ln(V - v_C) = \frac{-1}{RC} t + \mathrm{K} \tag{11.18}$$

where K is a constant of integration and is evaluated by applying initial conditions. The initial condition is that at $t = 0$, $v_C = 0$.

Substituting this into Eq. (11.18), we get

$$\ln(V) = \mathrm{K}$$

Therefore, Eq. (11.18) becomes,

$$\ln(V - v_C) = \frac{-1}{RC} t + \ln(V)$$

$$\ln\left(\frac{V - v_C}{V}\right) = \frac{-t}{RC}$$

$$\left(\frac{V - v_C}{V}\right) = e^{\frac{-t}{RC}}$$

$$V - v_C = V e^{\frac{-t}{RC}}$$

$$v_C = V - V e^{\frac{-t}{RC}}$$

or

$$v_C = V\left(1 - e^{\frac{-t}{RC}}\right) \text{ volts} \tag{11.19}$$

From Eq. (11.19), current $i(t)$ is obtained as:

$$i(t) = CV \frac{d}{dt}\left(1 - e^{\frac{-t}{RC}}\right) \qquad \left[\because \; i(t) = C\frac{dv_C}{dt}\right]$$

$$= CV\left[0 - \left(\frac{-1}{RC}\right) e^{\frac{-t}{RC}}\right]$$

$$i(t) = \frac{V}{R} e^{\frac{-t}{RC}} \tag{11.20}$$

In Eq. (11.20), term *V*/*R* is the initial value of the current and is denoted by *I*. Equation (11.20) can also be written as:

$$i(t) = \frac{V}{R} e^{-\frac{t}{\tau}} \tag{11.21}$$

where $\tau = RC$ is known as *time constant* of the circuit. It is defined as the time required for the current to fall 0.368 times of its initial value.

Let us examine the values of current $i(t)$ for different values of time t.

At $t = 0^+$, $$i(0^+) = \frac{V}{R} e^{-0} = \frac{V}{R} = I$$

At $t = 1\tau$, $$i(\tau) = \frac{V}{R} e^{-\frac{\tau}{\tau}} = \frac{V}{R} e^{-1} = 0.368 \frac{V}{R}$$

At $t = \infty$, $$i(\infty) = \frac{V}{R} e^{-\infty} = \frac{V}{R}(0) = 0$$

The variation of current $i(t)$ with respect to time is plotted in Fig. 11.8.

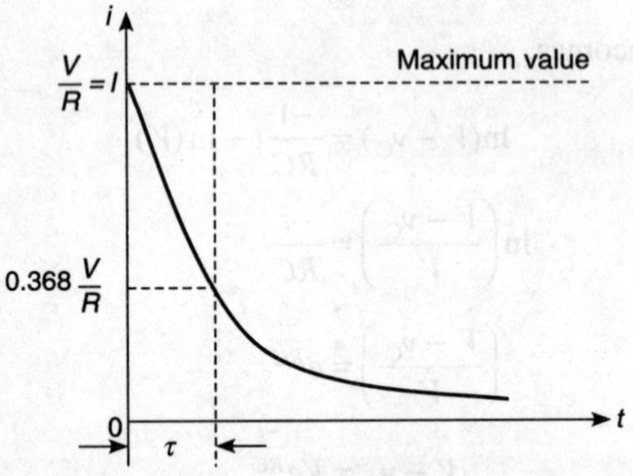

Fig. 11.8 Current decay in *RC* circuit.

Referring to Fig. 11.8, initially the current falls fast and at later stage the rate of fall reduces and the curve becomes almost flat. The current reaches zero value when t becomes infinity as the capacitor behaves like an open-circuit at $t = \infty$.

The voltage across resistor v_R is given by

$$v_R = iR$$
$$= Ve^{\frac{-t}{RC}} \tag{11.22}$$

From Eq. (11.19), the voltage across the capacitor is obtained as:

$$v_C = V\left(1 - e^{\frac{-t}{RC}}\right) \text{ volts} \tag{11.23}$$

The variation of v_R and v_C against time t is shown in Fig. 11.9.

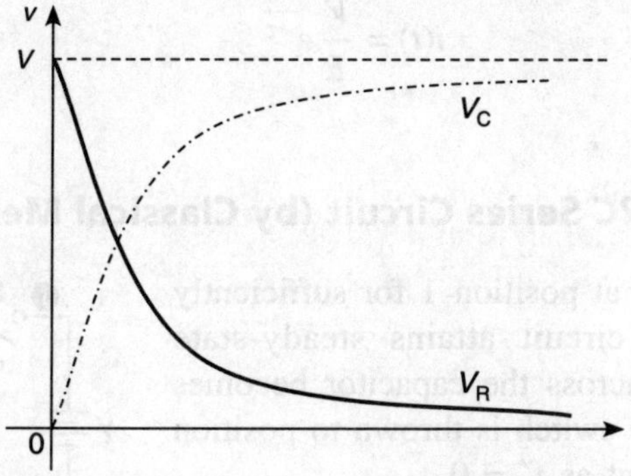

Fig. 11.9 Variation of v_R and v_C with respect to time t.

11.5.2 Charging in *RC* Series Circuit (by Laplace Transformation Method)

In view of Eq. (11.17),

$$Ri + v_C = V$$

$$Ri + \frac{1}{C}\int i dt = V \qquad \left[\because v_C = \frac{1}{C}\int i dt\right]$$

Taking Laplace transformation on both sides, we get

$$RI(s) + \frac{1}{C}\frac{I(s)}{s} = \frac{V}{s} \tag{11.24}$$

$$I(s)\left(R + \frac{1}{sC}\right) = \frac{V}{s}$$

$$I(s) = \frac{\left(\frac{V}{s}\right)}{\left(R + \frac{1}{sC}\right)} = \frac{\left(\frac{V}{s}\right)}{\left(\frac{sRC+1}{sC}\right)}$$

$$= \frac{VC}{(sRC+1)} = \frac{VC}{RC\left(s + \frac{1}{RC}\right)}$$

$$= \frac{V}{R\left(s + \frac{1}{RC}\right)}$$

Taking inverse Laplace transformation on both sides, we get

$$i(t) = \frac{V}{R}e^{-\frac{t}{RC}}$$

as obtained earlier in Eq. (11.20). The above equation can be written as:

$$i(t) = \frac{V}{R} e^{-\frac{t}{\tau}}$$

where $\tau = RC$.

11.5.3 Discharging in *RC* Series Circuit (by Classical Method)

Assume that switch S is kept at position 1 for sufficiently longer period so that the circuit attains steady-state condition. Now the voltage across the capacitor becomes V volts. After this instant, the switch is thrown to position 2. Let us take this time instant as $t' = 0$.

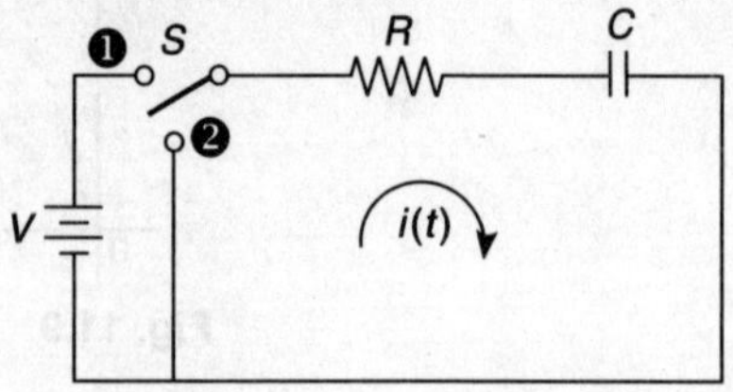

Fig. 11.10 RC series circuit.

Applying KVL at this instant (i.e. moving the switch from position 1 to position 2),

$$Ri + v_C = 0 \tag{11.25}$$

$$RC\frac{dv_C}{dt'} + v_C = 0$$

$$RC\frac{dv_C}{dt'} = -v_C$$

Separating the variables,

$$\frac{dv_C}{v_C} = -\frac{1}{RC}dt'$$

Integrating both sides, we get

$$\int \frac{dv_C}{v_C} = -\frac{1}{RC}dt'$$

$$\ln (v_C) = \frac{-1}{RC}t' + K' \tag{11.26}$$

The constant of integration K′ is evaluated by applying initial conditions. The initial condition is that at $t' = 0$, $v_C = V$

Substituting $t' = 0$ and $v_C = V$ into Eq. (11.26), we obtain

$$\ln (V) = 0 + K' \tag{11.27}$$

Substituting the value of K′ from Eq. (11.27) into Eq. (11.26), we obtain

$$\ln (v_C) = \frac{-1}{RC}t' + \ln (V)$$

$$\ln (v_C) - \ln (V) = \frac{-1}{RC}t'$$

$$\ln\left(\frac{v_C}{V}\right) = \frac{-1}{RC}t'$$

$$\left(\frac{v_C}{V}\right) = e^{\frac{-t'}{RC}}$$

$$v_C = Ve^{\frac{-t'}{RC}} \tag{11.28}$$

Current i is found as follows:

$$i = C\frac{dv_C}{dt'} = C\frac{d}{dt'}\left(Ve^{\frac{-t'}{RC}}\right)$$

$$= CV\left(\frac{-1}{RC}\right)e^{\frac{-t'}{RC}}$$

$$i = -\frac{V}{R}e^{\frac{-t'}{RC}} \tag{11.29}$$

or

$$i = -\frac{V}{R}e^{\frac{-t'}{\tau}} \tag{11.30}$$

where $\tau = RC$.

Let us examine the values of current $i(t')$ for different values of time t'.

At $t' = 0^+$, $\quad i(0^+) = -\frac{V}{R}e^0 = -\frac{V}{R}$

At $t' = 1\tau$, $\quad i(\tau) = -\frac{V}{R}e^{\frac{-\tau}{\tau}} = -\frac{V}{R}e^{-1} = -0.368\frac{V}{R}$

At $t' = \infty$, $\quad i(\infty) = -\frac{V}{R}e^{-\infty} = 0$

The variation of current with respect to time is shown in Fig. 11.11.

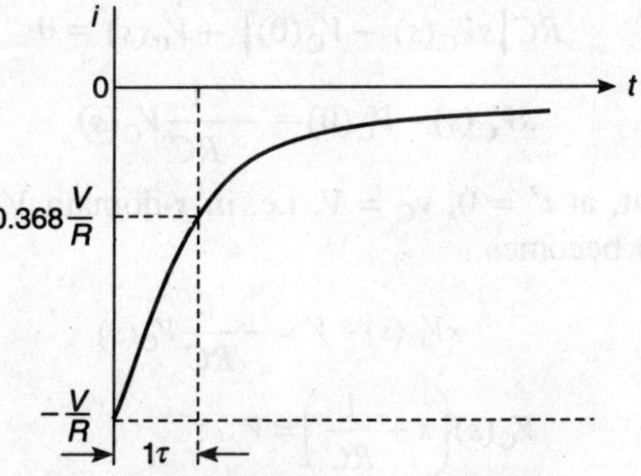

Fig. 11.11 Variation of i with time t'.

Referring to Fig. 11.11, current i rises exponentially from $-V/R$ to zero as t' increases from zero to infinity.

The voltage across resistor v_R is given by

$$v_R = iR$$

Substituting the value of i from Eq. (11.29) into the above, we get

$$v_R = -Ve^{\frac{-t'}{RC}} \tag{11.31}$$

Using Eqs. (11.28) and (11.31), the variation of v_R and v_C against time t' is shown in Fig. 11.12.

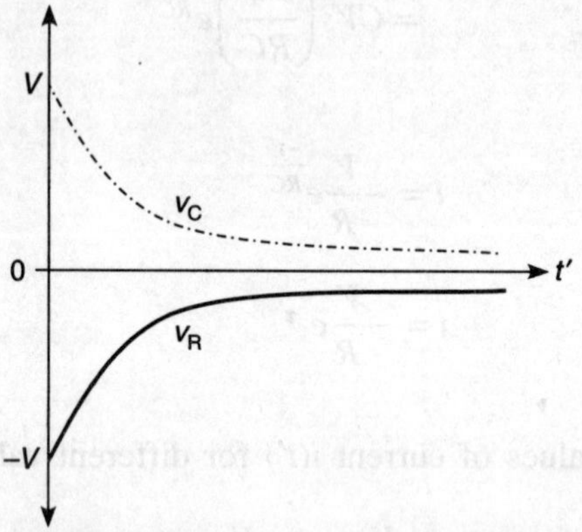

Fig. 11.12 Variation of v_R and v_C with time t'.

11.5.4 Discharging in *RC* Series Circuit (by Laplace Transformation Method)

In view of Eq. (11.25),

$$Ri + v_C = 0$$

$$RC\frac{dv_C}{dt'} + v_C = 0$$

Taking Laplace transformation on both sides, we get

$$RC[sV_C(s) - V_C(0)] + V_C(s) = 0$$

$$sV_C(s) - V_C(0) = -\frac{1}{RC}V_C(s) \tag{11.32}$$

The initial condition is that, at $t' = 0$, $v_C = V$, i.e. in s-domain $V_C(0) = V$.

Therefore, Eq. (11.32) becomes

$$sV_C(s) - V = -\frac{1}{RC}V_C(s)$$

$$V_C(s)\left(s + \frac{1}{RC}\right) = V$$

$$V_C(s) = \frac{V}{[s + (1/RC)]}$$

Taking inverse Laplace transformation on both sides, we get

$$v_C(t) = Ve^{\frac{-t'}{RC}}$$

as obtained earlier in Eq. (11.28).

11.6 *RLC* TRANSIENTS

11.6.1 Response by Classical Method

Let switch S be closed at $t = 0$, so that, the series *RLC* circuit shown in Fig. 11.13 is excited by the DC voltage V.

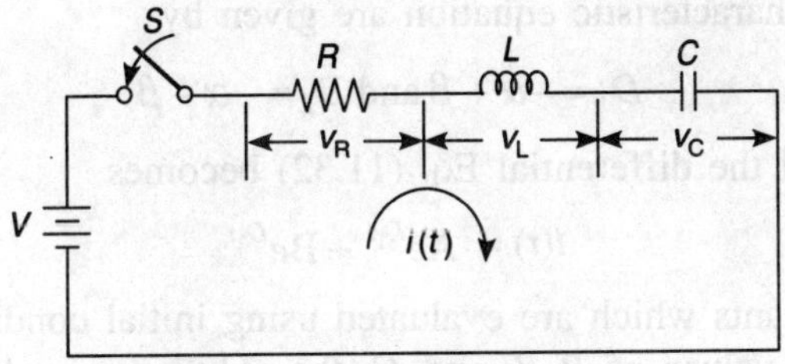

Fig. 11.13 Series *RLC* circuit.

Assuming no initial voltage across the capacitor and applying KVL to the circuit,

$$Ri + L\frac{di}{dt} + \frac{1}{C}\int i\,dt = V \tag{11.33}$$

Differentiating both sides and rearranging,

$$L\frac{d^2i}{dt^2} + R\frac{di}{dt} + \frac{1}{C}i = 0$$

To solve this linear homogeneous second-order differential equation, let

$$\left(LD^2 + RD + \frac{1}{C}\right)i = 0$$

where $D = \dfrac{d}{dt}$ and $D^2 = \dfrac{d^2}{dt^2}$

For the above, the auxiliary equation is written as:

$$LD^2 + RD + \frac{1}{C} = 0$$

$$D^2 + \frac{R}{L}D + \frac{1}{LC} = 0$$

The roots of the above equation are,

$$D = \frac{-\frac{R}{L} \pm \sqrt{\left(\frac{R}{L}\right)^2 - \frac{4}{LC}}}{2}$$

$$= -\frac{R}{2L} \pm \sqrt{\left(\frac{R}{2L}\right)^2 - \frac{1}{LC}}$$

$$= -\alpha + \beta$$

where $\alpha = \frac{R}{2L}$ and $\beta = \sqrt{\left(\frac{R}{2L}\right)^2 - \frac{1}{LC}}$

Hence, the roots of the characteristic equation are given by

$$D_1 = -\alpha + \beta \text{ and } D_2 = -\alpha - \beta \tag{11.34}$$

Therefore, the solution of the differential Eq. (11.32) becomes

$$i(t) = \text{A}e^{D_1 t} + \text{B}e^{D_2 t} \tag{11.35}$$

where A and B are constants which are evaluated using initial conditions.

Depending upon the values of R, L and C, the solution may be one of the three cases discussed below.

Case 1 When $\left(\frac{R}{2L}\right)^2 > \frac{1}{LC}$

In this case, β is positive and real. Hence, roots D_1 and D_2 are real and unequal, i.e.

$$D_1 = -\alpha + \beta \quad \text{and} \quad D_2 = -\alpha - \beta$$

So, Eq. (11.35) becomes

$$i(t) = \text{A}e^{(-\alpha+\beta)t} + \text{B}e^{(-\alpha-\beta)t}$$

$$i(t) = e^{-\alpha t}(\text{A}e^{\beta t} + \text{B}e^{-\beta t}) \tag{11.36}$$

In this case, the current response is over-damped and shown in Fig. 11.14.

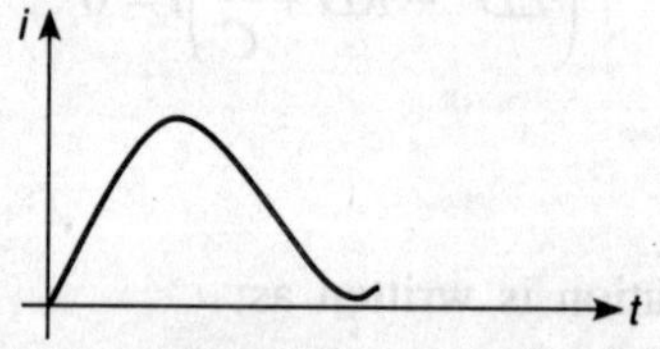

Fig. 11.14 Over-damped response of *RLC* circuit.

Case 2 When $\left(\frac{R}{2L}\right)^2 = \frac{1}{LC}$

In this case, β is zero. Hence, roots D_1 and D_2 are real and equal, i.e.

$$D_1 = D_2 = -\alpha$$

Hence, Eq. (11.35) becomes,

$$i(t) = Ae^{-\alpha t} + Bte^{-\alpha t} \tag{11.37}$$

In this case, the current response is critically damped and is shown in Fig. 11.15.

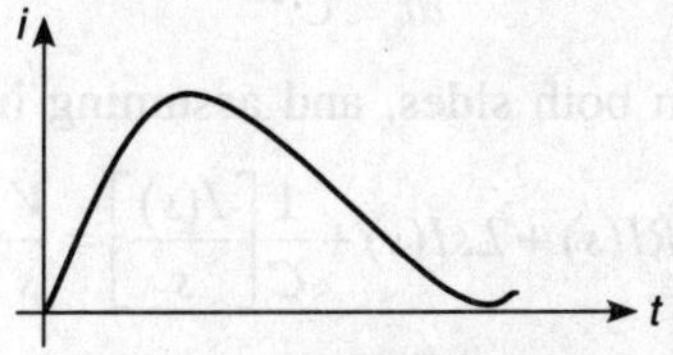

Fig. 11.15 Critically damped response of *RLC* circuit.

Case 3 When $\left(\frac{R}{2L}\right)^2 < \frac{1}{LC}$

In this case, β is imaginary and roots D_1 and D_2 are complex conjugates, i.e.

$$D_1 = -\alpha + j\beta \quad \text{and} \quad D_2 = -\alpha - j\beta$$

Hence, Eq. (11.35) becomes

$$i(t) = Ae^{(-\alpha + j\beta)t} + Be^{(-\alpha - j\beta)t}$$

$$i(t) = Ae^{-\alpha t}e^{j\beta t} + Be^{-\alpha t}e^{-j\beta t}$$

$$= e^{-\alpha t}(Ae^{j\beta t} + Be^{-j\beta t})$$

$$= e^{-\alpha t}[A(\cos\beta t + j\sin\beta t) + B(\cos\beta t - j\sin\beta t)]$$

$$= e^{-\alpha t}[(A + B)\cos\beta t + (A - B)j\sin\beta t]$$

$$= e^{-\alpha t}[C_1\cos\beta t + C_2\sin\beta t] \tag{11.38}$$

where $C_1 = (A + B)$ and $C_2 = j(A - B)$.

The solution has both sinusoidal and exponential components. The sinusoidal oscillations are damped by the exponential term $e^{-\alpha t}$. Hence, current $i(t)$ will have damped oscillations, whose amplitude decays to zero as t tends to infinity. This response is under-damped and is shown in Fig. 11.16.

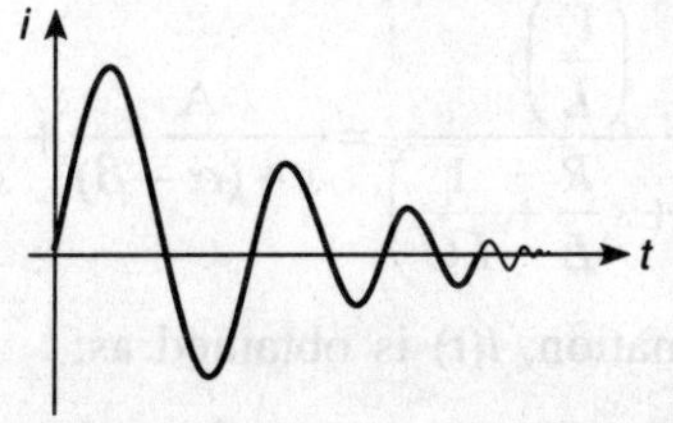

Fig. 11.16 Under-damped response of *RLC* circuit.

11.6.2 Response by Laplace Transformation Method

In view of Eq. (11.33)

$$Ri + L\frac{di}{dt} + \frac{1}{C}\int i\,dt = V$$

Taking Laplace transformation on both sides, and assuming initial conditions are zero, we get

$$RI(s) + LsI(s) + \frac{1}{C}\left[\frac{I(s)}{s}\right] = \frac{V}{s} \tag{11.39}$$

$$I(s)\left[R + sL + \frac{1}{sC}\right] = \frac{V}{s}$$

$$I(s) = \frac{V}{s\left(R + sL + \frac{1}{sC}\right)} = \frac{V}{\left(sR + s^2L + \frac{1}{C}\right)}$$

$$= \frac{V}{L\left(s^2 + s\frac{R}{L} + \frac{1}{LC}\right)} = \frac{\left(\frac{V}{L}\right)}{\left(s^2 + s\frac{R}{L} + \frac{1}{LC}\right)}$$

The roots of the denominator are

$$-\frac{R}{2L} \pm \sqrt{\left(\frac{R}{2L}\right)^2 - \frac{1}{LC}} = -\alpha \pm \beta$$

where $\qquad \alpha = \frac{R}{2L}$ and $\beta = \sqrt{\left(\frac{R}{2L}\right)^2 - \frac{1}{LC}}$

As before, depending upon the values of R, L and C, the solution may be one of the three cases.

Case 1 When $\left(\frac{R}{2L}\right)^2 > \frac{1}{LC}$

The roots are real and unequal. Hence,

$$I(s) = \frac{\left(\frac{V}{L}\right)}{\left(s^2 + s\frac{R}{L} + \frac{1}{LC}\right)} = \frac{\text{A}}{s + (\alpha - \beta)} + \frac{\text{B}}{s + (\alpha + \beta)}$$

Taking inverse Laplace transformation, $i(t)$ is obtained as:

$$i(t) = \text{A}e^{-(\alpha-\beta)t} + \text{B}e^{-(\alpha+\beta)t}$$

$$= e^{-\alpha t}(\text{A}e^{\beta t} + \text{B}e^{-\beta t})$$

In this case, the current response is over-damped and shown in Fig. 11.14.

Case 2 When $\left(\dfrac{R}{2L}\right)^2 = \dfrac{1}{LC}$

The roots are real and equal. Therefore,

$$i(t) = \frac{\left(\frac{V}{L}\right)}{\left(s^2 + s\frac{R}{L} + \frac{1}{LC}\right)} = \frac{\left(\frac{V}{L}\right)}{(s+\alpha)^2}$$

$$= \frac{\text{A}}{(s+\alpha)} + \frac{\text{B}}{(s+\alpha)^2}$$

Taking inverse Laplace transformation, $i(t)$ is obtained as:

$$\begin{aligned} i(t) &= \text{A}e^{-\alpha t} + \text{B}te^{-\alpha t} \\ &= e^{-\alpha t}(\text{A} + \text{B}t) \end{aligned}$$

In this case, the current response is critically damped and is shown in Fig. 11.15.

Case 3 When $\left(\dfrac{R}{2L}\right)^2 < \dfrac{1}{LC}$

Therefore, β is imaginary and the roots are $(-\alpha + j\beta)$ and $(-\alpha - j\beta)$.

So,
$$i(t) = \frac{\left(\frac{V}{L}\right)}{\left(s^2 + s\frac{R}{L} + \frac{1}{LC}\right)} = \frac{\text{A}}{s+(\alpha - j\beta)} + \frac{\text{B}}{s+(\alpha + j\beta)}$$

Taking inverse Laplace transformation on both sides,

$$\begin{aligned} i(t) &= \text{A}e^{-(\alpha - j\beta)t} + \text{B}^{-(\alpha + j\beta)t} \\ &= \text{A}e^{-\alpha t}e^{j\beta t} + \text{B}e^{-\alpha t}e^{-j\beta t} \\ &= e^{-\alpha t}(\text{A}e^{j\beta t} + \text{B}e^{-j\beta t}) \\ &= e^{-\alpha t}[\text{A}(\cos\beta t + j\sin\beta t) + \text{B}(\cos\beta t - j\sin\beta t)] \\ &= e^{-\alpha t}[(\text{A}+\text{B})\cos\beta t + (\text{A}-\text{B})j\sin\beta t] \\ &= e^{-\alpha t}[C_1\cos\beta t + C_2\sin\beta t] \end{aligned}$$

where C_1 = (A + B) and C_2 = j(A – B). The solution is oscillatory and at the same time the current decays in a short time (see Fig. 11.16).

EXAMPLE 11.1 A series *RL* circuit with R = 100 Ω and L = 20 H has a DC voltage of 200 V applied through a switch at t = 0 as shown in Fig. 11.17. Find the equation for the current.

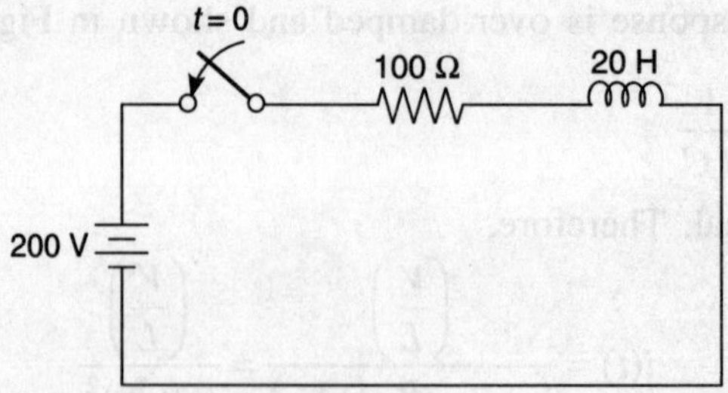

Fig. 11.17

Solution *By Classical method*

The differential equation for the circuit is

$$100i + 20\frac{di}{dt} = 200$$

$$20\frac{di}{dt} = 200 - 100i$$

Separating the variables,

$$\frac{di}{200 - 100i} = \frac{dt}{20}$$

Multiplying by '–100' on both sides,

$$\frac{-100\,di}{200 - 100\,i} = \frac{-100\,dt}{20}$$

Integrating both sides,

$$\int \frac{-100}{200 - 100i} di = \int \frac{-100}{20} dt$$

$$\ln(200 - 100i) = \frac{-100}{20} t + \mathrm{K} \quad \text{(i)}$$

The initial condition is that at $t = 0$, $i = 0$

Hence, $\ln 200 = \mathrm{K}$

Substituting in Eq. (i), we obtain

$$\ln(200 - 100i) = \frac{-100}{20} t + \ln 200$$

$$\ln\left(\frac{200 - 100i}{200}\right) = -5t$$

$$\left(\frac{200 - 100i}{200}\right) = e^{-5t}$$

$$1 - 0.5i = e^{-5t}$$

$$0.5i = 1 - e^{-5t}$$

Therefore, the equation for the current, $\boldsymbol{i(t) = 2(1 - e^{-5t})}$ **A**

By Laplace transformation method

The differential equation for the circuit is

$$100i + 20\frac{di}{dt} = 200 \quad \text{(ii)}$$

Taking Laplace transformation on both sides, we get

$$100I(s) + 20[sI(s) - I(0)] = \frac{200}{s} \quad \text{(iii)}$$

$I(0) = 0$, therefore, Eq. (iii) becomes

$$100I(s) + 20sI(s) = \frac{200}{s}$$

$$I(s)[100 + 20s] = \frac{200}{s}$$

$$I(s) = \frac{200}{s(20s + 100)}$$

$$= \frac{10}{s(s+5)}$$

By applying partial fraction technique,

$$\frac{10}{s(s+5)} = \frac{A}{s} + \frac{B}{(s+5)} \quad \text{(iv)}$$

Solving for constants A and B,

$$A = \left.\frac{10}{(s+5)}\right|_{s=0} = 2 \quad \text{and} \quad B = \left.\frac{10}{s}\right|_{s=-5} = -2$$

Substituting the values of A and B into Eq. (iv), we get

$$I(s) = \frac{2}{s} - \frac{2}{(s+5)}$$

Taking inverse Laplace transformation on both sides, we get

$$i(t) = 2 - 2e^{-5t}$$

or

$$\boldsymbol{i(t) = 2(1 - e^{-5t})\,\text{A}}$$

EXAMPLE 11.2 In the circuit shown in Fig. 11.18, find the equation for the current and sketch it. Also find the initial rate of growth of current.

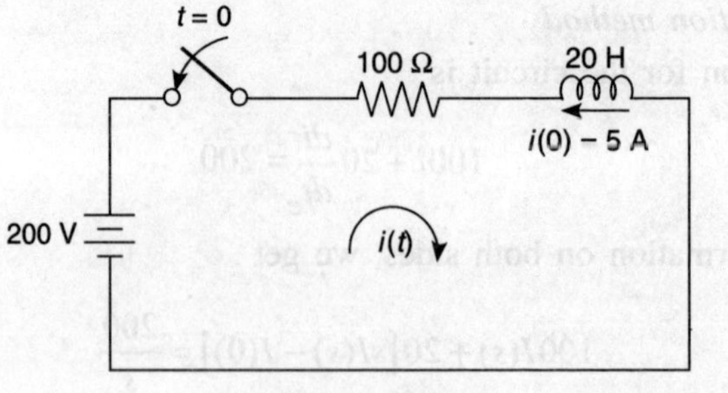

Fig. 11.18

Solution Determination of initial conditions

Since the initial current flows through the inductor assumes opposite direction to the circuit current $i(t)$, $i(0) = I(0) = -5$ A.

By Classical method

The differential equation for the circuit is

$$100i + 20\frac{di}{dt} = 200$$

$$20\frac{di}{dt} = 200 - 100i$$

Separating the variables,

$$\frac{di}{200 - 100i} = \frac{dt}{20}$$

Multiplying by '–100' on both sides,

$$\frac{-100\,di}{200 - 100\,i} = \frac{-100\,dt}{20}$$

Integrating both sides,

$$\int \frac{-100}{200 - 100\,i}di = \int \frac{-100}{20}dt$$

$$\ln(200 - 100i) = \frac{-100}{20}t + \text{K} \qquad \text{(i)}$$

The initial condition is that at $t = 0$, $i = -5$

$$\therefore \qquad \ln(200 + 500) = \text{K}$$

Substituting the value of K into Eq. (i),

$$\ln(200 - 100i) = \frac{-100}{20}t + \ln 700$$

$$\ln\left(\frac{200 - 100i}{700}\right) = -5t$$

$$\left(\frac{200-100i}{700}\right) = e^{-5t}$$

$$200 - 100i = 700e^{-5t}$$

or

$$2 - i = 7e^{-5t}$$

Hence, the equation for the current, $\boldsymbol{i(t) = 2 - 7e^{-5t}}$ **A**

By Laplace transformation method

The differential equation for the circuit is

$$100i + 20\frac{di}{dt} = 200 \tag{ii}$$

Taking Laplace transformation on both sides, we get

$$100I(s) + 20[sI(s) - I(0)] = \frac{200}{s} \tag{iii}$$

Since at $t = 0$ the current through the inductor is 5 A, $I(0) = -5$. Therefore, Eq. (iii) becomes

$$100I(s) + 20[sI(s) + 5] = \frac{200}{s}$$

$$I(s)\,[100 + 20s] + 100 = \frac{200}{s}$$

$$I(s) = \frac{\left(\frac{200}{s}\right) - 100}{(20s + 100)}$$

$$= \frac{200 - 100s}{s(20s+100)} = \frac{200-100s}{20s(s+5)} = \frac{10-5s}{s(s+5)}$$

By applying partial fraction technique,

$$I(s) = \frac{A}{s} + \frac{B}{(s+5)} \tag{iv}$$

Solving for constants A and B,

$$A = \left.\frac{10-5s}{(s+5)}\right|_{s=0} = 2 \quad \text{and} \quad B = \left.\frac{10-5s}{s}\right|_{s=-5} = -7$$

Substituting the values of A and B into Eq. (iv), we obtain

$$I(s) = \frac{2}{s} - \frac{7}{(s+5)}$$

Taking inverse Laplace transformation on both sides, we get

$$\boldsymbol{i(t) = 2 - 7e^{-5t}\ \text{A}} \tag{v}$$

The transient current is shown in Fig. 11.19.

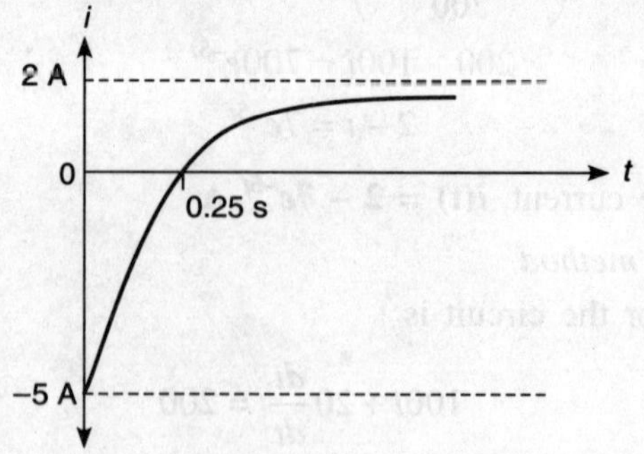

Fig. 11.19 Transient response of Example 11.2.

The time at which the response changes from negative value to positive value is obtained by equating Eq. (v) to zero, i.e.

$$i(t) = 2 - 7e^{-5t} = 0$$

$$e^{-5t} = \frac{2}{7} = 0.2857$$

$$-5t = \ln (0.2857) = -1.253$$

$$\therefore \qquad t = 0.25 \text{ s}$$

Rate of growth,

$$\frac{di}{dt} = \frac{d(2 - 7e^{-5t})}{dt} = (-7)(-5)e^{-5t} = 35e^{-5t}$$

Initial rate of growth,

$$\left.\frac{di}{dt}\right|_{t=0} = \mathbf{35\ A/s}$$

EXAMPLE 11.3 In the circuit shown in Fig. 11.20, switch S is in position 1 for a long time and brought to position 2 at time $t = 0$. Determine the circuit current.

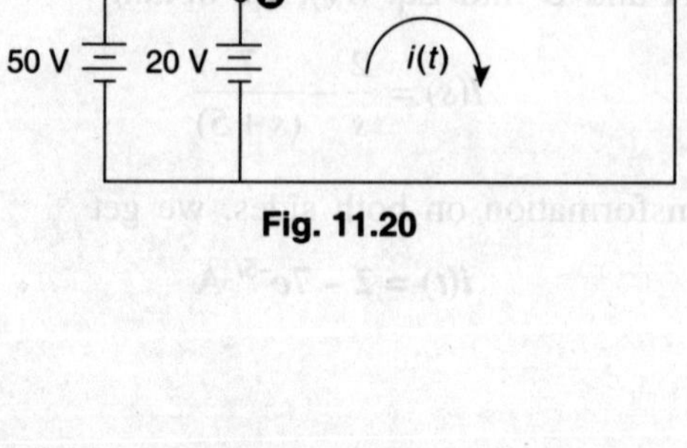

Fig. 11.20

Solution Determination of initial conditions

Since switch S is kept at position 1 for a long time, steady state is reached at this position and the steady-state current is 50/10 = 5 A.

By Classical method

At $t = 0$, the switch is moved to position 2 and the differential equation is

$$10i + 2\frac{di}{dt} = 20$$

$$2\frac{di}{dt} = 20 - 10i$$

Separating the variables,

$$\frac{di}{20-10i} = \frac{dt}{2}$$

Multiplying by '–10' on both sides,

$$\frac{-10di}{20-10i} = \frac{-10dt}{2}$$

Integrating both sides,

$$\int \frac{-10}{20-10i} di = \int \frac{-10}{2} dt$$

$$\ln(20-10i) = \frac{-10}{2}t + K \qquad \text{(i)}$$

The initial condition is that at $t = 0$, $i = 5$ A

Hence, $$\ln(20 - 50) = K$$

Substituting the value of K into Eq. (i),

$$\ln(20-10i) = \frac{-10}{2}t + \ln(-30)$$

$$\ln\left(\frac{20-10i}{-30}\right) = -5t$$

$$\left(\frac{20-10i}{-30}\right) = e^{-5t}$$

$$20 - 10i = -30e^{-5t}$$

$$2 - i = -3e^{-5t}$$

Therefore, the equation for the current, $\boldsymbol{i(t) = 2 + 3e^{-5t}}$ **A**

By Laplace transformation method

At $t = 0$, the switch is moved to position 2 and the differential equation is

$$10i + 2\frac{di}{dt} = 20$$

Taking Laplace transformation on both sides, we get

$$10I(s)+2\left[sI(s)-I(0)\right]=\frac{20}{s} \tag{ii}$$

At $t = 0$, $I(0) = 5$ A. Therefore Eq. (ii) becomes

$$10I(s)+2[sI(s)-5]=\frac{20}{s}$$

$$I(s)[10+2s]-10=\frac{20}{s}$$

$$I(s)=\frac{\left(\frac{20}{s}\right)+10}{(2s+10)}$$

$$=\frac{20+10s}{s(2s+10)}=\frac{20+10s}{2s(s+5)}$$

$$=\frac{10+5s}{s(s+5)}$$

By applying partial fraction technique,

$$I(s)=\frac{A}{s}+\frac{B}{(s+5)} \tag{iii}$$

Solving for constants A and B,

$$A=\left.\frac{10+5s}{(s+5)}\right|_{s=0}=2 \quad \text{and} \quad B=\left.\frac{10+5s}{s}\right|_{s=-5}=\frac{10-25}{-5}=3$$

Substituting the values of A and B into Eq. (iii),

$$I(s)=\frac{2}{s}+\frac{3}{(s+5)}$$

Taking inverse Laplace transformation on both sides, the current equation is obtained as:

$$i(t) = 2 + 3e^{-5t}\ \text{A}$$

EXAMPLE 11.4 In the circuit shown in Fig. 11.21, switch S is closed on position 1 at $t = 0$. At $t = 0.001$ s, the switch is moved to position 2. Obtain the equations for the current in both intervals and draw the current curves.

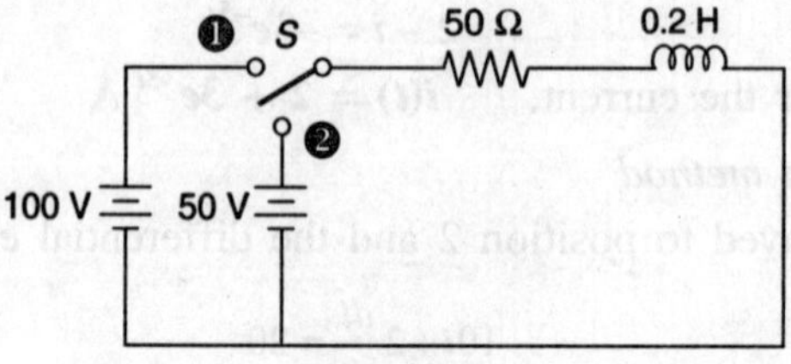

Fig. 11.21

Solution *By Classical method*

At $t = 0$, the switch is connected to position 1 and the differential equation is

$$50i + 0.2\frac{di}{dt} = 100$$

$$0.2\frac{di}{dt} = 100 - 50i$$

Separating the variables,

$$\frac{di}{100 - 50i} = \frac{dt}{0.2}$$

Multiplying by '–50' on both sides,

$$\frac{-50di}{100 - 50i} = \frac{-50dt}{0.2}$$

Integrating both sides,

$$\int \frac{-50}{100 - 50i} di = \int \frac{-50}{0.2} dt$$

$$\ln(100 - 50i) = \frac{-50}{0.2}t + \text{K} \qquad \text{(i)}$$

The initial condition is that at $t = 0$, $i = 0$

So, $\ln(100) = \text{K}$

Substituting into Eq. (i),

$$\ln(100 - 50i) = \frac{-50}{0.2}t + \ln(100)$$

$$\ln\left(\frac{100 - 50i}{100}\right) = -250t$$

$$\left(\frac{100 - 50i}{100}\right) = e^{-250t}$$

$$100 - 50i = 100e^{-250t}$$

$$2 - i = 2e^{-250t}$$

Hence, $$i(t) = 2 - 2e^{-250t} \text{ A} \qquad \text{(ii)}$$

The switch is kept at this position for 0.001 s. Therefore, at $t = 0.001$ s, the current through the circuit is determined by putting $t = 0.001$ in Eq. (ii), i.e.

$$\text{at } t = 0.001 \text{ s}, \; i(t) = 2 - 2e^{-250(0.001)} = 0.4424 \text{ A}$$

This value becomes initial current for the circuit when the switch is put in position 2. Now the switch is moved to position 2 at $t = 0.001$ s. Due to this action, 50 V source is connected in the circuit and 100 V source is disconnected. This moment ($t = 0.001$ s) can be redefined as $t' = 0$. Applying KVL,

$$50i + 0.2\frac{di}{dt'} = 50$$

$$0.2\frac{di}{dt'} = 50 - 50i$$

Separating the variables,

$$\frac{di}{50 - 50i} = \frac{dt'}{0.2}$$

Multiplying by '–50' on both sides,

$$\frac{-50di}{50 - 50i} = \frac{-50dt'}{0.2}$$

Integrating both sides,

$$\int \frac{-50}{50 - 50i}di = \int \frac{-50}{0.2}dt'$$

$$\ln(50 - 50i) = \frac{-50}{0.2}t' + \text{K} \qquad \text{(iii)}$$

The initial condition is that at $t' = 0$, $i = 0.4424$.

$$\ln(50 - 50 \times 0.4424) = \text{K}$$

Therefore, $\ln(27.88) = \text{K}$

Substituting the value of K into Eq. (iii),

$$\ln(50 - 50i) = \frac{-50}{0.2}t' + \ln(27.88)$$

$$\ln\left(\frac{50 - 50i}{27.88}\right) = -250t'$$

$$\left(\frac{50 - 50i}{27.88}\right) = e^{-250t'}$$

$$50 - 50i = 27.88e^{-250t'}$$

$$1 - i = 0.5576e^{-250t'}$$

$$\boldsymbol{i(t) = 1 - 0.5576e^{-250t'}}\ \textbf{A}$$

By Laplace transformation method

At $t = 0$, the switch is connected to position 1 and the differential equation is

$$50i + 0.2\frac{di}{dt} = 100$$

Taking Laplace transformation on both sides, we get

$$50I(s) + 0.2\left[sI(s) - I(0)\right] = \frac{100}{s} \tag{iv}$$

At $t = 0$, $i(t) = 0$, and in s-domain, $I(0) = 0$. Therefore, Eq. (iv) becomes

$$50I(s) + 0.2sI(s) = \frac{100}{s}$$

$$I(s)\,[50 + 0.2s] = \frac{100}{s}$$

$$I(s) = \frac{100}{s(0.2s + 50)}$$

$$= \frac{100}{0.2s(s + 250)} = \frac{500}{s(s + 250)}$$

By applying partial fraction technique,

$$= \frac{A}{s} + \frac{B}{(s + 250)} \tag{v}$$

Solving for constants A and B,

$$A = \left.\frac{500}{(s + 250)}\right|_{s=0} = 2 \quad \text{and} \quad B = \left.\frac{500}{s}\right|_{s=-250} = -2$$

Substituting the values of A and B into Eq. (v), we get

$$I(s) = \frac{2}{s} - \frac{2}{(s + 250)}$$

Taking inverse Laplace transformation on both sides, the current equation is obtained as:

$$\boldsymbol{i(t) = 2 - 2e^{-250t}\ \mathrm{A}} \tag{vi}$$

The switch is kept at this position for 0.001 s. Therefore, at $t = 0.001$ s, the current through the circuit is determined by putting $t = 0.001$ in Eq. (vi), i.e. at $t = 0.001$ s,

$$i(t) = 2 - 2e^{-250(0.001)} = 0.4424\ \mathrm{A}$$

This value becomes initial current for the circuit when the switch is put in position 2. Now the switch is moved to position 2 at $t = 0.001$ s. Due to this action, 50 V source is connected in the circuit and 100 V source is disconnected. This moment ($t = 0.001$ s) can be redefined as $t' = 0$.

Applying KVL,

$$50i + 0.2\frac{di}{dt'} = 50$$

Taking Laplace transformation on both sides, we get

$$50I(s) + 0.2[sI(s) - I(0)] = \frac{50}{s} \tag{vii}$$

At $t' = 0$, $I(0) = 0.4424$ A. Therefore, Eq. (vii) becomes

$$50I(s) + 0.2sI(s) - 0.08848 = \frac{50}{s}$$

$$I(s)[50 + 0.2s] = \frac{50}{s} + 0.08848$$

$$I(s) = \frac{50 + 0.08848s}{s(0.2s + 50)}$$

$$= \frac{50 + 0.08848s}{0.2s(s + 250)}$$

$$= \frac{250 + 0.4424s}{s(s + 250)}$$

By applying partial fraction technique,

$$I(s) = \frac{A}{s} + \frac{B}{(s + 250)} \qquad \text{(viii)}$$

Solving for constants A and B,

$$A = \left.\frac{250 + 0.4424s}{(s + 250)}\right|_{s=0} = 1 \quad \text{and} \quad B = \left.\frac{250 + 0.4424s}{s}\right|_{s=-250} = -0.5576$$

Substituting the values of A and B into Eq. (viii), we get

$$I(s) = \frac{1}{s} - \frac{0.5576}{(s + 250)}$$

Taking inverse Laplace transformation on both sides, the current equation is obtained as:

$$\mathbf{i(t) = 1 - 0.5576e^{-250t'}\ A}$$

or

$$\mathbf{= 1 - 0.5576e^{-250(t-0.001)}\ A}$$

Result:

$$\mathbf{i(t) = 2 - 2e^{-250t}\ A} \qquad \text{for } 0 < t < 0.001$$

$$\mathbf{= 1 - 0.5576e^{-250(t-0.001)}\ A} \qquad \text{for } t > 0.001$$

The transient current is shown in Fig. 11.22.

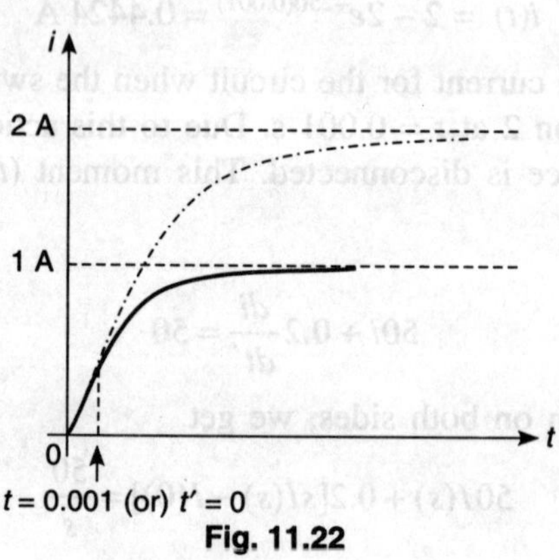

Fig. 11.22

EXAMPLE 11.5 In the circuit shown in Fig. 11.23, switch S is closed on position 1 at $t = 0$. At $t = 0.1$ s, the switch is moved to position 2. Obtain the equations for the current in both intervals and draw the current curve.

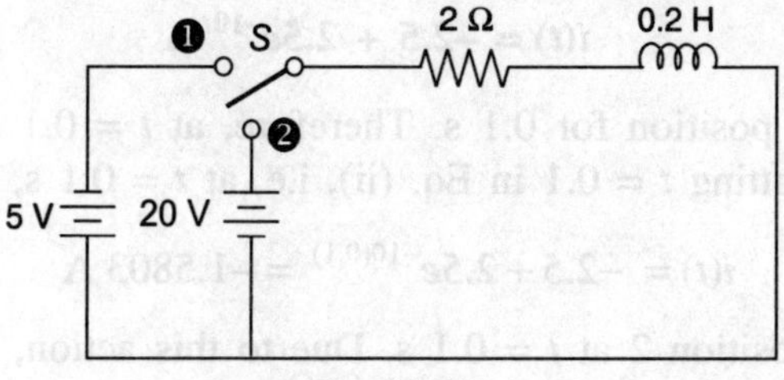

Fig. 11.23

Solution *By Classical method*

At $t = 0$, the switch is connected to position 1 and the differential equation is

$$2i + 0.2\frac{di}{dt} = -5$$

$$0.2\frac{di}{dt} = -5 - 2i$$

Separating the variables,

$$\frac{di}{-5-2i} = \frac{dt}{0.2}$$

Multiplying by '–2' on both sides,

$$\frac{-2di}{-5-2i} = \frac{-2dt}{0.2}$$

Integrating both sides,

$$\int \frac{-2}{-5-2i}di = \int \frac{-2}{0.2}dt$$

$$\ln(-5-2i) = \frac{-2}{0.2}t + \text{K} \qquad \text{(i)}$$

The initial condition is that at $t = 0$, $i = 0$

Therefore, $\ln(-5) = \text{K}$

Substituting into Eq. (i), we get

$$\ln(-5-2i) = \frac{-2}{0.2}t + \ln(-5)$$

$$\ln\left(\frac{-5-2i}{-5}\right) = -10t$$

$$\left(\frac{-5-2i}{-5}\right) = e^{-10t}$$

$$-5 - 2i = -5e^{-10t}$$

$$\boldsymbol{i(t) = -2.5 + 2.5e^{-10t}\ \text{A}} \qquad \text{(ii)}$$

The switch is kept at this position for 0.1 s. Therefore, at $t = 0.1$ s, the current through the circuit is determined by putting $t = 0.1$ in Eq. (ii), i.e. at $t = 0.1$ s,

$$i(t) = -2.5 + 2.5e^{-10(0.1)} = -1.5803\ \text{A}$$

The switch is moved to position 2 at $t = 0.1$ s. Due to this action, 20 V source is connected to the circuit and 5 V source is disconnected. This moment can be redefined as $t' = 0$. Applying KVL,

$$2i + 0.2\frac{di}{dt'} = 20$$

$$0.2\frac{di}{dt'} = 20 - 2i$$

Separating the variables,

$$\frac{di}{20-2i} = \frac{dt'}{0.2}$$

Multiplying by '–2' on both sides,

$$\frac{-2di}{20-2i} = \frac{-2dt'}{0.2}$$

Integrating both sides,

$$\int \frac{-2}{20-2i}di = \int \frac{-2}{0.2}dt'$$

$$\ln(20-2i) = \frac{-2}{0.2}t' + \text{K} \qquad \text{(iii)}$$

The initial condition is that at $t' = 0$, $i = -1.5803$.

$$\ln[20 + (2 \times 1.5803)] = \text{K}$$

Hence,

$$\ln(23.1606) = \text{K}$$

Substituting the value of K into Eq. (iii), we obtain

$$\ln(20-2i) = \frac{-2}{0.2}t' + \ln(23.1606)$$

$$\ln\left(\frac{20-2i}{23.1606}\right) = -10t'$$

$$\left(\frac{20-2i}{23.1606}\right) = e^{-10t'}$$

$$20 - 2i = 23.1606e^{-10t'}$$

$$10 - i = 11.5803e^{-10t'}$$

Therefore, $$\boldsymbol{i(t') = 10 - 11.5803e^{-10t'}\ \text{A}} \qquad \text{(iv)}$$

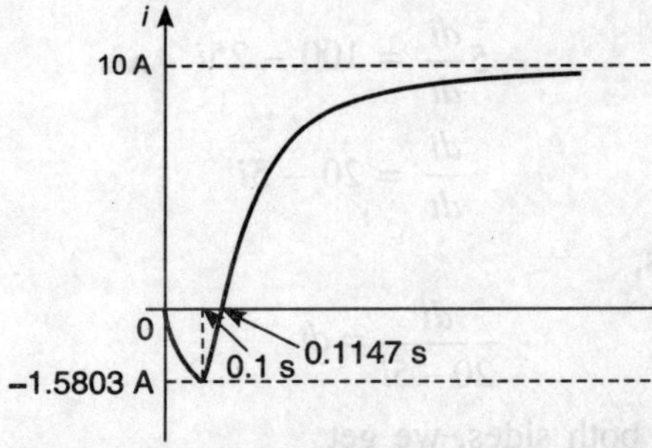

Fig. 11.24 Transient response of Example 11.5.

The time at which the transient curve crosses zero value t_0 is obtained as follows:
From Eq. (iv),

$$i(t') = 10 - 11.5803e^{-10t'} = 0$$

$$11.5803e^{-10t'} = 10$$

$$e^{-10t'} = \frac{10}{11.5803} = 0.8635$$

$$-10t' = \ln(0.8635) = -0.14676$$

$$t' = 0.014676$$

$$\therefore \quad t_0 = t' + 0.1 = 0.014676 + 0.1$$

$$t_0 = 0.1147 \text{ s}$$

The transient current is shown in Fig. 11.24.

Result:

$$\boldsymbol{i(t) = -2.5 + 2.5e^{-10t}\ \text{A}} \qquad \text{for } 0 < t < 0.1$$

$$\boldsymbol{= 10 - 11.5803e^{-10(t-0.1)}\ \text{A}} \qquad \text{for } t > 0.1$$

Note: The solution by Laplace transformation technique is left to the readers as exercise.

EXAMPLE 11.6 A series *RL* circuit has $R = 25\ \Omega$ and $L = 5$ H. A DC voltage of 100 V is applied at $t = 0$. Find (i) the equations for current, (ii) voltage across *R* and *L*, (iii) the current in the circuit after 0.5 s and (iv) the time at which the voltage drops across *R* and *L* are same.

Solution The switch is closed at $t = 0$. Therefore, at $t = 0^-$, the current through the circuit is zero and also $i(0) = 0$, since the inductor acts as open-circuit initially. This is the initial condition for this problem.

By Classical method

(i) The equation for current

The differential equation is written as

$$25i + 5\frac{di}{dt} = 100$$

$$5\frac{di}{dt} = 100 - 25i$$

$$\frac{di}{dt} = 20 - 5i$$

Separating the variables,

$$\frac{di}{20 - 5i} = dt$$

Multiplying by '–5' on both sides, we get

$$\frac{-5}{20 - 5i}di = -5dt$$

Integrating both sides, we get

$$\int \frac{-5}{20 - 5i}di = \int -5dt$$

$$\ln (20 - 5i) = -5t + K \qquad \text{(i)}$$

At $t = 0$, $i = 0$ A.

Hence Eq. (i) becomes, $\ln (20) = K$

Substituting the value of K into Eq. (i), we obtai

$$\ln (20 - 5i) = -5t + \ln (20)$$

$$\ln\left(\frac{20 - 5i}{20}\right) = -5t$$

$$\frac{20 - 5i}{20} = e^{-5t}$$

$$20 - 5i = 20e^{-5t}$$

Therefore, the equation for the current, $\boldsymbol{i(t) = 4(1 - e^{-5t})}$ **A**

(ii) Voltage across R and L

Voltage across R, $v_R = iR$

$$= 4(1 - e^{-5t}) \times 25$$

$$= \mathbf{100(1 - e^{-5t})\ V} \qquad \text{(ii)}$$

Voltage across L, $v_L = L\frac{di}{dt}$

$$= 5\frac{d}{dt}[4(1-e^{-5t})] = 20\frac{d}{dt}(1-e^{-5t})$$

$$= \mathbf{100e^{-5t}\ V}$$

(iii) Current in the circuit after 0.5 seconds

$$i = 4(1 - e^{-5t})$$

Substituting t = 0.5 into the above equation, we obtain

$$i = 4(1 - e^{-5\times 0.5})$$

$$= \mathbf{3.6717\ A}$$

(iv) Time at which the voltage across R and L are equal

This condition is satisfied only if v_R and v_L are equal to 50 V, as the applied voltage is 100 V.

∴ From Eq. (ii), voltage across resistor $v_R = 100(1 - e^{-5t}) = 50$

$$(1 - e^{-5t}) = \frac{50}{100}$$

$$e^{-5t} = 0.5$$

$$\mathbf{t = 0.1386\ s}$$

By Laplace transformation method

The differential equation for the circuit is

$$25i + 5\frac{di}{dt} = 100 \qquad \text{(iii)}$$

Taking Laplace transformation on both sides, we get

$$25I(s) + 5[sI(s) - I(0)] = \frac{100}{s} \qquad \text{(iv)}$$

$I(0)$ = 0, therefore Eq. (iv) becomes,

$$25I(s) + 5sI(s) = \frac{100}{s}$$

$$I(s)\ [25 + 5s] = \frac{100}{s}$$

$$I(s) = \frac{100}{s(5s+25)} = \frac{20}{s(s+5)}$$

By applying partial fraction technique,

$$I(s) = \frac{A}{s} + \frac{B}{(s+5)} \qquad \text{(v)}$$

Solving for constants A and B,

$$A = \left.\frac{20}{(s+5)}\right|_{s=0} = 4 \quad \text{and} \quad B = \left.\frac{20}{s}\right|_{s=-5} = -4$$

Substituting the values of A and B into Eq. (v), we get

$$I(s) = \frac{4}{s} - \frac{4}{(s+5)}$$

Taking inverse Laplace transformation on both sides, we get

$$i(t) = 4 - 4e^{-5t}$$

or $$\mathbf{i(t) = 4(1 - e^{-5t})\ A}$$

> ***Note:*** For the other sub questions from (ii) to (iv), the readers are requested to refer to the solutions described earlier under classical method.

EXAMPLE 11.7 In the circuit shown in Fig. 11.25, a DC voltage of 100 V is applied and switch S is open. It is closed at $t = 0$. Find the expression for the current.

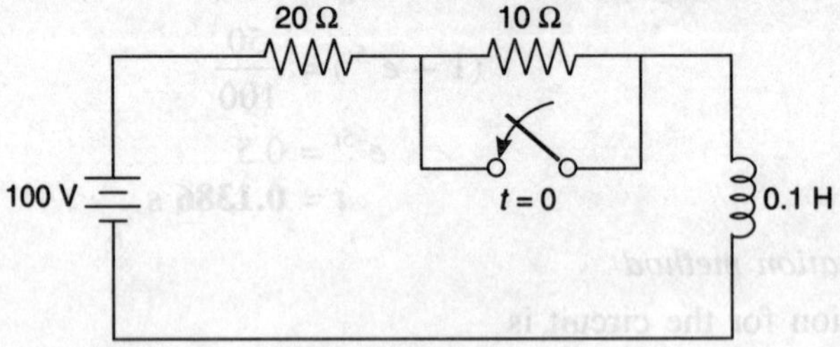

Fig. 11.25

Solution When the switch is open, the steady-state current through the circuit is 100/30 = 3.33 A. The switch is closed at $t = 0$, therefore $i(0) = 3.33$ A.

By classical method

The mesh equation, after closing the switch is written as:

$$20i + 0.1\frac{di}{dt} = 100$$

$$0.1\frac{di}{dt} = 100 - 20i$$

Separating the variables,

$$\frac{di}{100 - 20i} = \frac{dt}{0.1}$$

Multiplying by '–20' on both sides,

$$\frac{-20di}{100 - 20i} = \frac{-20dt}{0.1}$$

Integrating both sides, we get

$$\int \frac{-20}{100 - 20i}\, di = \int \frac{-20}{0.1}\, dt$$

$$\ln(100 - 20i) = -200t + K \quad \text{(i)}$$

At $t = 0$, $i = 3.33$ A. Hence,

$$\ln(100 - 20 \times 3.33) = K$$

$$\ln(33.4) = K$$

Substituting the value of K into Eq. (i), we get

$$\ln(100 - 20i) = -200t + \ln(33.4)$$

$$\ln\left(\frac{100 - 20i}{33.4}\right) = -200t$$

$$\left(\frac{100 - 20i}{33.4}\right) = e^{-200t}$$

$$100 - 20i = 33.4e^{-200t}$$

$$5 - i = 1.67e^{-200t}$$

$$i = 5 - 1.67e^{-200t}$$

Therefore, the equation for the current, $\boldsymbol{i(t) = 5 - 1.67e^{-200t}}$ **A**

By Laplace transformation method

The differential equation for the circuit after $t = 0$ is

$$20i + 0.1\frac{di}{dt} = 100 \quad \text{(ii)}$$

Taking Laplace transformation on both sides, we get

$$20I(s) + 0.1[sI(s) - I(0)] = \frac{100}{s} \quad \text{(iii)}$$

Since at $t = 0$, $I(0) = 3.33$. Therefore, Eq. (iii) becomes,

$$20I(s) + 0.1[sI(s) - 3.33] = \frac{100}{s}$$

$$20I(s) + 0.1sI(s) = \frac{100}{s} + 0.333$$

$$I(s)[20 + 0.1s] = \frac{100}{s} + 0.333$$

$$I(s) = \frac{\left(\frac{100}{s}\right) + 0.333}{(0.1s + 20)}$$

$$= \frac{100 + 0.333s}{0.1s(s + 200)} = \frac{1000 + 3.33s}{s(s + 200)}$$

By applying partial fraction technique,

$$I(s) = \frac{A}{s} + \frac{B}{(s+200)} \quad \text{(iv)}$$

Solving for constants A and B,

$$A = \left.\frac{1000+3.33s}{(s+200)}\right|_{s=0} = 5 \quad \text{and} \quad B = \left.\frac{1000+3.33s}{s}\right|_{s=-200} = -1.67$$

Substituting the values of A and B into Eq. (iv), we obtain

$$I(s) = \frac{5}{s} - \frac{1.67}{(s+200)}$$

Taking inverse Laplace transformation on both sides, the current is obtained as:

$$\boldsymbol{i(t) = 5 - 1.67e^{-200t}\ \text{A}}$$

EXAMPLE 11.8 In the circuit shown in Fig. 11.26, switch S is closed at $t = 0$. Find the circuit current.

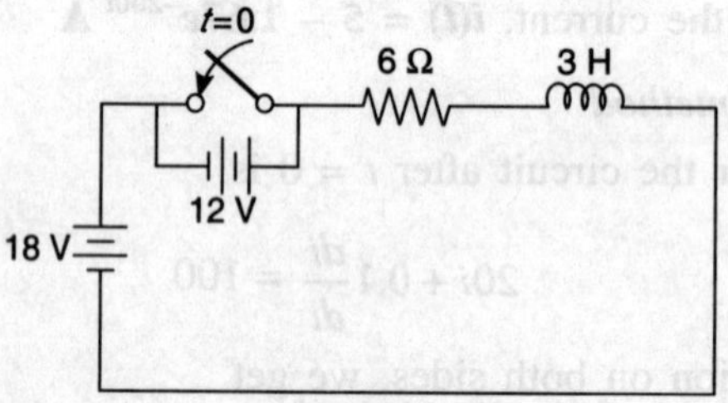

Fig. 11.26

Solution Determination of initial conditions

At $t = 0^-$, the switch is open, and the total voltage applied to the circuit is the sum of two sources, i.e. 30 V. Therefore, $i(0^-) = i(0^+) = \frac{30}{6} = 5$ A

By Classical method

At $t = 0$, the switch is closed and the 12 V source is short-circuited. Therefore, 18 V source is acting alone. The differential equation is

$$6i + 3\frac{di}{dt} = 18$$

$$3\frac{di}{dt} = 18 - 6i$$

Separating the variables,

$$\frac{di}{18-6i} = \frac{dt}{3}$$

Multiplying by '–6' on both sides,

$$\frac{-6di}{18-6i} = -\frac{-6dt}{3}$$

Integrating both sides,

$$\int\frac{-6}{18-6i}\,di = \int\frac{-6}{3}\,dt$$

$$\ln(18-6i) = \frac{-6}{3}t + \text{K} \qquad \text{(i)}$$

The initial condition is that at $t = 0$, $i = 5$ A.

Therefore, $\ln(18-30) = \text{K}$

and $\ln(-12) = \text{K}$

Substituting the value of K into Eq. (i), we obtain

$$\ln(18-6i) = \frac{-6}{3}t + \ln(-12)$$

$$\ln\left(\frac{18-6i}{-12}\right) = -2t$$

$$\left(\frac{18-6i}{-12}\right) = e^{-2t}$$

$$18-6i = -12e^{-2t}$$

$$3-i = -2e^{-2t}$$

Hence, the equation for the current, $\boldsymbol{i(t) = 3 + 2e^{-2t}}$ **A**

By Laplace transformation method

At $t = 0$, the switch is closed and the 12 V source is short-circuited. Therefore, 18 V source is acting alone. The differential equation is

$$6i + 3\frac{di}{dt} = 18 \qquad \text{(ii)}$$

Taking Laplace transformation on both sides, we get

$$6I(s) + 3[sI(s) - I(0)] = \frac{18}{s} \qquad \text{(iii)}$$

At $t = 0$, $i(t) = 5$ A, and in s-domain $I(0) = 5$ A. Therefore, Eq. (iii) becomes

$$6I(s) + 3[sI(s) - 5] = \frac{18}{s}$$

$$I(s)\,[6 + 3s] - 15 = \frac{18}{s}$$

$$I(s) = \frac{\left(\frac{18}{s}\right) + 15}{(3s+6)}$$

$$= \frac{18+15s}{s(3s+6)} = \frac{18+15s}{3s(s+2)}$$

$$= \frac{6+5s}{s(s+2)}$$

By applying partial fraction technique,

$$I(s) = \frac{A}{s} + \frac{B}{(s+2)} \qquad \text{(iv)}$$

Solving for constants A and B,

$$A = \left.\frac{6+5s}{(s+2)}\right|_{s=0} = 3 \quad \text{and} \quad B = \left.\frac{6+5s}{s}\right|_{s=-2} = \frac{6-10}{-2} = 2$$

Substituting the values of A and B into Eq. (iv), we get

$$I(s) = \frac{3}{s} + \frac{2}{(s+2)}$$

Taking inverse Laplace transformation on both sides, the current is obtained as:

$$\boldsymbol{i(t) = 3 + 2e^{-2t}\ \text{A}}$$

EXAMPLE 11.9 For the circuit shown in Fig. 11.27, find the current equation when the switch is opened at $t = 0$. Find the circuit current $i(t)$.

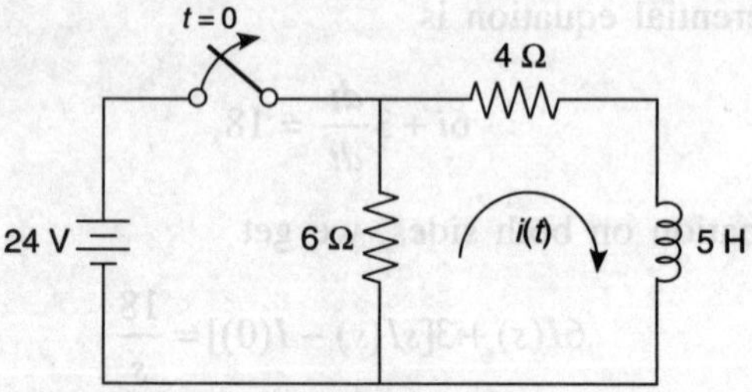

Fig. 11.27

Solution Determination of initial conditions

At $t = 0$, switch is opened. Therefore, prior to that, the current through the inductor is determined using the equivalent circuit at $t = 0^-$ as shown in Fig. 11.28.

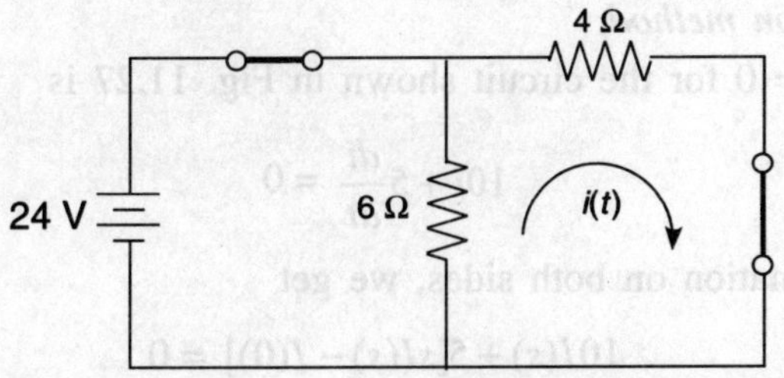

Fig. 11.28 Equivalent circuit of Fig. 11.27 at $t = 0^-$.

From Fig. 11.28, current supplied by the source = $\dfrac{24}{6\|4} = \dfrac{24}{\left(\dfrac{6\times 4}{6+4}\right)} = 10$ A. Using current division rule, current through the inductor = $10 \times \dfrac{6}{6+4} = 6$ A.

By Classical method

In Fig. 11.27, by using KVL, thc mesh equation at $t = 0$ is written as:

$$6i + 4i + 5\frac{di}{dt} = 0$$

$$10i + 5\frac{di}{dt} = 0$$

$$5\frac{di}{dt} = -10i$$

$$\frac{di}{dt} = -2i$$

Separating the variables,

$$\frac{di}{i} = -2dt$$

Integrating both sides, we get

$$\ln (i) = -2t + \text{K} \tag{i}$$

At $t = 0$, $i = 6$ A. Hence,

$$\ln (6) = \text{K}$$

Substituting the value of constant K into Eq. (i), we get

$$\ln (i) = -2t + \ln (6)$$

$$\ln\left(\frac{i}{6}\right) = -2t$$

$$\frac{i}{6} = e^{-2t}$$

Therefore, the equation for the current, $\boldsymbol{i(t) = 6e^{-2t}}$ **A**

By Laplace transformation method

The mesh equation at $t = 0$ for the circuit shown in Fig. 11.27 is

$$10i + 5\frac{di}{dt} = 0$$

Taking Laplace transformation on both sides, we get

$$10I(s) + 5[sI(s) - I(0)] = 0 \quad \text{(iii)}$$

At $t = 0$, $I(0) = 6$ A. Therefore, Eq. (iii) becomes

$$10I(s) + 5[sI(s) - 6] = 0$$

$$I(s)[10 + 5s] - 30 = 0$$

$$I(s)[10 + 5s] = 30$$

$$I(s) = \frac{30}{(5s + 10)}$$

$$= \frac{6}{(s + 2)}$$

Taking inverse Laplace transformation on both sides, the current equation is obtained as:

$$\mathbf{i(t) = 6e^{-2t}\ A}$$

EXAMPLE 11.10 For the circuit shown in Fig. 11.29, find the current equations for i_1 and i_2 when the switch is closed at $t = 0$.

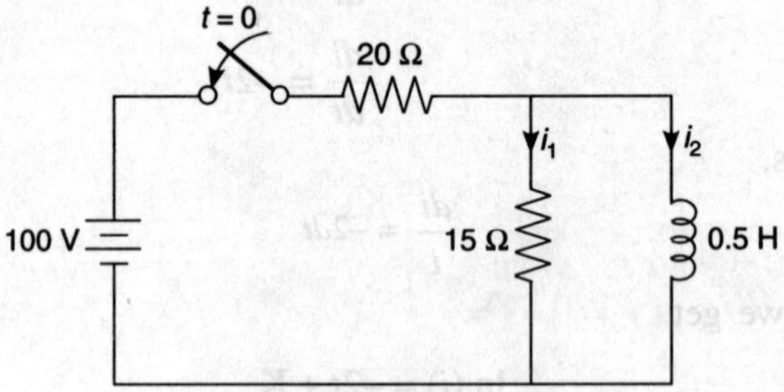

Fig. 11.29

Solution Determination of initial conditions

The switch is closed at $t = 0$. Therefore, before closing the switch (at $t = 0^-$), the current through the inductor was zero, i.e. $i_2(0) = 0$.

Solution by Classical method

Applying Kirchhoff's voltage laws to both loops, we get

$$20(i_1 + i_2) + 15i_1 = 100$$

or

$$35i_1 + 20i_2 = 100 \quad \text{(i)}$$

and

$$20(i_1 + i_2) + 0.5\frac{di_2}{dt} = 100 \quad \text{(ii)}$$

From Eq. (i), we have

$$35i_1 = 100 - 20i_2$$

or $$i_1 = 2.8571 - 0.5714i_2 \tag{iii}$$

Substituting i_1 into Eq. (ii), we have

$$20(2.8571 - 0.5714i_2) + 20i_2 + 0.5\frac{di_2}{dt} = 100$$

$$57.14 - 11.43i_2 + 20i_2 + 0.5\frac{di_2}{dt} = 100$$

$$8.57i_2 + 0.5\frac{di_2}{dt} = 42.86$$

or $$17.14i_2 + \frac{di_2}{dt} = 85.72$$

$$\frac{di_2}{dt} = 85.72 - 17.14i_2$$

Separating the variables,

$$\frac{di_2}{85.72 - 17.14i_2} = dt$$

Multiplying by '–17.14' on both sides,

$$\frac{-17.14}{85.72 - 17.14i_2} di_2 = -17.14\,dt$$

Integrating both sides,

$$\int \frac{-17.14}{85.72 - 17.14i_2} di_2 = \int -17.14\,dt$$

$$\ln(85.72 - 17.14i_2) = -17.14t + K \tag{iv}$$

The initial condition is that at $t = 0$, $i_2 = 0$

Hence, $$\ln(85.72) = K$$

Substituting the value of K into Eq. (iv), we get

$$\ln(85.72 - 17.14i_2) = -17.14t + \ln(85.72)$$

$$\ln\left(\frac{85.72 - 17.14i_2}{85.72}\right) = -17.14t$$

$$\left(\frac{85.72 - 17.14i_2}{85.72}\right) = e^{-17.14t}$$

$$1 - 0.2i_2 = e^{-17.14t}$$

$$0.2i_2 = 1 - e^{-17.14t}$$

So, $$\boldsymbol{i_2(t) = 5(1 - e^{-17.14t})\text{ A}} \tag{v}$$

Substituting the value of i_2 from Eq. (v) into Eq. (iii), we obtain

$$i_1 = 2.8571 - 0.5714\,(5 - 5e^{-17.14t})$$

or

$$\mathbf{i_1(t) = 2.8571e^{-17.14t}\ A}$$

Result:

$$\mathbf{i_1(t) = 2.8571e^{-17.14t}\ A}$$
$$\mathbf{i_2(t) = 5(1 - e^{-17.14t})\ A}$$

> ***Note:*** The solution by Laplace transformation technique is left to the readers as exercise.

EXAMPLE 11.11 In the circuit shown in Fig. 11.30, find the current through the inductor for all values of t.

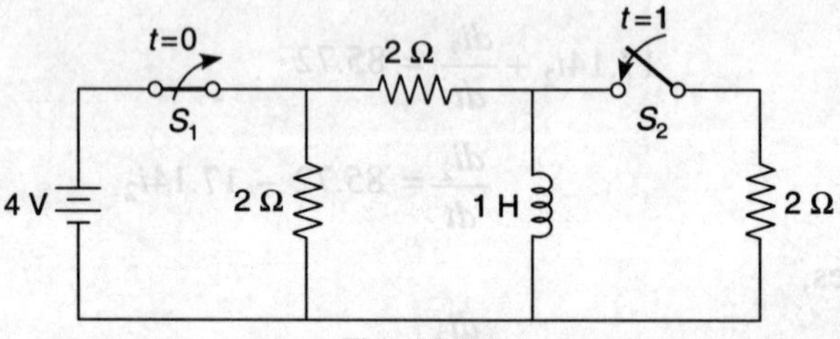

Fig. 11.30

Solution

Case 1 (for all values of $t < 0$)

At $t = 0^-$, switch S_1 is in closed state and S_2 is in open state. The circuit might have attained steady state. The equivalent circuit for this condition is shown in Fig. 11.31.

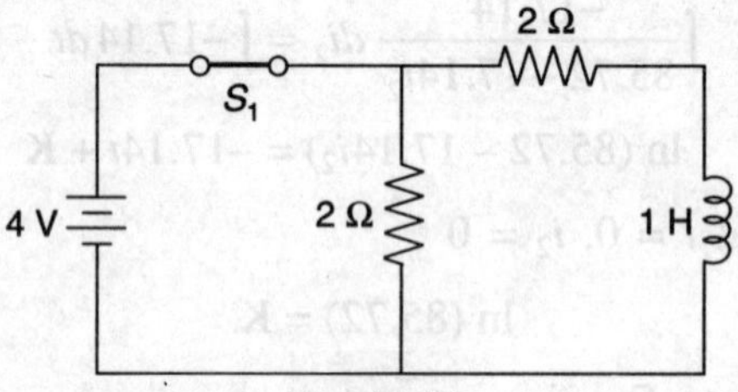

Fig. 11.31

Referring to Fig. 11.31, current through the inductor is $\left[4 \div (2\|2)\right] \times \dfrac{2}{(2+2)} = 2$ A, since in steady state inductor acts as a short-circuit. Therefore, for all values of $t < 0$, current through the inductor = **2 A**.

Case 2 (for all values of $0 < t < 1$)

In this interval, both S_1 and S_2 are in open state. The equivalent circuit for this case is shown in Fig. 11.32.

Applying KVL,

$$4i + \frac{di}{dt} = 0$$

$$\frac{di}{dt} = -4i$$

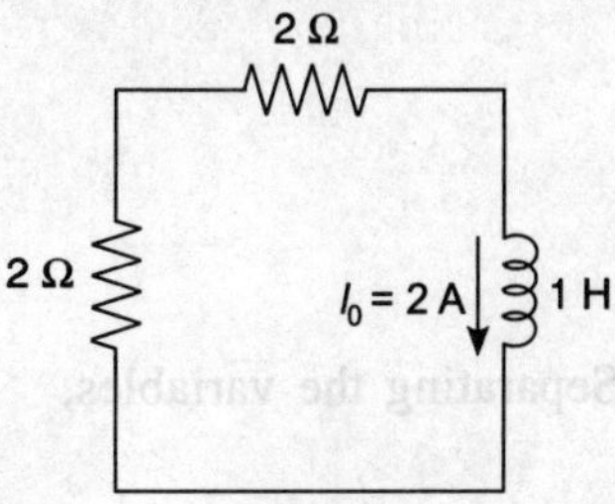

Fig. 11.32

Separating the variables,

$$\frac{di}{i} = -4dt$$

$$\ln(i) = -4t + K \quad \text{(i)}$$

Substituting the initial condition, i.e. at $t = 0$, $i = 2$ A, into the above equation, we get

$$\ln(2) = K$$

Substituting the value of K in Eq. (i), we get

$$\ln(i) = -4t + \ln(2)$$

$$\frac{i}{2} = e^{-4t}$$

$$\boldsymbol{i = 2e^{-4t}} \textbf{ A} \quad \text{(ii)}$$

Case 3 (for all values of $t > 1$)

At $t = 1$, switch S_2 is closed. Let us redefine the time $t = 1$ as $t' = 0$ and is taken as reference. The equivalent circuit for this case is shown in Fig. 11.33. Figures 11.34(a) and (b) show the reduction steps of circuit shown in Fig. 11.33.

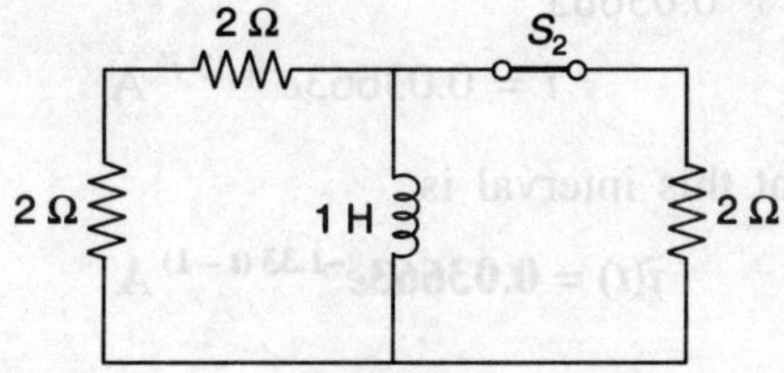

Fig. 11.33

Applying KVL to the circuit shown in Fig. 11.34(b),

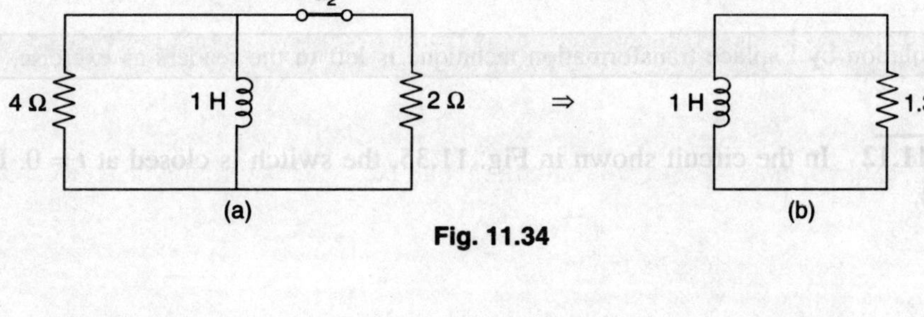

Fig. 11.34

$$1.33i + \frac{di}{dt'} = 0$$

$$\frac{di}{dt'} = -1.33i$$

Separating the variables,

$$\frac{di}{i} = -1.33dt'$$

$$\ln(i) = -1.33t' + K \qquad \text{(iii)}$$

The value of constant K is determined from initial conditions. Current at $t' = 0$ (i.e. $t = 1$) is found by substituting $t = 1$ into Eq. (ii), i.e.

$$i = 2e^{-4(1)} = 0.03663 \text{ A}$$

Substituting $t' = 0$ and $i = 0.03663$ in Eq. (iii), we get

$$\ln(0.03663) = K$$

Substituting the value of K found above into Eq. (iii), we obtain

$$\ln(i) = -1.33t' + \ln(0.03663)$$

$$\ln\left(\frac{i}{0.03663}\right) = -1.33t'$$

$$\frac{i}{0.03663} = e^{-1.33t'}$$

$$i = 0.03663e^{-1.33t'} \text{ A}$$

Hence, the current equation at this interval is

$$\mathbf{i(t) = 0.03663e^{-1.33(t-1)} \text{ A}}$$

Result:

$$\begin{aligned} \mathbf{i(t)} &= \mathbf{2\ A} && \text{for } t < 0 \\ &= \mathbf{2e^{-4t}\ A} && \text{for } 0 < t < 1 \\ &= \mathbf{0.03663e^{-1.33(t-1)}\ A} && \text{for } t > 1 \end{aligned}$$

Note: The solution by Laplace transformation technique is left to the readers as exercise.

EXAMPLE 11.12 In the circuit shown in Fig. 11.35, the switch is closed at $t = 0$. Determine $i_L(t)$ and $v_L(t)$.

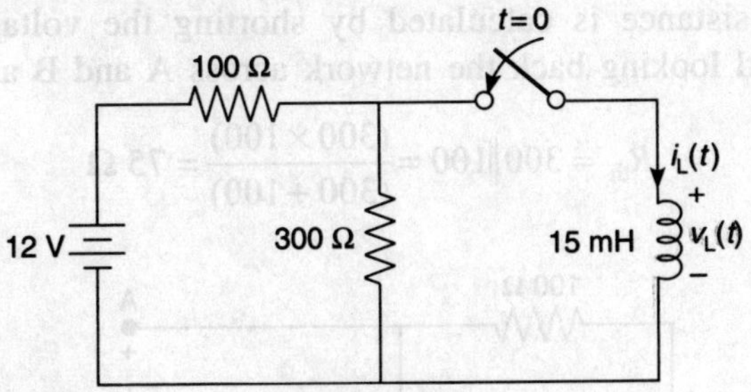

Fig. 11.35

Solution Determination of initial conditions

The switch is closed at $t = 0$. Therefore, at $t = 0^-$, the current through the inductor was zero, i.e. $i_L(0)^- = 0$. Since the current through the inductor cannot be changed instantaneously, $i_L(0)^+$ is also zero.

For all values of $t > 0$, the switch is closed and the circuit is drawn as shown in Fig. 11.36.

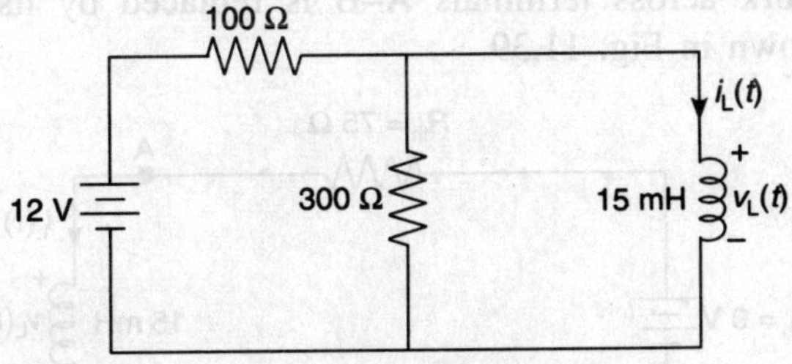

Fig. 11.36

Next, Thevenin's equivalent circuit is found as follows:

Step 1 The 15 mH inductor is removed from the circuit and the open circuit terminals are marked as A and B as shown in Fig. 11.37.

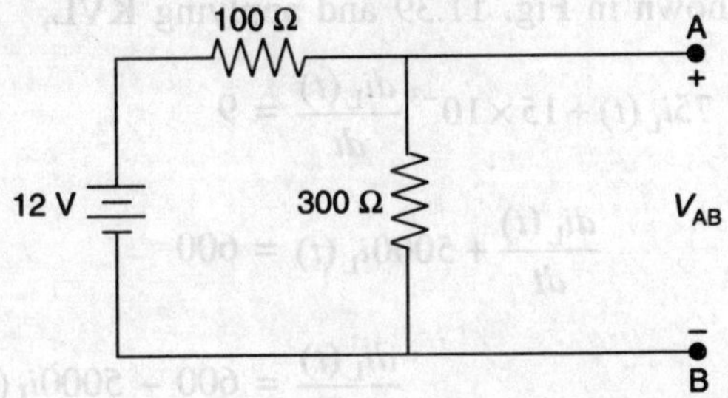

Fig. 11.37

Step 2 The open circuit voltage V_{AB} is calculated as $\dfrac{12}{(300+100)} \times 300 = 9$ V.

Step 3 Thevenin's resistance is calculated by shorting the voltage source as shown in Fig. 11.38. and looking back the network across A and B as:

$$R_{th} = 300 \| 100 = \frac{(300 \times 100)}{(300 + 100)} = 75\ \Omega$$

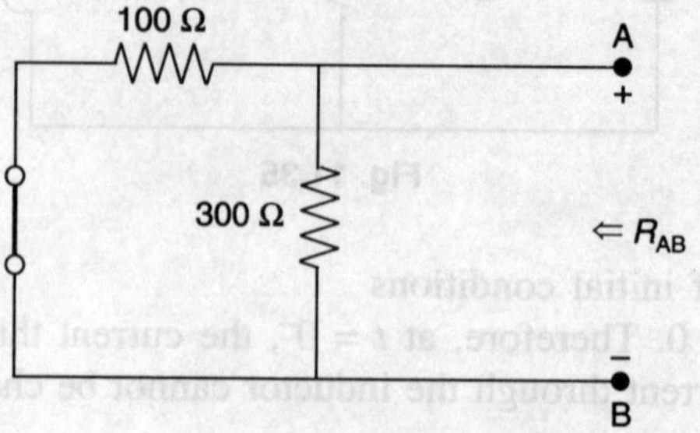

Fig. 11.38

Step 4 Original network across terminals A–B is replaced by its Thevenin's equivalent network as shown in Fig. 11.39.

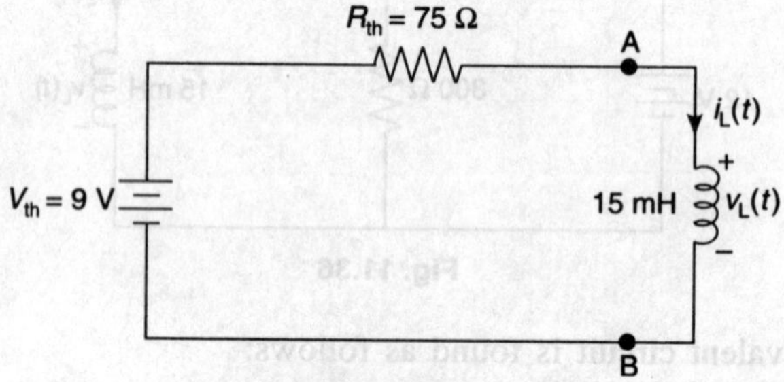

Fig. 11.39

By Classical method

Considering the network shown in Fig. 11.39 and applying KVL,

$$75 i_L(t) + 15 \times 10^{-3} \frac{di_L(t)}{dt} = 9$$

$$\frac{di_L(t)}{dt} + 5000 i_L(t) = 600$$

$$\frac{di_L(t)}{dt} = 600 - 5000 i_L(t)$$

Separating the variables,

$$\frac{di_L(t)}{600 - 5000 i_L(t)} = dt$$

Multiplying by '–5000' on both sides,

$$\frac{-5000di_L(t)}{600-5000i_L(t)}=-5000dt$$

Integrating both sides,

$$\int\frac{-5000di_L(t)}{600-5000i_L(t)}=\int -5000dt$$

$$\ln[600-5000i_L(t)]=-5000t+K \quad \text{(i)}$$

The initial condition is that at $t = 0$, $i_L(t) = 0$,

Hence, $\ln(600) = K$

Substituting the value of K into Eq. (i), we get

$$\ln[600-5000i_L(t)]=-5000t+\ln(600)$$

$$\ln\left(\frac{600-5000i_L(t)}{600}\right)=-5000t$$

$$\left(\frac{600-5000i_L(t)}{600}\right)=e^{-5000t}$$

$$600-5000i_L(t)=600e^{-5000t}$$

$$0.12-i_L(t)=0.12e^{-5000t}$$

$$\therefore \quad \mathbf{i_L(t)=0.12-0.12e^{-5000t}\ A}$$

By Laplace transformation method

Considering the network shown in Fig. 11.39 and applying KVL,

$$75i_L(t)+15\times10^{-3}\frac{di_L(t)}{dt}=9$$

Taking Laplace transformation on both sides, we get

$$75I_L(s)+15\times10^{-3}[sI_L(s)-I_L(0)]=\frac{9}{s} \quad \text{(ii)}$$

At $t = 0$, $I_L(0) = 0$. Therefore, Eq. (ii) becomes

$$75I_L(s)+15\times10^{-3}sI_L(s)=\frac{9}{s}$$

$$I_L(s)(75+15\times10^{-3}s)=\frac{9}{s}$$

$$I_L(s) = \frac{\left(\frac{9}{s}\right)}{(75 + 15\times10^{-3}s)}$$

$$= \frac{9}{15\times10^{-3}s\left(s + \frac{75}{15\times10^{-3}}\right)}$$

$$= \frac{600}{s(s+5000)}$$

By applying partial fraction technique

$$I_L(s) = \frac{A}{s} + \frac{B}{(s+5000)} \quad \text{(iii)}$$

Solving for constants A and B,

$$A = \left.\frac{600}{(s+5000)}\right|_{s=0} = 0.12 \quad \text{and} \quad B = \left.\frac{600}{s}\right|_{s=-5000} = -0.12$$

Substituting the values of A and B in Eq. (iii), we get

$$I_L(s) = \frac{0.12}{s} - \frac{0.12}{(s+5000)}$$

Taking inverse Laplace transformation on both sides, the current equation is obtained as:

$$\boldsymbol{i_L(t) = 0.12 - 0.12e^{-5000t}\ \text{A}}$$

Voltage across inductor, $v_L(t) = L\dfrac{di_L(t)}{dt}$

$$= 15\times10^{-3}\frac{d}{dt}(0.12 - 0.12e^{-5000t})$$

$$= 15 \times 10^{-3} \times [0 - (0.12) \times (-5000)e^{-5000t}]$$

$$\boldsymbol{v_L(t) = 9e^{-5000t}\ \text{V}}$$

Result

$$\boldsymbol{i_L(t) = 0.12 - 0.12e^{-5000t}\ \text{A}}$$

$$\boldsymbol{v_L(t) = 9e^{-5000t}\ \text{V}}$$

EXAMPLE 11.13 A series *RC* circuit (Fig. 11.40) with R = 100 Ω and C = 5000 μF has a DC voltage of 25 V applied through a switch at t = 0. Find the equation for the current, voltage across the capacitor and resistor. Assume that there is no initial charge on the capacitor.

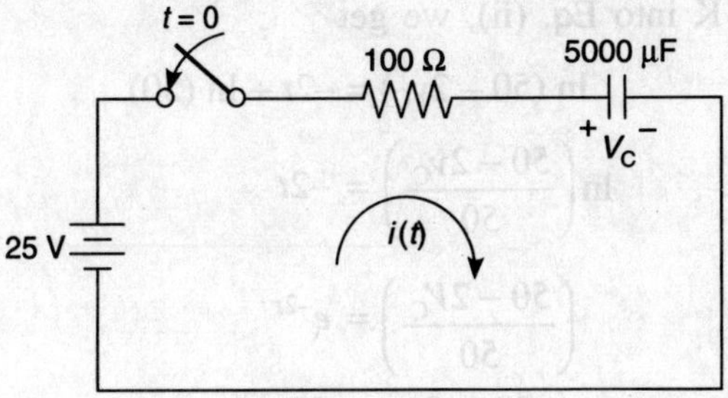

Fig. 11.40

Solution *By Classical method*

The differential equation for the circuit is

$$Ri + v_C = V$$

$$R\left(C\frac{dv_C}{dt}\right) + v_C = V \qquad \left[\because i = C\frac{dv_C}{dt}\right]$$

$$100\left(5000\times10^{-6}\frac{dv_C}{dt}\right) + v_C = 25$$

$$0.5\frac{dv_C}{dt} + v_C = 25 \qquad \text{(i)}$$

$$\frac{dv_C}{dt} + 2v_C = 50$$

$$\frac{dv_C}{dt} = 50 - 2v_C$$

Separating the variables,

$$\frac{dv_C}{50 - 2V_C} = dt$$

Multiplying by '–2' on both sides

$$\frac{-2}{50 - 2V_C}\,dv_C = -2dt$$

Integrating both sides,

$$\int\frac{-2}{50 - 2V_C}\,dv_C = \int -2dt$$

$$\ln(50 - 2v_C) = -2t + K \qquad \text{(ii)}$$

The initial condition is that at $t = 0$, $v_C = 0$ (since there is no initial charge on the capacitor)

$$\therefore \qquad \ln(50) = K$$

Substituting the value of K into Eq. (ii), we get

$$\ln(50 - 2v_C) = -2t + \ln(50)$$

$$\ln\left(\frac{50 - 2v_C}{50}\right) = -2t$$

$$\left(\frac{50 - 2V_C}{50}\right) = e^{-2t}$$

$$50 - 2v_C = 50e^{-2t}$$

Therefore, voltage across the capacitor, $\mathbf{v_C = 25 - 25e^{-2t}\ V}$

The current is given by

$$i(t) = C\frac{dv_C}{dt}$$

$$= 5000 \times 10^{-6}\frac{d}{dt}25(1 - e^{-2t})$$

$$= 0.125\frac{d}{dt}(1 - e^{-2t})$$

$\therefore$ $$\mathbf{i(t) = 0.25e^{-2t}\ A}$$

Voltage across resistor, $v_R = iR$

$$= 100 \times 0.25e^{-2t}$$

$$\mathbf{v_R = 25e^{-2t}\ V}$$

By Laplace transformation method

In view of Eq. (i),

$$0.5\frac{dv_C}{dt} + v_C = 25$$

Taking Laplace transformation on both sides, we get

$$V_C(s) + 0.5[sV_C(s) - V_C(0)] = \frac{25}{s} \quad \text{(iii)}$$

At $t = 0$, $V_C(0) = 0$ (since there is no initial charge on the capacitor); therefore, Eq. (iii) becomes

$$V_C(s) + 0.5sV_C(s) = \frac{25}{s}$$

$$V_C(s)[1 + 0.5s] = \frac{25}{s}$$

$$V_C(s) = \frac{25}{s(0.5s + 1)}$$

$$= \frac{25}{0.5s(s + 2)} = \frac{50}{s(s + 2)}$$

By applying partial fraction technique,

$$V_C(s) = \frac{A}{s} + \frac{B}{(s+2)} \qquad \text{(iv)}$$

Solving for constants A and B, we obtain

$$A = \left.\frac{50}{(s+2)}\right|_{s=0} = 25 \quad \text{and} \quad B = \left.\frac{50}{s}\right|_{s=-2} = -25$$

Substituting the values of A and B into Eq. (iv), we get

$$V_C(s) = \frac{25}{s} - \frac{25}{(s+2)}$$

Taking inverse Laplace transformation on both sides, we get

$$\boldsymbol{v_C(t) = 25 - 25e^{-2t}\ \text{V}}$$

The equations for current and voltage across resistor are as presented under classical method.

Result:

$$\boldsymbol{i(t) = 0.25e^{-2t}\ \text{A}}$$
$$\boldsymbol{v_C = 25(1 - e^{-2t})\ \text{V}}$$
$$\boldsymbol{v_R = 25e^{-2t}\ \text{V}}$$

EXAMPLE 11.14 A 4 μF capacitor with an initial voltage of 2 V is connected to a 12 V battery through resistor $R = 5\ \text{k}\Omega$ at $t = 0$ as shown in Fig. 11.41. Find the voltage across and the current through the capacitor for $t > 0$ and sketch v_R and v_C.

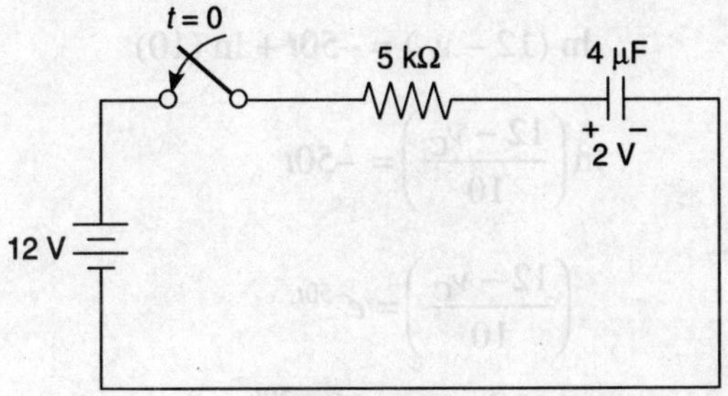

Fig. 11.41

Solution *By Classical method*

The differential equation for the circuit is

$$Ri + v_C = V$$

$$R\left(C\frac{dv_C}{dt}\right) + v_C = V$$

$$5000\left(4 \times 10^{-6}\frac{dv_C}{dt}\right) + v_C = 12$$

$$0.02\frac{dv_C}{dt} + v_C = 12 \tag{i}$$

$$0.02\frac{dv_C}{dt} = 12 - v_C$$

Separating the variables,

$$\frac{dv_C}{12 - v_C} = \frac{dt}{0.02}$$

Multiplying by '–1' on both sides

$$\frac{-1}{12 - v_C}dv_C = \frac{-1}{0.02}dt$$

Integrating both sides,

$$\int\frac{-1}{12 - v_C}dv_C = \int\frac{-1}{0.02}dt$$

$$\ln(12 - v_C) = -50t + K \tag{ii}$$

The initial condition is that at $t = 0$, $v_C = 2$ V

Therefore, $\ln(12 - 2) = \ln(10) = K$

Substituting the value of K into Eq. (ii), we get

$$\ln(12 - v_C) = -50t + \ln(10)$$

$$\ln\left(\frac{12 - v_C}{10}\right) = -50t$$

$$\left(\frac{12 - v_C}{10}\right) = e^{-50t}$$

$$12 - v_C = 10e^{-50t}$$

So, voltage across the capacitor, $\mathbf{v_C = 12 - 10e^{-50t}}$ **V**

The current is given by

$$i(t) = C\frac{dv_C}{dt} = 4\times10^{-6}\frac{d}{dt}(12 - 10e^{-50t})$$

$\therefore$ $\mathbf{i(t) = 0.002e^{-50t}}$ **A**

By Laplace transformation method
In view of Eq. (i),

$$0.02\frac{dv_C}{dt} + v_C = 12$$

Taking Laplace transformation on both sides, we get

$$V_C(s) + 0.02[sV_C(s) - V_C(0)] = \frac{12}{s} \tag{iii}$$

At $t = 0$, $v_C = 2$, i.e. $V_C(0) = 2$, therefore, Eq. (iii) becomes,

$$V_C(s) + 0.02[sV_C(s) - 2] = \frac{12}{s}$$

$$V_C(s) + 0.02sV_C(s) = \frac{12}{s} + 0.04$$

$$V_C(s)\,[1 + 0.02s] = \frac{12}{s} + 0.04$$

$$V_C(s) = \frac{(12/s) + 0.04}{(0.02s + 1)} = \frac{12 + 0.04s}{s(0.02s + 1)} = \frac{12 + 0.04s}{0.02s(s + 50)}$$

$$= \frac{12 + 0.04s}{0.02s(s + 50)}$$

$$= \frac{600 + 2s}{s(s + 50)}$$

By applying partial fraction technique,

$$V_C(s) = \frac{A}{s} + \frac{B}{(s + 50)} \tag{iv}$$

Solving for constants A and B,

$$A = \left.\frac{600 + 2s}{(s + 50)}\right|_{s=0} = 12 \quad \text{and} \quad B = \left.\frac{600 + 2s}{s}\right|_{s=-50} = -10$$

Substituting the values of A and B into Eq. (iv), we get

$$V_C(s) = \frac{12}{s} - \frac{10}{(s + 50)}$$

Taking inverse Laplace transformation on both sides, we get

$$\mathbf{v_C(t) = 12 - 10e^{-50t}\ V}$$

The equation for current is as presented under in classical method, i.e.

$$\mathbf{i(t) = 0.002e^{-50t}\ A}$$

The voltage across capacitor and current through it are shown in Figs. 11.42(a) and (b), respectively. The initial current through the capacitor is found as $\dfrac{12-2}{5\times10^3}=0.002$ A.

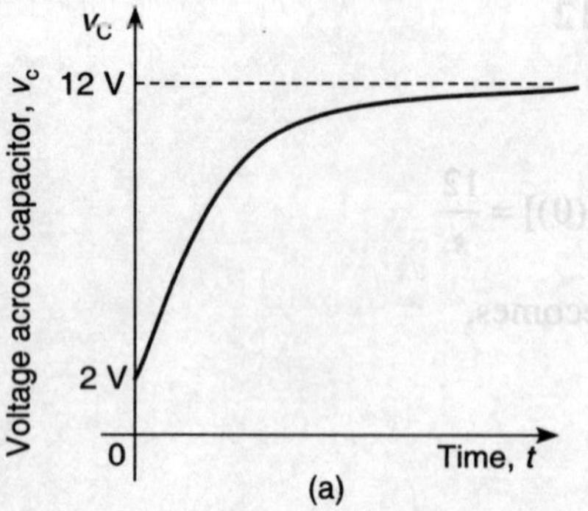

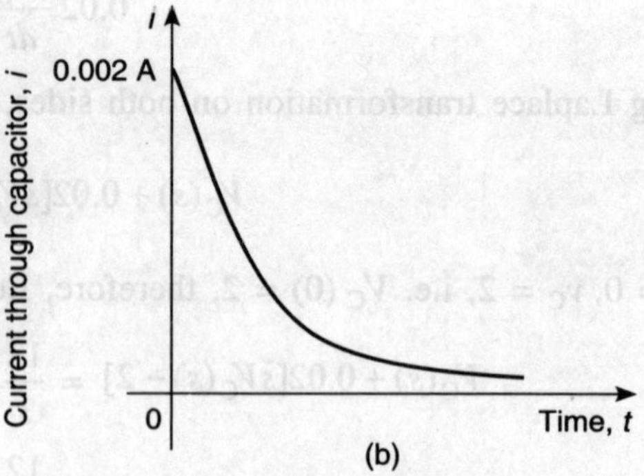

Fig. 11.42

EXAMPLE 11.15 The switch in the circuit shown in Fig. 11.43 is closed at $t = 0$, at which moment the capacitor has charge $Q_0 = 500$ μC, with the polarity indicated. Obtain v_C and q, for $t > 0$, and sketch the graph of q.

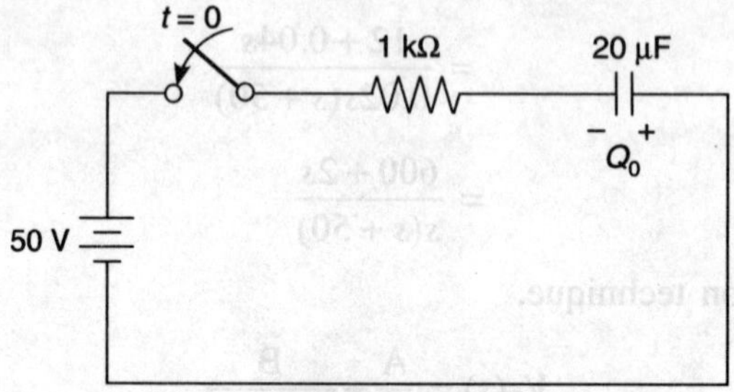

Fig. 11.43

Solution Determination of initial conditions

The corresponding voltage for the initial charge on the capacitor is

$$v_0=\frac{Q_0}{C}=\frac{500\times10^{-6}}{20\times10^{-6}}=25\text{ V}$$

and hence $v_C(0) = -25$ V.

By Classical method

The differential equation for the circuit is

$$Ri + v_C = V$$

$$R\left(C\frac{dv_C}{dt}\right)+v_C=V$$

$$1000\left(20\times10^{-6}\frac{dv_C}{dt}\right)+v_C=50$$

$$0.02\frac{dv_C}{dt}+v_C=50 \qquad \text{(i)}$$

$$0.02\frac{dv_C}{dt}=50-v_C$$

Separating the variables,

$$\frac{dv_C}{50-v_C}=\frac{dt}{0.02}$$

Multiplying by '–1' on both sides

$$\frac{-1}{50-v_C}dv_C=\frac{-1}{0.02}dt$$

Integrating both sides,

$$\int\frac{-1}{50-v_C}dv_C=\int\frac{-1}{0.02}dt$$

$$\ln(50-v_C)=-50t+K \qquad \text{(ii)}$$

The initial condition is that at $t = 0$, $v_C = -25$ V. Therefore,

$$\ln(50+25)=K$$

Substituting the value of K into Eq. (ii), we obtain

$$\ln(50-v_C)=-50t+\ln(75)$$

$$\ln\left(\frac{50-v_C}{75}\right)=-50t$$

$$\left(\frac{50-v_C}{75}\right)=e^{-50t}$$

$$50-v_C=75e^{-50t}$$

Hence, voltage across the capacitor, $\boldsymbol{v_C = 50 - 75e^{-50t}}$ **V**

By Laplace transformation method

In view of Eq. (i),

$$0.02\frac{dv_C}{dt}+v_C=50$$

Taking Laplace transformation on both sides, we get

$$V_C(s)+0.02[sV_C(s)-V_C(0)]=\frac{50}{s} \qquad \text{(iii)}$$

At $t = 0$, $V_C(0) = -25$; therefore, Eq. (iii) becomes

$$V_C(s) + 0.02[sV_C(s) - (-25)] = \frac{50}{s}$$

$$V_C(s) + 0.02sV_C(s) = \frac{50}{s} - 0.5$$

$$V_C(s)[1 + 0.02s] = \frac{50}{s} - 0.5$$

$$V_C(s) = \frac{\frac{50}{s} - 0.5}{(0.02s+1)} = \frac{50 - 0.5s}{s(0.02s+1)}$$

$$= \frac{2500 - 25s}{s(s+50)}$$

By applying partial fraction technique,

$$V_C(s) = \frac{A}{s} + \frac{B}{(s+50)} \tag{iv}$$

Solving for constants A and B,

$$A = \left.\frac{2500 - 25s}{(s+50)}\right|_{s=0} = 50 \quad \text{and} \quad B = \left.\frac{2500 - 25s}{s}\right|_{s=-50} = -75$$

Substituting the values of A and B into Eq. (iv), we get

$$V_C(s) = \frac{50}{s} - \frac{75}{(s+50)}$$

Taking inverse Laplace transformation on both sides, we get

$$\mathbf{v_C = 50 - 75e^{-50t}\ V}$$

The charge,

$$q = Cv_C = 20 \times 10^{-6}(50 - 75e^{-50t})$$

$$\mathbf{q = 1000 - 1500e^{-50t}\ \mu C} \tag{v}$$

The time at which q changes from negative to positive is obtained as follows: Equating Eq. (v) to zero,

$$1000 - 1500e^{-50t} = 0$$

$$e^{-50t} = 0.6666$$

Taking natural logarithm on both sides,

$$-50t = -0.4055$$

Hence,

$$t = 8.11 \text{ ms}$$

Figure 11.44 shows that the initial charge changes from –500 μC to 1000 μC when time t approaches infinity.

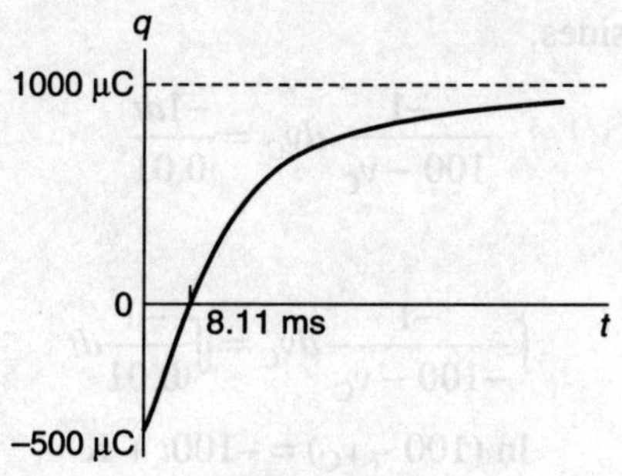

Fig. 11.44

EXAMPLE 11.16 In the circuit shown in Fig. 11.45, the switch is closed on position 1 at $t = 0$ and after one time constant is moved to position 2. Find the current before and after moving to position 2. Assume no initial charge on the capacitor. Sketch the transient current.

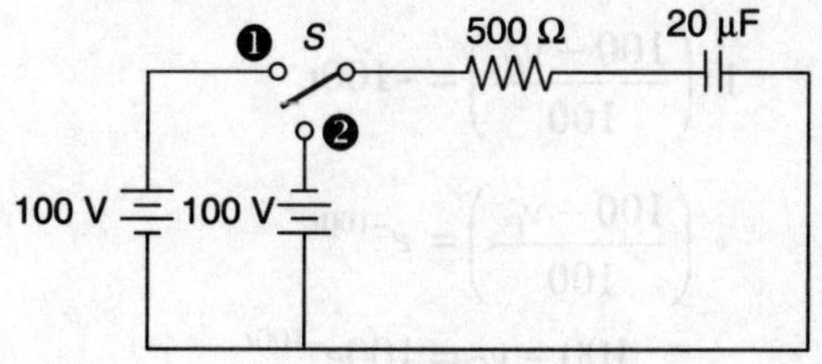

Fig. 11.45

Solution *By Classical method*

At $t = 0$, the switch is put on position 1. The differential equation for this condition is

$$Ri + v_C = V$$

$$R\left(C\frac{dv_C}{dt}\right) + v_C = V$$

$$500\left(20 \times 10^{-6}\frac{dv_C}{dt}\right) + v_C = 100$$

$$0.01\frac{dv_C}{dt} + v_C = 100 \qquad \text{(i)}$$

$$0.01\frac{dv_C}{dt} = 100 - v_C$$

Separating the variables,

$$\frac{dv_C}{100 - v_C} = \frac{dt}{0.01}$$

Multiplying by '–1' on both sides,

$$\frac{-1}{100-v_C}dv_C = \frac{-1dt}{0.01}$$

Integrating both sides,

$$\int\frac{-1}{-100-v_C}dv_C = \int\frac{-1}{0.01}dt$$

$$\ln(100 - v_C) = -100t + K \qquad \text{(ii)}$$

The initial condition is that at $t = 0$, $v_C = 0$

Hence, $\ln(100) = K$

Substituting the value of K into Eq. (ii), we get

$$\ln(100 - v_C) = -100t + \ln(100)$$

$$\ln\left(\frac{100-v_C}{100}\right) = -100t$$

$$\left(\frac{100-v_C}{100}\right) = e^{-100t}$$

$$100 - v_C = 100e^{-100t}$$

Therefore, voltage across the capacitor, $\boldsymbol{v_C = 100 - 100e^{-100t}}$ **V**

The current is found by

$$i(t) = C\frac{dv_C}{dt}$$

$$= 20\times10^{-6}\frac{d}{dt}(100-100e^{-100t})$$

Hence. $\boldsymbol{i(t) = 0.2e^{-100t}}$ **A**

This continues till one time constant, i.e. $RC = 500 \times 20 \times 10^{-6} = 0.01$ s. This time is redefined as t' and is taken as reference.

The current at $t = 0.01$ s ($t' = 0$),

$$i(t) = 0.2e^{-100(0.01)} = 0.0736 \text{ A}$$

The capacitor then has a voltage of

$$v_C = 100 - 100e^{-100(0.01)} = 63.212 \text{ V}$$

The differential equation at position 2 is written as:

$$Ri + v_C = V$$

$$R\left(C\frac{dv_C}{dt'}\right) + v_C = V$$

$$500\left(20\times10^{-6}\frac{dv_C}{dt'}\right)+v_C=-100$$

$$0.01\frac{dv_C}{dt'}+v_C=-100 \qquad \text{(iii)}$$

$$0.01\frac{dv_C}{dt'}=-100-v_C$$

Separating the variables,

$$\frac{dv_C}{-100-v_C}=\frac{dt'}{0.01}$$

Multiplying by '–1' on both sides,

$$\frac{-1}{-100-v_C}dv_C=\frac{-1dt'}{0.01}$$

Integrating both sides,

$$\int\frac{-1}{-100-v_C}dv_C=\int\frac{-1}{0.01}dt'$$

$$\ln(-100-v_C)=-100t'+K \qquad \text{(iv)}$$

The initial condition is that at $t' = 0$ ($t = 0.01$ s), $v_C = 63.212$ V

Hence, $\ln(-163.212) = K$

Substituting the value of K into Eq. (iv), we obtain

$$\ln(-100-v_C)=-100t'+\ln(-163.212)$$

$$\ln\left(\frac{-100-v_C}{-163.212}\right)=-100t'$$

$$\left(\frac{-100-v_C}{-163.212}\right)=e^{-100t'}$$

$$-100-v_C=-163.212e^{-100t'}$$

Therefore, voltage across the capacitor, $v_C = -100 + 163.212e^{-100t'}$ V

The current is found by

$$i(t)=C\frac{dv_C}{dt}=20\times10^{-6}\frac{d}{dt}(-100+163.212e^{-100t'})$$

So, $\boldsymbol{i(t) = -0.3264e^{-100t'}}$ **A**

By Laplace transformation method

In view of Eq. (i),

$$0.01\frac{dv_C}{dt}+v_C=100$$

Taking Laplace transformation on both sides, we get

$$V_C(s) + 0.01[sV_C(s) - V_C(0)] = \frac{100}{s} \qquad \text{(v)}$$

At t = 0, V_C (0) = 0; therefore, Eq. (v) becomes

$$V_C(s) + 0.01sV_C(s) = \frac{100}{s}$$

$$V_C(s)\,[1 + 0.01s] = \frac{100}{s}$$

$$V_C(s) = \frac{\frac{100}{s}}{(0.01s+1)} = \frac{100}{0.01s(s+100)}$$

$$= \frac{10000}{s(s+100)}$$

By applying partial fraction technique,

$$V_C(s) = \frac{A}{s} + \frac{B}{(s+100)} \qquad \text{(vi)}$$

Solving for constants A and B,

$$A = \left.\frac{10000}{(s+100)}\right|_{s=0} = 100 \quad \text{and} \quad B = \left.\frac{10000}{s}\right|_{s=-100} = -100$$

Substituting the values of A and B into Eq. (vi), we get

$$V_C(s) = \frac{100}{s} - \frac{100}{(s+100)}$$

Taking inverse Laplace transformation on both sides, we get

$$\mathbf{v_C = 100 - 100e^{-100t}\ V}$$

The current is found by

$$i(t) = C\frac{dv_C}{dt} = 20 \times 10^{-6}\frac{d}{dt}(100 - 100e^{-100t})$$

Hence, $\qquad \boldsymbol{i(t) = 0.2e^{-100t}}$ **A**

This continues till one time constant, i.e. $RC = 500 \times 20 \times 10^{-6} = 0.01$ s. This time is redefined as t' and is taken as reference.

The current at t = 0.01 s (t' = 0),

$$i(t) = 0.2e^{-100(0.01)} = 0.0736 \text{ A}$$

The capacitor then has a voltage of

$$v_C = 100 - 100e^{-100(0.01)} = 63.212 \text{ V}$$

From Eq. (iii), the differential equation at position 2 is

$$0.01\frac{dv_C}{dt'} + v_C = -100$$

Taking Laplace transformation on both sides, we get

$$V_C(s) + 0.01[sV_C(s) - V_C(0)] = -\frac{100}{s} \quad \text{(vii)}$$

At $t' = 0$, $V_C(0) = 63.212$, therefore, Eq. (vii) becomes

$$V_C(s) + 0.01[sV_C(s) - 63.212] = -\frac{100}{s}$$

$$V_C(s) + 0.01sV_C(s) = -\frac{100}{s} + 0.63212$$

$$V_C(s)\,[1 + 0.01s] = -\frac{100}{s} + 0.63212$$

$$V_C(s) = \frac{-\frac{100}{s} + 0.63212}{(0.01s + 1)} = \frac{-100 + 0.63212s}{0.01s(s + 100)}$$

$$= \frac{-10000 + 63.212s}{s(s + 100)}$$

By applying partial fraction technique,

$$V_C(s) = \frac{A}{s} + \frac{B}{(s + 100)} \quad \text{(viii)}$$

Solving for constants A and B,

$$A = \left.\frac{-10000 + 63.212s}{(s + 100)}\right|_{s=0} = -100 \quad \text{and} \quad B = \left.\frac{-10000 + 63.212s}{s}\right|_{s=-100} = 163.212$$

Substituting the values of A and B into Eq. (viii), we get

$$V_C(s) = \frac{-100}{s} + \frac{163.212}{(s + 100)}$$

Taking inverse Laplace transformation on both sides, we get

$$v_C(t') = -100 + 163.212e^{-100t'} \text{ V}$$

The current equation is as same as obtained under classical method. The transient current is shown in Fig. 11.46.

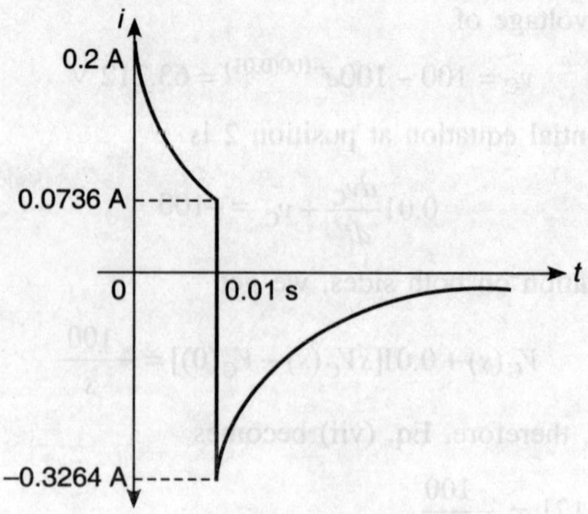

Fig. 11.46

Result

$$i(t) = 0.2e^{-100t} \text{ A} \qquad \text{for } 0 < t < 1 \text{ time constant}$$
$$= -0.3264e^{-100t'} \text{ A} \qquad \text{for } t > 1 \text{ time constant}$$

EXAMPLE 11.17 For the circuit shown in Fig. 11.47, find the current equation when the switch is opened at $t = 0$.

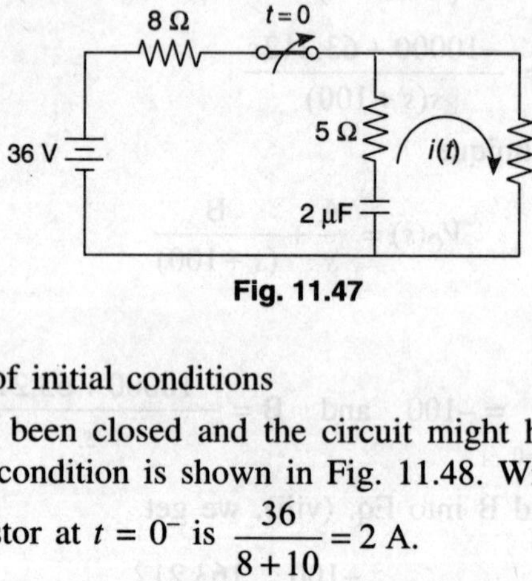

Fig. 11.47

Solution Determination of initial conditions

At $t = 0^-$, the switch had been closed and the circuit might have attained steady state. The equivalent circuit for this condition is shown in Fig. 11.48. With reference to Fig. 11.48, the current through 10 Ω resistor at $t = 0^-$ is $\dfrac{36}{8+10} = 2$ A.

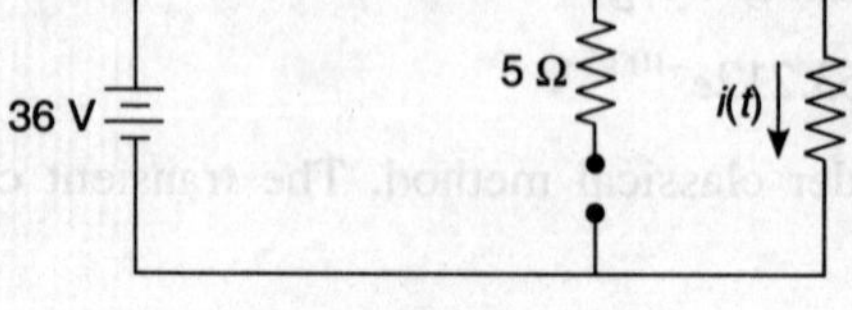

Fig. 11.48

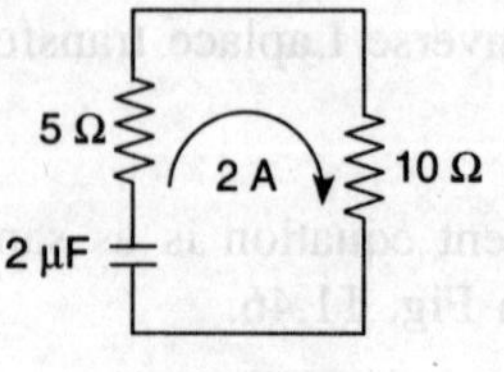

Fig. 11.49

By Classical method

The switch is opened at $t = 0$ and the equivalent circuit for this condition is shown in Fig. 11.49. By using Kirchhoff's law, the differential equation is given by

$$Ri + \frac{1}{C}\int i\,dt = V$$

$$15i + \frac{1}{2\times 10^{-6}}\int i\,dt = 0$$

$$15\frac{di}{dt} + (0.5\times 10^{6})i = 0$$

$$15\frac{di}{dt} = -(0.5 \times 10^{6})i$$

Separating the variables,

$$\frac{di}{i} = -33333.33t$$

Integrating both sides,

$$\ln(i) = -33333.33t + K \quad \text{(i)}$$

The initial condition is that at $t = 0$, $i = 2$

Hence, $\ln(2) = K$

Substituting the value of K into Eq. (i), we get

$$\ln(i) = -33333.33t + \ln(2)$$

$$\ln\left(\frac{i}{2}\right) = -33333.33t$$

$$\left(\frac{i}{2}\right) = e^{-33333.33t}$$

Therefore, the current, $i = 2e^{-33333.33t}$ A

Note: The solution by Laplace transformation technique is left to the readers as exercise.

EXAMPLE 11.18 In the circuit shown in Fig. 11.50, the switch is closed for a long time. At $t = 0$, the switch is opened. Find the voltage across the capacitor.

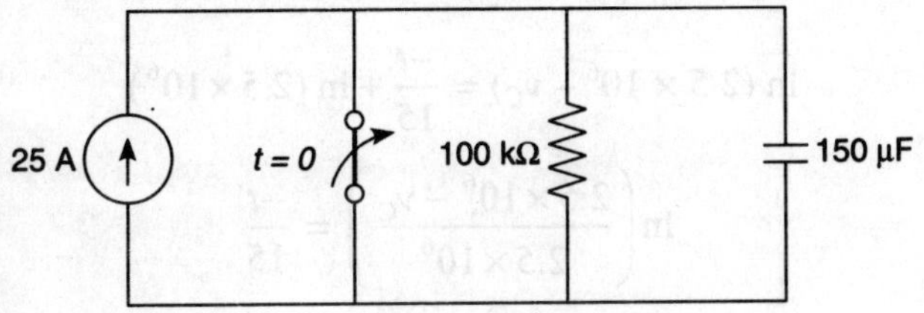

Fig. 11.50

Solution Determination of initial conditions

The switch is in closed state for a long period and during this period, the current neither flows through R nor C. The current supplied by the current source circulates through the switch only. Therefore, the voltage across the capacitor $v_C(0^-)$ is zero. $v_C(0)$ is also zero because the capacitor cannot change its voltage immediately.

By Classical method

At $t = 0$, the switch is opened. Applying Kirchhoff's current law, the current equation is given by

$$\frac{v_C}{R} + C\frac{dv_C}{dt} = I$$

$$\frac{v_C}{100\times10^3} + 150\times10^{-6}\frac{dv_C}{dt} = 25$$

$$v_C + 15\frac{dv_C}{dt} = 2.5\times10^6$$

$$15\frac{dv_C}{dt} = 2.5\times10^6 - v_C$$

Separating the variables,

$$\frac{dv_C}{2.5\times10^6 - v_C} = \frac{dt}{15}$$

Multiplying by '–1' on both sides,

$$\frac{-dv_C}{2.5\times10^6 - v_C} = \frac{-dt}{15}$$

Integrating both sides,

$$\int\frac{-dv_C}{2.5\times10^6 - v_C} = \int\frac{-dt}{15}$$

$$\ln(2.5\times10^6 - v_C) = \frac{-t}{15} + K \qquad \text{(i)}$$

The initial condition is that at $t = 0$, $v_C = 0$

Therefore, $\ln(2.5\times10^6) = K$

Substituting the value of K into Eq. (i), we get

$$\ln(2.5\times10^6 - v_C) = \frac{-t}{15} + \ln(2.5\times10^6)$$

$$\ln\left(\frac{2.5\times10^6 - v_C}{2.5\times10^6}\right) = \frac{-t}{15}$$

$$\left(\frac{2.5\times10^6 - v_C}{2.5\times10^6}\right) = e^{-\frac{t}{15}}$$

$$2.5\times10^6 - v_C = 2.5\times10^6 e^{-\frac{t}{15}}$$

So, voltage across the capacitor, $\boldsymbol{v_C = 2.5\times10^6\left(1-e^{-\frac{t}{15}}\right)}$ **V**

By Laplace transformation method

$$\frac{v_C}{R} + C\frac{dv_C}{dt} = I$$

Taking Laplace transformation on both sides, we get

$$\frac{1}{R}V_C(s) + C[sV_C(s) - V_C(0)] = \frac{I}{s}$$

The initial condition is that at $t = 0$, $v_C(0) = 0$. Therefore, in s-domain, $V_C(0) = 0$. Substituting this into the above equation and rearranging, we get

$$V_C(s)\left(\frac{1}{R} + Cs\right) = \frac{I}{s}$$

$$V_C(s) = \frac{I}{s\left(\frac{1}{R} + Cs\right)}$$

$$= \frac{(I/C)}{s\left(s + \frac{1}{RC}\right)} \qquad \text{(ii)}$$

Substituting the values of R, C and I into Eq. (ii), we get

$$= \frac{\left(\frac{25}{150\times10^{-6}}\right)}{s\left(s + \frac{1}{100\times10^3\times150\times10^{-6}}\right)} = \frac{\left(\frac{25}{150\times10^{-6}}\right)}{s\left(s+\frac{1}{15}\right)}$$

$$= \frac{A}{s} + \frac{B}{\left(s + \frac{1}{15}\right)} \qquad \text{(iii)}$$

Solving for constants A and B,

$$A = \left.\frac{\left(\frac{25}{150\times10^{-6}}\right)}{\left(s+\frac{1}{15}\right)}\right|_{s=0} = 2.5\times10^6 \quad \text{and} \quad B = \left.\frac{\left(\frac{25}{150\times10^{-6}}\right)}{s}\right|_{s=-\frac{1}{15}} = -2.5\times10^6$$

Substituting the values of A and B into Eq. (iii), we obtain

$$V_C(s) = \frac{2.5\times10^6}{s} - \frac{2.5\times10^6}{\left(s+\frac{1}{15}\right)}$$

Taking inverse Laplace transformation on both sides, we get

$$v_C(t) = 2.5\times10^6 - 2.5\times10^6 e^{-\frac{t}{15}}$$

or

$$\mathbf{v_C(t) = 2.5\times10^6 \left(1 - e^{-\frac{t}{15}}\right) V}$$

EXAMPLE 11.19 In the circuit shown in Fig. 11.51, the switch is kept open for a longer time. The capacitor attains a value of 80 V, 10 ms after closing the switch. Then the switch is kept closed for very long time. Again the switch is opened and after 0.5 s, the voltage across the capacitor becomes 90 V. Find the values of *R* and *C*.

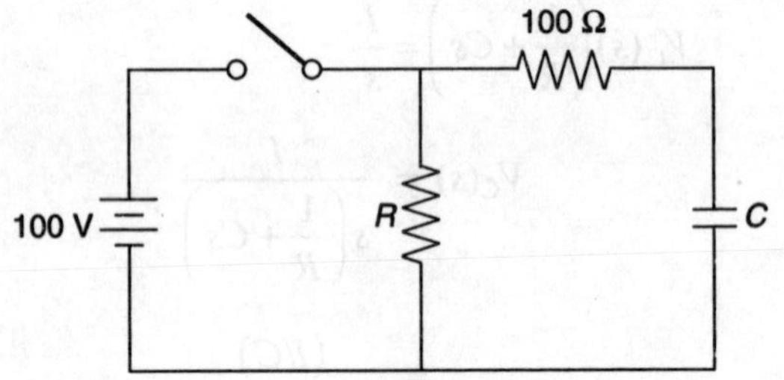

Fig. 11.51

Solution Determination of initial conditions

Initially the switch is kept open for a long period; therefore, the voltage across the capacitor is zero. This is the initial condition for the next transition where the switch is closed. Let us assume this happens at $t_1 = 0$ and the circuit for this position is drawn as shown in Fig. 11.52.

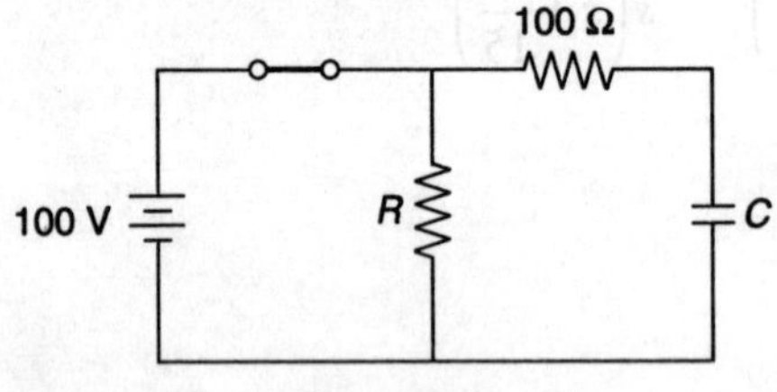

Fig. 11.52

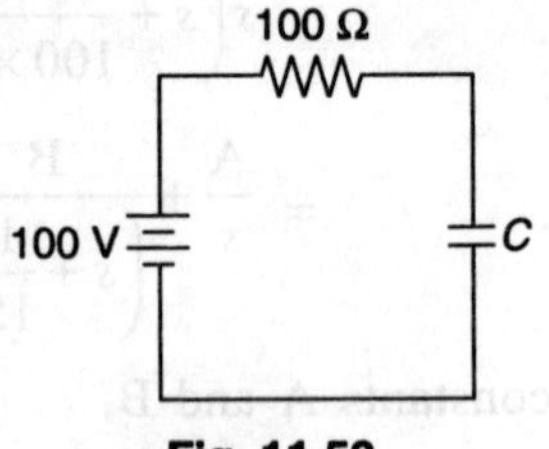

Fig. 11.53

In Fig. 11.52, the voltage across *R* should also be 100 V, since it is connected in parallel with the source. By using this property, the circuit in Fig. 11.52 is redrawn as shown in Fig. 11.53. Applying KVL, we can write

$$Ri + v_C = V$$

$$R\left(C\frac{dv_C}{dt_1}\right) + v_C = V$$

$$100C\frac{dv_C}{dt_1} + v_C = 100$$

$$100C\frac{dv_C}{dt_1} = 100 - v_C$$

Separating the variables,

$$\frac{dv_C}{100 - v_C} = \frac{dt_1}{100C}$$

Multiplying by '–1' on both sides

$$\frac{-dv_C}{100 - v_C} = \frac{-dt_1}{100C}$$

Integrating both sides,

$$\int \frac{-1}{100 - v_C} dv_C = \int \frac{-1}{100C} dt_1$$

$$\ln(100 - v_C) = \frac{-1}{100C} t_1 + K \qquad \text{(i)}$$

The initial condition is that at $t = 0$, $v_C = 0$

Hence, $\ln(100) = K$

Substituting the value of K into Eq. (i), we obtain

$$\ln(100 - v_C) = \frac{-1}{100C} t_1 + \ln(100)$$

$$\ln\left(\frac{100 - v_C}{100}\right) = \frac{-1}{100C} t_1$$

$$\left(\frac{100 - v_C}{100}\right) = e^{\frac{-1}{100C} t_1}$$

$$100 - v_C = 100 e^{\frac{-1}{100C} t_1}$$

Therefore, voltage across the capacitor, $v_C = 100\left(1 - e^{\frac{-1}{100C}t_1}\right)$ V. (ii)

Given that at t_1 = 10 ms, v_C = 80 V. Substituting this into Eq. (ii), we get

$$80 = 100\left(1 - e^{\frac{-1}{100C}\times\frac{10}{1000}}\right) \quad \text{(iii)}$$

Solving,

$$C = 62.1349\ \mu\text{F}$$

The circuit remains in this state (as shown in Fig. 11.53) for another long period and so attains steady state. Now the voltage across the capacitor becomes 100 V because at steady state the capacitor acts as an open-circuit. This is the initial condition for the next transition when the switch is opened. Let us assume this happens at t_2 = 0 and the circuit for this position is drawn as shown in Fig. 11.54.

Applying KVL,

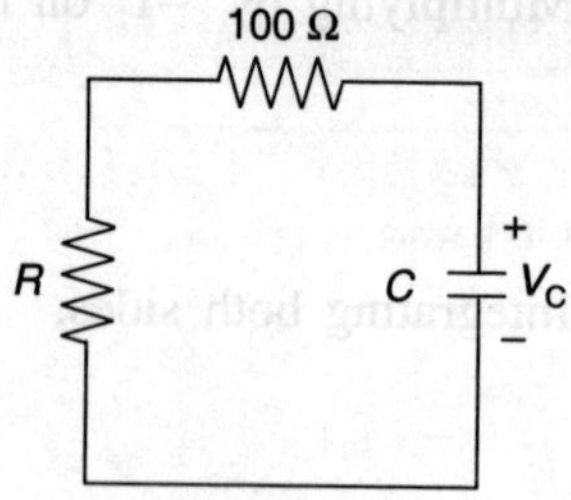

Fig. 11.54

$$Ri + 100i + v_C = 0$$

$$(R + 100)i = -v_C$$

$$(R+100)C\frac{dv_C}{dt_2} = -v_C$$

Separating the variables,

$$\frac{dv_C}{v_C} = \frac{-dt_2}{(R+100)C}$$

Integrating both sides,

$$\ln(v_C) = \frac{-t_2}{(R+100)C} + K \quad \text{(iv)}$$

The initial condition is that at t_2 = 0, v_C = 100

Hence, $\ln(100) = K$

Substituting the value of K into Eq. (iv), we get

$$\ln(v_C) = \frac{-t_2}{(R+100)C} + \ln(100)$$

$$\ln\left(\frac{v_C}{100}\right) = \frac{-t_2}{(R+100)C}$$

$$\left(\frac{v_C}{100}\right) = e^{\frac{-1}{(R+100)C}t_2}$$

$$v_C = 100e^{\frac{-1}{(R+100)C}t_2} \quad \text{(v)}$$

Substituting the value of C found into Eq. (iii), we obtain

$$v_C = 100e^{\frac{-1}{(R+100)(62.1349\times10^{-6})}t_2} = 100e^{\frac{-16094}{(R+100)}t_2}$$

Given that at $t_2 = 0.5$ s, $v_C = 90$ V. Substituting this into Eq. (v), we get

$$90 = 100e^{\frac{-16094}{(R+100)}(0.5)}$$

Solving,

$$R = 76276\ \Omega$$

Result:

$$R = 76276\ \Omega$$
$$C = 62.1349\ \mu F$$

Note: The solution by Laplace transformation technique is left to the readers as exercise.

EXAMPLE 11.20 In the circuit shown in Fig. 11.55, the switch is closed at $t = 0$. Determine $i_C(t)$ and $v_C(t)$.

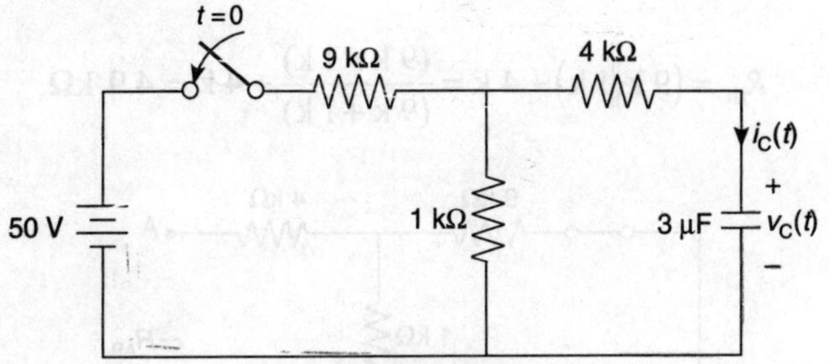

Fig. 11.55

Solution Determination of initial conditions

The switch is closed at $t = 0$. Therefore, at $t = 0^-$, the current through the capacitor was zero. Thus, the initial voltage across the capacitor, i.e. $v_C(0)^- = 0$. Since the voltage across the capacitor cannot be changed instantaneously, $v_C(0)^+$ is also zero.

For all values of $t > 0$, the switch is closed and the circuit is drawn as shown in Fig. 11.56.

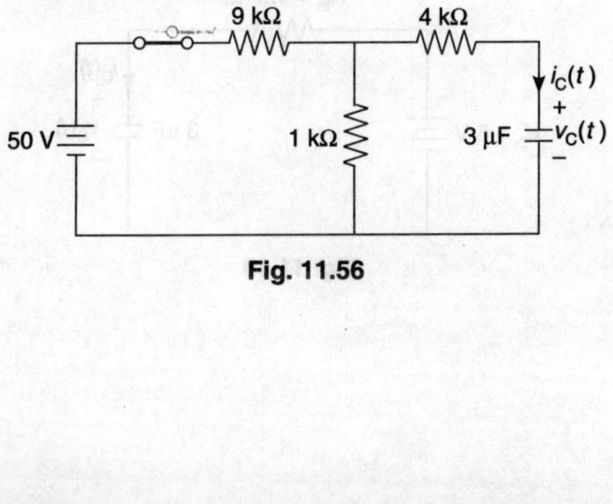

Fig. 11.56

Next, Thevenin's equivalent circuit is found as follows:

Step 1 The capacitor is removed from the circuit and the open circuit terminals are marked as A and B as shown in Fig. 11.57.

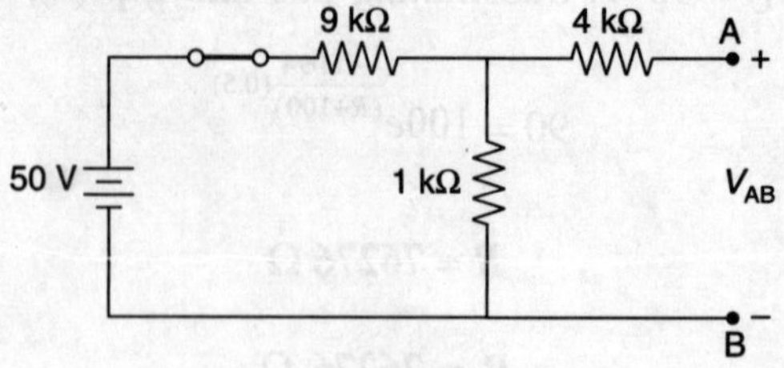

Fig. 11.57

Step 2 The open circuit voltage V_{AB} is calculated as $\dfrac{50}{(9000+1000)}\times 1000 = 5$ V.

Step 3 Thevenin's resistance is calculated by shorting the voltage source as shown in Fig. 11.58, and looking back the network across A and B as:

$$R_{th} = (9\,k \| 1\,k) + 4\,k = \frac{(9\,k \times 1\,k)}{(9\,k + 1\,k)} + 4\,k = 4.9\text{ k}\Omega.$$

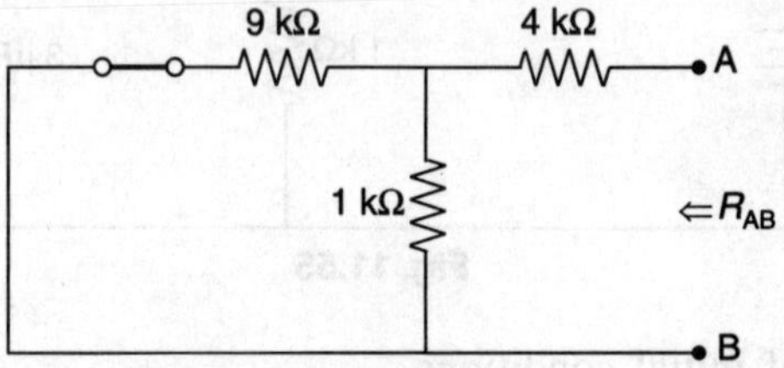

Fig. 11.58

Step 4 Original network across terminals A–B is replaced by its Thevenin's equivalent network as shown in Fig. 11.59.

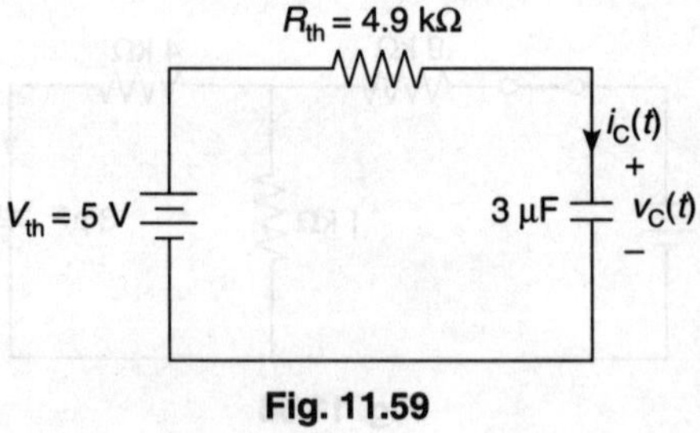

Fig. 11.59

By Classical method

Considering the network shown in Fig. 11.59 and applying KVL,

$$Ri + v_C = V$$

$$R\left(C\frac{dv_C}{dt}\right) + v_C = V$$

$$4900 \times 3 \times 10^{-6}\frac{dv_C}{dt} + v_C = 5$$

$$0.0147\frac{dv_C}{dt} = 5 - v_C$$

Separating the variables,

$$\frac{dv_C}{5 - v_C} = \frac{dt}{0.0147}$$

Multiplying by '–1' on both sides

$$\frac{-dv_C}{5 - v_C} = \frac{-dt}{0.0147}$$

Integrating both sides,

$$\int \frac{-1}{5 - v_C} dv_C = \int \frac{-1}{0.0147} dt$$

$$\ln(5 - v_C) = \frac{-1}{0.0147}t + K \qquad \text{(i)}$$

The initial condition is that, at $t = 0$, $v_C = 0$

Hence, $\ln(5) = K$

Substituting the value of K into Eq. (i), we get

$$\ln(5 - v_C) = -68.0272t + \ln(5)$$

$$\ln\left(\frac{5 - v_C}{5}\right) = -68.0272t$$

$$\left(\frac{5 - v_C}{5}\right) = e^{-68.0272t}$$

$$5 - v_C = 5e^{-68.0272t}$$

Therefore, voltage across the capacitor, $\mathbf{v_C = 5(1 - e^{-68.0272t})}$ **V**

The current is found by

$$i(t) = C\frac{dv_C}{dt} = 3 \times 10^{-6} \times 5\frac{d}{dt}(1 - e^{-68.0272t})$$

$$\mathbf{i(t) = 0.00102e^{-68.0272t}\ A}$$

Note: **The solution by Laplace transformation technique is left to the readers as exercise.**

EXAMPLE 11.21 A series *RLC* circuit shown in Fig. 11.60, has a DC voltage of 200 V applied at $t = 0$. Find the resulting current. Assume initial voltage on the capacitor is zero.

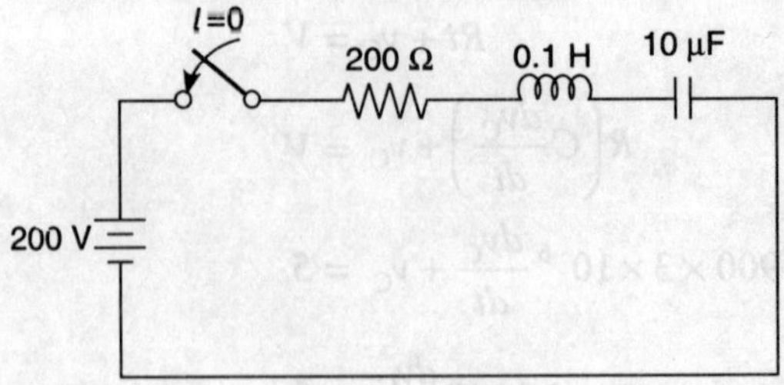

Fig. 11.60

Solution Determination of initial conditions

(i) The switch is closed at $t = 0$. Therefore, at $t = 0^-$, the current through the inductor is zero, i.e. $i(0^-) = 0$, which is also equal to $i(0^+)$, since the current through the inductor cannot be changed instantaneously. Hence, we can write

$$i(0^-) = 0 = i(0^+) \tag{i}$$

(ii) Initial voltage on the capacitor is zero, which results in

$$v_C(0^-) = 0 = v_C(0^+) \tag{ii}$$

as the voltage across the capacitor cannot be changed instantaneously.

By Classical method

Applying KVL after closing the switch

$$200i + 0.1\frac{di}{dt} + \frac{1}{10\times10^{-6}}\int idt = 200 \tag{iii}$$

At $t = 0^+$,

$$200i(0^+) + 0.1\frac{di(0^+)}{dt} + \frac{1}{10\times10^{-6}}\int i(0^+)\cdot dt = 200$$

$$200i(0^+) + v_L(0^+) + v_C(0^+) = 200 \tag{iv}$$

Substituting the values of $i(0^+)$ and $v_C(0^+)$ from Eqs. (i) and (ii) into Eq. (iv), we get

$$v_L(0^+) = 200$$

or

$$L\frac{di(0^+)}{dt} = 200$$

or

$$\frac{di(0^+)}{dt} = \frac{200}{L}$$

$$\frac{di(0^+)}{dt} = \frac{200}{0.1} = 2000 \text{ A/s} \tag{v}$$

Differentiating Eq. (iii) with respect to t,

$$200\frac{di}{dt}+0.1\frac{d^2i}{dt^2}+\frac{1}{10\times10^{-6}}i=0$$

or

$$\frac{d^2i}{dt^2}+2000\frac{di}{dt}+10^6 i=0$$

$$(D^2+2000D+10^6)\,i=0$$

where $D=\frac{d}{dt}$ and $D^2=\frac{d^2}{dt^2}$.

The auxiliary equation is

$$D^2+2000D+10^6=0$$

The roots of the above equation are

$$D_1, D_2=\frac{-2000\pm\sqrt{2000^2-(4\times10^6)}}{2}$$

$$=-1000,\ -1000$$

Roots D_1 and D_2 are real and equal and the solution is of the form (see Section 11.6.1):

$$i(t)=\text{A}e^{-D_1t}+\text{B}te^{-D_2t}$$

or

$$i(t)=\text{A}e^{-1000t}+\text{B}te^{-1000t} \tag{vi}$$

Constants A and B are determined by applying initial conditions as follows:
From Eq. (i), at $t = 0$, $i(t) = 0$. Hence, Eq. (vi) becomes,

$$0=\text{A}e^0+0$$

or

$$\text{A}=0 \tag{vii}$$

Substituting the value of A from Eq. (vii) into Eq. (vi), we get

$$i(t)=\text{B}te^{-1000t} \tag{viii}$$

Differentiating Eq. (viii) with respect to t,

$$\frac{di(t)}{dt}=\text{B}[te^{-1000t}(-1000)+e^{-1000t}] \tag{ix}$$

From Eq. (v), at $t = 0^+$,

$$\frac{di(0^+)}{dt}=2000$$

Hence, Eq. (ix) becomes

$$2000=\text{B} \tag{x}$$

Substituting the value of B from Eq. (x) into Eq. (viii), we get

$$\mathbf{i(t)=2000te^{-1000t}\ A}$$

By Laplace transformation method

In view of Eq. (iii),

$$200i + 0.1\frac{di}{dt} + \frac{1}{10\times 10^{-6}}\int i dt = 200$$

Taking Laplace transformation on both sides, we get

$$200I(s) + 0.1[sI(s) - I(0)] + \frac{1}{10\times 10^{-6}}\frac{I(s)}{s} = \frac{200}{s}$$

From Eq. (i), $i(0^+) = 0$, and in *s*-domain, $I(0) = 0$. Therefore, the above equation becomes

$$200I(s) + 0.1sI(s) + \frac{1}{10\times 10^{-6}}\frac{I(s)}{s} = \frac{200}{s}$$

$$I(s)\left[200 + 0.1s + \frac{1}{10\times 10^{-6}}\frac{1}{s}\right] = \frac{200}{s}$$

$$I(s)\left(\frac{0.002s + 1\times 10^{-6}s^2 + 1}{10\times 10^{-6}s}\right) = \frac{200}{s}$$

$$I(s) = \frac{200}{s}\left(\frac{10\times 10^{-6}s}{0.002s + 1\times 10^{-6}s^2 + 1}\right)$$

$$= \left(\frac{0.002}{0.002s + 1\times 10^{-6}s^2 + 1}\right)$$

$$= \left(\frac{2000}{s^2 + 2000s + 1\times 10^6}\right) \quad \text{(xi)}$$

The roots of the denominator are

$$= \frac{-2000 \pm \sqrt{2000^2 - (4\times 10^6)}}{2}$$

$$= -1000, -1000$$

Therefore, Eq. (xi) becomes

$$I(s) = \frac{2000}{(s+1000)^2}$$

Taking inverse Laplace transformation on both sides, we get

$$\boldsymbol{i(t) = 2000te^{-1000t}\ \text{A}}$$

EXAMPLE 11.22 Consider *RLC* series circuit shown in Fig. 11.61. The circuit is initially relaxed. The switch is closed at $t = 0$. Find the transient current, the voltage across resistor, the voltage across inductor and the voltage across the capacitor.

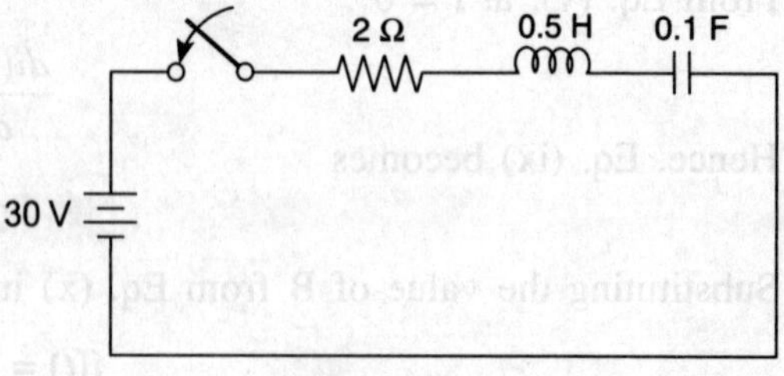

Fig. 11.61

Solution Determination of initial conditions

(i) The switch is closed at $t = 0$. Therefore, at $t = 0^-$, the current through the inductor is zero. i.e. $i(0^-) = 0$, which is also equal to $i(0^+)$, since the current through the inductor cannot be changed instantaneously. Hence, we can write

$$i(0^-) = 0 = i(0^+) \tag{i}$$

(ii) Initial voltage on the capacitor is zero, which results in

$$v_C(0^-) = 0 = v_C(0^+) \tag{ii}$$

as the voltage across the capacitor cannot be changed instantaneously.

By Classical method

Applying KVL after closing the switch

$$2i + 0.5\frac{di}{dt} + \frac{1}{0.1}\int idt = 30 \tag{iii}$$

At $t = 0^+$,

$$2i(0^+) + 0.5\frac{di(0^+)}{dt} + \frac{1}{0.1}\int i(0^+)dt = 30$$

$$2i(0^+) + v_L(0^+) + v_C(0^+) = 30 \tag{iv}$$

Substituting the values of $i(0^+)$ and $v_C(0^+)$ from Eqs. (i) and (ii) into Eq. (iv), we get

$$v_L(0^+) = 30$$

or

$$L\frac{di(0^+)}{dt} = 30$$

or

$$\frac{di(0^+)}{dt} = \frac{30}{L}$$

$$\frac{di(0^+)}{dt} = \frac{30}{0.5} = 60 \text{ A/s} \tag{v}$$

Differentiating Eq. (iii) with respect to t,

$$2\frac{di}{dt} + 0.5\frac{d^2i}{dt^2} + \frac{1}{0.1}i = 0$$

or

$$\frac{d^2i}{dt^2} + 4\frac{di}{dt} + 20i = 0$$

$$(D^2 + 4D + 20)i = 0$$

where $D = \dfrac{d}{dt}$ and $D^2 = \dfrac{d^2}{dt^2}$

The auxiliary equation is

$$D^2 + 4D + 20 = 0$$

The roots of the above equation are

$$D_1, D_2 = \frac{-4 \pm \sqrt{4^2 - (4 \times 20)}}{2}$$
$$= -2 + j4,\ -2 - j4$$

The roots are complex conjugates and the solution is of the form (see Section 11.6.1),

$$i(t) = \text{A}e^{D_1 t} + \text{B}e^{D_2 t}$$
$$= \text{A}e^{(-2+j4)t} + \text{B}e^{(-2-j4)t}$$
$$= e^{-2t}\left(\text{A}e^{j4t} + \text{B}e^{-j4t}\right)$$
$$= e^{-2t}\left[\text{A}(\cos 4t + j\sin 4t) + \text{B}(\cos 4t - j\sin 4t)\right]$$
$$= e^{-2t}\left[(\text{A}+\text{B})\cos 4t + j(\text{A}-\text{B})\sin 4t\right]$$

Let C_1 = (A + B) and $C_2 = j$(A − B), we get

$$i(t) = e^{-2t}[C_1 \cos 4t + C_2 \sin 4t] \qquad \text{(vi)}$$

Constants C_1 and C_2 are determined by applying initial conditions as follows:
From Eq. (i), at $t = 0$, $i(t) = 0$. Hence, Eq. (vi) becomes,

$$0 = e^{-0}[C_1 \cos 0 + C_2 \sin 0]$$

or

$$C_1 = 0 \qquad \text{(vii)}$$

Substituting the value of C_1 from Eq. (vii) into Eq. (vi), we get

$$i(t) = e^{-2t} C_2 \sin 4t \qquad \text{(viii)}$$

Differentiating Eq. (viii) with respect to t,

$$\frac{di}{dt} = C_2\,[e^{-2t}(-2)\sin 4t + e^{-2t}(4)\cos 4t] \qquad \text{(ix)}$$

From Eq. (v), at $t = 0^+$,

$$\frac{di(0^+)}{dt} = 60$$

So, Eq. (ix) becomes,

$$60 = C_2[e^{-0}(-2)\sin 0 + e^{-0}(4)\cos 0]$$
$$60 = 4C_2$$

or

$$C_2 = 15 \qquad \text{(x)}$$

Substituting the value C_2 from Eq. (x) into Eq. (viii), we get

$$i(t) = 15e^{-2t} \sin 4t \text{ A}$$

By Laplace transformation method

In view of Eq. (iii),

$$2i + 0.5\frac{di}{dt} + \frac{1}{0.1}\int i dt = 30$$

Taking Laplace transformation on both sides, we get

$$2I(s) + 0.5[sI(s) - I(0)] + \frac{1}{0.1}\frac{I(s)}{s} = \frac{30}{s}$$

From Eq. (i), $i(0^+) = 0$, and in s-domain, $I(0) = 0$. Therefore, the above equation becomes,

$$2I(s) + 0.5sI(s) + \frac{1}{0.1}\frac{I(s)}{s} = \frac{30}{s}$$

$$I(s)\left[2 + 0.5s + \frac{1}{0.1}\frac{1}{s}\right] = \frac{30}{s}$$

$$I(s)\left(\frac{0.2s + 0.05s^2 + 1}{0.1s}\right) = \frac{30}{s}$$

$$I(s) = \frac{30}{s}\left(\frac{0.1s}{0.2s + 0.05s^2 + 1}\right)$$

$$= \frac{3}{0.05(s^2 + 4s + 20)}$$

$$= \frac{60}{s^2 + 4s + 20} \quad \text{(xi)}$$

$$= \frac{60}{(s+2)^2 + 16} = \frac{60}{(s+2)^2 + 4^2}$$

$$= (15)\frac{4}{(s+2)^2 + 4^2}$$

Taking inverse Laplace transformation on both sides,

$$i(t) = 15e^{-2t} \sin 4t \text{ A}$$

Voltage across resistor,

$$v_R = Ri$$

$$= 2 \times 15e^{-2t} \sin 4t$$

$$v_R = 30e^{-2t} \sin 4t \text{ V} \quad \text{(xii)}$$

Voltage across inductor, $v_L = L\frac{di}{dt}$

$$= 0.5\frac{d}{dt}(15e^{-2t}\sin 4t)$$

$$= 7.5\,[e^{-2t}(-2)\sin 4t + e^{-2t}\cos 4t\,(4)]$$

$$= 7.5e^{-2t}\,[\cos 4t\,(4) - \sin 4t\,(2)] \qquad \text{(xiii)}$$

A right-angled triangle is constructed with 4 and 2 as sides as shown in Fig. 11.62. With reference to Fig. 11.62, we can write

$$\phi = \tan^{-1}\left(\frac{4}{2}\right) = 63.43°$$

$$\sin\phi = \sin 63.43° = \frac{4}{4.472}$$

or $$4 = 4.472\sin 63.43° \qquad \text{(xiv)}$$

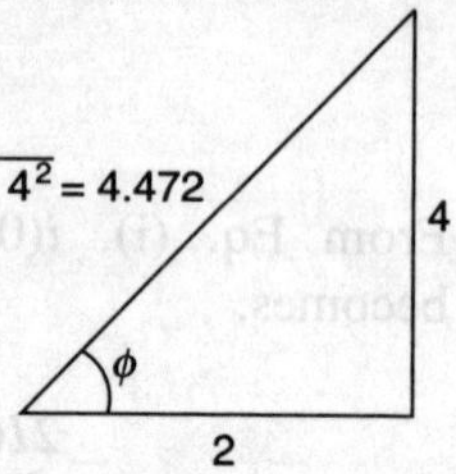

Fig. 11.62

Similarly,

$$\cos\phi = \cos 63.43° = \frac{2}{4.472}$$

or $$2 = 4.472\cos 63.43° \qquad \text{(xv)}$$

Substituting the values for 4 and 2 from Eqs. (xiv) and (xv) respectively into Eq. (xiii), we get

$$v_L = 7.5e^{-2t}\,[\cos 4t\,(4.472\sin 63.43°) - \sin 4t\,(4.472\cos 63.43°)]$$

$$= 7.5 \times 4.472e^{-2t}\,[\cos 4t\,(\sin 63.43°) - \sin 4t\,(\cos 63.43°)]$$

$$\mathbf{v_L = 33.54e^{-2t}\,[\sin(63.43° - 4t)]\ V} \qquad \text{(xvi)}$$

With reference to Fig. 11.61, by KVL we get

$$v_R + v_L + v_C = 30$$

So, $$v_C = 30 - v_R - v_L$$

Substituting the values of v_R and v_L respectively from Eqs. (xii) and (xiii) into above equation,

$$v_C = 30 - 30e^{-2t}\sin 4t - 7.5e^{-2t}\,[\cos 4t\,(4) - \sin 4t\,(2)]$$

$$= 30 - 30e^{-2t}\sin 4t - 30e^{-2t}\cos 4t + 15e^{-2t}\sin 4t$$

$$= 30 - 15e^{-2t}\sin 4t - 30e^{-2t}\cos 4t$$

$$= 30 - e^{-2t}\,[\sin 4t\,(15) + \cos 4t\,(30)] \qquad \text{(xvii)}$$

Let us again construct right-angled triangle with 15 and 30 as sides as shown in Fig. 11.63. With reference to Fig. 11.63, we can write

$$\phi = \tan^{-1}\left(\frac{30}{15}\right) = 63.43°$$

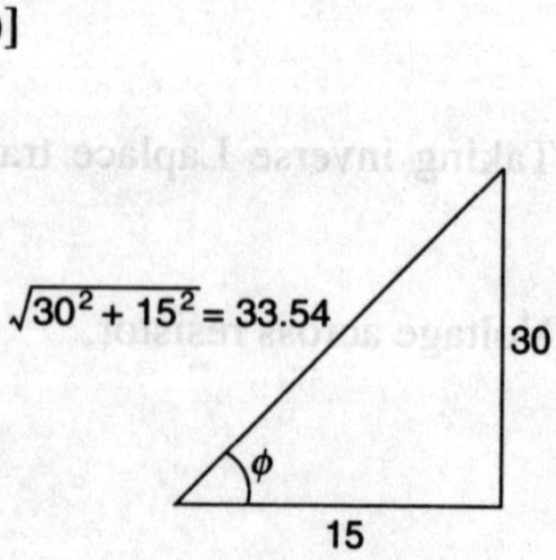

Fig. 11.63

$$\sin\phi = \frac{30}{33.54}$$

$$30 = 33.54 \sin\phi$$

or $$30 = 33.54 \sin 63.43° \qquad \text{(xviii)}$$

Similarly,

$$15 = 33.54 \cos 63.43° \qquad \text{(xix)}$$

Substituting the values of 30 and 15 from Eqs. (xviii) and (xix) respectively into Eq. (xvii), we get

$$v_C = 30 - e^{-2t}\,[\sin 4t\,(33.54 \cos 63.43°) + \cos 4t\,(33.54 \sin 63.43°)]$$
$$= 30 - 33.54e^{-2t}\,[\sin 4t\,(\cos 63.43°) + \cos 4t\,(\sin 63.43°)]$$
$$\mathbf{v_C = 30 - 33.54e^{-2t}\,[\sin\,(63.43° + 4t)]\ V}$$

EXAMPLE 11.23 In a series *RLC* circuit, $R = 5\ \Omega$, $L = 1$ H and $C = 1$ F. A DC voltage of 20 V is applied at $t = 0$. Obtain $i(t)$. Also find the maximum current.

Solution Determination of initial conditions

(i) The switch is closed at $t = 0$. Therefore, at $t = 0^-$, the current through the inductor was zero, i.e. $i(0^-) = 0$, which is also equal to $i(0^+)$, since the current through the inductor cannot be changed instantaneously. Hence, we can write

$$i(0^-) = 0 = i(0^+) \qquad \text{(i)}$$

(ii) Initial voltage on the capacitor is zero, which results in

$$v_C(0^-) = 0 = v_C(0^+) \qquad \text{(ii)}$$

as the voltage across the capacitor cannot be changed instantaneously.

By Classical method

Applying KVL after closing the switch,

$$5i + 1\frac{di}{dt} + \frac{1}{1}\int i\,dt = 20 \qquad \text{(iii)}$$

At $t = 0^+$,

$$5i(0^+) + \frac{di(0^+)}{dt} + \int i(0^+)dt = 20$$

$$5i(0^+) + v_L(0^+) + v_C(0^+) = 20 \qquad \text{(iv)}$$

Substituting the values of $i(0^+)$ and $v_C(0^+)$ from Eqs. (i) and (ii) into Eq. (iv), we get

$$v_L(0^+) = 20$$

or $$L\frac{di(0^+)}{dt} = 20$$

or
$$\frac{di(0^+)}{dt} = \frac{20}{L}$$

$$\frac{di(0^+)}{dt} = \frac{20}{1} = 20 \text{ A/s} \quad \text{(v)}$$

Differentiating Eq. (iii) with respect to t,

$$5\frac{di}{dt} + \frac{d^2i}{dt^2} + i = 0$$

or
$$(D^2 + 5D + 1)i = 0$$

where $D = \frac{d}{dt}$ and $D^2 = \frac{d^2}{dt^2}$

The auxiliary equation is

$$D^2 + 5D + 1 = 0$$

The roots of the above equation are

$$D_1, D_2 = \frac{-5 \pm \sqrt{5^2 - 4}}{2}$$
$$= -0.21, -4.79$$

The roots are real and unequal and the solution is of the form (see Section 11.6.1),

$$i(t) = \text{A}e^{D_1 t} + \text{B}e^{D_2 t}$$
$$= \text{A}e^{-0.21t} + \text{B}e^{-4.79t} \quad \text{(vi)}$$

Constants A and B are determined by applying initial conditions as follows:
From Eq. (i), at $t = 0$, $i(t) = 0$. So, Eq. (vi) becomes

$$0 = \text{A}e^0 + \text{B}e^0$$
$$\text{A} + \text{B} = 0$$

or
$$\text{A} = -\text{B} \quad \text{(vii)}$$

Differentiating Eq. (vi) with respect to t,

$$\frac{di}{dt} = \text{A}(-0.21)e^{-0.21t} + \text{B}(-4.79)e^{-4.79t} \quad \text{(viii)}$$

From Eq. (v), at $t = 0^+$,

$$\frac{di(0^+)}{dt} = 20$$

Hence, Eq. (viii) becomes

$$20 = A(-0.21)e^{-0} + B(-4.79)e^{-0}$$

$$20 = -0.21\text{ A} - 4.79\text{ B}$$

or

$$20 = -0.21\text{ A} + 4.79\text{ A} \quad \text{[from Eq. (vii)]}$$

So,

$$A = 4.37 \text{ and } B = -4.37 \tag{ix}$$

Substituting the value A and B from Eq. (ix) into Eq. (vi), we have

$$\mathbf{i(t) = 4.37e^{-0.21t} - 4.37e^{-4.79t}\ A}$$

By Laplace transformation method

From Eq. (iii),

$$5i + 1\frac{di}{dt} + \frac{1}{1}\int idt = 20$$

Taking Laplace transformation on both sides, we get

$$5I(s) + [sI(s) - I(0)] + \frac{I(s)}{s} = \frac{20}{s}$$

From Eq. (i), $i(0^+) = 0$, and in s-domain, $I(0) = 0$. Therefore, the above equation becomes

$$5I(s) + sI(s) + \frac{I(s)}{s} = \frac{20}{s}$$

$$I(s)\left(5 + s + \frac{1}{s}\right) = \frac{20}{s}$$

$$I(s)\left(\frac{5s + s^2 + 1}{s}\right) = \frac{20}{s}$$

$$I(s) = \frac{20}{s}\left(\frac{s}{5s + s^2 + 1}\right)$$

$$= \frac{20}{s^2 + 5s + 1} = \frac{20}{(s + 0.21)(s + 4.79)}$$

By applying partial fraction technique,

$$I(s) = \frac{A}{(s + 0.21)} + \frac{B}{(s + 4.79)} \tag{x}$$

Solving for constants A and B,

$$A = \left.\frac{20}{(s + 4.79)}\right|_{s=-0.21} = 4.37 \quad \text{and} \quad B = \left.\frac{20}{(s + 0.21)}\right|_{s=-4.79} = -4.37$$

Substituting the values of A and B into Eq. (x), we obtain

$$= \frac{4.37}{(s+0.21)} - \frac{4.37}{(s+4.79)}$$

Taking inverse Laplace transformation on both sides, we get

$$\mathbf{i(t) = 4.37e^{-0.21t} - 4.37e^{-4.79t}\ A} \tag{xi}$$

To find the maximum current

When the current is maximum, $\frac{di}{dt} = 0$

Hence, $$\frac{di}{dt} = \frac{d}{dt}(4.37e^{-0.21t} - 4.37e^{-4.79t}) = 0$$

$$4.37[e^{-0.21t}(-0.21) - e^{-4.79t}(-4.79)] = 0$$

$$4.79e^{-4.79t} = 0.21e^{-0.21t}$$

$$22.81 = e^{4.58t}$$

Taking natural logarithm on both sides, we get

$$3.1272 = 4.58t$$

$$\therefore \qquad t = 0.6828 \text{ s}$$

Substituting $t = 0.6828$ into Eq. (xi), we get

$$i(t) = 4.37e^{-0.21t} - 4.37e^{-4.79t}$$

$$= 4.37e^{-0.21(0.6828)} - 4.37e^{-4.79(0.6828)}$$

$$= \mathbf{3.6203\ A}$$

EXAMPLE 11.24 A step voltage of 100 $u(t)$ is applied to a series *RLC* circuit at $t = 0$ as shown in Fig. 11.64. The initial current in the circuit is zero but there is an initial voltage of 50 V on the capacitor in a direction which opposes the applied source. Find the expression for the current in the circuit.

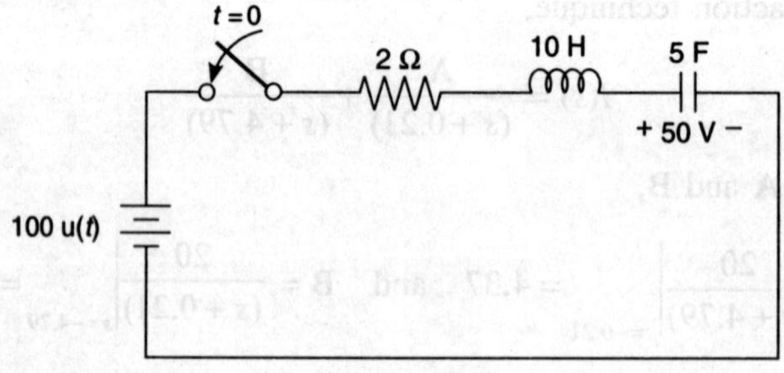

Fig. 11.64

Solution Determination of initial conditions

(i) The switch is closed at $t = 0$. Therefore, at $t = 0^-$, the current through the inductor was zero, i.e. $i(0^-) = 0$, which is also equal to $i(0^+)$, since the current through the inductor cannot be changed instantaneously. Hence, we can write

$$i(0^-) = 0 = i(0^+) \quad \text{(i)}$$

(ii) Initial voltage on the capacitor is 50 V, i.e.

$$v_C(0^-) = 50 = v_C(0^+) \quad \text{(ii)}$$

as the voltage across the capacitor cannot be changed instantaneously.

By Classical method

Applying KVL after closing the switch,

$$2i + 10\frac{di}{dt} + \frac{1}{5}\int i\,dt = 100 \quad \text{(iii)}$$

At $t = 0^+$,

$$2i(0^+) + 10\frac{di(0^+)}{dt} + \frac{1}{5}\int i(0^+)dt = 100$$

$$2i(0^+) + v_L(0^+) + v_C(0^+) = 100 \quad \text{(iv)}$$

Substituting the values of $i(0^+)$ and $v_C(0^+)$ from Eqs. (i) and (ii) into Eq. (iv), we get

$$0 + v_L(0^+) + 50 = 100$$

$$v_L(0^+) = 50$$

or

$$L\frac{di(0^+)}{dt} = 50$$

or

$$\frac{di(0^+)}{dt} = \frac{50}{L}$$

$$\frac{di(0^+)}{dt} = \frac{50}{10} = 5 \text{ A/s} \quad \text{(v)}$$

Differentiating Eq. (iii) with respect to t,

$$2\frac{di}{dt} + 10\frac{d^2i}{dt^2} + \frac{1}{5}i = 0$$

or

$$\frac{d^2i}{dt^2} + 0.2\frac{di}{dt} + 0.02i = 0$$

$$(D^2 + 0.2D + 0.02)i = 0$$

where $D = \frac{d}{dt}$ and $D^2 = \frac{d^2}{dt^2}$

The auxiliary equation is

$$D^2 + 0.2D + 0.02 = 0$$

The roots of the above equation are

$$D_1, D_2 = \frac{-0.2 \pm \sqrt{0.2^2 - (4 \times 0.02)}}{2}$$

$$= -0.1 + j0.1, -0.1 - j0.1$$

The roots are complex conjugates and the solution is of the form (see Section 11.6.1):

$$i(t) = \mathrm{A}e^{D_1 t} + \mathrm{B}e^{D_2 t}$$

$$= \mathrm{A}e^{(-0.1+j0.1)t} + \mathrm{B}e^{(-0.1-j0.1)t}$$

$$= e^{-0.1t}(\mathrm{A}e^{j0.1t} + \mathrm{B}e^{-j0.1t})$$

$$= e^{-0.1t}\left[\mathrm{A}(\cos 0.1t + j\sin 0.1t) + \mathrm{B}(\cos 0.1t - j\sin 0.1t)\right]$$

$$= e^{-0.1t}\left[(\mathrm{A}+\mathrm{B})\cos 0.1t + j(\mathrm{A}-\mathrm{B})\sin 0.1t\right]$$

Let C_1 = (A + B) and $C_2 = j$(A – B), we get

$$i(t) = e^{-0.1t}[C_1 \cos 0.1t + C_2 \sin 0.1t] \qquad \text{(vi)}$$

Constants C_1 and C_2 are determined by applying initial conditions as follows:
From Eq. (i), at t = 0, $i(t)$ = 0. Hence, Eq. (vi) becomes

$$0 = e^{-0}[C_1 \cos 0 + C_2 \sin 0]$$

or

$$C_1 = 0 \qquad \text{(vii)}$$

Substituting the value of C_1 from Eq. (vii) into Eq. (vi), we get

$$i(t) = e^{-0.1t}\, C_2 \sin 0.1t \qquad \text{(viii)}$$

Differentiating Eq. (viii) with respect to t,

$$\frac{di}{dt} = C_2[e^{-0.1t}(-0.1)\sin 0.1t + e^{-0.1t}(0.1)\cos 0.1t] \qquad \text{(ix)}$$

From Eq. (v), at $t = 0^+$,

$$\frac{di(0^+)}{dt} = 5$$

Hence, Eq. (ix) becomes

$$5 = C_2[e^{-0}(-0.1)\sin 0 + e^{-0}(0.1)\cos 0]$$

$$5 = 0.1C_2$$

or

$$C_2 = 50 \qquad \text{(x)}$$

Substituting the value C_2 from Eq. (x) into Eq. (viii), we obtain

$$i(t) = 50e^{-0.1t} \sin 0.1t \text{ A}$$

By Laplace transformation method

In view of Eq. (iii),

$$2i + 10\frac{di}{dt} + \frac{1}{5}\int i dt + 50 = 100$$

50 V is being the initial voltage on the capacitor.

$$2i + 10\frac{di}{dt} + \frac{1}{5}\int i dt = 50$$

Taking Laplace transformation on both sides, we get

$$2I(s) + 10[sI(s) - I(0)] + \frac{1}{5}\frac{I(s)}{s} = \frac{50}{s}$$

From Eq. (i), $i(0^+) = 0$, in s-domain, $I(0) = 0$. Therefore, the above equation becomes

$$2I(s) + 10sI(s) + \frac{1}{5}\frac{I(s)}{s} = \frac{50}{s}$$

$$I(s)\left[2 + 10s + \frac{1}{5}\frac{1}{s}\right] = \frac{50}{s}$$

$$I(s)\left(\frac{10s + 50s^2 + 1}{5s}\right) = \frac{50}{s}$$

$$I(s) = \frac{50}{s}\left(\frac{5s}{10s + 50s^2 + 1}\right)$$

$$= \frac{250}{50(s^2 + 0.2s + 0.02)}$$

$$= \frac{5}{s^2 + 0.2s + 0.02}$$

$$= \frac{5}{(s + 0.1)^2 + 0.01} = \frac{5}{(s + 0.1)^2 + 0.1^2}$$

$$= \left(\frac{5}{0.1}\right)\frac{0.1}{(s + 0.1)^2 + 0.1^2}$$

Taking inverse Laplace transformation on both sides,

$$i(t) = 50e^{-0.1t} \sin 0.1t \text{ A}$$

EXAMPLE 11.25 Consider *RLC* series circuit in Fig. 11.65. The switch is closed to position 1 for a long time. The switch is moved to position 2 at $t = 0$. Find the transient current.

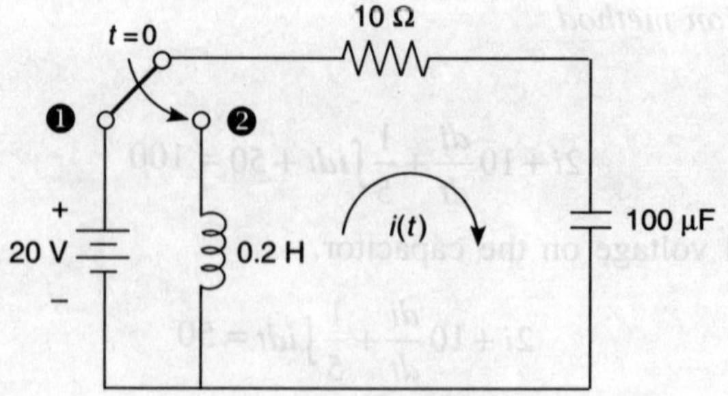

Fig. 11.65

Solution Determination of initial conditions

(i) The switch is closed to position 1 long before $t = 0$. At this position, the circuit should have reached steady state. We know that in steady state, the capacitor would get charged fully and act as an open-circuit. Hence, the voltage across the capacitor will be equal to and opposite of supply voltage, i.e.

$$v_C(0^-) = -20 = v_C(0^+) \qquad \text{(i)}$$

as the voltage across the capacitor cannot be changed instantaneously. The steady state condition of the given circuit is shown in Fig. 11.66.

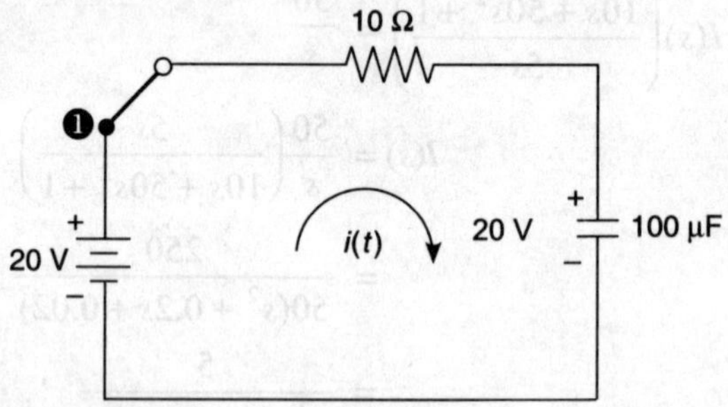

Fig. 11.66

(ii) At $t = 0^-$, the current through the inductor is zero. Though the inductor is connected with the circuit at $t = 0$, the current through it cannot be changed instantaneously. Hence, we can write

$$i(0^-) = 0 = i(0^+) \qquad \text{(ii)}$$

Let $i(t)$ be the current through the circuit when the switch is moved to position 2. The equivalent of the circuit for this position is shown in Fig. 11.67.

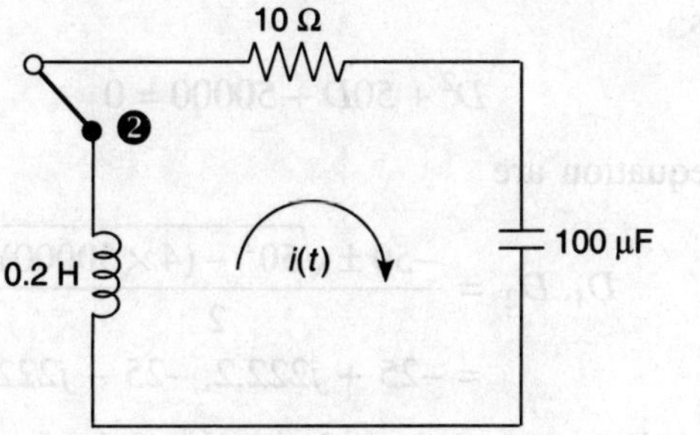

Fig. 11.67

By Classical method

Applying KVL to the circuit shown in Fig. 11.67,

$$10i + 0.2\frac{di}{dt} + \frac{1}{100\times10^{-6}}\int i dt = 0 \qquad \text{(iii)}$$

At $t = 0^+$,

$$10i(0^+) + 0.2\frac{di(0^+)}{dt} + \frac{1}{100\times10^{-6}}\int i(0^+)dt = 0$$

$$10i(0^+) + v_L(0^+) + v_C(0^+) = 0 \qquad \text{(iv)}$$

Substituting the values of $v_C(0^+)$ and $i(0^+)$ from Eqs. (i) and (ii) into Eq. (iv), we get

$$v_L(0^+) = 20$$

or

$$L\frac{di(0^+)}{dt} = 20$$

or

$$\frac{di(0^+)}{dt} = \frac{20}{L}$$

$$\frac{di(0^+)}{dt} = \frac{20}{0.2} = 100 \text{ A/s} \qquad \text{(v)}$$

Differentiating Eq. (iii) with respect to t,

$$10\frac{di}{dt} + 0.2\frac{d^2i}{dt^2} + \frac{1}{100\times10^{-6}}i = 0$$

or

$$\frac{d^2i}{dt^2} + 50\frac{di}{dt} + 50000i = 0$$

$$(D^2 + 50D + 50000)i = 0$$

where $D = \frac{d}{dt}$ and $D^2 = \frac{d^2}{dt^2}$

The auxiliary equation is

$$D^2 + 50D + 50000 = 0$$

The roots of the above equation are

$$D_1, D_2 = \frac{-50 \pm \sqrt{50^2 - (4 \times 50000)}}{2}$$

$$= -25 + j222.2, -25 - j222.2$$

The roots are complex conjugates and the solution is of the form (see Section 11.6.1):

$$\begin{aligned} i(t) &= \text{A}e^{D_1 t} + \text{B}e^{D_2 t} \\ &= \text{A}e^{(-25+j222.2)t} + \text{B}e^{(-25-j222.2)t} \\ &= e^{-25t}(\text{A}e^{j222.2t} + \text{B}e^{-j222.2t}) \\ &= e^{-25t}[\text{A}(\cos 222.2t + j\sin 222.2t) + \text{B}(\cos 222.2t - j\sin 222.2t)] \\ &= e^{-25t}[(\text{A}+\text{B})\cos 222.2t + j(\text{A}-\text{B})\sin 222.2t] \end{aligned}$$

Let C_1 = (A + B) and C_2 = j(A – B), we get

$$i(t) = e^{-25t}[C_1 \cos 222.2t + C_2 \sin 222.2t] \quad \text{(vi)}$$

Constants C_1 and C_2 are determined by applying initial conditions as follows:
From Eq. (ii), at $t = 0$, $i(0) = 0$. Hence, Eq. (vi) becomes

$$0 = e^{-0}[C_1 \cos 0 + C_2 \sin 0]$$

or

$$C_1 = 0 \quad \text{(vii)}$$

Substituting the value of C_1 from Eq. (vii) into Eq. (vi), we obtain

$$i(t) = e^{-25t}\, C_2 \sin 222.2t \quad \text{(viii)}$$

Differentiating Eq. (viii) with respect to t,

$$\frac{di}{dt} = C_2[e^{-25t}(-25)\sin 222.2t + e^{-25t}(222.2)\cos 222.2t] \quad \text{(ix)}$$

From Eq. (v), at $t = 0^+$,

$$\frac{di(0^+)}{dt} = 100$$

Therefore, Eq. (ix) becomes

$$100 = C_2[e^{-0}(-25)\sin 0 + e^{-0}(222.2)\cos 0]$$

$$100 = 222.2C_2$$

or

$$C_2 = 0.45 \quad \text{(x)}$$

Substituting the value C_2 from Eq. (x) into Eq. (viii), we get

$$i(t) = 0.45e^{-25t} \sin 222.2t \text{ A}$$

By Laplace transformation method

With reference to Fig. 11.67, by applying KVL we can write

$$10i + 0.2\frac{di}{dt} + \frac{1}{100 \times 10^{-6}}\int i\,dt - 20 = 0$$

20 V is being the initial voltage on the capacitor.

$$10i + 0.2\frac{di}{dt} + \frac{1}{100 \times 10^{-6}}\int i\,dt = 20$$

Taking Laplace transformation on both sides, we get

$$10I(s) + 0.2[sI(s) - I(0)] + \frac{1}{100 \times 10^{-6}}\frac{I(s)}{s} = \frac{20}{s}$$

From Eq. (ii), $i(0^+) = 0$ and in s-domain, $I(0) = 0$. Therefore, the above equation becomes

$$10I(s) + 0.2sI(s) + \frac{1}{100 \times 10^{-6}}\frac{I(s)}{s} = \frac{20}{s}$$

$$I(s)\left[10 + 0.2s + \frac{1}{100 \times 10^{-6}}\frac{1}{s}\right] = \frac{20}{s}$$

$$I(s)\left(\frac{1000 \times 10^{-6}s + 20 \times 10^{-6}s^2 + 1}{100 \times 10^{-6}s}\right) = \frac{20}{s}$$

$$I(s) = \frac{20}{s}\left(\frac{100 \times 10^{-6}s}{1000 \times 10^{-6}s + 20 \times 10^{-6}s^2 + 1}\right)$$

$$= \frac{20 \times 100 \times 10^{-6}}{20 \times 10^{-6}(s^2 + 50s + 50000)}$$

$$= \frac{100}{s^2 + 50s + 50000}$$

$$= \frac{100}{(s+25)^2 + 49375} = \frac{100}{(s+25)^2 + (222.2)^2}$$

$$= \left(\frac{100}{222.2}\right)\frac{222.2}{(s+25)^2 + (222.2)^2}$$

$$= (0.45)\frac{222.2}{(s+25)^2 + (222.2)^2}$$

Taking inverse Laplace transformation on both sides, we get

$$i(t) = 0.45e^{-25t} \sin 222.2t \text{ A}$$

SHORT QUESTIONS WITH ANSWERS

1. **What is transient?**

   When a circuit is switched from one state to another, either by a change in the applied source or a change in the circuit elements, there is a brief period during which the branch currents and voltage drops change from their previous values to new values. This brief period is called the transient.

2. **What is transient response?**

   When a circuit is switched from one state to another, either by a change in the applied source or a change in the circuit elements, there is a brief period during which the branch currents and voltage drops change from their previous values to new values. This brief period is called the transient. The value of current/voltage during the transient period is called transient response.

3. **Why transient occurs in electric circuits?**

   The inductance does not allow sudden change in current and the capacitance does not allow sudden change in voltage. Hence, in the circuits which have inductances and capacitances, transient occurs during switching operations.

4. **Does transient occur in purely resistive circuits?**

   No. In purely resistive circuits the transition from one state to another state occurs in zero time.

5. **What is free and forced response?**

   The response of a circuit due to the internal energy stored in the elements is called *free response*, and the response of a circuit due to the application of external energy source is called *forced response*.

6. **What is complementary function?**

   The part of the solution or response which becomes zero as t tends to infinity is called *complementary function*. It is the transient part of the solution.

7. **What is particular solution?**

   The part of the solution or response which attains a steady value as t tends to infinity is called complementary function. It is the steady-state part of the solution.

8. **Define time constant of *RL* circuit.**

   The time constant of *RL* circuit is defined as the time taken by the current through the inductance to reach steady value if initial rate of rise is maintained.

9. **What is the time constant of *RL* circuit with R = 100 Ω and L = 200 mH?**

   Time constant, $$\tau = \frac{L}{R} = \frac{200 \times 10^{-3}}{100} = 2 \text{ ms}$$

10. Define time constant of *RC* circuit.

The time constant of *RC* circuit is defined as the time taken by the voltage across the capacitance to reach steady value if initial rate of rise is maintained.

11. What is the time constant of *RC* circuit with $R = 100\ \Omega$ and $C = 2000\ \mu F$?

Time constant, $\tau = RC = 100 \times 2000 \times 10^{-6} = 0.2$ s

UNSOLVED PROBLEMS

1. In the *RL* circuit in Fig. U11.1, the switch is closed at $t = 0$. Find current $i(t)$ and the voltage across resistance and inductance.

[Ans. $i(t) = 2(1 - e^{-10t})$; $10(1 - e^{-10t})$; $10e^{-10t}$]

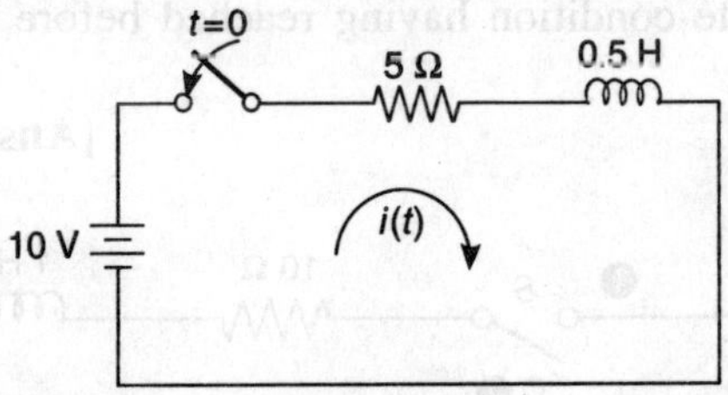

Fig. U11.1

2. In the given network in Fig. U11.2, the switch is closed at $t = 0$. With no initial current through inductor, find i, $\frac{di}{dt}$, $\frac{d^2i}{dt^2}$ at $t = 0^+$. **[Ans.** 0; 100 A/s; –1000 A/s^2]

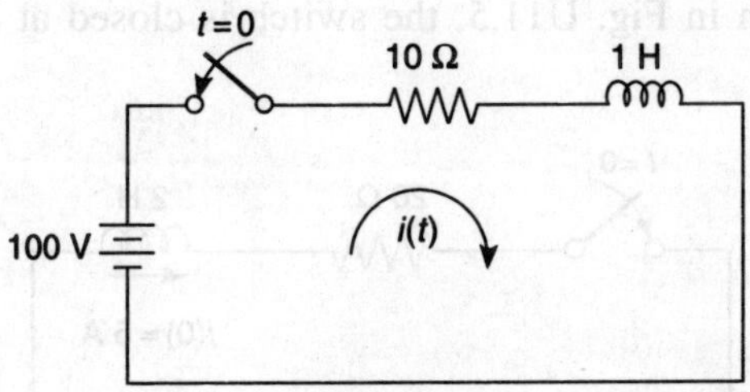

Fig. U11.2

3. In the circuit shown in Fig. U11.3, switch *S* is closed at position 1 for a long time and moved to position 2 at time $t = 0$. Obtain the equation for the current.

[Ans. $10 - 20e^{-2.5t}$]

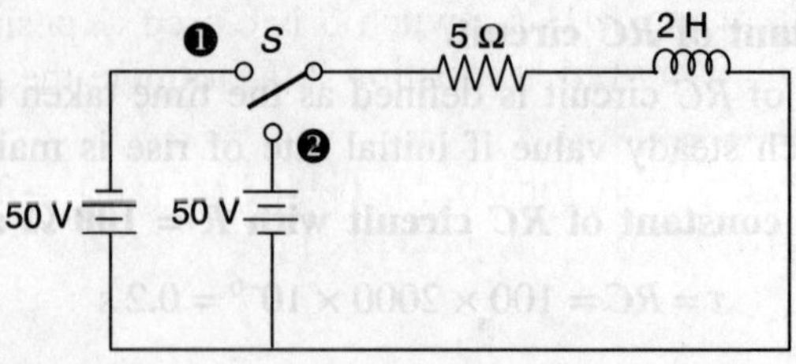

Fig. U11.3

4. A 5 Ω resistor and 20 H inductance are connected in series across a battery. At the instant of closing the switch, the current increases at the rate of 4 A/s. Find (i) supply voltage and (ii) the rate of growth of current when 5 A flows in the circuit.

[**Ans.** (i) 80 V (ii) 2.75 A/s]

5. In the network shown in Fig. U11.4, the switch is changed from position 1 to position 2 at $t = 0$, steady-state condition having reached before switching. Find i, $\frac{di}{dt}, \frac{d^2i}{dt^2}$ at $t = 0^+$.

[**Ans.** 2 A; –60 A/s; 1800 A/s^2]

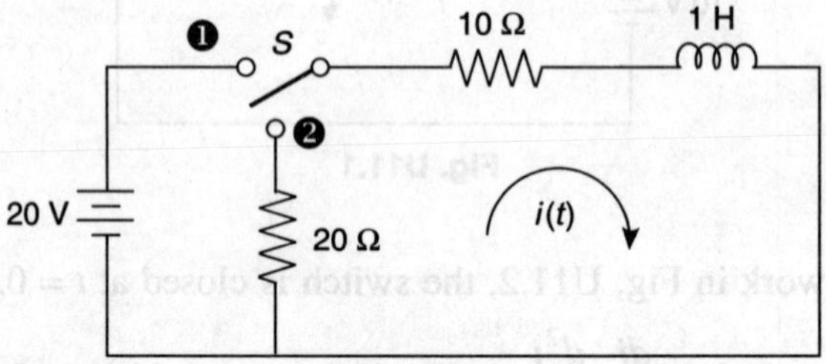

Fig. U11.4

6. In the circuit shown in Fig. U11.5, the switch is closed at $t = 0$, find the equation for the current. [**Ans.** $5e^{-10t}$ A]

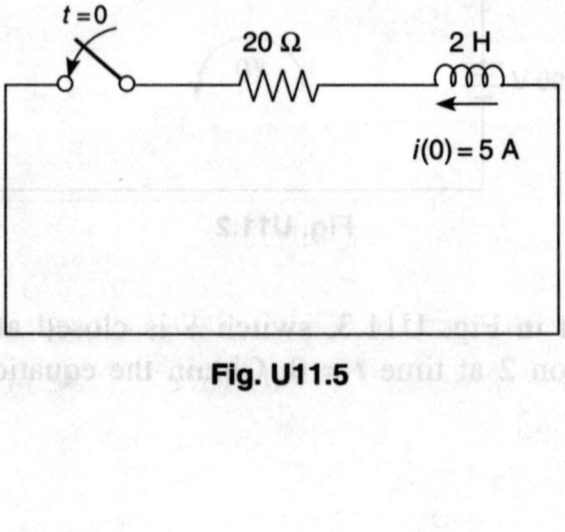

Fig. U11.5

7. In the circuit shown in Fig. U11.6, switch S is closed at position 1 at $t = 0$. Then, at $t = 0.24$ s, the switch is moved to position 2. Determine the response $i_L(t)$. Also find the time at which $i_L(t)$ is zero.
[**Ans.** $i_L(t) = -1 + e^{-8.333t}$ A for $0 \le t \le 0.24$; $i_L(t) = 0.1 - 0.9647e^{-(t-0.24)/0.06}$ A for $t \ge 0.24$; $i_L(t) = 0$ when $t = 0.376$ s]

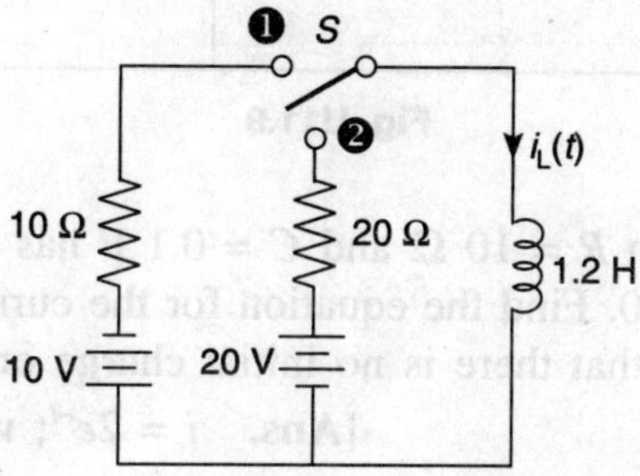

Fig. U11.6

8. In the circuit shown in Fig. U11.7, a DC voltage of 100 V is applied and switch S is open. It is closed at $t = 0$. Find the expression for the current. [**Ans.** $1 - 0.8e^{-25t}$]

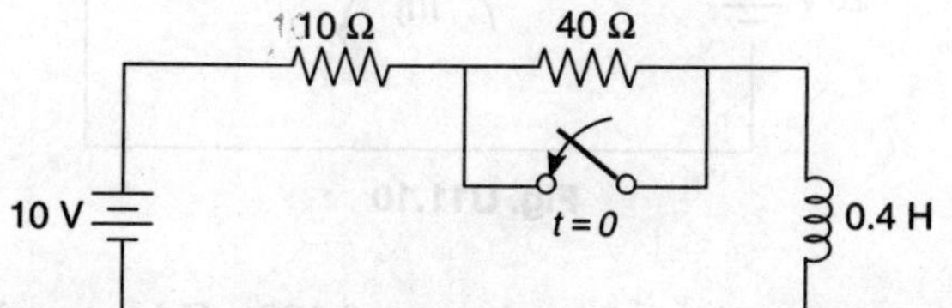

Fig. U11.7

9. In the circuit shown in Fig. U11.8, the switch remains closed for a long time. At $t = 0$, the switch is opened. Find $i_2(t)$. [**Ans.** $0.4e^{-10t}$]

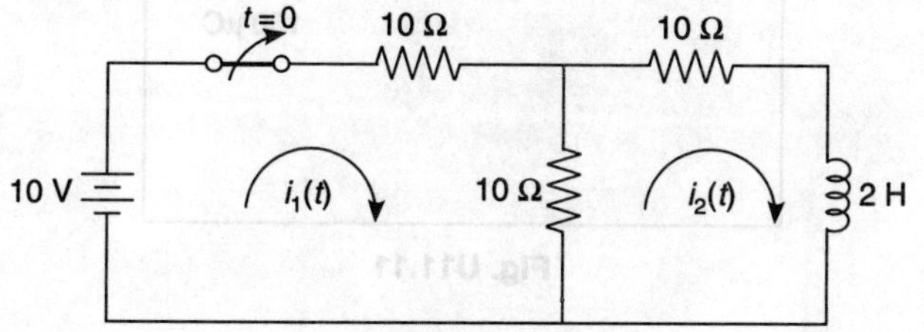

Fig. U11.8

10. In the circuit shown in Fig. U11.9, the switch is closed at $t = 0$. Find $i_1(t)$ and $i_2(t)$.
[**Ans.** $i_1(t) = 0.7778 - 0.4611e^{-2.45t} - 0.3167e^{-12.21t}$ A; $i_2(t) = 0.5556 - 0.6955e^{-2.45t} + 0.14e^{-12.21t}$ A]

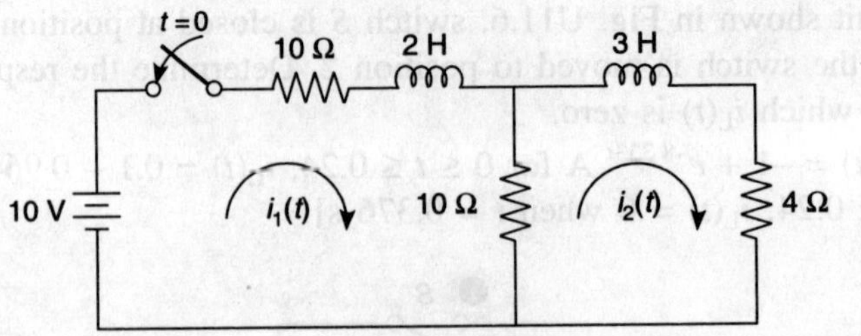

Fig. U11.9

11. A series *RC* circuit with $R = 10\ \Omega$ and $C = 0.1$ F has a DC voltage of 20 V applied through a switch at $t = 0$. Find the equation for the current, voltage across the resistor and capacitor. Assume that there is no initial charge on the capacitor.

[**Ans.** $i = 2e^{-t}$; $v_R = 20e^{-t}$; $v_C = 20\ (1 - e^{-t})$ V]

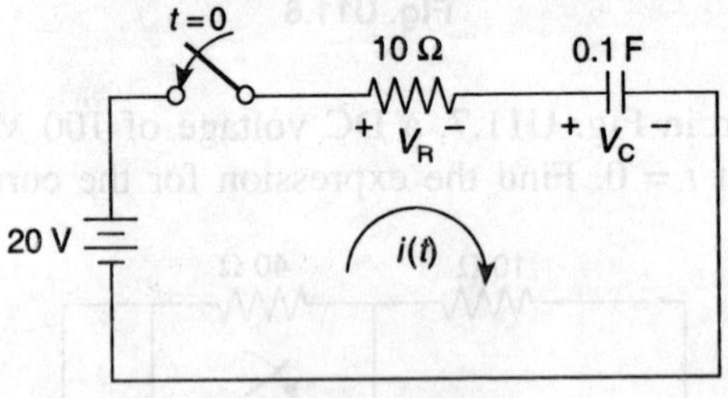

Fig. U11.10

12. In the *RC* circuit of Fig. U11.11, a charge of 100 μC is stored in the capacitor before closing the switch. At $t = 0$, the switch is closed to discharge the charge through 10 kΩ resistance. Determine the voltage across the capacitor. [**Ans.** $4e^{-4t}$]

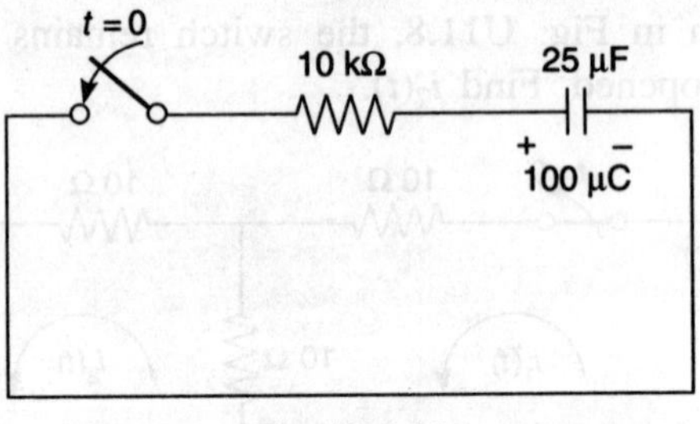

Fig. U11.11

13. In the circuit shown in Fig. U11.12, capacitor *C* has an initial charge of 12×10^{-4} C. Find the current through *RC* circuit if the switch is closed at $t = 0$.

[**Ans.** $0.001e^{-2.5t}$ A]

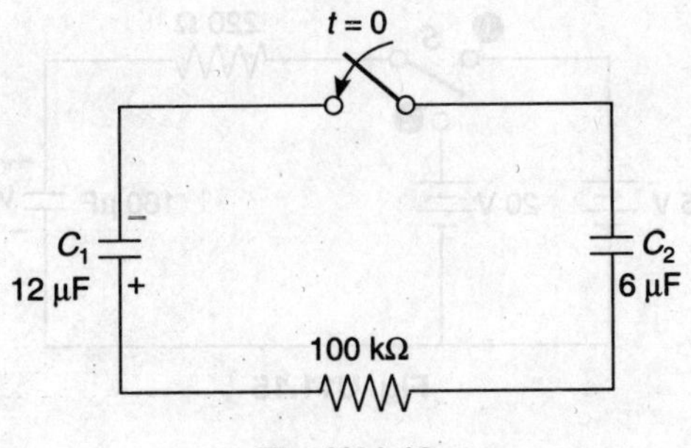

Fig. U11.12

14. In the given network in Fig. U11.13, the switch is closed at $t = 0$. With no initial charge on the capacitor, find i, $\dfrac{di}{dt}, \dfrac{d^2i}{dt^2}$ at $t = 0^+$.

[Ans. 0.1 A; –100 A/s; 10^5 A/s^2]

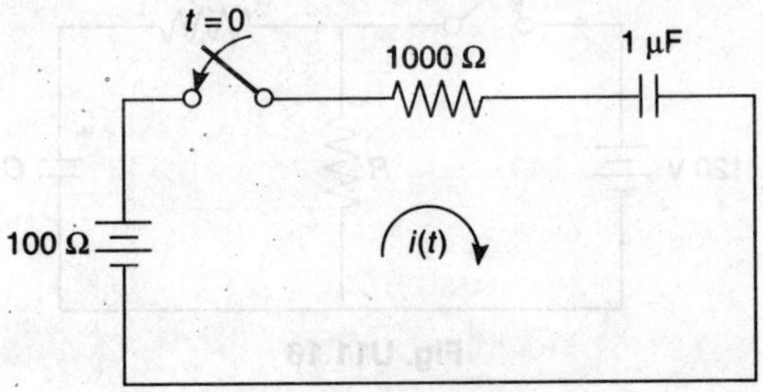

Fig. U11.13

15. In the given network shown in Fig. U11.14, the switch is changed from position 1 to position 2 at $t = 0$, steady-state condition having reached before switching. Find i, $\dfrac{di}{dt}, \dfrac{d^2i}{dt^2}$ at $t = 0^+$. **[Ans.** 0.1 A; 0.33 × 10^5 A/s; –1.1 × 10^9 A/s^2]

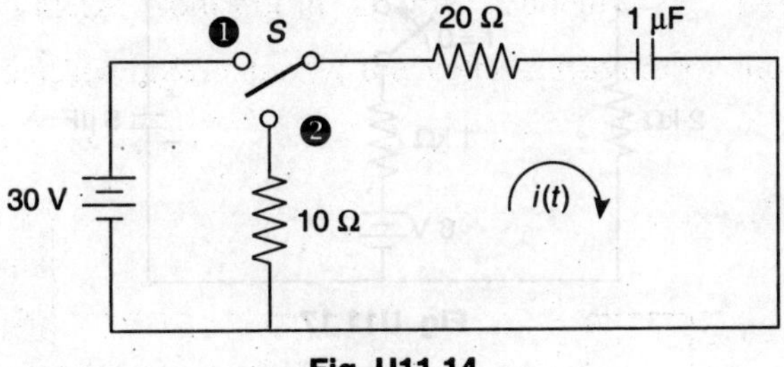

Fig. U11.14

16. In the *RC* circuit shown in Fig. U11.15, the switch is closed to position 1 for a long time. At time $t = 0$, the switch is moved to position 2. Determine the voltage across capacitance for $t \geq 0$. **[Ans.** $20 - 15e^{-28.41t}$ V]

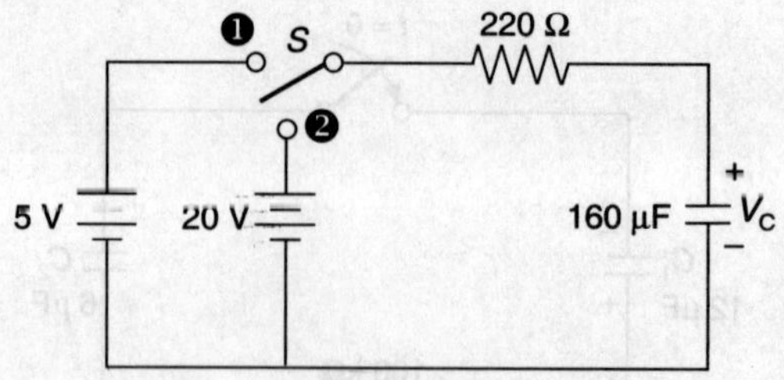

Fig. U11.15

17. In the circuit shown Fig. U11.16, the switch is open for a long time. On closing the switch at $t = 0$, the capacitor voltage raises to 70 V in 10 ms. After the steady state is reached, the switch is opened again and found that the capacitor voltage is 90 V in 0.5 s. Find the values of R and C. **[Ans.** 18.159 kΩ; 95.1837 μF]

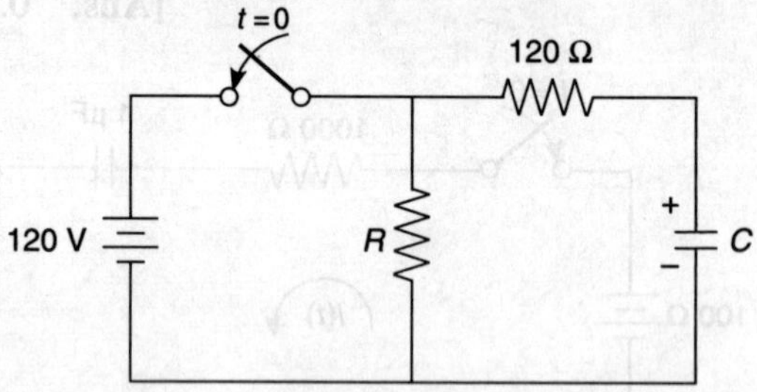

Fig. U11.16

18. In the network shown in Fig. U11.17, the steady state is reached with switch open. At $t = 0$, the switch is closed. Find the voltage across the capacitor for $t \geq 0$.

[Ans. $5e^{-20t}$ V]

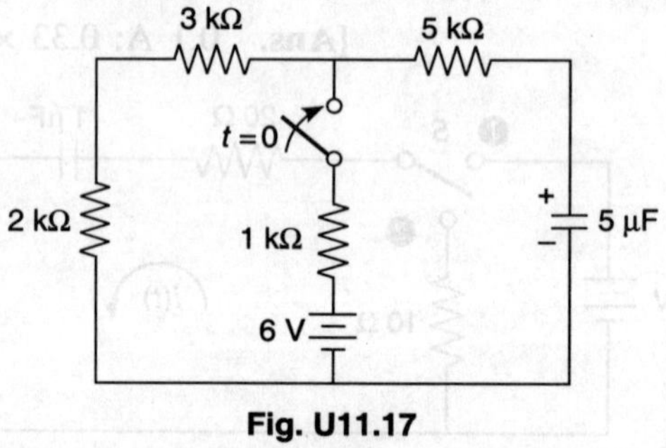

Fig. U11.17

19. In the network shown in Fig. U11.18, switch is closed at $t = 0$. Obtain current $i_C(t)$.

[Ans. $i_C(t) = 5e^{-100000t}$ A]

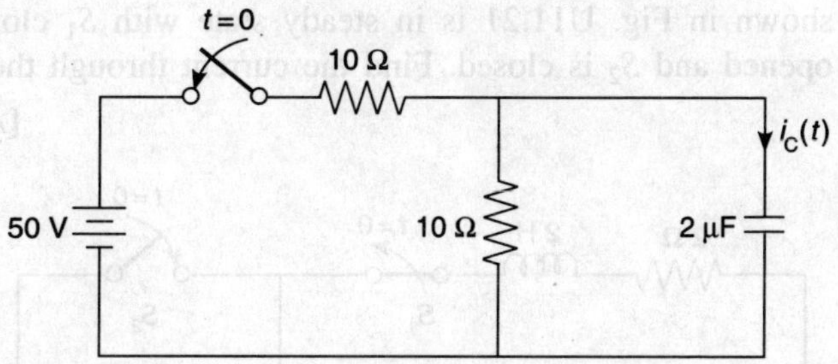

Fig. U11.18

20. A series *RLC* circuit of $R = 60\ \Omega$, $L = 0.2$ H and $C = 50\ \mu$F is excited by a DC source of 100 V through a switch at $t = 0$. Find an expression for current and voltage in the elements of the circuit.

[**Ans.** $i(t) = 1.796e^{-150t} \sin 278.4t$ A; $v_R(t) = 107.76e^{-150t} \sin 278.4t$ V; $v_L(t) = 113.592e^{-150t} \sin (278.4t + 118.3°$ V; $v_C(t) = 100 - 113.592e^{-150t} \sin (278.4t + 61.7°)$ V]

21. In the *RLC* circuit in Fig. U11.19, the capacitor has an initial voltage of 40 V, when the switch is closed at $t = 0$. Find an expression for current $i(t)$.

[**Ans.** $i(t) = 48.9796e^{-0.1t} \sin 0.1225t$ A]

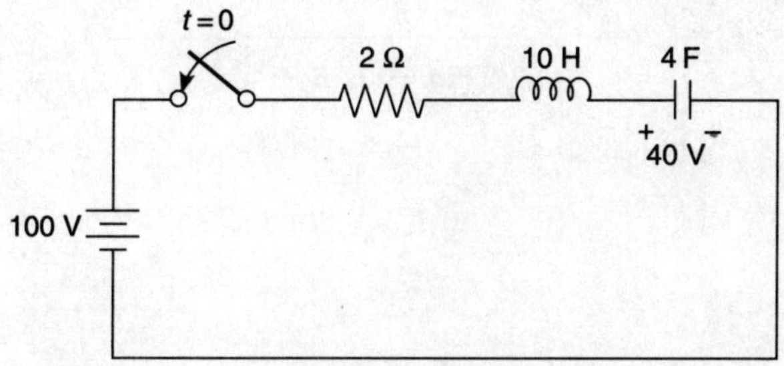

Fig. U11.19

22. In the *RLC* circuit in Fig. U11.20, the inductor has an initial current of 10 A, when the switch is closed at $t = 0$. Find an expression for current $i(t)$.

[**Ans.** $i(t) = 10.2459\ (1.976e^{-0.4033t} - e^{-49.5968t})$ A]

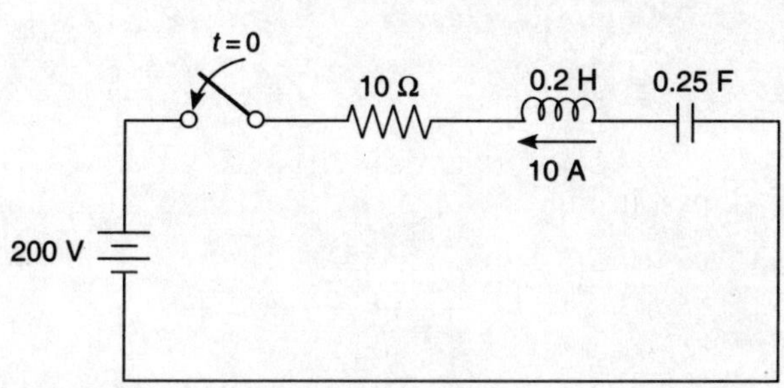

Fig. U11.20

23. The circuit shown in Fig. U11.21 is in steady state with S_1 closed and S_2 open. At $t = 0$, S_1 is opened and S_2 is closed. Find the current through the capacitor.

[Ans. 5 cos 577*t* A]

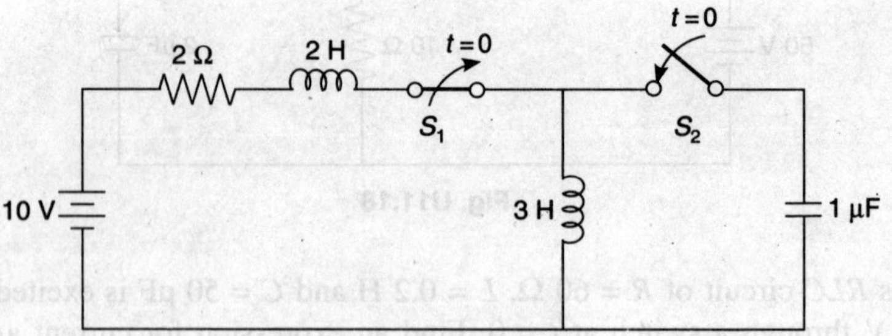

Fig. U11.21

CHAPTER 12

AC Transients

12.1 *RL* TRANSIENTS WITH SINUSOIDAL EXCITATION

12.1.1 Response by Classical Method

Consider the *RL* circuit shown in Fig. 12.1. Assume switch S is closed at $t = 0$, and the circuit is excited by source $V_m \cos(\omega t + \theta)$, where V_m and θ are constants. Let us assume that there is no initial stored energy in the inductor.

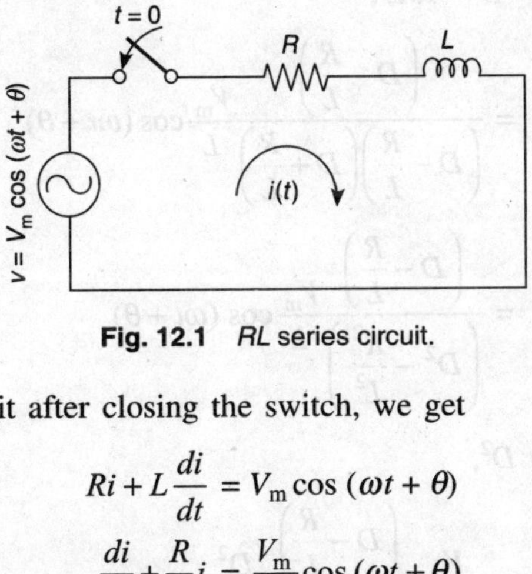

Fig. 12.1 *RL* series circuit.

Applying KVL to the circuit after closing the switch, we get

$$Ri + L\frac{di}{dt} = V_m \cos(\omega t + \theta) \tag{12.1}$$

$$\frac{di}{dt} + \frac{R}{L}i = \frac{V_m}{L}\cos(\omega t + \theta) \tag{12.2}$$

The solution of the above equation contains two parts, *viz.* complementary function (i_c) and particular integral (i_p).

i.e. the total current, $i = i_c + i_p$

Equation (12.2) can be written as:

$$\left(D+\frac{R}{L}\right)i = \frac{V_m}{L}\cos(\omega t+\theta) \tag{12.3}$$

where $D = \dfrac{d}{dt}$.

The auxiliary equation is written by making the forcing function as zero, i.e.

$$\left(D+\frac{R}{L}\right) = 0$$

or

$$D = -\frac{R}{L}$$

Therefore, the complementary function of the solution i is given by

$$i_c = Ke^{-\frac{R}{L}t} \tag{12.4}$$

where K is a constant and is evaluated using initial conditions.

Particular integral i_p is found as follows:

From Eq. (12.3),

$$i_p = \frac{\dfrac{V_m}{L}\cos(\omega t+\theta)}{\left(D+\dfrac{R}{L}\right)}$$

Multiplying and dividing by '$D - R/L$',

$$= \frac{\left(D-\dfrac{R}{L}\right)}{\left(D-\dfrac{R}{L}\right)\left(D+\dfrac{R}{L}\right)}\frac{V_m}{L}\cos(\omega t+\theta)$$

$$= \frac{\left(D-\dfrac{R}{L}\right)}{\left(D^2-\dfrac{R^2}{L^2}\right)}\frac{V_m}{L}\cos(\omega t+\theta) \tag{12.5}$$

Multiplying and dividing by D^2,

$$= \frac{V_m}{L}\frac{\left(D-\dfrac{R}{L}\right)}{\left(D^2-\dfrac{R^2}{L^2}\right)}\frac{D^2}{D^2}\cos(\omega t+\theta) \tag{12.6}$$

Consider the term '$D^2 \cos(\omega t + \theta)$' only:

$$D^2 \cos(\omega t+\theta) = \frac{d^2}{dt^2}(\cos(\omega t+\theta))$$

$$= \frac{d}{dt}\left[\frac{d}{dt}(\cos(\omega t+\theta))\right]$$

$$= \frac{d}{dt}[(-\omega)\sin(\omega t+\theta)]$$

$$= \left[(-\omega^2)\cos(\omega t+\theta)\right]$$

Comparing both sides, $D^2 = -\omega^2$ (12.7)

Substituting $D^2 = -\omega^2$ into Eq. (12.6), we get

$$i_p = \frac{V_m}{L}\frac{\left(D-\frac{R}{L}\right)}{\left(-\omega^2-\frac{R^2}{L^2}\right)}\cos(\omega t+\theta)$$

$$= -\frac{V_m}{L}\frac{\left(D-\frac{R}{L}\right)}{\left(\omega^2+\frac{R^2}{L^2}\right)}\cos(\omega t+\theta)$$

$$= -\frac{V_m L}{(\omega^2L^2+R^2)}\left(D-\frac{R}{L}\right)\cos(\omega t+\theta)$$

$$= -\frac{V_m L}{(\omega^2L^2+R^2)}\left(D[\cos(\omega t+\theta)]-\frac{R}{L}\cos(\omega t+\theta)\right)$$

$$= -\frac{V_m L}{(\omega^2L^2+R^2)}\left(\frac{d}{dt}[\cos(\omega t+\theta)]-\frac{R}{L}\cos(\omega t+\theta)\right)$$

$$= -\frac{V_m L}{(\omega^2L^2+R^2)}\left(-\omega[\sin(\omega t+\theta)]-\frac{R}{L}\cos(\omega t+\theta)\right)$$

$$= \frac{V_m}{(\omega^2L^2+R^2)}[\omega L\sin(\omega t+\theta)+R\cos(\omega t+\theta)]$$

$$= \frac{V_m}{\sqrt{\omega^2L^2+R^2}}\left[\frac{\omega L}{\sqrt{\omega^2L^2+R^2}}\sin(\omega t+\theta)+\frac{R}{\sqrt{\omega^2L^2+R^2}}\cos(\omega t+\theta)\right] \quad (12.8)$$

Let us construct impedance triangle as shown in Fig. 12.2.

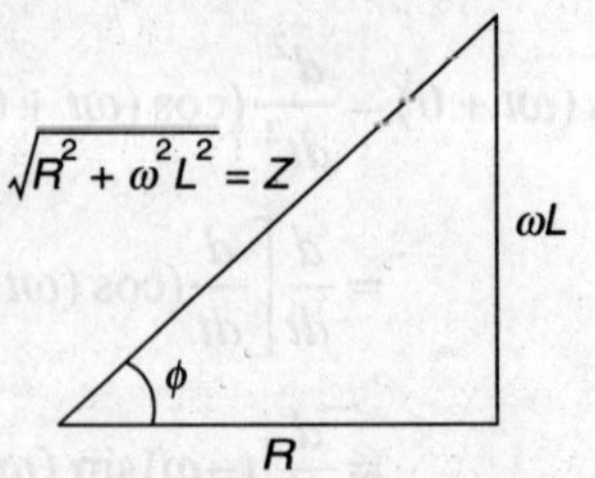

Fig 12.2

Referring to Fig. 12.2, Eq. (12.8) can be written as:

$$= \frac{V_m}{\sqrt{\omega^2 L^2 + R^2}}[\sin\phi \sin(\omega t + \theta) + \cos\phi \cos(\omega t + \theta)]$$

$$i_p = \frac{V_m}{\sqrt{\omega^2 L^2 + R^2}}[\cos(\omega t + \theta - \phi)] \tag{12.9}$$

By combining Eqs. (12.4) and (12.9), the complete solution for i is written as:

$$i = Ke^{-\frac{R}{L}t} + \frac{V_m}{\sqrt{\omega^2 L^2 + R^2}}[\cos(\omega t + \theta - \phi)] \tag{12.10}$$

Constant K is evaluated using initial conditions.
Since the inductor does not allow sudden change in current, at $t = 0$, $i = 0$, i.e.

$$0 = Ke^{-0} + \frac{V_m}{\sqrt{\omega^2 L^2 + R^2}}[\cos(0 + \theta - \phi)]$$

$$\therefore \quad K = -\frac{V_m}{\sqrt{\omega^2 L^2 + R^2}}[\cos(\theta - \phi)] \tag{12.11}$$

Substituting the value of K into Eq. (12.10) and the complete solution for i is obtained as:

$$i = -\frac{V_m}{\sqrt{\omega^2 L^2 + R^2}}[\cos(\theta - \phi)]e^{-\frac{R}{L}t} + \frac{V_m}{\sqrt{\omega^2 L^2 + R^2}}[\cos(\omega t + \theta - \phi)] \tag{12.12}$$

Similarly, the response for a source of $V_m \sin(\omega t + \theta)$ could be obtained as:

$$i = -\frac{V_m}{\sqrt{\omega^2 L^2 + R^2}}[\sin(\theta - \phi)]e^{-\frac{R}{L}t} + \frac{V_m}{\sqrt{\omega^2 L^2 + R^2}}[\sin(\omega t + \theta - \phi)] \tag{12.13}$$

Note: Derivation of Eq. (12.13) is left to the readers as exercise.

12.1.2 Response by Laplace Transformation Method

In view of Eq. (12.1),

$$Ri + L\frac{di}{dt} = V_m \cos(\omega t + \theta)$$

Taking Laplace transformation on both sides,

$$RI(s) + sLI(s) = V_m \frac{(s\cos\theta - \omega\sin\theta)}{(s^2 + \omega^2)}$$

$$I(s) = V_m \frac{(s\cos\theta - \omega\sin\theta)}{(s^2 + \omega^2)} \frac{1}{(R + sL)}$$

$$= \frac{V_m}{L\left(s + \frac{R}{L}\right)} \frac{(s\cos\theta - \omega\sin\theta)}{(s^2 + \omega^2)}$$

$$= \frac{\left(\frac{V_m}{L}\right)}{\left(s + \frac{R}{L}\right)} \frac{(s\cos\theta - \omega\sin\theta)}{(s + j\omega)(s - j\omega)}$$

By partial fraction technique,

$$I(s) = \frac{A}{\left(s + \frac{R}{L}\right)} + \frac{B}{(s + j\omega)} + \frac{C}{(s - j\omega)} \tag{12.14}$$

Solving for constants A, B and C

$$A = \left(\frac{V_m}{L}\right) \frac{(s\cos\theta - \omega\sin\theta)}{(s + j\omega)(s - j\omega)}\bigg|_{s=-\frac{R}{L}}$$

$$= \left(\frac{V_m}{L}\right) \frac{\left(-\frac{R}{L}\cos\theta - \omega\sin\theta\right)}{\left(\frac{R^2}{L^2} + \omega^2\right)}$$

$$= \frac{V_m L}{(R^2 + \omega^2 L^2)}\left(-\frac{R}{L}\cos\theta - \omega\sin\theta\right)$$

$$= -\frac{V_m}{\sqrt{(R^2 + \omega^2 L^2)}}\left(\frac{R}{\sqrt{(R^2 + \omega^2 L^2)}}\cos\theta + \frac{\omega L}{\sqrt{(R^2 + \omega^2 L^2)}}\sin\theta\right)$$

$$= -\frac{V_m}{\sqrt{(R^2+\omega^2L^2)}}(\cos\phi\cos\theta+\sin\phi\sin\theta) \qquad \text{[From Fig. 12.2]}$$

$$\therefore \qquad A = -\frac{V_m}{\sqrt{(R^2+\omega^2L^2)}}\cos(\theta-\phi) \tag{12.15}$$

Constant B is evaluated as follows:

$$B = \frac{\left(\dfrac{V_m}{L}\right)}{\left(s+\dfrac{R}{L}\right)}\frac{(s\cos\theta-\omega\sin\theta)}{(s-j\omega)}\Bigg|_{s=-j\omega}$$

$$= \frac{\left(\dfrac{V_m}{L}\right)}{\left(-j\omega+\dfrac{R}{L}\right)}\frac{(-j\omega\cos\theta-\omega\sin\theta)}{(-j\omega-j\omega)}$$

$$= \frac{\left(\dfrac{V_m}{L}\right)}{\left(\dfrac{R}{L}-j\omega\right)}\frac{(-j\omega\cos\theta-\omega\sin\theta)}{(-2j\omega)} = \left(\frac{V_m}{R-j\omega L}\right)\frac{(-j\cos\theta-\sin\theta)}{(-j2)}$$

$$= \frac{V_m}{2}\left(\frac{1}{R-j\omega L}\right)\frac{(-j\cos\theta-\sin\theta)}{\left(\dfrac{1}{j}\right)}$$

$$= \frac{V_m}{2}\left(\frac{1}{R-j\omega L}\right)j(-j\cos\theta-\sin\theta) = \frac{V_m}{2}\left(\frac{1}{R-j\omega L}\right)(\cos\theta-j\sin\theta)$$

$$= \left(\frac{V_m}{2}\right)\frac{1}{(R-j\omega L)}\frac{(R+j\omega L)}{(R+j\omega L)}(\cos\theta-j\sin\theta)$$

$$= \left(\frac{V_m}{2}\right)\frac{(R+j\omega L)}{(R^2+\omega^2L^2)}(\cos\theta-j\sin\theta)$$

$$= \left(\frac{V_m}{2\sqrt{(R^2+\omega^2L^2)}}\right)\frac{(R+j\omega L)}{\sqrt{(R^2+\omega^2L^2)}}(\cos\theta-j\sin\theta)$$

$$= \left(\frac{V_m}{2\sqrt{(R^2+\omega^2L^2)}}\right)\left(\frac{R}{\sqrt{(R^2+\omega^2L^2)}}+\frac{j\omega L}{\sqrt{(R^2+\omega^2L^2)}}\right)(\cos\theta-j\sin\theta)$$

With reference to Fig. 12.2,

$$= \left(\frac{V_{\mathrm{m}}}{2\sqrt{(R^2+\omega^2L^2)}}\right)(\cos\phi + j\sin\phi)(\cos\theta - j\sin\theta)$$

$$= \left(\frac{V_{\mathrm{m}}}{2\sqrt{(R^2+\omega^2L^2)}}\right)(\cos\phi\cos\theta - j\cos\phi\sin\theta + j\sin\phi\cos\theta + \sin\phi\sin\theta)$$

$$= \left(\frac{V_{\mathrm{m}}}{2\sqrt{(R^2+\omega^2L^2)}}\right)[(\cos\phi\cos\theta + \sin\phi\sin\theta) - j(\cos\phi\sin\theta - \sin\phi\cos\theta)]$$

or
$$\mathrm{B} = \left(\frac{V_{\mathrm{m}}}{2\sqrt{(R^2+\omega^2L^2)}}\right)[\cos(\theta-\phi) - j\sin(\theta-\phi)] \qquad (12.16)$$

Constant C is nothing but the complex conjugate of B.

i.e. $\mathrm{C} = \mathrm{B}^*$

Hence,
$$\mathrm{C} = \left(\frac{V_{\mathrm{m}}}{2\sqrt{(R^2+\omega^2L^2)}}\right)[\cos(\theta-\phi) + j\sin(\theta-\phi)] \qquad (12.17)$$

Substituting the values of A, B and C respectively found in Eqs. (12.15), (12.16) and (12.17) into Eq. (12.14), we get

$$I(s) = \left\{\frac{\dfrac{-V_{\mathrm{m}}}{\sqrt{(R^2+\omega^2L^2)}}\cos(\theta-\phi)}{\left(s+\dfrac{R}{L}\right)} + \frac{\dfrac{V_{\mathrm{m}}}{2\sqrt{(R^2+\omega^2L^2)}}[\cos(\theta-\phi) - j\sin(\theta-\phi)]}{(s+j\omega)} + \frac{\dfrac{V_{\mathrm{m}}}{2\sqrt{(R^2+\omega^2L^2)}}[\cos(\theta-\phi) + j\sin(\theta-\phi)]}{(s-j\omega)}\right\}$$

Taking inverse Laplace transformation on both sides, we get

$$i(t) = \left\{\frac{-V_{\mathrm{m}}}{\sqrt{(R^2+\omega^2L^2)}}\cos(\theta-\phi)e^{-\frac{R}{L}t} + \left(\frac{V_{\mathrm{m}}}{2\sqrt{(R^2+\omega^2L^2)}}\right)e^{-j(\theta-\phi)}e^{-j\omega t} + \left(\frac{V_{\mathrm{m}}}{2\sqrt{(R^2+\omega^2L^2)}}\right)e^{j(\theta-\phi)}e^{j\omega t}\right\}$$

$$= \left\{ \frac{-V_m}{\sqrt{(R^2+\omega^2L^2)}} \cos(\theta-\phi) e^{-\frac{R}{L}t} + \left(\frac{V_m}{2\sqrt{(R^2+\omega^2L^2)}} \right) e^{-j(\omega t+\theta-\phi)} + \left(\frac{V_m}{2\sqrt{(R^2+\omega^2L^2)}} \right) e^{j(\omega t+\theta-\phi)} \right\}$$

$$= \left\{ \frac{-V_m}{\sqrt{(R^2+\omega^2L^2)}} \cos(\theta-\phi) e^{-\frac{R}{L}t} + \left(\frac{V_m}{\sqrt{(R^2+\omega^2L^2)}} \right) \left(\frac{e^{-j(\omega t+\theta-\phi)} + e^{j(\omega t+\theta-\phi)}}{2} \right) \right\}$$

$$i(t) = \left\{ \frac{-V_m}{\sqrt{(R^2+\omega^2L^2)}} \cos(\theta-\phi) e^{-\frac{R}{L}t} + \left(\frac{V_m}{\sqrt{(R^2+\omega^2L^2)}} \right) \cos(\omega t+\theta-\phi) \right\} \quad (12.18)$$

as obtained earlier in Eq. (12.12) using classical method.

12.2 *RC* TRANSIENTS WITH SINUSOIDAL EXCITATION

12.2.1 Response by Classical Method

Consider the *RC* circuit shown in Fig. 12.3. Assume switch S is closed at $t = 0$, and the circuit is excited by source $V_m \cos(\omega t + \theta)$, where V_m and θ are constants. Let us assume that there is no initial charge on the capacitor.

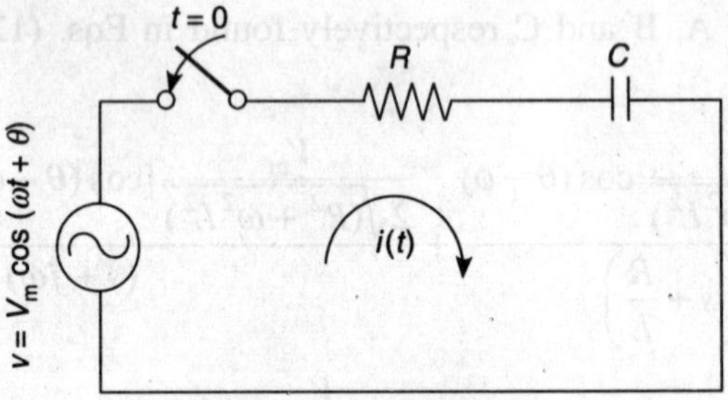

Fig. 12.3 *RC* series circuit.

Applying KVL to the circuit after closing the switch, we get

$$Ri + \frac{1}{C}\int i\,dt = V_m \cos(\omega t + \theta) \quad (12.19)$$

$$R\frac{di}{dt} + \frac{1}{C}i = -\omega V_m \sin(\omega t + \theta) \quad (12.20)$$

$$\frac{di}{dt} + \frac{1}{RC}i = -\omega \frac{V_m}{R} \sin(\omega t + \theta) \quad (12.21)$$

The solution of the above equation contains two parts, *viz.* complementary function (i_c) and particular integral (i_p).

i.e. the total current, $i = i_c + i_p$

Equation (12.21) can be written as:

$$\left(D+\frac{1}{RC}\right)i=-\omega\frac{V_{\mathrm{m}}}{R}\sin(\omega t+\theta) \tag{12.22}$$

where $D = \dfrac{d}{dt}$.

The auxiliary equation is written by making the forcing function as zero, i.e.

$$\left(D+\frac{1}{RC}\right)=0$$

or

$$D=-\frac{1}{RC}$$

Hence, the complementary function of the solution i is given by

$$i_{\mathrm{c}} = \mathrm{K}e^{-\frac{1}{RC}t} \tag{12.23}$$

where K is a constant and is evaluated using initial conditions.

Particular integral, i_{p} is found as follows:

From Eq. (12.22),

$$i_{\mathrm{p}}=\frac{(-\omega)\dfrac{V_{\mathrm{m}}}{R}\sin(\omega t+\theta)}{\left(D+\dfrac{1}{RC}\right)}$$

Multiplying and dividing by '$D - 1/RC$',

$$=\frac{\left(D-\dfrac{1}{RC}\right)}{\left(D-\dfrac{1}{RC}\right)\left(D+\dfrac{1}{RC}\right)}(-\omega)\frac{V_{\mathrm{m}}}{R}\sin(\omega t+\theta)$$

$$=\left(-\omega\frac{V_{\mathrm{m}}}{R}\right)\frac{\left(D-\dfrac{1}{RC}\right)}{\left(D^2-\dfrac{1}{R^2C^2}\right)}\sin(\omega t+\theta) \tag{12.24}$$

From Eq. (12.7), substituting $D^2 = -\omega^2$ into Eq. (12.24), we get

$$i_{\mathrm{p}}=\left(-\omega\frac{V_{\mathrm{m}}}{R}\right)\frac{\left(D-\dfrac{1}{RC}\right)}{\left(-\omega^2-\dfrac{1}{R^2C^2}\right)}\sin(\omega t+\theta)$$

$$= \left(\omega \frac{V_m}{R}\right) \frac{\left(D - \frac{1}{RC}\right)}{\left(\frac{\omega^2 R^2 C^2 + 1}{R^2 C^2}\right)} \sin(\omega t + \theta)$$

$$= \left(\omega \frac{V_m}{R}\right) \frac{R^2 C^2}{(\omega^2 R^2 C^2 + 1)} \left(D - \frac{1}{RC}\right) \sin(\omega t + \theta)$$

$$= \left(\omega \frac{V_m}{R}\right) \frac{R^2 C^2}{\omega^2 C^2 \left(R^2 + \frac{1}{\omega^2 C^2}\right)} \left(D - \frac{1}{RC}\right) \sin(\omega t + \theta)$$

$$= \left(\frac{V_m R}{\omega}\right) \frac{1}{\left(R^2 + \frac{1}{\omega^2 C^2}\right)} \left(D - \frac{1}{RC}\right) \sin(\omega t + \theta)$$

$$= \left(\frac{V_m R}{\omega}\right) \frac{1}{\left(R^2 + \frac{1}{\omega^2 C^2}\right)} \left(D \sin(\omega t + \theta) - \frac{1}{RC} \sin(\omega t + \theta)\right)$$

$$= \left(\frac{V_m R}{\omega}\right) \frac{1}{\left(R^2 + \frac{1}{\omega^2 C^2}\right)} \left(\frac{d}{dt} \sin(\omega t + \theta) - \frac{1}{RC} \sin(\omega t + \theta)\right)$$

$$= \left(\frac{V_m R}{\omega}\right) \frac{1}{\left(R^2 + \frac{1}{\omega^2 C^2}\right)} \left(\omega \cos(\omega t + \theta) - \frac{1}{RC} \sin(\omega t + \theta)\right)$$

$$= V_m \frac{1}{\left(R^2 + \frac{1}{\omega^2 C^2}\right)} \left(R \cos(\omega t + \theta) - \frac{1}{\omega C} \sin(\omega t + \theta)\right)$$

$$= \frac{V_m}{\sqrt{\left(R^2 + \frac{1}{\omega^2 C^2}\right)}} \left(\frac{R}{\sqrt{\left(R^2 + \frac{1}{\omega^2 C^2}\right)}} \cos(\omega t + \theta) - \frac{1/\omega C}{\sqrt{\left(R^2 + \frac{1}{\omega^2 C^2}\right)}} \sin(\omega t + \theta)\right) \quad (12.25)$$

Let us construct impedance triangle as shown in Fig. 12.4.

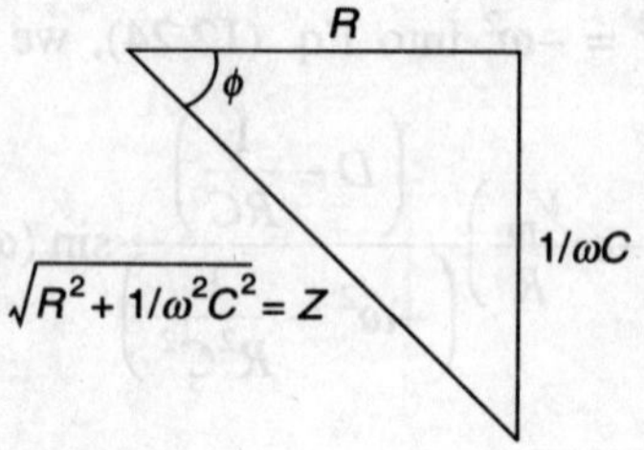

Fig. 12.4 Impedance triangle.

Referring to Fig. 12.4, Eq. (12.25) can be written as:

$$= \frac{V_m}{\sqrt{\left(R^2 + \frac{1}{\omega^2 C^2}\right)}}[\cos\phi \cos(\omega t + \theta) - \sin\phi \sin(\omega t + \theta)]$$

or

$$i_p = \frac{V_m}{\sqrt{\left(R^2 + \frac{1}{\omega^2 C^2}\right)}} \cos(\omega t + \theta + \phi) \tag{12.26}$$

By combining Eqs. (12.23) and (12.26), the complete solution for i is written as:

$$i = Ke^{-\frac{1}{RC}t} + \frac{V_m}{\sqrt{\left(R^2 + \frac{1}{\omega^2 C^2}\right)}}[\cos(\omega t + \theta + \phi)] \tag{12.27}$$

Constant K is evaluated using initial conditions.

Since the capacitor does not allow sudden change in voltage, at $t = 0$, $q = 0$ and $i = i_0$ Equation (12.19) is rewritten as:

$$Ri + \frac{1}{C}q = V_m \cos(\omega t + \theta) \qquad \left[\because \int i\,dt = q\right]$$

Upon substitution of initial conditions,

$$Ri_0 = V_m \cos\theta$$

Hence

$$i_0 = \frac{V_m}{R}\cos\theta$$

Substituting the value of i_0 into Eq. (12.27) and putting $t = 0$, we get

$$\frac{V_m}{R}\cos\theta = Ke^{-0} + \frac{V_m}{\sqrt{\left(R^2 + \frac{1}{\omega^2 C^2}\right)}}[\cos(0 + \theta + \phi)]$$

or

$$K = \frac{V_m}{R}\cos\theta - \frac{V_m}{\sqrt{\left(R^2 + \frac{1}{\omega^2 C^2}\right)}}[\cos(\theta + \phi)]$$

Substituting the value of K into Eq. (12.27), we get

$$i = \left[\frac{V_m}{R}\cos\theta - \frac{V_m}{\sqrt{\left(R^2 + \frac{1}{\omega^2 C^2}\right)}}\cos(\theta + \phi)\right]e^{-\frac{1}{RC}t} + \frac{V_m}{\sqrt{\left(R^2 + \frac{1}{\omega^2 C^2}\right)}}[\cos(\omega t + \theta + \phi)] \tag{12.28}$$

Similarly, the response for a source of $V_m \sin(\omega t + \theta)$ can also be obtained as:

$$i = \left[\frac{V_m}{R}\sin\theta - \frac{V_m}{\sqrt{\left(R^2 + \frac{1}{\omega^2 C^2}\right)}}\sin(\theta + \phi) \right] e^{-\frac{1}{RC}t} + \frac{V_m}{\sqrt{\left(R^2 + \frac{1}{\omega^2 C^2}\right)}}[\sin(\omega t + \theta + \phi)] \quad (12.29)$$

Note: Derivation of Eq. (12.29) is left to the readers as exercise.

12.2.2 Response by Laplace Transformation Method

In view of Eq. (12.19),

$$Ri + \frac{1}{C}\int i\,dt = V_m \cos(\omega t + \theta)$$

Taking Laplace transformation on both sides,

$$RI(s) + \frac{1}{sC}I(s) = V_m \frac{(s\cos\theta - \omega\sin\theta)}{(s^2 + \omega^2)}$$

$$I(s) = V_m \frac{(s\cos\theta - \omega\sin\theta)}{(s^2 + \omega^2)} \frac{1}{\left(R + \frac{1}{sC}\right)} = V_m \frac{(s\cos\theta - \omega\sin\theta)}{(s^2 + \omega^2)} \frac{sC}{(sCR + 1)}$$

$$= V_m \frac{(s\cos\theta - \omega\sin\theta)}{(s^2 + \omega^2)} \frac{sC}{CR\left(s + \frac{1}{CR}\right)}$$

$$= \frac{\left(\frac{V_m}{R}\right)}{\left(s + \frac{1}{CR}\right)} \frac{s(s\cos\theta - \omega\sin\theta)}{(s^2 + \omega^2)}$$

$$= \frac{\left(\frac{V_m}{R}\right)}{\left(s + \frac{1}{CR}\right)} \frac{s(s\cos\theta - \omega\sin\theta)}{(s + j\omega)(s - j\omega)}$$

By partial fraction technique,

$$I(s) = \frac{A}{\left(s + \frac{1}{CR}\right)} + \frac{B}{(s + j\omega)} + \frac{C}{(s - j\omega)} \quad (12.30)$$

Solving for constants, A, B and C, we obtain

$$A = \left(\frac{V_m}{R}\right)\frac{s(s\cos\theta - \omega\sin\theta)}{(s+j\omega)(s-j\omega)}\bigg|_{s=\frac{-1}{CR}}$$

$$= \left(\frac{V_m}{R}\right)\frac{\left(\frac{-1}{CR}\right)\left(\frac{-1}{CR}\cos\theta - \omega\sin\theta\right)}{\left(\frac{1}{C^2R^2}+\omega^2\right)}$$

$$= \left(\frac{V_m}{R}\right)\frac{\left(\frac{1}{C^2R^2}\right)(\cos\theta + \omega CR\sin\theta)}{\left(\frac{1}{C^2R^2}\right)(1+\omega^2C^2R^2)}$$

$$= \left(\frac{V_m}{R}\right)\frac{(\cos\theta + \omega CR\sin\theta)}{(1+\omega^2C^2R^2)} \qquad (12.31)$$

$$= \frac{\left(\frac{V_m}{R}\cos\theta\right)}{(1+\omega^2C^2R^2)} + \frac{(V_m\omega C\sin\theta)}{(1+\omega^2C^2R^2)}$$

$$= \frac{\left(\frac{V_m}{R}\cos\theta\right)(1+\omega^2C^2R^2 - \omega^2C^2R^2)}{(1+\omega^2C^2R^2)} + \frac{(V_m\omega C\sin\theta)}{(1+\omega^2C^2R^2)}$$

$$= \frac{\left(\frac{V_m}{R}\cos\theta\right)(1+\omega^2C^2R^2) - (V_m\omega^2C^2R\cos\theta)}{(1+\omega^2C^2R^2)} + \frac{(V_m\omega C\sin\theta)}{(1+\omega^2C^2R^2)}$$

$$= \frac{V_m}{R}\cos\theta - \frac{(V_m\omega^2C^2R\cos\theta)}{(1+\omega^2C^2R^2)} + \frac{(V_m\omega C\sin\theta)}{(1+\omega^2C^2R^2)}$$

$$= \frac{V_m}{R}\cos\theta - \frac{(V_m\omega^2C^2)\left(R\cos\theta - \frac{1}{\omega C}\sin\theta\right)}{(1+\omega^2C^2R^2)}$$

$$= \frac{V_m}{R}\cos\theta - \frac{(V_m\omega^2C^2)}{\omega^2C^2\left(R^2+\frac{1}{\omega^2C^2}\right)}\left(R\cos\theta - \frac{1}{\omega C}\sin\theta\right)$$

$$= \frac{V_m}{R}\cos\theta - \frac{V_m}{\sqrt{\left(R^2+\frac{1}{\omega^2C^2}\right)}}\left(\frac{R}{\sqrt{\left(R^2+\frac{1}{\omega^2C^2}\right)}}\cos\theta - \frac{1}{\omega C}\frac{1}{\sqrt{\left(R^2+\frac{1}{\omega^2C^2}\right)}}\sin\theta\right) \qquad (12.32)$$

With reference to Fig. 12.4, Eq. (12.32) can be written as:

$$= \frac{V_m}{R}\cos\theta - \frac{V_m}{\sqrt{\left(R^2 + \frac{1}{\omega^2 C^2}\right)}}(\cos\phi\cos\theta - \sin\phi\sin\theta)$$

so,

$$A = \frac{V_m}{R}\cos\theta - \frac{V_m}{\sqrt{\left(R^2 + \frac{1}{\omega^2 C^2}\right)}}\cos(\theta + \phi) \tag{12.33}$$

Constant B is evaluated as follows:

$$B = \left.\frac{\left(\frac{V_m}{R}\right)}{\left(s + \frac{1}{CR}\right)}\frac{s(s\cos\theta - \omega\sin\theta)}{(s - j\omega)}\right|_{s=-j\omega}$$

$$= \frac{\left(\frac{V_m}{R}\right)}{\left(-j\omega + \frac{1}{CR}\right)}\frac{(-j\omega)(-j\omega\cos\theta - \omega\sin\theta)}{(-j\omega - j\omega)}$$

$$= \frac{\left(\frac{V_m}{R}\right)}{\left(-j\omega + \frac{1}{CR}\right)}\frac{(j\omega)(j\omega\cos\theta + \omega\sin\theta)}{(-2j\omega)}$$

$$= \left(-\frac{V_m}{2R}\right)\frac{(j\omega\cos\theta + \omega\sin\theta)}{\left(-j\omega + \frac{1}{CR}\right)}$$

$$= \left(-\frac{V_m}{2R}\right)\frac{(j\omega\cos\theta + \omega\sin\theta)}{\left(\frac{(-j\omega CR + 1)}{CR}\right)}$$

$$= \left(-\frac{V_m\omega C}{2}\right)\frac{(j\cos\theta + \sin\theta)}{(-j\omega CR + 1)}$$

$$= \left(-\frac{V_m\omega C}{2}\right)\frac{(j\cos\theta + \sin\theta)}{(-j\omega CR + 1)} \times \frac{j}{j}$$

$$= \left(-\frac{V_m\omega C}{2}\right)\frac{(-\cos\theta + j\sin\theta)}{(\omega CR + j1)} \times \frac{(\omega CR - j1)}{(\omega CR - j1)}$$

$$= \left(-\frac{V_m\omega C}{2}\right)\frac{(\cos\theta - j\sin\theta)(j1 - \omega CR)}{(1 + \omega^2 C^2 R^2)}$$

$$= \left(-\frac{V_m \omega C}{2}\right) \frac{e^{-j\theta}(j1 - \omega CR)}{\omega^2 C^2 \left(R^2 + \frac{1}{\omega^2 C^2}\right)}$$

$$= \left(-\frac{V_m}{2}\right)\left(\frac{1}{\omega C}\right) \frac{(j1 - \omega CR)}{\left(R^2 + \frac{1}{\omega^2 C^2}\right)} e^{-j\theta}$$

$$= \left(-\frac{V_m}{2}\right) \frac{\left(j\frac{1}{\omega C} - R\right)}{\left(R^2 + \frac{1}{\omega^2 C^2}\right)} e^{-j\theta}$$

$$= \left(\frac{-\frac{V_m}{2}}{\sqrt{\left(R^2 + \frac{1}{\omega^2 C^2}\right)}}\right)\left(j\frac{1}{\omega C}\frac{1}{\sqrt{\left(R^2 + \frac{1}{\omega^2 C^2}\right)}} - R\frac{1}{\sqrt{\left(R^2 + \frac{1}{\omega^2 C^2}\right)}}\right) e^{-j\theta}$$

With reference to Fig. 12.4, we can write

$$= \frac{\left(-\frac{V_m}{2}\right)}{\sqrt{\left(R^2 + \frac{1}{\omega^2 C^2}\right)}} (j\sin\phi - \cos\phi) e^{-j\theta}$$

$$= \frac{\left(\frac{V_m}{2}\right)}{\sqrt{\left(R^2 + \frac{1}{\omega^2 C^2}\right)}} (\cos\phi - j\sin\phi) e^{-j\theta}$$

$$= \frac{\left(\frac{V_m}{2}\right)}{\sqrt{\left(R^2 + \frac{1}{\omega^2 C^2}\right)}} e^{-j\phi} e^{-j\theta}$$

or

$$B = \frac{V_m}{2} \frac{1}{\sqrt{\left(R^2 + \frac{1}{\omega^2 C^2}\right)}} e^{-j(\theta+\phi)} \qquad (12.34)$$

Constant C is nothing but the complex conjugate of B.

i.e.

$$C = B^*$$

Hence,

$$C = \frac{V_m}{2} \frac{1}{\sqrt{\left(R^2 + \frac{1}{\omega^2 C^2}\right)}} e^{j(\theta+\phi)} \tag{12.35}$$

Substituting the values of A, B and C respectively from Eqs. (12.33), (12.34) and (12.35) into Eq. (12.30), we get

$$I(s) = \left\{ \frac{\frac{V_m}{R}\cos\theta - \frac{V_m}{\sqrt{\left(R^2 + \frac{1}{\omega^2 C^2}\right)}}\cos(\theta+\phi)}{\left(s + \frac{1}{RC}\right)} + \frac{\frac{V_m}{2}\frac{1}{\sqrt{\left(R^2 + \frac{1}{\omega^2 C^2}\right)}} e^{-j(\theta+\phi)}}{(s + j\omega)} + \frac{\frac{V_m}{2}\frac{1}{\sqrt{\left(R^2 + \frac{1}{\omega^2 C^2}\right)}} e^{j(\theta+\phi)}}{(s - j\omega)} \right\}$$

Taking inverse Laplace transformation on both sides,

$$i(t) = \left\{ \frac{V_m}{R}\cos\theta e^{-\frac{1}{RC}t} - \frac{V_m}{\sqrt{\left(R^2 + \frac{1}{\omega^2 C^2}\right)}}\cos(\theta+\phi) e^{-\frac{1}{RC}t} + \frac{V_m}{2}\frac{1}{\sqrt{\left(R^2 + \frac{1}{\omega^2 C^2}\right)}} e^{-j(\theta+\phi)} e^{-j\omega t} + \frac{V_m}{2}\frac{1}{\sqrt{\left(R^2 + \frac{1}{\omega^2 C^2}\right)}} e^{j(\theta+\phi)} e^{j\omega t} \right\}$$

$$= \left\{ \frac{V_m}{R}\cos\theta e^{-\frac{1}{RC}t} - \frac{V_m}{\sqrt{\left(R^2 + \frac{1}{\omega^2 C^2}\right)}}\cos(\theta+\phi) e^{-\frac{1}{RC}t} + \frac{V_m}{2}\frac{1}{\sqrt{\left(R^2 + \frac{1}{\omega^2 C^2}\right)}} e^{-j(\omega t+\theta+\phi)} + \frac{V_m}{2}\frac{1}{\sqrt{\left(R^2 + \frac{1}{\omega^2 C^2}\right)}} e^{j(\omega t+\theta+\phi)} \right\}$$

$$= \left\{ \frac{V_m}{R} \cos\theta e^{-\frac{1}{RC}t} - \frac{V_m}{\sqrt{\left(R^2 + \frac{1}{\omega^2 C^2}\right)}} \cos(\theta+\phi) e^{-\frac{1}{RC}t} + \frac{V_m}{\sqrt{\left(R^2 + \frac{1}{\omega^2 C^2}\right)}} \frac{e^{-j(\omega t+\theta+\phi)} + e^{j(\omega t+\theta+\phi)}}{2} \right\}$$

$$= \left\{ \frac{V_m}{R} \cos\theta e^{-\frac{1}{RC}t} - \frac{V_m}{\sqrt{\left(R^2 + \frac{1}{\omega^2 C^2}\right)}} \cos(\theta+\phi) e^{-\frac{1}{RC}t} + \frac{V_m}{\sqrt{\left(R^2 + \frac{1}{\omega^2 C^2}\right)}} \cos(\omega t+\theta+\phi) \right\}$$

Hence,

$$i = \left[\frac{V_m}{R} \cos\theta - \frac{V_m}{\sqrt{\left(R^2 + \frac{1}{\omega^2 C^2}\right)}} \cos(\theta+\phi) \right] e^{-\frac{1}{RC}t} + \frac{V_m}{\sqrt{\left(R^2 + \frac{1}{\omega^2 C^2}\right)}} [\cos(\omega t+\theta+\phi)] \quad (12.36)$$

as obtained earlier in Eq. (12.28) using classical method.

12.3 *RLC* TRANSIENTS WITH SINUSOIDAL EXCITATION

12.3.1 Response by Classical Method

Consider the *RLC* circuit shown in Fig. 12.5. Assume switch S is closed at $t = 0$, and the circuit is excited by source $V_m \cos(\omega t + \theta)$, where V_m and θ are constants. Let us assume that there is no initial stored energy in the inductor and capacitor.

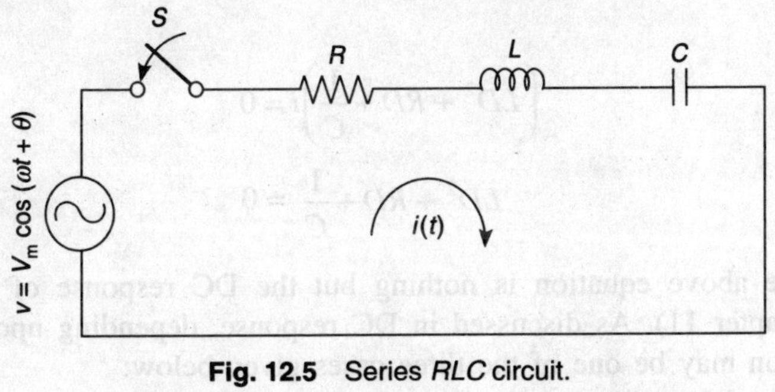

Fig. 12.5 Series *RLC* circuit.

Applying KVL to the circuit after closing the switch, we get

$$Ri + L\frac{di}{dt} + \frac{1}{C}\int i\,dt = V_m \cos(\omega t + \theta) \quad (12.37)$$

The solution of the above equation contains two parts, viz. complementary function (i_c) and particular integral (i_p).

i.e. the total current, $i = i_c + i_p$

Particular integral i_p is found as follows:
By Ohm's law,

$$i = \frac{v}{Z}$$

So, $$i_p = \frac{V_m \cos(\omega t + \theta - \phi)}{Z} \tag{12.38}$$

where $\phi = \tan^{-1}\left(\frac{X_L - X_C}{R}\right) = \tan^{-1}\left(\frac{\omega L - \frac{1}{\omega C}}{R}\right)$

Thus, $$i_p = \frac{V_m}{\sqrt{R^2 + \left(\omega L - \frac{1}{\omega C}\right)^2}} \cos(\omega t + \theta - \phi)$$

Complementary function i_c is found as follows:
Differentiating Eq. (12.37) w.r.t. 't', and rearranging, we get

$$L\frac{d^2 i}{dt^2} + R\frac{di}{dt} + \frac{i}{C} = -\omega V_m \sin(\omega t + \theta) \tag{12.39}$$

Equation (12.39) can be written as:

$$\left(LD^2 + RD + \frac{1}{C}\right)i = -\omega V_m \sin(\omega t + \theta) \tag{12.40}$$

Complementary function is found by making the forcing function in the above equation to be zero.

i.e. $$\left(LD^2 + RD + \frac{1}{C}\right)i = 0$$

or $$LD^2 + RD + \frac{1}{C} = 0$$

The solution of the above equation is nothing but the DC response of *RLC* circuit (refer Section 11.6 in Chapter 11). As discussed in DC response, depending upon the values of *R*, *L* and *C*, the solution may be one of the three cases given below:

Case 1 When $\left(\frac{R}{2L}\right)^2 > \frac{1}{LC}$ (over-damped case)

$$i = e^{-\alpha t}(Ae^{\beta t} + Be^{-\beta t}) + \frac{V_m}{\sqrt{R^2 + \left(\omega L - \frac{1}{\omega C}\right)^2}} \cos(\omega t + \theta - \phi) \tag{12.41}$$

Case 2 When $\left(\frac{R}{2L}\right)^2 = \frac{1}{LC}$ (critically-damped case)

$$i = Ae^{-\alpha t} + Bte^{-\alpha t} + \frac{V_m}{\sqrt{R^2 + \left(\omega L - \frac{1}{\omega C}\right)^2}} \cos(\omega t + \theta - \phi) \tag{12.42}$$

Case 3 When $\left(\frac{R}{2L}\right)^2 < \frac{1}{LC}$ (under-damped case)

$$i = e^{-\alpha t}(C_1 \cos \beta t + C_2 \sin \beta t) + \frac{V_m}{\sqrt{R^2 + \left(\omega L - \frac{1}{\omega C}\right)^2}} \cos(\omega t + \theta - \phi) \tag{12.43}$$

12.3.2 Response by Laplace Transformation Method

In view of Eq. (12.37),

$$Ri + L\frac{di}{dt} + \frac{1}{C}\int i dt = V_m \cos(\omega t + \theta)$$

Taking Laplace transformation on both sides,

$$RI(s) + sLI(s) + \frac{1}{sC}I(s) = V_m \frac{(s\cos\theta - \omega\sin\theta)}{(s^2 + \omega^2)}$$

$$\left(R + sL + \frac{1}{sC}\right)I(s) = V_m \frac{(s\cos\theta - \omega\sin\theta)}{(s^2 + \omega^2)}$$

$$I(s) = V_m \frac{(s\cos\theta - \omega\sin\theta)}{\left(R + sL + \frac{1}{sC}\right)(s^2 + \omega^2)}$$

$$= V_m \frac{s(s\cos\theta - \omega\sin\theta)}{\left(sR + s^2L + \frac{1}{C}\right)(s^2 + \omega^2)}$$

$$= \frac{V_m}{L} \frac{s(s\cos\theta - \omega\sin\theta)}{\left(s^2 + \frac{R}{L}s + \frac{1}{LC}\right)(s^2 + \omega^2)} \tag{12.44}$$

As discussed in DC response, depending upon the values of R, L and C, the solution may be one of the three cases given below.

Case 1 When $\left(\frac{R}{2L}\right)^2 > \frac{1}{LC}$ (over-damped case)

The roots of the part $\left(s^2 + \frac{R}{L}s + \frac{1}{LC}\right)$ are

$$-\frac{R}{2L} \pm \sqrt{\left(\frac{R}{2L}\right)^2 - \frac{1}{LC}} = -\alpha \pm \beta$$

where $\alpha = \frac{R}{2L}$ and $\beta = \sqrt{\left(\frac{R}{2L}\right)^2 - \frac{1}{LC}}$

Hence, Eq. (12.44) can be written as:

$$I(s) = \frac{V_m}{L}\frac{s(s\cos\theta - \omega\sin\theta)}{(s+\alpha-\beta)(s+\alpha+\beta)(s^2+\omega^2)} \tag{12.45}$$

Applying partial fraction technique, Eq. (12.45) is written as:

$$= \frac{A}{s+(\alpha-\beta)} + \frac{B}{s+(\alpha+\beta)} + \frac{Cs+D}{(s^2+\omega^2)}$$

Taking inverse Laplace transformation on both sides, we get

$$i = e^{-\alpha t}(Ae^{\beta t} + Be^{-\beta t}) + \frac{V_m}{\sqrt{R^2 + \left(\omega L - \frac{1}{\omega C}\right)^2}}\cos(\omega t + \theta - \phi) \tag{12.46}$$

Similarly, for the other two cases the response could be obtained as follows:

Case 2 When $\left(\frac{R}{2L}\right)^2 = \frac{1}{LC}$ (critically-damped case)

$$i = Ae^{-\alpha t} + Bte^{-\alpha t} + \frac{V_m}{\sqrt{R^2 + \left(\omega L - \frac{1}{\omega C}\right)^2}}\cos(\omega t + \theta - \phi) \tag{12.47}$$

Case 3 When $\left(\frac{R}{2L}\right)^2 < \frac{1}{LC}$ (under-damped case)

$$i = e^{-\alpha t}\left[C_1\cos\beta t + C_2\sin\beta t\right] + \frac{V_m}{\sqrt{R^2 + \left(\omega L - \frac{1}{\omega C}\right)^2}}\cos(\omega t + \theta - \phi) \tag{12.48}$$

Constants A, B, C_1 and C_2 are evaluated using initial conditions [refer Section 11.6 in Chapter 11].

EXAMPLE 12.1 An *RL* series-circuit shown in Fig. 12.6 is excited by a sinusoidal source $v = 10 \sin 100t$ volts, by closing the switch at $t = 0$. Take $R = 5\ \Omega$ and $L = 0.1$ H. Determine current $i(t)$ flowing through the *RL* circuit.

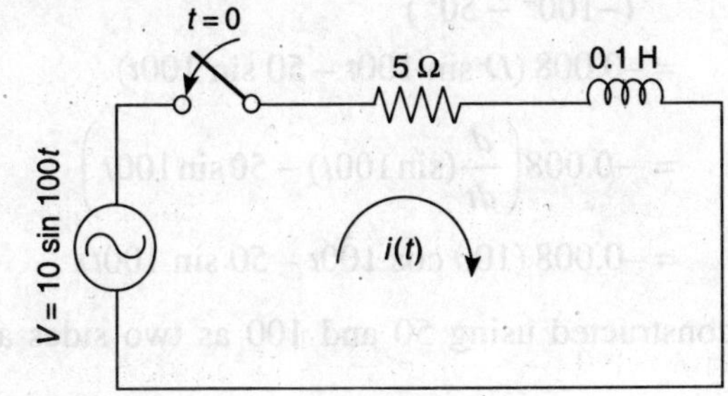

Fig. 12.6

Solution *By Classical method*

Applying KVL to the circuit after closing the switch, we get

$$5i + 0.1\frac{di}{dt} = 10\sin 100t \qquad \text{(i)}$$

$$50i + \frac{di}{dt} = 100\sin 100t \qquad \text{(ii)}$$

The solution of the above equation contains two parts, *viz.* complementary function (i_c) and particular integral (i_p), i.e.

$$\text{The total current, } i = i_c + i_p \qquad \text{(iii)}$$

Equation (ii) can be written as:

$$(D + 50)i = 100 \sin 100t \qquad \text{(iv)}$$

The auxiliary equation is $(D + 50)i = 0$

or $D = -50$ (v)

Hence, the complementary function of the solution of i is given by

$$i_c = Ke^{-50t} \qquad \text{(vi)}$$

The particular integral of i is obtained as follows:

From Eq. (iv),

$$i_p = \frac{100}{(D+50)}\sin 100t$$

$$= \frac{(D-50)100}{(D-50)(D+50)}\sin 100t$$

$$= \frac{(D-50)100}{(D^2-50^2)}\sin 100t \qquad \text{(vii)}$$

Substituting $D^2 = -100^2$ into Eq. (vii), we get $[\because \omega = 100]$

$$i_p = \frac{(D-50)100}{(-100^2 - 50^2)} \sin 100t$$

$$= -0.008\,(D \sin 100t - 50 \sin 100t)$$

$$= -0.008\left(\frac{d}{dt}(\sin 100t) - 50 \sin 100t\right)$$

$$= -0.008\,(100 \cos 100t - 50 \sin 100t)$$

A right-angled triangle is constructed using 50 and 100 as two sides as shown in Fig. 12.7. Referring to Fig. 12.7,

$$\phi = \tan^{-1}\left(\frac{100}{50}\right) = 63.43°$$

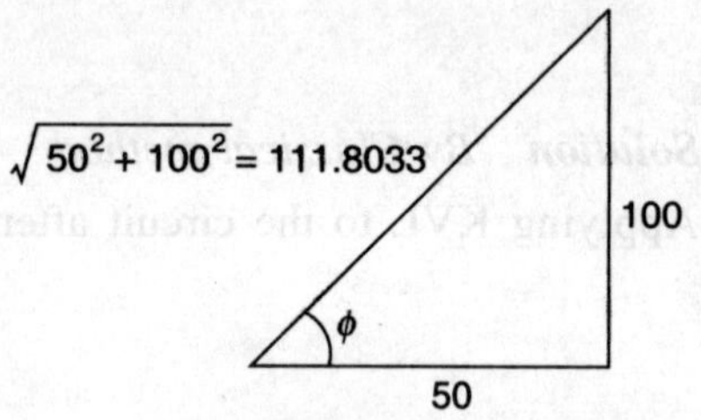

Fig. 12.7

Also, $$\cos\phi = \frac{50}{111.8033}$$

or $$50 = 111.8033 \cos\phi = 111.8033 \cos 63.43°$$

Similarly,

$$100 = 111.8033 \sin\phi = 111.8033 \sin 63.43°$$

So, $$i_p = -0.008\,(111.8033 \sin 63.43° \cos 100t - 111.8033 \cos 63.43° \sin 100t)$$

$$= 0.8944\,(\cos 63.43° \sin 100t - \sin 63.43° \cos 100t)$$

$$= 0.8944 \sin (100t - 63.43°) \qquad \text{(viii)}$$

Substituting the values of i_c and i_p into Eq. (iii), we obtain

$$i = \text{K}e^{-50t} + 0.8944 \sin (100t - 63.43°) \qquad \text{(ix)}$$

The value of constant K is determined by using initial conditions.

At $t = 0$, $i = 0$. Putting these values in Eq. (ix),

$$0 = \text{K}e^{-0} + 0.8944 \sin (0 - 63.43°)$$

or $$\text{K} = 0.8$$

Substituting the value of K into Eq. (ix), we get

$$i = \underbrace{0.8e^{-50t}}_{\text{Transient part}} + \underbrace{0.8944 \sin (100t - 63.43°)\ \text{A}}_{\text{Steady-state part}}$$

Laplace transformation method

In view of Eq. (i),

$$5i + 0.1\frac{di}{dt} = 10 \sin 100t$$

Taking Laplace transformation on both sides,

$$5I(s) + 0.1sI(s) = 10 \times \frac{100}{s^2 + 100^2}$$

$$I(s)[5 + 0.1s] = \frac{1000}{s^2 + 100^2}$$

$$I(s) = \frac{1}{(5 + 0.1s)} \frac{1000}{(s^2 + 100^2)}$$

$$I(s) = \frac{1}{0.1(s + 50)} \frac{1000}{(s^2 + 100^2)}$$

$$I(s) = \frac{10000}{(s + 50)(s^2 + 100^2)} = \frac{A}{(s + 50)} + \frac{Bs + C}{(s^2 + 100^2)} \quad \text{(x)}$$

$$A = \left.\frac{10000}{(s^2 + 100^2)}\right|_{s=-50} = \frac{10000}{(-50)^2 + 100^2} = 0.8$$

On cross-multiplying Eq. (x), we get

$$10000 = A(s^2 + 100^2) + (Bs + C)(s + 50)$$

$$10000 = As^2 + 10000A + Bs^2 + 50Bs + Cs + 50C$$

$$10000 = s^2(A + B) + s(50B + C) + (10000A + 50C) \quad \text{(xi)}$$

On equating the coefficients of s^2 of Eq. (xi), we get

$$A + B = 0$$

or

$$B = -A = -0.8$$

On equating the coefficients of s of Eq. (xi), we get

$$50B + C = 0$$

Hence, $C = 40$

So,

$$I(s) = \frac{0.8}{(s + 50)} + \frac{-0.8s + 40}{(s^2 + 100^2)}$$

$$= \frac{0.8}{(s + 50)} - \frac{0.8s}{(s^2 + 100^2)} + \frac{40}{(s^2 + 100^2)}$$

$$= \frac{0.8}{(s + 50)} - \frac{0.8s}{(s^2 + 100^2)} + \frac{40}{100} \frac{100}{(s^2 + 100^2)}$$

Taking inverse Laplace transformation on both sides, we get

$$i(t) = 0.8e^{-50t} - 0.8 \cos 100t + 0.4 \sin 100t$$

$$= 0.8e^{-50t} + (0.4 \sin 100t - 0.8 \cos 100t)$$

A right-angled triangle is constructed using 0.4 and 0.8 as two sides as shown in Fig. 12.7. Referring to Fig. 12.8,

$$\phi = \tan^{-1}\left(\frac{0.8}{0.4}\right) = 63.43°$$

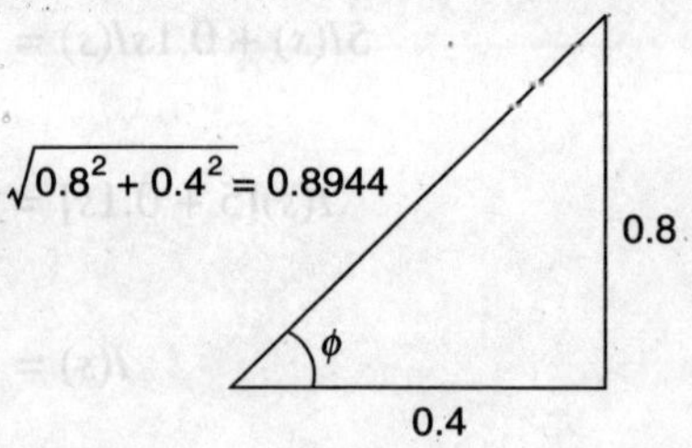

Fig. 12.8

Also, $\cos\phi = \dfrac{0.4}{0.8944}$

or $0.4 = 0.8944 \cos \phi = 0.8944 \cos 63.43°$

Similarly,

$$0.8 = 0.8944 \sin \phi = 0.8944 \sin 63.43°$$

Hence, $i(t) = 0.8e^{-50t} + (0.8944 \cos 63.43° \sin 100t - 0.8944 \sin 63.43° \cos 100t)$

$$= 0.8e^{-50t} + 0.8944 (\cos 63.43° \sin 100t - \sin 63.43° \cos 100t)$$

$$i(t) = \underbrace{0.8e^{-50t}}_{\text{Transient part}} + \underbrace{0.8944 \sin(100t - 63.43°)}_{\text{Steady-state part}}$$

EXAMPLE 12.2 In the circuit shown in Fig. 12.9, the switch remains in position 1 until steady-state is reached. At time $t = 0$, the switch is changed to position 2. Find $i(t)$.

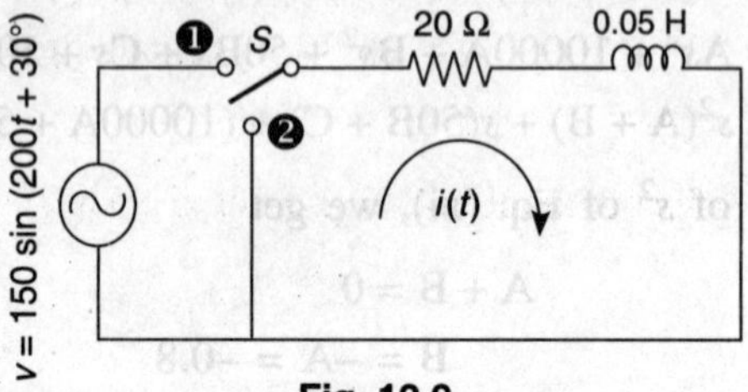

Fig. 12.9

Solution Determination of initial conditions

In position 1, the circuit has already reached steady-state. The standard form of writing instantaneous value v in steady-state analysis is

$$v = v_m \sin(\omega t + \theta).$$

Upon comparing with the value of source,

$$V_m = 150, \omega = 200, \theta = 30°$$

At position 1, the steady-state current flowing through the circuit is

$$I_1 = \frac{V_{\text{rms}}}{\sqrt{R^2 + \omega^2 L^2}} = \frac{\left(\frac{150}{\sqrt{2}}\right)}{\sqrt{20^2 + (200^2)(0.05^2)}} = 4.7434 \text{ A}$$

This steady-state current I_1 is the initial current when the switch is thrown to position 2.

By Classical method

Applying KVL to the circuit after closing the switch at position 2, we get

$$20i + 0.05\frac{di}{dt} = 0 \qquad \text{(i)}$$

$$400i + \frac{di}{dt} = 0 \qquad \text{(ii)}$$

The solution of the above equation contains complementary function (transient part) only, because the steady-state current will be obviously zero, as there is no source connected in the circuit when the switch is put on position 2.

Equation (ii) can be written as:

$$(D + 400)i = 0$$

or

$$D = -400 \qquad \text{(iii)}$$

Hence, the solution of i is given by

$$i = \text{K}e^{-400t} \qquad \text{(iv)}$$

The value of K is calculated by using initial conditions.

At t = 0, i = 4.7434 A. Substituting this value into Eq. (iv), we get

$$\boldsymbol{i = 4.7434e^{-400t}} \textbf{ A}$$

By Laplace transformation method

In view of Eq. (i),

$$20i + 0.05\frac{di}{dt} = 0$$

Taking Laplace transformation on both sides, we get

$$20I(s) + 0.05[sI(s) - I(0)] = 0$$

$$20I(s) + 0.05[sI(s) - 4.7434] = 0 \qquad [\because I(0) = 4.7434]$$

$$I(s)\ [20 + 0.05s] - 0.23717 = 0$$

$$I(s) = \frac{0.23717}{20 + 0.05s}$$

$$= \frac{0.23717}{0.05(s + 400)}$$

$$I(s) = \frac{4.7434}{(s + 400)}$$

Taking inverse Laplace transformation on both sides, we get

$$\boldsymbol{i(t) = 4.7434e^{-400t}} \textbf{ A}$$

EXAMPLE 12.3 The switch in the circuit in Fig. 12.10 is closed at position 1 for a long time. At time $t = 0$, the switch is moved to position 2. Find $i(t)$ for $t \geq 0$.

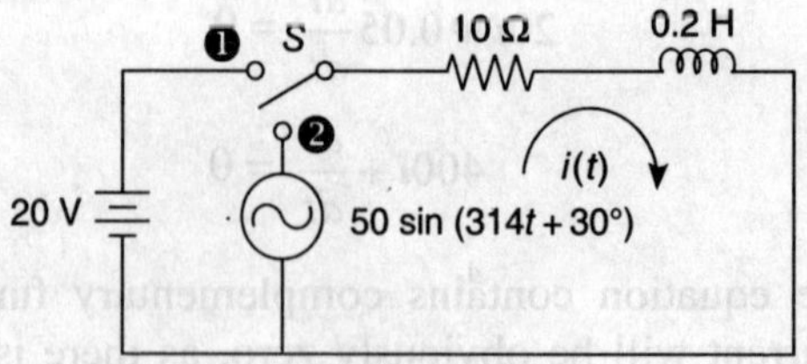

Fig. 12.10

Solution Determination of initial conditions

The switch is closed to position 1 long before $t = 0$. At this position, the circuit should have reached steady-state. We know that in steady-state the inductor would act as a short-circuit. Hence, the current through the inductor is 20/10 = 2 A.

By Classical method

Applying KVL to the circuit after closing the switch at $t = 0$, we get

$$10i + 0.2\frac{di}{dt} = 50 \sin (314t + 30°) \quad \text{(i)}$$

$$50i + \frac{di}{dt} = 250 \sin (314t + 30°) \quad \text{(ii)}$$

The solution of the above equation contains two parts, viz. complementary function (i_c) and particular integral (i_p), i.e.

$$\text{The total current, } i = i_c + i_p \quad \text{(iii)}$$

Equation (ii) can be written as

$$(D + 50)i = 250 \sin (314t + 30°) \quad \text{(iv)}$$

The auxiliary equation is $(D + 50)i = 0$

or

$$D = -50 \quad \text{(v)}$$

Hence, the complementary function of the solution of i is given by

$$i_c = Ke^{-50t} \quad \text{(vi)}$$

The particular integral of i is obtained as follows:
From Eq. (iv),

$$i_p = \frac{250}{(D+50)} \sin (314t + 30°)$$

$$= \frac{(D-50)250}{(D-50)(D+50)} \sin (314t + 30°)$$

$$= \frac{(D-50)250}{(D^2 - 50^2)} \sin (314t + 30°) \quad \text{(vii)}$$

Substituting $D^2 = -314^2$ into Eq. (vii), we obtain $[\because \quad \omega = 314]$

$$i_p = \frac{(D-50)250}{(-314^2 - 50^2)} \sin(314t + 30^\circ)$$

$$= -0.002473\,(D - 50)\sin(314t + 30^\circ)$$

$$= -0.002473\left(\frac{d}{dt}\sin(314t+30^\circ) - 50\sin(314t+30^\circ)\right)$$

$$= -0.002473\,[314\cos(314t + 30^\circ) - 50\sin(314t + 30^\circ)]$$

A right-angled triangle is constructed using 50 and 314 as two sides as shown in Fig. 12.11. From Fig. 12.11,

$$\phi = \tan^{-1}\left(\frac{314}{50}\right) = 80.95^\circ$$

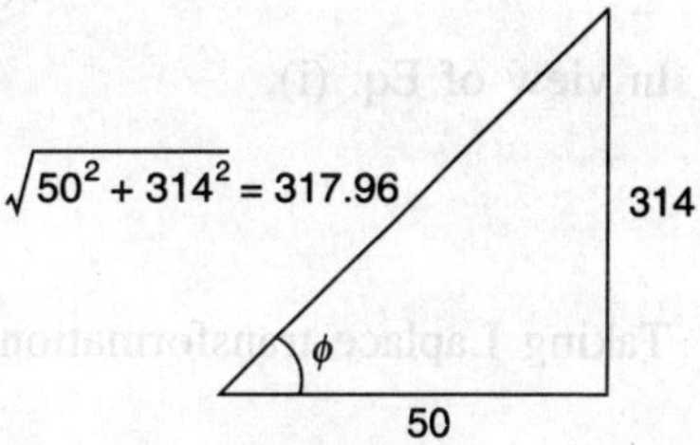

Fig. 12.11

Also, $$\cos\phi = \frac{50}{317.96}$$

or $$50 = 317.96\cos\phi = 317.96\cos 80.95^\circ$$

Similarly,

$$314 = 317.96\sin\phi = 317.96\sin 80.95^\circ$$

$$i_p = -0.002473\,[317.96\sin 80.95^\circ \cos(314t + 30^\circ) - 317.96\cos 80.95^\circ \sin(314t + 30^\circ)$$

$$= 0.7863\,[\cos 80.95^\circ \sin(314t + 30^\circ) - \sin 80.95^\circ \cos(314t + 30^\circ)]$$

$$= 0.7863\sin(314t - 50.95^\circ) \qquad \text{(viii)}$$

Substituting the values of i_c and i_p into Eq. (iii), we get

$$i = \mathrm{K}e^{-50t} + 0.7863\sin(314t - 50.95^\circ) \qquad \text{(ix)}$$

The value of constant K is determined by using initial conditions.
At $t = 0$, $i = 2$ A. Hence Eq. (ix) becomes

$$2 = \mathrm{K}e^{-0} + 0.7863\sin(0 - 50.95^\circ)$$

or $$\mathrm{K} = 2.6106$$

Substituting the value of K into Eq. (ix), we get

$$i = \underbrace{2.6106e^{-50t}}_{\text{Transient part}} + \underbrace{0.7863\sin(314t - 50.95^\circ)\text{ A}}_{\text{Steady-state part}}$$

Laplace transformation method

The source voltage $\quad v(t) = 50\sin(314t + 30^\circ)$

$$= 50\,(\sin 314t\cos 30^\circ + \cos 314t\sin 30^\circ)$$

$$= 50\left(\sin 314t\frac{\sqrt{3}}{2} + \cos 314t\frac{1}{2}\right)$$

$$= 25\sqrt{3}\sin 314t + 25\cos 314t$$

Taking Laplace transformation on both sides,

$$V(s) = 25\sqrt{3}\frac{314}{s^2+314^2} + 25\frac{s}{s^2+314^2}$$

$$= \frac{13597}{s^2+314^2} + \frac{25s}{s^2+314^2}$$

$$V(s) = \frac{25s+13597}{s^2+314^2} \qquad \text{(x)}$$

In view of Eq. (i),

$$10i + 0.2\frac{di}{dt} = 50\sin(314t + 30°)$$

Taking Laplace transformation on both sides,

$$10I(s) + 0.2[sI(s) - I(0)] = \frac{25s+13597}{s^2+314^2} \qquad \text{[From Eq. (x)]}$$

From the initial conditions, substituting the value $I(0) = 2$,

$$10I(s) + 0.2[sI(s) - 2] = \frac{25s+13597}{s^2+314^2}$$

$$10I(s) + 0.2sI(s) = \frac{25s+13597}{s^2+314^2} + 0.4$$

$$I(s)[10+0.2s] = \frac{25s+13597}{s^2+314^2} + 0.4$$

$$I(s) = \frac{25s+13597}{(10+0.2s)(s^2+314^2)} + \frac{0.4}{(10+0.2s)}$$

$$= \frac{\left(\frac{25}{0.2}\right)s + \left(\frac{13597}{0.2}\right)}{(s+50)(s^2+314^2)} + \frac{\left(\frac{0.4}{0.2}\right)}{(s+50)}$$

$$= \frac{125s+67985}{(s+50)(s^2+314^2)} + \frac{2}{(s+50)}$$

By partial fraction technique,

$$I(s) = \frac{125s+67985}{(s+50)(s^2+314^2)} + \frac{2}{(s+50)} = \frac{A}{(s+50)} + \frac{Bs+C}{(s^2+314^2)} + \frac{2}{(s+50)} \qquad \text{(xi)}$$

Solving for the values of A, B and C, we get

$$A = 0.6107$$

$$B = -A = -0.6107$$

and $$C = 155.535$$

Substituting the values of A, B and C into Eq. (xi), we obtain

$$= \frac{0.6107}{(s+50)} + \frac{-0.6107s + 155.535}{(s^2 + 314^2)} + \frac{2}{(s+50)}$$

$$= \frac{2.6107}{(s+50)} - \frac{0.6107s}{(s^2 + 314^2)} + \frac{155.535}{(s^2 + 314^2)}$$

$$= \frac{2.6107}{(s+50)} - \frac{0.6107s}{(s^2 + 314^2)} + \frac{155.535}{314} \times \frac{314}{(s^2 + 314^2)}$$

Taking inverse Laplace transformation on both sides, we get

$$i(t) = 2.6107e^{-50t} - 0.6107 \cos 314t + 0.4953 \sin 314t$$

A right-angled triangle is constructed using 0.4953 and 0.6107 as two sides as shown in Fig. 12.12. Referring to Fig. 12.12,

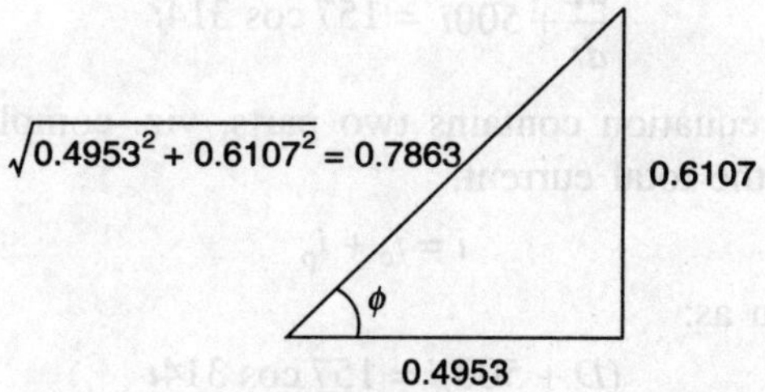

Fig. 12.12

$$\phi = \tan^{-1}\left(\frac{0.6107}{0.4953}\right) = 50.96°$$

Also, $$\cos\phi = \frac{0.4953}{0.7863}$$

or $$0.4953 = 0.7863 \cos\phi = 0.7863 \cos 50.96°$$

Similarly,

$$0.6107 = 0.7863 \sin\phi = 0.7863 \sin 50.96°$$

$\therefore$ $$i(t) = 2.6107e^{-50t} + (0.7863 \cos 50.96° \sin 314t - 0.7863 \sin 50.96° \cos 314t)$$

$$= 2.6107e^{-50t} + 0.7863 (\cos 50.96° \sin 314t - \sin 50.96° \cos 100t)$$

$$i(t) = \underbrace{2.6107e^{-50t}}_{\text{Transient part}} + \underbrace{0.7863 \sin(314t - 50.96°)\ \text{A}}_{\text{Steady-state part}}$$

EXAMPLE 12.4 An *RC* circuit shown in Fig. 12.13 is excited by a sinusoidal voltage source of $v = 50 \sin 314t$ at $t = 0$. Find current $i(t)$, after $t > 0$.

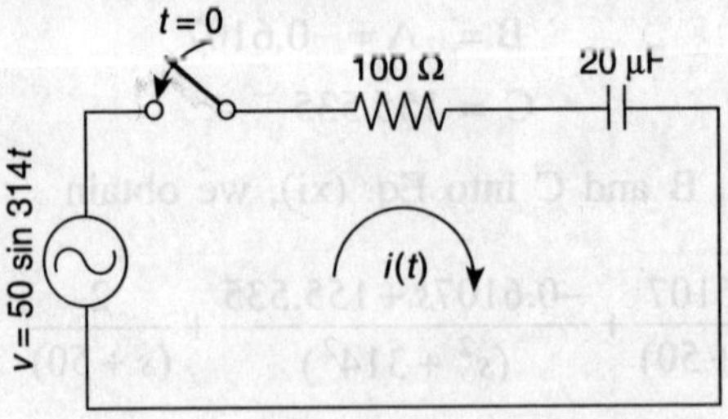

Fig. 12.13

Solution *By Classical method*

Applying KVL to the circuit after closing the switch, we get

$$100i + \frac{1}{20 \times 10^{-6}} \int i dt = 50 \sin 314t \qquad \text{(i)}$$

Differentiating both sides,

$$100 \frac{di}{dt} + 50000i = 50\,(314) \cos 314t$$

$$\frac{di}{dt} + 500i = 157 \cos 314t \qquad \text{(ii)}$$

The solution of the above equation contains two parts, viz. complementary function (i_c) and particular integral (i_p), i.e. the total current,

$$i = i_c + i_p \qquad \text{(iii)}$$

Equation (ii) can be written as:

$$(D + 500)i = 157 \cos 314t \qquad \text{(iv)}$$

The auxiliary equation is $(D + 500)i = 0$

or

$$D = -500 \qquad \text{(v)}$$

Hence, the complementary function of the solution of i is given by

$$i_c = Ke^{-500t} \qquad \text{(vi)}$$

The particular integral of i is obtained as follows:

From Eq. (iv),

$$i_p = \frac{157}{(D + 500)} \cos 314t$$

$$= \frac{(D - 500)157}{(D - 500)(D + 500)} \cos 314t$$

$$= \frac{(D - 500)157}{(D^2 - 500^2)} \cos 314t \qquad \text{(vii)}$$

Substituting $D^2 = -314^2$ into Eq. (vii), we obtain $[\because \quad \omega = 314]$

$$i_p = \frac{(D-500)157}{(-314^2 - 500^2)} \cos 314t$$
$$= -0.00045\,(D - 500)\cos 314t$$
$$= -0.00045\left(\frac{d}{dt}\cos 314t - 500\cos 314t\right)$$
$$= -0.00045\,[-(314)\sin 314t - 500\cos 314t]$$
$$= 0.00045\,[314\sin 314t + 500\cos 314t]$$

A right-angled triangle is constructed using 314 and 500 as two sides as shown in Fig. 12.14. From Fig. 12.14,

$$\phi = \tan^{-1}\left(\frac{500}{314}\right) = 57.87°$$

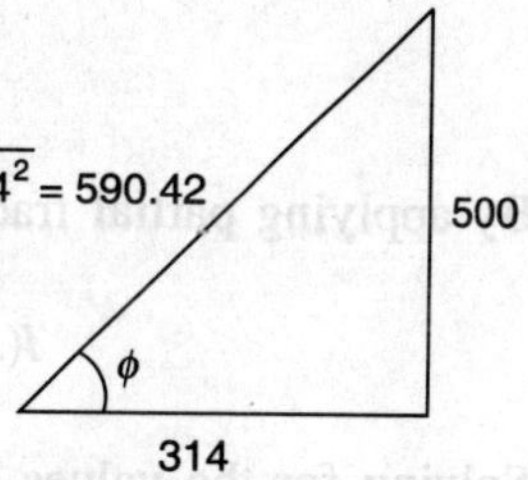

Fig. 12.14

Also, $\cos\phi = \dfrac{314}{590.42}$

or $314 = 590.42\cos\phi = 590.42\cos 57.87°$

Similarly,

$$500 = 590.42\sin\phi = 590.42\sin 57.87°$$
$$i_p = 0.00045\,(590.42\cos 57.87°\sin 314t + 590.42\sin 57.87°\cos 314t)$$
$$= 0.2657\,(\cos 57.87°\sin 314t + \sin 57.87°\cos 314t)$$
$$= 0.2657\sin(314t + 57.87°) \qquad \text{(viii)}$$

Substituting the values of i_c and i_p into Eq. (iii), we get

$$i = Ke^{-500t} + 0.2657\sin(314t + 57.87°) \qquad \text{(ix)}$$

The value of constant K is determined by using initial conditions.

At $t = 0$, $i = 0$. Hence, Eq. (ix) becomes

$$0 = Ke^{-0} + 0.2657\sin(0 + 57.87°)$$

or $$K = -0.225$$

Substituting the value of K into Eq. (ix), we get

$$i = \underbrace{-0.225e^{-500t}}_{\text{Transient part}} + \underbrace{0.2657\sin(314t + 57.87°)\text{ A}}_{\text{Steady-state part}}$$

Laplace transformation method

In view of Eq. (i),

$$100i + \frac{1}{20\times 10^{-6}}\int i\,dt = 50\sin 314t$$

Taking Laplace transformation on both sides,

$$100I(s) + 50000\left[\frac{1}{s}I(s)\right] = 50 \times \frac{314}{s^2 + 314^2}$$

$$I(s)\left(100 + \frac{50000}{s}\right) = \frac{15700}{s^2 + 314^2}$$

$$I(s) = \left(\frac{s}{100s + 50000}\right)\left(\frac{15700}{s^2 + 314^2}\right)$$

$$= \left(\frac{s}{100(s + 500)}\right)\left(\frac{15700}{s^2 + 314^2}\right)$$

$$= \frac{157s}{(s + 500)(s^2 + 314^2)}$$

By applying partial fraction expansion technique,

$$I(s) = \frac{157s}{(s + 500)(s^2 + 314^2)} = \frac{A}{(s + 500)} + \frac{Bs + C}{(s^2 + 314^2)} \qquad \text{(x)}$$

Solving for the values of A, B and C, we get

$$A = -0.2252$$

$$B = -A = 0.2252$$

and

$$C = 44.4$$

Substituting the values of A, B and C into Eq. (x), we get

$$= \frac{-0.2252}{(s + 500)} + \frac{0.2252s + 44.4}{(s^2 + 314^2)}$$

$$= \frac{-0.2252}{(s + 500)} + \frac{0.2252s}{(s^2 + 314^2)} + \frac{44.4}{(s^2 + 314^2)}$$

$$= -\frac{0.2252}{(s + 500)} + \frac{0.2252s}{(s^2 + 314^2)} + \frac{44.4}{314} \times \frac{314}{(s^2 + 314^2)}$$

Taking inverse Laplace transformation on both sides, we get

$$i(t) = -0.2252e^{-500t} + 0.2252 \cos 314t + 0.1414 \sin 314t$$

A right-angled triangle is constructed using 0.1414 and 0.2252 as two sides as shown in Fig. 12.15.

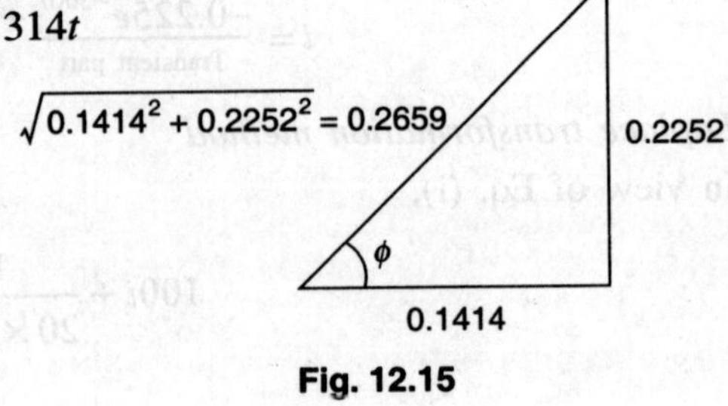

Fig. 12.15

Referring to Fig. 12.15,

$$\phi = \tan^{-1}\left(\frac{0.2252}{0.1414}\right) = 57.87°$$

Also, $$\cos\phi = \frac{0.1414}{0.2659}$$

or $$0.1414 = 0.2659 \cos\phi = 0.2659 \cos 57.87°$$

Similarly,

$$0.2252 = 0.2659 \sin\phi = 0.2659 \sin 57.87°$$

$$\therefore \quad i(t) = -0.2252e^{-500t} + 0.2659 \sin 57.87° \cos 314t + 0.2659 \cos 57.87° \sin 314t$$

$$= -0.2252e^{-500t} + 0.2659 (\sin 57.87° \cos 314t + \cos 57.87° \sin 314t)$$

$$i(t) = \underbrace{-0.2252e^{-500t}}_{\text{Transient part}} + \underbrace{0.2659 \sin(314t + 57.87°)\text{ A}}_{\text{Steady-state part}}$$

EXAMPLE 12.5 An *RC* circuit shown in Fig. 12.16 is excited by a sinusoidal voltage source of $v = 200 \sin(2000t + 90°)$ at $t = 0$. Find current $i(t)$ after $t > 0$. Assume the initial charge q_0 on the capacitor as 1250 μC.

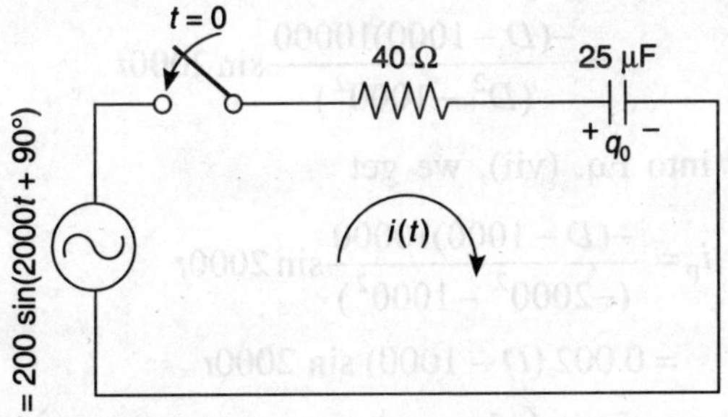

Fig. 12.16

Solution *By Classical method*

Applying KVL to the circuit after closing the switch, we get

$$40i + \frac{1}{25 \times 10^{-6}} \int i\,dt = 200 \sin(2000t + 90°) \quad \text{(i)}$$

$$40i + 40000 \int i\,dt = 200 \cos 2000t$$

Differentiating both sides,

$$40\frac{di}{dt} + 40000i = -200(2000) \sin 2000t$$

or $$\frac{di}{dt} + 1000i = -10000 \sin 2000t \quad \text{(ii)}$$

The solution of the above equation contains two parts, viz. complementary function (i_c) and particular integral (i_p), i.e. the total current,

$$i = i_c + i_p \quad \text{(iii)}$$

Equation (ii) can be written as:

$$(D + 1000)i = -10000 \sin 2000t \quad \text{(iv)}$$

The auxiliary equation is $(D + 1000)i = 0$

or

$$D = -1000 \quad \text{(v)}$$

Hence, the complementary function of the solution of i is given by

$$i_c = Ke^{-1000t} \quad \text{(vi)}$$

The particular integral of i is given as follows:
From Eq. (iv),

$$i_p = \frac{-10000}{(D+1000)} \sin 2000t$$

$$= \frac{-(D-1000)10000}{(D-1000)(D+1000)} \sin 2000t$$

$$= \frac{-(D-1000)10000}{(D^2-1000^2)} \sin 2000t \quad \text{(vii)}$$

Substituting $D^2 = -2000^2$ into Eq. (vii), we get [$\because \quad \omega = 2000$]

$$i_p = \frac{-(D-1000)10000}{(-2000^2-1000^2)} \sin 2000t$$

$$= 0.002\,(D - 1000) \sin 2000t$$

$$= 0.002\left(\frac{d}{dt} \sin 2000t - 1000 \sin 2000t\right)$$

$$= 0.002\,(2000 \cos 2000t - 1000 \sin 2000t)$$

A right-angled triangle is constructed using 2000 and 1000 as two sides as shown in Fig. 12.17.
From Fig. 12.17,

$$\phi = \tan^{-1}\left(\frac{1000}{2000}\right) = 26.57°$$

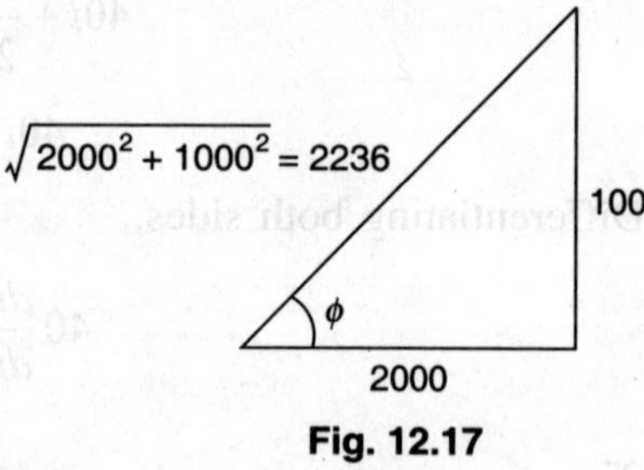

Fig. 12.17

Also, $\cos\phi = \dfrac{2000}{2236}$

or $2000 = 2236 \cos \phi = 2236 \cos 26.57°$

Similarly,

$$1000 = 2236 \sin \phi = 2236 \sin 26.57°$$

$$i_p = 0.002\,(2236 \cos 26.57° \cos 2000t - 2236 \sin 26.57° \sin 2000t)$$

$$= 4.472\,(\cos 26.57° \cos 2000t - \sin 26.57° \sin 2000t)$$

$$= 4.472 \cos (2000t + 26.57°)$$

$$= 4.472 \sin (2000t + 26.57° + 90°)$$

$$= 4.472 \sin (2000t + 116.57°) \quad \text{(viii)}$$

Substituting the values of i_c and i_p into Eq. (iii), we obtain

$$i = Ke^{-1000t} + 4.472 \sin (2000t + 116.57°) \quad \text{(ix)}$$

The value of constant K is determined by using initial conditions.

At $t = 0$, $i = i_0$ $\qquad \int idt = q_0 = 1250 \times 10^{-6}$

Substituting into Eq. (i), we get

$$40i_0 + 40000\,(1250 \times 10^{-6}) = 200$$

or $\qquad i_0 = 3.75$

Therefore, when $t = 0$, $\qquad i = i_0 = 3.75$

Substituting this value into Eq. (ix), we get

$$3.75 = Ke^{-0} + 4.472 \sin (0 + 116.57°)$$

or $\qquad K = -0.25$

Substituting the value of K into Eq. (ix), we get

$$i = \underbrace{-0.25e^{-1000t}}_{\text{Transient part}} + \underbrace{4.472 \sin (2000t + 116.57°)\text{ A}}_{\text{Steady-state part}}$$

Laplace transformation method

In view of Eq. (i),

$$40i + \frac{1}{25 \times 10^{-6}} \int idt = 200 \sin (2000t + 90°)$$

The above equation is rewritten to include the initial charge on the capacitor as:

$$40i + \frac{1}{25 \times 10^{-6}} \left[\int_{-\infty}^{0} idt + \int_{0}^{t} idt \right] = 200 \sin (2000t + 90°)$$

$$40i + 40000 \left[1250 \times 10^{-6} + \int_{0}^{t} idt \right] = 200 \cos 2000t \qquad \left[\because \int_{-\infty}^{0} idt = q_0 = 1250 \times 10^{-6} \right]$$

$$40i + 50 + 40000 \int_{0}^{t} idt = 200 \cos 2000t$$

$$40i + 40000 \int_{0}^{t} idt = 200 \cos 2000t - 50$$

Taking Laplace transformation on both sides,

$$40I(s) + 40000 \left[\frac{1}{s} I(s) \right] = 200 \times \frac{s}{s^2 + 2000^2} - \frac{50}{s}$$

$$I(s)\left(40+\frac{40000}{s}\right)=\frac{200s}{s^2+2000^2}-\frac{50}{s}$$

$$I(s)=\left(\frac{s}{40s+40000}\right)\left(\frac{200s}{s^2+2000^2}-\frac{50}{s}\right)$$

$$=\left(\frac{s}{40(s+1000)}\right)\left(\frac{200s}{s^2+2000^2}-\frac{50}{s}\right)$$

$$=\frac{5s^2}{(s+1000)(s^2+2000^2)}-\frac{1.25}{(s+1000)}$$

By applying partial fraction expansion technique,

$$\frac{5s^2}{(s+1000)(s^2+2000^2)}-\frac{1.25}{(s+1000)}=\frac{\text{A}}{(s+1000)}+\frac{\text{B}}{(s+j2000)}+\frac{\text{C}}{(s-j2000)}-\frac{1.25}{(s+1000)} \quad \text{(x)}$$

Solving for the values of A, B and C, we get

$$\text{A}=1;\ \text{B}=2-j1;\ \text{C}=2+j1$$

Substituting the values of A, B and C into Eq. (x), we get

$$I(s)=\frac{1}{(s+1000)}+\frac{2-j1}{(s+j2000)}+\frac{2+j1}{(s-j2000)}-\frac{1.25}{(s+1000)}$$

Taking inverse Laplace transformation,

$$i(t) = 1e^{-1000t} + (2 - j1)\, e^{-j2000t} + (2 + j1)e^{j2000t} - 1.25e^{-1000t}$$

$$= -0.25e^{-1000t} + 2e^{-j2000t} - j1e^{-j2000t} + 2e^{j2000t} + j1e^{j2000t}$$

$$= -0.25e^{-1000t} + 4 \cos 2000t - 2 \sin 2000t \quad \text{(xi)}$$

A right-angled triangle is constructed using 4 and 2 as two sides as shown in Fig. 12.18. From Fig. 12.18,

$$\phi = \tan^{-1}\left(\frac{2}{4}\right) = 26.57°$$

Also, $\cos\phi = \dfrac{4}{4.472}$

or $4 = 4.472 \cos \phi = 4.472 \cos 26.57°$

Similarly, $2 = 4.472 \sin \phi = 4.472 \sin 26.57°$

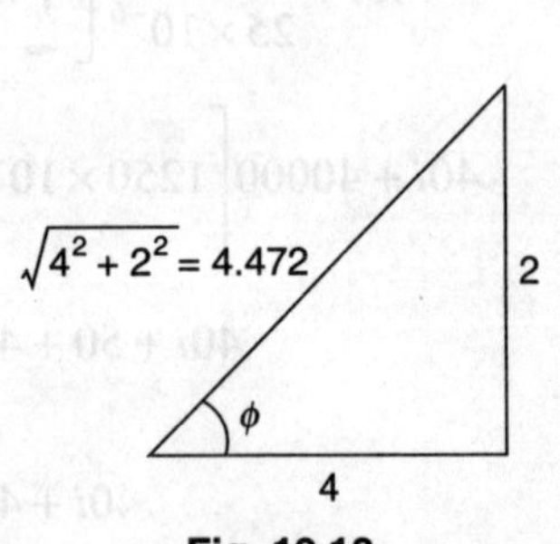

Fig. 12.18

Substituting these values into Eq. (xi), we get

$$i(t) = -0.25e^{-1000t} + 4.472 \cos 26.57° \cos 2000t - 4.472 \sin 26.57° \sin 2000t$$

$$= -0.25e^{-1000t} + 4.472 (\cos 26.57° \cos 2000t - \sin 26.57° \sin 2000t)$$

$$= -0.25e^{-1000t} + 4.472 \cos (2000t + 26.57°)$$

$$i(t) = \underbrace{-0.25e^{-1000t}}_{\text{Transient part}} + \underbrace{4.472 \sin (2000t + 116.57°) \text{ A}}_{\text{Steady-state part}}$$

EXAMPLE 12.6 In the circuit shown in Fig. 12.19, the switch remains in position 1 until steady state is reached. At time $t = 0$, the switch is changed to position 2. Find $i(t)$.

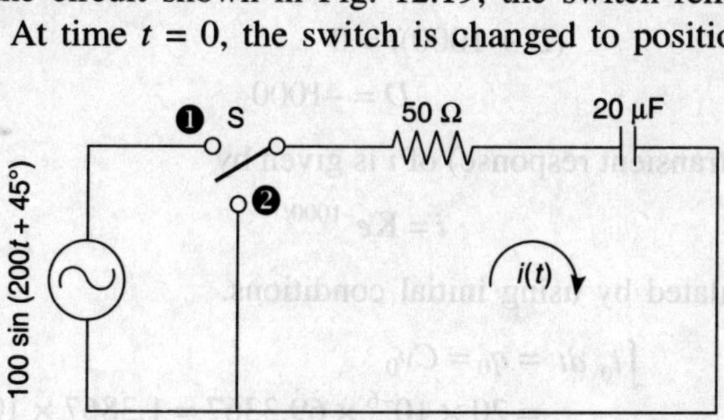

Fig. 12.19

Solution Determination of initial conditions

In position 1, the circuit has already reached steady state. Comparing 100 sin (200t + 45°) with $v_m \sin(\omega t + \theta)$,

$$v_m = 100,\ \omega = 200,\ \theta = 45°$$

At position 1, the steady state voltage across the capacitor v_0 is found by voltage division rule as:

$$v_0 = V_{rms} \times \frac{-jX_C}{R - jX_C}$$

$$= \frac{100\angle 45°}{\sqrt{2}} \times \frac{-j250}{50 - j250}$$

$$= 69.3367\angle 33.7°$$

Hence, the magnitude of the voltage v_0 = 69.3367 V.

This steady-state voltage v_0 is the initial condition when the switch is thrown to position 2.

By Classical method

Applying KVL to the circuit after closing the switch at position 2, we get

$$50i + \frac{1}{20 \times 10^{-6}} \int i dt = 0 \qquad \text{(i)}$$

$$50\frac{di}{dt} + 50000i = 0$$

or

$$\frac{di}{dt} + 1000i = 0 \qquad \text{(ii)}$$

The solution of the above equation contains complementary function (transient part) only because the steady-state current will be obviously zero, as there is no source connected in the circuit after the switch is put at position 2.

Equation (ii) can be written as:

$$(D + 1000)i = 0$$

or

$$D = -1000 \quad \text{(iii)}$$

Therefore, the solution (transient response) of i is given by

$$i = \text{K}e^{-1000t} \quad \text{(iv)}$$

The value of K is calculated by using initial conditions.

At $t = 0$, $i = i_0$,

$$\int i_0\, dt = q_0 = Cv_0$$

$$= 20 \times 10^{-6} \times 69.3367 = 1.3867 \times 10^{-3}$$

Upon substitution into Eq. (i), we get

$$50i_0 + \frac{1}{20 \times 10^{-6}}(1.3867 \times 10^{-3}) = 0$$

$$50i_0 + 69.335 = 0$$

$$i_0 = -1.3867$$

Hence, at $t = 0$,

$$i = i_0 = -1.3867$$

Substituting the above values into Eq. (iv), we obtain

$$-1.3867 = \text{K}$$

Substituting the value of K into Eq. (iv), we get

$$\boldsymbol{i = -1.3867e^{-1000t}\ \text{A}}$$

By Laplace transformation method

In view of Eq. (i),

$$50i + \frac{1}{20 \times 10^{-6}} \int i dt = 0$$

The above equation can be rewritten to include the initial condition as:

$$50i + \frac{1}{20 \times 10^{-6}} \left[\int_{-\infty}^{0} i dt + \int_{0}^{t} i dt \right] = 0 \quad \text{(v)}$$

Here,

$$\int_{-\infty}^{0} i dt = q_0 = Cv_0$$

$$= 20 \times 10^{-6} \times 69.3367 = 1.3867 \times 10^{-3}$$

Upon substitution, Eq. (v) becomes

$$50i + \frac{1}{20 \times 10^{-6}} \left(1.3867 \times 10^{-3} + \int_{0}^{t} i dt \right) = 0$$

$$50i + 69.3367 + \frac{1}{20\times10^{-6}}\left(\int_0^t i dt\right) = 0$$

$$50i + 69.3367 + 50000\int_0^t i dt = 0$$

$$50i + 50000\int_0^t i dt = -69.3367$$

Taking Laplace transformation on both sides,

$$50I(s) + 50000\left[\frac{I(s)}{s}\right] = \frac{-69.3367}{s}$$

$$I(s)\left[50 + \frac{50000}{s}\right] = \frac{-69.3367}{s}$$

$$I(s) = \frac{-69.3367}{(50s + 50000)}$$

$$= \frac{-1.3867}{(s+1000)}$$

Taking inverse Laplace transformation on both sides, we get

$$\boldsymbol{i(t) = -1.3867e^{-1000t}\ \text{A}}$$

EXAMPLE 12.7 The *RLC* circuit shown in Fig. 12.20 has a sinusoidal voltage source v = 100 sin (500t + ϕ) volts. If the switch is closed when ϕ = 90°, find the current flowing through the circuit assuming zero initial charge on the capacitor.

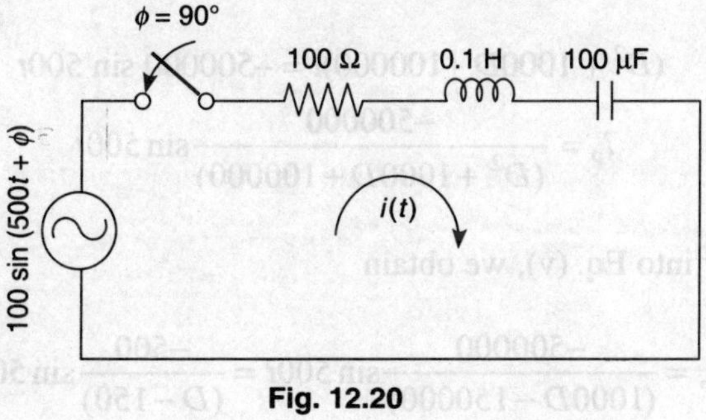

Fig. 12.20

Solution *By Classical method*

Applying KVL to the circuit after closing the switch, we get

$$100i + 0.1\frac{di}{dt} + \frac{1}{100\times10^{-6}}\int i\, dt = 100\sin(500t + 90°) \qquad \text{(i)}$$

$$100i + 0.1\frac{di}{dt} + \frac{1}{100 \times 10^{-6}}\int i\,dt = 100 \cos 500t$$

Differentiating both sides,

$$100\frac{di}{dt} + 0.1\frac{d^2i}{dt^2} + \frac{1}{100 \times 10^{-6}} i = -100\,(500) \sin 500t$$

$$\frac{d^2i}{dt^2} + 1000\frac{di}{dt} + 100000i = -500000 \sin 500t \qquad \text{(ii)}$$

The solution of the above equation contains two parts, viz. complementary function (i_c) and particular integral (i_p), i.e. the total current,

$$i = i_c + i_p \qquad \text{(iii)}$$

Equation (ii) can be written as:

$$(D^2 + 1000D + 100000)i = -500000 \sin 500t \qquad \text{(iv)}$$

The auxiliary equation is $\quad (D^2 + 1000D + 100000)i = 0$

Therefore, $\quad D = \dfrac{-1000 \pm \sqrt{1000^2 - 400000}}{2}$

The two roots –887.3 and –112.7 are real and distinct. So, the complementary function is of the form:

$$i_c = K_1 e^{D1t} + K_2 e^{D2t}$$

Hence, $\quad i_c = K_1 e^{-887.3t} + K_2 e^{-112.7t}$

The particular integral of i is given as follows:
From Eq. (iv),

$$(D^2 + 1000D + 100000)i = -500000 \sin 500t$$

$$i_p = \frac{-500000}{(D^2 + 1000D + 100000)} \sin 500t \qquad \text{(v)}$$

Substituting $D^2 = -500^2$ into Eq. (v), we obtain $\qquad [\because \quad \omega = 500]$

$$i_p = \frac{-500000}{(1000D - 150000)} \sin 500t = \frac{-500}{(D - 150)} \sin 500t$$

$$= \frac{-500(D + 150)}{(D + 150)(D - 150)} \sin 500t$$

$$= \frac{-500(D + 150)}{(D^2 - 22500)} \sin 500t$$

Again substituting $D^2 = -500^2$,

$$i_p = \frac{-500(D+150)}{-272500}\sin 500t$$

$$= 0.001835\,(D \sin 500t + 150 \sin 500t)$$

$$= 0.001835\left(\frac{d}{dt}\sin 500t + 150\sin 500t\right)$$

$$= 0.001835\,[500 \cos 500t + 150 \sin 500t] \qquad \text{(vi)}$$

A right-angled triangle is constructed using 500 and 150 as two sides as shown in Fig. 12.21. From Fig. 12.21,

$$\phi = \tan^{-1}\left(\frac{500}{150}\right) = 73.3°$$

Also, $\cos\phi = \dfrac{150}{522.0153}$

or $150 = 522.0153 \cos \phi = 522.0153 \cos 73.3°$

Similarly, $500 = 522.0153 \sin \phi = 522.0153 \sin 73.3°$

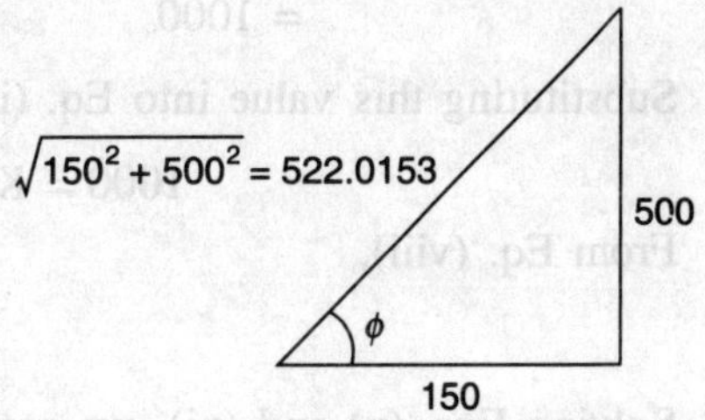

Fig. 12.21

Therefore, Eq. (vi) becomes

$$i_p = 0.001835\,(522.0153 \sin 73.3° \cos 500t + 522.0153 \cos 73.3° \sin 500t)$$
$$= 0.9579\,(\sin 73.3° \cos 500t + \cos 73.3° \sin 500t)$$
$$= 0.9579 \sin (500t + 73.3°)$$

Substituting the values of i_c and i_p into Eq. (iii), we get

$$i = K_1 e^{-887.3t} + K_2 e^{-112.7t} + 0.9579 \sin (500t + 73.3°) \qquad \text{(vii)}$$

Constants K_1 and K_2 are determined by using the initial conditions. Therefore, at $t = 0$, $i = 0$.
Substituting this value into Eq. (vii), we get

$$0 = K_1 + K_2 + 0.9175$$

$$K_1 + K_2 = -0.9175 \qquad \text{(viii)}$$

Differentiating Eq. (vii) we get

$$\frac{di}{dt} = K_1(-887.3)e^{-887.3t} + K_2(-112.7)e^{-112.7t} + 0.9579(500)\cos(500t + 73.3°)$$

$$\left.\frac{di}{dt}\right|_{t=0} = K_1(-887.3) + K_2(-112.7) + 137.6313 \qquad \text{(ix)}$$

Also, from Eq. (i),

$$100i + 0.1\frac{di}{dt} + \frac{1}{100\times 10^{-6}}\int i dt = 100\sin(500t + 90°)$$

$$1000i + \frac{di}{dt} + \frac{10}{100\times 10^{-6}}\int i dt = 1000\sin(500t + 90°)$$

$$\frac{di}{dt} = 1000\sin(500t + 90°) - 1000i - 100000\int i\, dt$$

$$\left.\frac{di}{dt}\right|_{t=0} = 1000\sin(0 + 90°) - 1000(0) - 100000\int (0)\, dt \qquad [\because \text{ at } t = 0,\ i = 0]$$

$$= 1000$$

Substituting this value into Eq. (ix), we obtain

$$1000 = K_1(-887.3) + K_2(-112.7) + 137.6313 \tag{x}$$

From Eq. (viii),

$$K_2 = -0.9175 - K_1 \tag{xi}$$

Solving Eqs. (x) and (xi), we get

$$K_1 = -0.98$$

$$K_2 = 0.0624$$

Substituting these values into Eq. (vii), we get

$$i = \underbrace{-0.98e^{-887.3t} + 0.0624e^{-112.7t}}_{\text{Transient part}} + \underbrace{0.9579\sin(500t + 73.3°)\text{ A}}_{\text{Steady-state part}}$$

Laplace transformation method

In view of Eq. (i),

$$100i + 0.1\frac{di}{dt} + \frac{1}{100\times 10^{-6}}\int i dt = 100\sin(500t + 90°)$$

$$100i + 0.1\frac{di}{dt} + \frac{1}{100\times 10^{-6}}\int i dt = 100\cos 500t$$

Taking Laplace transformation on both sides,

$$100I(s) + 0.1sI(s) + 10000\frac{I(s)}{s} = 100\times\frac{s}{s^2 + 500^2}$$

$$I(s)\left(100 + 0.1s + \frac{10000}{s}\right) = \frac{100s}{s^2 + 500^2}$$

$$I(s)\left(\frac{100s + 0.1s^2 + 10000}{s}\right) = \frac{100s}{s^2 + 500^2}$$

$$I(s) = \frac{100s^2}{(100s + 0.1s^2 + 10000)(s^2 + 500^2)}$$

$$= \frac{1000s^2}{(s^2 + 1000s + 100000)(s^2 + 500^2)}$$

$$= \frac{1000s^2}{(s + 887.3)(s + 112.7)(s^2 + 500^2)}$$

By applying partial fraction expansion technique,

$$\frac{1000s^2}{(s + 887.3)(s + 112.7)(s^2 + 500^2)} = \frac{A}{(s + 887.3)} + \frac{B}{(s + 112.7)} + \frac{Cs + D}{(s^2 + 500^2)} \quad \text{(xii)}$$

Constants A, B, C and D are evaluated as follows:

$$A = \left.\frac{1000s^2}{(s + 112.7)(s^2 + 500^2)}\right|_{s=-887.3} = -0.98 \quad \text{(xiii)}$$

$$B = \left.\frac{1000s^2}{(s + 887.3)(s^2 + 500^2)}\right|_{s=-112.7} = 0.0624 \quad \text{(xiv)}$$

On cross-multiplying Eq. (xii), we get

$$1000s^2 = A(s + 112.7)(s^2 + 500^2) + B(s + 887.3)(s^2 + 500^2) + (Cs + D)(s + 112.7)(s + 887.3)$$

$$= As^3 + 500^2As + 112.7As^2 + (112.7)(500)^2A + Bs^3 + 500^2Bs + 887.3Bs^2$$
$$+ (887.3)(500)^2B + Cs^3 + 1000Cs^2 + 10000Cs + Ds^2 + 1000Ds + 10000D$$

$$= s^3(A + B + C) + s^2(112.7A + 887.3B + 1000C + D)$$
$$+ s(500^2A + 500^2B + 10000C + 1000D) + (112.7)(500)^2A + (887.3)(500)^2B + 10000D \quad \text{(xv)}$$

On comparing the coefficients of 's^3' in Eq. (xv),

$$0 = A + B + C$$

Substituting the values of A and B respectively from Eqs. (xiii) and (xiv),

$$C = 0.9176 \quad \text{(xvi)}$$

On comparing the coefficients of 's^2' in Eq. (xv),

$$1000 = 112.7A + 887.3B + 1000C + D \quad \text{(xvii)}$$

Substituting the values of A, B and C respectively from Eqs. (xiii), (xiv) and (xvi) into Eq. (xvii), we get

$$D = 137.478 \quad \text{(xviii)}$$

Putting the values of A, B, C and D in Eq. (xii),

$$I(s) = \frac{-0.98}{(s+887.3)} + \frac{0.0624}{(s+112.7)} + \frac{0.9176s+137.478}{(s^2+500^2)}$$

$$= \frac{-0.98}{(s+887.3)} + \frac{0.0624}{(s+112.7)} + \frac{0.9176s}{(s^2+500^2)} + \frac{137.478}{(s^2+500^2)}$$

$$= \frac{-0.98}{(s+887.3)} + \frac{0.0624}{(s+112.7)} + \frac{0.9176s}{(s^2+500^2)} + \frac{137.478}{500}\frac{500}{(s^2+500^2)}$$

Taking inverse Laplace transformation on both sides,

$$i(t) = -0.98e^{-887.3t} + 0.0624e^{-112.7t} + 0.9176 \cos 500t + 0.2749 \sin 500t \qquad \text{(xix)}$$

A right-angled triangle is constructed with 0.9176 and 0.2749 as two sides as shown in Fig. 12.22.

Referring to Fig. 12.22,

$$\phi = \tan^{-1}\left(\frac{0.9176}{0.2749}\right) = 73.32^\circ$$

Also, $\cos\phi = \dfrac{0.2749}{0.9579}$

or $\quad 0.2749 = 0.9579 \cos \phi = 0.9579 \cos 73.32^\circ$

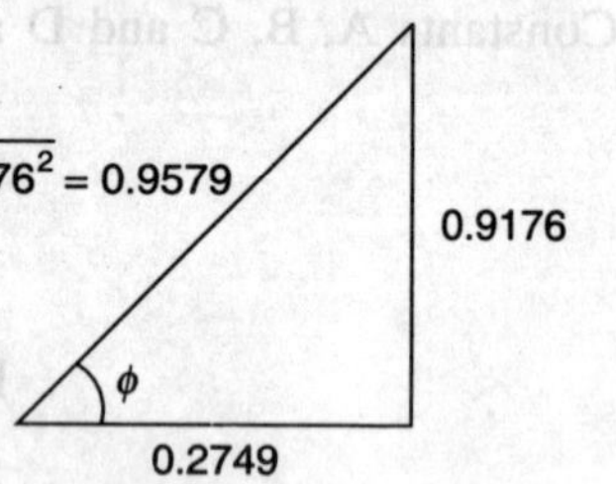

Fig. 12.22

Similarly, $\quad 0.9176 = 0.9579 \sin \phi = 0.9579 \sin 73.32^\circ$

Substituting these values into Eq. (xix), we get

$$i(t) = -0.98e^{-887.3t} + 0.0624e^{-112.7t} + 0.9579 \sin 73.32^\circ \cos 500t + 0.9579 \cos 73.32^\circ \sin 500t$$

$$= -0.98e^{-887.3t} + 0.0624e^{-112.7t} + 0.9579 \sin (500t + 73.32^\circ)$$

$$i(t) = \underbrace{-0.98e^{-887.3t} + 0.0624e^{-112.7t}}_{\text{Transient part}} + \underbrace{0.9579 \sin(500t + 73.32^\circ)\text{ A}}_{\text{Steady-state part}}$$

EXAMPLE 12.8 Find the complete solution for the current in the network shown in Fig. 12.23 for $t > 0$, if the switch closes at $t = 0$. The state of the network just before switching is given as, $i(0^-) = 1$ A and $V_C(0^-) = 2$ V.

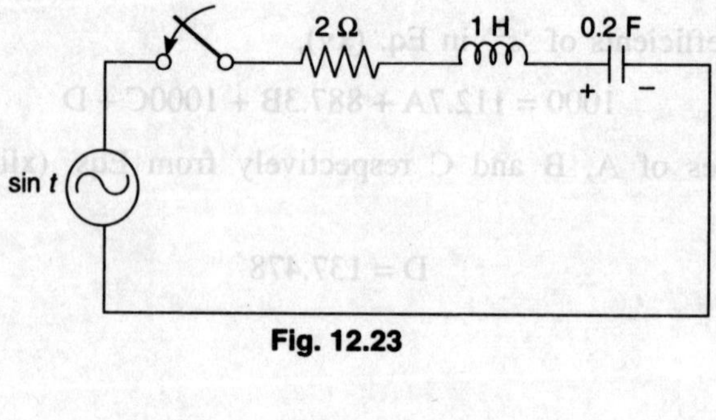

Fig. 12.23

Solution *By Classical method*

Applying KVL to the circuit after closing the switch, we get

$$2i + 1\frac{di}{dt} + \frac{1}{0.2}\int_{-\infty}^{t} i\,dt \qquad \text{(i)}$$

Equation (i) is rewritten to include the initial voltage across the capacitor as:

$$2i + \frac{di}{dt} + \frac{1}{0.2}\left[\int_{-\infty}^{0} i\,dt + \int_{0}^{t} i\,dt\right] = \sin t$$

$$2i + \frac{di}{dt} + \frac{1}{0.2}\left[0.4 + \int_{0}^{t} i\,dt\right] = \sin t \qquad \left[\because \int_{-\infty}^{0} i\,dt = q_0 = CV_0 = 0.2 \times 2 = 0.4\right]$$

$$2i + \frac{di}{dt} + 2 + \frac{1}{0.2}\int_{0}^{t} i\,dt = \sin t$$

Differentiating both sides,

$$\frac{d^2i}{dt^2} + 2\frac{di}{dt} + 5i = \cos t \qquad \text{(ii)}$$

The solution of the above equation contains two parts, *viz.* complementary function (i_c) and particular integral (i_p), i.e. the total current,

$$i = i_c + i_p \qquad \text{(iii)}$$

Equation (ii) can be written as:

$$(D^2 + 2D + 5)i = \cos t \qquad \text{(iv)}$$

The auxiliary equation is $(D^2 + 2D + 5)i = 0$

Hence, $$D = \frac{-2 \pm \sqrt{4 - 20}}{2} = \frac{-2 \pm j4}{2}$$

Thus, the two roots $-1 + j2$ and $-1 - j2$ are complex and distinct. So, the complementary function is of the form:

$$i_c = K_1 e^{D1t} + K_2 e^{D2t}$$

Hence, $$i_c = K_1 e^{(-1 + j2)t} + K_2 e^{(-1 - j2)t}$$

The particular integral of i is given as follows:

From Eq. (iv),

$$i_p = \frac{1}{(D^2 + 2D + 5)}\cos t \qquad \text{(v)}$$

Substituting $D^2 = -1^2$ (because $\omega = 1$) into Eq. (v), we obtain

$$i_p = \frac{1}{(2D+4)}\cos t$$

$$= \frac{(2D-4)}{(2D+4)(2D-4)}\cos t$$

$$= \frac{(2D-4)}{(4D^2-16)}\cos t$$

Again substituting $D^2 = -1^2$,

$$i_p = \frac{2(D-2)}{-20}\cos t$$

$$= -0.1(D\cos t - 2\cos t)$$

$$= -0.1\left(\frac{d}{dt}\cos t - 2\cos t\right)$$

$$i_p = 0.1\sin t + 0.2\cos t$$

Substituting the values of i_c and i_p into Eq. (iii), we get

$$i = K_1 e^{(-1+j2)t} + K_2 e^{(-1-j2)t} + 0.1\sin t + 0.2\cos t \quad \text{(vi)}$$

Constants K_1 and K_2 are determined by using the initial conditions.

Since inductor does not permit any sudden change in current, $i(0^-) = i(0^+) = 1$ A. Therefore, at $t = 0$, $i = 1$.

Substituting this value into Eq. (vi), we get

$$1 = K_1 + K_2 + 0 + 0.2$$

$$K_1 + K_2 = 0.8 \quad \text{(vii)}$$

Differentiating Eq. (vi), we get

$$\frac{di}{dt} = K_1(-1+j2)e^{(-1+j2)t} + K_2(-1-j2)e^{(-1-j2)t} + 0.1\cos t - 0.2\sin t$$

$$\left.\frac{di}{dt}\right|_{t=0} = K_1(-1+j2) + K_2(-1-j2) + 0.1 \quad \text{(viii)}$$

Also, from Eq. (i),

$$2i + \frac{di}{dt} + \frac{1}{0.2}\int_{-\infty}^{t} i\,dt = \sin t$$

$$\frac{di}{dt} = \sin t - 2i - \frac{1}{0.2}\int_{-\infty}^{t} i\,dt$$

$$\left.\frac{di}{dt}\right|_{t=0} = \sin 0 - 2i_0 - V_C(0^+)$$

$$= 0 - 2(1) - 2 = -4$$

Substituting this value into Eq. (viii), we obtain

$$-4 = K_1(-1 + j2) + K_2(-1 - j2) + 0.1 \qquad \text{(ix)}$$

From Eq. (vii),

$$K_2 = 0.8 - K_1 \qquad \text{(x)}$$

Solving Eqs. (ix) and (x), we get

$$K_1 = 0.4 + j0.825$$
$$K_2 = 0.4 - j0.825$$

Substituting these values into Eq. (vi), we obtain

$$i = (0.4 + j0.825)e^{(-1+j2)t} + (0.4 - j0.825)e^{(-1-j2)t} + 0.1 \sin t + 0.2 \cos t \qquad \text{(xi)}$$

$$= e^{-t}[(0.4 + j0.825)e^{j2t} + (0.4 - j0.825)e^{-j2t}] + 0.1 \sin t + 0.2 \cos t$$

$$= e^{-t}[(0.4 + j0.825)(\cos 2t + j \sin 2t) + (0.4 - j0.825)(\cos 2t - j \sin 2t)] + 0.1 \sin t + 0.2 \cos t$$

$$= e^{-t}[0.8 \cos 2t - 1.65 \sin 2t] + 0.1 \sin t + 0.2 \cos t$$

A right-angled triangle is constructed using 0.8 and 1.65 as two sides as shown in Fig. 12.24.

From Fig. 12.24,

$$\phi = \tan^{-1}\left(\frac{1.65}{0.8}\right) = 64.13°$$

Also, $\cos\phi = \dfrac{0.8}{1.8337}$

or $0.8 = 1.8337 \cos\phi = 1.8337 \cos 64.13°$

Similarly, $1.65 = 1.8337 \sin\phi = 1.8337 \sin 64.13°$

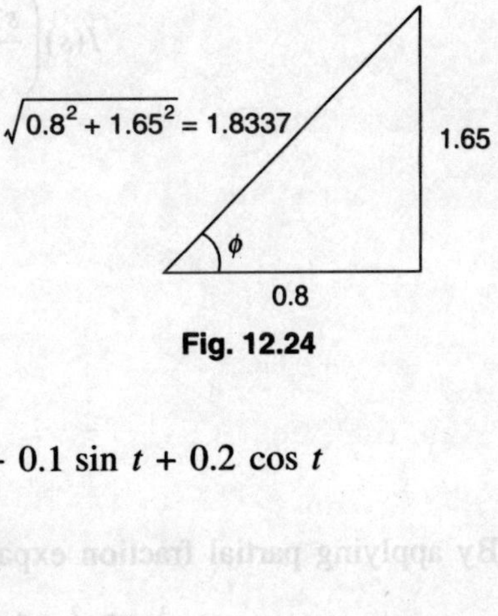

Fig. 12.24

$i = e^{-t}[1.8337 \cos 64.13° \cos 2t - 1.8337 \sin 64.13° \sin 2t] + 0.1 \sin t + 0.2 \cos t$

$$= 1.8337e^{-t}[\cos 64.13° \cos 2t - \sin 64.13° \sin 2t] + 0.1 \sin t + 0.2 \cos t$$

$$i = \underbrace{1.8337e^{-t}\cos(2t + 64.13°)}_{\text{Transient part}} + \underbrace{0.1 \sin t + 0.2 \cos t}_{\text{Steady-state part}} \text{ A}$$

Laplace transformation method

In view of Eq. (i),

$$2i + 1\frac{di}{dt} + \frac{1}{0.2}\int_{-\infty}^{t} i\,dt = \sin t$$

$$2i + \frac{di}{dt} + \frac{1}{0.2}\left[\int_{-\infty}^{0} i\,dt + \int_{0}^{t} i\,dt\right] = \sin t$$

$$2i + \frac{di}{dt} + \frac{1}{0.2}\left[0.4 + \int_{0}^{t} i\,dt\right] = \sin t \qquad \left[\because \int_{-\infty}^{0} idt = q_0 = CV_0 = 0.2\times 2 = 0.4\right]$$

$$2i + \frac{di}{dt} + 2 + \frac{1}{0.2}\int_{0}^{t} i\,dt = \sin t$$

Taking Laplace transformation on both sides,

$$2I(s) + sI(s) - I(0) + \frac{2}{s} + 5\frac{I(s)}{s} = \frac{1}{s^2+1} \tag{xii}$$

Since inductor does not permit any sudden change in current, $i(0^-) = i(0^+) = I(0) = 1$ A. So, Eq. (xii) becomes

$$2I(s) + sI(s) - 1 + \frac{2}{s} + 5\frac{I(s)}{s} = \frac{1}{s^2+1}$$

$$I(s)\left(2 + s + \frac{5}{s}\right) = \frac{1}{s^2+1} - \frac{2}{s} + 1$$

$$I(s)\left(\frac{s^2+2s+5}{s}\right) = \frac{s - 2(s^2+1) + s(s^2+1)}{s(s^2+1)}$$

$$= \frac{s^3 - 2s^2 + 2s - 2}{s(s^2+1)}$$

$$I(s) = \left(\frac{s}{s^2+2s+5}\right)\left(\frac{s^3 - 2s^2 + 2s - 2}{s(s^2+1)}\right)$$

$$= \frac{s^3 - 2s^2 + 2s - 2}{(s+1-j2)(s+1+j2)(s^2+1)}$$

By applying partial fraction expansion technique,

$$\frac{s^3 - 2s^2 + 2s - 2}{(s+1-j2)(s+1+j2)(s^2+1)} = \frac{A}{(s+1-j2)} + \frac{B}{(s+1+j2)} + \frac{Cs + D}{(s^2+1)} \tag{xiii}$$

Constants A, B, C and D are evaluated as follows:

$$A = \left.\frac{s^3 - 2s^2 + 2s - 2}{(s+1+j2)(s^2+1)}\right|_{s=-1+j2} = 0.4 + j0.825 \quad \text{(xiv)}$$

$$B = A^* = 0.4 - j0.825 \quad \text{(xv)}$$

On cross multiplying Eq. (xiii), we get

$$s^3 - 2s^2 + 2s - 2 = A(s+1+j2)(s^2+1) + B(s+1-j2)(s^2+1) + (Cs+D)(s+1-j2)(s+1+j2)$$

$$= A(s^3 + s + s^2 + 1 + j2s^2 + j2) + B(s^3 + s + s^2 + 1 - j2s^2 - j2) + (Cs + D)(s^2 + 2s + 5)$$

$$= As^3 + As + As^2 + A + j2As^2 + j2A + Bs^3 + Bs + Bs^2 + B - j2Bs^2 - j2B + Cs^3 + 2Cs^2 + 5Cs + Ds^2 + 2Ds + 5D$$

$$= s^3[A + B + C] + s^2[(A + B + 2C + D) + j(2A - 2B)] + s[A + B + 5C + 2D] + j[2A - 2B] \quad \text{(xvi)}$$

Comparing the coefficients of 's^3' in Eq. (xvi),

$$1 = A + B + C$$

Putting the values of A and B respectively from Eqs. (xiv) and (xv) in the above, we get

$$C = 0.2 \quad \text{(xvii)}$$

Comparing the coefficients of 's^2' in Eq. (xvi),

$$-2 = (A + B + 2C + D) + j(2A - 2B)$$

Putting the values of A, B and C respectively from Eqs. (xiv), (xv) and Eq. (xvii), in the above, we get

$$D = 0.1 \quad \text{(xviii)}$$

Substituting the values of A, B, C and D into Eq. (xiii),

$$I(s) = \frac{0.4 + j0.825}{(s+1-j2)} + \frac{0.4 - j0.825}{(s+1+j2)} + \frac{0.2s + 0.1}{(s^2+1)}$$

$$= \frac{0.4 + j0.825}{(s+1-j2)} + \frac{0.4 - j0.825}{(s+1+j2)} + \frac{0.2s}{(s^2+1)} + \frac{0.1}{(s^2+1)}$$

$$= \frac{0.4 + j0.825}{(s+1-j2)} + \frac{0.4 - j0.825}{(s+1+j2)} + \frac{0.2s}{(s^2+1)} + \frac{0.1}{1}\frac{1}{(s^2+1)}$$

Taking inverse Laplace transformation on both sides, we get

$$= (0.4 + j0.825)e^{(-1+j2)t} + (0.4 - j0.825)e^{(-1-j2)t} + 0.2\cos t + 0.1\sin t \quad \text{(xix)}$$

as obtained earlier in Eq. (xi).

Further simplification as given in classical method leads to the following equation for current:

$$i = \underbrace{1.8337e^{-t}\cos(2t + 64.13^\circ)}_{\text{Transient part}} + \underbrace{0.1\sin t + 0.2\cos t}_{\text{Steady-state part}} \text{ A}$$

SHORT QUESTIONS WITH ANSWERS

1. **What is damping ratio?**

   The ratio of resistance of the circuit to resistance for critical damping is called damping ratio.

2. **What is critical damping?**

   The critical damping is the condition of the circuit at which the oscillations in the response are just eliminated. This is possible by increasing the value of resistance in the circuit.

3. **What is critical resistance?**

   The critical resistance is the value of the resistance of the circuit to achieve critical damping.

4. **What is natural frequency?**

   The response of a circuit is completely oscillatory with a frequency ω_n without resistance, and the frequency ω_n is called natural frequency.

5. **What is damped frequency?**

   The response of under-damped circuit is oscillatory with a frequency of ω_d and these oscillations are damped as t tends to infinity. The frequency of damped oscillatory response is called damped frequency.

6. **Write the condition for over damping in RLC series circuit?**

   The condition for over-damping is $\left(\dfrac{R}{2L}\right)^2 > \dfrac{1}{LC}$.

7. **Write the condition for under damping in RLC series circuit?**

   The condition for under-damping is $\left(\dfrac{R}{2L}\right)^2 < \dfrac{1}{LC}$.

8. **Write the condition for critical damping in RLC series circuit?**

   The condition for critical-damping is $\left(\dfrac{R}{2L}\right)^2 = \dfrac{1}{LC}$.

UNSOLVED PROBLEMS

1. An RL series circuit excited by a sinusoidal source $v = 10 \sin 100t$ volts, by closing the switch at $t = 0$. Take $R = 10\ \Omega$ and $L = 0.1$ H. Determine current $i(t)$ flowing through the RL circuit. **[Ans.** $i(t) = 0.5e^{-100t} + 0.7071 \sin (100t - 45°)$ A]

2. In the *RL* circuit shown in Fig. U12.1, the switch is closed at position 1 at $t = 0$. Then, at $t = 0.24$ second, the switch is moved to position 2. Determine the response $i(t)$ after $t = 0.24$ second.

$$\left[\textbf{Ans.}\quad i(t) = 0.1 - 0.9647e^{-\frac{(t-0.24)}{0.06}}\ \text{A}\right]$$

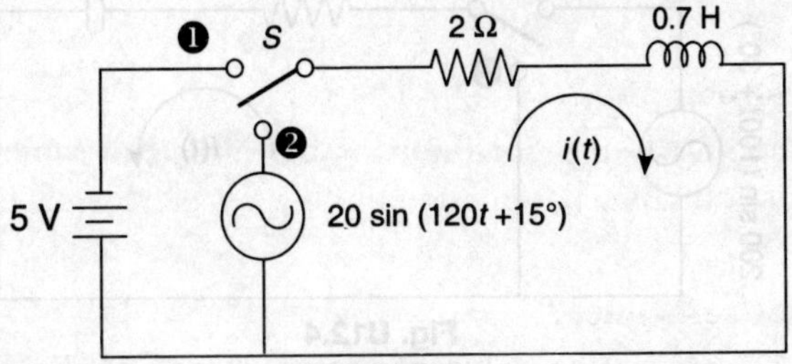

Fig. U12.1

3. In the circuit shown in Fig. U12.2, the switch remains in position 1 until steady state is reached. At time $t = 0$, the switch is changed to position 2. Find $i(t)$.

[**Ans.** $12.5e^{-100t}$ A]

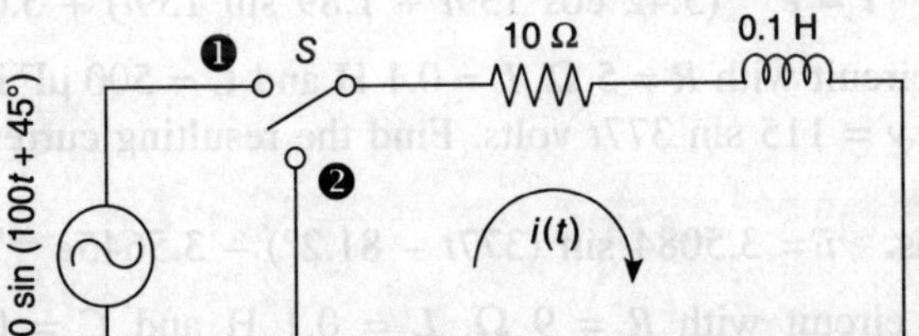

Fig. U12.2

4. An *RC* series circuit with $R = 100\ \Omega$ and $C = 2.5\ \mu F$ is excited by a sinusoidal voltage source of $v = 250 \sin 500t$ at $t = 0$. Find current $i(t)$, after $t > 0$ assuming that there is no initial charge on the capacitor.

[**Ans.** $i(t) = -1.22e^{-400t} + 1.95 \sin (500t + 38.7°)$ A]

5. An *RC* circuit shown in Fig. U12.3 is excited by a sinusoidal voltage source of $v = 100 \sin (200t + 40°)$ V at $t = 0$. Determine an expression for voltage across the capacitor. Assume the initial charge on the capacitor, $q_0 = 250 \times 10^{-6}$ C.

[**Ans.** $11.1628e^{-200t} + 70.7107 \sin (200t - 5°)$ V]

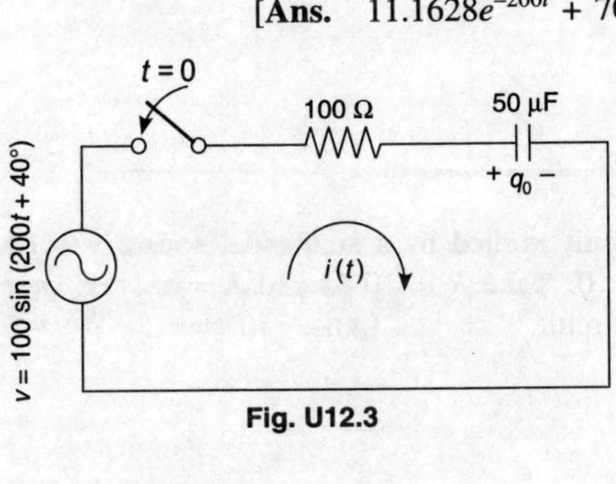

Fig. U12.3

6. In the circuit shown in Fig. U12.4, the switch remains in position 1 until steady state is reached. At time $t = 0$, the switch is changed to position 2. Find $i(t)$.

[Ans. $-28.282e^{-1000t}$ A]

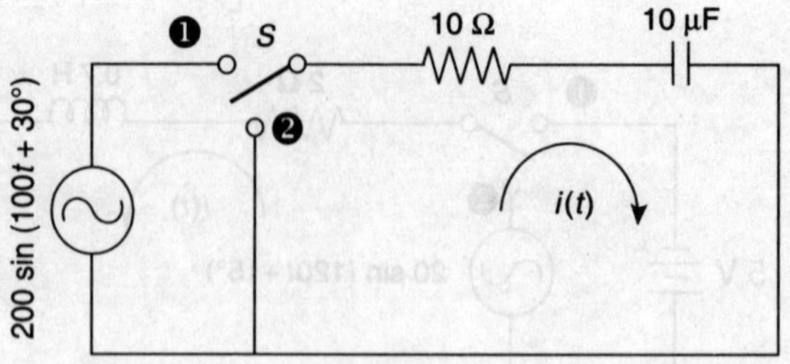

Fig. U12.4

7. A series *RLC* circuit with $R = 5\ \Omega$, $L = 0.1$ H and $C = 500\ \mu$F has a sinusoidal voltage source $v = 100 \sin(250t + \phi)$ volts. Find the resulting current if the switch is closed when $\phi = 0°$.

[Ans. $i = e^{-25t}(5.42 \cos 139t + 1.89 \sin 139t) + 5.65 \sin(250t - 73.6°)$ A]

8. A series *RLC* circuit with $R = 5\ \Omega$, $L = 0.1$ H and $C = 500\ \mu$F is excited by a sinusoidal voltage source $v = 115 \sin 377t$ volts. Find the resulting current if the switch is closed at $t = 0°$.

[Ans. $i = 3.5084 \sin(377t - 81.2°) - 3.5645e^{-25t} \sin(139.2t - 76.6°)$ A]

9. A series *RLC* circuit with $R = 9\ \Omega$, $L = 0.1$ H and $C = 0.005$ F is excited by a sinusoidal voltage source $v = 200 \sin(314t + 40°)$ V. Find an expression for voltage across the inductance.

[Ans. $v_L(t) = 68.61e^{-25t} - 103.0696e^{-25t} + 195.928 \sin(314t + 56.3°)$ V]

CHAPTER 13

Two-Port Networks

13.1 INTRODUCTION

A pair of terminals at which an electrical signal enters or leaves a network is called *port*. The port is required to supply energy or to withdraw energy. It is also used for measuring network variables. Generally, the minimum number of terminals required for a network is two. One such network is called *one-port network*, having only one pair of terminals.

A network with two pair of terminals is called *two-port network*. Examples of two-port network are transmission line, transformer, electronic amplifier, communication circuits, and so on. Figure 13.1 shows a diagrammatic representation of two-port network. Terminals 1–1′ constitute input port and this port is connected to input. So, it is also known as driving point of the network. Terminals 2–2′ constitute output port and this port is connected to load.

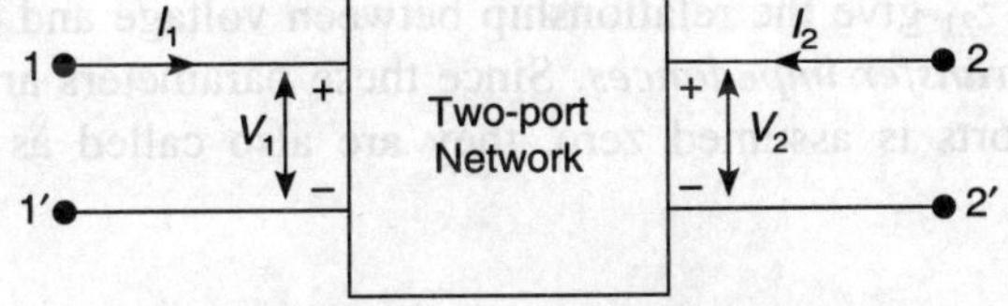

Fig. 13.1 Two-port network.

The following assumptions are quite useful as regard to the analysis of two-port networks:

1. The network contains only the linear elements along with dependent sources.
2. The initial conditions on capacitor and inductor present in the network are zero.
3. The voltages and currents inside the network are not available for measurements.
4. Currents I_1 and I_2 are entering into the network.

The voltage and current at the input port are V_1 and I_1, whereas those at the output port are V_2 and I_2. Among the four variables V_1, I_1, V_2, I_2, two variables are independent and the other two are dependent. There are six possible ways of obtaining two independent variables out of four variables. The different set of parameters such as impedance (z), admittance (y), hybrid (h), inverse hybrid (g), transmission ($ABCD$) and inverse transmission parameters ($A'B'C'D'$) are thus described.

13.2 z-PARAMETERS

The z-parameters or impedance parameters are obtained by expressing voltages at two ports in terms of currents at two ports. Therefore, currents I_1 and I_2 are independent variables, whereas V_1 and V_2 are dependent variables. The performance equations thus are

$$V_1 = z_{11}I_1 + z_{12}I_2 \tag{13.1}$$

$$V_2 = z_{21}I_1 + z_{22}I_2 \tag{13.2}$$

In matrix form, the above equations are written as

$$\begin{bmatrix} V_1 \\ V_2 \end{bmatrix} = \begin{bmatrix} z_{11} & z_{12} \\ z_{21} & z_{22} \end{bmatrix} \begin{bmatrix} I_1 \\ I_2 \end{bmatrix}$$

The 2×2 matrix containing the z-parameters z_{11}, z_{12}, z_{21} and z_{22} is known as *impedance matrix*. The z-parameters are defined as follows:

$$z_{11} = \left.\frac{V_1}{I_1}\right|_{I_2=0} \qquad z_{12} = \left.\frac{V_1}{I_2}\right|_{I_1=0}$$

$$z_{21} = \left.\frac{V_2}{I_1}\right|_{I_2=0} \qquad z_{22} = \left.\frac{V_2}{I_2}\right|_{I_1=0}$$

From the above equations, z_{11} and z_{22} are called *driving point impedances* in port 1 and port 2, respectively as they give the relationship between voltage and current at the respective port. Parameters z_{12} and z_{21} give the relationship between voltage and current at different ports and they are known as *transfer impedances*. Since these parameters are defined only when the current in one of the ports is assumed zero, they are also called as *open-circuit impedance parameters*.

Thus,

z_{11} = open-circuit driving point input impedance
z_{22} = open-circuit output impedance
z_{12} = open-circuit forward transfer impedance
z_{21} = open-circuit reverse transfer impedance

Figure 13.2 shows an equivalent circuit to represent z-parameters. Equations (13.1) and (13.2) can be verified by applying KVL around two meshes.

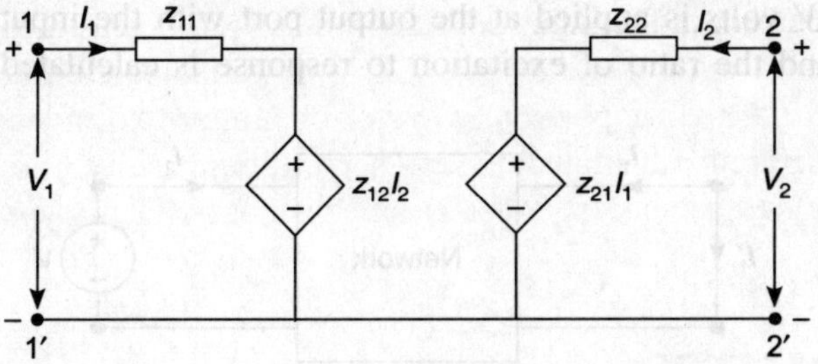

Fig. 13.2 Equivalent circuit of a two-port network in terms of *z*-parameters.

Condition for reciprocity

A network is said to be reciprocal if the ratio of excitation at one port to response at the other port is same if excitation and response are interchanged. A voltage of V volts is applied at the input port with the output port short-circuited as shown in Fig. 13.3 and the ratio of excitation to response is calculated as follows.

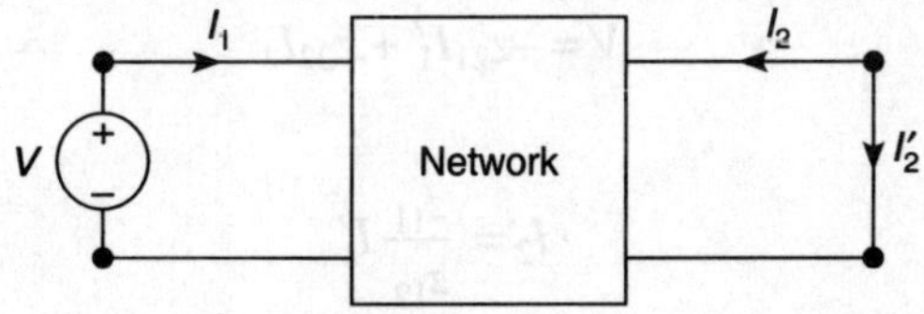

Fig. 13.3 Input port excited and output port shorted.

From Fig. 13.3,

$$V_1 = V;\ V_2 = 0 \text{ and } I_2 = -I_2'$$

From the defining equations of z-parameters,

$$V = z_{11}I_1 - z_{12}I_2' \tag{13.3}$$

$$0 = z_{21}I_1 - z_{22}I_2' \tag{13.4}$$

From Eq. (13.4),

$$I_1 = \frac{z_{22}}{z_{21}} I_2' \tag{13.5}$$

Substituting Eq. (13.5) into Eq. (13.3), we obtain

$$V = z_{11}\frac{z_{22}}{z_{21}}I_2' - z_{12}I_2'$$

$$= I_2'\left(\frac{z_{11}z_{22} - z_{12}z_{21}}{z_{21}}\right)$$

$$\frac{V}{I_2'} = \left(\frac{z_{11}z_{22} - z_{12}z_{21}}{z_{21}}\right) \tag{13.6}$$

Next, the voltage of V volts is applied at the output port with the input port short-circuited as shown in Fig. 13.4 and the ratio of excitation to response is calculated as follows.

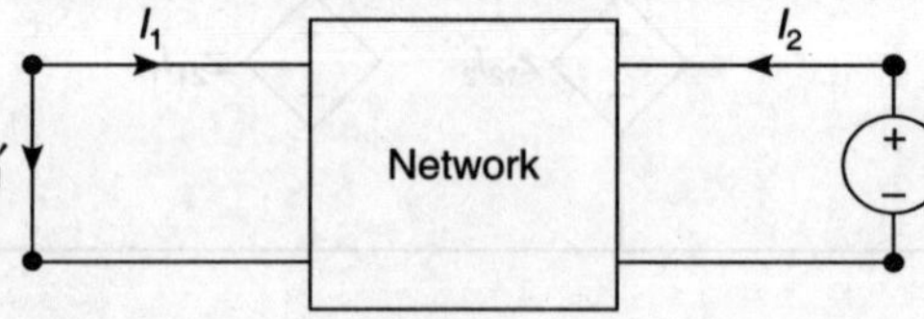

Fig. 13.4 Input port shorted and output port excited.

From Fig. 13.4,

$$V_2 = V;\ V_1 = 0 \text{ and } I_1 = I_1'$$

From the defining equations of z-parameters,

$$0 = -z_{11}I_1' + z_{12}I_2 \tag{13.7}$$

$$V = -z_{21}I_1' + z_{22}I_2 \tag{13.8}$$

From Eq. (13.7),

$$I_2 = \frac{z_{11}}{z_{12}}I_1' \tag{13.9}$$

Substituting Eq. (13.9) into Eq. (13.8), we get

$$V = -z_{21}I_1' + z_{22}\frac{z_{11}}{z_{12}}I_1'$$

$$= I_1'\left(\frac{-z_{21}z_{12} + z_{22}z_{11}}{z_{12}}\right)$$

$$\frac{V}{I_1'} = \left(\frac{z_{11}z_{22} - z_{21}z_{12}}{z_{21}}\right) \tag{13.10}$$

It is found from Eqs. (13.6) and (13.10), for the network to be reciprocal,

$$\frac{V}{I_1'} = \frac{V}{I_2'}$$

and

$$z_{12} = z_{21}$$

Condition for symmetry

A network is said to be symmetrical, if the voltage-to-current ratios at both input and output ports are same with one of the ports is open-circuited. When the output port is open-circuited, I_2 is zero. Thus, from the defining equations of z-parameters,

$$V = z_{11}I_1$$

or voltage-to-current ratio, $\dfrac{V}{I_1} = z_{11}$

Similarly, if the input port is open-circuited I_1 is zero. Thus, from the defining equations of z-parameters,

$$V = z_{22}I_2$$

or voltage-to-current ratio, $\dfrac{V}{I_2} = z_{22}$

Hence, for the network to be symmetrical,

$$z_{11} = z_{22}$$

13.3 *y*-PARAMETERS

The y-parameters or admittance parameters are obtained by expressing currents at two ports in terms of voltages at two ports. Therefore, voltages V_1 and V_2 are independent variables, whereas I_1 and I_2 are dependent variables. The performance equations thus are

$$I_1 = y_{11}V_1 + y_{12}V_2 \quad (13.11)$$

$$I_2 = y_{21}V_1 + y_{22}V_2 \quad (13.12)$$

In matrix form, the above equations are written as:

$$\begin{bmatrix} I_1 \\ I_2 \end{bmatrix} = \begin{bmatrix} y_{11} & y_{12} \\ y_{21} & y_{22} \end{bmatrix} \begin{bmatrix} V_1 \\ V_2 \end{bmatrix}$$

The 2 × 2 matrix containing the y-parameters y_{11}, y_{12}, y_{21} and y_{22} is known as *admittance matrix*. The y-parameters are defined as follows:

$$y_{11} = \left.\frac{I_1}{V_1}\right|_{V_2=0} \qquad y_{12} = \left.\frac{I_1}{V_2}\right|_{V_1=0}$$

$$y_{21} = \left.\frac{I_2}{V_1}\right|_{V_2=0} \qquad y_{22} = \left.\frac{I_2}{V_2}\right|_{V_1=0}$$

From the above equations, y_{11} and y_{22} are called *driving point admittances* in port 1 and port 2, respectively as they give the relationship between current and voltage at the respective port. Parameters y_{12} and y_{21} give the relationship between current and voltage at different ports and they are known as *transfer admittances*. Since these parameters are defined only when the voltage in one of the ports is assumed zero, they are also called as *short-circuit admittance parameters*.

Thus,

y_{11} = short-circuit driving point input admittance

y_{22} = short-circuit output admittance
y_{12} = short-circuit forward transfer admittance
y_{21} = short-circuit reverse transfer admittance

Figure 13.5 shows an equivalent circuit to represent y-parameters. Equations (13.11) and (13.12) can be verified by applying KCL at two nodes.

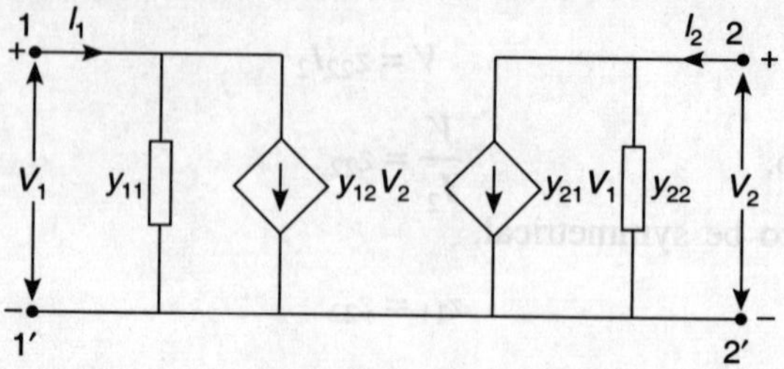

Fig. 13.5 Equivalent circuit of two-port network in terms of y-parameters.

Condition for reciprocity

A voltage of V volts is applied at the input port with the output port short-circuited as shown in Fig. 13.6 and the ratio of excitation to response is calculated as follows:

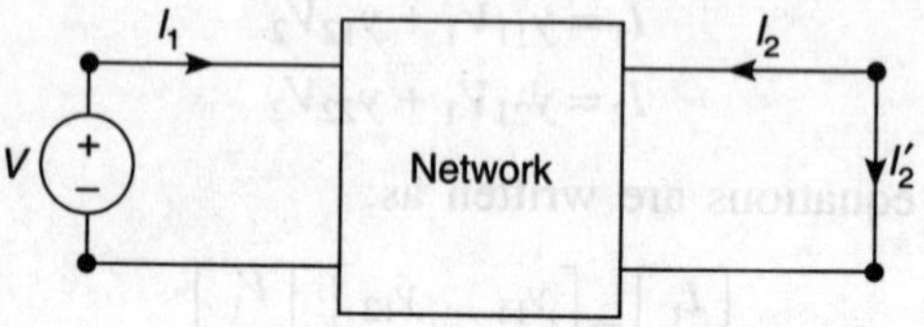

Fig. 13.6 Input port excited and output port shorted.

From Fig. 13.6,

$$V_1 = V;\ V_2 = 0 \text{ and } I_2 = -I_2'$$

From the defining equations of y-parameters,

$$I_1 = y_{11}V \tag{13.13}$$

$$-I_2' = y_{21}V \tag{13.14}$$

From Eq. (13.14),

$$-y_{21} = \frac{I_2'}{V} \tag{13.15}$$

Next, the voltage of V volts is applied at the output port with the input port short-circuited as shown in Fig. 13.7 and the ratio of excitation to response is calculated as follows:

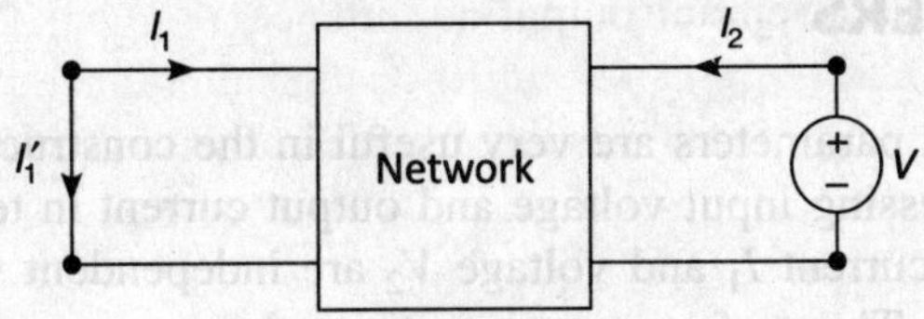

Fig. 13.7 Input port shorted and output port excited.

From Fig. 13.7,

$$V_2 = V\,;\ V_1 = 0 \text{ and } I_1 = -I_1'$$

From the defining equations of *y*-parameters,

$$-I_1' = y_{12}\,V \tag{13.16}$$

$$I_2 = y_{22}\,V \tag{13.17}$$

From Eq. (13.16),

$$-y_{12} = \frac{I_1'}{V} \tag{13.18}$$

Hence, for the network to be reciprocal,

$$\frac{I_2'}{V} = \frac{I_1'}{V}$$

That is

$$y_{12} = y_{21}$$

Condition for symmetry

When the output port is short-circuited, V_2 is zero. Thus, from the defining equations of *y*-parameters,

$$I_1 = y_{11}V$$

or voltage-to-current ratio, $\dfrac{V}{I_1} = y_{11}$

Similarly, if the input port is short-circuited, V_1 is zero. Thus, from the defining equations of *y*-parameters,

$$I_2 = y_{22}V$$

or voltage-to-current ratio, $\dfrac{V}{I_2} = y_{22}$

Hence, for the network to be symmetrical,

$$\frac{V}{I_1} = \frac{V}{I_2}$$

That is

$$y_{11} = y_{22}$$

13.4 *h*-PARAMETERS

The *h*-parameters or hybrid parameters are very useful in the construction of transistor models. They are obtained by expressing input voltage and output current in terms of input current and output voltage. Therefore, current I_1 and voltage V_2 are independent variables, whereas I_2 and V_1 are dependent variables. The performance equations thus are

$$V_1 = h_{11}I_1 + h_{12}V_2 \tag{13.19}$$

$$I_2 = h_{21}I_1 + h_{22}V_2 \tag{13.20}$$

In matrix form, the above equations are written as:

$$\begin{bmatrix} V_1 \\ I_2 \end{bmatrix} = \begin{bmatrix} h_{11} & h_{12} \\ h_{21} & h_{22} \end{bmatrix} \begin{bmatrix} I_1 \\ V_2 \end{bmatrix}$$

The *h*-parameters are defined as follows:

$$h_{11} = \left.\frac{V_1}{I_1}\right|_{V_2=0}$$, is called short-circuit input impedance

$$h_{12} = \left.\frac{V_1}{V_2}\right|_{I_1=0}$$, is called reverse open-circuit voltage gain

$$h_{21} = \left.\frac{I_2}{I_1}\right|_{V_2=0}$$, is called forward short-circuit current gain

$$h_{22} = \left.\frac{I_2}{V_2}\right|_{I_1=0}$$, is called open-circuit output admittance

Since all the above parameters do have different units such as 'ohm' and 'mho', they are called as *hybrid parameters*. Figure 13.8 shows an equivalent circuit to represent *h*-parameters. Equations (13.19) and (13.20) can be verified by applying KVL around input mesh and KCL at output node.

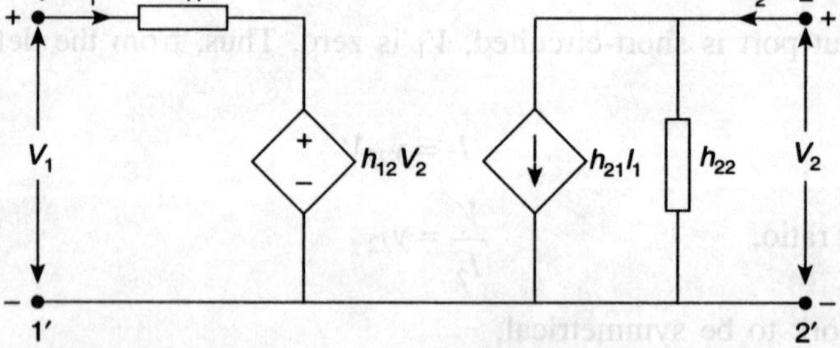

Fig. 13.8 Equivalent circuit of a two-port network in terms of *h*-parameters.

Condition for reciprocity

A voltage of V volts is applied at the input port with the output port short-circuited as shown in Fig. 13.9 and the ratio of excitation to response is calculated as follows:

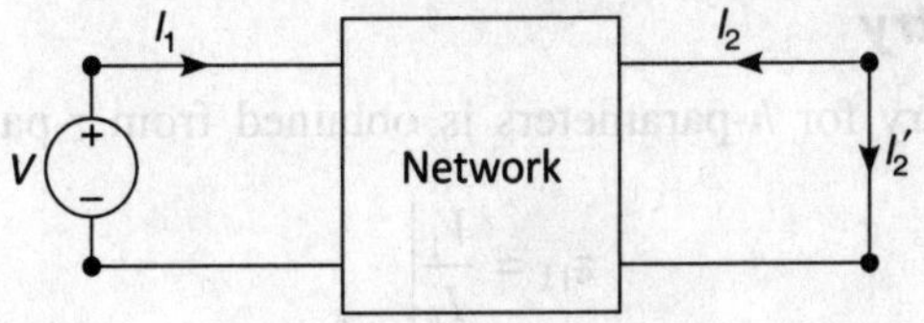

Fig. 13.9 Input port excited and output port shorted.

From Fig. 13.9,

$$V_1 = V;\ V_2 = 0 \text{ and } I_2 = -I_2'$$

From the defining equations of *h*-parameters,

$$V_1 = h_{11}I_1 \tag{13.21}$$

$$-I_2' = h_{21}I_1 \tag{13.22}$$

From Eqs. (13.21) and (13.22),

$$\frac{V_1}{I_2'} = -\frac{h_{11}}{h_{21}} \tag{13.23}$$

Next, the voltage of *V* volts is applied at the output port with the input port short-circuited as shown in Fig. 13.10 and the ratio of excitation to response is calculated as follows.

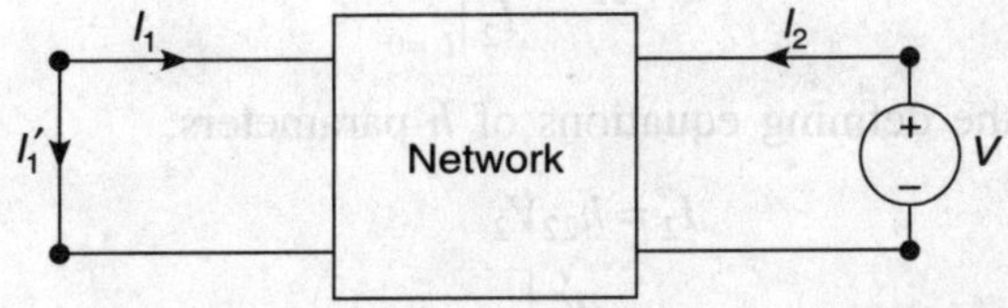

Fig. 13.10 Input port shorted and output port excited.

From Fig. 13.10,

$$V_2 = V;\ V_1 = 0 \text{ and } I_1 = -I_1'$$

From the defining equations of *h*-parameters,

$$0 = -h_{11}I_1' + h_{12}V \tag{13.24}$$

$$I_2 = h_{21}I_1 + h_{22}V \tag{13.25}$$

From Eq. (13.24),

$$\frac{V}{I_1'} = \frac{h_{11}}{h_{12}}$$

Hence, for the network to be reciprocal,

$$\frac{V}{I_2'} = \frac{V}{I_1'}$$

That is,
$$h_{21} = -h_{12}$$

Condition for symmetry

The condition for symmetry for h-parameters is obtained from z-parameters.

$$z_{11} = \left.\frac{V_1}{I_1}\right|_{I_2=0}$$

From Eq. (13.19),

$$= \left.\frac{h_{11}I_1 + h_{12}V_2}{I_1}\right|_{I_2=0}$$

Since $I_2 = 0$, and from the defining equations of h-parameters,

$$0 = h_{21}I_1 + h_{22}V_2$$

$$z_{11} = h_{11} - \frac{h_{12}h_{21}}{h_{22}} = \frac{h_{11}h_{22} - h_{12}h_{21}}{h_{22}}$$

$$= \frac{\Delta h}{h_{22}}$$

where $\Delta h = h_{11}h_{22} - h_{12}h_{21}$

Similarly,

$$z_{22} = \left.\frac{V_2}{I_2}\right|_{I_1=0}$$

Since $I_1 = 0$, and from the defining equations of h-parameters,

$$I_2 = h_{22}V_2$$

$$z_{22} = \left.\frac{V_2}{I_2}\right|_{I_1=0} = \frac{1}{h_{22}}$$

For a symmetrical network, $z_{11} = z_{22}$.

or $$\frac{\Delta h}{h_{22}} = \frac{1}{h_{22}}$$

or $$\Delta h = 1$$

or $$h_{11}h_{22} - h_{12}h_{21} = 1$$

13.5 *g*-PARAMETERS

The g-parameters are otherwise called as *inverse hybrid parameters*. They are obtained by expressing input current and output voltage in terms of input voltage and output current. Therefore, current I_2 and voltage V_1 are independent variables, whereas I_1 and V_2 are dependent variables. The performance equations thus are

$$I_1 = g_{11}V_1 + g_{12}I_2 \quad (13.26)$$

$$V_2 = g_{21}V_1 + g_{22}I_2 \quad (13.27)$$

In matrix form, the above equations are written as:

$$\begin{bmatrix} I_1 \\ V_2 \end{bmatrix} = \begin{bmatrix} g_{11} & g_{12} \\ g_{21} & g_{22} \end{bmatrix} \begin{bmatrix} V_1 \\ I_2 \end{bmatrix}$$

The *g*-parameters are defined as follows:

$$g_{11} = \left.\frac{I_1}{V_1}\right|_{I_2=0}$$, is called open-circuit input admittance

$$g_{12} = \left.\frac{I_1}{I_2}\right|_{V_1=0}$$, is called short-circuit reverse current gain

$$g_{21} = \left.\frac{V_2}{V_1}\right|_{I_2=0}$$, is called open-circuit voltage gain

$$g_{22} = \left.\frac{V_2}{I_2}\right|_{V_1=0}$$, is called short-circuit output impedance

Figure 13.11 shows an equivalent circuit to represent *g*-parameters. Equations (13.26) and (13.27) can be verified by applying KCL at input node and KVL around output mesh.

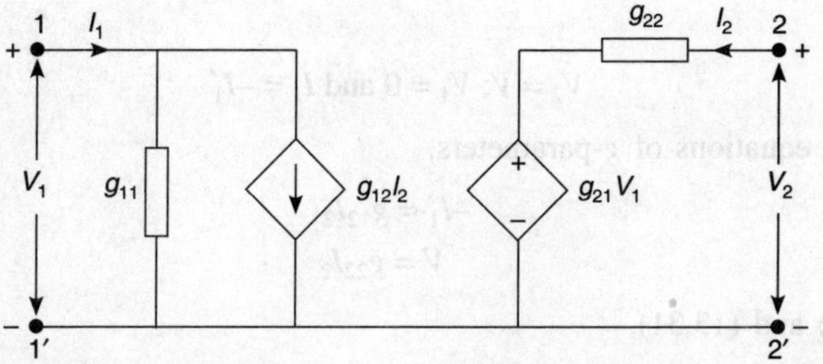

Fig. 13.11 Equivalent circuit of a two-port network in terms of g-parameters.

Condition for reciprocity

A voltage of *V* volts is applied at the input port with the output port short-circuited as shown in Fig. 13.12 and the ratio of excitation to response is calculated as follows:

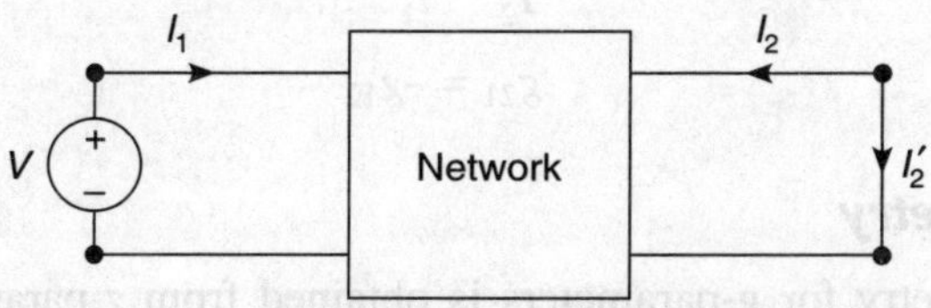

Fig. 13.12 Input port excited and output port shorted.

From Fig. 13.12,

$$V_1 = V;\ V_2 = 0 \text{ and } I_2 = -I_2'$$

From the defining equations of g-parameters,

$$I_1 = g_{11}V - g_{12}I_2' \quad (13.28)$$

$$0 = g_{21}V - g_{22}I_2' \quad (13.29)$$

From Eq. (13.29),

$$g_{21}V = g_{22}I_2'$$

$$\frac{V}{I_2'} = \frac{g_{22}}{g_{21}}$$

Next, the voltage of V volts is applied at the output port with the input port short-circuited as shown in Fig. 13.13 and the ratio of excitation to response is calculated as follows:

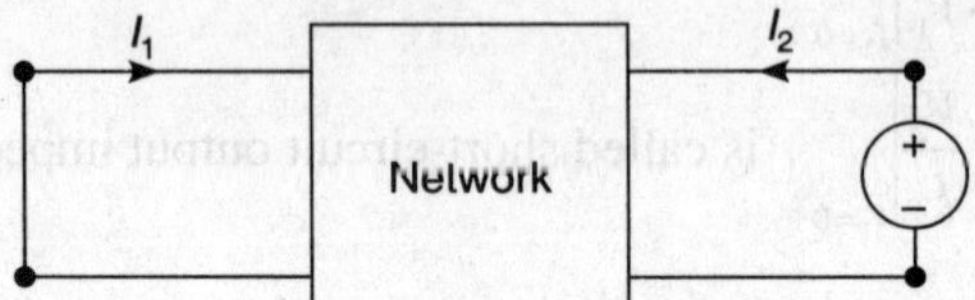

Fig. 13.13 Input port shorted and output port excited.

From Fig. 13.13,

$$V_2 = V;\ V_1 = 0 \text{ and } I_1 = -I_1'$$

From the defining equations of g-parameters,

$$-I_1' = g_{12}I_2 \quad (13.30)$$

$$V = g_{22}I_2 \quad (13.31)$$

From Eqs. (13.30) and (13.31),

$$\frac{V}{I_1'} = -\frac{g_{22}}{g_{12}}$$

Hence, for the network to be reciprocal,

$$\frac{V}{I_2'} = \frac{V}{I_1'}$$

That is

$$g_{21} = -g_{12}$$

Condition for symmetry

The condition for symmetry for g-parameters is obtained from z-parameters.

$$z_{11} = \left.\frac{V_1}{I_1}\right|_{I_2=0} = \frac{1}{g_{11}}$$

Similarly,

$$z_{22} = \left.\frac{V_2}{I_2}\right|_{I_1=0}$$

Since $I_1 = 0$, and from the defining equations of g-parameters,

$$0 = g_{11}V_1 + g_{12}I_2$$

$$V_1 = -\frac{g_{12}}{g_{11}} I_2$$

and $$V_2 = g_{21}\left(-\frac{g_{12}}{g_{11}}\right)I_2 + g_{22}I_2$$

$$z_{22} = \left.\frac{V_2}{I_2}\right|_{I_1=0} = \frac{g_{11}g_{22} - g_{12}g_{21}}{g_{11}}$$

For a symmetrical network, $z_{11} = z_{22}$

or $$\frac{1}{g_{11}} = \frac{g_{11}g_{22} - g_{12}g_{21}}{g_{11}}$$

or $$g_{11}g_{22} - g_{12}g_{21} = 1$$

13.6 *ABCD*-PARAMETERS

ABCD (or) transmission parameters are generally used in power transmission circuits. Here, the input port is called as *sending end* and output port is called as *receiving end*. *ABCD* parameters are obtained by expressing input voltage V_1 and input current I_1 in terms of output voltage V_2 and output current I_2. Therefore, V_2 and I_2 are independent variables, whereas V_1 and I_1 are dependent variables. The performance equations thus are

$$V_1 = AV_2 + B(-I_2) \tag{13.32}$$

$$I_1 = CV_2 + D(-I_2) \tag{13.33}$$

In matrix form, the above equations are written as:

$$\begin{bmatrix} V_1 \\ I_1 \end{bmatrix} = \begin{bmatrix} A & B \\ C & D \end{bmatrix}\begin{bmatrix} V_2 \\ -I_2 \end{bmatrix}$$

The 2 × 2 matrix containing the *ABCD*-parameters is known as *transmission matrix*. The *ABCD*-parameters are defined as follows:

$$A = \left.\frac{V_1}{V_2}\right|_{I_2=0}$$, is called open-circuit reverse voltage gain

$$B = \left.\frac{V_1}{-I_2}\right|_{V_2=0}$$, is called short-circuit transfer impedance

$$C = \left.\frac{I_1}{V_2}\right|_{I_2=0}$$, is called open-circuit transfer admittance

$$D = \left.\frac{I_1}{-I_2}\right|_{V_2=0}$$, is called short-circuit reverse current gain

These parameters are also called as *general circuit parameters* and *chain parameters*.

Condition for reciprocity

A voltage of V volts is applied at the input port with the output port short-circuited as shown in Fig. 13.14 and the ratio of excitation to response is calculated as follows:

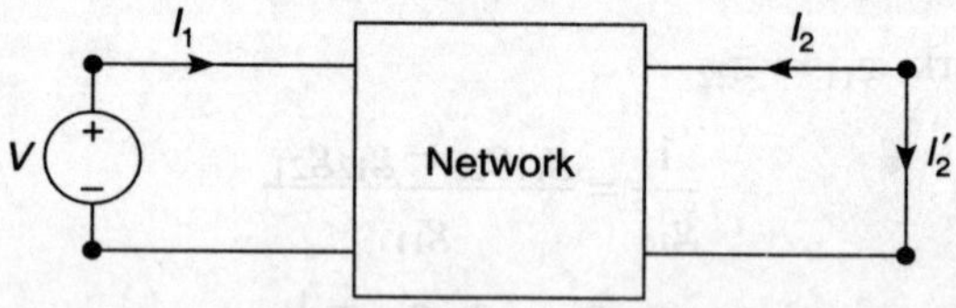

Fig. 13.14 Input port excited and output port shorted.

From Fig. 13.14,

$$V_1 = V;\ V_2 = 0 \text{ and } I_2 = -I_2'$$

From the defining equations of *ABCD*-parameters,

$$V = BI_2' \tag{13.34}$$

$$I_1 = DI_2' \tag{13.35}$$

From Eq. (13.34),

$$\frac{V}{I_2'} = B \tag{13.36}$$

Next, the voltage of V volts is applied at the output port with the input port short-circuited as shown in Fig. 13.15 and the ratio of excitation to response is calculated as follows.

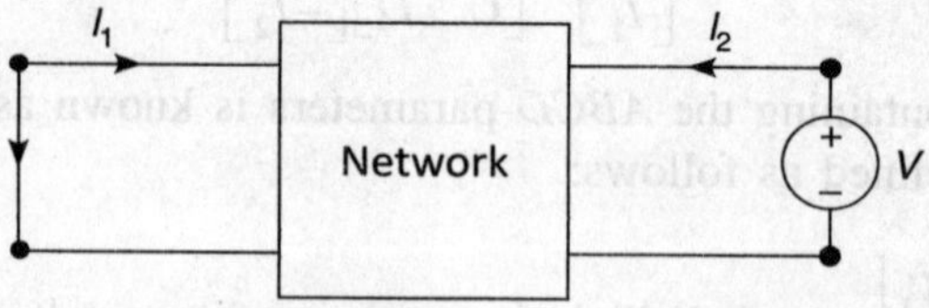

Fig. 13.15 Input port shorted and output port excited.

From Fig. 13.15,

$$V_2 = V;\ V_1 = 0 \text{ and } I_1 = -I_1'$$

From the defining equations of *ABCD*-parameters,

$$0 = AV - BI_2 \tag{13.37}$$

$$-I_1' = CV - DI_2 \tag{13.38}$$

From Eq. (13.37),

$$I_2 = \frac{A}{B}V \tag{13.39}$$

Substituting Eq. (13.39) into Eq. (13.38), we obtain

$$-I_1' = CV - D\frac{A}{B}V$$

$$= V\left(\frac{BC - AD}{B}\right)$$

$$\therefore \quad \frac{V}{I_1'} = \frac{B}{AD - BC} \tag{13.40}$$

Hence, for the network to be reciprocal,

$$\frac{V}{I_2'} = \frac{V}{I_1'}$$

From Eqs. (13.36) and (13.40),

$$B = \frac{B}{AD - BC}$$

or

$$AD - BC = 1$$

Condition for symmetry

When the output port is open-circuited, I_2 is zero. Thus, from the defining equations of *ABCD*-parameters,

$$V = AV_2$$

$$I_1 = CV_2$$

or voltage-to-current ratio,

$$\frac{V}{I_1} = \frac{A}{C}$$

Similarly, if the input port is open-circuited, I_1 is zero. Thus, from the defining equations of *ABCD*-parameters,

$$CV = DI_2$$

or voltage-to-current ratio, $$\frac{V}{I_2} = \frac{D}{C}$$

Hence, for the network to be symmetrical,

$$\frac{V}{I_1} = \frac{V}{I_2}$$

That is $$\frac{A}{C} = \frac{D}{C}$$

or $$A = D$$

13.7 *A′B′C′D′*-PARAMETERS

Generally *ABCD*-parameters are defined by expressing input port quantities (current and voltage) in terms of output port quantities. It is also possible to express output voltage and current in terms of input port voltage and current, respectively. Such expression gives inverse transmission or *A′B′C′D′* parameters. Here, V_1 and I_1 are independent variables, whereas V_2 and I_2 are dependent variables. The performance equations thus are

$$V_2 = A'V_1 + B'(-I_1) \quad (13.41)$$

$$I_2 = C'V_1 + D'(-I_1) \quad (13.42)$$

In matrix form, the above equations are written as:

$$\begin{bmatrix} V_2 \\ I_2 \end{bmatrix} = \begin{bmatrix} A' & B' \\ C' & D' \end{bmatrix} \begin{bmatrix} V_1 \\ -I_1 \end{bmatrix}$$

The *A′B′C′D′*-parameters are defined as follows:

$$A' = \left.\frac{V_2}{V_1}\right|_{I_1=0}$$, is called open-circuit voltage gain

$$B' = \left.\frac{V_2}{-I_1}\right|_{V_1=0}$$, is called short-circuit transfer impedance

$$C' = \left.\frac{I_2}{V_1}\right|_{I_1=0}$$, is called open-circuit transfer admittance

$$D' = \left.\frac{I_2}{-I_1}\right|_{V_1=0}$$, is called short-circuit current gain

Condition for reciprocity

A voltage of V volts is applied at the input port with the output port short-circuited as shown in Fig. 13.16 and the ratio of excitation to response is calculated as follows:

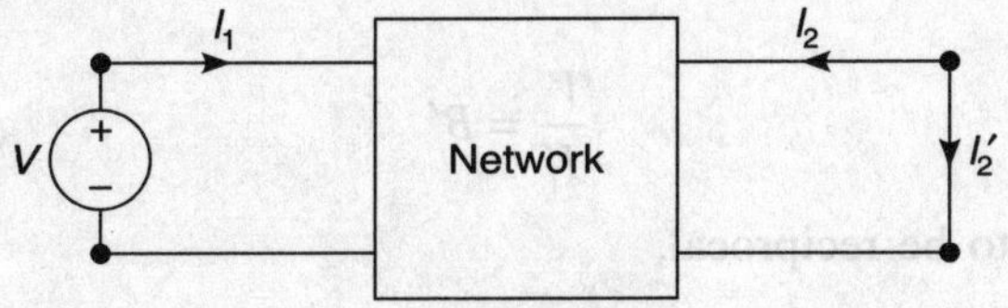

Fig. 13.16 Input port excited and output port shorted.

From Fig. 13.16,

$$V_1 = V;\ V_2 = 0 \text{ and } I_2 = -I_2'$$

From the defining equations of $A'B'C'D'$-parameters,

$$0 = A'V - B'I_1 \tag{13.43}$$

$$-I_2' = C'V - D'I_1 \tag{13.44}$$

From Eq. (13.43),

$$I_1 = \frac{A'}{B'}V$$

$$-I_2' = C'V - D'\frac{A'}{B'}V$$

or

$$-I_2' = V\left(\frac{B'C' - A'D'}{B'}\right)$$

Hence,

$$\frac{V}{I_2'} = \frac{B'}{A'D' - B'C'} \tag{13.45}$$

Next, the voltage of V volts is applied at the output port with the input port short-circuited as shown in Fig. 13.17 and the ratio of excitation to response is calculated as follows:

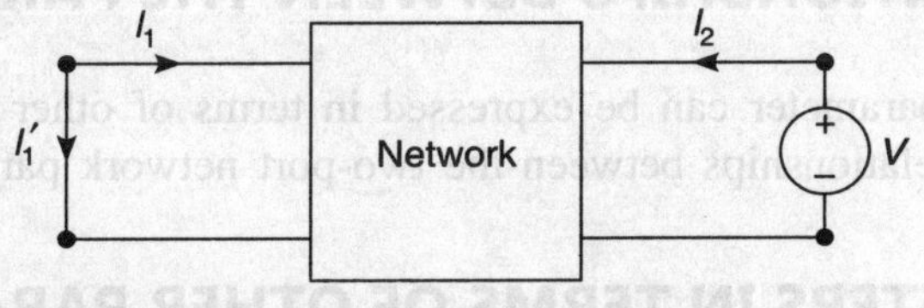

Fig. 13.17 Input port shorted and output port excited.

From Fig. 13.17,

$$V_2 = V;\ V_1 = 0 \text{ and } I_1 = -I_1'$$

From the defining equations of $A'B'C'D'$-parameters,

$$V = B'I_1' \tag{13.46}$$

$$I_2 = D'I_1' \tag{13.47}$$

From Eq. (13.46),

$$\frac{V}{I_1'} = B' \tag{13.48}$$

Hence, for the network to be reciprocal,

$$\frac{V}{I_2'} = \frac{V}{I_1'}$$

From Eqs. (13.45) and (13.48),

$$\frac{B'}{A'D' - B'C'} = B'$$

or

$$A'D' - B'C' = 1$$

Condition for symmetry

The condition for symmetry is obtained from z-parameters.

$$z_{11} = \left.\frac{V_1}{I_1}\right|_{I_2=0} = \frac{D'}{C'}$$

Similarly,

$$z_{22} = \left.\frac{V_2}{I_2}\right|_{I_1=0} = \frac{A'}{C'}$$

Hence, for the network to be symmetrical,

$$z_{11} = z_{22}$$

or

$$A' = D'$$

13.8 INTERRELATIONSHIPS BETWEEN THE PARAMETERS

Any two-port network parameter can be expressed in terms of other parameter. This section discusses all such interrelationships between the two-port network parameters.

13.9 *z*-PARAMETERS IN TERMS OF OTHER PARAMETERS

The defining equations of z-parameters [Eqs. (13.1) and (13.2)] are as follows:

$$V_1 = z_{11}I_1 + z_{12}I_2$$

$$V_2 = z_{21}I_1 + z_{22}I_2$$

13.9.1 *z*-Parameters in Terms of *y*-Parameters

The equations for *y*-parameters [Eqs. (13.11) and (13.12)] are as follows:

$$I_1 = y_{11}V_1 + y_{12}V_2$$
$$I_2 = y_{21}V_1 + y_{22}V_2$$

Putting the above equations in matrix form,

$$\begin{bmatrix} I_1 \\ I_2 \end{bmatrix} = \begin{bmatrix} y_{11} & y_{12} \\ y_{21} & y_{22} \end{bmatrix} \begin{bmatrix} V_1 \\ V_2 \end{bmatrix}$$

By applying Cramer's rule,

$$V_1 = \frac{\begin{vmatrix} I_1 & y_{12} \\ I_2 & y_{22} \end{vmatrix}}{\begin{vmatrix} y_{11} & y_{12} \\ y_{21} & y_{22} \end{vmatrix}} = \frac{y_{22}I_1 - y_{12}I_2}{y_{11}y_{22} - y_{12}y_{21}}$$

$$= \frac{y_{22}}{y_{11}y_{22} - y_{12}y_{21}} I_1 - \frac{y_{12}}{y_{11}y_{22} - y_{12}y_{21}} I_2 \tag{13.49}$$

Similarly,

$$V_2 = \frac{\begin{vmatrix} y_{11} & I_1 \\ y_{21} & I_2 \end{vmatrix}}{\begin{vmatrix} y_{11} & y_{12} \\ y_{21} & y_{22} \end{vmatrix}} = \frac{y_{11}I_2 - y_{21}I_1}{y_{11}y_{22} - y_{12}y_{21}}$$

$$= \frac{-y_{21}}{y_{11}y_{22} - y_{12}y_{21}} I_1 + \frac{y_{11}}{y_{11}y_{22} - y_{12}y_{21}} I_2 \tag{13.50}$$

Let $y_{11}y_{22} - y_{12}y_{21} = \Delta_y$

Rewriting Eqs. (13.49) and (13.50), we have

$$V_1 = \left[\frac{y_{22}}{\Delta_y}\right] I_1 + \left[\frac{-y_{12}}{\Delta_y}\right] I_2 \tag{13.51}$$

$$V_2 = \left[\frac{-y_{21}}{\Delta_y}\right] I_1 + \left[\frac{y_{11}}{\Delta_y}\right] I_2 \tag{13.52}$$

Comparing Eqs. (13.51) and (13.52) with Eqs. (13.1) and (13.2), respectively, we have

$$z_{11} = \frac{y_{22}}{\Delta_y} \qquad z_{21} = \frac{-y_{21}}{\Delta_y}$$

$$z_{12} = \frac{-y_{12}}{\Delta_y} \qquad z_{22} = \frac{y_{11}}{\Delta_y}$$

In matrix form, z-parameters can be written as:

$$[z] = \begin{bmatrix} \dfrac{y_{22}}{\Delta_y} & \dfrac{-y_{12}}{\Delta_y} \\ \dfrac{-y_{21}}{\Delta_y} & \dfrac{y_{11}}{\Delta_y} \end{bmatrix}$$

or

$$[z] = \begin{bmatrix} \dfrac{y_{22}}{y_{11}y_{22} - y_{12}y_{21}} & \dfrac{-y_{12}}{y_{11}y_{22} - y_{12}y_{21}} \\ \dfrac{-y_{21}}{y_{11}y_{22} - y_{12}y_{21}} & \dfrac{y_{11}}{y_{11}y_{22} - y_{12}y_{21}} \end{bmatrix}$$

13.9.2 z-Parameters in Terms of h-Parameters

The defining equations of h-parameters [Eqs. (13.19) and (13.20)] are as follows:

$$V_1 = h_{11}I_1 + h_{12}V_2$$
$$I_2 = h_{21}I_1 + h_{22}V_2$$

From Eq. (13.20), we can write

$$h_{22}V_2 = -h_{21}I_1 + I_2$$

$$V_2 = \left[\frac{-h_{21}}{h_{22}}\right]I_1 + \left[\frac{1}{h_{22}}\right]I_2 \tag{13.53}$$

Substituting the value of V_2 into Eq. (13.19), we have

$$V_1 = h_{11}I_1 + h_{12}\left[\frac{-h_{21}}{h_{22}}I_1 + \frac{1}{h_{22}}I_2\right]$$

$$V_1 = \left[h_{11} - \frac{h_{12}h_{21}}{h_{22}}\right]I_1 + \frac{h_{12}}{h_{22}}I_2$$

or

$$V_1 = \left[\frac{h_{11}h_{22} - h_{12}h_{21}}{h_{22}}\right]I_1 + \frac{h_{12}}{h_{22}}I_2 \tag{13.54}$$

Comparing Eqs. (13.53) and (13.54) with Eqs. (13.2) and (13.1), respectively, we have

$$z_{11} = \frac{h_{11}h_{22} - h_{12}h_{21}}{h_{22}} \qquad z_{21} = \frac{-h_{21}}{h_{22}}$$

$$z_{12} = \frac{h_{12}}{h_{22}} \qquad z_{22} = \frac{1}{h_{22}}$$

In matrix form, z-parameters can be written as:

$$[z] = \begin{bmatrix} \dfrac{h_{11}h_{22} - h_{12}h_{21}}{h_{22}} & \dfrac{h_{12}}{h_{22}} \\ \dfrac{-h_{21}}{h_{22}} & \dfrac{1}{h_{22}} \end{bmatrix}$$

13.9.3 *z*-Parameters in Terms of Transmission (*ABCD*) Parameters

The defining equations of transmission parameters [Eqs. (13.32) and (13.33)] are as follows:

$$V_1 = AV_2 + B(-I_2)$$
$$I_1 = CV_2 + D(-I_2)$$

We can rewrite Eq. (13.33) as follows:

$$CV_2 = I_1 + DI_2$$

Therefore,

$$V_2 = \left[\frac{1}{C}\right] I_1 + \left[\frac{D}{C}\right] I_2 \tag{13.55}$$

Substituting the value of V_2 into Eq. (13.32), we have

$$V_1 = A\left[\left(\frac{1}{C}\right) I_1 + \left(\frac{D}{C}\right) I_2\right] - BI_2$$

$$V_1 = \left[\frac{A}{C}\right] I_1 + \left[\frac{AD}{C} - B\right] I_2$$

or

$$V_1 = \left[\frac{A}{C}\right] I_1 + \left[\frac{AD - BC}{C}\right] I_2 \tag{13.56}$$

Comparing Eqs. (13.56) and (13.55) with the Eqs. (13.1) and (13.2), respectively, we have

$$z_{11} = \frac{A}{C} \qquad z_{21} = \frac{1}{C}$$

$$z_{12} = \frac{AD - BC}{C} \qquad z_{22} = \frac{D}{C}$$

In matrix form, the z-parameters can be written as:

$$[z] = \begin{bmatrix} \dfrac{A}{C} & \dfrac{AD - BC}{C} \\ \dfrac{1}{C} & \dfrac{D}{C} \end{bmatrix}$$

13.10 *y*-PARAMETERS IN TERMS OF OTHER PARAMETERS

The defining equations of *y*-parameters [Eqs. (13.11) and (13.12)] are as follows:

$$I_1 = y_{11}V_1 + y_{12}V_2$$
$$I_2 = y_{21}V_1 + y_{22}V_2$$

13.10.1 *y*-Parameters in Terms of *z*-Parameters

The equations for *z*-parameters [Eqs. (13.1) and (13.2)] are as follows:

$$V_1 = z_{11}I_1 + z_{12}I_2$$
$$V_2 = z_{21}I_1 + z_{22}I_2$$

Putting the above equations in matrix form,

$$\begin{bmatrix} V_1 \\ V_2 \end{bmatrix} = \begin{bmatrix} z_{11} & z_{12} \\ z_{21} & z_{22} \end{bmatrix} \begin{bmatrix} I_1 \\ I_2 \end{bmatrix}$$

By applying Cramer's rule,

$$I_1 = \frac{\begin{vmatrix} V_1 & z_{12} \\ V_2 & z_{22} \end{vmatrix}}{\begin{vmatrix} z_{11} & z_{12} \\ z_{21} & z_{22} \end{vmatrix}} = \frac{z_{22}V_1 - z_{12}V_2}{z_{11}z_{22} - z_{12}z_{21}}$$

$$= \frac{z_{22}}{z_{11}z_{22} - z_{12}z_{21}}V_1 + \frac{-z_{12}}{z_{11}z_{22} - z_{12}z_{21}}V_2 \tag{13.57}$$

Similarly,

$$I_2 = \frac{\begin{vmatrix} z_{11} & V_1 \\ z_{21} & V_2 \end{vmatrix}}{\begin{vmatrix} z_{11} & z_{12} \\ z_{21} & z_{22} \end{vmatrix}} = \frac{z_{11}V_2 - z_{21}V_1}{z_{11}z_{22} - z_{12}z_{21}}$$

$$= \frac{-z_{21}}{z_{11}z_{22} - z_{12}z_{21}}V_1 + \frac{z_{11}}{z_{11}z_{22} - z_{12}z_{21}}V_2 \tag{13.58}$$

Let, $\Delta_z = z_{11}z_{22} - z_{12}z_{21}$

Rewriting Eqs. (13.57) and (13.58),

$$I_1 = \left[\frac{z_{22}}{\Delta_z}\right]V_1 + \left[\frac{-z_{12}}{\Delta_z}\right]V_2 \tag{13.59}$$

and

$$I_2 = \left[\frac{-z_{21}}{\Delta_z}\right]V_1 + \left[\frac{z_{11}}{\Delta_z}\right]V_2 \tag{13.60}$$

Comparing Eqs. (13.59) and (13.60) with Eqs. (13.11) and (13.12), respectively, we have

$$y_{11} = \frac{z_{22}}{\Delta_z} \qquad y_{21} = \frac{-z_{21}}{\Delta_z}$$

$$y_{12} = \frac{-z_{12}}{\Delta_z} \qquad y_{22} = \frac{z_{11}}{\Delta_z}$$

In matrix form, *y*-parameters can be written as:

$$[y] = \begin{bmatrix} \dfrac{z_{22}}{\Delta_z} & \dfrac{-z_{12}}{\Delta_z} \\[2ex] \dfrac{-z_{21}}{\Delta_z} & \dfrac{z_{11}}{\Delta_z} \end{bmatrix}$$

or

$$[y] = \begin{bmatrix} \dfrac{z_{22}}{z_{11}z_{22} - z_{12}z_{21}} & \dfrac{-z_{12}}{z_{11}z_{22} - z_{12}z_{21}} \\[2ex] \dfrac{-z_{21}}{z_{11}z_{22} - z_{12}z_{21}} & \dfrac{z_{11}}{z_{11}z_{22} - z_{12}z_{21}} \end{bmatrix}$$

13.10.2 *y*-Parameters in Terms of *h*-Parameters

The defining equations of *h*-parameters [Eqs. (13.19) and (13.20)] are as follows:

$$V_1 = h_{11}I_1 + h_{12}V_2$$

$$I_2 = h_{21}I_1 + h_{22}V_2$$

We can rewrite Eq. (13.19) as follows:

$$h_{11}I_1 = V_1 - h_{12}V_2$$

$$I_1 = \left[\frac{1}{h_{11}}\right]V_1 + \left[\frac{-h_{12}}{h_{11}}\right]V_2 \tag{13.61}$$

Substituting Eq. (13.61) into Eq. (13.20), we have

$$I_2 = h_{21}\left[\frac{1}{h_{11}}V_1 - \frac{h_{12}}{h_{11}}V_2\right] + h_{22}V_2$$

$$I_2 = \left[\frac{h_{21}}{h_{11}}\right]V_1 + \left[h_{22} - \frac{h_{12}h_{21}}{h_{11}}\right]V_2$$

or

$$I_2 = \left[\frac{h_{21}}{h_{11}}\right]V_1 + \left[\frac{h_{11}h_{22} - h_{12}h_{21}}{h_{11}}\right]V_2 \tag{13.62}$$

Comparing Eqs. (13.61) and (13.62) with Eqs. (13.11) and (13.12), respectively, we have

$$y_{11} = \frac{1}{h_{11}} \qquad y_{21} = \frac{h_{21}}{h_{11}}$$

$$y_{12} = \frac{-h_{12}}{h_{11}} \qquad y_{22} = \frac{h_{11}h_{22} - h_{12}h_{21}}{h_{11}}$$

In matrix form, y-parameters can be written as:

$$[y] = \begin{bmatrix} \dfrac{1}{h_{11}} & \dfrac{-h_{12}}{h_{11}} \\ \dfrac{h_{21}}{h_{11}} & \dfrac{h_{11}h_{22} - h_{12}h_{21}}{h_{11}} \end{bmatrix}$$

13.10.3 *y*-Parameters in Terms of Transmission (*ABCD*) Parameters

The defining equations of transmission parameters [Eqs. (13.32) and (13.33)] are:

$$V_1 = AV_2 + B(-I_2)$$

$$I_1 = CV_2 + D(-I_2)$$

We can rewrite Eq. (13.32) as follows:

$$-BI_2 = V_1 - AV_2$$

$$I_2 = \left[-\frac{1}{B}\right]V_1 + \left[\frac{A}{B}\right]V_2 \qquad (13.63)$$

Substituting the value of I_2 into Eq. (13.33), we have

$$I_1 = CV_2 + D\left[\frac{1}{B}V_1 - \frac{A}{B}V_2\right]$$

$$I_1 = \left[\frac{D}{B}\right]V_1 + \left[C - \frac{AD}{B}\right]V_2$$

or

$$I_1 = \left[\frac{D}{B}\right]V_1 + \left[\frac{BC - AD}{B}\right]V_2 \qquad (13.64)$$

Comparing Eqs. (13.64) and (13.63) with Eqs. (13.11) and (13.12), respectively, we have

$$y_{11} = \frac{D}{B} \qquad y_{21} = \frac{-1}{B}$$

$$y_{12} = \frac{BC - AD}{B} \qquad y_{22} = \frac{A}{B}$$

In matrix form, y-parameters can be written as:

$$[y] = \begin{bmatrix} \dfrac{D}{B} & \dfrac{BC - AD}{B} \\ \dfrac{-1}{B} & \dfrac{A}{B} \end{bmatrix}.$$

13.11 *h*-PARAMETERS IN TERMS OF OTHER PARAMETERS

The defining equations for h-parameters [Eqs. (13.19) and (13.20)] are as follows:

$$V_1 = h_{11}I_1 + h_{12}V_2$$
$$I_2 = h_{21}I_1 + h_{22}V_2$$

13.11.1 *h*-Parameters in Terms of *z*-Parameters

The defining equations for z-parameters [Eqs. (13.1) and (13.2)] are as follows:

$$V_1 = z_{11}I_1 + z_{12}I_2$$
$$V_2 = z_{21}I_1 + z_{22}I_2$$

We can rewrite Eq. (13.2) as follows:

$$z_{22}I_2 = V_2 - z_{21}I_1$$

$$I_2 = \left[\frac{-z_{21}}{z_{22}}\right]I_1 + \left[\frac{1}{z_{22}}\right]V_2 \tag{13.65}$$

Substituting the value of I_2 into Eq. (13.1), we have

$$V_1 = z_{11}I_1 + z_{12}\left[\frac{-z_{21}}{z_{22}}I_1 + \frac{1}{z_{22}}V_2\right]$$

$$V_1 = \left[z_{11} + \frac{(z_{12})(-z_{21})}{z_{22}}\right]I_1 + \frac{z_{12}}{z_{22}}V_2$$

or

$$V_1 = \left[\frac{z_{11}z_{22} - z_{12}z_{21}}{z_{22}}\right]I_1 + \left[\frac{z_{12}}{z_{22}}\right]V_2 \tag{13.66}$$

Comparing Eqs. (13.65) and (13.66) with Eqs. (13.20) and (13.19), respectively, we have

$$h_{11} = \frac{z_{11}z_{22} - z_{12}z_{21}}{z_{22}} \qquad h_{21} = \frac{-z_{21}}{z_{22}}$$

$$h_{12} = \frac{z_{12}}{z_{22}} \qquad h_{22} = \frac{1}{z_{22}}$$

In matrix form, h-parameters can be written as:

$$[h] = \begin{bmatrix} \dfrac{z_{11}z_{22} - z_{12}z_{21}}{z_{22}} & \dfrac{z_{12}}{z_{22}} \\ \dfrac{-z_{21}}{z_{22}} & \dfrac{1}{z_{22}} \end{bmatrix}$$

13.11.2 h-Parameters in Terms of y-Parameters

The defining equations of y-parameters [Eqs. (13.11) and (13.12)] are as follows:

$$I_1 = y_{11}V_1 + y_{12}V_2$$
$$I_2 = y_{21}V_1 + y_{22}V_2$$

We can rewrite Eq. (13.11) as follows:

$$y_{11}V_1 = I_1 - y_{12}V_2$$

$$V_1 = \left[\frac{1}{y_{11}}\right]I_1 + \left[\frac{-y_{12}}{y_{11}}\right]V_2 \tag{13.67}$$

Substituting the value of V_1 into Eq. (13.12), we have

$$I_2 = y_{21}\left[\frac{1}{y_{11}}I_1 - \frac{y_{12}}{y_{11}}V_2\right] + y_{22}V_2$$

$$I_2 = \left[\frac{y_{21}}{y_{11}}\right]I_1 + \left[y_{22} - \frac{y_{12}y_{21}}{y_{11}}\right]V_2$$

or

$$I_2 = \left[\frac{y_{21}}{y_{11}}\right]I_1 + \left[\frac{y_{11}y_{22} - y_{12}y_{21}}{y_{11}}\right]V_2 \tag{13.68}$$

Comparing Eqs. (13.67) and (13.68) with Eqs. (13.19) and (13.20), respectively, we have

$$h_{11} = \frac{1}{y_{11}} \qquad h_{21} = \frac{y_{21}}{y_{11}}$$

$$h_{12} = \frac{-y_{12}}{y_{11}} \qquad h_{22} = \frac{y_{11}y_{22} - y_{12}y_{21}}{y_{11}}$$

In matrix form, the h-parameters can be written as:

$$[h] = \begin{bmatrix} \dfrac{1}{y_{11}} & \dfrac{-y_{12}}{y_{11}} \\ \dfrac{y_{21}}{y_{11}} & \dfrac{y_{11}y_{22} - y_{12}y_{21}}{y_{11}} \end{bmatrix}$$

13.11.3 *h*-Parameters in Terms of Transmission (*ABCD*) Parameters

The defining equations of transmission parameters [Eqs. (13.32) and (13.33)] are as follows:

$$V_1 = AV_2 + B(-I_2)$$
$$I_1 = CV_2 + D(-I_2)$$

We can rewrite Eq. (13.33) as follows:

$$DI_2 = -I_1 + CV_2$$

$$I_2 = \left[-\frac{1}{D}\right]I_1 + \left[\frac{C}{D}\right]V_2 \tag{13.69}$$

Substituting the value of I_2 into Eq. (13.32), we have

$$V_1 = AV_2 + B\left[\frac{1}{D}I_1 - \frac{C}{D}V_2\right]$$

$$V_1 = \left[\frac{B}{D}\right]I_1 + \left[A - \frac{BC}{D}\right]V_2$$

or

$$V_1 = \left[\frac{B}{D}\right]I_1 + \left[\frac{AD - BC}{D}\right]V_2 \tag{13.70}$$

Comparing Eqs. (13.70) and (13.69) with Eqs. (13.19) and (13.20), respectively, we have

$$h_{11} = \frac{B}{D} \qquad h_{21} = \frac{-1}{D}$$

$$h_{12} = \frac{AD - BC}{D} \qquad h_{22} = \frac{C}{D}$$

In matrix form, the h-parameters can be written as:

$$[h] = \begin{bmatrix} \dfrac{B}{D} & \dfrac{AD - BC}{D} \\ \dfrac{-1}{D} & \dfrac{C}{D} \end{bmatrix}$$

13.12 TRANSMISSION (*ABCD*) PARAMETERS IN TERMS OF OTHER PARAMETERS

The defining equations of transmission parameters [Eqs. (13.32) and (13.33)] are as follows:

$$V_1 = AV_2 + B(-I_2)$$
$$I_1 = CV_2 + D(-I_2)$$

13.12.1 *ABCD*-Parameters in Terms of *z*-Parameters

The defining equations for *z*-parameters [Eqs. (13.1) and (13.2)] are as follows:

$$V_1 = z_{11}I_1 + z_{12}I_2$$
$$V_2 = z_{21}I_1 + z_{22}I_2$$

We can rewrite Eq. (13.2) as follows:

$$-z_{21}I_1 = -V_2 + z_{22}I_2 \tag{13.71}$$

or

$$I_1 = \left[\frac{1}{z_{21}}\right]V_2 + \left[\frac{z_{22}}{z_{21}}\right](-I_2) \tag{13.72}$$

Substituting the value of I_1 into Eq. (13.1), we have

$$V_1 = z_{11}\left[\left(\frac{1}{z_{21}}\right)V_2 + \left(\frac{-z_{22}}{z_{21}}\right)I_2\right] + z_{12}I_2$$

$$V_1 = \left[\frac{z_{11}}{z_{21}}\right]V_2 + \left[\frac{-z_{11}z_{22}}{z_{21}} + z_{12}\right]I_2$$

or

$$V_1 = \left[\frac{z_{11}}{z_{21}}\right]V_2 + \left[\frac{z_{11}z_{22} - z_{12}z_{21}}{z_{21}}\right](-I_2) \tag{13.73}$$

Comparing Eqs. (13.72) and (13.73) with Eqs. (13.33) and (13.32), respectively, we have

$$A = \frac{z_{11}}{z_{21}} \qquad C = \frac{1}{z_{21}}$$

$$B = \frac{z_{11}z_{22} - z_{12}z_{21}}{z_{21}} \qquad D = \frac{z_{22}}{z_{21}}$$

In matrix form, the transmission parameters can be written as:

$$[ABCD] = \begin{bmatrix} \dfrac{z_{11}}{z_{21}} & \dfrac{z_{11}z_{22} - z_{12}z_{21}}{z_{21}} \\ \dfrac{1}{z_{21}} & \dfrac{z_{22}}{z_{21}} \end{bmatrix}.$$

13.12.2 *ABCD*-Parameters in Terms of *y*-Parameters

The defining equations of *y*-parameters [Eqs. (13.11) and (13.12)] are as follows:

$$I_1 = y_{11}V_1 + y_{12}V_2$$
$$I_2 = y_{21}V_1 + y_{22}V_2$$

We can rewrite Eq. (13.12) as follows:

$$y_{21}V_1 = -y_{22}V_2 + I_2$$

$$V_1 = \left[\frac{-y_{22}}{y_{21}}\right]V_2 + \left[\frac{-1}{y_{21}}\right](-I_2) \tag{13.74}$$

Substituting the value of V_1 into Eq. (13.11), we have

$$I_1 = y_{11}\left[\frac{-y_{22}}{y_{21}}V_2 + \frac{1}{y_{21}}I_2\right] + y_{12}V_2$$

$$I_1 = \left[y_{12} - \frac{y_{11}y_{22}}{y_{21}}\right]V_2 + \left[\frac{-y_{11}}{y_{21}}\right](-I_2)$$

or

$$I_1 = \left[\frac{y_{12}y_{21} - y_{11}y_{22}}{y_{21}}\right]V_2 + \left[\frac{-y_{11}}{y_{21}}\right](-I_2) \tag{13.75}$$

Comparing Eqs. (13.74) and (13.75) with Eqs. (13.32) and (13.33), respectively, we have

$$A = \frac{-y_{22}}{y_{21}} \qquad C = \frac{y_{12}y_{21} - y_{11}y_{22}}{y_{21}}$$

$$B = \frac{-1}{y_{21}} \qquad D = \frac{-y_{11}}{y_{21}}$$

In matrix form, the transmission parameters can be written as:

$$[ABCD] = \begin{bmatrix} \dfrac{-y_{22}}{y_{21}} & \dfrac{-1}{y_{21}} \\ \dfrac{y_{12}y_{21} - y_{11}y_{22}}{y_{21}} & \dfrac{-y_{11}}{y_{21}} \end{bmatrix}$$

13.12.3 *ABCD*-Parameters in Terms of *h*-Parameters

The defining equations for *h*-parameters [Eqs. (13.19) and (13.20)] are as follows:

$$V_1 = h_{11}I_1 + h_{12}V_2$$

$$I_2 = h_{21}I_1 + h_{22}V_2$$

We can rewrite Eq. (13.20) as follows:

$$h_{21}I_1 = I_2 - h_{22}V_2$$

$$I_1 = \left[\frac{-h_{22}}{h_{21}}\right]V_2 + \left[\frac{-1}{h_{21}}\right](-I_2) \tag{13.76}$$

Substituting the value of I_1 into Eq. (13.19), we have

$$V_1 = h_{11}\left[\frac{-h_{22}}{h_{21}}V_2 + \frac{1}{h_{21}}I_2\right] + h_{12}V_2$$

$$V_1 = \left[\frac{-h_{11}h_{22}}{h_{21}} + h_{12}\right]V_2 + \left[\frac{h_{11}}{h_{21}}\right]I_2$$

or

$$V_1 = \left[\frac{h_{12}h_{21} - h_{11}h_{22}}{h_{21}}\right]V_2 + \left[\frac{-h_{11}}{h_{21}}\right](-I_2) \qquad (13.77)$$

Comparing Eqs. (13.77) and (13.76) with Eqs. (13.32) and (13.33), respectively, we have

$$A = \frac{h_{12}h_{21} - h_{11}h_{22}}{h_{21}} \qquad C = \frac{-h_{22}}{h_{21}}$$

$$B = \frac{-h_{11}}{h_{21}} \qquad D = \frac{-1}{h_{21}}$$

In matrix form, the transmission parameters can be written as:

$$[ABCD] = \begin{bmatrix} \dfrac{h_{12}h_{21} - h_{11}h_{22}}{h_{21}} & \dfrac{-h_{11}}{h_{21}} \\ \dfrac{-h_{22}}{h_{21}} & \dfrac{-1}{h_{21}} \end{bmatrix}$$

EXAMPLE 13.1 Determine the z-parameters for the network shown in Fig. 13.18.

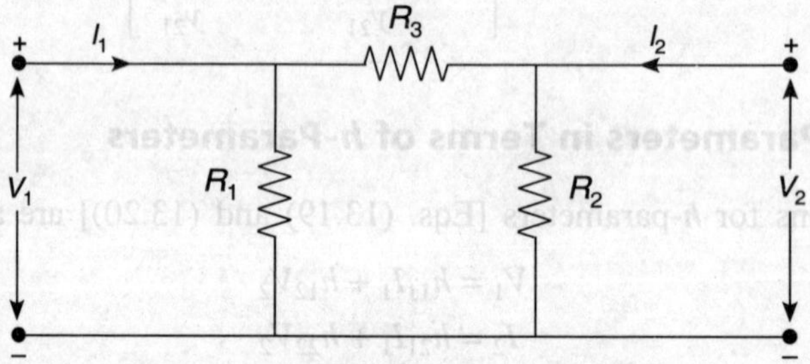

Fig. 13.18

Solution The defining equations of z-parameters are:

$$V_1 = z_{11}I_1 + z_{12}I_2$$
$$V_2 = z_{21}I_1 + z_{22}I_2$$

Let us assume the mesh equations in the directions as shown in Fig. 13.19.

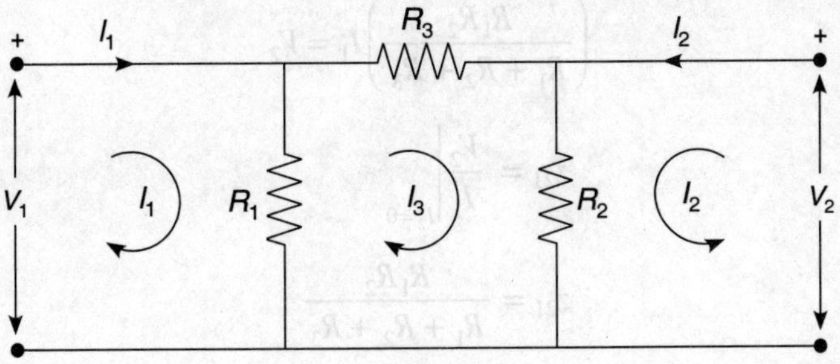

Fig. 13.19

The mesh equations by inspection are written as:

$$\begin{bmatrix} R_1 & 0 & -R_1 \\ 0 & R_2 & R_2 \\ -R_1 & R_2 & R_1+R_2+R_3 \end{bmatrix} \begin{bmatrix} I_1 \\ I_2 \\ I_3 \end{bmatrix} = \begin{bmatrix} V_1 \\ V_2 \\ 0 \end{bmatrix}$$

The z-parameters z_{11} and z_{21} are determined by taking $I_2 = 0$. By expanding the above matrix equation and substituting $I_2 = 0$, we get

$$R_1 I_1 - R_1 I_3 = V_1 \quad \text{(i)}$$

$$R_2 I_3 = V_2 \quad \text{(ii)}$$

$$-R_1 I_1 + (R_1 + R_2 + R_3) I_3 = 0 \quad \text{(iii)}$$

From Eq. (iii),

$$I_3 = \left(\frac{R_1}{R_1 + R_2 + R_3} \right) I_1 \quad \text{(iv)}$$

Substituting Eq. (iv) into Eq. (i), we obtain

$$R_1 I_1 - \left(\frac{R_1^2}{R_1 + R_2 + R_3} \right) I_1 = V_1$$

or

$$\left(R_1 - \frac{R_1^2}{R_1 + R_2 + R_3} \right) I_1 = V_1$$

By definition,

$$z_{11} = \left. \frac{V_1}{I_1} \right|_{I_2=0}$$

Hence,

$$z_{11} = \left(R_1 - \frac{R_1^2}{R_1 + R_2 + R_3} \right) = \frac{R_1 R_2 + R_1 R_3}{R_1 + R_2 + R_3}$$

Using Eqs. (ii) and (iv), we can write

$$\left(\frac{R_1 R_2}{R_1 + R_2 + R_3}\right) I_1 = V_2$$

By definition,

$$z_{21} = \left.\frac{V_2}{I_1}\right|_{I_2=0}$$

$\therefore$

$$z_{21} = \frac{R_1 R_2}{R_1 + R_2 + R_3}$$

The z-parameters z_{12} and z_{22} are determined by taking $I_1 = 0$. By expanding the matrix equation and substituting $I_1 = 0$, we get

$$-R_1 I_3 = V_1 \quad \text{(v)}$$

$$R_2 I_2 + R_2 I_3 = V_2 \quad \text{(vi)}$$

$$R_2 I_2 + (R_1 + R_2 + R_3) I_3 = 0 \quad \text{(vii)}$$

From Eq. (vii),

$$I_3 = \left(-\frac{R_2}{R_1 + R_2 + R_3}\right) I_2 \quad \text{(viii)}$$

Substituting Eq. (viii) into Eq. (vi), we get

$$R_2 I_2 - \left(\frac{R_2^2}{R_1 + R_2 + R_3}\right) I_2 = V_2$$

or

$$\left(R_2 - \frac{R_2^2}{R_1 + R_2 + R_3}\right) I_2 = V_2$$

By definition,

$$z_{22} = \left.\frac{V_2}{I_2}\right|_{I_1=0}$$

So,

$$z_{22} = \left(R_2 - \frac{R_2^2}{R_1 + R_2 + R_3}\right) = \frac{R_1 R_2 + R_2 R_3}{R_1 + R_2 + R_3}$$

Using Eqs. (v) and (viii), we can write

$$\left(\frac{R_1 R_2}{R_1 + R_2 + R_3}\right) I_2 = V_1$$

By definition,

$$z_{12} = \left.\frac{V_1}{I_2}\right|_{I_1=0}$$

So, $$z_{12} = \frac{R_1R_2}{R_1 + R_2 + R_3}$$

Thus, $$[z] = \begin{bmatrix} \dfrac{R_1R_2 + R_1R_3}{R_1 + R_2 + R_3} & \dfrac{R_1R_2}{R_1 + R_2 + R_3} \\ \dfrac{R_1R_2}{R_1 + R_2 + R_3} & \dfrac{R_1R_2 + R_2R_3}{R_1 + R_2 + R_3} \end{bmatrix}$$

EXAMPLE 13.2 Find the z-parameters for the network shown in Fig. 13.20.

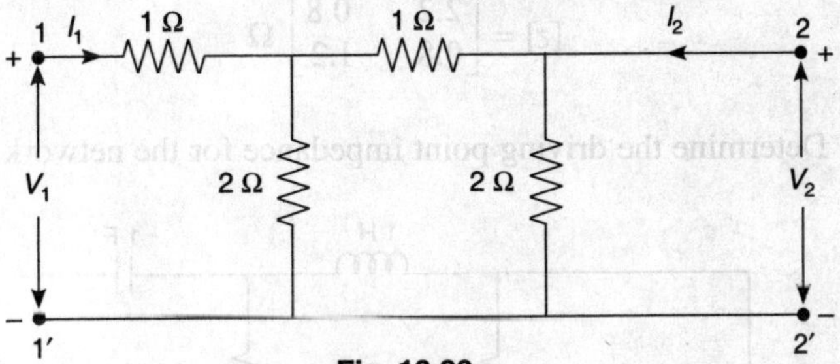

Fig. 13.20

Solution The mesh currents are assumed in the directions as shown in Fig. 13.21.

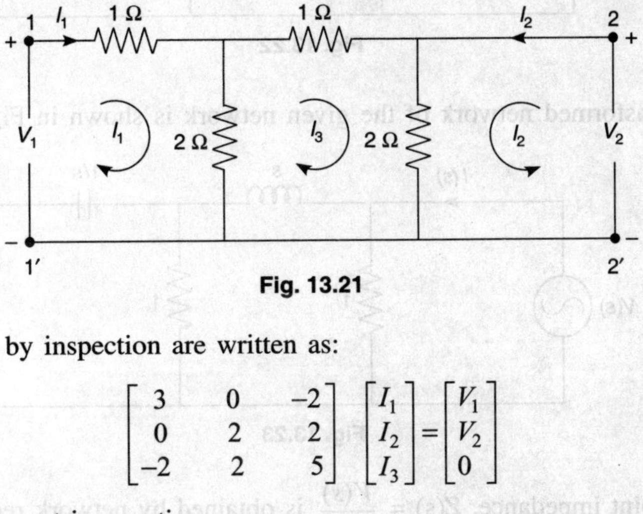

Fig. 13.21

The mesh equations by inspection are written as:

$$\begin{bmatrix} 3 & 0 & -2 \\ 0 & 2 & 2 \\ -2 & 2 & 5 \end{bmatrix} \begin{bmatrix} I_1 \\ I_2 \\ I_3 \end{bmatrix} = \begin{bmatrix} V_1 \\ V_2 \\ 0 \end{bmatrix}$$

Expanding the above matrix equation,

$$3I_1 - 2I_3 = V_1 \quad (i)$$

$$2I_2 + 2I_3 = V_2 \quad \text{(ii)}$$

$$-2I_1 + 2I_2 + 5I_3 = 0 \quad \text{(iii)}$$

From Eq. (iii),

$$I_3 = \frac{2I_1 - 2I_2}{5} = 0.4I_1 - 0.4I_2 \tag{iv}$$

Substituting the value of I_3 found above into Eqs. (i) and (ii), we have

$$V_1 = 2.2I_1 + 0.8I_2 \tag{v}$$

$$V_2 = 0.8I_1 + 1.2I_2 \tag{vi}$$

Comparing Eqs. (v) and (vi) with the defining equations of z-parameters, we get

$$[z] = \begin{bmatrix} 2.2 & 0.8 \\ 0.8 & 1.2 \end{bmatrix} \Omega$$

EXAMPLE 13.3 Determine the driving point impedance for the network shown in Fig. 13.22.

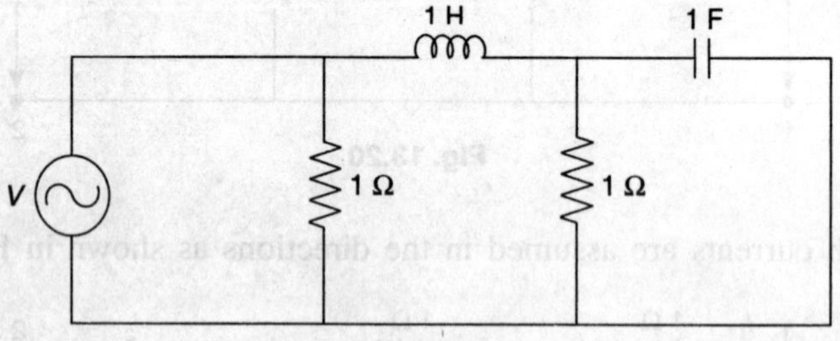

Fig. 13.22

Solution The transformed network of the given network is shown in Fig. 13.23.

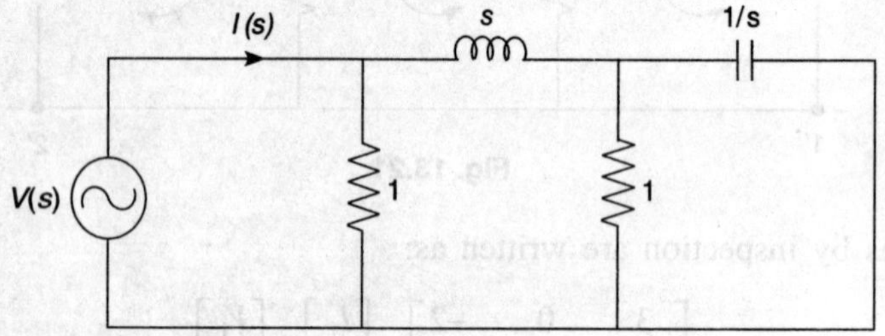

Fig. 13.23

The driving point impedance, $Z(s) = \dfrac{V(s)}{I(s)}$ is obtained by network reduction technique as follows:

$(1||1/s)$ gives $\dfrac{1}{(s+1)}$, and it results in the circuit shown in Fig. 13.24. It is further simplified as shown in Fig. 13.25.

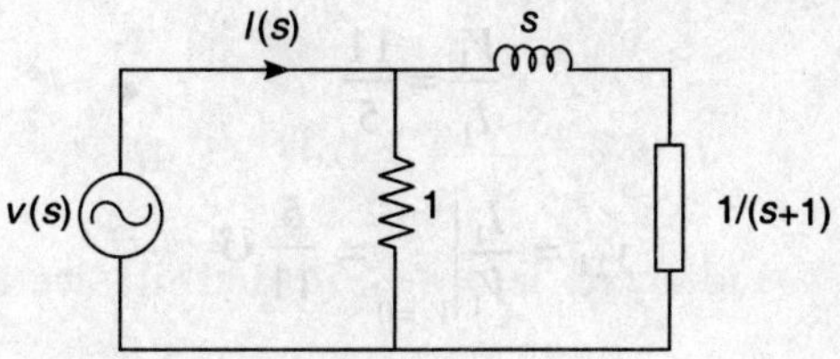

Fig. 13.24

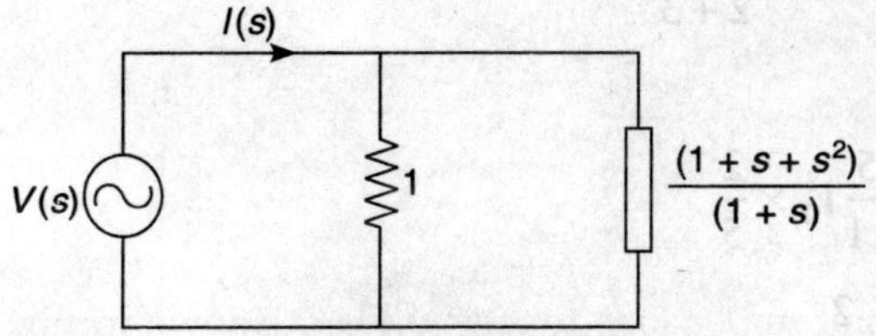

Fig. 13.25

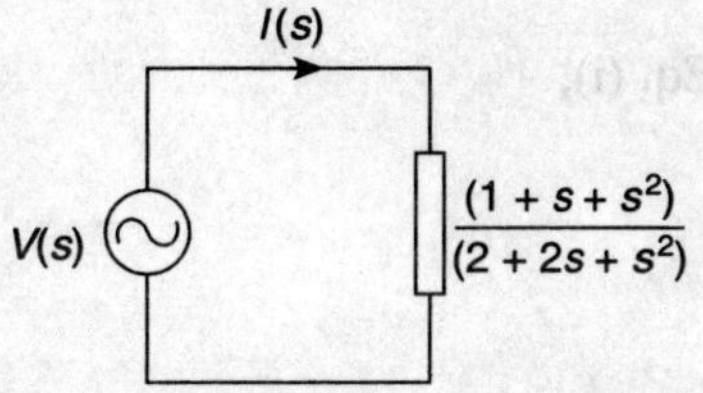

Fig. 13.26

As observed from Fig. 13.26, the driving point impedance, $Z(s) = \dfrac{V(s)}{I(s)} = \dfrac{(1+s+s^2)}{(2+2s+s^2)}$.

EXAMPLE 13.4 Find the *y*-parameters for the network shown in Fig. 13.27.

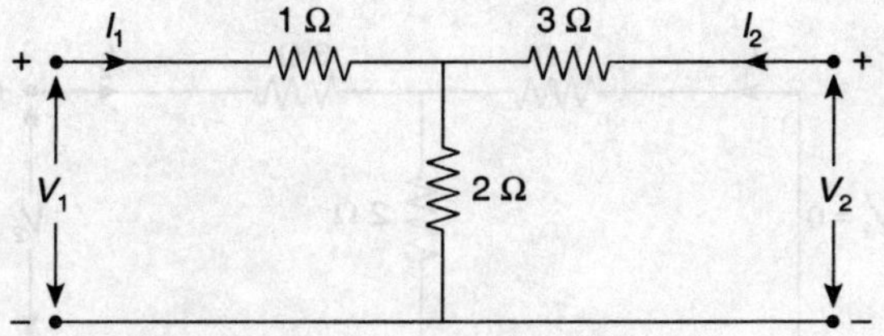

Fig. 13.27

Solution First let us assume that the output port is short-circuited so that $V_2 = 0$, and the circuit may be redrawn as shown in Fig. 13.28.

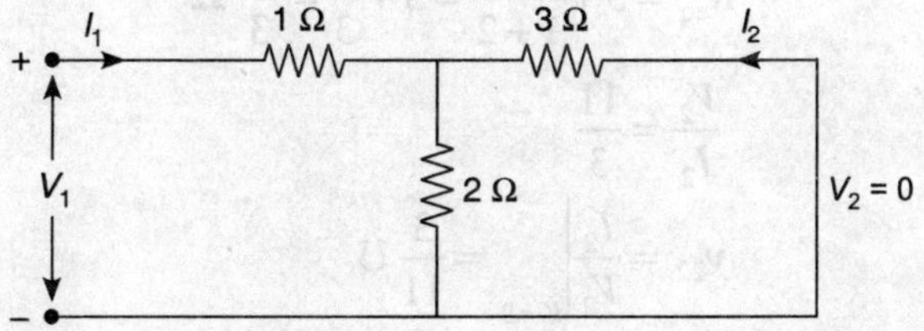

Fig. 13.28

Equivalent resistance seen from input port,

$$R_{eq} = 1 + \frac{2 \times 3}{2+3} = 1 + \frac{6}{5} = \frac{11}{5}\ \Omega$$

or
$$\frac{V_1}{I_1} = \frac{11}{5} \quad \text{(i)}$$

Hence,
$$y_{11} = \left.\frac{I_1}{V_1}\right|_{V_2=0} = \frac{5}{11}\ ℧$$

Using current division rule,
$$-I_2 = I_1 \times \frac{2}{2+3}$$

From Eq. (i),
$$-I_2 = \frac{5}{11}V_1 \times \frac{2}{5}$$

or
$$\frac{I_2}{V_1} = -\frac{2}{11}$$

Therefore,
$$y_{21} = \left.\frac{I_2}{V_1}\right|_{V_2=0} = -\frac{2}{11}\ ℧$$

Next, let us short-circuit the input port so that $V_1 = 0$ and the circuit may be redrawn as shown in Fig. 13.29.

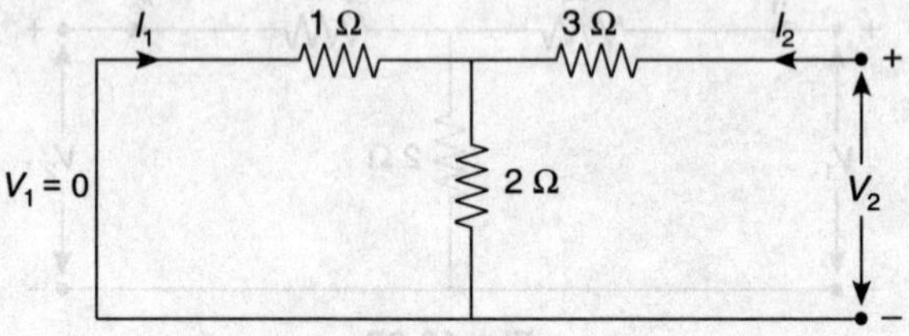

Fig. 13.29

Equivalent resistance seen from output port,
$$R_{eq} = 3 + \frac{1 \times 2}{1+2} = 3 + \frac{2}{3} = \frac{11}{3}\ \Omega$$

or
$$\frac{V_2}{I_2} = \frac{11}{3} \quad \text{(ii)}$$

So,
$$y_{22} = \left.\frac{I_2}{V_2}\right|_{V_1=0} = \frac{3}{11}\ ℧$$

Using current division rule,
$$-I_1 = I_2 \times \frac{2}{2+1}$$

From Eq. (ii),

$$-I_1 = \frac{3}{11}V_2 \times \frac{2}{3}$$

or

$$\frac{I_1}{V_2} = -\frac{2}{11}$$

Hence,

$$y_{12} = \left.\frac{I_1}{V_2}\right|_{V_1=0} = -\frac{2}{11}\ \mho$$

Thus,

$$[y] = \begin{bmatrix} \frac{5}{11} & -\frac{2}{11} \\ -\frac{2}{11} & \frac{3}{11} \end{bmatrix} \mho$$

EXAMPLE 13.5 Determine the *y*-parameters for the network shown in Fig. 13.30.

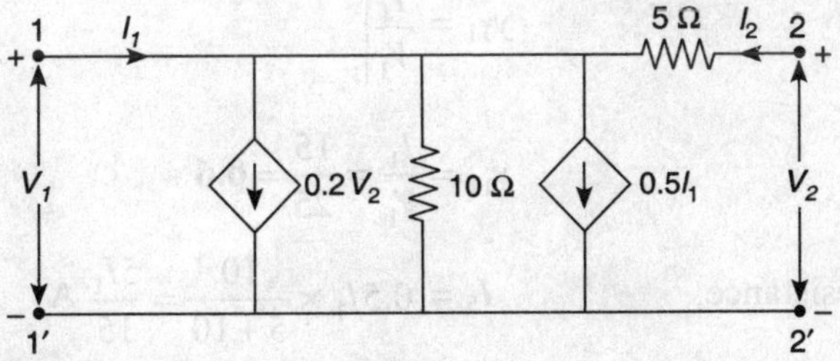

Fig. 13.30

Solution The defining equations of *y*-parameters are

$$I_1 = y_{11}V_1 + y_{12}V_2$$
$$I_2 = y_{21}V_1 + y_{22}V_2$$

First let us assume that the output port is short-circuited so that $V_2 = 0$, and the circuit may be redrawn as shown in Fig. 13.31.

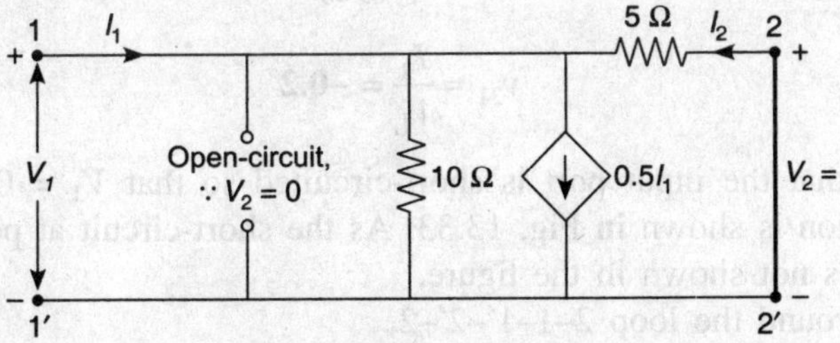

Fig. 13.31

For simplicity, the circuit is again redrawn by interchanging the positions of 10 Ω resistance and $0.5I_1$ current source as shown in Fig. 13.32.

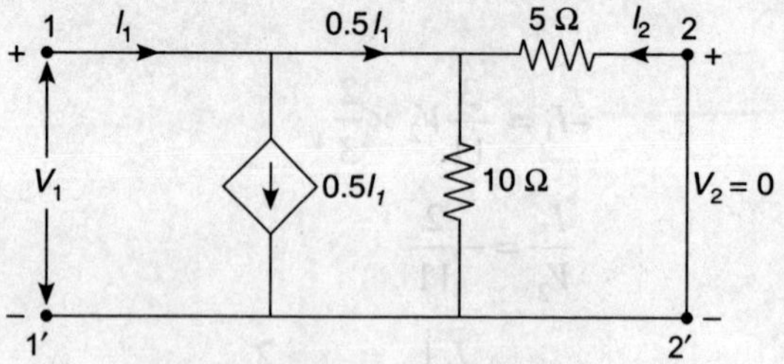

Fig. 13.32

By current division rule, the current through 10 Ω resistance,

$$I_{10} = 0.5I_1 \times \frac{5}{5+10} = \frac{2.5I_1}{15}\text{ A}$$

Therefore, voltage across 10 Ω resistance, $V_1 = 10 \times \frac{2.5I_1}{15} = \frac{25I_1}{15}$ V (i)

By definition, $$y_{11} = \left.\frac{I_1}{V_1}\right|_{V_2=0}$$

So, from Eq. (i), $$y_{11} = \frac{I_1}{V_1} = \frac{15}{25} = \mathbf{0.6}$$

Current through 5 Ω resistance, $$I_5 = 0.5I_1 \times \frac{10}{5+10} = \frac{5I_1}{15}\text{ A} \qquad \text{(ii)}$$

Also, $$I_2 = -I_5 = -\frac{5I_1}{15} \qquad \text{(iii)}$$

Substituting the value of I_1 from Eq. (i) into Eq. (iii), we obtain

$$I_2 = -\frac{5}{15}\left(\frac{15V_1}{25}\right)$$

By definition, $$y_{21} = \left.\frac{I_2}{V_1}\right|_{V_2=0}$$

So, $$y_{21} = \frac{I_2}{V_1} = \mathbf{-0.2}$$

Next, let us assume that the input port is short-circuited so that $V_1 = 0$, and the equivalent circuit for this condition is shown in Fig. 13.33. As the short-circuit at port 1–1′ makes 10 Ω resistance useless, it is not shown in the figure.

Applying KVL around the loop 2–1–1′–2′–2,

$$V_2 = 5I_2 \qquad \text{(iv)}$$

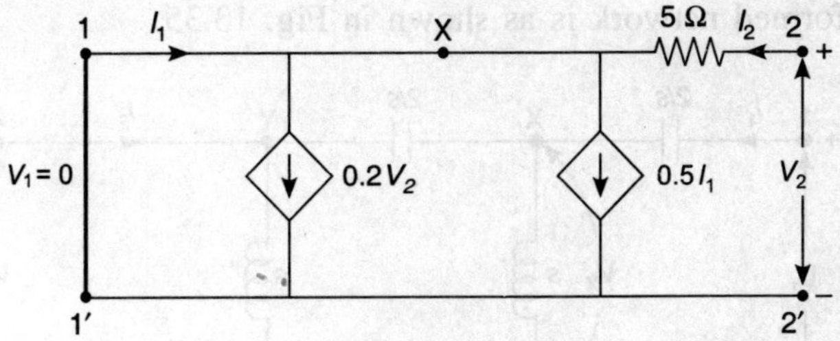

Fig. 13.33

By definition,

$$y_{22} = \left.\frac{I_2}{V_2}\right|_{V_1=0}$$

Hence,

$$y_{22} = \frac{I_2}{V_2} = \mathbf{0.2}$$

Applying KCL at node X,

$$-I_1 + 0.2V_2 + 0.5I_1 - I_2 = 0$$

From Eq. (iv), putting the value of $V_2 = 5I_2$,

$$-I_1 + I_2 + 0.5I_1 - I_2 = 0$$

or

$$I_1 = 0$$

So,

$$y_{12} = \frac{I_1}{V_2} = \mathbf{0}$$

Thus,

$$y = \begin{bmatrix} 0.6 & 0 \\ -0.2 & 0.2 \end{bmatrix} \mho$$

EXAMPLE 13.6 Determine y-parameters for the network shown in Fig. 13.34.

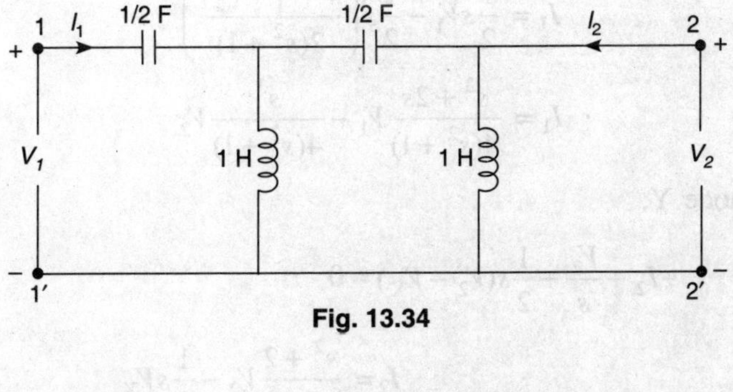

Fig. 13.34

Solution The transformed network is as shown in Fig. 13.35.

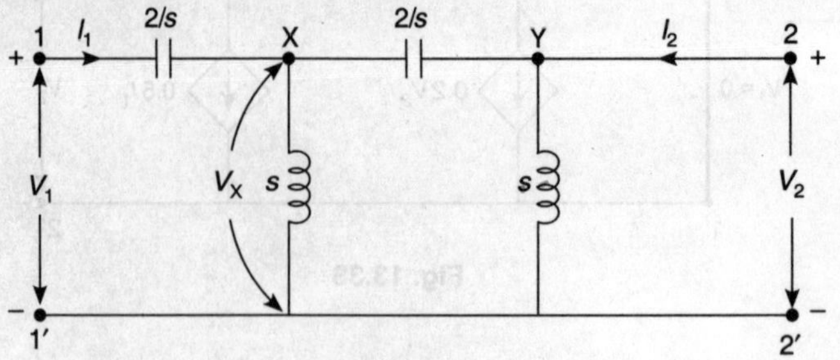

Fig. 13.35

From Fig. 13.35,

$$I_1 = \frac{V_1 - V_X}{\left(\frac{2}{s}\right)}$$

$$= \frac{s(V_1 - V_X)}{2} \quad \text{(i)}$$

Applying KCL at node X,

$$-\frac{V_1 - V_X}{\left(\frac{2}{s}\right)} + \frac{V_X}{s} + \frac{V_X - V_2}{\left(\frac{2}{s}\right)} = 0$$

$$\frac{1}{2}sV_X + \frac{1}{s}V_X + \frac{1}{2}sV_X = \frac{1}{2}sV_1 + \frac{1}{2}sV_2$$

$$V_X = \frac{s^2(V_1 + V_2)}{2(s^2 + 1)} \quad \text{(ii)}$$

Substituting the value of V_X found above into Eq. (i), we get

$$I_1 = \frac{1}{2}sV_1 - \frac{1}{2}s\left(\frac{s^2(V_1 + V_2)}{2(s^2 + 1)}\right)$$

or

$$I_1 = \frac{s^3 + 2s}{4(s^2 + 1)}V_1 - \frac{s^3}{4(s^2 + 1)}V_2 \quad \text{(iii)}$$

Applying KCL at node Y,

$$-I_2 + \frac{V_2}{s} + \frac{1}{2}s(V_2 - V_X) = 0$$

$$I_2 = \frac{s^2 + 2}{2s}V_2 - \frac{1}{2}sV_X \quad \text{(iv)}$$

Substituting Eq. (ii) into Eq. (iv), we get

$$I_2 = \frac{s^2+2}{2s}V_2 - \frac{1}{2}s\left[\frac{s^2(V_1+V_2)}{2(s^2+1)}\right]$$

$$= \frac{-s^3}{4(s^2+1)}V_1 + \left[\frac{s^2+2}{2s} - \frac{s^3}{4(s^2+1)}\right]V_2$$

$$I_2 = \frac{-s^3}{4(s^2+1)}V_1 + \frac{s^4+6s^2+4}{4s(s^2+1)}V_2 \quad \text{(v)}$$

Comparing Eqs. (iii) and (v) with the defining equations of y-parameters, we get

$$[y] = \begin{bmatrix} \dfrac{s^3+2s}{4(s^2+1)} & \dfrac{-s^3}{4(s^2+1)} \\ \dfrac{-s^3}{4(s^2+1)} & \dfrac{s^4+6s^2+4}{4s(s^2+1)} \end{bmatrix} \mho$$

EXAMPLE 13.7 Determine the y-parameters of the network shown in Fig. 13.36.

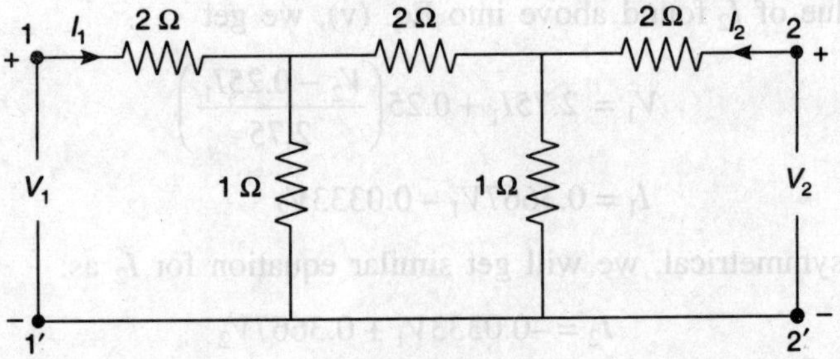

Fig. 13.36

Solution The mesh currents are assumed in the directions as shown in Fig. 13.37.

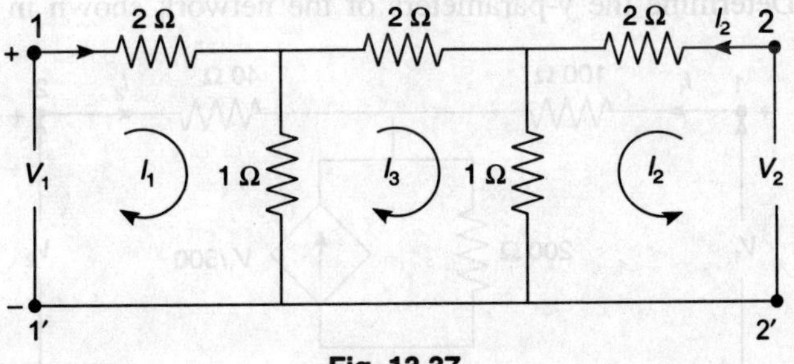

Fig. 13.37

The mesh equations by inspection are written as:

$$\begin{bmatrix} 3 & 0 & -1 \\ 0 & 3 & 1 \\ -1 & 1 & 4 \end{bmatrix} \begin{bmatrix} I_1 \\ I_2 \\ I_3 \end{bmatrix} = \begin{bmatrix} V_1 \\ V_2 \\ 0 \end{bmatrix}$$

Expanding the above matrix equation,

$$3I_1 - I_3 = V_1 \tag{i}$$

$$3I_2 + I_3 = V_2 \tag{ii}$$

$$-I_1 + I_2 + 4I_3 = 0 \tag{iii}$$

From Eq. (iii),

$$I_3 = \frac{I_1 - I_2}{4} = 0.25I_1 - 0.25I_2 \tag{iv}$$

Substituting the value of I_3 found above into Eqs. (i) and (ii), we get

$$V_1 = 2.75I_1 + 0.25I_2 \tag{v}$$

$$V_2 = 0.25I_1 + 2.75I_2 \tag{vi}$$

From Eq. (vi),

$$I_2 = \frac{V_2 - 0.25I_1}{2.75} \tag{vii}$$

Substituting the value of I_2 found above into Eq. (v), we get

$$V_1 = 2.75I_1 + 0.25\left(\frac{V_2 - 0.25I_1}{2.75}\right) \tag{viii}$$

Solving,

$$I_1 = 0.3667V_1 - 0.0333V_2$$

As the network is symmetrical, we will get similar equation for I_2 as:

$$I_2 = -0.0333V_1 + 0.3667V_2$$

Hence,

$$[y] = \begin{bmatrix} 0.3667 & -0.0333 \\ -0.0333 & 0.3667 \end{bmatrix} \mho$$

EXAMPLE 13.8 Determine the y-parameters of the network shown in Fig. 13.38.

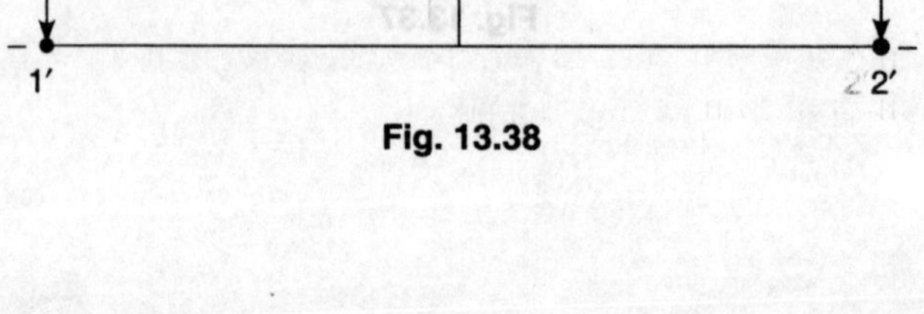

Fig. 13.38

Solution First let us assume that the output port is short-circuited so that $V_2 = 0$ and the equivalent circuit for this condition is shown in Fig. 13.39.

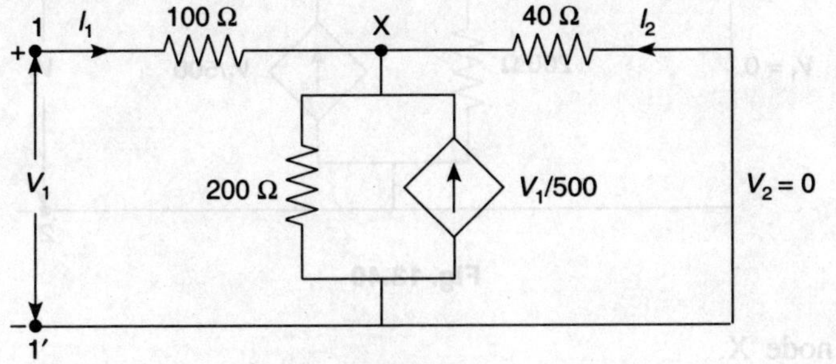

Fig. 13.39

Applying KCL at node X,

$$\frac{V_1 - V_X}{100} - \frac{V_X}{200} + \frac{V_1}{500} - \frac{V_X}{40} = 0$$

Solving,

$$V_X = 0.3V_1$$

Also,

$$I_1 = \frac{V_1 - V_X}{100}$$

$$= \frac{V_1 - 0.3V_1}{100}$$

$$I_1 = 7 \times 10^{-3} V_1$$

By definition,

$$y_{11} = \left.\frac{I_1}{V_1}\right|_{V_2=0}$$

So,

$$y_{11} = \frac{I_1}{V_1} = 7 \times 10^{-3}\ \mho$$

Current,

$$I_2 = \frac{-V_X}{40} = \frac{-0.3V_1}{40}$$

$$I_2 = -7.5 \times 10^{-3} V_1$$

By definition,

$$y_{21} = \left.\frac{I_2}{V_1}\right|_{V_2=0}$$

So,

$$y_{21} = \frac{I_2}{V_1} = -7.5 \times 10^{-3}\ \mho$$

Next, let us assume that the input port is short-circuited so that $V_1 = 0$, and the equivalent circuit for this condition is shown in Fig. 13.40.

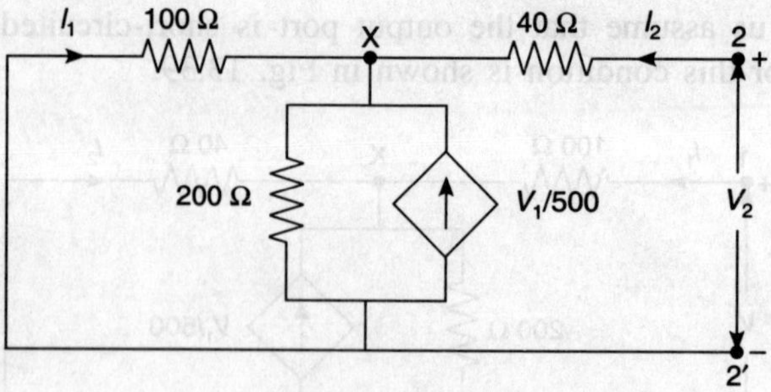

Fig. 13.40

Applying KCL at node X,

$$-\frac{V_X}{100}-\frac{V_X}{200}+\frac{V_2-V_X}{40}=0$$

Solving,

$$V_X = 0.625V_2$$

Also,

$$I_2 = \frac{V_2 - V_X}{40} = \frac{V_2 - 0.625V_2}{40}$$

$$I_2 = 9.375 \times 10^{-3} V_2$$

By definition,

$$y_{22} = \left.\frac{I_2}{V_2}\right|_{V_1=0}$$

∴

$$y_{22} = \frac{I_2}{V_2} = 9.375 \times 10^{-3}\ \mho$$

Current,

$$I_1 = \frac{-V_X}{100} = \frac{-0.625V_2}{100}$$

$$I_1 = -6.25 \times 10^{-3} V_2$$

By definition,

$$y_{21} = \left.\frac{I_1}{V_2}\right|_{V_2=0}$$

Hence,

$$y_{12} = \frac{I_1}{V_2} = -6.25 \times 10^{-3}\ \mho$$

Thus,

$$[y] = \begin{bmatrix} 7 & -6.25 \\ -7.5 & 9.375 \end{bmatrix} \text{m}\mho$$

EXAMPLE 13.9 Determine the driving point admittance for the network shown in Fig. 13.41.

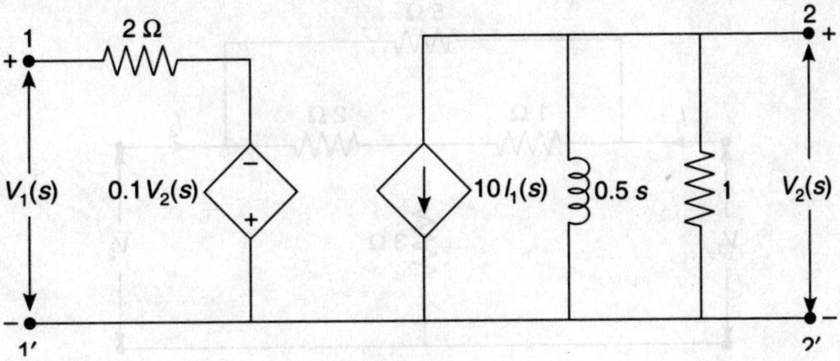

Fig. 13.41

Solution Applying KVL in the input port side loop,

$$-V_1(s) + 2I_1(s) - 0.1V_2(s) = 0$$

or

$$I_1(s) = \frac{V_1(s) + 0.1V_2(s)}{2} \quad \text{(i)}$$

Applying KCL at node 2,

$$10I_1(s) + \frac{V_2(s)}{0.5s} + \frac{V_2(s)}{1} = 0$$

or

$$10I_1(s) + \frac{2V_2(s)}{s} + V_2(s) = 0$$

$$V_2(s)\left[\frac{2}{s} + 1\right] = -10I_1(s)$$

$$V_2(s) = -10\left[\frac{s}{s+2}\right]I_1(s)$$

Substituting the value of $V_2(s)$ found above into Eq. (i), we get

$$I_1(s) = \frac{V_1(s) + 0.1\left[\frac{-10s}{s+2}\right]I_1(s)}{2}$$

Simplifying,

Driving point admittance,

$$\frac{I_1(s)}{V_1(s)} = \frac{s+2}{3s+4}$$

EXAMPLE 13.10 Find the z-parameters and y-parameters of the network shown in Fig. 13.42.

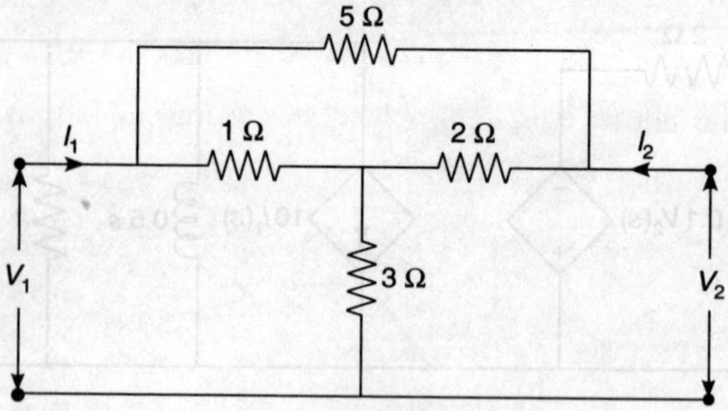

Fig. 13.42

Solution Mesh currents I_1, I_2 and I_3 are assumed in the directions as shown in Fig. 13.43.

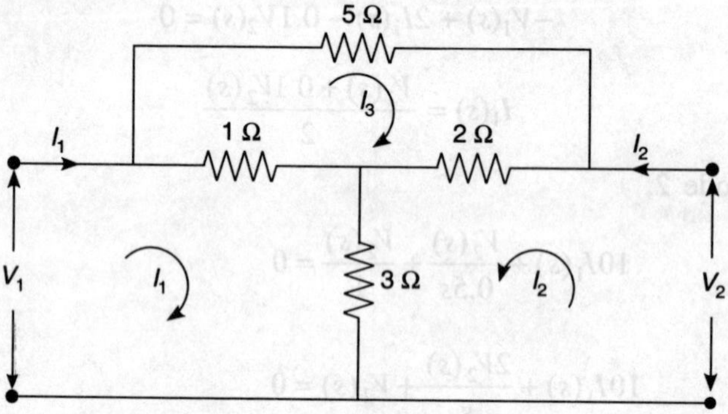

Fig. 13.43

Applying mesh analysis and by inspection, we can write

$$\begin{bmatrix} 4 & 3 & -1 \\ 3 & 5 & 2 \\ -1 & 2 & 8 \end{bmatrix} \begin{bmatrix} I_1 \\ I_2 \\ I_3 \end{bmatrix} = \begin{bmatrix} V_1 \\ V_2 \\ 0 \end{bmatrix}$$

Expanding the above matrix equation,

$$4I_1 + 3I_2 - I_3 = V_1 \qquad \text{(i)}$$

$$3I_1 + 5I_2 + 2I_3 = V_2 \qquad \text{(ii)}$$

$$-I_1 + 2I_2 + 8I_3 = 0 \qquad \text{(iii)}$$

From Eq. (iii),

$$I_3 = \frac{I_1 - 2I_2}{8} = 0.125I_1 - 0.25I_2 \qquad \text{(iv)}$$

Substituting the value of I_3 from Eq. (iv) into Eqs. (i) and (ii), we have

$$V_1 = 3.875I_1 + 3.25I_2 \quad \text{(v)}$$

$$V_2 = 3.25I_1 + 4.5I_2 \quad \text{(vi)}$$

Upon comparison of the above equations with the defining equations of z-parameters

$$z_{11} = 3.875\ \Omega;\ z_{12} = 3.25\ \Omega;\ z_{21} = 3.25\ \Omega \text{ and } z_{22} = 4.5\ \Omega.$$

Hence, the z-matrix is

$$[z] = \begin{bmatrix} 3.875 & 3.25 \\ 3.25 & 4.5 \end{bmatrix} \Omega$$

$$[y] = [z]^{-1}$$

Determinant of $[z]$, $\Delta z = (3.875 \times 4.5) - (3.25 \times 3.25) = 6.875$

$$[y] = [z]^{-1} = \frac{1}{6.875}\begin{bmatrix} 4.5 & -3.25 \\ -3.25 & 3.875 \end{bmatrix}$$

Therefore, the y-matrix is

$$[y] = \begin{bmatrix} 0.6545 & -0.4727 \\ -0.4727 & 0.5636 \end{bmatrix} \mho$$

EXAMPLE 13.11 Find the z-parameters and y-parameters for the network shown in Fig. 13.44.

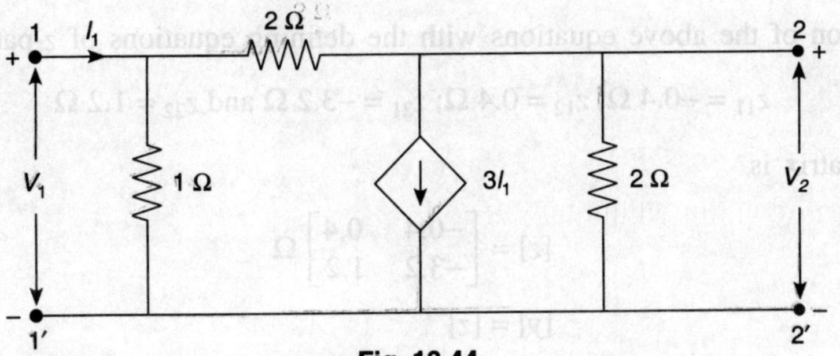

Fig. 13.44

Solution The branch currents flowing through various branches may be marked as shown in Fig. 13.45.

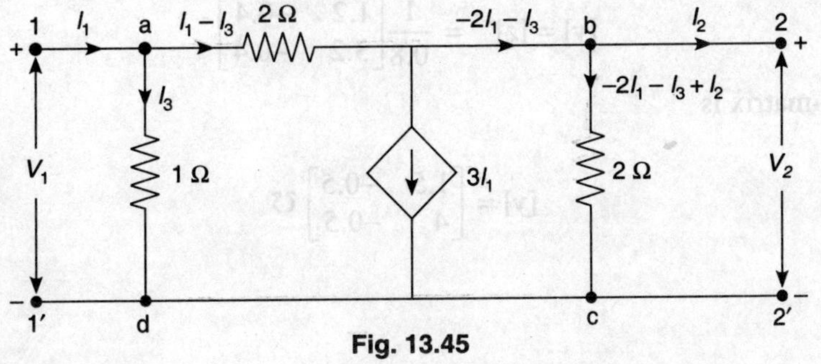

Fig. 13.45

KVL equations are:

Loop 1–a–d–1′–1:

$$-V_1 + I_3 = 0$$

or $$V_1 = I_3 \qquad \text{(i)}$$

Loop a–b–c–d–a:

$$2(I_1 - I_3) + 2(-2I_1 - I_3 + I_2) - I_3 = 0$$

or $$-2I_1 + 2I_2 - 5I_3 = 0 \qquad \text{(ii)}$$

Loop 2–b–c–2′–2:

$$-4I_1 + 2I_2 - 2I_3 - V_2 = 0$$

or $$V_2 = -4I_1 + 2I_2 - 2I_3 \qquad \text{(iii)}$$

From Eq. (ii),

$$I_3 = \frac{-2I_1 + 2I_2}{5} = -0.4I_1 + 0.4I_2 \qquad \text{(iv)}$$

Substituting the value of I_3 from Eq. (iv) into Eqs. (i) and (iii), we get

$$V_1 = -0.4I_1 + 0.4I_2$$
$$V_2 = -3.2I_1 + 1.2I_2$$

Upon comparison of the above equations with the defining equations of *z*-parameters

$$z_{11} = -0.4\ \Omega;\ z_{12} = 0.4\ \Omega;\ z_{21} = -3.2\ \Omega \text{ and } z_{22} = 1.2\ \Omega$$

Hence, the *z*-matrix is

$$[z] = \begin{bmatrix} -0.4 & 0.4 \\ -3.2 & 1.2 \end{bmatrix} \Omega$$

$$[y] = [z]^{-1}$$

Determinant of [*z*],

$$\Delta z = (-0.4 \times 1.2) - (-3.2 \times 0.4) = 0.8$$

$$[y] = [z]^{-1} = \frac{1}{0.8}\begin{bmatrix} 1.2 & -0.4 \\ 3.2 & -0.4 \end{bmatrix}$$

Therefore, the *y*-matrix is

$$[y] = \begin{bmatrix} 1.5 & -0.5 \\ 4 & -0.5 \end{bmatrix} \mho$$

EXAMPLE 13.12 Determine the z and y parameters of the network shown in Fig. 13.46.

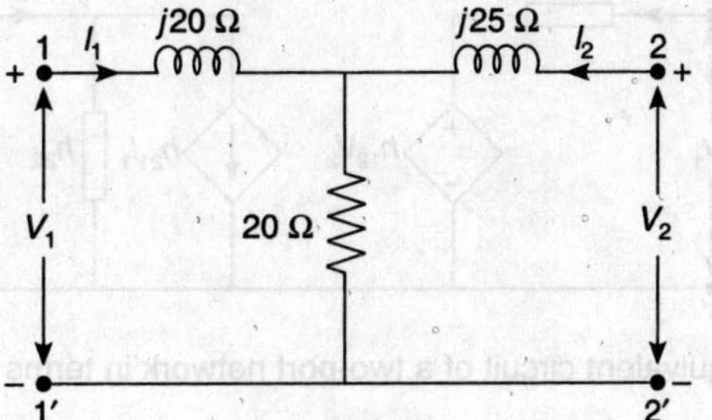

Fig. 13.46

Solution The defining equations of z-parameters are

$$V_1 = z_{11}I_1 + z_{12}I_2$$
$$V_2 = z_{21}I_1 + z_{22}I_2$$

Applying KVL for the two meshes, we get

$$V_1 = (20 + j20)I_1 + 20I_2 \quad \text{(i)}$$
$$V_2 = 20I_1 + (20 + j25)I_2 \quad \text{(ii)}$$

Comparing Eqs. (i) and (ii) with the defining equations of z-parameters, we get

$$[z] = \begin{bmatrix} 20 + j20 & 20 \\ 20 & 20 + j25 \end{bmatrix} \Omega$$

$$[y] = [z]^{-1}$$

Determinant of $[z]$, $$\Delta z = (20 + j20)\,(20 + j25) - 20^2$$
$$= (-500 + j900)$$

$$[y] = [z]^{-1} = \frac{1}{(-500 + j900)} \begin{bmatrix} 20 + j25 & -20 \\ -20 & 20 + j20 \end{bmatrix}$$

Therefore, the y-matrix is

$$[y] = \begin{bmatrix} 0.0118 - j0.0288 & 0.0094 + j0.0169 \\ 0.0094 + j0.0169 & 0.0075 - j0.0264 \end{bmatrix} \mho$$

EXAMPLE 13.13 For the hybrid equivalent circuit shown in Fig. 13.47, determine (i) h-parameters (ii) current gain (iii) voltage gain and (iv) input impedance.

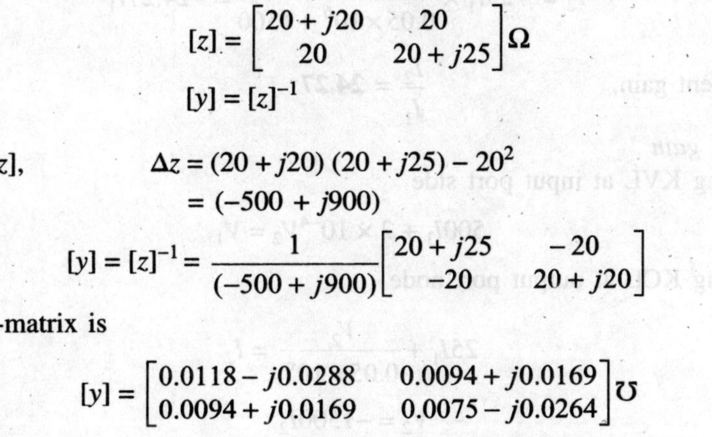

Fig. 13.47

Solution

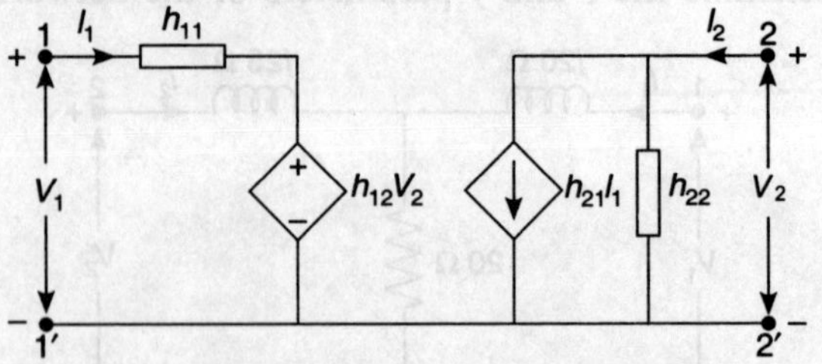

Fig. 13.48 Equivalent circuit of a two-port network in terms of *h*-parameters.

(i) *Hybrid parameters*

Comparing Figs. 13.47 and 13.48, we can write

$$h_{11} = \mathbf{500\ \Omega}$$
$$h_{12} = \mathbf{2 \times 10^{-4}}$$
$$h_{21} = \mathbf{25}$$
$$h_{22} = \frac{1}{0.05 \times 10^6} = \mathbf{20 \times 10^{-6}\ ℧}$$

(ii) *Current gain*

$$-I_2 = -25I_1 \times \frac{0.05 \times 10^6}{0.05 \times 10^6 + 1500} = -24.27I_1$$

∴ Current gain, $\quad \dfrac{I_2}{I_1} = \mathbf{24.27}$

(iii) *Voltage gain*

Applying KVL at input port side

$$500I_1 + 2 \times 10^{-4}V_2 = V_1 \tag{i}$$

Applying KCL at output port node

$$25I_1 + \frac{V_2}{0.05 \times 10^6} = I_2 \tag{ii}$$

Also, $\quad V_2 = -1500I_2 \quad$ (iii)

Substituting the value of I_2 from Eq. (iii) into Eq. (ii), we get

$$25I_1 + \frac{V_2}{0.05 \times 10^6} = -\frac{V_2}{1500}$$

or $\quad I_1 = -2.7467 \times 10^{-5}V_2 \quad$ (iv)

Substituting the value of I_1 found above into Eq. (i), we get

$$500\,(-2.7467 \times 10^{-5}V_2) + 2 \times 10^{-4}V_2 = V_1$$

From which, $\quad$ voltage gain, $\dfrac{V_2}{V_1} = \mathbf{-73.89} \quad$ (v)

(iv) *Input impedance*

From Eq. (iv),

$$I_1 = -2.7467 \times 10^{-5} V_2$$

or $$I_1 = (-2.7467 \times 10^{-5}) \times (-73.89 V_1) \quad \text{[from Eq. (v)]}$$

Hence, $$\text{input impedance, } \frac{V_1}{I_1} = \mathbf{492.72\ \Omega}$$

EXAMPLE 13.14 In the two-port network shown in Fig. 13.49, determine *h*-parameters from the following data.

(i) With the output port short-circuited: $V_1 = 20$ V, $I_1 = 1$ A, $I_2 = 2$ A

(ii) With the input port open-circuited: $V_1 = 10$ V, $V_2 = 25$ V, $I_2 = 2$ A

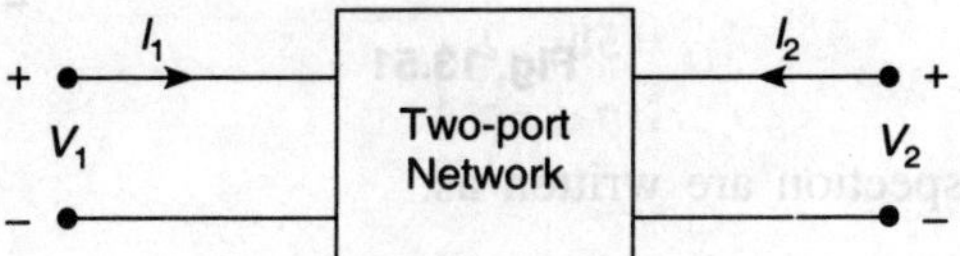

Fig. 13.49

Solution

$$h_{11} = \left.\frac{V_1}{I_1}\right|_{V_2=0} = \frac{20}{1} = 20\ \Omega$$

$$h_{21} = \left.\frac{I_2}{I_1}\right|_{V_2=0} = \frac{2}{1} = 2$$

$$h_{12} = \left.\frac{V_1}{V_2}\right|_{I_1=0} = \frac{10}{25} = 0.4$$

$$h_{22} = \left.\frac{I_2}{V_2}\right|_{I_1=0} = \frac{2}{25} = 0.08\ ℧$$

Hence, $$[h] = \begin{bmatrix} 20 & 0.4 \\ 2 & 0.08 \end{bmatrix}$$

EXAMPLE 13.15 Find the *h*-parameters for the network shown in Fig. 13.50.

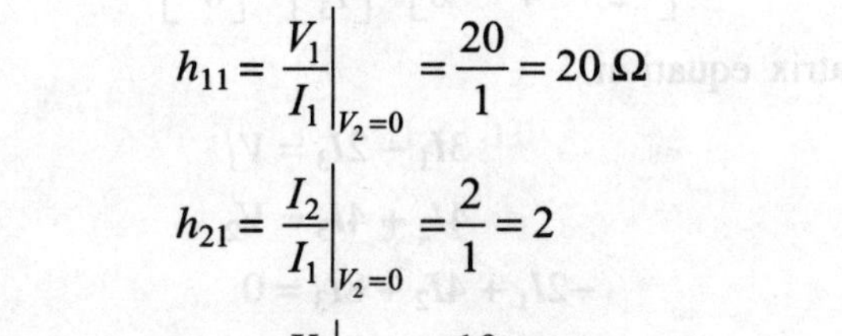

Fig. 13.50

Solution The defining equations of h-parameters are

$$V_1 = h_{11}I_1 + h_{12}V_2$$
$$I_2 = h_{21}I_1 + h_{22}V_2$$

The mesh currents are assumed in the directions as shown in Fig. 13.51.

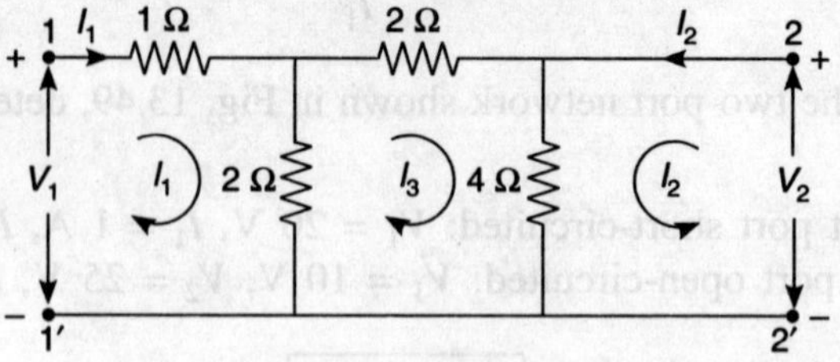

Fig. 13.51

The mesh equations by inspection are written as:

$$\begin{bmatrix} 3 & 0 & -2 \\ 0 & 4 & 4 \\ -2 & 4 & 8 \end{bmatrix} \begin{bmatrix} I_1 \\ I_2 \\ I_3 \end{bmatrix} = \begin{bmatrix} V_1 \\ V_2 \\ 0 \end{bmatrix}$$

Expanding the above matrix equation,

$$3I_1 - 2I_3 = V_1 \qquad \text{(i)}$$
$$4I_2 + 4I_3 = V_2 \qquad \text{(ii)}$$
$$-2I_1 + 4I_2 + 8I_3 = 0 \qquad \text{(iii)}$$

From Eq. (iii),

$$I_3 = \frac{2I_1 - 4I_2}{8} = 0.25I_1 - 0.5I_2 \qquad \text{(iv)}$$

Substituting the value of I_3 found above into Eqs. (i) and (ii), we get

$$V_1 = 2.5I_1 + I_2 \qquad \text{(v)}$$
$$V_2 = I_1 + 2I_2 \qquad \text{(vi)}$$

From Eq. (vi),

$$I_2 = \frac{V_2 - I_1}{2} \qquad \text{(vii)}$$

or

$$I_2 = -0.5I_1 + 0.5V_2 \qquad \text{(viii)}$$

Substituting I_2 from Eq. (viii) into Eq. (v), we obtain

$$V_1 = 2.5I_1 - 0.5I_1 + 0.5V_2$$

or

$$V_1 = 2I_1 + 0.5V_2 \qquad \text{(ix)}$$

On comparison of Eqs. (ix) and (viii) with the defining equations of h-parameters, we get

$$[h] = \begin{bmatrix} 2 & 0.5 \\ -0.5 & 0.5 \end{bmatrix}$$

EXAMPLE 13.16 Determine the h-parameters for the network shown in Fig. 13.52.

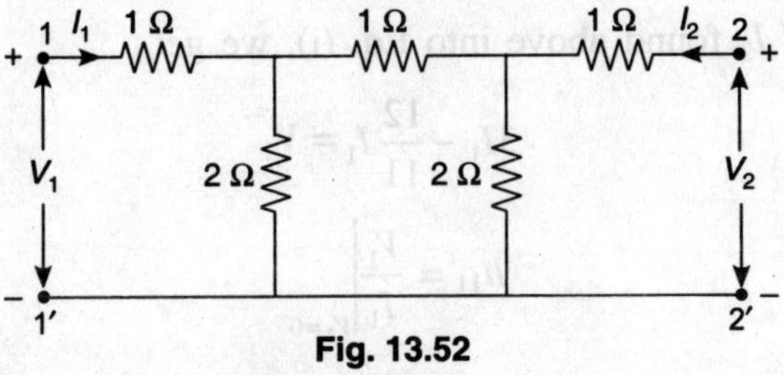

Fig. 13.52

Solution The defining equations of h-parameters are

$$V_1 = h_{11}I_1 + h_{12}V_2$$

$$I_2 = h_{21}I_1 + h_{22}V_2$$

The mesh currents are assumed in the directions as shown in Fig. 13.53.

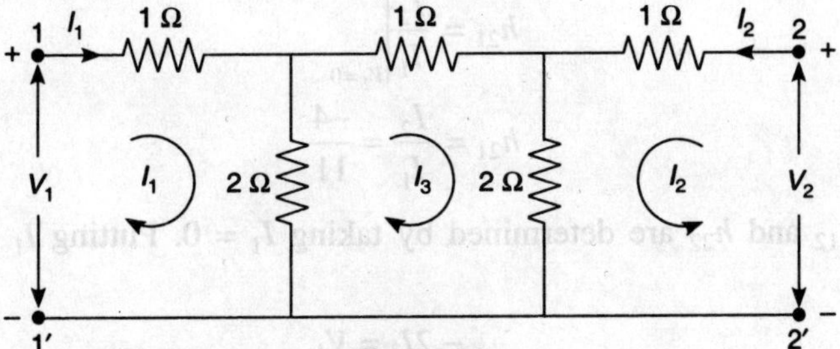

Fig. 13.53

The mesh equations by inspection are written as:

$$\begin{bmatrix} 3 & 0 & -2 \\ 0 & 3 & 2 \\ -2 & 2 & 5 \end{bmatrix} \begin{bmatrix} I_1 \\ I_2 \\ I_3 \end{bmatrix} = \begin{bmatrix} V_1 \\ V_2 \\ 0 \end{bmatrix}$$

By expanding the above matrix equation,

$$3I_1 - 2I_3 = V_1 \qquad \text{(i)}$$

$$3I_2 + 2I_3 = V_2 \qquad \text{(ii)}$$

$$-2I_1 + 2I_2 + 5I_3 = 0 \qquad \text{(iii)}$$

The h-parameters h_{11} and h_{21} are determined by taking $V_2 = 0$.

From Eq. (ii),

$$I_2 = -\frac{2}{3}I_3 \qquad [\because V_2 = 0] \qquad \text{(iv)}$$

Substitution of Eq. (iv) into Eq. (iii) gives

$$-2I_1 + \frac{11}{3}I_3 = 0$$

or $$I_3 = \frac{6}{11}I_1 \qquad \text{(v)}$$

Substituting the value of I_3 found above into Eq. (i), we get

$$3I_1 - \frac{12}{11}I_1 = V_1$$

By definition, $$h_{11} = \left.\frac{V_1}{I_1}\right|_{V_2=0}$$

Hence, $$h_{11} = \frac{V_1}{I_1} = \frac{21}{11}$$

Substitution of Eq. (v) into Eq. (iv) gives

$$I_2 = -\frac{2}{3}\left(\frac{6}{11}I_1\right) = -\frac{4}{11}I_1$$

By definition, $$h_{21} = \left.\frac{I_2}{I_1}\right|_{V_2=0}$$

So, $$h_{21} = \frac{I_2}{I_1} = \frac{-4}{11}$$

The *h*-parameters h_{12} and h_{22} are determined by taking $I_1 = 0$. Putting $I_1 = 0$ in Eqs. (i), (ii) and (iii), we get

$$-2I_3 = V_1 \qquad \text{(vi)}$$

$$3I_2 + 2I_3 = V_2 \qquad \text{(vii)}$$

$$2I_2 + 5I_3 = 0 \qquad \text{(viii)}$$

Eliminating I_3 from Eqs. (vii) and (viii), we get

$$11I_2 = 5V_2$$

By definition, $$h_{22} = \left.\frac{I_2}{V_2}\right|_{I_1=0}$$

Hence, $$h_{22} = \frac{I_2}{V_2} = \frac{5}{11}$$

Eliminating I_2 and I_3 from Eqs. (vi), (vii) and (viii), we get

$$\frac{11}{4}V_1 = V_2$$

By definition, $$h_{12} = \left.\frac{V_1}{V_2}\right|_{I_1=0}$$

Therefore, $$h_{12} = \frac{V_1}{V_2} = \frac{4}{11}$$

Thus, $$[h] = \begin{bmatrix} \frac{21}{11} & \frac{4}{11} \\ -\frac{4}{11} & \frac{5}{11} \end{bmatrix}$$

EXAMPLE 13.17 Determine *h*-parameters for the network shown in Fig. 13.54.

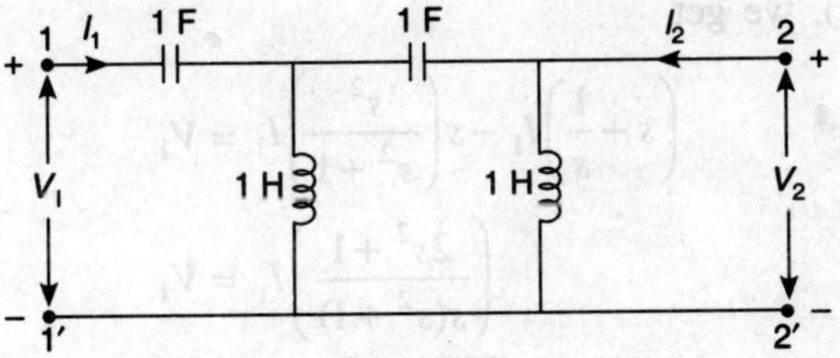

Fig. 13.54

Solution The transformed network is as shown in Fig. 13.55. Mesh currents with assumed directions are shown as well.

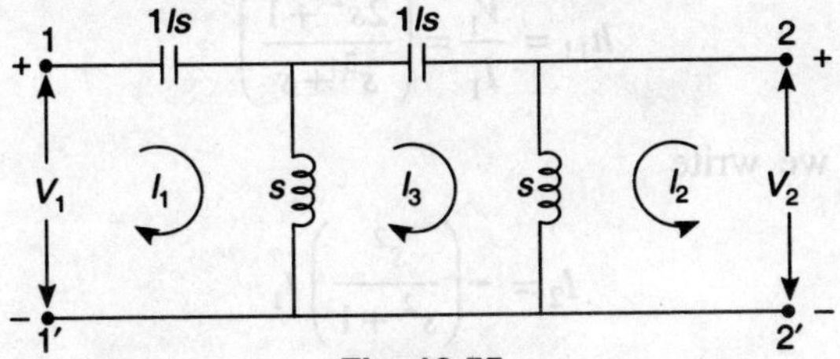

Fig. 13.55

The KVL equations by inspection are written as:

$$\begin{bmatrix} s+\frac{1}{s} & 0 & -s \\ 0 & s & s \\ -s & s & 2s+\frac{1}{s} \end{bmatrix} \begin{bmatrix} I_1 \\ I_2 \\ I_3 \end{bmatrix} = \begin{bmatrix} V_1 \\ V_2 \\ 0 \end{bmatrix}$$

The *h*-parameters h_{11} and h_{21} are determined by taking $V_2 = 0$. Expanding the above matrix equation and substituting $V_2 = 0$, we obtain

$$\left(s+\frac{1}{s}\right)I_1 - sI_3 = V_1 \qquad \text{(i)}$$

$$sI_2 + sI_3 = 0 \qquad [\because V_2 = 0] \qquad \text{(ii)}$$

$$-sI_1 + sI_2 + \left(2s + \frac{1}{s}\right)I_3 = 0 \qquad \text{(iii)}$$

From Eq. (ii),

$$I_2 = -I_3 \qquad \text{(iv)}$$

Substitution of Eq. (iv) into Eq. (iii) gives

$$-sI_1 - sI_3 + \left(2s + \frac{1}{s}\right)I_3 = 0$$

or

$$I_3 = \left(\frac{s^2}{s^2+1}\right)I_1 \qquad \text{(v)}$$

Putting Eq. (v) in Eq. (i), we get

$$\left(s + \frac{1}{s}\right)I_1 - s\left(\frac{s^2}{s^2+1}\right)I_1 = V_1$$

$$\left(\frac{2s^2+1}{s(s^2+1)}\right)I_1 = V_1$$

By definition,

$$h_{11} = \left.\frac{V_1}{I_1}\right|_{V_2=0}$$

Hence,

$$h_{11} = \frac{V_1}{I_1} = \left(\frac{2s^2+1}{s^3+s}\right)$$

Using Eqs. (iv) and (v), we write

$$I_2 = -\left(\frac{s^2}{s^2+1}\right)I_1$$

By definition,

$$h_{21} = \left.\frac{I_2}{I_1}\right|_{V_2=0}$$

So,

$$h_{21} = \frac{I_2}{I_1} = -\left(\frac{s^2}{s^2+1}\right)$$

Expanding the matrix equations and substituting $I_1 = 0$, we get

$$-sI_3 = V_1 \qquad \text{(vi)}$$

$$sI_2 + sI_3 = V_2 \qquad \text{(vii)}$$

$$sI_2 + \left(2s + \frac{1}{s}\right)I_3 = 0 \qquad \text{(viii)}$$

Eliminating I_2 from Eqs. (vii) and (viii),

$$-\left(s+\frac{1}{s}\right)I_3 = V_2 \tag{ix}$$

Substituting Eq. (vi) into Eq. (ix), we get

$$-\left(s+\frac{1}{s}\right)\left(-\frac{1}{s}\right)V_1 = V_2$$

By definition,

$$h_{12} = \left.\frac{V_1}{V_2}\right|_{I_1=0}$$

Hence,

$$h_{12} = \frac{V_1}{V_2} = \left(\frac{s^2}{s^2+1}\right)$$

Eliminating I_3 from Eqs. (vii) and (viii), we get

$$(s^2+1)I_2 = \left(2s+\frac{1}{s}\right)V_2$$

By definition,

$$h_{22} = \left.\frac{I_2}{V_2}\right|_{I_1=0}$$

So,

$$h_{22} = \frac{I_2}{V_2} = \frac{\left(\frac{2s^2+1}{s}\right)}{(s^2+1)} = \left(\frac{2s^2+1}{s(s^2+1)}\right)$$

Thus,

$$[h] = \begin{bmatrix} \frac{2s^2+1}{s^3+s} & \frac{s^2}{s^2+1} \\ -\frac{s^2}{s^2+1} & \frac{2s^2+1}{s(s^2+1)} \end{bmatrix}$$

EXAMPLE 13.18 Determine the *ABCD*-parameters for the network shown in Fig. 13.56.

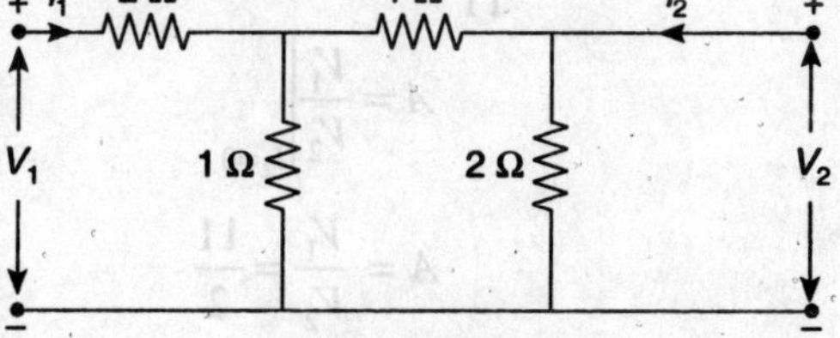

Fig. 13.56

Solution The defining equations of *ABCD*-parameters are

$$V_1 = AV_2 + B(-I_2)$$
$$I_1 = CV_2 + D(-I_2)$$

The mesh currents are assumed in the directions as shown in Fig. 13.57.

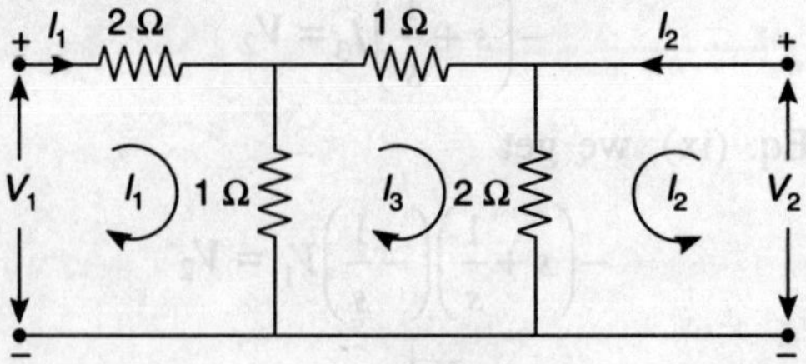

Fig. 13.57

The mesh equations by inspection are written as:

$$\begin{bmatrix} 3 & 0 & -1 \\ 0 & 2 & 2 \\ -1 & 2 & 4 \end{bmatrix} \begin{bmatrix} I_1 \\ I_2 \\ I_3 \end{bmatrix} = \begin{bmatrix} V_1 \\ V_2 \\ 0 \end{bmatrix}$$

Parameters A and C are determined by taking $I_2 = 0$. By expanding the matrix equation and substituting $I_2 = 0$, we get

$$3I_1 - I_3 = V_1 \quad \text{(i)}$$

$$2I_3 = V_2 \quad \text{(ii)}$$

$$-I_1 + 4I_3 = 0 \quad \text{(iii)}$$

From Eq. (iii),

$$I_1 = 4I_3 \quad \text{(iv)}$$

Substituting Eq. (iv) into Eq. (i), we get

$$11I_3 = V_1$$

or

$$I_3 = \frac{1}{11}V_1 \quad \text{(v)}$$

Substituting Eq. (v) into Eq. (ii), we have

$$\frac{2}{11}V_1 = V_2$$

By definition,

$$A = \left.\frac{V_1}{V_2}\right|_{I_2=0}$$

So,

$$A = \frac{V_1}{V_2} = \frac{11}{2}$$

Using Eqs. (iv) and (ii), we can write

$$\frac{1}{2}I_1 = V_2$$

By definition, $C = \left.\dfrac{I_1}{V_2}\right|_{I_2=0}$

Hence $C = \dfrac{I_1}{V_2} = 2$

Parameters B and D are determined by taking $V_2 = 0$. By expanding the above matrix equation and substituting $V_2 = 0$, we get

$$3I_1 - I_3 = V_1 \quad \text{(vi)}$$

$$2I_2 + 2I_3 = 0 \quad \text{(vii)}$$

$$-I_1 + 2I_2 + 4I_3 = 0 \quad \text{(viii)}$$

Eliminating I_3 from Eqs. (vi) and (vii),

$$6I_1 + 2I_2 = 2V_1 \quad \text{(ix)}$$

Eliminating I_3 from Eqs. (vii) and (viii),

$$I_1 + 2I_2 = 0 \quad \text{(x)}$$

Eliminating I_1 from Eqs. (ix) and (x),

$$-10I_2 = 2V_1$$

By definition, $B = -\left.\dfrac{V_1}{I_2}\right|_{V_2=0}$

So, $B = -\dfrac{V_1}{I_2} = 5$

From Eq. (x)

$$I_1 = -2I_2$$

By definition, $D = -\left.\dfrac{I_1}{I_2}\right|_{V_2=0}$

So, $D = -\dfrac{I_1}{I_2} = 2$

Thus, $[ABCD] = \begin{bmatrix} \dfrac{11}{2} & 5 \\ 2 & 2 \end{bmatrix}$

Check:

$$AD - BC = (11/2 \times 2) - (5 \times 2) = 1$$

EXAMPLE 13.19 Determine the *ABCD*-parameters for the network shown in Fig. 13.58.

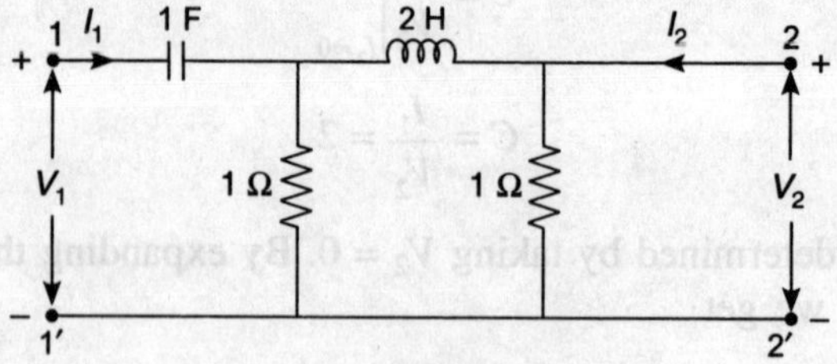

Fig. 13.58

Solution The defining equations of *ABCD*-parameters are

$$V_1 = AV_2 + B(-I_2)$$
$$I_1 = CV_2 + D(-I_2)$$

The mesh currents are assumed in the directions as shown in Fig. 13.59.

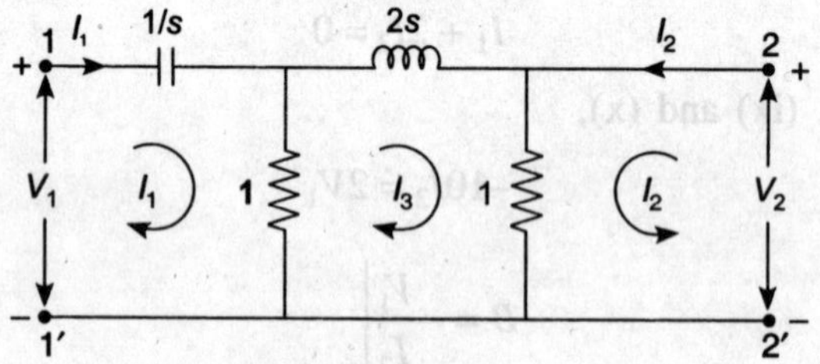

Fig. 13.59

The mesh equations by inspection are written as:

$$\begin{bmatrix} 1+\dfrac{1}{s} & 0 & -1 \\ 0 & 1 & 1 \\ -1 & 1 & 2+2s \end{bmatrix} \begin{bmatrix} I_1 \\ I_2 \\ I_3 \end{bmatrix} = \begin{bmatrix} V_1 \\ V_2 \\ 0 \end{bmatrix}$$

Parameters *A* and *C* are determined by taking $I_2 = 0$. By expanding the above matrix equation and substituting $I_2 = 0$, we get

$$\left(1+\frac{1}{s}\right)I_1 - I_3 = V_1 \quad \text{(i)}$$

$$I_3 = V_2 \quad \text{(ii)}$$

$$-I_1 + (2+2s)I_3 = 0 \quad \text{(iii)}$$

From Eqs. (ii) and (iii), we get

$$I_1 = (2+2s)V_2 \quad \text{(iv)}$$

Substituting Eqs. (ii) and (iv) into Eq. (i), we get

$$\left(1+\frac{1}{s}\right)(2+2s)V_2 - V_2 = V_1$$

$$\left(2s+\frac{2}{s}+3\right)V_2 = V_1$$

By definition,

$$A = \left.\frac{V_1}{V_2}\right|_{I_2=0}$$

Hence, $$A = \frac{V_1}{V_2} = \left(\frac{2s^2+3s+2}{s}\right)$$

By definition,

$$C = \left.\frac{I_1}{V_2}\right|_{I_2=0}$$

Therefore, from Eq. (iv), $$C = 2 + 2s$$

Parameters B and D are determined by taking $V_2 = 0$. By expanding the matrix equation and substituting $V_2 = 0$, we get

$$\left(1+\frac{1}{s}\right)I_1 - I_3 = V_1 \quad \text{(v)}$$

$$I_2 + I_3 = 0 \quad \text{(vi)}$$

$$-I_1 + I_2 + (2+2s)I_3 = 0 \quad \text{(vii)}$$

Substituting Eq. (vii) into Eq. (v), we have

$$\left(1+\frac{1}{s}\right)[I_2 + (2+2s)I_3] - I_3 = V_1 \quad \text{(viii)}$$

From Eq. (vi), $$I_2 = -I_3 \quad \text{(ix)}$$

Thus, Eq. (viii) becomes

$$I_2\left(\frac{-2s^2-2s-1}{s}\right) = V_1$$

By definition,

$$B = -\left.\frac{V_1}{I_2}\right|_{V_2=0}$$

$$\therefore \quad B = -\frac{V_1}{I_2} = \left(\frac{2s^2+2s+1}{s}\right)$$

Using Eq. (ix), Eq. (vii) is written as:

$$-I_1 + I_2 - (2 + 2s)I_2 = 0$$

or

$$I_1 = (-1 - 2s)I_2$$

By definition,

$$D = -\frac{I_1}{I_2}\bigg|_{V_2=0}$$

Hence,

$$D = -\frac{I_1}{I_2} = (2s+1)$$

Thus,

$$[ABCD] = \begin{bmatrix} \dfrac{2s^2+3s+2}{s} & \dfrac{2s^2+2s+1}{s} \\ 2+2s & 2s+1 \end{bmatrix}$$

SHORT QUESTIONS WITH ANSWERS

1. What is meant by two-port networks?

A pair of terminals at which an electrical signal enters or leaves a network is called port. A network which has two pairs of terminals is called a two-port network. One pair forms the input terminals is called input port and the other pair forms the output terminals is called output port.

2. What are the assumptions made in the analysis of two-port networks?

1. The network contains only the linear elements along with dependent sources.
2. The initial conditions on capacitor and inductor present in the network are zero.
3. The voltages and currents inside the network are not available for measurements.
4. The two currents I_1 and I_2 are entering into the network.

3. Define reciprocal network.

A network is said to be reciprocal if the ratio of excitation at one port to response at the other port is same if excitation and response are interchanged.

4. Define symmetrical network.

A network is said to be symmetrical, if the voltage-to-current ratios at both input and output ports are same with one of the ports open-circuited.

5. Write the performance equations of *z*-parameters.

$$V_1 = z_{11}I_1 + z_{12}I_2$$
$$V_2 = z_{21}I_1 + z_{22}I_2$$

6. Write the performance equations of *y*-parameters.

$$I_1 = y_{11}V_1 + y_{12}V_2$$
$$I_2 = y_{21}V_1 + y_{22}V_2$$

7. Write the performance equations of *h*-parameters.

$$V_1 = h_{11}I_1 + h_{12}V_2$$
$$I_2 = h_{21}I_1 + h_{22}V_2$$

8. Write the performance equations of *g*-parameters.

$$I_1 = g_{11}V_1 + g_{12}I_2$$
$$V_2 = g_{21}V_1 + g_{22}I_2$$

9. Write the performance equations of *ABCD*-parameters.

$$V_1 = AV_2 + B(-I_2)$$
$$I_1 = CV_2 + D(-I_2)$$

10 Write the performance equations of *A′B′C′D′*-parameters.

$$V_2 = A'V_1 + B'(-I_1)$$
$$I_2 = C'V_1 + D'(-I_1)$$

11. Write *z*-parameters in terms of *y*-parameters.

$$[z] = \begin{bmatrix} \dfrac{y_{22}}{y_{11}y_{22} - y_{12}y_{21}} & \dfrac{-y_{12}}{y_{11}y_{22} - y_{12}y_{21}} \\ \dfrac{-y_{21}}{y_{11}y_{22} - y_{12}y_{21}} & \dfrac{Y_{11}}{y_{11}y_{22} - y_{12}y_{21}} \end{bmatrix}$$

12. Write *z*-parameters in terms of *h*-parameters.

$$[z] = \begin{bmatrix} \dfrac{h_{11}h_{22} - h_{12}h_{21}}{h_{22}} & \dfrac{h_{12}}{h_{22}} \\ \dfrac{-h_{21}}{h_{22}} & \dfrac{1}{h_{22}} \end{bmatrix}$$

13. Write *z*-parameters in terms of *ABCD*-parameters.

$$[z] = \begin{bmatrix} \dfrac{A}{C} & \dfrac{AD - BC}{C} \\ \dfrac{1}{C} & \dfrac{D}{C} \end{bmatrix}$$

14. Write *y*-parameters in terms of *z*-parameters.

$$[y] = \begin{bmatrix} \dfrac{z_{22}}{z_{11}z_{22} - z_{12}z_{21}} & \dfrac{-z_{12}}{z_{11}z_{22} - z_{12}z_{21}} \\ \dfrac{-z_{21}}{z_{11}z_{22} - z_{12}z_{21}} & \dfrac{z_{11}}{z_{11}z_{22} - z_{12}z_{21}} \end{bmatrix}$$

15. Write *y*-parameters in terms of *h*-parameters.

$$[y] = \begin{bmatrix} \dfrac{1}{h_{11}} & \dfrac{-h_{12}}{h_{11}} \\ \dfrac{h_{21}}{h_{11}} & \dfrac{h_{11}h_{22} - h_{12}h_{21}}{h_{11}} \end{bmatrix}$$

16. Write *y*-parameters in terms of *ABCD*-parameters.

$$[y] = \begin{bmatrix} \dfrac{D}{B} & \dfrac{BC - AD}{B} \\ \dfrac{-1}{B} & \dfrac{A}{B} \end{bmatrix}$$

17. Write *h*-parameters in terms of *z*-parameters.

$$[h] = \begin{bmatrix} \dfrac{z_{11}z_{22} - z_{12}z_{21}}{z_{22}} & \dfrac{z_{12}}{z_{22}} \\ \dfrac{-z_{21}}{z_{22}} & \dfrac{1}{z_{22}} \end{bmatrix}$$

18. Write *h*-parameters in terms of *y*-parameters.

$$[h] = \begin{bmatrix} \dfrac{1}{y_{11}} & \dfrac{-y_{12}}{y_{11}} \\ \dfrac{y_{21}}{y_{11}} & \dfrac{y_{11}y_{22} - y_{12}y_{21}}{y_{11}} \end{bmatrix}$$

19. Write *h*-parameters in terms of *ABCD*-parameters.

$$[h] = \begin{bmatrix} \dfrac{B}{D} & \dfrac{AD - BC}{D} \\ \dfrac{-1}{D} & \dfrac{C}{D} \end{bmatrix}$$

20. Write *ABCD*-parameters in terms of *z*-parameters.

$$[ABCD] = \begin{bmatrix} \dfrac{z_{11}}{z_{21}} & \dfrac{z_{11}z_{22} - z_{12}z_{21}}{z_{21}} \\ \dfrac{1}{z_{21}} & \dfrac{z_{22}}{z_{21}} \end{bmatrix}$$

21. Write *ABCD*-parameters in terms of *y*-parameters.

$$[ABCD] = \begin{bmatrix} \dfrac{-y_{22}}{y_{21}} & \dfrac{-1}{y_{21}} \\ \dfrac{y_{12}y_{21} - y_{11}y_{22}}{y_{21}} & \dfrac{-y_{11}}{y_{21}} \end{bmatrix}$$

22. Write *ABCD*-parameters in terms of *h*-parameters.

$$[ABCD] = \begin{bmatrix} \dfrac{h_{12}h_{21} - h_{11}h_{22}}{h_{21}} & \dfrac{-h_{11}}{h_{21}} \\ \dfrac{-h_{22}}{h_{21}} & \dfrac{-1}{h_{21}} \end{bmatrix}$$

23. Define input impedance of a two-port network.

The ratio of input voltage to input current in a transformed two-port network is called input impedance.

24. Define input admittance of a two-port network.

The ratio of input current to input voltage in a transformed two-port network is called input admittance.

25. Define output impedance of a two-port network.

The ratio of output voltage to output current in a transformed two-port network is called input impedance.

26. Define output admittance of a two-port network.

The ratio of output current to output voltage in a transformed two-port network is called input admittance.

27. Define voltage gain.

The ratio of output voltage to input voltage in a network is called voltage gain.

28. Define current gain.

The ratio of output current to input current in a network is called current gain.

UNSOLVED PROBLEMS

1. Calculate the output impedance of the network shown in Fig. U13.1 with a source admittance of 1 ℧ at the input port. [Ans. 1.7143 ℧]

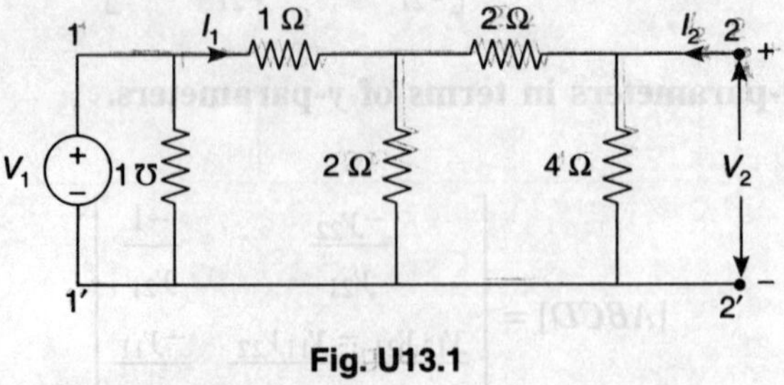

Fig. U13.1

2. Find the z-parameters for the network shown in Fig. U13.2.
[Ans. z_{11} = 17 Ω, z_{12} = 12 Ω, z_{21} = 12 Ω, z_{22} = 15 Ω]

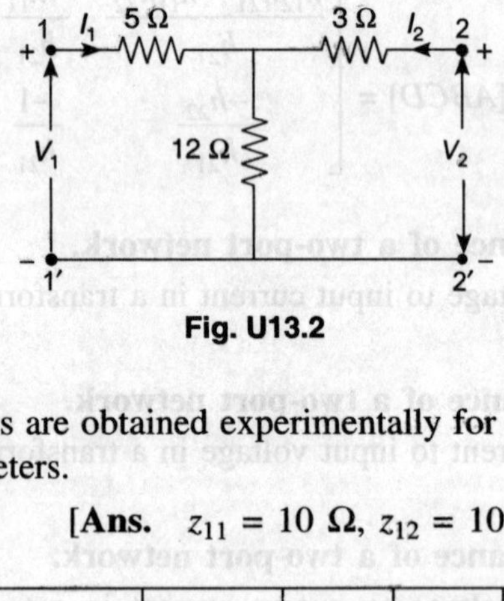

Fig. U13.2

3. The following readings are obtained experimentally for an unknown two-port network. Compute the z-parameters.

[Ans. z_{11} = 10 Ω, z_{12} = 10 Ω, z_{21} = 6 Ω, z_{22} = 13.33 Ω]

	V_1	V_2	I_1	I_2
Output open	100 V	60 V	10 A	0
Input open	30 V	40 V	0	3 A

4. The following measurements were obtained from a two-port network.
   (i) With port 2 short-circuited, a voltage of 15 V applied to port 1 results in I_1 = 2 A and I_2 = –1 A.
   (ii) With port 1 short-circuited, a voltage of 12 V applied to port 2 results in I_1 = –3 A and I_2 = 1.5 A.

   Find the y-parameters of the network.

   [Ans. y_{11} = 0.133 ℧, y_{12} = –0.25 ℧, y_{21} = –0.067 ℧, y_{22} = 0.125 ℧]

5. Find the z- and y-parameters of the following resistive network (Fig. U13.3).
[**Ans.** z_{11} = 2.375 Ω, z_{12} = 0.25 Ω, z_{21} = 0.25 Ω, z_{22} = 0.5 Ω; y_{11} = 0.444 ℧, y_{12} = –0.22 ℧, y_{21} = –0.22 ℧, y_{22} = 2.11 ℧]

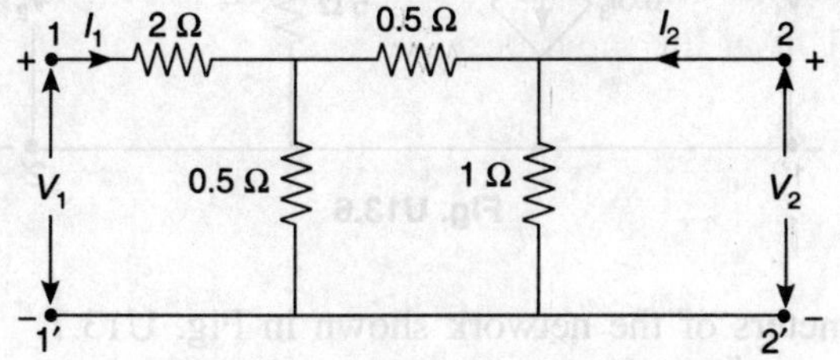

Fig. U13.3

6. For the circuit shown in Fig. U13.4, (i) find the z-parameters if the circuit is terminated by a load resistance of 300 Ω at port 2 and (ii) find input resistance.
[**Ans.** (i) z_{11} = 166.42∠33.68° Ω, z_{12} = 55.47∠33.69° Ω, z_{21} = 55.47∠33.69° Ω, z_{22} = 78.43∠78.69° Ω; (ii) 157.5∠32.52° Ω]

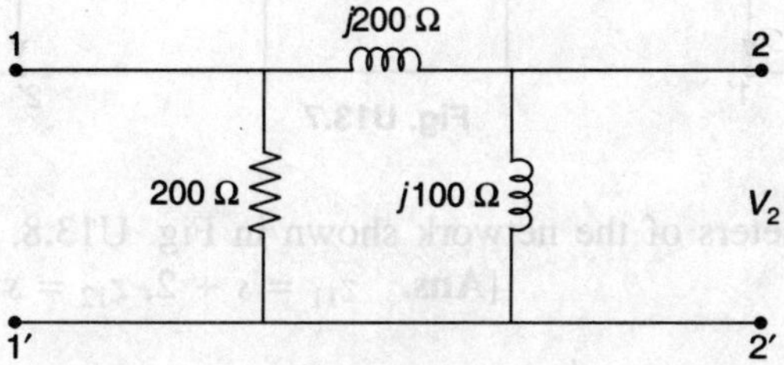

Fig. U13.4

7. Determine the z-parameters of the network shown in Fig. U13.5.
[**Ans.** z_{11} = 20 Ω, z_{12} = 5 Ω, z_{21} = 20 Ω, z_{22} = 15 Ω]

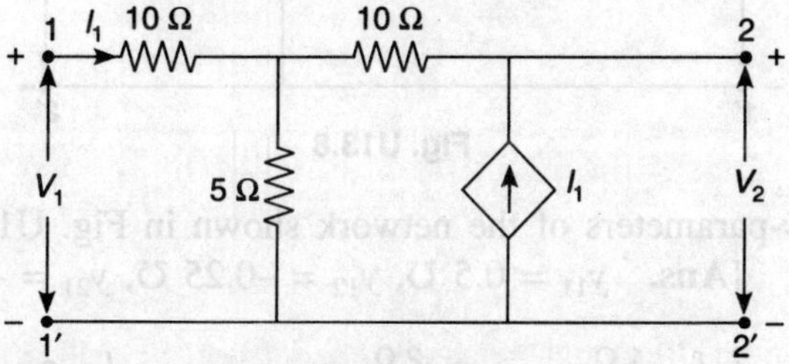

Fig. U13.5

8. Determine the z-parameters of the network shown in Fig. U13.6.
[**Ans.** z_{11} = 8.67 Ω, z_{12} = 0.67 Ω, z_{21} = 3.33 Ω, z_{22} = 3.33 Ω]

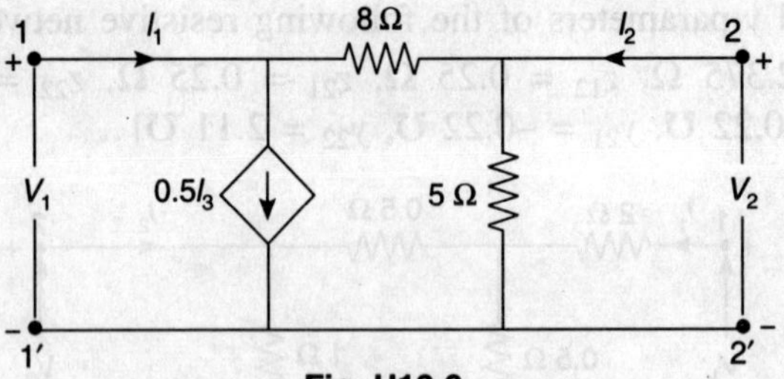

Fig. U13.6

9. Find the z-parameters of the network shown in Fig. U13.7.

$$\left[\textbf{Ans.}\quad z_{11} = \frac{s^2+3s+1}{2s+1};\ z_{12} = \frac{s^2}{2s+1};\ z_{21} = \frac{s^2}{2s+1};\ z_{22} = \frac{s^2+s}{2s+1}\right]$$

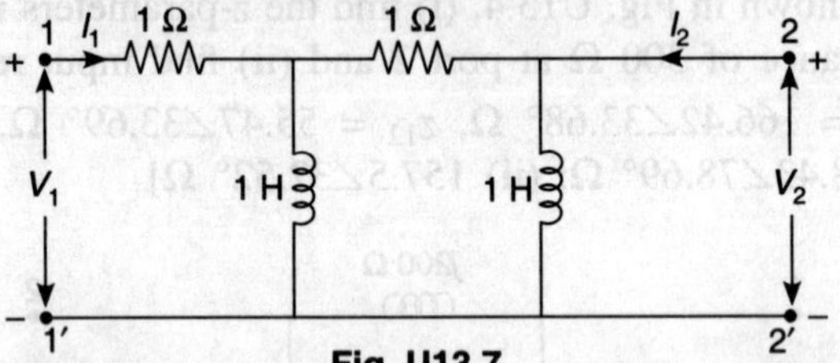

Fig. U13.7

10. Find the z-parameters of the network shown in Fig. U13.8.

[**Ans.** $z_{11} = s + 2$, $z_{12} = s - 1$, $z_{21} = s$, $z_{22} = s + 3$]

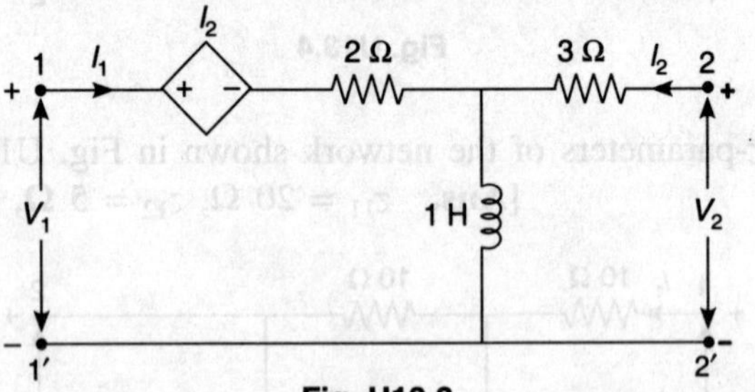

Fig. U13.8

11. Determine the y-parameters of the network shown in Fig. U13.9.

[**Ans.** $y_{11} = 0.5$ ℧, $y_{12} = -0.25$ ℧, $y_{21} = -0.25$ ℧, $y_{22} = 0.625$ ℧]

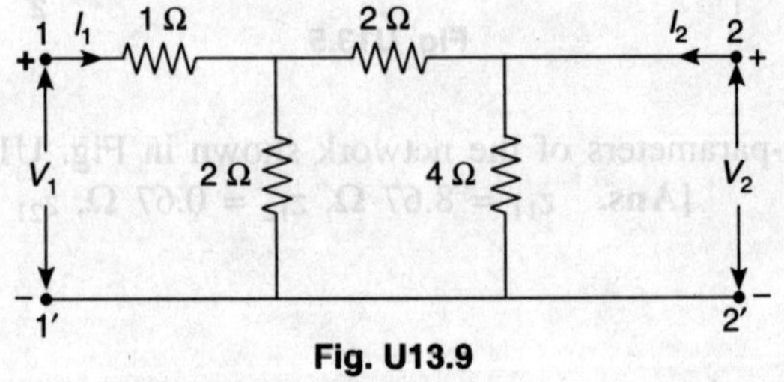

Fig. U13.9

12. Find the *y*-parameters of the network shown in Fig. U13.10.

$$\left[\textbf{Ans.}\quad y_{11} = \frac{s+1}{s+2},\ y_{12} = \frac{-1}{s+2},\ y_{21} = \frac{-1}{s+2},\ y_{22} = \frac{s^2+3s+1}{s+2}\right]$$

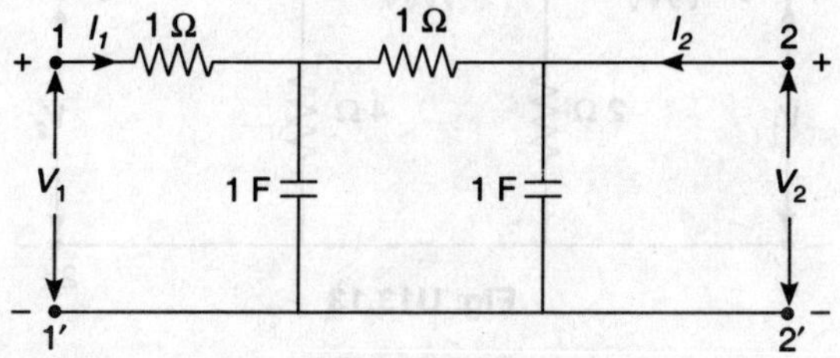

Fig. U13.10

13. Find the *y*-parameters of the network shown in Fig. U13.11.

$$\left[\textbf{Ans.}\quad y_{11} = \frac{s^2+3s+1}{s+2},\ y_{12} = \frac{-(s^2+2s+1)}{s+2},\ y_{21} = \frac{-(s^2+2s+1)}{s+2},\ y_{22} = \frac{s^2+3s+1}{s+2}\right]$$

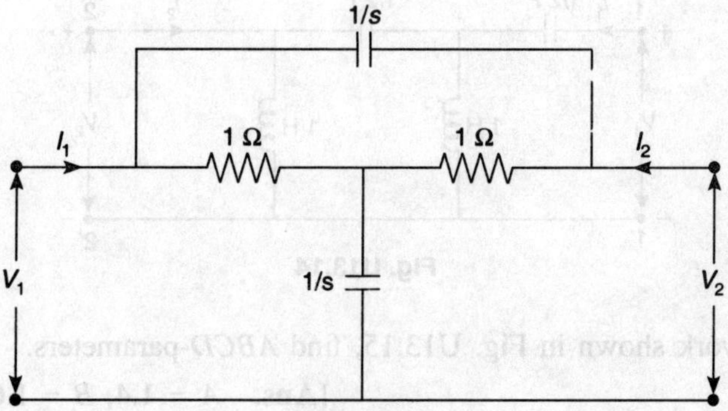

Fig. U13.11

14. Find *h*-parameters for the network shown in Fig. U13.12.

[**Ans.** h_{11} = 5.33, h_{12} = 0.33, h_{21} = 0.33, h_{22} = 0.33]

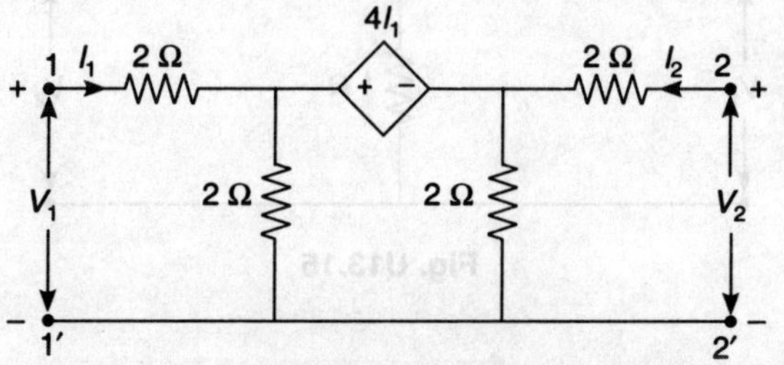

Fig. U13.12

15. Determine the h-parameters of the network shown in Fig. U13.13.

[**Ans.** $h_{11} = 2\ \Omega$, $h_{12} = 0.5\ \Omega$, $h_{21} = -0.5\ \Omega$, $h_{22} = 0.5\ \Omega$]

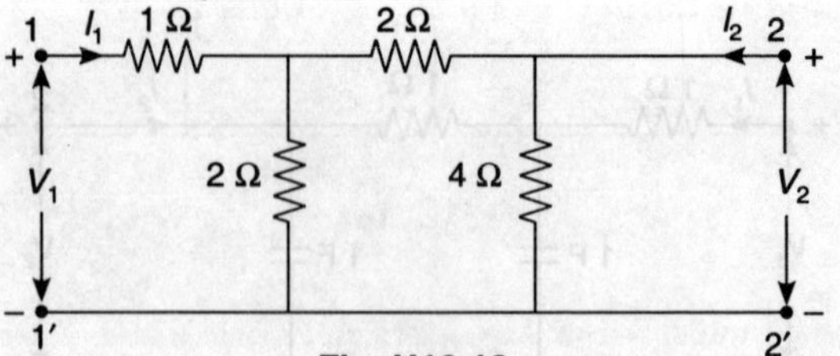

Fig. U13.13

16. Determine h-parameters for the network shown in Fig. U13.14.

$$\left[\textbf{Ans.}\quad h_{11} = \frac{4s^2+4}{s^3+2s},\ h_{12} = \frac{s^2}{s^2+2},\ h_{21} = \frac{-s^2}{s^2+2},\ h_{22} = \frac{2s^2+2}{s^3+2s}\right]$$

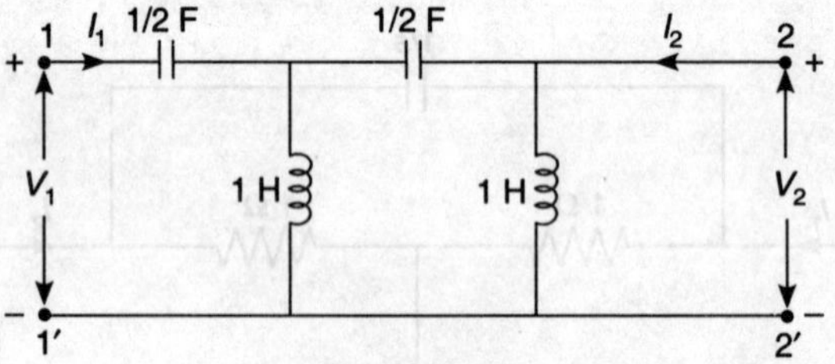

Fig. U13.14

17. For the network shown in Fig. U13.15, find $ABCD$-parameters.

[**Ans.** $A = 1.4$, $B = 1.6$, $C = 0.6$, $D = 1.4$]

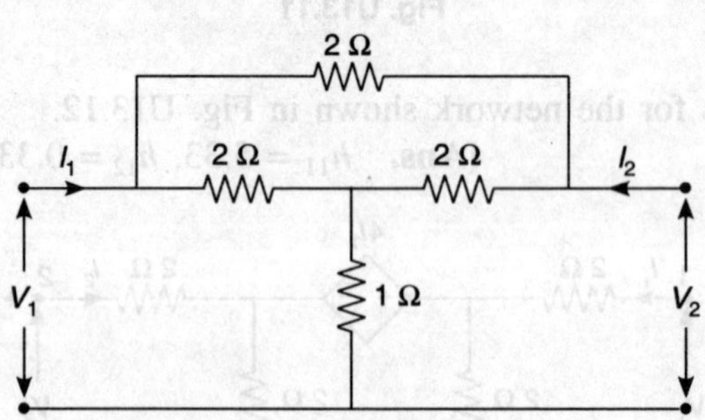

Fig. U13.15

18. Determine $ABCD$-parameters for the network shown in Fig. U13.16.
[**Ans.** $A = 8s^3 + 10s^2 + 2s + 2$; $B = 8s^2 + 2s + 2$; $C = 4s^3 + 4s^2 + 3s + 1$; $D = 4s^2 + 1$]

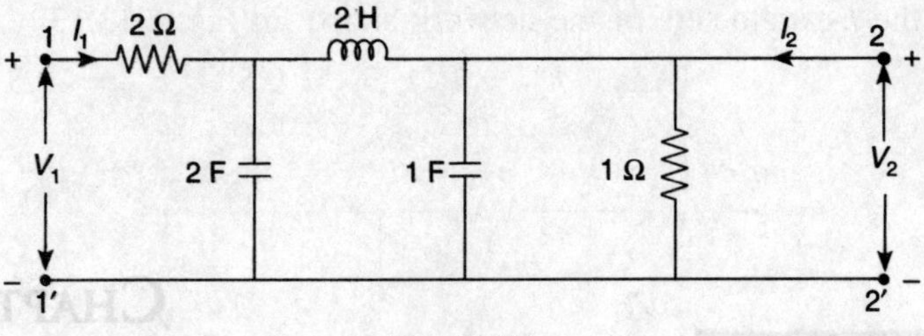

Fig. 13.16

19. The impedance parameters of a two-port network are z_{11} = 6 Ω, z_{12} = 3 Ω, z_{21} = 3 Ω, z_{22} = 4 Ω. Compute the *y*-parameters and *ABCD*-parameters.

[**Ans.** y_{11} = 6 ℧, y_{12} = 3 ℧, y_{21} = 3 ℧, y_{22} = 4 ℧; A = 2; B = 5; C = 0.3333; D = 1.3333]

20. Find the transmission parameters for the circuit shown in Fig. U13.17.

[**Ans.** A = 2; B = 2; C = 1.5; D = 2]

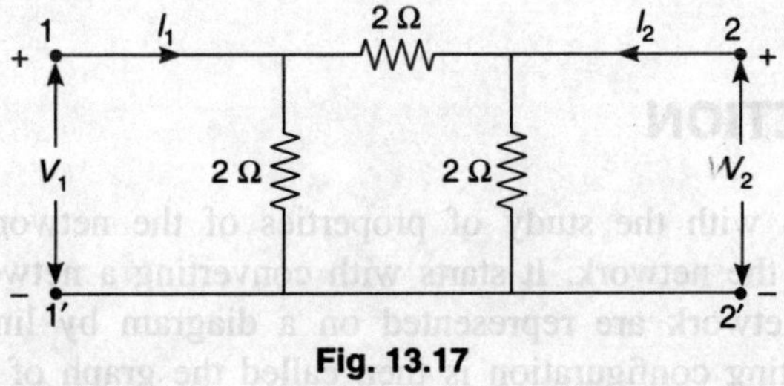

Fig. 13.17

CHAPTER 14

Network Topology

14.1 INTRODUCTION

Network topology deals with the study of properties of the network that arise out of the geometrical structure of the network. It starts with converting a network into a *graph.* When all the elements of a network are represented on a diagram by lines irrespective of their characteristics, the resulting configuration is then called the graph of the network. The graph is developed by replacing every two-terminal element by a line segment called a *branch* and each end point of the element by a dot called *node*. In constructing a graph, all sources are set as zero so that a current source becomes an open-circuit and a voltage source becomes a short-circuit. When arrows are placed on the branches of graph, it is called *oriented graph.* The arrow indicates the direction of branch current. A typical network, its graph and oriented graph are shown in Figs. 14.1 to 14.3, respectively.

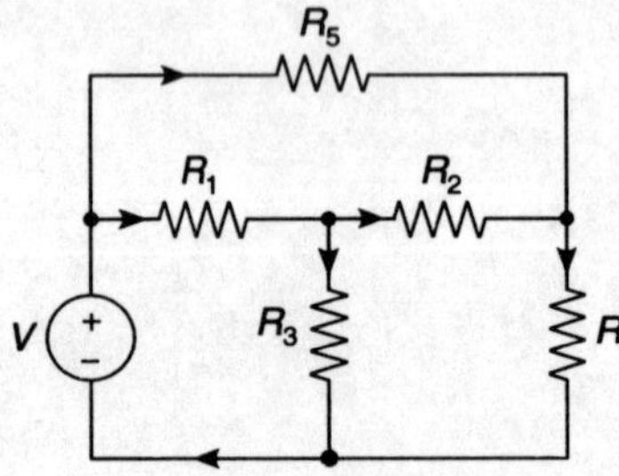

Fig. 14.1 Network.

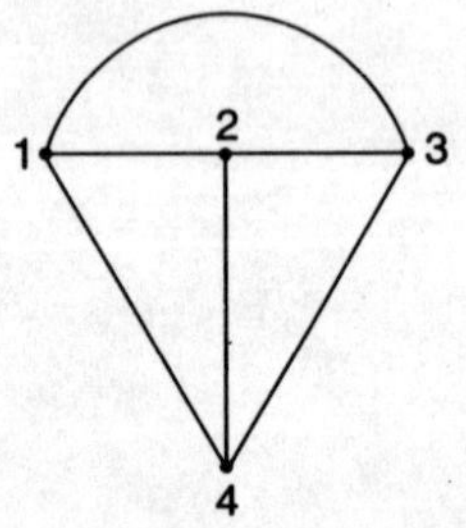

Fig. 14.2 Graph.

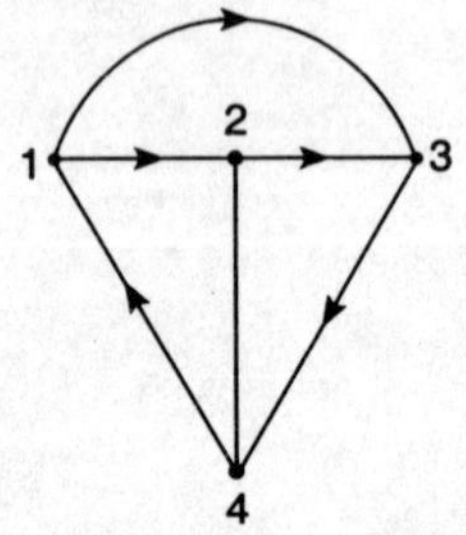

Fig. 14.3 Oriented graph.

A sequence of branches traversed while going from one node to another node is called a *path.* A graph is said to be *connected graph* if there exists at least one path from each node to every other node of the graph.

A *planar graph* is a graph drawn on a two-dimensional plane so that no two branches cross each other as shown in Fig. 14.4. A *non-planar graph* is a graph drawn on a two-dimensional plane such that two or more branches cross at a point other than a node. The graphs shown in Fig. 14.5 are non-planar graphs.

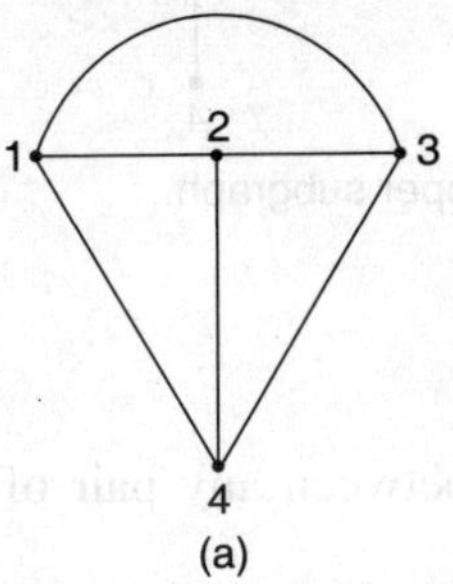

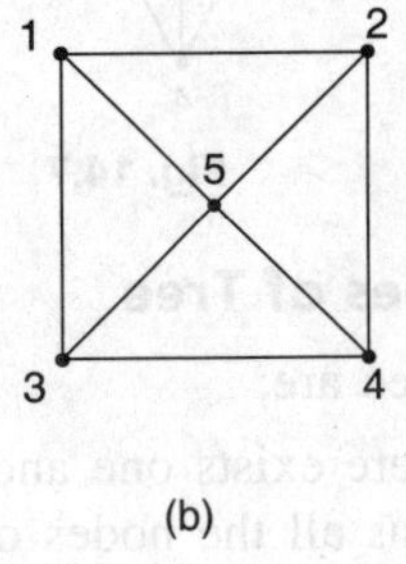

Fig. 14.4 Planar graphs.

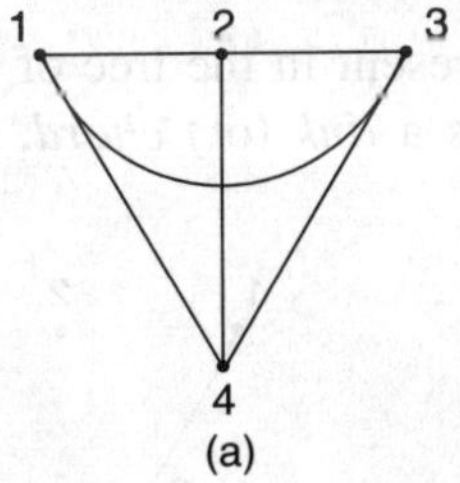

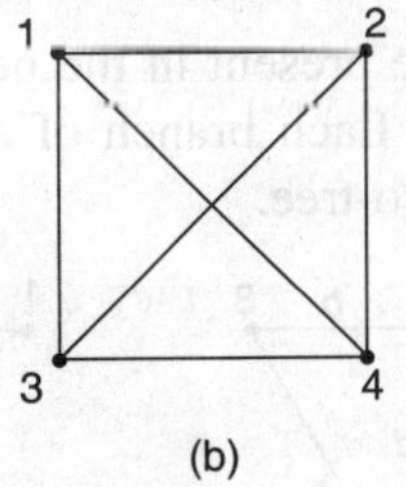

Fig. 14.5 Non-planar graphs.

When some of the branches in an original graph are removed, the resultant graph is called *subgraph.* If the subgraph contains lesser number of nodes than the original graph, it is called *proper subgraph.* On the other hand, if it contains all the nodes of the graph, it is called *improper subgraph.* Proper subgraph and improper subgraph along with the original graph are shown in Figs. 14.6 and 14.7.

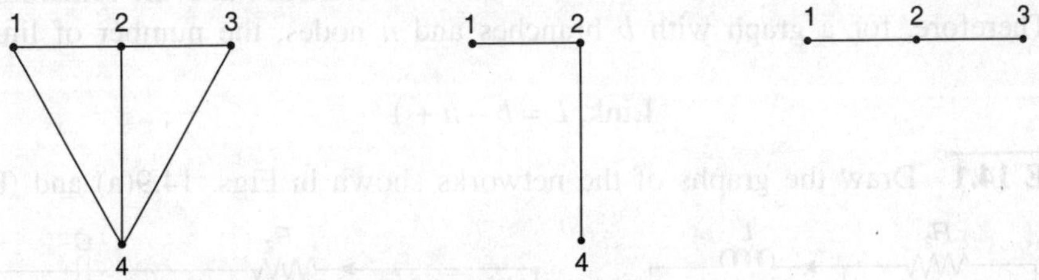

Fig. 14.6 Graph and its proper subgraphs.

The *tree* is a subgraph which is obtained by removing some branches such that the subgraph includes all the nodes of the original graph, but does not have any closed paths. For every graph, there may be more than one possible tree. Each branch of a tree is called as a *twig.*

Fig. 14.7 Graph and its improper subgraph.

14.1.1 Properties of Tree

The properties of tree are:

1. In a tree there exists one and only one path between any pair of nodes.
2. Tree contains all the nodes of graph.
3. Tree does not have a closed path or loop.
4. Every connected graph has at least one tree.
5. Each tree has (n – 1) branches, where n is the number of nodes of the graph.

All branches that are present in the network but not present in the tree of the network together constitute a *co-tree*. Each branch of a tree is called as a *link* (or) *chord*. Figure 14.8 shows a graph, its tree and co-tree.

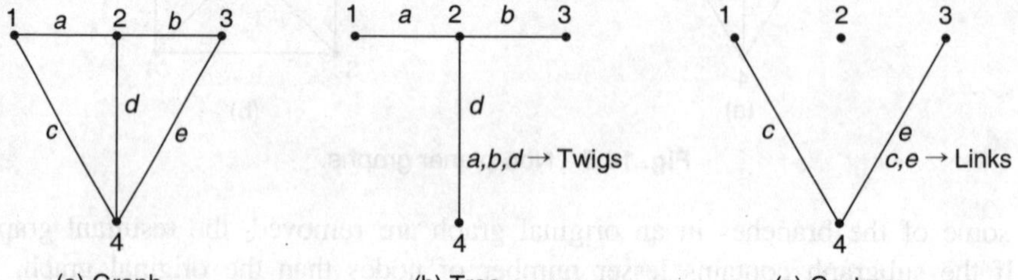

(a) Graph (b) Tree (c) Co-tree

Fig. 14.8

A definite relationship exists between number of nodes and branches in a tree. Any tree of the graph with b branches and n nodes consists of n – 1 branches and the remaining branches are links. Therefore, for a graph with b branches and n nodes, the number of links (chords) is given by

$$\text{Link, } L = b - n + 1$$

EXAMPLE 14.1 Draw the graphs of the networks shown in Figs. 14.9(a) and (b).

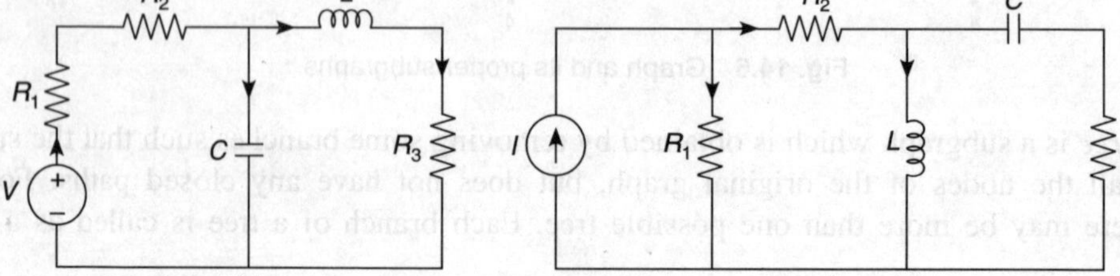

(a) (b)

Fig. 14.9

Solution Voltage source in Fig. 14.9(a) is replaced by short-circuit and the current source in Fig. 14.9(b) is replaced by open-circuit. The networks shown in Fig. 14.9(a) and (b) are converted into connected graphs and they are shown in Fig. 14.10(a) and (b), respectively.

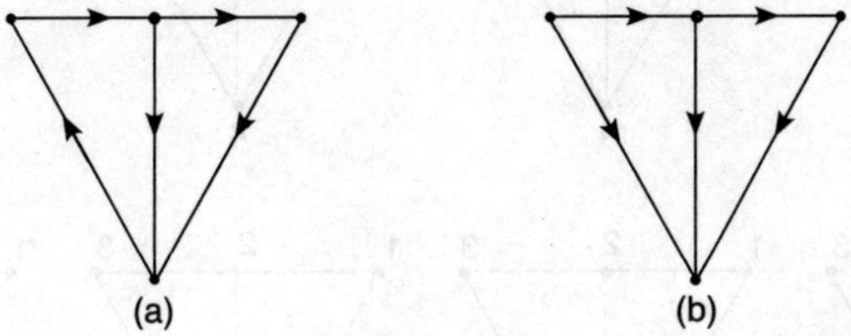

Fig. 14.10

EXAMPLE 14.2 Show the tree, twigs and links of the graph shown in Fig. 14.11.

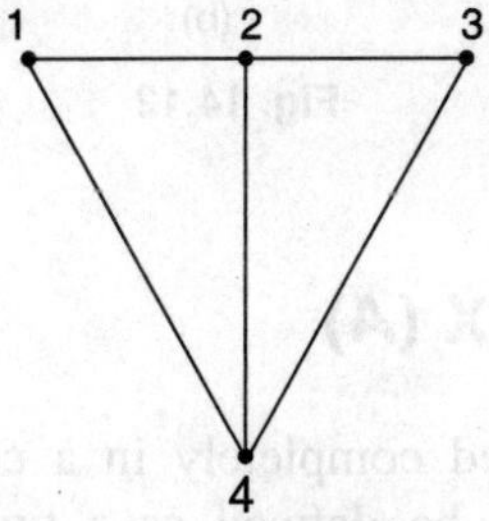

Fig. 14.11

Solution The possible trees for the given graph are shown in Fig. 14.12(a). The twigs and links are shown by bold and dotted lines respectively in Fig. 14.12(b) for each of the trees shown in Fig. 14.12(a).

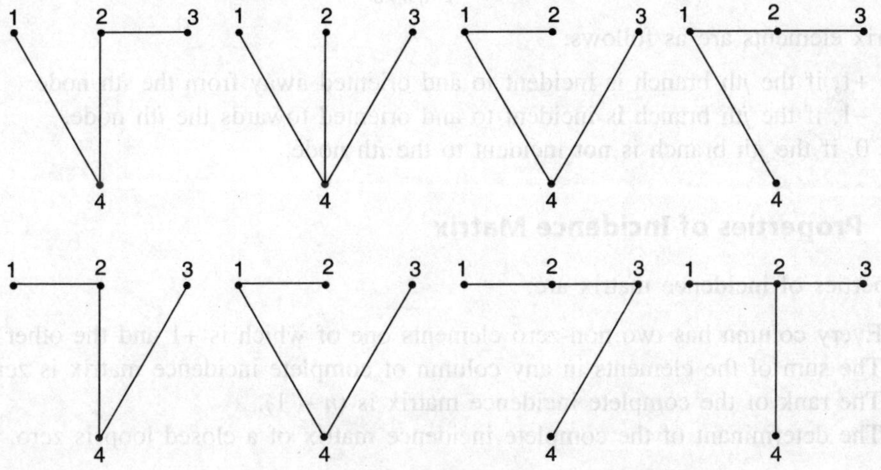

(a)

Fig. 14.12 (*Contd.*)

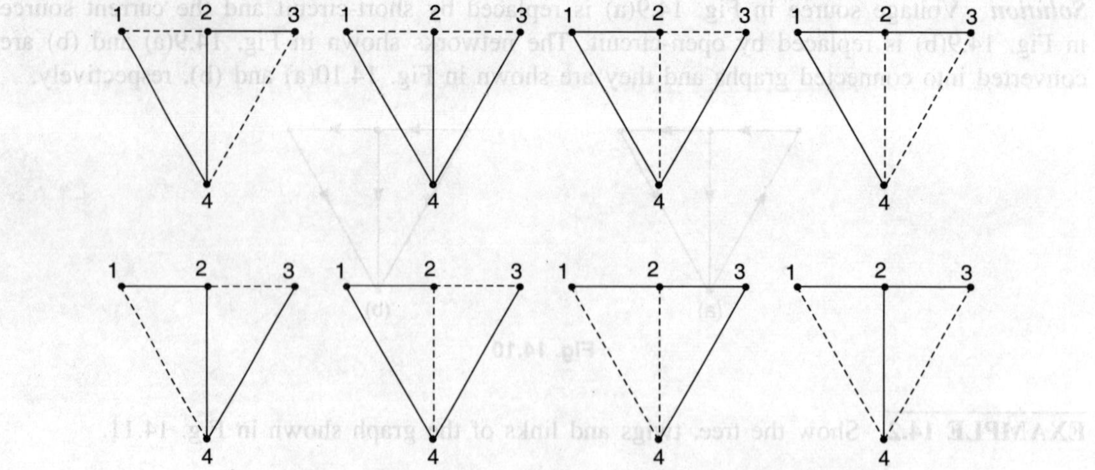

(b)

Fig. 14.12

14.2 INCIDENCE MATRIX (*A*)

An oriented graph may be described completely in a compact and convenient matrix form called *incidence matrix* (*A*). It can be defined as a two-dimensional array which provides information regarding which branches are incident at which node and what is the orientation relative to the nodes.

Let *G* be a graph with *n* nodes, *b* branches and no self loops, the complete incidence matrix is

$$A = [a_{ij}]_{n\times b}$$

The matrix elements are as follows:

$a_{ij} = +1$, if the *j*th branch is incident to and oriented away from the *i*th node.
$a_{ij} = -1$, if the *j*th branch is incident to and oriented towards the *i*th node.
$a_{ij} = 0$, if the *j*th branch is not incident to the *i*th node.

14.2.1 Properties of Incidence Matrix

The properties of incidence matrix are:

1. Every column has two non-zero elements one of which is +1 and the other is –1.
2. The sum of the elements in any column of complete incidence matrix is zero.
3. The rank of the complete incidence matrix is $(n - 1)$.
4. The determinant of the complete incidence matrix of a closed loop is zero.

EXAMPLE 14.3 Obtain the incidence matrix of the graph shown in Fig. 14.13.

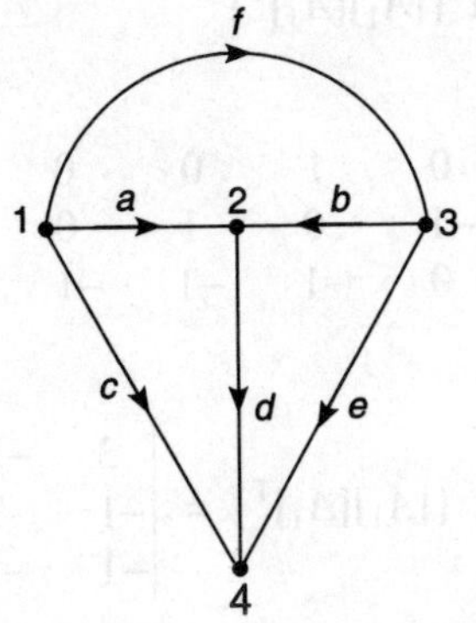

Fig. 14.13

Solution It is observed that branch a is incident to node 1 as well as node 2. It oriented away from node 1 and towards node 2. Thus, $a_{11} = 1$, but $a_{21} = -1$. Branch a is not incident to nodes 3 and 4. Hence, $a_{31} = 0$; $a_{41} = 0$. Similarly, other a_{ij} elements are found and the complete incidence matrix is shown below.

Branches

Nodes	*a*	*b*	*c*	*d*	*e*	*f*
1	1	0	1	0	0	1
2	–1	–1	0	1	0	0
3	0	1	0	0	1	–1
4	0	0	–1	–1	–1	0

$$A = \begin{bmatrix} 1 & 0 & 1 & 0 & 0 & 1 \\ -1 & -1 & 0 & 1 & 0 & 0 \\ 0 & 1 & 0 & 0 & 1 & -1 \\ 0 & 0 & -1 & -1 & -1 & 0 \end{bmatrix}$$

14.3 REDUCED INCIDENCE MATRIX (A_1)

It is possible to remove one row from A without sacrificing any information in A. When one row is deleted from the complete incidence matrix A, the resultant matrix is called a *reduced incidence matrix* A_1. Thus, for a graph when there are n nodes and b branches, the order of the incidence matrix is $[(n - 1) \times b]$.

In the above example, if the 3rd row is eliminated, then reduced incidence matrix is given as:

Branches

Nodes	*a*	*b*	*c*	*d*	*e*	*f*
1	1	0	1	0	0	1
2	–1	–1	0	1	0	0
4	0	0	–1	–1	–1	0

$$A_1 = \begin{bmatrix} 1 & 0 & 1 & 0 & 0 & 1 \\ -1 & -1 & 0 & 1 & 0 & 0 \\ 0 & 0 & -1 & -1 & -1 & 0 \end{bmatrix}$$

The number of possible trees is given by the following relation:

Number of possible trees = det $\{[A_1][A_1]^T\}$

$$[A_1][A_1]^T = \begin{bmatrix} 1 & 0 & 1 & 0 & 0 & 1 \\ -1 & -1 & 0 & 1 & 0 & 0 \\ 0 & 0 & -1 & -1 & -1 & 0 \end{bmatrix} \begin{bmatrix} 1 & -1 & 0 \\ 0 & -1 & 0 \\ 1 & 0 & -1 \\ 0 & 1 & -1 \\ 0 & 0 & -1 \\ 1 & 0 & 0 \end{bmatrix}$$

$$\text{Number of possible trees} = \det\{[A_1][A_1]^T\} = \begin{vmatrix} 3 & -1 & -1 \\ -1 & 3 & -1 \\ -1 & -1 & 3 \end{vmatrix}$$

$$= (3)\,[9 - 1] - (-1)[-3 - 1] - 1[1 + 3]$$

$$= 24 - 4 - 4 = 16$$

14.4 TIE-SET

A *tie-set* is a set of branches that form a closed path in a graph such that the closed path contains one link and remainders are twigs. The closed path is also known as loop.

A tree of a graph does not have any closed path. Upon adding a link to the tree, a closed path or loop is created. Therefore, on adding each link we get a loop and so the number of loops in a graph will be equal to number of links. The number of tie-sets will be equal to the number of links. For a graph with b branches and n nodes, the possible tie-sets are given by $(b - n + 1)$. The tie-set is also called as *fundamental circuit* or *f-circuit*. Therefore, a fundamental tie-set is a tie-set formed by one and only one link and a set of twigs.

Consider a tree of graph as shown in Fig. 14.14(a). The shown graph has six branches and four nodes.

Number of links, $L = b - n + 1 = 6 - 4 + 1 = 3$

Hence, three links are removed from the graph, and the tree thus formed is shown in Fig. 14.14(b). As shown in Fig. 14.14(b), the removed branches c, e and f are links and a, b and d are twigs. The graph has three links and so three tie-sets can be formed by involving one link in each tie-set as shown in Fig. 14.14(c), (d) and (e).

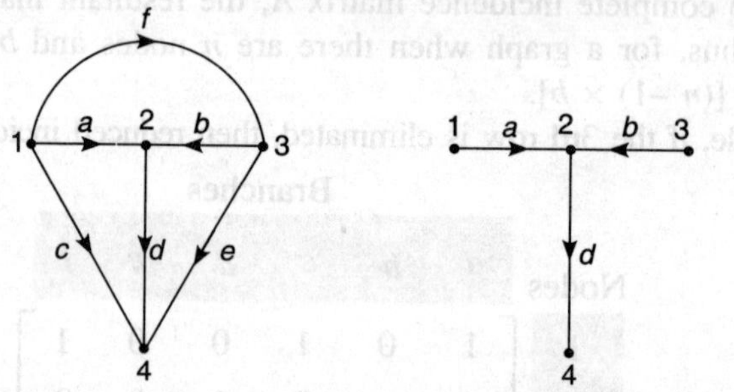

(a) Graph (b) Tree of the graph in (a)

Fig. 14.14 (*Contd.*)

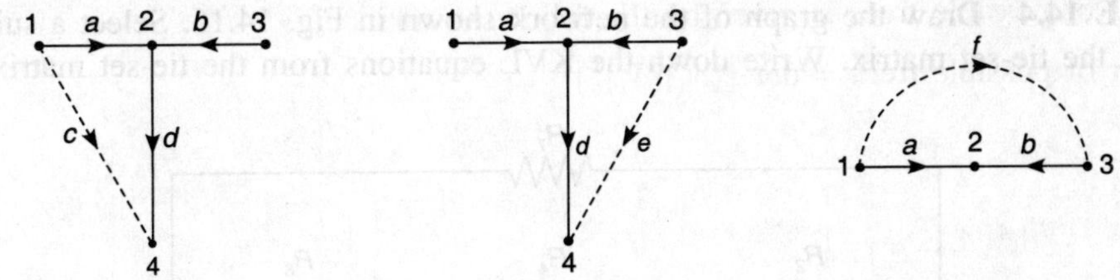

(c) Tie-set 1: [*a*, *c*, *d*] (d) Tie-set 2: [*b*, *d*, *e*] (e) Tie-set 3: [*a*, *b*, *f*]

Fig. 14.14

14.4.1 Tie-Set Matrix

Consider the graph shown in Fig. 14.15. There are three fundamental loops corresponding to the links *c*, *e* and *f*. If V_a, V_b,..., V_f are the branch voltages, the KVL equations for the three tie-sets can be written as:

$$\left.\begin{aligned} V_a + V_d - V_c &= 0 \\ z - V_b + V_e - V_d &= 0 \\ V_f - V_a + V_b &= 0 \end{aligned}\right\}$$

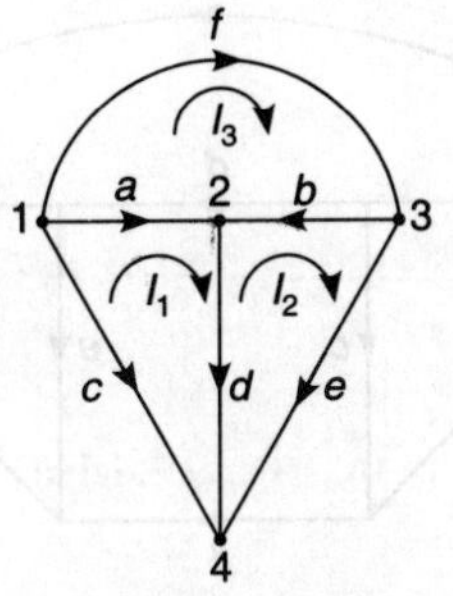

Fig. 14.15 Graph.

The above equation can be written in matrix form as:

Branches

$$\begin{array}{cccccc} a & b & c & d & e & f \end{array}$$

$$\begin{bmatrix} 1 & 0 & -1 & 1 & 0 & 0 \\ 0 & -1 & 0 & -1 & 1 & 0 \\ -1 & 1 & 0 & 0 & 0 & 1 \end{bmatrix} \begin{bmatrix} V_a \\ V_b \\ V_c \\ V_d \\ V_e \\ V_f \end{bmatrix} = \begin{bmatrix} 0 \\ 0 \\ 0 \end{bmatrix} \quad (14.1)$$

$$\boldsymbol{B}\boldsymbol{V}_{\text{branch}} = 0 \quad (14.2)$$

where B is a $L \times b$ matrix, called tie-set matrix or fundamental loop matrix and V_{branch} is a column vector of branch voltages.

EXAMPLE 14.4 Draw the graph of the network shown in Fig. 14.16. Select a suitable tree and obtain the tie-set matrix. Write down the KVL equations from the tie-set matrix.

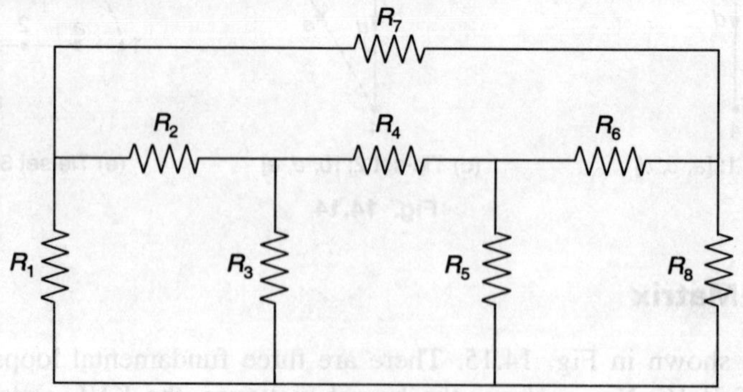

Fig. 14.16

Solution The graph of the given network is shown in Fig. 14.17. The orientations are chosen arbitrarily.

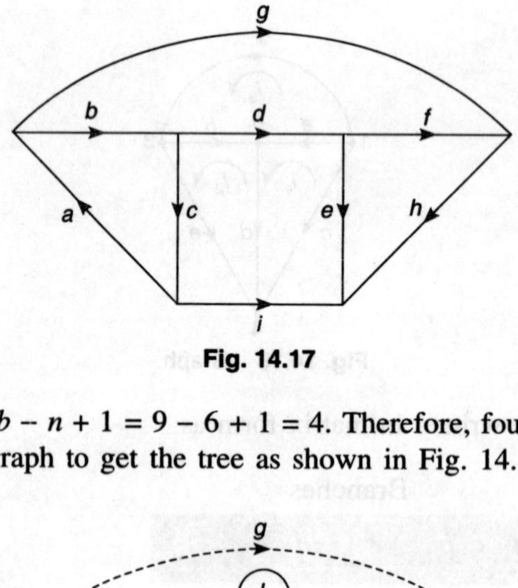

Fig. 14.17

Number of links, $L = b - n + 1 = 9 - 6 + 1 = 4$. Therefore, four links have been removed (shown dotted) from the graph to get the tree as shown in Fig. 14.18.

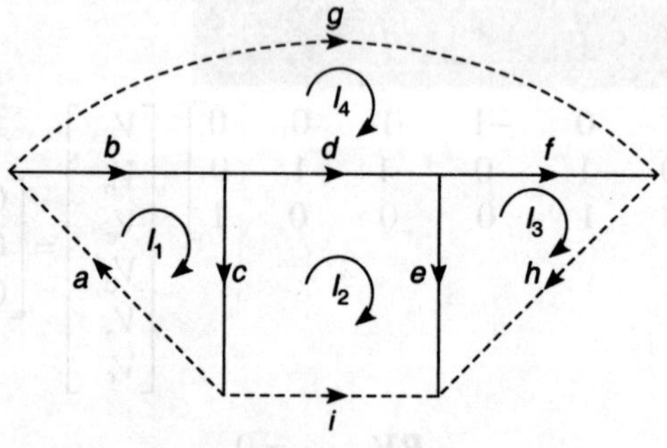

Fig. 14.18

The tie-set matrix formed is shown below:

Loop currents	Branches a	b	c	d	e	f	g	h	i
I_1	1	1	1	0	0	0	0	0	0
I_2	0	0	–1	1	1	0	0	0	1
I_3	0	0	0	0	–1	1	0	1	0
I_4	0	–1	0	–1	0	–1	1	0	0

From the tie-set matrix, the KVL equations are written as follows:

$$V_a + V_b + V_c = 0$$
$$-V_c + V_d + V_e + V_i = 0$$
$$-V_e + V_f + V_h = 0$$
$$-V_b - V_d - V_f + V_g = 0$$

where V_a, V_b, V_c,..., V_i are corresponding branch voltages.

14.4.2 Equilibrium Equations Using Tie-Sets

If I_B and I_L represent the branch current matrix and loop current matrix, respectively and B is the tie-set matrix, then

$$[I_B] = [B^T]\,[I_L] \tag{14.3}$$

where $[B^T]$ is the transpose of the matrix $[B]$. Equation (14.3) is known as *link current transformation equation.*

Consider the tie-set matrix in Fig. 14.15.

$$B = \begin{bmatrix} 1 & 0 & -1 & 1 & 0 & 0 \\ 0 & -1 & 0 & -1 & 1 & 0 \\ -1 & 1 & 0 & 0 & 0 & 1 \end{bmatrix}$$

The branch current vector $[I_B]$ is a column vector.

$$I_B = \begin{bmatrix} I_a \\ I_b \\ I_c \\ I_d \\ I_e \\ I_f \end{bmatrix}$$

The loop current vector $[I_L]$ is a column vector.

$$I_L = \begin{bmatrix} I_1 \\ I_2 \\ I_3 \end{bmatrix}$$

From Eq. (14.3), the link current information is given by $[I_B] = [B^T][I_L]$, i.e.

$$\begin{bmatrix} I_a \\ I_b \\ I_c \\ I_d \\ I_e \\ I_f \end{bmatrix} = \begin{bmatrix} 1 & 0 & -1 \\ 0 & -1 & 1 \\ -1 & 0 & 0 \\ 1 & -1 & 0 \\ 0 & 1 & 0 \\ 0 & 0 & 1 \end{bmatrix} \begin{bmatrix} I_1 \\ I_2 \\ I_3 \end{bmatrix}$$

The branch currents are

$$\left.\begin{aligned} I_a &= I_1 - I_3 \\ I_b &= -I_2 + I_3 \\ I_c &= -I_1 \\ I_d &= I_1 - I_2 \\ I_e &= I_2 \\ I_f &= I_3 \end{aligned}\right\} \tag{14.4}$$

Multiplying Eq. (14.1), we get

$$\left.\begin{aligned} V_a - V_c + V_d &= 0 \\ -V_b - V_d + V_e &= 0 \\ -V_a + V_b + V_f &= 0 \end{aligned}\right\} \tag{14.5}$$

By Ohm's law, the branch voltage can be expressed as a product of current and resistance (or impedance) of the branch.

Therefore, the branch voltages can be expressed as:

$$V_a = I_aR_a; \quad V_b = I_bR_b; \quad V_c = I_cR_c; \quad V_d = I_dR_d; \quad V_e = I_eR_e; \quad V_f = I_fR_f$$

where R_a, R_b, R_c, R_d, R_e and R_f are resistances of the branches *a*, *b*, *c*, *d*, *e* and *f*, respectively.

Now, Eq. (14.5) can be written as:

$$\left.\begin{aligned} I_aR_a - I_cR_c + I_dR_d &= 0 \\ -I_bR_b - I_dR_d + I_eR_e &= 0 \\ -I_aR_a + I_bR_b + I_fR_f &= 0 \end{aligned}\right\} \tag{14.6}$$

Using Eq. (14.4), Eq. (14.6) can be written as:

$$(I_1 - I_3)R_a + I_1R_c + (I_1 - I_2)R_d = 0$$
$$(I_2 - I_3)R_b + (-I_1 + I_2)R_d + I_2R_e = 0$$
$$(-I_1 + I_3)R_a + (-I_2 + I_3)R_b + I_3R_f = 0$$

Rearranging,

$$I_1(R_a + R_c + R_d) + I_2(-R_d) + I_3(-R_a) = 0$$
$$I_1(-R_d) + I_2(R_b + R_d + R_e) + I_3(-R_b) = 0$$
$$I_1(-R_a) + I_2(-R_b) + I_3(R_a + R_b + R_f) = 0$$

$$\begin{bmatrix} R_a + R_c + R_d & -R_d & -R_a \\ -R_d & R_b + R_d + R_e & -R_b \\ -R_a & -R_b & R_a + R_b + R_f \end{bmatrix} \begin{bmatrix} I_1 \\ I_2 \\ I_3 \end{bmatrix} = \begin{bmatrix} 0 \\ 0 \\ 0 \end{bmatrix} \tag{14.7}$$

$$\boldsymbol{RI}_{\mathrm{L}} = \boldsymbol{V}$$

where V is voltage source matrix.

Matrix R formed by branch resistances in Eq. (14.7) is called *parameter matrix.*

EXAMPLE 14.5 Determine the branch currents of the circuit shown in Fig. 14.19 by using tie-set schedule.

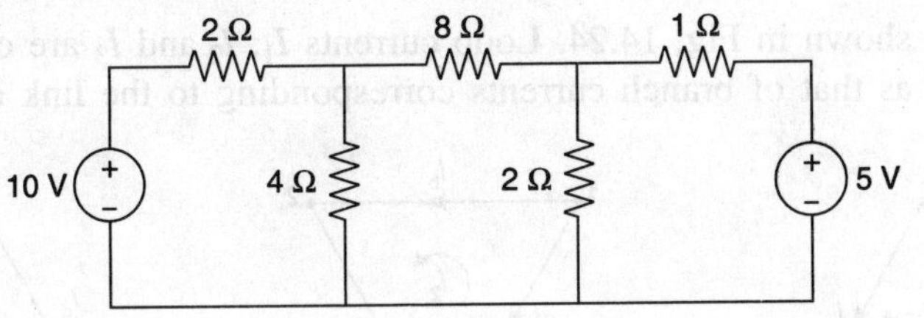

Fig. 14.19

Solution The voltage sources are short-circuited and the given circuit is redrawn as shown in Fig. 14.20.

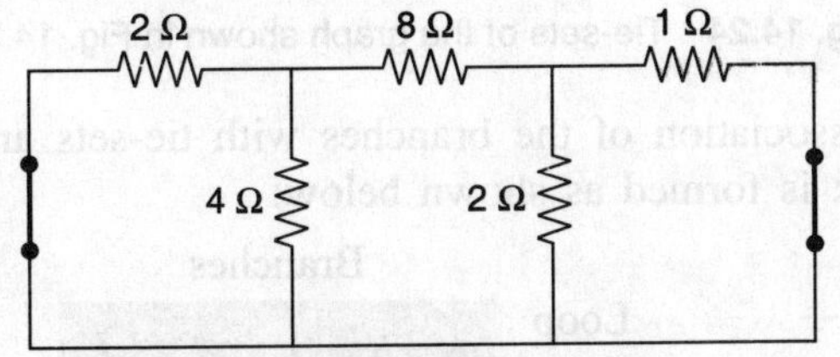

Fig. 14.20

The graph of the network is drawn by representing the nodes as dots and the elements connected between the nodes by lines as shown in Fig. 14.21. Arrows are placed on the branches arbitrarily and the oriented graph is obtained as shown in Fig. 14.22. The directions of arrow heads indicate the direction of currents attached to each branch.

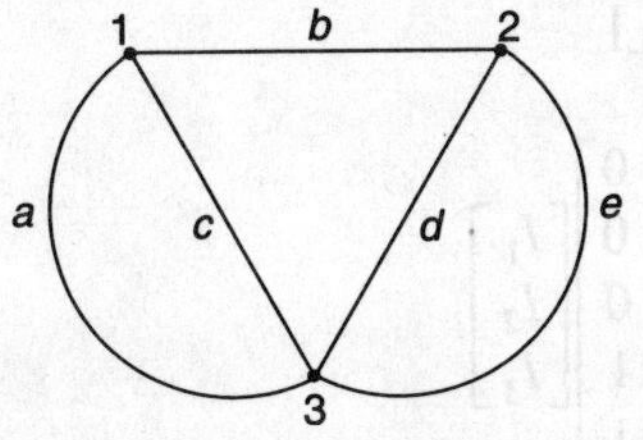

Fig. 14.21 Graph.

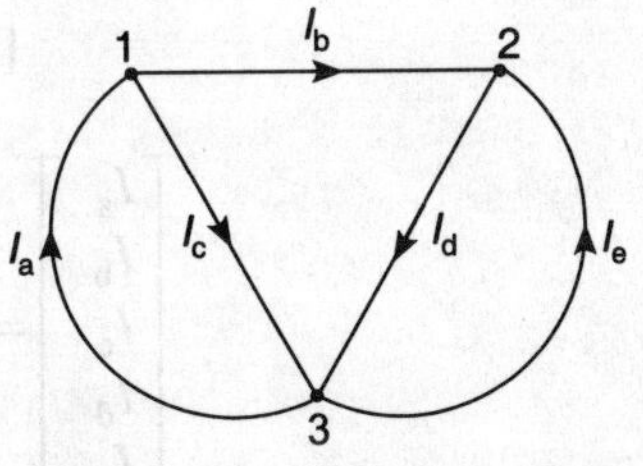

Fig. 14.22 Oriented graph.

By removing three branches (number of links, $L = b - n + 1 = 5 - 3 + 1 = 3$), a tree of the graph is formed as shown in Fig. 14.23 by removing branches a, b and e.

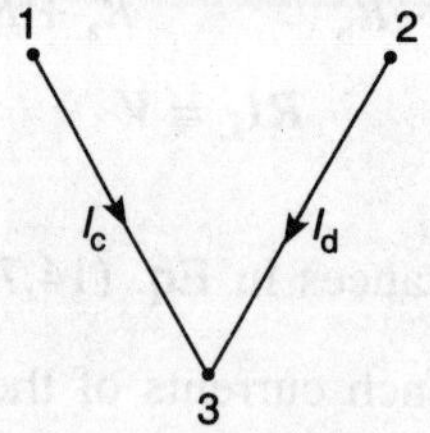

Fig. 14.23 Tree.

The tie-sets are shown in Fig. 14.24. Loop currents I_1, I_2 and I_3 are chosen such that their directions are same as that of branch currents corresponding to the link associated with.

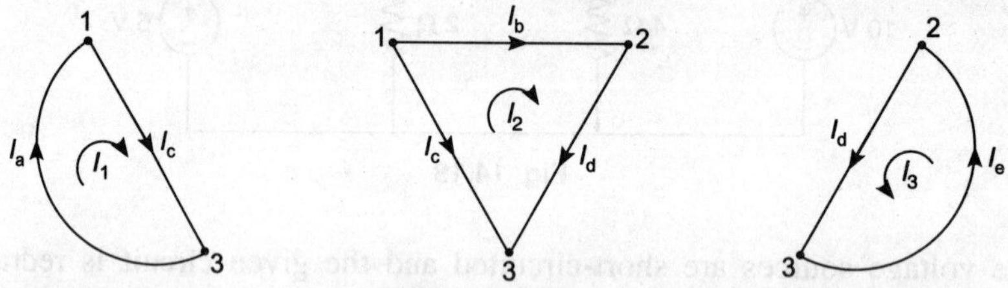

Fig. 14.24 Tie-sets of the graph shown in Fig. 14.21.

By considering the association of the branches with tie-sets and the direction of branch currents, the tie-set matrix is formed as shown below:

Loop currents	a	b	c	d	e
1	1	0	1	0	0
2	0	1	−1	1	0
3	0	0	0	1	1

(Branches)

$$B = \begin{bmatrix} 1 & 0 & 1 & 0 & 0 \\ 0 & 1 & -1 & 1 & 0 \\ 0 & 0 & 0 & 1 & 1 \end{bmatrix} \quad \text{(i)}$$

According to link current transformation equation,

$$[I_B] = [B^T]\,[I_L]$$

Therefore,

$$\begin{bmatrix} I_a \\ I_b \\ I_c \\ I_d \\ I_e \end{bmatrix} = \begin{bmatrix} 1 & 0 & 0 \\ 0 & 1 & 0 \\ 1 & -1 & 0 \\ 0 & 1 & 1 \\ 0 & 0 & 1 \end{bmatrix} \begin{bmatrix} I_1 \\ I_2 \\ I_3 \end{bmatrix} \quad \text{(ii)}$$

On multiplying Eq. (ii),

$$\begin{bmatrix} I_a \\ I_b \\ I_c \\ I_d \\ I_e \end{bmatrix} = \begin{bmatrix} I_1 \\ I_2 \\ I_1 - I_2 \\ I_2 + I_3 \\ I_3 \end{bmatrix} \quad \text{(iii)}$$

According to Eq. (14.2),

$$\boldsymbol{BV}_{\textbf{branch}} = 0$$

Hence,

$$\begin{bmatrix} 1 & 0 & 1 & 0 & 0 \\ 0 & 1 & -1 & 1 & 0 \\ 0 & 0 & 0 & 1 & 1 \end{bmatrix} \begin{bmatrix} V_a \\ V_b \\ V_c \\ V_d \\ V_e \end{bmatrix} = 0 \quad \text{(iv)}$$

On multiplying Eq. (iv),

$$\begin{bmatrix} V_a + V_c \\ V_b - V_c + V_d \\ V_d + V_e \end{bmatrix} = 0 \quad \text{(v)}$$

From Eq. (v), we can write

$$\begin{aligned} V_a + V_c &= 0 \\ V_b - V_c + V_d &= 0 \\ V_d + V_e &= 0 \end{aligned} \quad \text{(vi)}$$

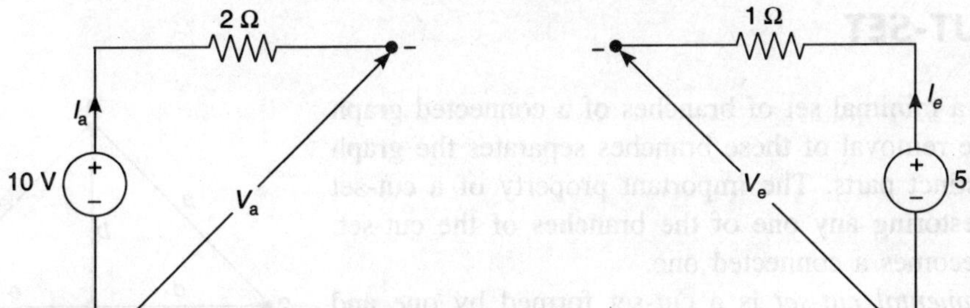

Fig. 14.25

The branches *a* and *e* are having voltage sources. By applying KVL to the circuits shown in Fig. 14.25, we can write

$$\begin{aligned} V_a &= -10 + 2I_a \\ V_e &= -5 + I_e \end{aligned} \quad \text{(vii)}$$

The branches that do not possess voltage sources are *b*, *c* and *d*. Their respective voltages V_b, V_c and V_d are expressed as a product of branch current and resistance of the branch as follows:

$$\begin{aligned} V_b &= I_b \times 8 \\ V_c &= I_c \times 4 \\ V_d &= I_d \times 2 \end{aligned} \qquad \text{(viii)}$$

Using Eqs. (vii) and (viii), Eq. (vi) can be written as:

$$\begin{aligned} -10 + 2I_a + 4I_c &= 0 \\ 8I_b - 4I_c + 2I_d &= 0 \\ 2I_d - 5 + I_e &= 0 \end{aligned} \qquad \text{(ix)}$$

Using Eq. (iii), the above equation is written as:

$$\begin{aligned} -10 + 2I_1 + 4(I_1 - I_2) &= 0 \\ 8I_2 - 4(I_1 - I_2) + 2(I_2 + I_3) &= 0 \\ 2(I_2 + I_3) - 5 + I_3 &= 0 \end{aligned} \qquad \text{(x)}$$

Equation (x) is simplified as:

$$\begin{aligned} 6I_1 - 4I_2 &= 10 \\ -4I_1 + 14I_2 + 2I_3 &= 0 \\ 2I_2 + 3I_3 &= 5 \end{aligned} \qquad \text{(xi)}$$

Solving Eq. (xi), we get

$$\begin{aligned} I_1 &= 1.8889 \text{ A} \\ I_2 &= 0.3333 \text{ A} \\ I_3 &= 1.4444 \text{ A} \end{aligned}$$

14.5 CUT-SET

A *cut-set* is a minimal set of branches of a connected graph such that the removal of these branches separates the graph into two distinct parts. The important property of a cut-set is that by restoring any one of the branches of the cut-set, the graph becomes a connected one.

A *fundamental cut-set* is a cut-set formed by one and only one twig and a set of links. Thus, in a graph, for each twig of a tree, there would be a fundamental cut-set. For a graph with *n* nodes, there will be $(n - 1)$ number of fundamental cut-sets.

Consider the graph shown in Fig. 14.26. Branches and nodes are marked respectively using alphabets and numerals as shown in Fig. 14.26.

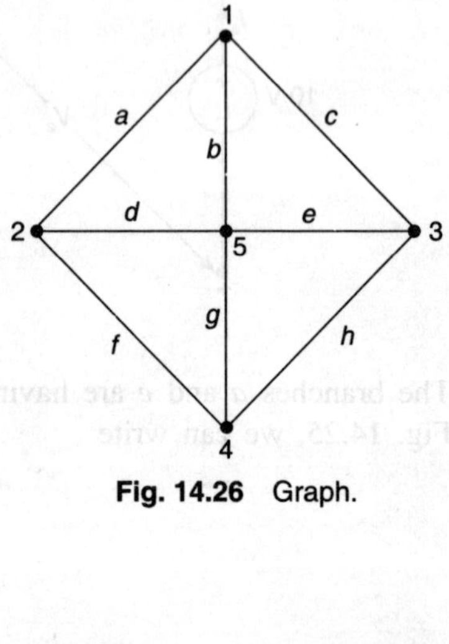

Fig. 14.26 Graph.

Number of links, $L = b - n + 1$

$$= 8 - 5 + 1 = 4$$

Therefore, the four branches, namely *a*, *c*, *f* and *h* may be removed and the resulting tree is shown in Fig. 14.27.

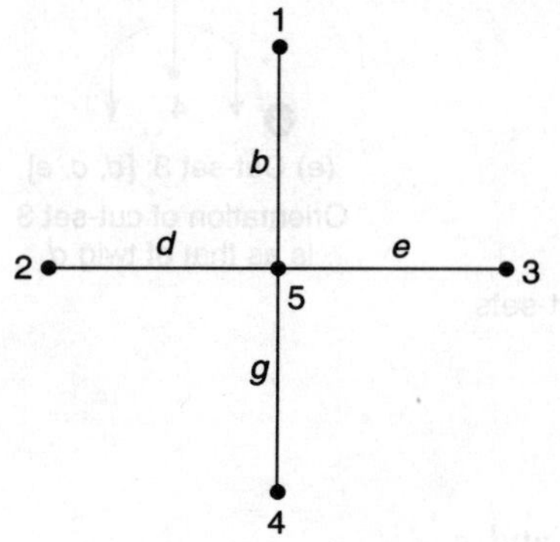

Fig. 14.27 Tree.

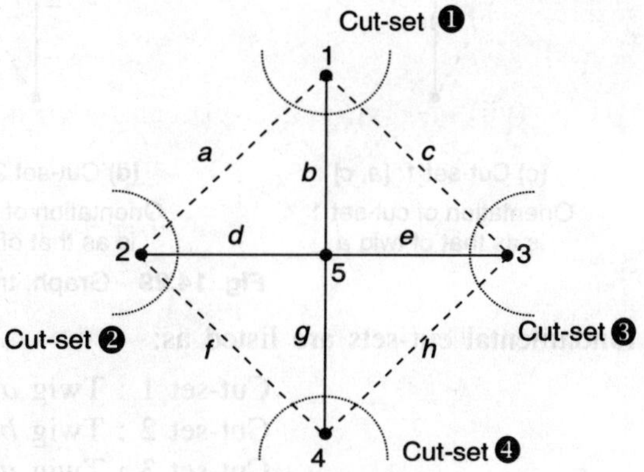

Fig. 14.28 Cut-sets.

The following four cut-sets are possible and their orientations are shown in Fig. 14.28.

Cut-set 1 : Twig *b*, links *a* and *c*
Cut-set 2 : Twig *d*, links *a* and *f*.
Cut-set 3 : Twig *e*, links *c* and *h*
Cut-set 4 : Twig *g*, links *f* and *h*

14.5.1 Cut-Set Matrix

Consider an oriented graph as shown in Fig. 14.29(a). Let us a select a tree arbitrarily as shown in Fig. 14.29(b). The twigs are shown by solid lines and links are shown by dotted lines. The fundamental cut-sets are given by the relation $(n - 1)$, i.e. $(4 - 1) = 3$. The fundamental cut-sets are shown in Fig. 14.29(c), (d) and (e). The directions of the cut-sets are assumed in the same direction as that of concerned twig.

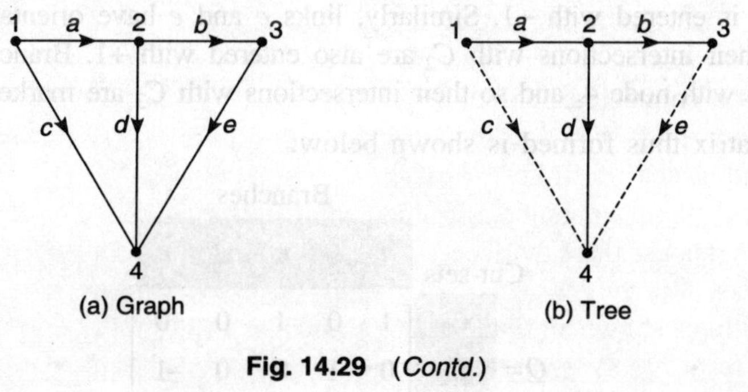

Fig. 14.29 (*Contd.*)

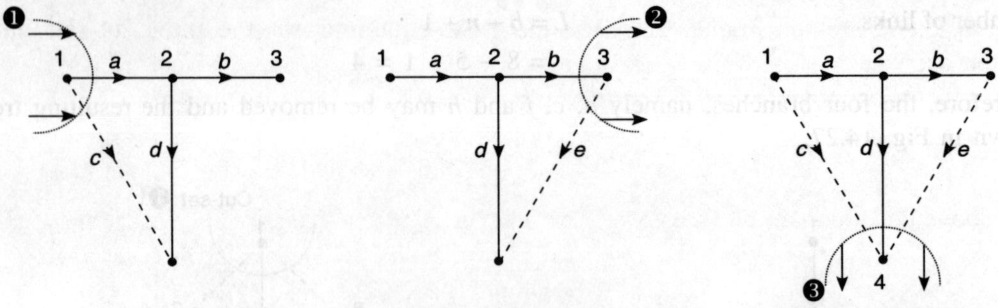

(c) Cut-set 1: [*a*, *c*]
Orientation of cut-set 1 is as that of twig *a*

(d) Cut-set 2: [*b*, *e*]
Orientation of cut-set 2 is as that of twig *b*

(e) Cut-set 3: [*d*, *c*, *e*]
Orientation of cut-set 3 is as that of twig *d*

Fig. 14.29 Graph, tree and cut-sets.

The fundamental cut-sets are listed as:

Cut-set 1 : Twig *a*, link *c*
Cut-set 2 : Twig *b*, link *e*
Cut-set 3 : Twig *d*, links *c* and *e*

The procedure to form the cut-set matrix is as follows:

(i) As cut-set 1 is associated with twig *a*, its orientation is taken as same as that of twig *a* as shown in Fig. 14.29(c). The direction of branch *a* is away from node 1, and therefore, the intersection of branch *a* and C_1 is entered with +1. The direction of link *c* is also away from node 1, its intersection with C_1 is entered with +1 as well in the matrix given below. The branches *b*, *d* and *e* are no way connected with node 1, and so their intersections with C_1 are marked with 0.

(ii) Cut-set 2 is associated with twig *b*, and its orientation is taken as same as that of twig *b* as shown in Fig. 14.29(d). The direction of branch *b* is towards node 3, and therefore, the intersection of branch *b* and C_2 is entered with +1. However, the direction of link *e* is away from node 3, its intersection with C_2 is entered with –1. Branches *a*, *c* and *d* are no way connected with node 3, and so their intersections with C_1 are marked with 0.

(iii) Cut-set 3 is associated with twig *d*, its orientation is taken as same as that of twig *d*, i.e. towards node 4, which is shown in Fig. 14.29(e). Therefore, the intersection of branch *d* and C_3 is entered with +1. Similarly, links *c* and *e* have orientations of towards the node 4, their intersections with C_2 are also entered with +1. Branches *a* and *b* are not connected with node 4, and so their intersections with C_3 are marked with 0.

The cut-set matrix thus formed is shown below:

Cut-sets \ Branches	*a*	*b*	*c*	*d*	*e*
C_1	1	0	1	0	0
C_2	0	1	0	0	–1
C_3	0	0	1	1	1

$$Q = \begin{bmatrix} 1 & 0 & 1 & 0 & 0 \\ 0 & 1 & 0 & 0 & -1 \\ 0 & 0 & 1 & 1 & 1 \end{bmatrix}$$

If I_a, I_b,..., I_e are the branch currents, the KCL equations for the three cut-sets can be written as:

$$\left.\begin{aligned} I_a + I_c &= 0 \\ I_b - I_e &= 0 \\ I_c + I_d + I_e &= 0 \end{aligned}\right\}$$

The above equation can be written in matrix form as:

$$\begin{array}{c} \text{Branches} \\ \begin{array}{ccccc} a & b & c & d & e \end{array} \end{array}$$

$$\begin{bmatrix} 1 & 0 & 1 & 0 & 0 \\ 0 & 1 & 0 & 0 & -1 \\ 0 & 0 & 1 & 1 & 1 \end{bmatrix} \begin{bmatrix} I_a \\ I_b \\ I_c \\ I_d \\ I_e \end{bmatrix} = \begin{bmatrix} 0 \\ 0 \\ 0 \end{bmatrix}$$

i.e.

$$QI_{\text{branch}} = 0 \tag{14.8}$$

where Q is called cut-set matrix or fundamental loop matrix and I_{branch} is a column vector of branch currents.

EXAMPLE 14.6 Draw the graph of the network shown in Fig. 14.30. Select a suitable tree and obtain the cut-set matrix. Write down the KCL equations from the cut-set matrix.

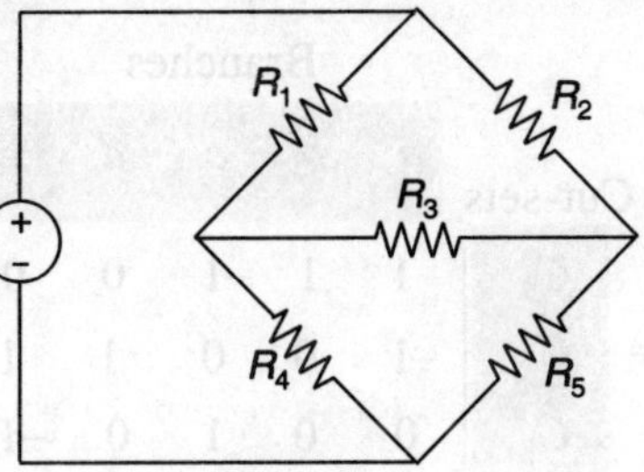

Fig. 14.30

Solution The graph for the given network is shown in Fig. 14.31. The orientations are chosen arbitrarily.

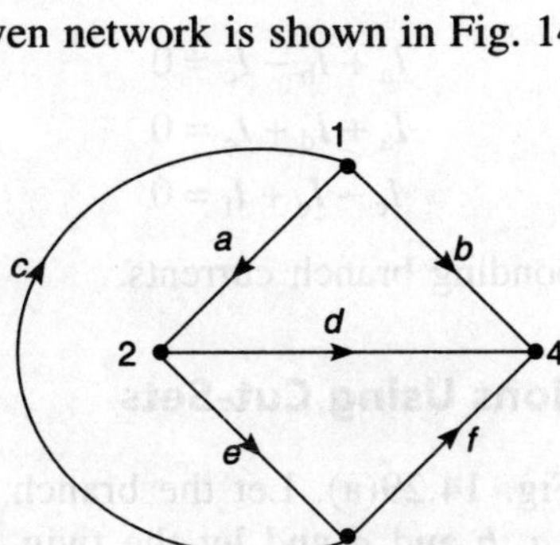

Fig. 14.31

Number of links, $L = b - n + 1 = 6 - 4 + 1 = 3$

As the number of links is three, branches a, c and e may be removed and a tree is formed as shown in Fig. 14.32.

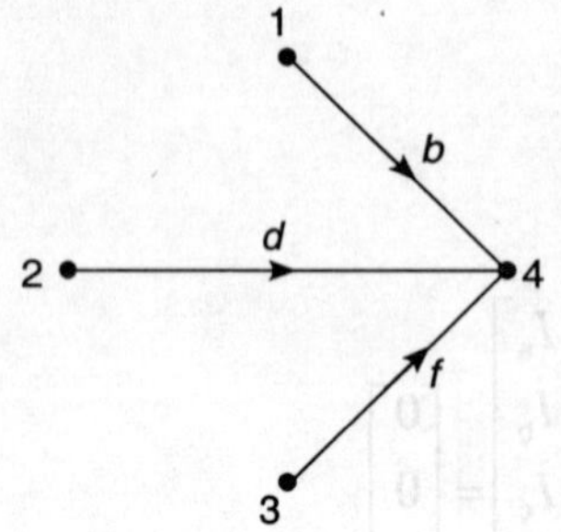

Fig. 14.32

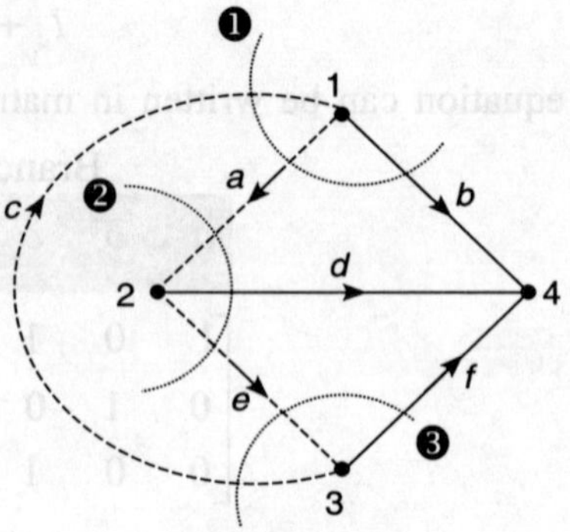

Fig. 14.33

The fundamental cut-sets are shown in Fig. 14.33. The directions of the cut-sets are assumed in the same direction as that of the concerned twig.

The following three cut-sets are possible with node 4 chosen as reference node.

Cut-set 1 : Twig b, links a and c
Cut-set 2 : Twig d, links a and e
Cut-set 3 : Twig f, links c and e

The cut-set matrix formed is shown below:

$$Q = \begin{array}{c} \\ C_1 \\ C_2 \\ C_3 \end{array} \begin{array}{c} \text{Branches} \\ \begin{matrix} a & b & c & d & e & f \end{matrix} \\ \begin{bmatrix} 1 & 1 & -1 & 0 & 0 & 0 \\ -1 & 0 & 0 & 1 & 1 & 0 \\ 0 & 0 & 1 & 0 & -1 & 1 \end{bmatrix} \end{array}$$

(Rows: Cut-sets)

From the cut-set matrix, the KCL equations are written as follows:

$$I_a + I_b - I_c = 0$$
$$-I_a + I_d + I_e = 0$$
$$I_c - I_e + I_f = 0$$

where I_a, I_b, I_c,..., I_f are corresponding branch currents.

14.5.2 Equilibrium Equations Using Cut-Sets

Consider the graph shown in Fig. 14.29(a). Let the branch voltages be V_a, V_b, V_c, V_d and V_e. There are three twigs, viz. a, b and d and let the twig voltages be V_A, V_B and V_D for twigs a, b and d, respectively. The branch voltages are expressed in terms of twig voltages as follows:

$$\begin{aligned} V_a &= V_A \\ V_b &= V_B \\ V_c &= V_A + V_D \\ V_d &= V_D \\ V_e &= -V_B + V_D \end{aligned} \tag{14.9}$$

The above equations are written in matrix form as:

$$\begin{bmatrix} V_a \\ V_b \\ V_c \\ V_d \\ V_e \end{bmatrix} = \begin{bmatrix} 1 & 0 & 0 \\ 0 & 1 & 0 \\ 1 & 0 & 1 \\ 0 & 0 & 1 \\ 0 & -1 & 1 \end{bmatrix} \begin{bmatrix} V_A \\ V_B \\ V_D \end{bmatrix}$$

or $$\mathbf{V}_{\text{branch}} = \mathbf{Q}^T \mathbf{V}_{\text{node}} \tag{14.10}$$

From Eq. (14.8),

$$QI_B = \begin{bmatrix} 1 & 0 & 1 & 0 & 0 \\ 0 & 1 & 0 & 0 & -1 \\ 0 & 0 & 1 & 1 & 1 \end{bmatrix} \begin{bmatrix} I_a \\ I_b \\ I_c \\ I_d \\ I_e \end{bmatrix} = 0 \tag{14.11}$$

Upon multiplying,

$$\begin{aligned} I_a + I_c &= 0 \\ I_b - I_e &= 0 \\ I_c + I_d + I_e &= 0 \end{aligned} \tag{14.12}$$

By Ohm's law, the branch current can be expressed as a product of voltage and conductance (or admittance) of the branch.

Therefore the branch currents can be expressed as:

$$I_a = V_a G_a;\quad I_b = V_b G_b;\quad I_c = V_c G_c;\quad I_d = V_d G_d;\quad I_e = V_e G_e$$

where G_a, G_b, G_c, G_d and G_e are conductances of branches a, b, c, d and e, respectively. Now, Eq. (14.12) can be written as:

$$\begin{aligned} V_a G_a + V_c G_c &= 0 \\ V_b G_b - V_e G_e &= 0 \\ V_c G_c + V_d G_d + V_e G_e &= 0 \end{aligned} \tag{14.13}$$

Using Eq. (14.9), Eq. (14.13) can be written as:

$$\begin{aligned} V_A G_a + (V_A + V_D)G_c &= 0 \\ V_B G_b + (V_B - V_D)G_e &= 0 \\ (V_A + V_D)G_c + V_D G_d + (-V_B + V_D)G_e &= 0 \end{aligned} \tag{14.14}$$

Rearranging,

$$\begin{aligned} V_A(G_a + G_c) + V_D(G_c) &= 0 \\ V_B(G_b + G_e) + V_D(-G_e) &= 0 \\ V_A(G_c) + V_B(-G_e) + V_D(G_c + G_d + G_e) &= 0 \end{aligned} \tag{14.15}$$

$$\begin{bmatrix} G_a + G_c & 0 & G_c \\ 0 & G_b + G_e & -G_e \\ G_c & -G_e & G_c + G_d + G_e \end{bmatrix} \begin{bmatrix} V_A \\ V_B \\ V_D \end{bmatrix} = \begin{bmatrix} 0 \\ 0 \\ 0 \end{bmatrix} \tag{14.16}$$

$$GV_N = I$$

where I is current source matrix. Matrix G formed by branch resistances in Eq. (14.16) is called *parameter matrix.*

EXAMPLE 14.7 Determine the node voltages of the circuit shown in Fig. 14.34 by using cut-set schedule.

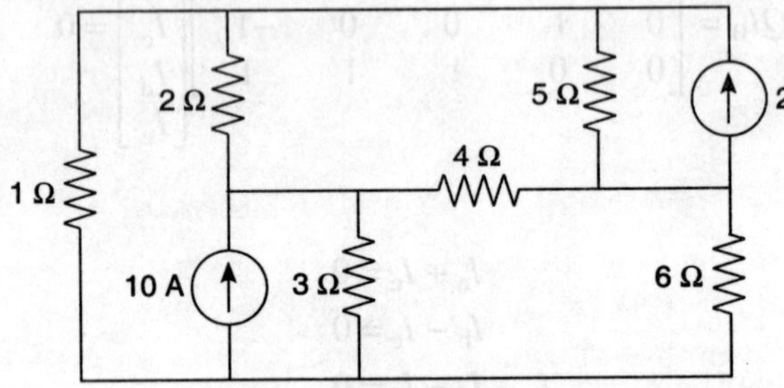

Fig. 14.34

Solution The current sources in the circuit are replaced with open-circuits and the circuit is redrawn as shown in Fig. 14.35 and the graph for the same is shown in Fig. 14.36. The orientations are chosen arbitrarily.

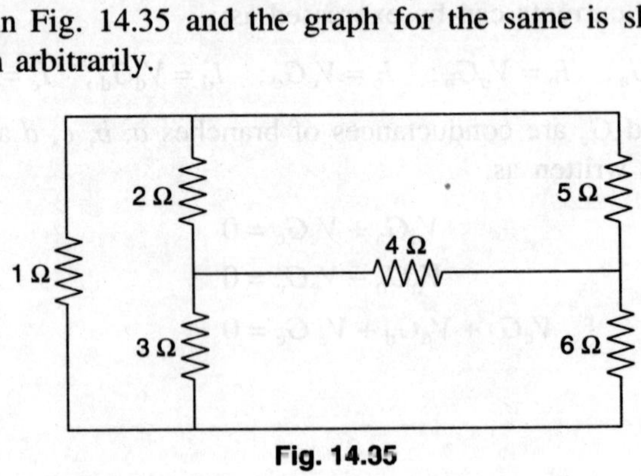

Fig. 14.35

Number of links, $L = b - n + 1 = 6 - 4 + 1 = 3$

As the number of links is three, branches b, c and d are removed and a tree is formed as shown in Fig. 14.37.

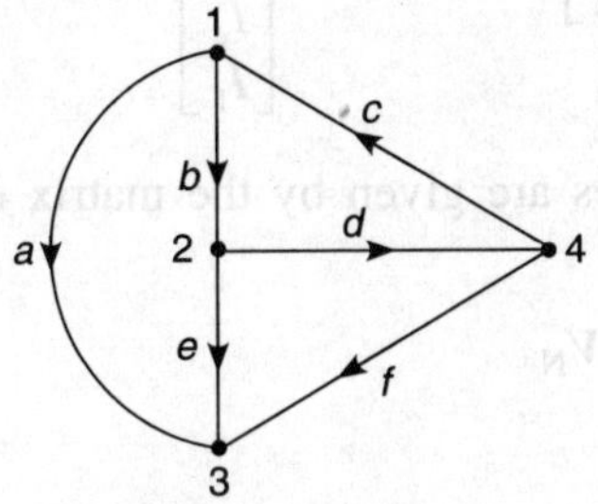

Fig. 14.36 Graph.

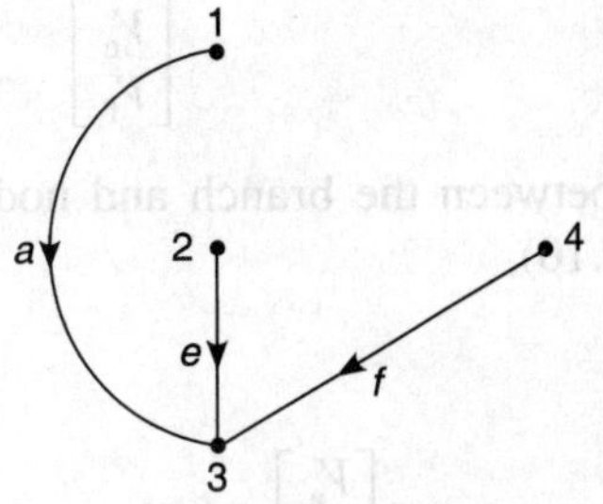

Fig. 14.37 Tree.

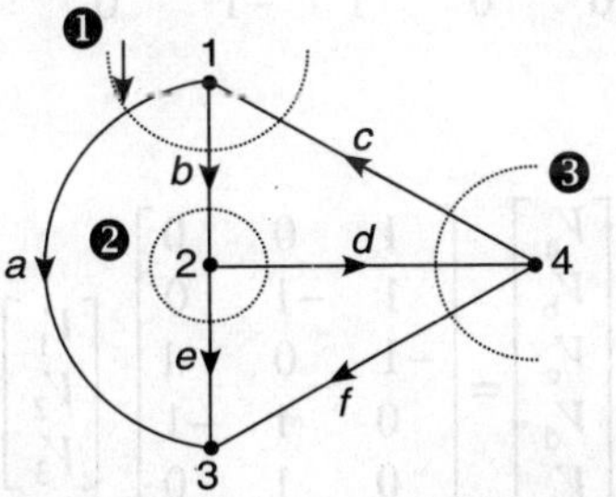

Fig. 14.38 Cut-sets of the graph in Fig. 14.36.

The following three cut-sets are possible with node 4 chosen as reference node.

Cut-set 1 : Twig a, links b and c
Cut-set 2 : Twig e, links b and d
Cut-set 3 : Twig f, links c and d

The cut-set matrix formed is shown below:

Cut-sets	Branches a	b	c	d	e	f
C_1	1	1	–1	0	0	0
C_2	0	–1	0	1	1	0
C_3	0	0	1	–1	0	1

$$Q = \begin{bmatrix} 1 & 1 & -1 & 0 & 0 & 0 \\ 0 & -1 & 0 & 1 & 1 & 0 \\ 0 & 0 & 1 & -1 & 0 & 1 \end{bmatrix}$$

Let V_1, V_2 and V_3 be the node voltages and V_a, V_b, V_c, V_d, V_e and V_f be the branch voltages. Similarly, I_a, I_b, I_c, I_d, I_e and I_f are the respective branch currents. They are given in a matrix form as shown below:

$$\boldsymbol{V}_{\text{branch}} = \begin{bmatrix} V_a \\ V_b \\ V_c \\ V_d \\ V_e \\ V_f \end{bmatrix}; \boldsymbol{V}_{\text{node}} = \begin{bmatrix} V_1 \\ V_2 \\ V_3 \end{bmatrix} \text{ and } \boldsymbol{I}_{\text{branch}} = \begin{bmatrix} I_a \\ I_b \\ I_c \\ I_d \\ I_e \\ I_f \end{bmatrix}$$

The relation between the branch and node voltages are given by the matrix equation:
From Eq. (14.10),

$$\boldsymbol{V}_B = \boldsymbol{Q}^T \boldsymbol{V}_N$$

Therefore,

$$\begin{bmatrix} V_a \\ V_b \\ V_c \\ V_d \\ V_e \\ V_f \end{bmatrix} = \begin{bmatrix} 1 & 1 & -1 & 0 & 0 & 0 \\ 0 & -1 & 0 & 1 & 1 & 0 \\ 0 & 0 & 1 & -1 & 0 & 1 \end{bmatrix}^T \begin{bmatrix} V_1 \\ V_2 \\ V_3 \end{bmatrix}$$

$$\begin{bmatrix} V_a \\ V_b \\ V_c \\ V_d \\ V_e \\ V_f \end{bmatrix} = \begin{bmatrix} 1 & 0 & 0 \\ 1 & -1 & 0 \\ -1 & 0 & 1 \\ 0 & 1 & -1 \\ 0 & 1 & 0 \\ 0 & 0 & 1 \end{bmatrix} \begin{bmatrix} V_1 \\ V_2 \\ V_3 \end{bmatrix} \tag{i}$$

On multiplying Eq. (i), we get

$$\begin{bmatrix} V_a \\ V_b \\ V_c \\ V_d \\ V_e \\ V_f \end{bmatrix} = \begin{bmatrix} V_1 \\ V_1 - V_2 \\ -V_1 + V_3 \\ V_2 - V_3 \\ V_2 \\ V_3 \end{bmatrix} \tag{ii}$$

From Eq. (ii),

$$\left.\begin{array}{ll} V_a = V_1 \quad ; & V_d = V_2 - V_3 \\ V_b = V_1 - V_2 \quad ; & V_e = V_2 \\ V_c = -V_1 + V_3 \,; & V_f = V_3 \end{array}\right\} \tag{iii}$$

From Eq. (14.8)

$$\boldsymbol{Q}\boldsymbol{I}_{\text{branch}} = 0$$

$$\begin{bmatrix} 1 & 1 & -1 & 0 & 0 & 0 \\ 0 & -1 & 0 & 1 & 1 & 0 \\ 0 & 0 & 1 & -1 & 0 & 1 \end{bmatrix} \begin{bmatrix} I_a \\ I_b \\ I_c \\ I_d \\ I_e \\ I_f \end{bmatrix} = 0 \qquad \text{(iv)}$$

On multiplying Eq. (iv), we get

$$\left.\begin{aligned} I_a + I_b - I_c &= 0 \\ -I_b + I_d + I_e &= 0 \\ I_c - I_d + I_f &= 0 \end{aligned}\right\} \qquad \text{(v)}$$

For the branches that do not have a parallel current source, the branch currents are expressed as follows:

$$\left.\begin{aligned} I_a &= \frac{V_a}{1} = V_a \\ I_b &= \frac{V_b}{2} = 0.5V_b \\ I_d &= \frac{V_d}{4} = 0.25V_d \\ I_f &= \frac{V_f}{6} = 0.1667V_f \end{aligned}\right\} \qquad \text{(vi)}$$

For the branches *c* and *e*, since they have a parallel current source each with them, the branch currents are expressed by applying KCL as follows:

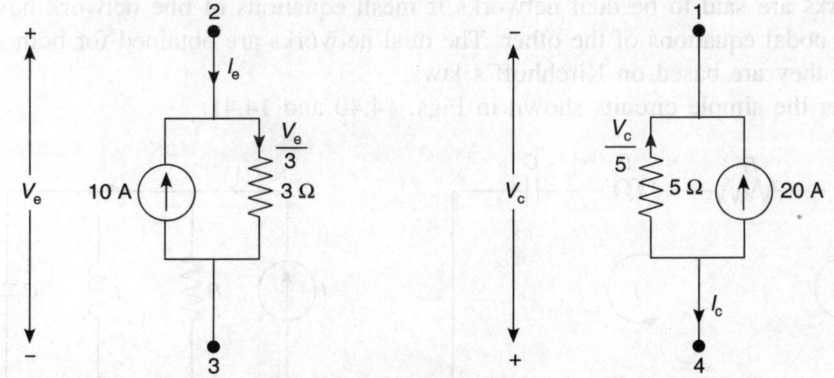

Fig. 14.39

With reference to Fig. 14.39, by applying KCL,

$$\left.\begin{aligned} I_e &= \frac{1}{3}V_e - 10 \\ I_c &= \frac{1}{5}V_c + 20 \end{aligned}\right\} \qquad \text{(vii)}$$

Using Eqs. (vi) and (vii), Eq. (v) can be written as:

$$\left.\begin{array}{r} V_a + 0.5V_b - 0.2V_c - 20 = 0 \\ -0.5V_b + 0.25V_d + 0.33V_e - 10 = 0 \\ 0.2V_c + 20 - 0.25V_d + 0.1667V_f = 0 \end{array}\right\} \quad \text{(viii)}$$

Using Eq. (iii) the branch voltages in Eq. (viii) are written in terms of node voltages as follows:

$$\left.\begin{array}{r} V_1 + 0.5(V_1 - V_2) - 0.2(-V_1 + V_3) - 20 = 0 \\ -0.5(V_1 - V_2) + 0.25(V_2 - V_3) + 0.33V_2 - 10 = 0 \\ 0.2(-V_1 + V_3) + 20 - 0.25(V_2 - V_3) + 0.1667V_3 = 0 \end{array}\right\} \quad \text{(ix)}$$

Rearranging,

$$\left.\begin{array}{r} 1.7V_1 - 0.5V_2 - 0.2V_3 = 20 \\ -0.5V_1 + 1.08V_2 - 0.25V_3 = 10 \\ -0.2V_1 - 0.25V_2 + 0.6167V_3 = -20 \end{array}\right\}$$

Solving,

$$V_1 = 11.3380 \text{ V}$$
$$V_2 = 8.6582 \text{ V}$$
$$V_3 = -25.2725 \text{ V}$$

14.6 DUALITY

Two networks are said to be dual networks if mesh equations of one network have the same form as the nodal equations of the other. The dual networks are obtained for both AC and DC circuits and they are based on Kirchhoff's laws.

Consider the simple circuits shown in Figs. 14.40 and 14.41.

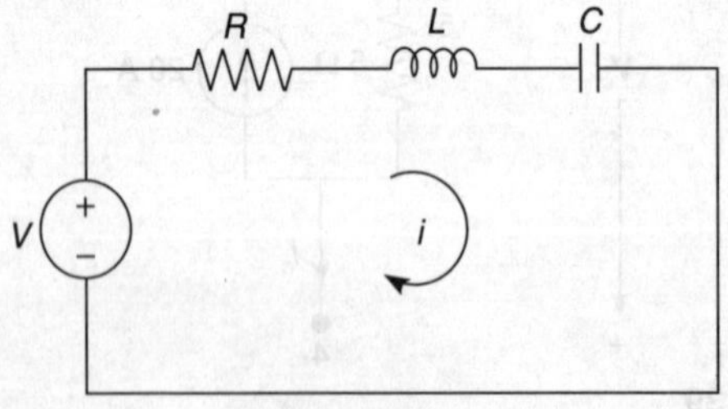

Fig. 14.40

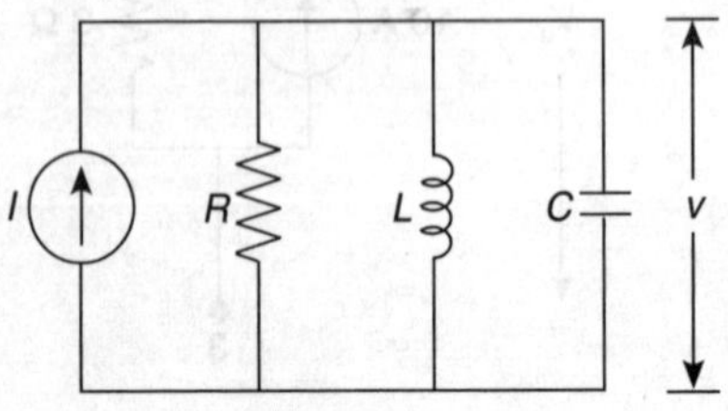

Fig. 14.41

Applying KVL in the circuit shown in Fig. 14.40, we get

$$iR + L\frac{di}{dt} + \frac{1}{C}\int i\,dt = V \quad (14.17)$$

Applying KCL in the circuit shown in Fig. 14.41, we get

$$vG + C\frac{dv}{dt} + \frac{1}{L}\int v dt = I \tag{14.18}$$

Equations (14.17) and (14.18) are seemed to be identical and the circuits shown in Figs. 14.40 and 14.41 are said to be dual circuits. Upon comparison, it is noticed that the following changes are made between the two equations:

$$i \to v;\ R \to G;\ L \to C;\ C \to L;\ V \to I$$

Dual circuits are obtained in planar networks. A planar network is a network laid on a two-dimensional plane without any branches cross each other. The duals of the elements and the variables are listed in Table. 14.1.

Table 14.1 Dual elements and variables

Original network	*Dual network*
Current	Voltage
Voltage	Current
Resistance	Conductance
Inductance	Capacitance
Capacitance	Inductance
Closed switch	Open switch
Loop	Node

For any given network, the dual network is easily obtained by the dot method as follows:

Step 1 A dot is placed in each mesh of the original network. The dots thus placed inside the meshes correspond to nodes in the dual network.

Step 2 A dot is placed outside the original network corresponds to the reference node of the dual network.

Step 3 All the dots are connected by dotted lines crossing all the branches one at a time. For each element in the original network, the dual element is formed between the corresponding nodes in the dual network.

Step 4 Voltage rise in the direction of clockwise mesh current is replaced by a current flowing towards the node and vice versa.

The above procedure is illustrated in Figs. 14.42 and 14.43. The given network (Fig. 14.42) has two meshes. Two dots are placed inside each mesh as shown in Fig. 14.42 the reference node "0" is placed outside the network as well. There are five elements and as many number of lines are drawn between nodes such that each line crossing only one element at a time. These dotted lines represent branches with dual elements in the dual network. The dual network is thus obtained and is shown in Fig. 14.43.

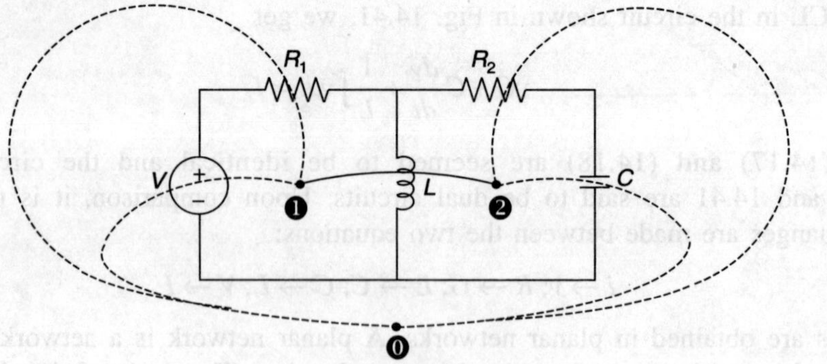

Fig. 14.42

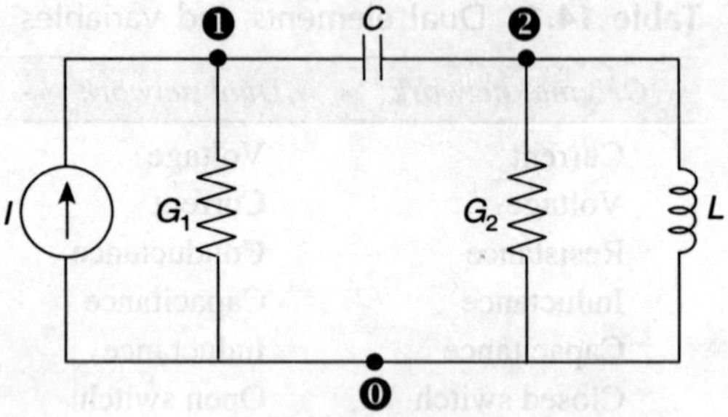

Fig. 14.43

EXAMPLE 14.8 Determine the dual of the network shown in Fig. 14.44.

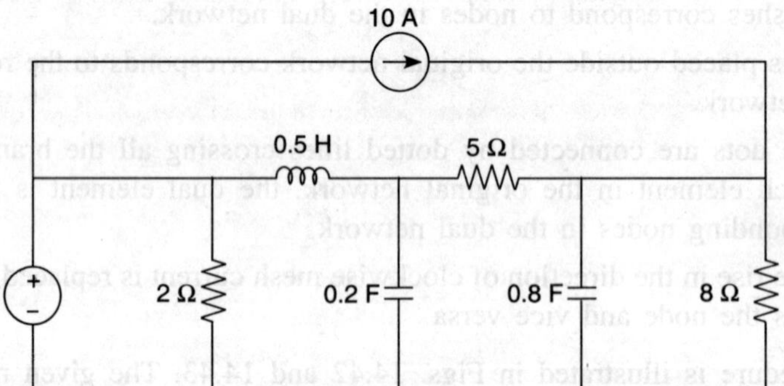

Fig. 14.44

Solution As illustrated in Fig. 14.45, the dots are connected crossing all elements. The dual network thus obtained is shown in Fig. 14.46.

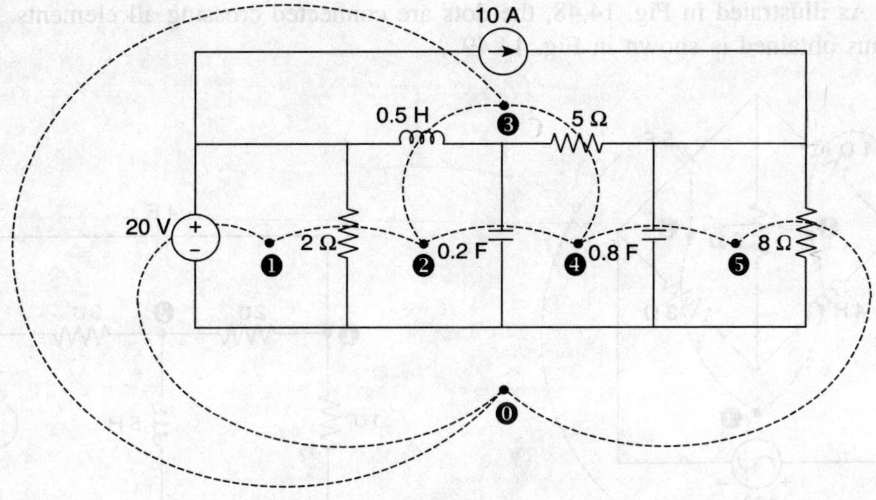

Fig. 14.45

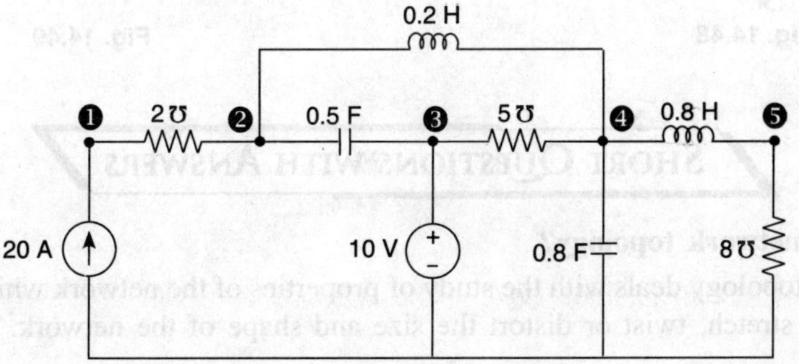

Fig. 14.46

EXAMPLE 14.9 Determine the dual of the network shown in Fig. 14.47.

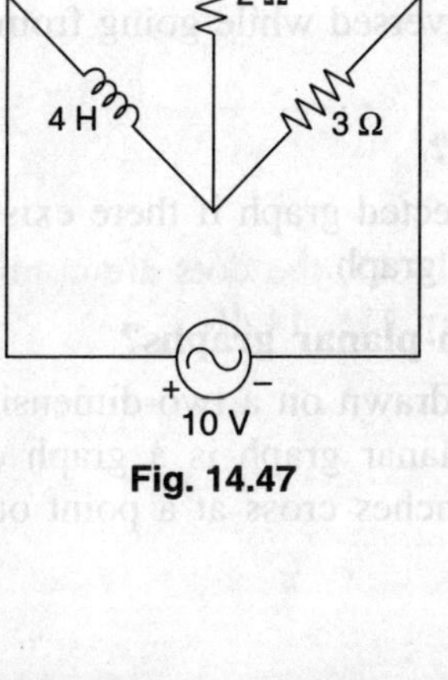

Fig. 14.47

Solution As illustrated in Fig. 14.48, the dots are connected crossing all elements. The dual network thus obtained is shown in Fig. 14.49.

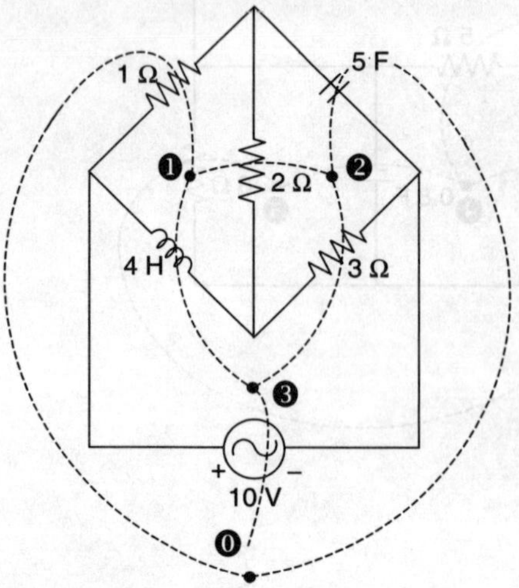

Fig. 14.48

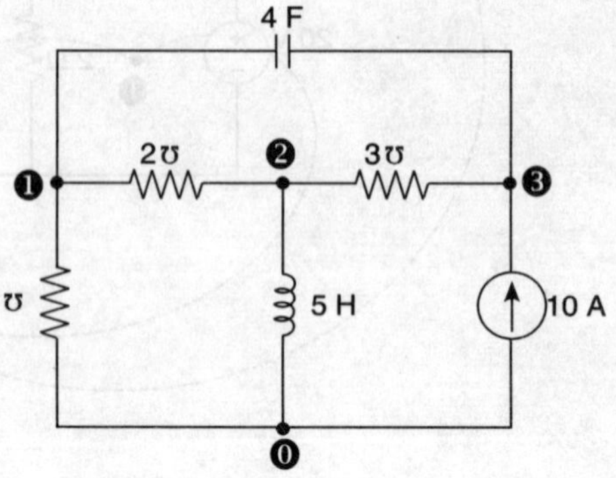

Fig. 14.49

Short Questions with Answers

1. What is network topology?

Network topology deals with the study of properties of the network which are unaffected when we stretch, twist or distort the size and shape of the network.

2. Define graph of a network.

When all the elements of a network are represented on a diagram by lines irrespective of their characteristics, the resulting configuration is then called the graph of the network.

3. What is oriented graph?

A graph is called oriented graph when arrows are placed on the branches of graph. The arrow indicates the direction of branch current.

4. What is a path in a graph?

A sequence of branches traversed while going from one node to another node is called a path.

5. What is connected graph?

A graph is said to be connected graph if there exists at least one path from each node to every other node of the graph.

6. What are planar and non-planar graphs?

A planar graph is a graph drawn on a two-dimensional plane so that no two branches cross each other. A non-planar graph is a graph drawn on a two-dimensional plane such that two or more branches cross at a point other than a node.

7. Define subgraph and what are its types?

When some of the branches in an original graph are removed, the resultant graph is called subgraph. If the subgraph contains lesser number of nodes than the original graph, it is called proper subgraph. On the other hand, if it contains all the nodes of the graph, it is called improper subgraph.

8. Define a tree.

The tree is a subgraph which is obtained by removing some branches such that the subgraph includes all the nodes of the original graph, but does not have any closed paths.

9. What are the properties of tree?

(i) In a tree there exists one and only one path between any pair of nodes.
(ii) Tree contains all the nodes of graph.
(iii) Tree does not have a closed path or loop.
(iv) Every connected graph has at least one tree.
(v) Each tree has $(n - 1)$ branches, where n is the number of nodes of the graph.

10. What is a co-tree?

All branches that are present in the network but not present in the tree of the network together constitute a co-tree.

11. What are links and twigs?

The branches of a graph which are removed to form a tree are called links or chords. The branches of a tree are called twigs.

In a graph with b branches and n nodes,

Number of links, $L = b - n + 1$

and number of twigs $n - 1$

12. Define incidence matrix.

Incidence matrix is defined as a two-dimensional array which provides information regarding which branches are incident at which node and what is the orientation relative to the nodes.

13. What are the properties of incidence matrix?

(i) Every column has two non-zero elements one of which is +1 and the other is –1.
(ii) The sum of the elements in any column of complete incidence matrix is zero.
(iii) The rank of the complete incidence matrix is $(n - 1)$.
(iv) The determinant of the complete incidence matrix of a closed loop is zero.

14. What is reduced incidence matrix?

When one row is deleted from the complete incidence matrix A, the resultant matrix is called a reduced incidence matrix A_1.

15. What is tie-set?

A tie-set is a set of branches that form a closed path in a graph such that the closed path contains one link and remainders are twigs. The closed path is also known as loop.

16. What is tie-set matrix?

The tie-set matrix is a two-dimensional array in which the relations between the loop currents and branch currents are tabulated.

17. What is cut-set?

A *cut-set* is a minimal set of branches of a connected graph such that the removal of these branches separates the graph into two distinct parts. The important property of a cut-set is that by restoring any one of the branches of the cut-set, the graph becomes a connected one.

18. What is cut-set matrix?

The tie-set matrix is a two-dimensional array in which the relation between the node voltages and branch voltages are tabulated.

19. Define duality in circuits.

The duality in circuits is about forming a voltage basis circuit for a given current basis circuit or vice versa.

20. Define dual networks.

Two networks are said to be dual networks if mesh equations of one network have the same form as the nodal equations of the other. The dual networks are obtained for both AC and DC circuits and they are based on Kirchhoff's laws.

21. List out few pairs or dual quantities.

Original network	*Dual network*
Current	Voltage
Voltage	Current
Resistance	Conductance
Inductance	Capacitance
Capacitance	Inductance
Closed switch	Open switch
Loop	Node

UNSOLVED PROBLEMS

1. Draw the graph of the circuit shown in Fig. U14.1, and hence determine the incidence matrix. Also draw any three possible trees of the graph.

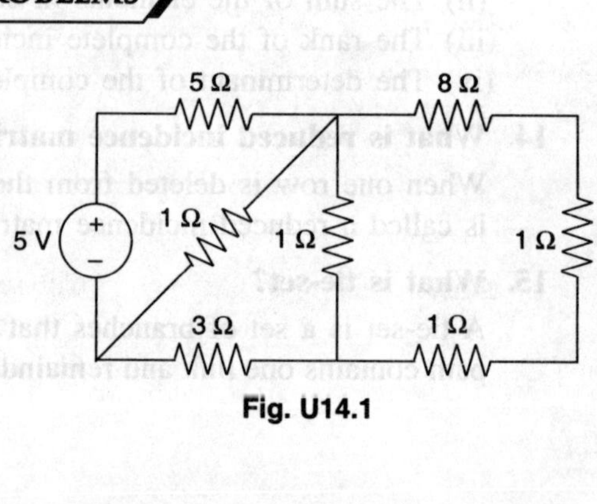

Fig. U14.1

Ans.

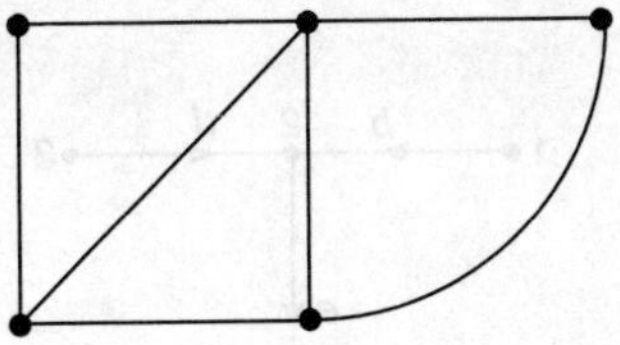

Fig. U14.2 Graph.

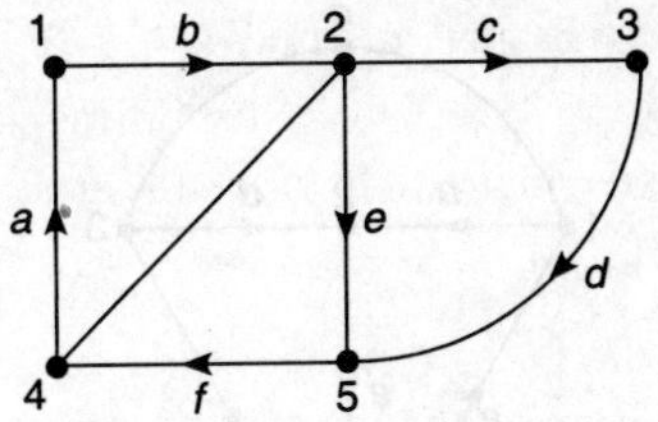

Fig. U14.3 Oriented graph.

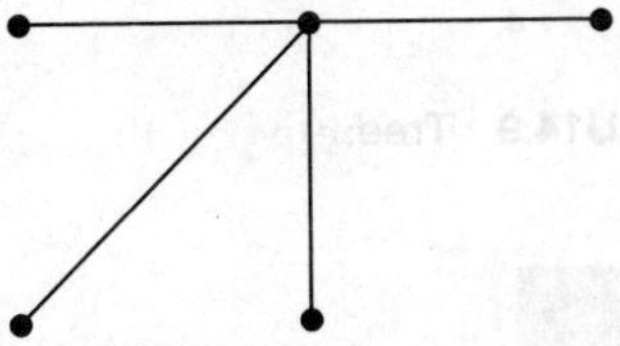

Fig. U14.4 Tree.

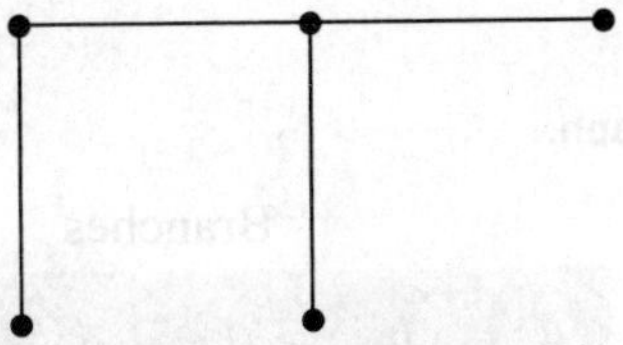

Fig. U14.5 Tree.

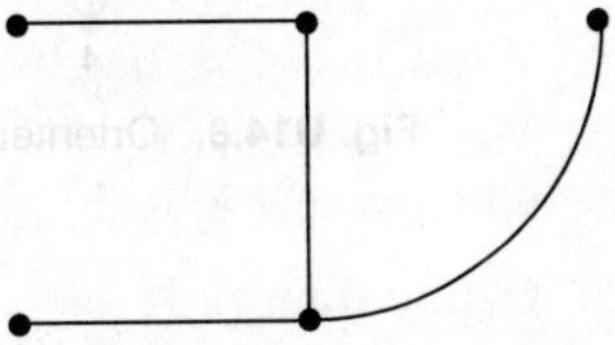

Fig. U14.6 Tree.

$$A = \begin{array}{c|ccccccc} & \multicolumn{7}{c}{\text{Branches}} \\ \text{Nodes} & a & b & c & d & e & f & g \\ \hline 1 & +1 & +1 & 0 & 0 & 0 & 0 & 0 \\ 2 & 0 & -1 & +1 & 0 & +1 & 0 & +1 \\ 3 & 0 & 0 & -1 & +1 & 0 & 0 & 0 \\ 4 & -1 & 0 & 0 & 0 & 0 & -1 & -1 \\ 5 & 0 & 0 & 0 & -1 & -1 & +1 & 0 \end{array}$$

2. For the circuit shown in Fig. U14.7, (i) draw its graph, (ii) draw its tree and (iii) write the fundamental cut-set matrix.

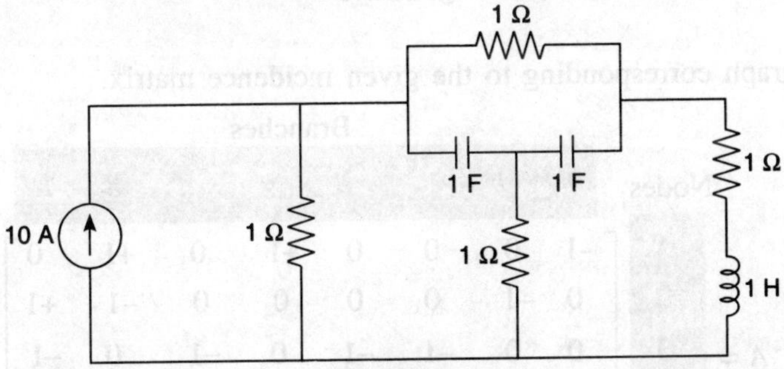

Fig. U14.7

Ans.

Cut-set 1 : Twig b, links a and c
Cut-set 2 : Twig d, links c and f
Cut-set 3 : Twig e, links a and f

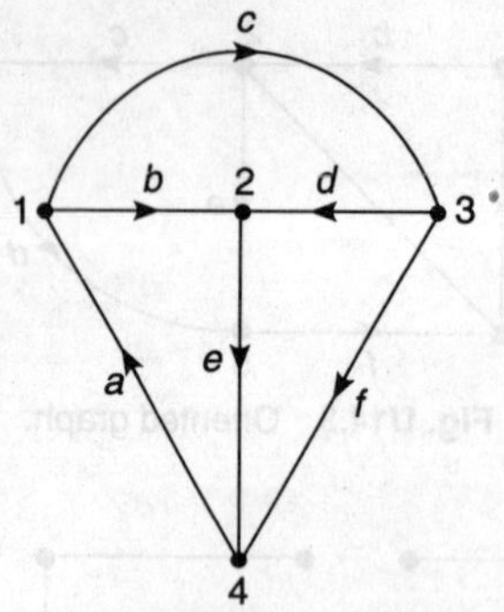

Fig. U14.8 Oriented graph.

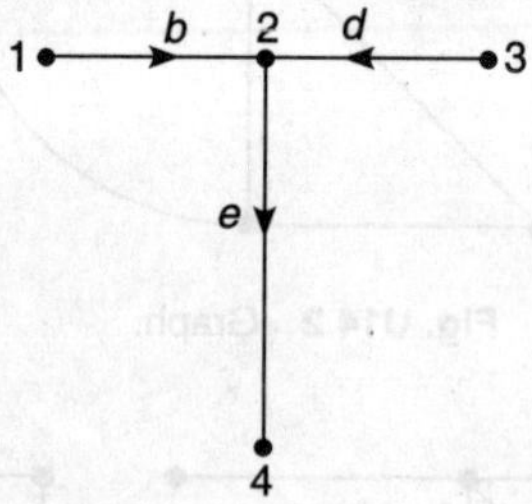

Fig. U14.9 Tree.

Branches

$Q =$ Cut-sets	a	b	c	d	e	f
C_1	+1	+1	+1	0	0	0
C_2	0	0	−1	+1	0	+1
C_3	+1	0	0	0	+1	+1

3. For the circuit shown in Fig. U14.10, how many trees are possible? **[Ans.** 5]

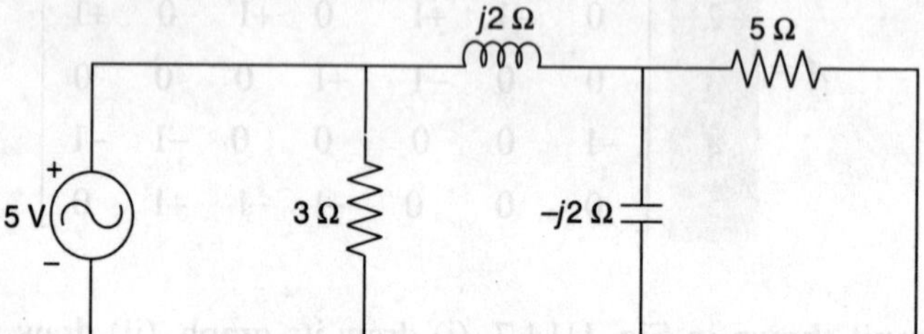

Fig. U14.10

4. Draw the graph corresponding to the given incidence matrix.

Branches

$A =$ Nodes	a	b	c	d	e	f	g	h
1	−1	0	0	0	+1	0	+1	0
2	0	−1	0	0	0	0	−1	+1
3	0	0	−1	−1	0	−1	0	−1
4	0	0	0	0	−1	+1	0	0
5	+1	+1	+1	+1	0	0	0	0

Ans.

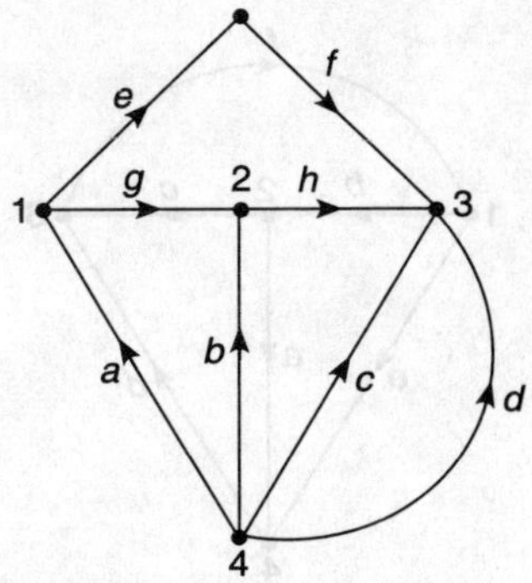

Fig. U14.11 Oriented graph.

5. Obtain the incidence matrix from the following reduced incidence matrix and draw its graph.

$A =$

Nodes \ Branches	a	b	c	d	e	f	g
1	−1	+1	0	0	0	0	0
2	0	−1	+1	+1	0	0	0
3	0	0	0	−1	+1	0	0
4	0	0	0	0	−1	+1	0
5	0	0	−1	0	0	−1	+1

Ans.

$A =$

Nodes \ Branches	a	b	c	d	e	f	g
1	−1	+1	0	0	0	0	0
2	0	−1	+1	+1	0	0	0
3	0	0	0	−1	+1	0	0
4	0	0	0	0	−1	+1	0
5	0	0	−1	0	0	−1	+1
6	+1	0	0	0	0	0	−1

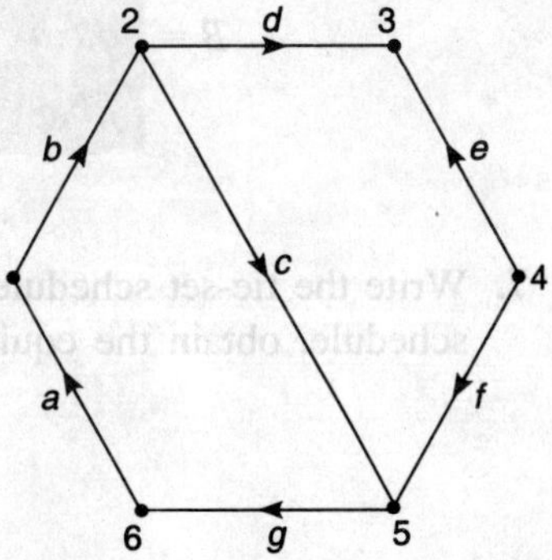

Fig. U14.12 Oriented graph.

6. In the following graph of a network, select a tree with branches a, b and c as the tree branches and write the tie-set schedule.

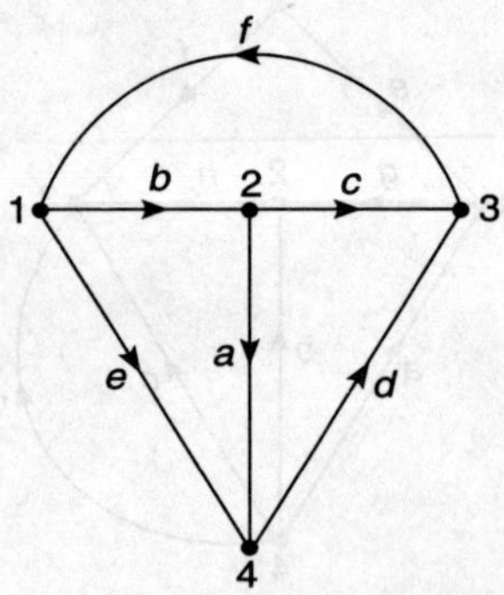

Fig. U14.13 Oriented graph.

Ans.

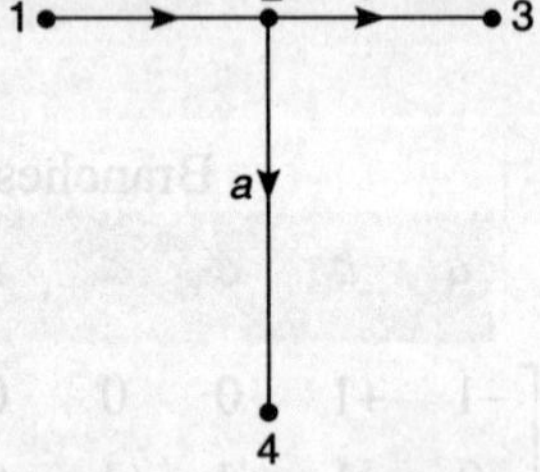

Fig. U14.14 Tree.

$B =$

Tie-sets \ Branches	*a*	*b*	*c*	*d*	*e*	*f*
e,a,b	–1	–1	0	0	+1	0
d,c,a	1	0	–1	+1	0	0
f,b,c	0	+1	+1	0	0	+1

Fig. U14.15 Tie-set matrix.

7. Write the tie-set schedule for the network shown in Fig. U14.16, and using the tie-set schedule, obtain the equilibrium equations on loop current basis.

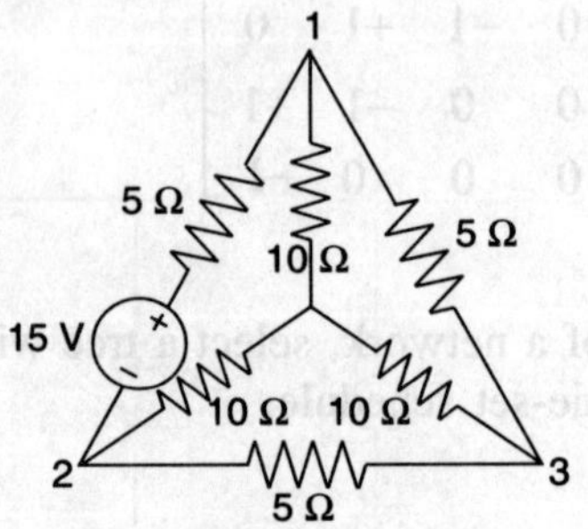

Fig. U14.16

Ans.

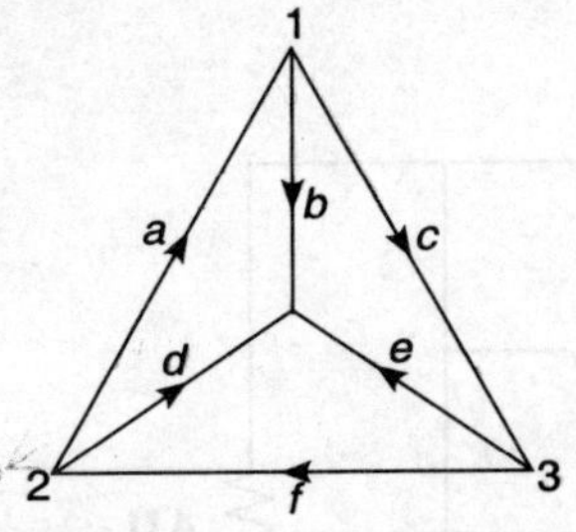

Fig. U14.17 Oriented graph.

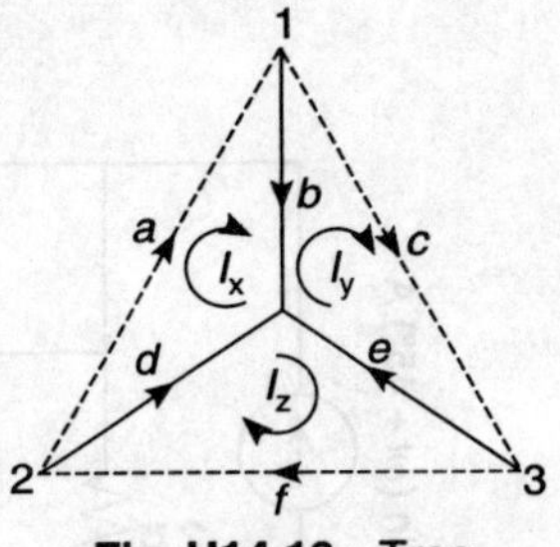

Fig. U14.18 Tree.

Branches

Tie-sets	a	b	c	d	e	f
a,b,d	+1	+1	0	−1	0	0
c,e,b	0	−1	+1	0	+1	0
f,d,e	0	0	0	+1	−1	+1

$$B = \begin{bmatrix} +1 & +1 & 0 & -1 & 0 & 0 \\ 0 & -1 & +1 & 0 & +1 & 0 \\ 0 & 0 & 0 & +1 & -1 & +1 \end{bmatrix}$$

Equilibrium equations:

$$25I_x - 10\,I_y - 10I_z = 15$$
$$-10I_x + 25I_y - 10I_z = 0$$
$$-10I_x - 10I_y + 25I_z = 15$$

8. Determine the dual of the network shown in Fig. U14.19.

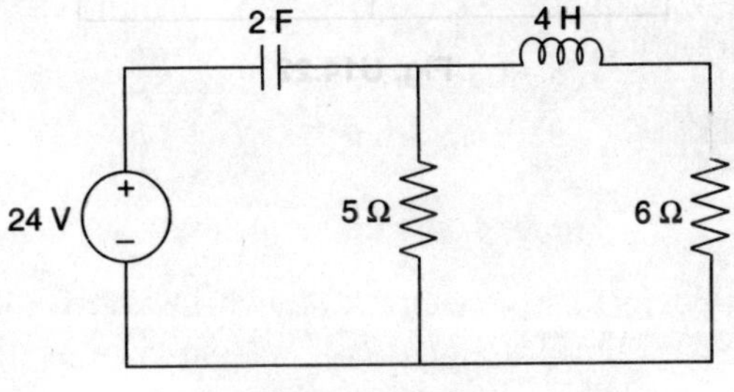

Fig. U14.19

Ans.

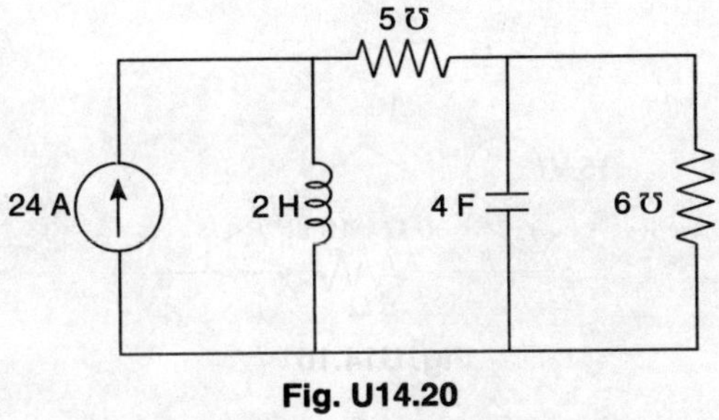

Fig. U14.20

9. Draw the dual network for the given network shown in Fig. U14.21.

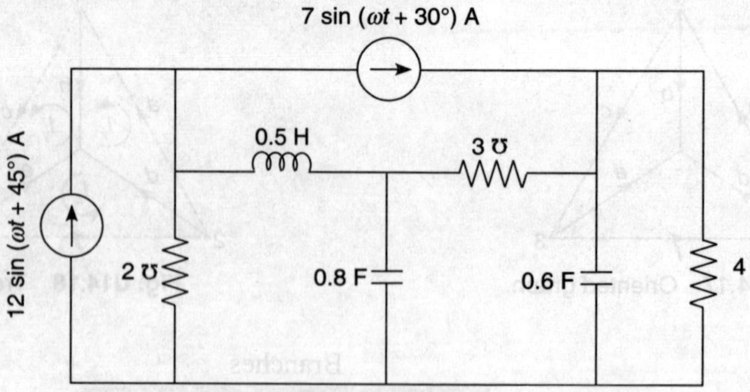

Fig. U14.21

Ans.

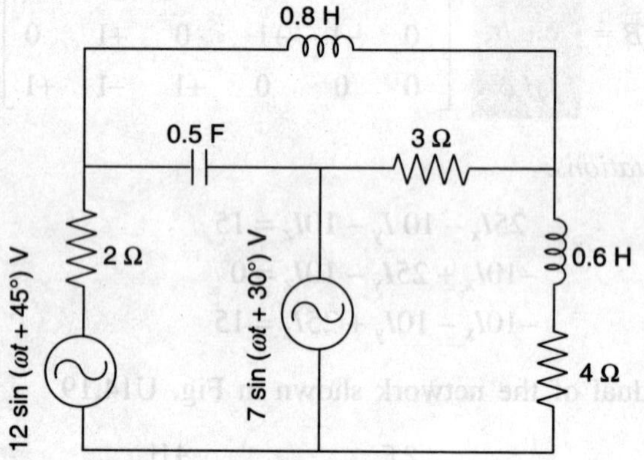

Fig. U14.22

Index